AF181284

FESTKÖRPERPROBLEME

ADVANCES IN SOLID STATE PHYSICS 38

FESTKÖRPER PROBLEME

ADVANCES IN SOLID STATE PHYSICS 38

Edited by
Bernhard Kramer

vieweg

Editor:

Prof. Dr. Bernhard Kramer
I. Institut für Theoretische Physik
Universität Hamburg
Jungusistr. 9
D-20355 Hamburg

All rights reserved
© Springer Fachmedien Wiesbaden 1999

Originally published by Friedr. Vieweg & Sohn Verlagsgesellschaft mbH, Braunschweig/Wiesbaden in 1999.

No part of this publication may be reproduced, stored in a retrieval system or transmitted, mechanical, photocopying or otherwise, without prior permission of the copyright holder.

http://www.vieweg.de

Cover design: Barbara Seebohm, Braunschweig
Printed on acid-free paper

ISSN 0430-3393

ISBN 978-3-662-16127-2 ISBN 978-3-540-44558-6 (eBook)
DOI 10.1007/978-3-540-44558-6

Preface

This volume contains the invited papers of the Spring meeting of the "Arbeitskreis Festkörperphysik" of the "Deutsche Physikalische Gesellschaft", which has been held in Regensburg in the period March 23–27, 1998.

The meeting has been attended by more than 2800 participants from all areas of solid state physics and other areas of physics as well since this time it has been organized together with the yearly general meeting.

The invited and plenary talks as well as the numerous symposia on selected topics reflected most recent developments in physics. We hope that this book covers the current status of the fields and reflects the excitement which has been felt by the speakers when presenting their results.

Hamburg, November 1998

Bernhard Kramer

Contents

Present Status of InGaN-Based UV/Blue/Green LEDs and Laser Diodes

Shuji Nakamura

Department of Research and Development, Nichia Chemical Industries, Ltd., 491 Oka, Kaminaka, Anan, Tokushima 774, Japan

Abstract: InGaN quantum well structure light emitting diodes operating with external quantum efficiencies of 7.5% at 371 nm (UV), 11.2% at 468 nm (blue) and 11.6% at green were developed. The blue and green InGaN LEDs with luminous efficiencies of 5 lm/W and 30 lm/W can be used for fabricating white LEDs with a luminous efficiency of 30 lm/W which is almost identical to that of conventional incandescent bulb lamps. Epitaxially laterally overgrown GaN on sapphire was used to reduce the number of threading dislocations originating from the interface of the GaN epilayer with the sapphire substrate. The GaN layer above the SiO_2 mask area surrounding the window and corresponding to the lateral overgrowth, was nearly free of the threading dislocations. InGaN multi-quantum well structure laser diodes grown on pure GaN substrates, which were fabricated by removing the sapphire substrate, were demonstrated. The laser diodes with an output power of 5 mW exhibited a lifetime of more than 290 hours. The far-field pattern of the laser diodes with a cleaved mirror facet revealed single mode emission without any interference effects.

1 Introduction

Gallium Nitride and other III-V Nitride-based semiconductors have a direct band gap energy structure that is suitable for blue light emitting devices. The band gap energy of Aluminium Gallium Indium Nitride (AlInGaN) varies between 6.2 and 2.0 eV, depending on its composition at room temperature. Thus, using these semiconductors, red to UV emitting devices are fabricated. Light emitting devices based on semiconductors have had a huge impact on modern technologies. The brightness and durability of light emitting diodes (LEDs) makes them ideal for displays, while semiconductor laser diodes (LDs) have been widely used from optical communications systems to compact disk players. But these applications have been limited by the lack of materials that can emit blue light efficiently. Full color displays, for example, require at least three primary colors, usually red, green and blue, to produce any visible color. Such a combination is also needed to make a white-light emitting device that would be more durable and less power consuming than conventional incandescent bulbs or fluorescent lamps.

The shorter wavelength means that the light can be focused more sharply, which would increase the storage capacity of magnetic and optical disks. Digital versatile disks which came onto the market in 1996, rely on red AlInGaP semiconductor lasers and have a data capacity of about 4.7 Gbytes, compared to 0.65 Gbytes for compact disks. By moving to violet wavelengths using III-V Nitride-based semiconductors, the capacity could be increased to 15 Gbytes. The violet LDs could also improve the performance of laser printers and underwater optical communications. At the end of 1995, the first current injection III-V Ntride-based LD was reported by the group of present author [1,2]. Since that time, 8 groups have reported pulsed or continuous-wave (CW) operation [3–10] of III-V Nitride-based LDs with an active layer of InGaN multi quantum well (MQW) structure at room temperature (RT). Thus, considerable progess has been achieved in the field of InGaN-based LEDs and LDs recently. In this paper, the present status and progress of the InGaN-based LEDs and LDs are described.

2 Ultraviolet, Blue, and Green LEDs

2.1 InGaN-Based LEDs

Major developments in wide-gap III-V Nitride semiconductors have recently led to the commercial production of high-power LEDs, operating in the UV, blue, and green regimes [1,11,12] and to the demonstration of violet laser light emission at RT in InGaN/GaN/-AlGaN-based heterostructures under pulsed and CW operations [1–10]. All of these light emitting devices use an InGaN active layer instead of a GaN active layer because it is difficult to fabricate a highly effcient light emitting device using a GaN active layer, the reason for which is yet unknown in detail. InGaN/AlGaN double heterostructure LEDs with an output power of 1 mW and an emission wavelength of 400 nm (UV) were reported using an InGaN active layer [13–15]. Only Akasaki et al. [16], and Akasaki and Amano [17] have reported high-effciency UV GaN/AlGaN double heterostructure LEDs with an external quantum efficiency of 1.5% and an emission wavelength of 370 nm using a GaN active layer. Here, the recent performance of InGaN QW structure LEDs operating in the UV, blue, and green regimes are described.

2.2 Structure of the LEDs

The III-V Nitride films were grown by the two-flow metalorganic chemical vapor deposition (MOCVD) method. Details of the two-flow MOCVD are described elsewhere [1]. The growth was conducted at atmospheric pressure. Sapphire with (0001) orientation (C face), of two inch diameter, was used as a substrate. The green LED device structures consist of a 300 Å GaN buffer layer grown at a low temperature (550 °C), a 4 μm thick layer of n-type GaN:Si, a 30 Å thick active layer of undoped $In_{0.45}Ga_{0.55}N$, a 1 000 Å thick layer of p-type $Al_{0.2}Ga_{0.8}N$:Mg, and a 0.5 μm thick layer of p-type GaN:Mg, as shown in Fig. 1 [11]. The active region forms a single quantum well (SQW) structure consisting of a 30 Å $In_{0.45}Ga_{0.55}N$ well layer sandwiched by 4 μm n-type GaN and 1 000 Å p-type $Al_{0.2}Ga_{0.8}N$ barrier layers. The In composition of the $In_{0.45}Ga_{0.55}N$ well layer was changed to nearly 0 and 0.2 for UV and blue LEDs, respectively [11,12]. Also, for UV LEDs, $Al_{0.2}Ga_{0.8}N$ barrier layer was used for both of n- and p-type barrier layers. Fabrication of LED chips was accomplished as follows. The surface of the p-type GaN layer was partially etched until the n-type GaN layer was exposed. Next, an Ni/Au contact was evaporated onto the p-type GaN layer and a Ti/Al contact onto the n-type GaN layer, as shown in Fig. 1. The wafer was cut into rectangles (350 μm × 350 μm). These chips were set on a lead frame, and were then molded. The characteristics of LEDs were measured under direct current (DC)-biased conditions at RT.

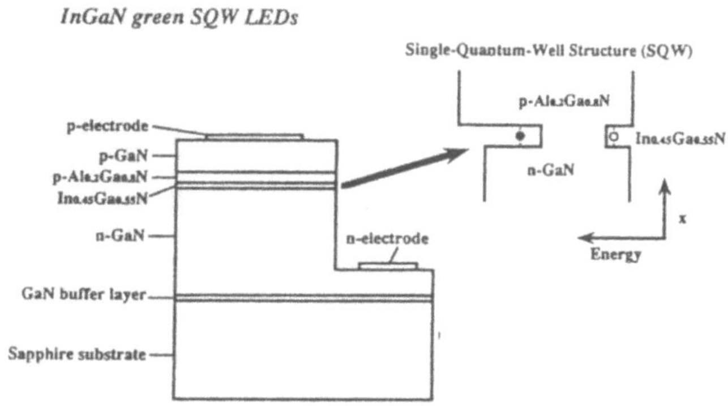

Figure 1 The structure of green InGaN SQW LEDs.

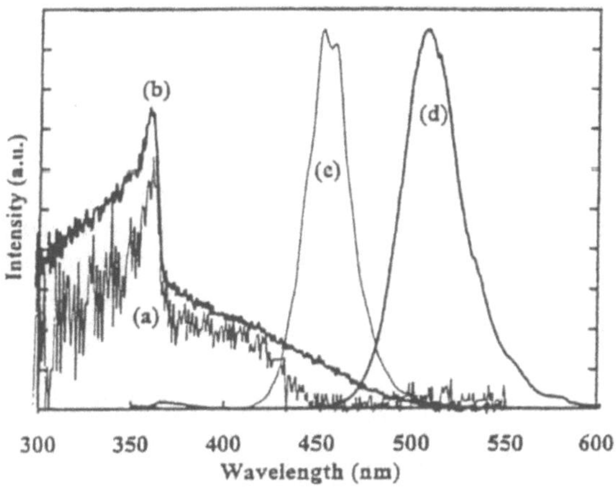

Figure 2 PC and EL spectra of the blue and green InGaN SQW LEDs. (a) PC of the blue, (b) PC of the green, (c) EL of the blue and (d) EL of the green InGaN SQW LEDs. The EL was observed at a forward current of 20 mA at RT.

2.3 Characteristics of UV/Blue/Green LEDs

The photocurrent (PC) spectra of the blue/green InGaN SQW structure LEDs were measured [18]. Monochromatic light was focused on the epitaxial wafer of the device in order to determine the PC spectrum. Figure 2 shows the PC and electro luminescence (EL) spectra of the blue and green InGaN SQW LEDs. The EL of the LEDs was measured at a current of 20 mA at room temperature. The peak wavelengths of the EL of the blue and green InGaN SQW LEDs were 453 nm (2.74 eV) and 520 nm (2.39 eV), respectively, as shown in spectra (c) and (d). Both, the PC spectra (a) and (b) show a strong peak at a wavelength around 360 nm (3.44 eV) which is due to absorption in the thick GaN layers. In the PC spectra there are also shoulder-like peaks at 410 nm (3.03 eV) for the blue (a) and 420 nm (2.96 eV) for the green SQW LEDs (b), respectively. These shoulder-like peaks are probably due to the absorption at the $n = 1$ electron heavy-hole excitonic transition of the quantum energy level in the InGaN well layer [19–22]. The Stokes shifts of energy differences between the absorption due to the $n = 1$ quantum energy state and the emission of the blue and green SQW LEDs were 290 meV and 570 meV, respectively. When the In-composition ratio of InGaN well layer increases, In-composition fluctuation

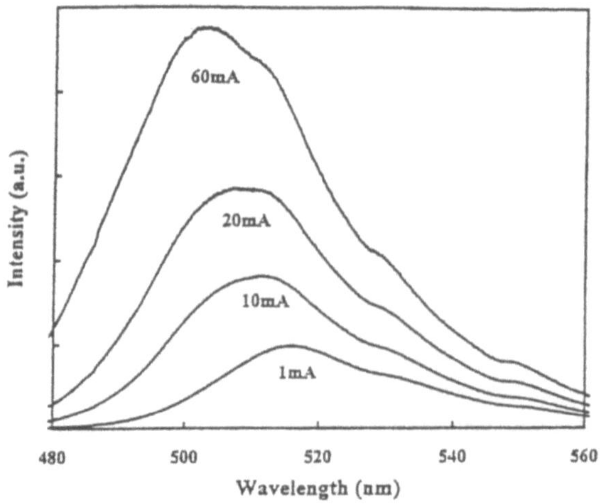

Figure 3
EL of green SQW LEDs with various forward currents.

increases also and a low energy tail of the absorption spectra is formed due to an InGaN-phase separation during the growth [23]. Then, the width of the PC spectrum of green LEDs (b) becomes broader and the peak wavelength is almost the same for the blue and green LEDs. All of the EL appears at this low energy tail of the absorption spectra. These results indicate that the EL of blue and green LEDs originates from a carrier recombination into strongly localized states with energies of 290 meV and 570 meV, respectively. The blue shift of the EL of the green SQW LEDs with increasing forward current (Fig. 3) is probably due to a band-filling effect of the local potential minima of the potential fluctuation.

Because of the low symmetry, wurtzite systems such as GaN-based materials display pyroelectric and piezoelectric behaviors [24–27]. The macroscopic polarization in the material comprising the active region of the SQW or MQW gives rise to a net electric field perpendicular to the plane of the well. This field, if strong enough, will induce a spatial separation of the electron and hole wave functions in the well. As the charge density concentrates near the walls of the well, the wave function overlap decreases and the interband recombination rate is reduced (Stark effect). The piezoelectric field strength in 1% strained InGaN QW was estimated to be of the order of MV/cm [27], and the Stark shift in 3 nm thick QWs was estimated to be 26 meV and 274 meV for fields of 1 MV/cm and 4 MV/cm [19], respectively. In Figs. 3, the blue shifts of the ELs of the green SQW LEDs with increasing forward current may be explained only by the Stark effects resulting from the piezoelectric field induced by the residual strain of 1%. However, the EL spectra showed spectral broadening with a blueshift when increasing the forward currents. This phenomenon cannot be explained by the Stark effect. Therefore, the blueshift in the EL peak energy may be due to a band-filling of the local potential minima of the potential fluctuation. Also, the higher efficiency of the LEDs when increasing the strain of the SQW, which is due to increasing the In content of the wells, was obsezved, as mentioned below. The localization induced by the composition fluctuations overcome the intrinsic limitations due to the piezoelectric field [26].

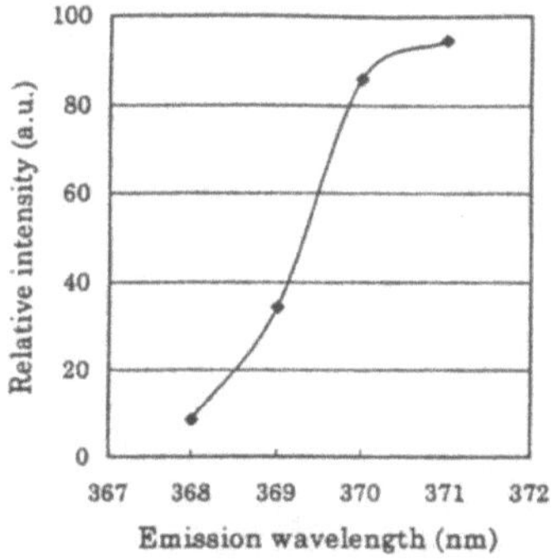

Figure 4
The output power of the UV LEDs as a function of the emission wavelength with a different In mole fraction in the active layer.

The typical output power and external quantum effciency of the UV LEDs were 5 mW and 7.5%, which were about 5 times higher than those of previously reported UV InGaN/AlGaN double heterostructure LEDs at a current of 20 mA [12]. The peak wavelength and the full-width at half-maximum (FWHM) of the emission spectra were 371 nm and 8.6 nm, respectively. Figure 4 shows the output power as a function of the emission wavelength [12]. The output power of the UV LEDs containing a small amount of In in the active layer, with the emission wavelength of 371 nm was about 10 times higher than that containing no In with the emission wavelength of 368 nm. Thus, high-power UV LEDs can be obtained only when using the InGaN active layer instead of the GaN active layer. This is probably related to the eregetically deep localized states which were caused by the In composition fluctuations of the InGaN active layer due to a phase separation during growth [19–22]. Chichibu et al. [19,22] and Narukawa and co-workers [20,21] observed a large Stokes shift of energy differences between the absorption and emission energies of the InGaN quantum energy levels, using the photovoltage (PV) method, on the InGaN SQW LEDs and MQW structure LDs. They attributed the high efficiency of InGaN-based LEDs and LDs, in spite of a large number of dislocation densities of $1 - 10 \times 10^{10}$ cm^{-2}, to the energetically deep localized states in the InGaN well layers which acted like quantum dot energy states. Without In in the active layer, there are no In composition fluctuations that form the deep localized states in the InGaN active layer. The Stark effect resulting from the piezoelectric field due to the strain becomes dominant without In in the active layer. This field, if strong enough, will induce a spatial separation of the electron and hole wave functions in the well. As the charge density concentrates near the walls of the well, the wave function overlap decreases and the interband recombination rate is reduced in the GaN well layer. Thus, the efficiency of the UV LEDs have to become extremely low when the active layer is GaN. In spite of these results, Akasaki et al. [16], and Akasaki and Amano [17] reported high-efficiency UV GaN/AlGaN double heterostructure LEDs with an external quantum efficiency of 1.5%. The spectral width of the emission observed by them was as broad as 30 nm at a peak wavelength of 370 nm.

Figure 5 shows the external quantum efficiencies of UV, blue and green InGaN SQW LEDs as a function of the emission wavelength. The external quantum efficiencies were 7.5% at 371 nm (UV), 11.2% at 468 nm (blue) and 11.6% at 520 nm (green), respectively, which were the highest values ever reported for the LEDs with those emission wavelengths. The luminous efficiencies of blue and green LEDs were 5 lm/W and 30 lm/W, respectively. The luminous efficiency of the red AlInGaP LEDs are around 20–30 lm/W. The luminous efficiency of the white conventional incandescent bulb lamp is around 10 lm/W.

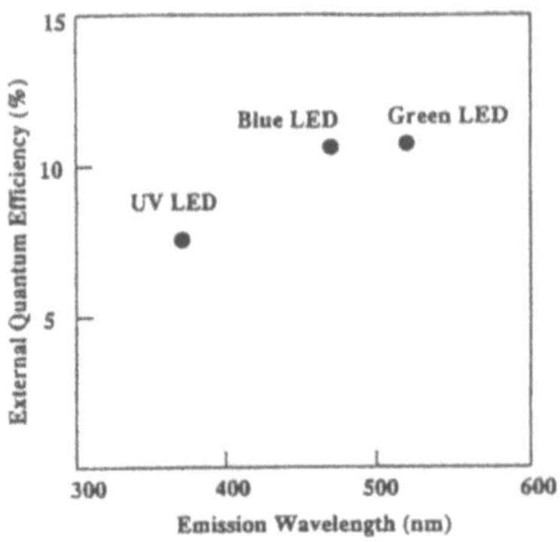

Figure 5
The external quantum efficiencies of UV, blue and green InGaN SQW LEDs as a function of the emission wavelengths.

By combining the blue, green and red LEDs, we can fabricate white LEDs with a luminous efficiency of 30 lm/W which is almost same as that of conventional incandescent bulb lamps. The lifetime of the LEDs is more than 100 000 hours which is much longer than that of the bulb lamps. Thus, we have to replace the conventional bulb lamps with the Nitride-based blue, green LEDs and AlInGaP-based red LEDs considering saving energies and resources.

3 InGaN/GaN/AlGaN-Based Violet Laser Diodes

3.1 Dislocation Free GaN Layers

The lifetimes of the III-V nitride-based LDs with an active layer of InGaN MQW sucture have recently improved to 10 000 hours under conditions of CW operation at room temperature [28]. Before this achievement, the lifetime of the LDs with the structure of InGaN-MQW/GaN/AlGaN separate confinement heterostructure (SCH) was only 300 hours [29]. The improvement of the lifetime of the LDs were performed using new structures and substrates for the LDs [28,30] . One of the problems was that it was difficult to grow the thick AlGaN cladding layer required for optical confinement, due to the formnation of cracks during growth. These cracks are due to the stress introduced in the AlGaN cladding layers caused by lattice mismatch, and the difference in thermal expansion coefficients between the AlGaN cladding layer and GaN layers. In the case of a thin AlGaN layer, the elastic strain is not relieved by the formadon of cracks and dislocations, thus the crystal quality of the AlGaN cladding layer improves. Here, the InGaN MQW-structure LDs, which have GaN/AlGaN modulation-doped strained-layer superlattices as cladding layers within the range of critical thickness instead of thick AlGaN layers are described. As a substrate, the epitaxially laterally overgrown GaN (ELOG) on sapphire was used to reduce the number of threading dislocations of the GaN epilayer.

First, selective growth of GaN was performed on a 2 μm thick GaN layer grown on a (0001) C-face sapphire substrate under a low pressure of 100 Torr. The 2 μm thick SiO_2 mask was patterned to form 3 μm wide stripe windows with a periodicity of 13 μm in the GaN $\langle 1\bar{1}00 \rangle$ direction. Following the 20 μm thick GaN growth on the SiO_2 mask pattern, the coalescence of the selectively grown GaN enabled to achieve a flat GaN surface over

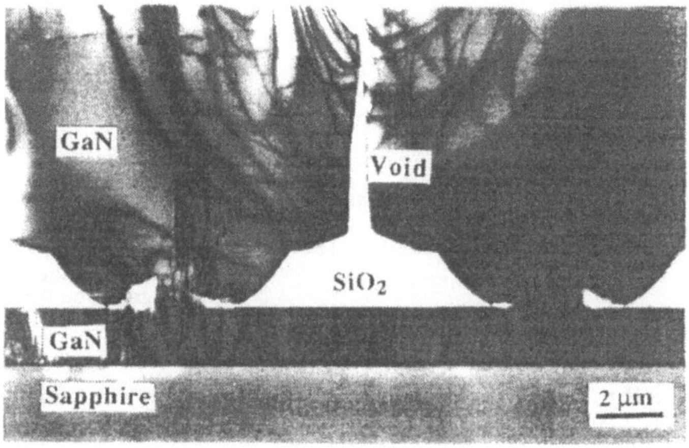

Figure 6
Cross-sectional transmission electron micrographs (TEM) of the laterally overgrown GaN layer on a SiO_2 mask and window area.

the entire substrate. This coalesced GaN was denoted as epitaxically laterally overgrown GaN.

Figure 6 shows the cross-sectional transmission electron microscopy (TEM) images of the ELOG substrate. Threading dislocations, originating from the GaN/sapphire interface, propagate to the regrown GaN layer within the window regions of the mask. It is important to note that the dislocations extend to only just above the window areas. In contrast, there were no observable threading dislocations in the overgrown layer. However, very few short edge-on dislocation segments parallel to the interface plane were observed in the GaN layer on the SiO_2 mask area. These dislocations were parallel to the (0001) plane via the extension of the vertical threading dislocations after a 90° bend in the regrown region. These dislocations did not subsequently propagate to the surface of the overgrown GaN layers. We examined the defect density by plan-view TEM observation of the surface of the ELOG substrates. The concentration of dislocations on the SiO_2 mask area was almost zero within the area of 10 μm $\times$ 10 μm, and that on the window area was approximately 1×10^7 /cm^2.

3.2 Structure of InGaN/GaN/AlGaN-Based Violet Laser Diodes

The InGaN MQW-structure LD was grown on the above mentioned ELOG substrate [28,30]. The InGaN MQW structure LD consisted of a 3 μm thick layer of n-type GaN:Si, a 0.1 μm thick n-type $In_{0.1}Ga_{0.9}N$:Si, a $Al_{0.14}Ga_{0.86}N$/GaN MD-SLS cladding layer consisting of 120 25 Å thick Si-doped GaN separated by 25 Å thick undoped $Al_{0.14}Ga_{0.86}N$ layers, a 0.1 μm thick layer of Si-doped GaN, an $In_{0.15}Ga_{0.85}N/In_{0.02}Ga_{0.98}N$ MQW structure consisting of four 35 Å thick Si-doped $In_{0.15}Ga_{0.85}N$ well layers forming a gain medium separated by 105 Å thick Si-doped $In_{0.02}Ga_{0.98}N$ barrier layers, a 200 Å thick layer of p-type $Al_{0.2}Ga_{0.8}N$:Mg, a 0.1 μm thick layer of Mg-doped GaN, a $Al_{0.14}Ga_{0.86}N$/-GaN MD-SLS cladding layer consisting of 120 25 Å thick Mg-doped GaN separated by 25 Å thick undoped $Al_{0.14}Ga_{0.86}N$ layers and a 0.05 μm thick layer of p-type GaN:Mg. The 0.1 μm thick n-type and p-type GaN layers were light-guiding layers. The n-type and p-type $Al_{0.14}Ga_{0.86}N$/GaN MD-SLS layers acted as cladding layers for confinement of the carriers and the light emitted from the active region of the InGaN MQW structure. The 0.1 μm thick n-type $In_{0.1}Ga_{0.9}N$ served as a buffer layer for the thick AlGaN film growth to prevent cracking. The structure of the ridge geometry InGaN MQW LD was almost the same as that described previously [1]. The LD structure grown on the ELOG substrate

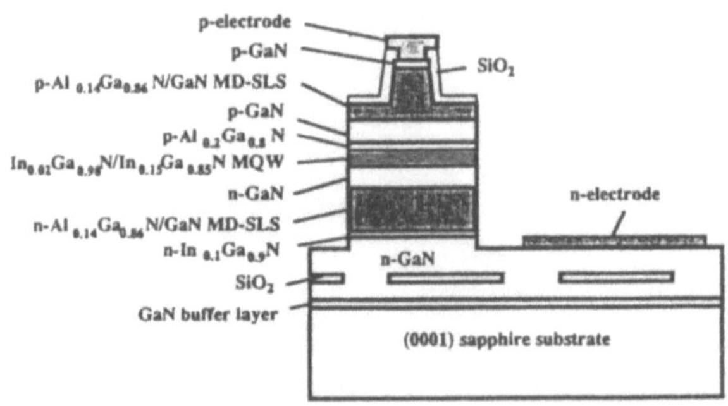

Figure 7
The structure of the InGaN MQW-structure LDs with MD-SLS cladding layers grown on the ELOG substrate.

is shown schematically in Fig. 7. Next, the surface of the p-type GaN layer was partially etched until the n-type GaN layer and the p-type $Al_{0.14}Ga_{0.86}N$/GaN MD-SLS cladding layer were exposed, to form a ridge geometry LD [1]. The laser cavity was formed parallel to the direction of the SiO_2 stripe. The ridge geometry LDs were fabricated with an area of $4 \times 450\,\mu m$. A mirror facet was also formed by dry etching. High-reflection facet coatings (50%) consisting of 2 pairs of quarter wave TiO_2/SiO_2 dielectric multilayers were used to reduce the threshold current. A Ni/Au contact was evaporated onto the p-type GaN layer, and a Ti/Al contact was evaporated onto the n-type GaN layer. The electrical characterisbcs of the LDs fabricated in this manner were measured under DC conditions at room temperature.

3.3 Characteristics of the LDs

Figure 8 shows typical voltage-current (V-I) characteristics and the light output power per coated facet of the LD as a function of the forward DC current (L-I) at room temperature. No stimulated emission was observed up to a threshold current of 53 mA, which corresponded to a current density of 3 kA/cm^{2}2.

Figure 9 shows the results of a lifetime test for CW-operated LDs carried out at 20 °C. The operating current is shown as a function of time for a constant output power of 2 mW per facet controlled using an autopower controller (APC). After 2 200 hours of operation, the operating current is still increasing gradually with increasing operating time. The LDs is still surviving after 2 200 hours of operation. The lifetime of some of the LDs was estimated to be longer than 10 000 hours from the degradation speed. The degradation speed

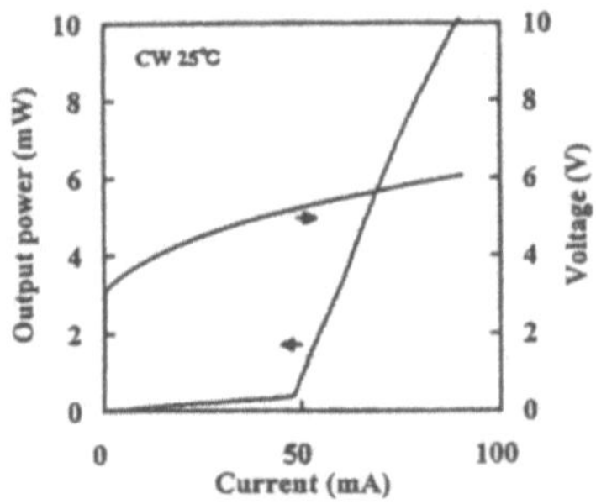

Figure 8
Typical L-I and V-I characteristics of InGaN MQW LDs measured under CW operation at RT.

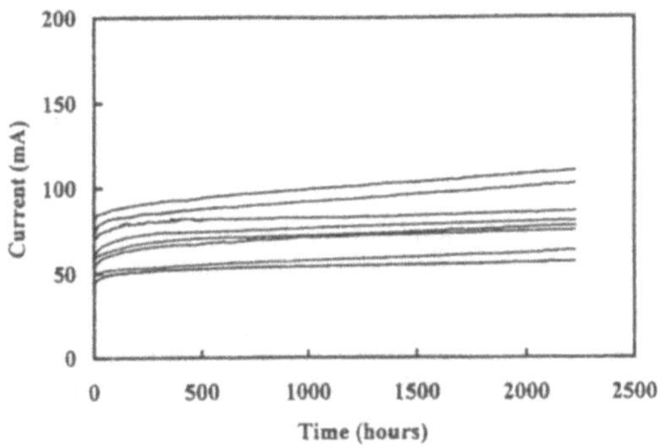

Figure 9
Operating current as a function of time under a constant output power of 2 mW per facet controlled using an autopower controller. The InGaN MQW LDs with MD-SLS cladding layers grown on the ELOG substrate were operated under DC at 20 °C.

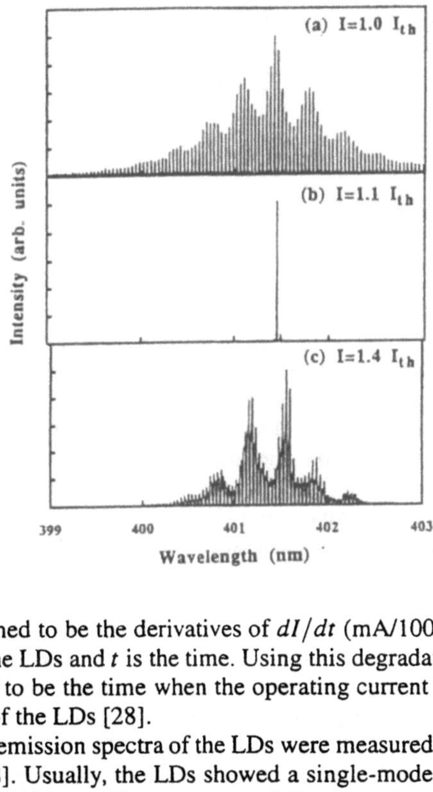

Figure 10
Laser emission spectra measured under RT CW operation with currents of (a) $I = 1.0I_{th}$ (b) $I = 1.1I_{th}$, and (c) $I = 1.4I_{th}$.

was defined to be the derivatives of dI/dt (mA/100 hours), where I is the operating current of the LDs and t is the time. Using this degradation speed, the estimated lifetime was obtained to be the time when the operating current became twice of the initial operating current of the LDs [28].

Next, emission spectra of the LDs were measured under CW operation at room temperature [28]. Usually, the LDs showed a single-mode stimulated emission at a wavelength of around 400 nm. However, some LDs showed a self pulsation with increasing the DC forward currents, as shown in Fig. 10. At $I = 1.0I_{th}$, where I is the operating current and I_{th} is the threshold current, longitudinal modes with many sharp peaks with a peak separation of 0.04 nm were observed (Fig 10a). At $I = 1.1I_{th}$, the single-mode emission at a wavelength of 401.4 nm was observed (Fig. 10b). At currents between $I = 1.3I_{th}$ and $1.7I_{th}$, however, the multi-mode emission was observed due to a self-pulsation with wide linewidths (Figs. 10c and Fig. 11) [28].

Next, this InGaN MQW-structure LD was modulated using a pulsed current (pulse width: 15.5 ns, pulse duty ratio: 1%) [28]. Figure 11 shows the modulation current (a) and the corresponding self pulsation (b) of the lasing light output modulated by a pulsed

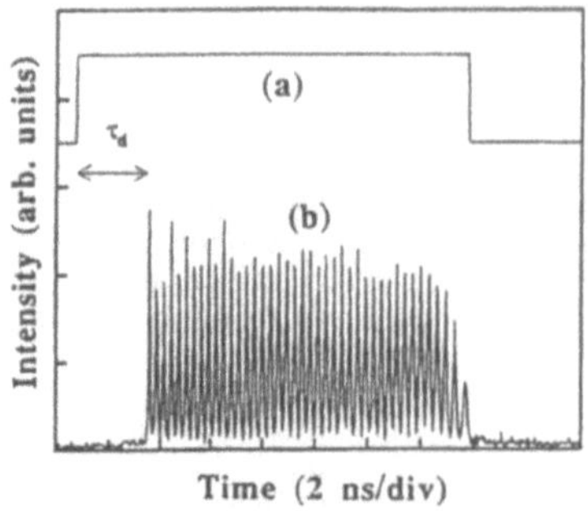

Figure 11
The modulation current (a) and the corresponding self-pulsation (b) of the lasing light output modulated by a pulsed current with a current of $I = 1.4I_{th}$. The τ_d shows the delay time of the laser emission just after the supply of the pulsed current.

current with a current of $I = 1.4I_{th}$. In the figure, the delay time τ_d of the laser emission just after the supply of the pulsed current is also shown. The self-pulsation is due to a saturable absorber in the wave guide region of the LDs. The frequency of the self-pulsation was measured as 3.5 GHz, (Fig. 11b).

4 Laser Diodes Grown on GaN Substrate

4.1 GaN Substrate

Now LDs grown on GaN substrate are described. These are based on homoepitaxy of GaN. The estimated lifetimes of the InGaN MQW LDs have been increased to more than 10 000 hours under CW operation at room temperature as mentioned above [28]. When sapphire substrates were used, it is diffcult to obtain cleaved mirror facets which were used for the cavides of conventional LDs. In addition, the thermal conductivity of sapphire (0.5 W/(cm·K)) is not as high as that of GaN (1.3 W/(cm·K)) for the heat dissipation generated by the LDs [31]. Here, the LDs grown on pure GaN substrates which are easily cleaved and have a high thermal conductivity, are discussed.

First, 20 μm thick ELOG was grown on the sapphire substrate as mentioned above. After obtaining the 20 μm thick ELOG substrate, the GaN growth was continued up to 100 μm thickness. After the 100 μm thick GaN growth, the sapphire substrate was removed by polishing in order to obtain a pure GaN substrate with a thickness of approximately 80 μm. The InGaN MQW LD structure was then grown on the surface of the 80 μm thick GaN substrate. The InGaN MQW LD structure was identical to that described in 3.2 except for the n-type $Al_{0.14}Ga_{0.86}N$/GaN MD-SLS cladding layer. The thickness of the n-type cladding layer was increased from 0.6 μm to 1.2 μm in order to reduce the penetration of the laser light to the 3 μm thick n-type GaN and the GaN substrate, which causes extra multi-modes in the far-field patterns (FFP) by coupling with the laser waveguide [32,33].

4.2 Characteristics of Laser Diodes Grown on GaN Substrates

The LDs were fabricated on the ELOG on sapphire and on the pure GaN substrate for a comparison. Between both LDs, there were no big differences on the *V*-*I* and *L*-*I* curves.

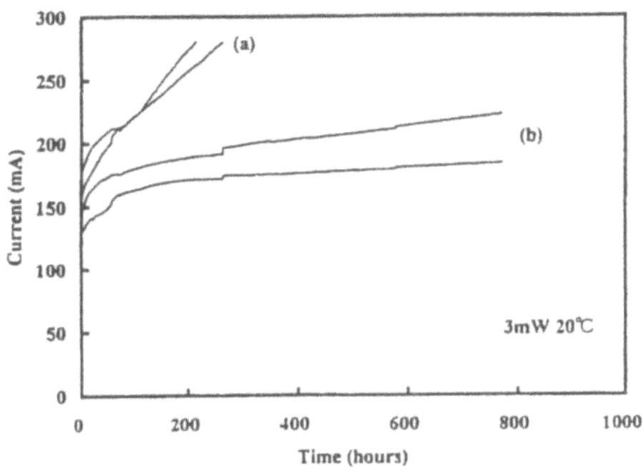

Figure 12 Operating current as a function of time under a constant output power of 3 mW per facet controlled using an autopower controller. The LDs grown on the sapphire substrate (a) and the GaN substrate (b) were operated under DC at RT.

No stimulated emission was observed up to a threshold current of 110 mA, which corresponds to a current density of 7 kA/cm^2. The operating voltage at the threshold current was 6 V. The operating voltage and threshold current density were relatively high since the growth conditions on an 80 μm thick GaN substrate have not been optimized. Figure 12 shows the results of a lifetime test of CW-operated LDs carried out at room temperature, in which the operating current is shown as a function of time under a constant output power of 3 mW per facet. A small increase in the operating current is observed with increasing operating time. Figure 12a shows the LDs grown on the 10 μm thick ELOG on the sapphire substrate with the same laser structure and the same threshold current density. The thickness of the sapphire was approximately 70 μm. The lifetime of the LDs was approximately 200 hours. However, the lifetime of the LDs grown on the GaN substrate was longer than 780 hours despite the large threshold current density of 7 kA/cm^2. The thermal resistance of the LDs was estimated by measuring the change in the threshold voltage under the pulsed current. The thermal resistance of the LDs grown on the GaN substrate was as low as 30 °C/W due to the high thermal conductivity of GaN (1.3 W/(cm·K)) [31]. The thermal resistance of the LDs grown on the ELOG on the sapphire substrate was about 60 °C/W due to the small thermal conductivity of the sapphire (0.5 W/(cm·K)) [31]. Probably, these differences in thermal resistance caused the difference of the lifetime of the LDs.

The best LD showed a threshold current of 70 mA, which corresponded to a threshold current density of 5 kA/cm^2, that was the lowest value ever obtained on the GaN substrate. The operating voltage at the threshold current was 6 V. The slope effciency was 0.3 W/A per facet. Figure 13 shows the results of a lifetime test of this CW-operated LD carried out at room temperature, in which the operating current is shown as a function of time under a constant output power of 5 mW per facet. After 100 hours of operation, the operating current was almost constant with increasing operating time. The LDs still survive after 290 hours of operation. From the degradation speed, the lifetime was estimated to be as long as approximately 10 000 hours.

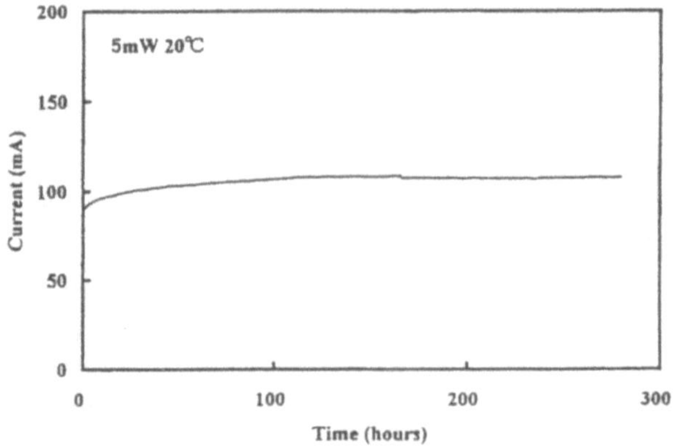

Figure 13
Operating current as a funcdon of time under a constant output power of 5 mW per facet controlled using an autopower controller. The LD grown on the GaN substrate was operated under DC at RT.

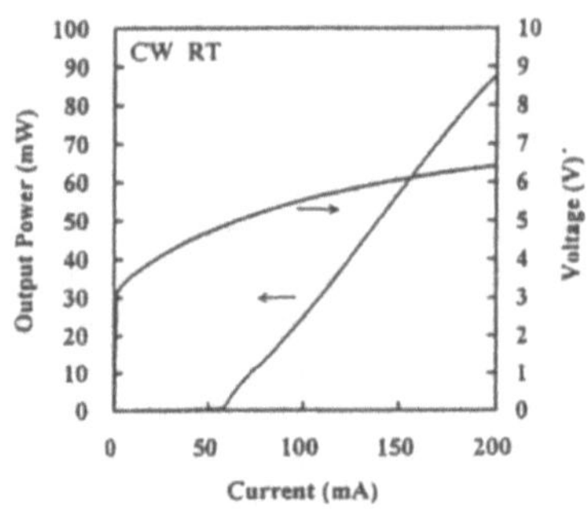

Figure 14
L-I and *V-I* characteristics of InGaN MQW LDs with a cleaved mirror facet measured under CW operation at RT. The output power was measured from the uncoated cleaved mirror facet.

When we cleaved one facet along $\{1\bar{1}00\}$ of the LDs grown on the GaN substrate, the cleaved mirror facet was obtained easily. The cleaved facet was much smoother than the etched mirror facet which was used for the previous LDs [1,2]. The facet at the opposite side was kept unchanged with a coated etched facet (50%). Figure 14 shows the typical *V-I* and *L-I* characteristics per uncoated cleaved mirror facet of the LD grown on the GaN substrate under CW operation at room temperature. The slope efficiency was as high as 0.6 W/A per facet because the output power was measured from the cleaved mirror facet with a low reflectivity. At a current of 200 mA, an output power of 90 mW was obtained, which was the highest value ever reported for III-V nitride-based LDs under RT-CW operation.

The measurement of the FFP was performed using an FFP measuring system (A3267: HAMAMATSU) which is composed of a charge coupled device (CCD) camera with an f-Θ lens, field lens and relay lens in front of the laser facet. The FFP of the LDs with the cleaved mirror facets was measured in the planes parallel and perpendicular to the junction, as shown in Fig. 15. At an output power of 5 mW, the FFP in the direction parallel to the epitaxial layers collapsed to 8°; in the other direction, the FFP extended to 31°. The aspect ratio was approximately 4.

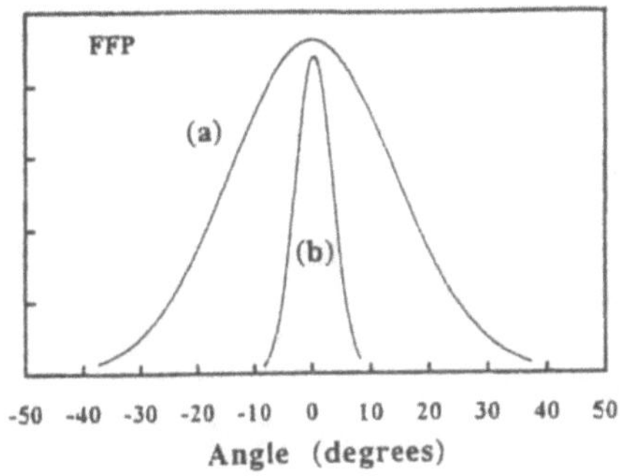

Figure 15
(a) Perpendicular and (b) parallel FFPs of the InGaN MQW LDs grown on the GaN substrate at an output power of 5 mW under RT-CW operation.

5 Conclusions

UV/blue/green InGaN QW structure LEDs with external quantum efficiencies of 7.5% at 371 nm (UV), 11.2% at 468 nm (blue) and 11.6% at green were developed. The blue and green InGaN QW structure LEDs with luminous efficiencies of 5 lm/W and 30 lm/W can be used for fabricating white LEDs with a luminous efficiency of 30 lm/W which is almost identical to that of conventional incandescent bulb lamps. The localized energy states caused by In composition fluctuation in the InGaN active layer are related to the high efficiency of the InGaN-based LEDs. InGaN MQW LDs grown on a GaN substrate were demonstrated for the first time. The LDs with an output power of 5 mW exhibited a lifetime of more than 290 hours and an estimated lifetime of 10 000 hours despite a relatively large threshold current density. The FFP of the LDs with a cleaved mirror facet revealed a single mode emission without any interference effects. If the threshold current density of the LDs grown on the GaN substrate could be reduced to approximately 1–4 kA/cm^2, the characteristics of the LDs would be improved further. The best substrate for GaN growth is GaN.

Bibliography

[1] For a review, see S. Nakamura and G. Fasol, The blue laser diode (Springer-Verlag, Heidelberg, Germany, 1997) 1st ed.

[2] S. Nakamura, M. Senoh, S. Nagahama, N. Iwasa, T. Yamada, T. Matsushita, H. Kiyoku, and Y. Sugimoto, Jpn. J. Appl. Phys. **35**, L74 (1996).

[3] S. Nakamura, M. Senoh, S. Nagahama, N. Iwasa, T. Yamada, T. Matsushita, Y. Sugimoto, and H. Kiyoku, Appl. Phys. Lett. **69**, L4056 (1996).

[4] K. Itaya, M. Onomura, J. Nishino, L. Sugiura, S. Saito, M. Suzuki, J. Rennie, S. Nunoue, M. Yamamoto, H. Fujimoto, Y. Kokubun, Y. Ohba, G. Hatakoshi, and M. Ishikawa, Jpn. J. Appl. Phys. **35**, L1315 (1996).

[5] G. E. Bulman, K. Doverspike, S. T. Sheppard, T. W. Weeks, H. S. Kong, H. M. Dieringer, J. A. Edmond, J. D. Brown, J. T. Swindell, and J. F. Schetzena, Electron. Lett. **33**, 1556 (1997).

[6] M. P. Mack, A. Abare, M. Aizcorbe, P. Kozodoy, S. Keller, U. K. Mishra, L. Coldren, and S. DenBaars, MRS Internet J. Nitride Semicond. Res. **2**, 41 (1997). (Available from http://nsr.mij.mrs.org/2/41/)

[7] A. Kuramata, K. Domen, R. Soejima, K. Horino, S. Kubota, and T. Tanahashi, Jpn J. Appl. Phys. **36**, L1130 (1997).

[8] F. Nakamura, T. Kobayashi, T. Asatsuma, K. Funato, K. Yanashima, S. Hashimoto, K. Naganuma, S. Tomioka, T. Miyajima, E. Morita, H. Kawai, and M. Ikeda, J. Crystal Growth **189/190**, 841 (1998).

[9] M.Kneissl, D.P. Bour, N.M. Johnson, L.T. Romano, D.S. Kruser, R. Donaldson, J. Walker, and C.D. Dunnrowicz, Appl. Phys. Lett **72**, 1539 (1998).

[10] H. Katoh, T. Takeuchi, C. Anbe, R. Mizumoto, S. Yamaguchi, C. Wetzel, H. Amano, I. Akasaki, Y. Kaneko, and N. Yamada, Jpn. J. Appl. Phys **37**, L444 (1998).

[11] S. Nakamura, M. Senoh, N. Iwasa, S. Nagahama, T. Yamada and T. Mukai, Jpn. J. Appl. Phys. **34**, L1332 (1995).

[12] T. Mukai, D. Morita, and S. Nakamura, J. Crystal Growth **189/190**, 778 (1998).

[13] S. Nakamura, M. Senoh and T. Mukai, Jpn. J. Appl. Phys. **32**, L8 (1993).

[14] S. Nakamura, M. Senoh and T. Mukai, Appl. Phys. Lett. **62**, 2390 (1993).

[15] S. Nakamura, Microelectron. J. **25**, 651 (1994).

[16] I. Akasaki, H. Amano, K. Itoh, N. Koide and K. Manabe, Inst. Phys. Conf. Ser. **129**, 851 (1992).

[17] I. Akasaki and H. Amano, Jpn. J. Appl. Phys. **36**, 5393 (1997).

[18] S. Nakamura, IEEE J. Selective Topics in Quantum Electron. **3**, 712 (1997).

[19] S. Chichibu, T. Azuhata, T. Sota, and S. Nakamura, Appl. Phys. Lett. **69**, 4188 (1996).

[20] Y. Narukawa, Y. Kawakami, Sz. Fujita, Sg. Fujita, and S. Nakamura, Phys. Rev. **B55**, 1938R (1997).

[21] Y. Narukawa, Y. Kawakami, M. Funato, Sz. Fujita, Sg. Fujita, and S. Nakamura, Appl. Phys. Lett. **70**, 981 (1997).

[22] S. Chichibu, K. Wada, and S. Nakamura, Appl. Phys. Lett. **71**, 2346 (1997).

[23] I. Ho and G. B. Stringfellow, Appl. Phys. Lett. **69**, 2701 (1996).

[24] D. L. Smith and C. Mailhiot, Phys. Rev. Lett. **58**, 1264 (1987).

[25] S. Chichibu, T. Azuhata, T. Sota, and S. Nakamura, Appl. Phys. Lett. **70**, 2822 (1997).

[26] M. B. Nardelli, K. Rapcewicz and J. Bernholc. Appl. Phys. Lett. **71**, 3135 (1997).

[27] T. Takeuchi, H. Takeuchi, S. Sota, H. Sakai, H. Amano, and I. Akasaki, Jpn. J. Appl. Phys. **36**, L177 (1997).

[28] S. Nakamura, M. Senoh, S. Nagahama, N. Iwasa, T. Yamada, T. Matsushita, H. Kiyoku, Y. Sugimoto, T. Kozaki, H. Umernoto, M. Sano, and K. Chocho, Jpn. J. Appl. Phys. **36**, L1568 (1997).

[29] S. Nakamura, M. Senoh, S. Nagahama, N. Iwasa, T. Yamada, T. Matsushita, Y. Sugimoto, and H. Kiyoku, Jpn. J. Appl. Phys. **36**, L1059 (1997).

[30] S. Nakamura, M. Senoh, S. Nagahama, N. Iwasa, T. Yamada, T. Matsushita, H. Kiyoku, Y. Sugimoto, T. Kozaki, H. Umemoto, M. Sano, and K. Chocho, Appl. Phys. Lett. **72**, 211 (1998).

[31] S. Strite and H. Morkos, J. Vac. Sci. & Technol. **B10**, 1237 (1992).

[32] D. Hofstetter, D. P. Bour, R. L. Thornton, and N. M. Johnson, Appl. Phys. Lett. **70**, 1650 (1997).

[33] I. A. Avrutsky, R. Gordon, R. Clayton, and J. M. Xu, IEEE J. Quantum Electron. **33**, 1801 (1997).

Group III-Nitride Based Blue Emitters

H. Obloh, K.-H. Bachem, D. Behr, U. Kaufmann, M. Kunzer, A. Ramakrishnan, P. Schlotter, M. Seelmann-Eggebert, J. Wagner

Fraunhofer-Institut für Angewandte Festkœrperphysik
Tullastrasse 72, D-79108 Freiburg, Germany

Abstract: (AlGaIn)N-based heterostructures are of considerable current interest for the realization of short-wavelength light-emitting semiconductor devices, covering the UV-to-blue spectral region. Recent advances in group III-nitride layer growth using MBE and, in particular, MOCVD allow the preparation of high-quality n- and p-type conducting GaN layers as well as GaN/(AlGa)N/(InGa)N hetero- and quantum-structures, in spite of the lack of lattice matched substrates. This progress in growth technology enabled the realization, and commercialization, of high-efficiency blue light emitting diodes and the demonstration of a diode laser emitting in the violet spectral range. Despite this tremendous technological progress, important questions regarding basic material properties and growth related issues remain yet to be answered. In the present paper we shall discuss the effect of GaN surface termination on layer quality and surface morphology, the mechanisms limiting the maximum obtainable hole concentration in GaN:Mg, and the effect of compositional inhomogeneity in low In-content (InGa)N on the photoluminescence, reflection, and dielectric function spectrum.

1 Introduction

Research on GaN started already in the early thirties of this century, when this material was synthesized in powder form for the first time [1]. During the following five decades, progress in the preparation of group III-nitrides was slow. Epitaxial layers were first realized using hydride vapor phase epitaxy (HVPE) in the late sixties [2]. On the basis of these layers, many basic physical and material properties of GaN have been established [3]. Since these early studies, GaN and its related compounds (AlGa)N and (InGa)N have been known to be promising materials for applications in optoelectronic devices, operating in the UV-to-blue spectral range. Nevertheless, by the end of the eighties only a few groups worldwide were still active in research on wide band gap group III-nitride semiconductors. The lack of a lattice matched substrate material and the failure to grow p-type conducting GaN prevented major progress. During the eighties, two remarkable breakthroughs in growing device quality group III-nitride semiconductors were achieved. One was the use of a thin AlN nucleation layer deposited at low temperature prior to the growth of GaN on sapphire substrates, which led to a substantial improvement in surface morphology of GaN layers [4,5]. The other was the achievement of p-type conductivity via post-growth treatment of Mg-doped GaN layers by low-energy electron-beam irradiation (LEEBI), as reported by the group of Akasaki [6]. In 1992 Nakamura and coworkers showed that p-type conductivity can also be achieved by post-growth thermal annealing

[7]. Within a few years, this group succeeded in growing high quality n- and p-type GaN layers as well as (InGa)N layers suitable for device fabrication [8–12]. Despite of this enormous progress the scientific community was highly surprised when the group led by Nakamura presented the first GaN-based high-brightness blue light-emitting diode (LED) in 1994 [13]. LEDs with a luminous intensity exceeding 1 cd were generally believed to be impossible with a material like heteroepitaxial GaN, which exhibits a dislocation density of the order of 10^8 cm^{-2} to 10^{10} cm^{-2}. The successes of the Japanese groups revived intensive research activities into group III-nitride materials. Today, only four years after the first generation of high-brightness blue LEDs became commercially available, lasers based on group III-nitrides emitting in the violet have been demonstrated by an increasing number of research groups.

Bulk GaN substrates are still rare and of small dimensions, but thick (> 10 μm) GaN quasi-substrates grown by HVPE on patterned sapphire or SiC substrates have been realized, using the technique of epitaxial lateral overgrowth [14–16]. These quasi-substrates show a strongly reduced defect density as compared to GaN layers grown directly on bare sapphire substrates.

This breath-taking technological progress tends to obscure the fact that many basic physical and material properties of group III-nitrides are still poorly understood. This applies in particular to the role of surface polarity in MOCVD growth of GaN, self-compensation in Mg-doped p-type GaN, and the incorporation of In in (InGa)N films. In the following sections these points will be addressed focusing on wurtzite group III-nitrides, and finally results on GaN-based LEDs will be presented.

2 Growth of GaN

Currently, metal-organic chemical vapour deposition (MOCVD) and molecular beam epitaxy (MBE) are the most promising techniques for group III-nitride growth. The best results reported so far for group III-nitride layers, LEDs, and laser diodes have been achieved by MOCVD on sapphire. The mechanism of heteroepitaxy of GaN on sapphire substrates has been investigated in detail [18–20]. A thin (30–50 nm) AlN or GaN nucleation layer deposited prior to the growth of GaN was found to be crucial in order to overcome the problems associated with both the large lattice mismatch between (00.1) *c*-plane sapphire and wurtzite GaN ($\approx 13.8\%$), and the difference in the thermal expansion coefficients between these two materials.

A key parameter for the growth of wurtzite GaN *with smooth surfaces* is the termination of the surface (by Ga or by N), i.e. the surface polarity. The (00.1) (Ga-terminated) and the $(00.\bar{1})$ (N-terminated) surface are inequivalent and thus behave differently during growth and also differ in chemical stability. A technique proven to be able to determine the polarity of compound semiconductor surfaces is hemispherically scanned X-ray photoelectron diffraction (HSXPD) [21]. Stereographic HSXPD patterns of two different layers 1 and 2, recorded using Ga 3d and N 1s photoelectrons, are shown in Fig 1. The surface of sample 1 (left side) is found to be Ga-terminated and has a smooth appearance, while the N-terminated surface of sample 2 (right side) shows a significantly increased roughness. Extensive atomic force microscope and HSXPD studies revealed a clear correlation between surface morphology and surface termination [22,23]. Surfaces with a mean roughness on an atomic scale (below 1 nm) were found to be typically Ga-terminated. This conclusion is in line with results obtained by transmission electron microscopy [24].

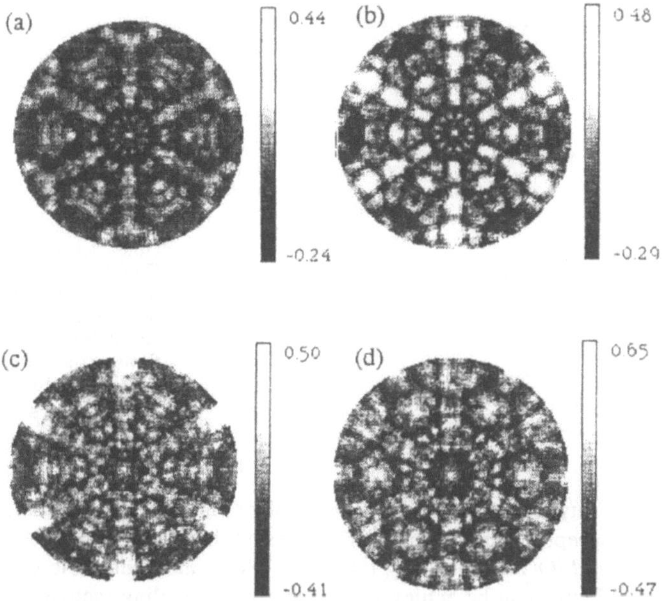

Figure 1 Stereographic HSXPD patterns as measured with Mg Kα excitation for two different GaN epilayers. (a) Ga 3d photoelectrons of sample 1, (b) Ga 3d of sample 2, (c) N 1s photoelectrons of sample 1, (d) N 1s of sample 2. The surface of sample 1 (2) has a smooth (rough) appearance.

According to our experience, an effective means to establish a Ga-terminated surface in the initial growth phase is heating of the sapphire substrate in a hydrogen flow to temperatures around 1 100 °C, prior to the deposition of the low temperature nucleation layer. On the other hand, nitridation of the sapphire surface by exposing it to ammonia at high temperature induces the unfavourable N-polarity. As reported recently [25], a reversal from N- to Ga-termination, and subsequent growth of layers with smooth surfaces, is only possible at the cost of a relatively thick (140 nm) nucleation layer. Therefore, one has to conclude that nitridation of the sapphire substrate is incompatible with the use of a thin ($\leq$ 50 nm) GaN or AlN nucleation layer, required for the growth of device grade active layers. Taken together, to obtain smooth GaN films by MOCVD it is necessary to ensure that Ga-termination is established during the initial stages of layer formation.

As a quick procedure to analyze the surface polarity and thus to distinguish between the two polar orientations of the (00.1)-type wurtzite GaN surface, a KOH based etching process can be used. Solutions of KOH or NaOH selectively attack GaN surfaces with N-polarity, whereas surfaces with Ga-polarity remain completely unaffected even by a longterm exposure to this chemical agent [22].

3 Doping of GaN with Mg for p-Type Conductivity

While n-type doping of GaN with Si is comparatively straightforward, the preparation of p-conducting films is considerably more difficult. So far, Mg is still the only acceptor of

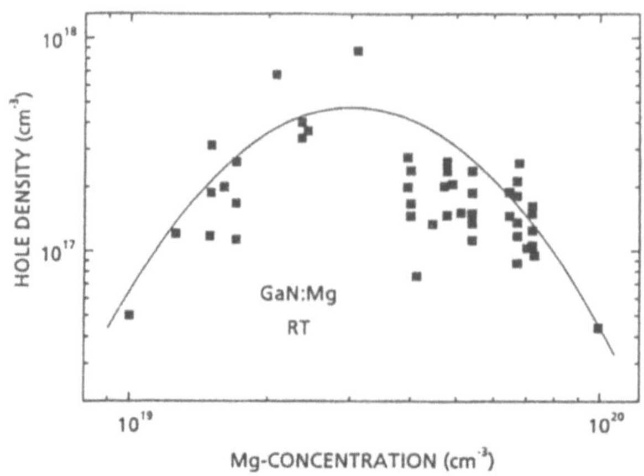

Figure 2
Hole density in GaN:Mg films as measured by Hall effect versus the Mg concentrations of the films as measured by SIMS. The full line is drawn as a guide to the eye.

practical importance for p-type doping of GaN. The problem in achieving p-conduction is the activation of the acceptors, and it has been suggested that they are passivated by hydrogen in the as-grown MOCVD GaN:Mg layers [26]. The maximum hole densities at room temperature (RT) that can be achieved after thermal annealing (activation), are roughly one percent of the Mg concentrations because of the relatively large ionization energy of about 200 meV for the Mg acceptor. For hole densities in the lower 10^{17} cm^{-3} range the mobilities are typically around 10 cm^2/Vs. In Fig. 2 the hole densities in p-type GaN:Mg layers measured by Hall effect after thermal activation of the acceptors, are plotted versus the Mg-concentration as determined by secondary ion mass spectroscopy (SIMS). The data points show that initially the hole density increases with increasing Mg-concentration, reaching maximum values of 5 to 6×10^{17} cm^{-3} at a Mg-concentration of about 3×10^{19} cm^{-3}. For higher Mg-concentrations the hole-density decreases again. This finding is in line with previous observations [9,27] where GaN:Mg layers with the highest Mg-doping levels remain highly resistive even after LEEBI or thermal annealing treatment.

The saturation, and eventual decrease, of the hole-concentration with increasing Mg-content is accompanied by a change in the RT photoluminescence (PL) spectrum. At RT Mg-doped GaN films show two broad PL bands centered around 3.2 eV and 2.8 eV. Which one of these dominates in intensity depends on the Mg concentration and on excitation power density (P_D). Fig. 3 shows normalized PL spectra recorded at high P_D (10 W/mm^2) for layers with Mg contents between 1.5×10^{19} cm^{-3} and 4.8×10^{19} cm^{-3}. A sudden change from the 3.2 eV to the 2.8 eV band occurs at a Mg content of 2.3×10^{19} cm^{-3}. The 3.2 eV band has been assigned to e-A^0 recombination of free electrons with holes bound to isolated Mg_{Ga} acceptors [28]. The assignment of the 2.8 eV band is less straightforward. One of the distinguishing features of this band is its peak shift to higher energy with increasing P_D. This is a characteristic typical for donor-acceptor pair (D-A) recombination involving electrons and holes trapped at distant donors and acceptors, respectively. The 2.8 eV band has therefore been assigned to D-A-recombination involving isolated Mg acceptors and deep donors D_d induced by self-compensation [28]. Based on the results of theory [29] and the analogy with II-VI compounds [30], it appears likely that D_d consists of Mg_{Ga} associated with a nearest-neighbour nitrogen vacancy (V_N). The recombination model shown in Fig. 4 provides a qualitative explanation for the switching between the two Mg related bands in Fig. 3. As long as the Mg content is moderate the concentration

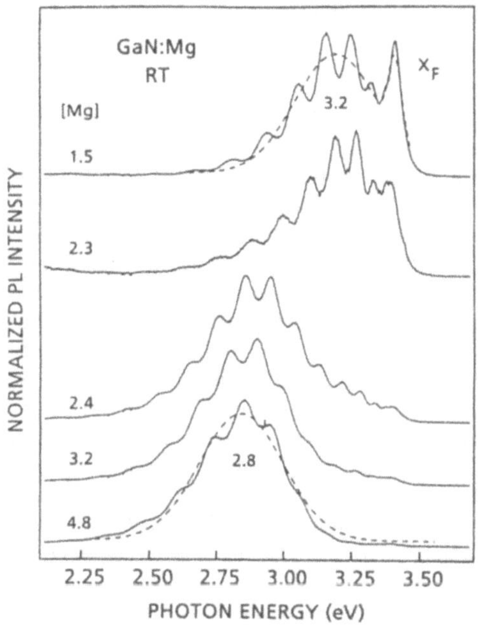

Figure 3
PL spectra at RT for various GaN:Mg films. The numbers to the left give the Mg concentrations in units of 10^{19} cm^{-3}. The spectra reveal the free exciton X_F and the broad Mg induced bands peaked at 3.2 and 2.8 eV. The oscillations superimposed on the broad bands are Fabry-Perot interferences.

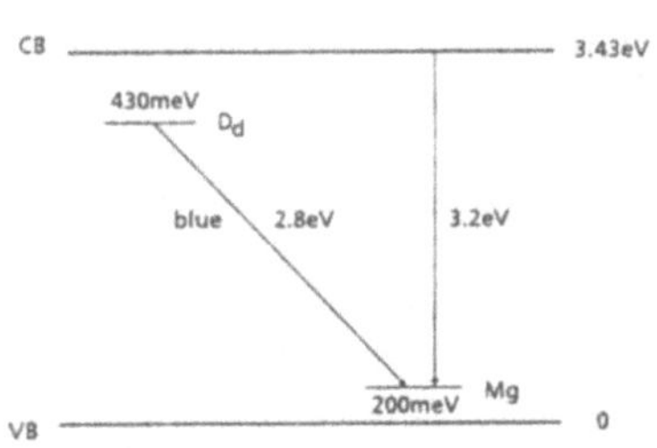

Figure 4
Recombination model for the RT Mg induced PL bands in GaN:Mg. The deep donor D_d is suggested to be formed by self-compensation.

of deep donors D_d (Mg_{Ga}-V_N associates) is low, and the D_d-Mg_{Ga} intrapair separations are relatively large. Consequently, the probability of recombination for trapped electrons and holes is low because of their small wavefunction overlap. Therefore, free electrons will preferentially recombine directly with holes on Mg^0. However, with increasing Mg concentration, the Fermi level moves towards the valence band edge thus increasing the V_N concentration [29]. Therefore the D_d concentration increases and consequently the D_d-Mg_{Ga} intrapair distances decrease. Correspondingly the pair recombination probability increases and eventually exceeds that of the direct e-A^0 recombination process such that the 2.8 eV band dominates. The results presented in Figs. 2 and 3 provide strong evidence for self-compensation in the case of Mg-doping and they suggest that hydrogen passivation is not the only mechanism limiting p-conductivity in GaN:Mg.

4 Growth and Characterization of (InGa)N

The growth of (InGa)N is complicated by the fact that the equilibrium vapor pressure of nitrogen over InN is several orders of magnitude higher compared to that over GaN and AlN. Therefore, high quality (InGa)N layers can only be grown in a nitrogen atmosphere at lower temperatures (typically below 800 °C) than GaN and (AlGa)N, and with large V/III ratios. For growth of (InGa)N layers by MOCVD, TMIn and TMGa are mostly used as precursors. For a given TMIn/(TMGa + TMIn) ratio in the gas phase the In incorporation into the solid increases strongly with decreasing growth temperature. This behavior does not change when the Ga precursor TMGa is replaced by TEGa, which has a lower decomposition temperature. Further difficulties in the growth of the (InGa)N alloy arise from the existence of a solid phase miscibility gap [31], which can give rise to compositional inhomogeneity, and thus local variations in the fundamental gap energy. Direct evidence for a phase separation in $In_xGa_{1-x}N$ layers with $0.28 < x < 0.5$ has been obtained from X-ray diffraction in combination with transmission electron microscopy and selected area diffraction [32]. But also for $In_xGa_{1-x}N$ layers with a lower In content of $x \leq 0.1$, less direct experimental evidence has been obtained for compositional fluctuations from optical spectroscopy, such as a combination of PL, electroluminescence, PL excitation, and electro-absorption spectroscopy [33] and by selectively enhanced Raman scattering [34].

In view of the above mentioned difficulties encountered in the epitaxial growth of (InGa)N layers, it is evident that reliable characterization techniques are required to assess layer quality and In content. High-resolution X-ray diffraction (HRXRD), which is routinely used for the analysis of conventional III/V alloys, has to rely for the analysis of the alloy composition on the applicability of Vegard's law, i.e. on the assumption that the lattice parameter scales linearly with alloy composition x. When dealing with pseudomorphically strained or only partially relaxed epitaxial layers, also the elastic parameters of the material under study have to be known with sufficient accuracy. This condition is certainly not fulfilled for (InGa)N, because the elastic constants of InN are only poorly known. The HRXRD analysis of pseudomorphically strained or partially relaxed layers requires further the measurement of symmetric *and* asymmetric reflections with respect to the growth axis, to confirm pseudomorphic growth in the former and to determine the degree of strain relaxation in the latter case. There is an ongoing controversy regarding the critical layer thickness for pseudomorphic growth of (InGa)N on GaN [35,36], with one report on fully strained pseudomorphic (InGa)N layers even for layer thicknesses exceeding 100 nm [36].

In the present work, the In content x of thin ($\leq$ 60 nm) $In_xGa_{1-x}N$ layers on GaN was independently determined by SIMS and by HRXRD respectively. The SIMS measurements were calibrated by energy dispersive X-ray (EDX) analysis on thick ($\approx$ 400 nm) (InGa)N layers. The HRXRD results were analyzed assuming either complete strain relaxation of the (InGa)N layers or fully strained pseudomorphic growth. Good agreement was found between the In content determined by SIMS and that derived from HRXRD assuming pseudomorphic (InGa)N films. Thus, the present results indicate that (InGa)N layers with thicknesses up to at least 60 nm grow pseudomorphically on GaN which is in line with the findings reported in [36].

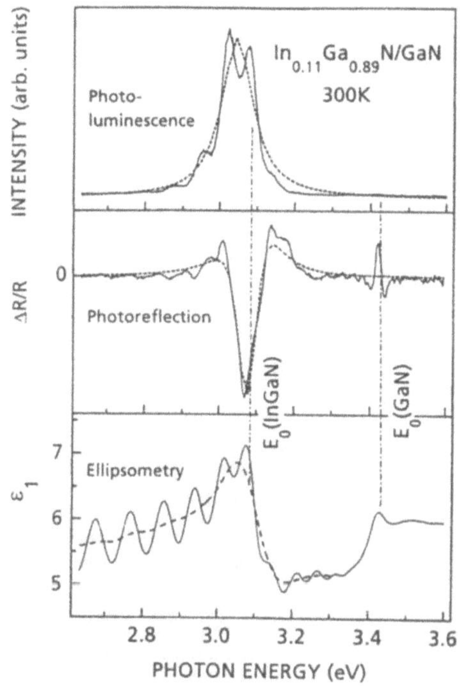

Figure 5
RT PL (top), photoreflection (middle), and pseudo-dielectric function (ε_1) spectra (bottom) for an (InGa)N/GaN heterostructure composed of a 60 nm thick $In_{0.11}Ga_{0.89}N$ layer grown on top of a GaN buffer layer. Short dashed lines indicate fits to the PL and photoreflection spectra not including interference oscillations. The long dashed line shows the pseudo-dielectric function spectrum in the transparency region of GaN after removal of the interference oscillations by Fourier transform filtering.

PL spectroscopy seems to be a promising alternative for the compositional analysis of, in particular thin, (InGa)N layers, once the composition dependence of the fundamental gap energy of (InGa)N is known. Fig. 5 shows a direct comparison of room-temperature PL, photoreflection, and pseudo-dielectric function spectra, recorded on the same sample consisting of a 60 nm thick $In_{0.11}Ga_{0.89}N$ film grown on top of a thick GaN layer. Photoreflection [37] and spectroscopic ellipsometry [38] allow a direct determination of the fundamental band gap energy of the (InGa)N. As shown in Fig. 5, good agreement is obtained between the fundamental gap energy deduced from a fit to the photoreflection spectrum, which exhibits a derivative-like line-shape, and the maximum in the spectrum of the real part of the pseudodielectric function (ε_1). The near-band-edge PL peak, which is modulated in its intensity due to interference oscillations, is shifted to lower energies by 40 meV with respect to the fundamental gap energy.

This Stokes-shift is a consequence of the before mentioned local variations in the fundamental gap energy caused by compositional inhomogeneity. PL spectroscopy probes preferentially regions with lower gap energy, and thus higher In content, to which photogenerated carriers relax prior to recombination. Photoreflection and spectroscopic ellipsometry, in contrast, probe the full range of optical interband transitions averaging over regions with higher and lower In content, which results in a broadening of the fundamental gap resonance. The compositional fluctuations, and thus the broadening of the fundamental gap resonance in, e.g., the photoreflection spectrum, increase with In content as seen from Fig. 6. There, photoreflection spectra are shown for samples with various In contents ranging from $x = 0.025$ to $x = 0.13$. Besides a low-energy shift of the (InGa)N fundamental gap resonance due to the reduction of the gap energy with increasing In content, a significant increase in the width of the (InGa)N-related photoreflection signal

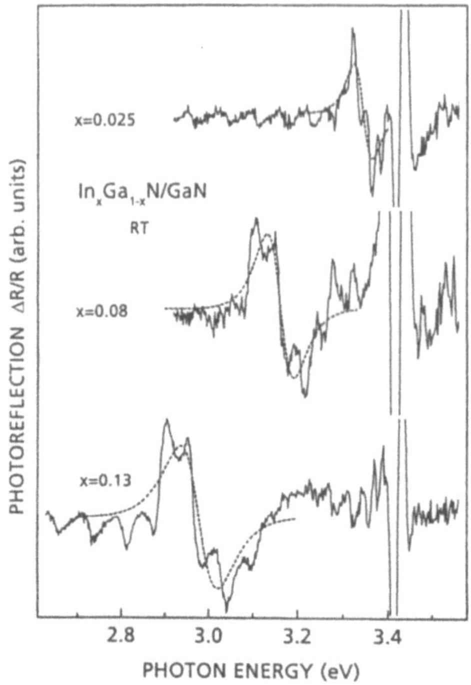

Figure 6
RT photoreflection spectra of 30 nm thick $In_xGa_{1-x}N$ layers on GaN with different In contents x as shown. Short dashed lines indicate fits to the (InGa)N fundamental gap resonance in the photoreflection spectra. Fitted spectra do not include interference oscillations.

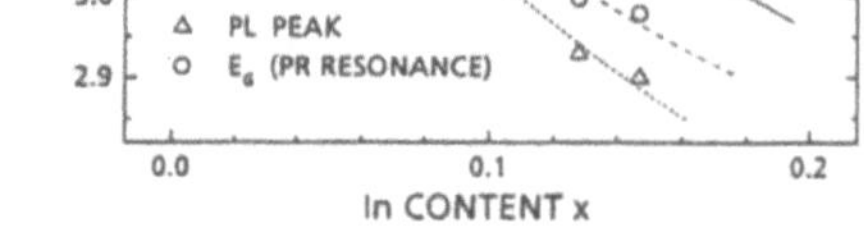

Figure 7
Fundamental gap energy E_G (•) and PL peak position (■) of pseudomorphically strained (≲ 60 nm) $In_xGa_{1-x}N$ layers on GaN versus In content x, as determined from RT photoreflection and PL measurements. The In content was determined by SIMS. The full curve indicates a linear fit to the band gap data. The open circles (∘) indicate band gap energies for unstrained (InGa)N obtained by correcting for the misfit strain induced gap shift. The dashed curve is a quadratic fit to these data (see Eq. (4.12)).

is observed. Fig. 7 shows the fundamental gap energy determined by photoreflection as well as the corresponding PL peak position plotted versus the In content x. The latter was determined by SIMS as discussed before.

The present PR band gap data for pseudomorphically strained (InGa)N films can be described by the linear relation $E_G = 3.43 - 3.28x$ (eV), as shown by the full curve in Fig. 7. After correction for the band gap shift induced by the biaxial compressive misfit strain [39] band gap energies valid for unstrained (InGa)N, indicated in Fig. 7 by open circles, are obtained. These data can be fitted by the following expression

$$E_G = 3.43 - (1.54 + b)x + bx^2 \text{ (eV)} \tag{4.1}$$

using a bowing parameter $b = 3.2$ eV, and RT band gap energies of GaN (3.43 eV) and InN (1.89 eV; [40]). The bowing parameter thus obtained is in good agreement with recent data by McClusky et al., which suggest a b-value of 3.5 eV. There is a clear increase in the Stokes shift between the fundamental gap energy and the PL peak position with increasing In content. Thus, PL spectroscopy is of only limited use for the determination of the In content of a given (InGa)N layer, unless the magnitude of the Stokes shift is known and constant for a given In concentration. Photoreflection and spectroscopic ellipsometry, in contrast, do not suffer from this limitation and can thus readily be used for determining the In content of (InGa)N films via the composition dependence of the fundamental gap energy.

5 (InGaAl)N Heterostructures and Light Emitting Devices

The first p-n-LEDs realized on the basis of group III-nitrides were GaN homojunction diodes emitting in the UV-to-blue spectral range [41]. However, the electroluminescence efficiency of these devices was low, in part due to the self-absorption of the emitted photons in the GaN contact layers. To improve the efficiency and to increase the emission wavelength into the visible blue or even green spectral region, GaN/(InGa)N/(AlGa)N double-heterostructure LEDs were developed employing a typically 50 nm thick (InGa)N layer as the active region [13]. To achieve emission in the blue spectral range while maintaining a moderately low In content of $x = 0.06$ in the $In_xGa_{1-x}N$ active layer, this layer was doped with Zn which presumably acts as a deep acceptor in (InGa)N [13].

Room-temperature electroluminescence spectra of a GaN/(InGa)N/(AlGa)N double-heterostructure LED emitting in the blue spectral range, which was fabricated in our laboratory (IAF), are shown in Fig. 8 for different injection currents. The epitaxial layer sequence used for the fabrication of the LED consisted of a 2.4 μm thick lower GaN:Si n-contact layer, a 30 nm thick (InGa)N active layer co-doped with Si and Zn, followed by a 85 nm wide (AlGa)N:Mg barrier layer inserted to prevent electrons from being injected into the 0.24 μm thick GaN:Mg top p-contact layer. The band edge profile of this LED structure is shown in Fig. 9. The layer sequence was grown on sapphire using low-pressure MOCVD. As can be seen from the electroluminescence spectra displayed in Fig. 8, the

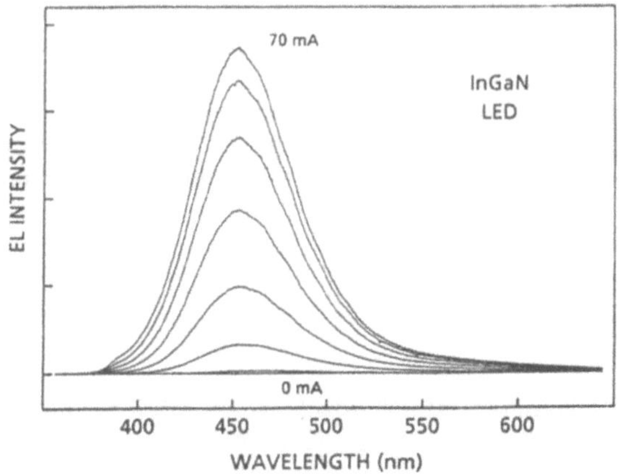

Figure 8
Electroluminescence spectra recorded at different injection currents of a GaN/(InGa)N/(AlGa)N double-heterostructure LED co-doped with Si and Zn in the (InGa)N active region.

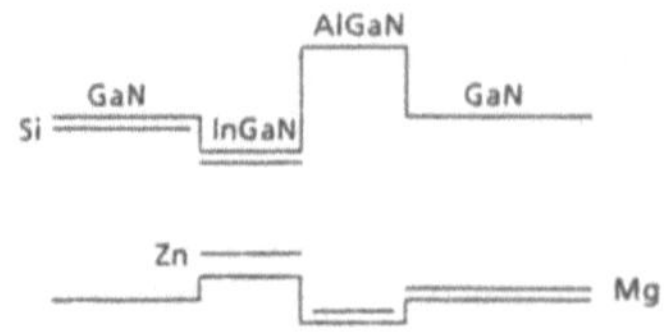

Figure 9
Band edge profile of a GaN/(InGa)N/(AlGa)N double-heterostructure LED.

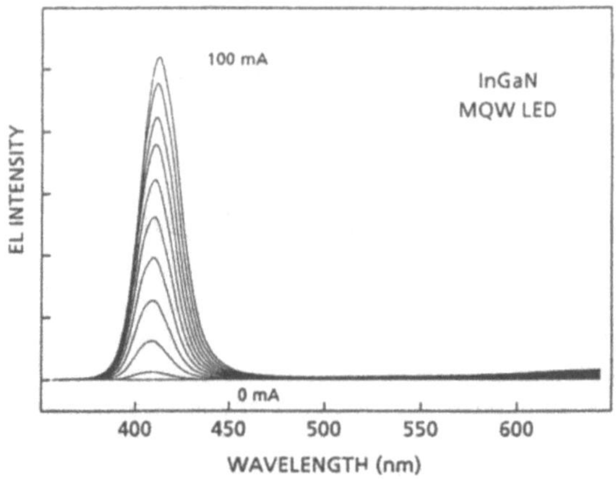

Figure 10
Electroluminescence spectra at different injection currents of a GaN/(InGa)N/(AlGa)N multiple-QW with an active region composed of three Si-doped (InGa)N quantum wells.

emission spectrum arising from band-to-acceptor recombination is rather broad with a full-width-at-half-maximum (FWHM) of 70 nm. This large spectral width is a general feature of LEDs with co-doped active layers [13,42]. Another feature specific to LEDs based on impurity assisted recombination is the occurrence of band-to-band emission at shorter wavelengths under high-injection conditions [43].

Much narrower emission spectra can be obtained when undoped, or only Si-doped, (InGa)N quantum wells (QW) are used as the active region of the LED [44]. Based on this approach high-efficiency single QW LEDs have been realized covering emission wavelengths from the violet to the yellow spectral region [44,45]. Injection current density dependent electroluminescence spectra are shown in Fig. 10 for a QW LED, fabricated at the IAF, containing three Si-doped (InGa)N QWs as its active region. The device emits in the violet spectral range with a rather narrow peak width of 20 nm FWHM. The shape of the emission spectrum remains unchanged when the injection current density is varied over a wide range, which is in contrast to what is typically observed for double-heterostructure LEDs with co-doped active regions [43]. The realization of high-brightness GaN-based QW LEDs was an important prerequisite for the subsequent development of a diode laser emitting in the violet spectral range [46,47]. Color photographs of violet, blue, and green emitting double-heterostructure and multiple QW LEDs fabricated at the IAF are shown in Fig. 11.

6 Summary and Outlook

In spite of the rapid development in (AlGaIn)N growth and the fabrication of group III-nitride based LEDs and diode lasers, important issues related to the growth of (InGaAl)N are unresolved and fundamental material parameters are still unknown. As outlined in

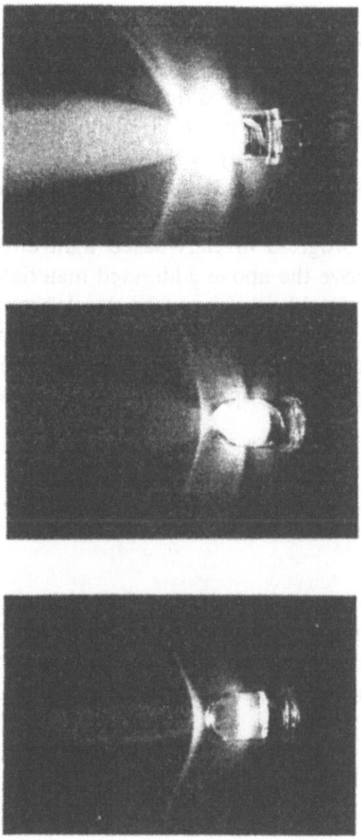

Figure 11
Violet, blue, and green emitting group III-nitride based LEDs fabricated at the IAF.

some detail above, this applies in particular to the issues of p-type doping of GaN and (AlGa)N and to the growth of (InGa)N. For Mg, the commonly used p-type dopant in GaN, it has been shown that self-compensation is the mechanism which limits the maximum obtainable concentration of electrically active dopant atoms even after post-growth activation of hydrogen passivated acceptors. Open questions regarding the preparation of (InGa)N are whether In segregation and phase separation can be suppressed to achieve homogeneous (InGa)N layers even for higher In contents. Dealing at present with band gap fluctuations induced by compositional inhomogeneity, there is an ongoing discussion whether the presence of such potential fluctuation, which lead to a localization of injected carriers, is beneficial for the luminescence yield of (InGa)N films. This is of particular importance for the understanding of the lasing mechanism in (InGa)N based diode lasers.

For GaN-based LEDs a further increase in internal efficiency should be possible via a reduction of the concentration of misfit dislocations and small angle grain boundaries, both induced by the growth on lattice mismatched substrates. Epitaxial lateral overgrowth (ELOG) seems to be a promising technique to realize GaN quasi-substrates with a reduced defect density [48,49]. As demonstrated by the Nichia research team GaN-ELOG substrates allow the growth of nitride lasers having superior device performances [50]. At the end of 1997 this team presented a blue laser diode with an extrapolated lifetime of more than 10 000 hours in cw-mode operation.

High-brightness GaN-based LEDs emitting in the blue, and in the future also in the UV spectral range, also form the basis of luminescence converting LEDs (LUCOLED). In the LUCOLED the emission of the LED is converted into light of longer wavelengths by absorption and reemission in either an organic dye or an inorganic phosphor [51]. When several dyes or phosphors are mixed, or when only part of the primary radiation is converted into light at longer wavelengths, also mixed colors including white can be generated by a single-chip LED [51].

In view of the already seen tremendous progress in GaN-based light emitting devices, combined with the possibility to improve the above addressed materials science and growth issues, it is easy to foresee further rapid progress in the development of both incoherent and coherent emitters employing group III-nitride heterostructures as the active region. This further progress will be accompanied by an expansion of the areas in which these devices will be used, such as high-density optical data storage, reprography, and indicator and lighting systems.

Acknowledgement

We are grateful to T. Fuchs, H. Güllich, N. Herres, Ch. Hofmann, M. Maier, Ch. Manz, S. Müller, W. Pletschen and R. Schmidt for their support with AFM, EDX, PL, XRD and SIMS measurements as well as with LED technology. We thank G. Weimann for continuous support and encouragement and the German Ministery for Research and Education for financial support.

Bibliography

[1] W.C Johnson, J.B. Parsons and M.C. Crew, J. Phys. Chem. **36**, 2561 (1932).

[2] H.P. Maruska and Y.Y. Tietyen, Appl. Phys. Lett. **15**, 327 (1969).

[3] see, e.g., R. Dingle, D.D. Sell, S.E. Stokowski and M. Ilegems, Phys. Rev. B **4**, 1211 (1971); R. Dingle and M. Ilegems, Solid State Commun. **9**, 175 (1971).

[4] S. Yoshioda, S. Misawa and S. Gonda, Appl. Phys. Lett. **42**, 427 (1983).

[5] H. Amano, N. Sawaki, I. Akasaki and Y. Toyoda, Appl. Phys. Lett. **48**, 353 (1986).

[6] H. Amano, M. Kito, K. Hiramatsu and I. Akasaki, Jpn. J. Appl. Phys. **28**, L2112 (1989).

[7] S. Nakamura, T. Mukai, M. Senoh and N. Iwasa, Jpn. J. Appl. Phys. **31**, L139 (1992).

[8] S. Nakamura, Jpn. J. Appl. Phys. **30**, L1705 (1991).

[9] S. Nakamura, Jpn. J. Appl. Phys. **30**, L1708 (1991).

[10] S. Nakamura, T. Mukai and M. Senoh, Jpn. J. Appl. Phys. **31**, 2883 (1992).

[11] S. Nakamura and T. Mukai, Jpn. J. Appl. Phys. **31**, L1457 (1992).

[12] S. Nakamura, T. Mukai and M. Senoh, Jpn. J. Appl. Phys. **32**, L16 (1993).

[13] S. Nakamura, T. Mukai and M. Senoh, Appl. Phys. Lett. **64**, 1687 (1994).

[14] Y. Kato, S. Kitamura, K. Hiramatsu and N. Sawaki, J. Cryst. Growth **144**, 133 (1994).

[15] T. Shibata, H. Sone, K. Yahashi, M. Yamaguchi, K. Hiramatsu, N. Sawaki and N. Itoh, Proceedings of the 2nd Int. Conf. On Nitride Semiconductors ICNS'97, 154 (1997), Tokushima Jpn.- to be published in J. Cryst. Growth

[16] R.P. Vaudo and V.M. Phanse, Proceedings of the 2nd Int. Conf. On Nitride Semiconductors ICNS'97, 442 (1997), Tokushima Jpn. - to be published in J. Cryst. Growth

[17] S. Nakamura, M. Senoh, S. Nagahama, N. Iwasa, T. Yamada, T. Matsushita, H. Kiyoku, Y. Sugimoto, T. Kozaki, H. Umemoto, M. Sano and K. Chocho, Proceedings of the 2nd Int. Conf. On Nitride Semiconductors ICNS'97, 444 (1997), Tokushima Jpn. - to be published in J. Cryst. Growth

[18] K. Hiramatsu, S. Itoh, H. Amano, I. Akasaki, N. Kuwano, T. Shiraishi, K. Oki, J. Cryst. Growth **115**, 628 (1991).

[19] S. Nakamura, T. Mukai and M. Senoh, J. Appl. Phys. **71**, 5543 (1992).

[20] I. Akasaki, H. Amano, H. Murakami, M. Sassa, H. Kato and K. Manabe, J. Cryst. Growth **128**, 379 (1993).

[21] M. Seelmann-Eggebert and H.J. Richter, J. Vac. Sci. Technol. B **9**, 1861 (1991).

[22] M. Seelmann-Eggebert, J.L. Weyher, H. Obloh, H. Zimmermann, A. Rar and S. Porowski, Appl. Phys. Lett. **71**, 2635 (1997).

[23] M. Seelmann-Eggebert, H. Zimmermann, H. Obloh, R. Niebuhr and B. Wachtendorf, Mater. Res. Symp. Proc. **468**, 193 (1997).

[24] M. Arlery, Ph. D. Thesis, University of Grenoble, France, 1998.

[25] S. Fuke, H. Teshigawara, K. Kuwahara, Y. Takano, T. Ito, M. Yanagihara and K. Ohtsuka, J. Appl. Phys. **83**, 764 (1998).

[26] S. Nakamura, N. Iwasa, M. Senoh and T. Mukai, Jpn. J. Appl. Phys. **31**, L139 (1992).

[27] D.P. Bour, H.F. Chung, W. Gœtz, L. Romano, B.S. Krusor, D. Hofstetter, S. Rudaz, C.P. Kuo, F.A. Ponce, N.M. Johnson, M.G. Craford and R.D. Brin gans, Mat. Res. Soc. Symp. Proc. Vol **449**, 509 (1997).

[28] U. Kaufmann, M. Kunzer, M. Maier, H. Obloh, A. Ramakrishnan, B. Santic and P. Schlotter, Appl. Phys. Lett. **72**, 1 (1998).

[29] J. Neugebauer and C. Van de Walle, Mat. Res. Soc. Symp. Proc. Vol. **395**, 645 (1996).

[30] J.W. Allen, Semicond. Sci. Technol. **10**, 1049 (1995). S.H. de Bruin, J. Dieleman and C.Z. van Doorn, Acta Physica Polonica **26**, 579 (1964).

[31] I. Ho and G.B. Stringfellow, Appl. Phys. Lett. **69**, 2701 (1996).

[32] N.A. El-Masry, E.L. Piner, S.X. Liu and S.M. Bedair, Appl. Phys. Lett. **72**, 40 (1998).

[33] S. Chichibu, T. Azuhata, T. Sota and S. Nakamura, Appl. Phys. Lett. **70**, 2822 (1997).

[34] D. Behr, J. Wagner, A. Ramakrishnan, H. Obloh and K.-H. Bachem, Appl. Phys. Lett. **73**, 241 (1998).

[35] S. Nakamura, J. Vac. Sci. Technol. A**13**, 705 (1995).

[36] H. Amano, T. Takeuchi, S. Sota, H. Sakai, and I. Akasaki, Mat. Res. Symp. Proc. Vol. **449**, 1143 (1997).

[37] W. Shan, B.D. Little, J.J. Song, Z.C. Feng, M. Schurmann and R.A. Stall, Appl. Phys. Lett. **69**, 3315 (1996).

[38] J. Wagner, A. Ramakrishnan, D. Behr, H. Obloh, M. Kunzer and K.-H. Bachem, to appear inAppl. Phys. Lett.

[39] M.D. McClusky, C.G. Van de Walle, C.P. Master, L.T. Romano, and N.M. Johnson, Appl. Phys. Lett. **72**, 2725 (1998).

[40] T.L. Tansley and C.P. Foley, J. Appl. Phys. **59**, 3241 (1986).

[41] H. Amano, M. Kitoh, K. Hiramatsu and I. Akasaki, Inst. Phys. Conf. Ser. **106**, 725 (1990).

[42] S. Nakamura, Solid State Commun. **102**, 237 (1997).

[43] J.P. Basur, F.S. Choa, P.-L. Liu, J. Sipior, G. Rao, G.M. Carter and Y.J. Chen, Appl. Phys. Lett. **71**, 1385 (1997).

[44] S. Nakamura, M. Senoh, N. Iwasa and S. Nagahama, Appl. Phys. Lett. **67**, 1868 (1995).

[45] S. Nakamura, M. Senoh, N. Iwasa and S. Nagahama, Jpn. J. Appl. Phys. **34**, L797 (1995).

[46] S. Nakamura, M. Senoh, S. Nagahama, N. Iwasa, T. Yamada, T. Matsushita, H. Kiyoku and Y. Sugimoto, Jpn. J. Appl. Phys. **35**, L74 (1996).

[47] S. Nakamura, M. Senoh, S. Nagahama, N. Iwasa, T. Yamada, T. Matsushita, Y. Sugimoto and H. Kiyoku, Appl. Phys. Lett. **70**, 868 (1997).

[48] A. Usui, H. Sunakawa, A. Sakai and A.A. Yamaguchi, Jpn. J. Appl. Phys. **36**, L899 (1997).

[49] O.-H. Nam, M.D. Bremser, T.S. Zheleva and R.F. Davis, Appl. Phys. Lett. **71**, 2638 (1997).

[50] S. Nakamura, M. Senoh, S.-I. Nagahama, N. Iwasa, T. Yamada, T. Matsushita, H. Kiyoku, Y. Sugimoto, T. Kozaki, H. Umemoto, M. Sano and K. Chocho, Appl. Phys. Lett. **72**, 211 (1998).

[51] For a review see: J. Baur, P. Schlotter and J. Schneider, in Festkœrperprobleme/Advances in Solid State Physics, Vol. **37**, p. 67, edited by R. Helbig (Vieweg, Braunschweig/Wiesbaden, 1998).

Laser Processes in Group-III Nitrides

Axel Hoffmann, Ludger Eckey and Jens Holst

Institut für Festkörperphysik der Technischen Universität Berlin,
Sekr. PN 5–1, Hardenbergstraße 36, D-10623 Berlin, Germany

Abstract: Wide-gap group-III nitrides have enormous potential in device applications, particularly as blue light emitters and in high-power and high-temperature electronics. The recent first realization of an InGaN multi-quantum-well laser diode is a great progress in the development of ultraviolet optoelectronic devices. In spite of this and other successes, there is a continuing discussion about the processes responsible for the laser emission. The present knowledge about stimulated emission processes which are observed in group-III nitrides will be summarized in this paper.

We report on high-excitation luminescence investigations and optical gain measurements of highly excited GaN epilayers in the cubic and hexagonal phases as well as of AlGaN epilayers.

For the hexagonal GaN layers at moderate excitation densities, the decay of biexcitons is observed in the low-temperature stimulated emission spectra up to 60 K. At higher temperatures inelastic scattering processes of excitons, phonon-assisted annihilation of excitons and free carriers dominate. At high excitation levels and high temperatures the electron-hole plasma (EHP) contributes to the gain in thin epitaxial layers. A similar behavior is observed for cubic GaN layers. AlGaN samples exhibit smaller gain values due to an increase of defects during the incorporation of Al.

Spatially resolved experiments enable us to analyze the inhomogeneity and the compositional content of In or Al in the ternary group-III nitrides. This is an important tool for investigating the influence of localized exciton states on the gain in the ternary compounds.

Finally, we will review investigations on the role of the quantum confined Stark effect induced by the piezoelectric field in group-III nitride based multi-quantum well structures. The consequences of these results for device applications are discussed.

1 Introduction

The realization of the first blue laser diode (LD) [1] is a remarkable achievement in the research of group-III nitrides as basic materials for light emitting devices. A brief summary of the research activities on lasing in these materials will be given. The first demonstration of stimulated emission from GaN single crystals was reported by Dingle et al. [2] in the early 70's. The first report on mechanisms responsible for the optical gain in GaN single crystals was given by Dai et al. [3] In 1986 Amano et al. [4] observed room temperature stimulated emission by optical pumping of high-quality GaN epilayers. Khan et al. [5] measured optical gain in AlGaN/GaN double heterostructures by optical pumping at room temperature (RT). Since then there has been a number of investigations on optically pumped stimulated emission and laser action in GaN and related heterostructures [6–19]. Akasaki et al. [20] observed the onset of stimulated emission by current injection for the first time. Nakamura et al. [21] fabricated the first CW operating LD in the ultra-violet spectral range at room temperature with an estimated life time of 10 000 hours [22].

In the present paper we give a survey about the optical gain processes in thin GaN epilayers, double heterostructures of group-III nitrides and quantum well structures of InGaN and AlGaN at high excitation levels. Despite the recent development of the first multilayer nitride laser structure [1] a thorough understanding has not been achieved of the physical processes causing stimulated emission in ternary InGaN multi-quantum well (MQW) structures. Localized excitons or biexcitons [23] and the electron-hole plasma (EHP) [24] have been claimed responsible for the laser mechanism in group-III-nitride heterostructures. These will be reviewed in the last section.

2 Excitonic Fine Structure

An important precondition to investigate the physical mechanisms involved in the stimulated emission are samples of good quality. Typical low-temperature photoluminescence spectra of high-quality GaN layers are often dominated by a strong emission due to excitons bound to an intrinsic donor having a halfwidth of less than 1 meV. This is shown in Fig. 1 [25] for the case of a high-quality GaN epitaxial layer grown on sapphire. The inset of Fig. 1(b) gives the luminescence on a linear intensity scale. The fine structure in the range of the free exciton is seen much more clearly if plotted on a logarithmic scale as

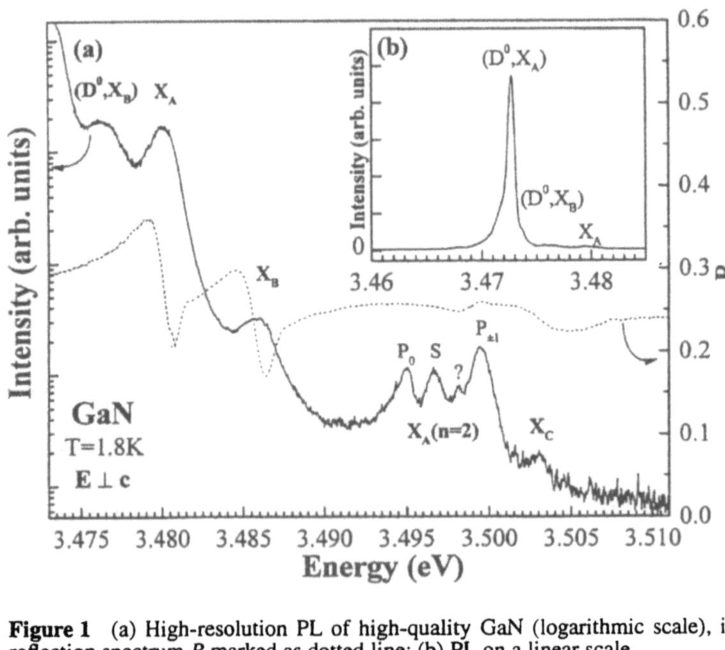

Figure 1 (a) High-resolution PL of high-quality GaN (logarithmic scale), in comparison reflection spectrum R marked as dotted line; (b) PL on a linear scale

in Fig. 1(a). The reflection spectrum of the sample shown for comparison can be used to identify the emissions due to the decay of A-, B-, and C-excitons. Additional fine structure with high contrast is observed in the range of the excited state $n = 2$ of the A-exciton. On the basis of recent polarized and magneto-optical measurements we identify these structures as due to S- and P-states as shown in the figure. From the energy difference between the S-states of the $n = 1$ and $n = 2$ A-exciton it is possible to determine the binding energy of the A-exciton to be 26.5 meV.

3 High-Density Effects in Group-III Nitrides

The optical properties of III-V nitride bulk samples under high-excitation density will be discussed in this section. Previously [26], we investigated lasing mechanisms of cubic and hexagonal GaN epilayers. We demonstrated [19] that in the region of intermediate excitation density up to 5 MW/cm^2 the radiative decay of biexcitons (M-band) can be the dominating low-temperature emission and gain process in GaN. This luminescence band M is shown in Fig. 2 as a function of excitation intensity for a GaN/SiC epilayer. It exhibits a slight red shift with increasing excitation density which is caused by the

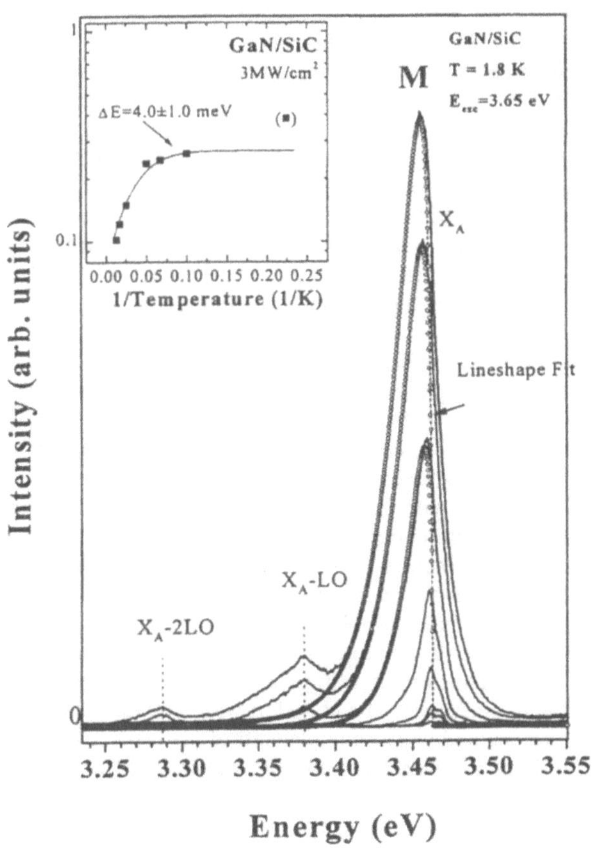

Figure 2
Stimulated emission of GaN/SiC as a function of excitation density, the lineshape fit according to the model of a biexciton decay is shown as dotted line; inset: Arrhenius plot of the temperature dependent PL measurements, an activation energy of 4 meV is observed.

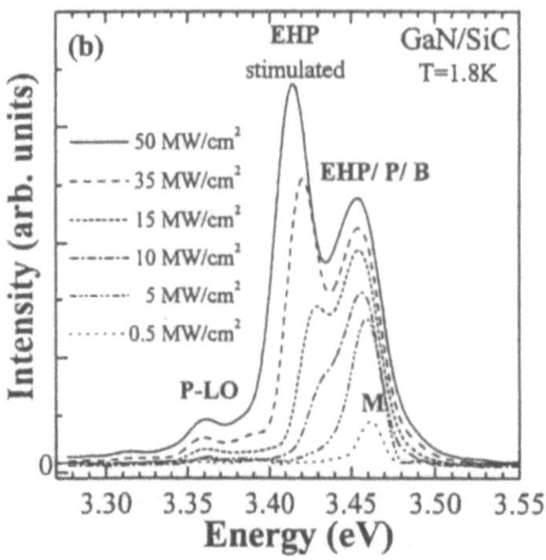

Figure 3
Emission spectra for various excitation densities up to 50 MW/cm^2.

enhanced kinetic energy the biexcitons. From the onset of the emission threshold in our lineshape fits we determined the biexciton binding energy in GaN to be 3.7 meV. This value is confirmed by temperature dependent measurements, shown in the inset of Fig. 2.

The emission spectrum of the same sample at high excitation densities up to 50 MW/cm^2 is given in Fig. 3 [28]. Here, the biexciton emission gives way to a superposition of different luminescences caused by inelastic exciton scattering (P) and the electron-hole-plasma (EHP). The most striking observation is the appearance of a new emission band around 3.43 eV at 10 MW/cm^2. It exhibits a superlinear growth of emission intensity while strongly shifting its peak position to lower energies. This behavior is typical for stimulated emission from an electron-hole plasma. Similar observations in GaN at low temperatures were made before [29]. The gain measurements presented below support this interpretation. Up to room temperature the integrated intensity of the EHP peak follows the predicted T^{-3}-rule [30]. At 270 K the measured peak energy of 3.221 eV is virtually equal to those often reported in the literature for stimulated emission from GaN at RT [8].

Fig. 4 [26] displays a series of gain spectra taken at various excitation densities for a hexagonal GaN epilayer. The spectra presented were smoothed to enhance the visibility of the observed spectral features. In the density series of Fig. 4 a broadening of the region of optical gain with increasing pump intensity is observed. An additional structure EHP appears on both the high- and low-energy shoulders of the main peak M. Its relative strengths with respect to M increase with increasing excitation density. The main peak M is due to biexciton decay at excitation densities up to 5 MW/cm^2. Inelastic exciton-exciton scattering P is likely to contribute at 3.44 eV. The low-energy peak EHP can be identified as due to the electron-hole plasma by its energy position and shift characteristics which are identical to those of the stimulated EHP emission shown in Fig. 4. At 30 MW/cm^2 the EHP represents the dominating low-temperature gain mechanism with a peak gain value of 250 cm^{-1}. The zero-crossing energy due to this process shifts to higher values at increased pump intensities. The blue shift with increasing pump intensity is typical for this gain mechanism and is caused by a shift of the respective quasi-Fermi levels of holes and electrons into the valence and conduction bands [31].

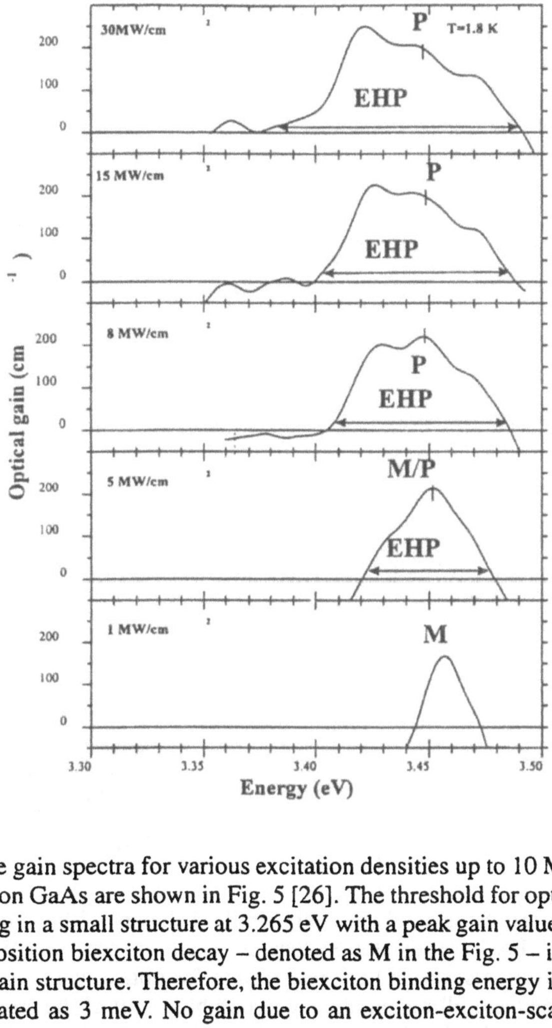

Figure 4
Intensity dependent gain spectra of hexagonal GaN for excitation intensities up to 30 MW/cm^2.

The gain spectra for various excitation densities up to 10 MW/cm^2 in a cubic GaN epilayer on GaAs are shown in Fig. 5 [26]. The threshold for optical gain is 150 kW/cm^2, resulting in a small structure at 3.265 eV with a peak gain value of 10 cm^{-1}. From the spectral position biexciton decay – denoted as M in the Fig. 5 – is likely to be responsible for this gain structure. Therefore, the biexciton binding energy in cubic GaN can be roughly estimated as 3 meV. No gain due to an exciton-exciton-scattering process, expected at around 3.24 eV, is found. For increasing excitation densities of 500 kW/cm^2 and more, the peak energy of the gainstructure shifts to energies higher than that of the bandgap of c-GaN. The peak position is 3.274 eV at 500 kW/cm^2 and 3.280 eV at 1.5 MW/cm^2. The gain above the GaN band edge originates from band filling processes, where the blue shift of the crossover gain-absorption is caused by a shift of the respective quasi-Fermi levels of holes and electrons into the valence and conduction bands [31]. At 3.219 eV another gain peak is observed, labeled EHP. With increasing excitation density this gain peak dominates the spectrum. Its peak position shifts to lower energies with increasing excitation density (3.210 eV at 5 MW/cm^2). From the energy position and the intensity dependent behavior we attribute this to an electron-hole-plasma. This process provides the highest gain values up to 210 cm^{-1}.

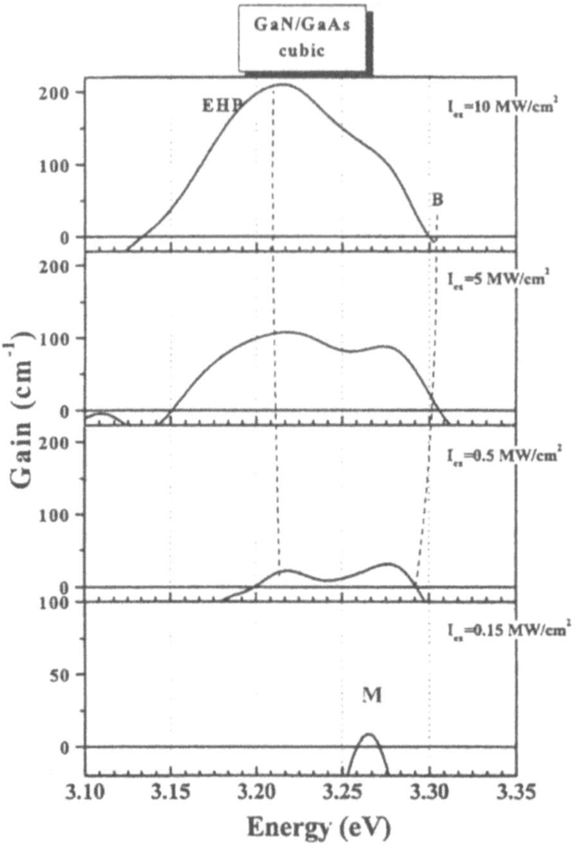

Figure 5
Intensity dependent gain spectra of cubic GaN/GaAs up to 10 MW/cm^2

By comparison of the gain processes involved a very similar behavior of the cubic and hexagonal phases is observed. For lower excitation densities excitonic processes like the biexciton decay or exciton-exciton-scattering are responsible for optical gain. With increasing excitation density the band filling results in a strong blue shift of the crossover gain-absorption. At highest excitation levels an electron-hole-plasma dominates the gain spectra providing high gain values up to 250 cm^{-1}. As the main difference the contribution of excitonic processes like the biexciton decay or the exciton-exciton-scattering to optical gain up to high excitation densities in h-GaN is observed while it is not detected in c-GaN. For an excitation density of 10 MW/cm^2 these excitonic processes provide optical gain in hexagonal GaN with gain values of 220 cm^{-1} whereas in cubic GaN many-particle effects are the dominant mechanisms for optical amplification. Our observations of gain caused by plasma, free carriers, and excitonic processes at the highest excitation levels do not imply that they all occur simultaneously. In our time-integrated measurements we detect all processes which produce optical gain during and after the excitation pulse which in our experiments had a duration of 15 ns. Time-resolved measurements of the optical gain in CdS showed that carrier diffusion as well as the formation of biexcitons and excitons contribute to the ultrashort decay of the plasma within 100–200 ps [32]. It is well known that free carriers at high densities as well as biexcitons and inelastic excitonic scattering

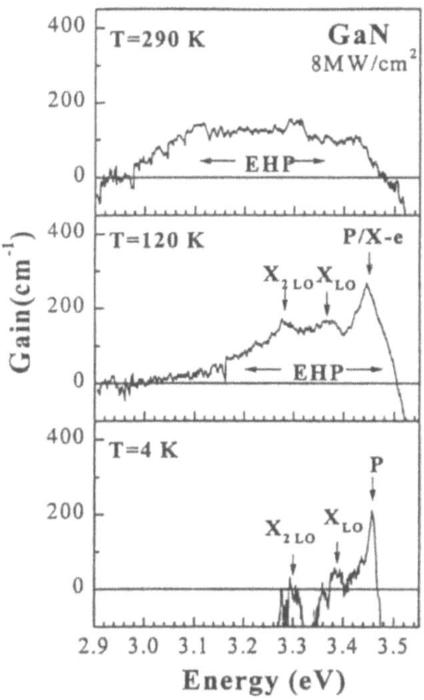

Figure 6
Temperature dependent gain spectra of hexagonal GaN at 8 MW/cm^2

processes give rise to optical gain themselves. Thus, our observations are in very good agreement with the typical behavior observed in other direct wide-gap semiconductors [32,33].

The temperature-dependent gain measurements of Fig. 6 [27] taken at a fixed excitation density of 8 MW/cm^2 show that excitonic processes like exciton-exciton scattering or the A-LO transition are responsible for optical gain at low and intermediate temperatures. At 290 K the gain band is typically broad, indicating that electron-hole-plasma is the dominating gain mechanism at higher temperatures.

The gain spectrum of an $Al_{0.23}Ga_{0.77}N$ epilayer is displayed in Fig. 7 [34]. As the main gain process an electron-hole plasma around 3.88 eV is observed. The weak low-energy gain structure around 3.5 eV is found 1 LO phonon energy below the free exciton transition energy and strongly increases in intensity with increasing pump power. Therefore, and from the energy position we assign the A-LO transition to be the responsible gain mechanism. The luminescence and optical-gain properties of AlGaN epilayers at high excitation densities are dominated by many-particle effects. While the gain values are somewhat lower than in GaN our results show the possibility to fabricate laser structures with active AlGaN layers operating in the ultraviolet spectral range. With increasing Al content the efficiency of optical amplification is reduced. This is explained by the increased defect density during the incorporation of Al.

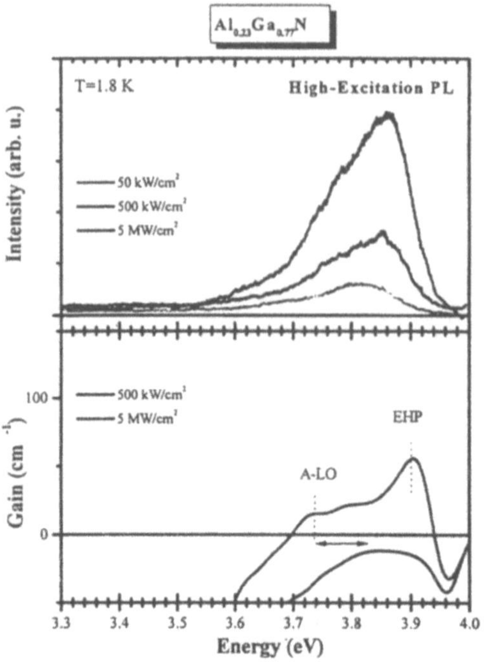

Figure 7
(upper part) High-excitation PL for various excitation densities; (lower part) gain spectra for various excitation densities

4 Indium Fluctuations in InGaN-Structures and Localized Excitons or Quantum Dots in Group-III Nitrides

Many authors report on the phase-separating nature [35–37] due to the mismatch of thermodynamical and chemical stabilities between GaN and InN. Therefore, it was one of the most important issues in recent years to understand the influence of the microstructural fluctuation of the In content on the lasing mechanism in the ternary InGaN epilayers and MQW structures. The spatial distribution of the In composition was often evaluated by energy-dispersive x-ray (EDX) analysis using a field-emission scanning transmission electron microscope (STEM) (Fig. 8) [38]. Here, the In fluctuations were monitored at various positions of the investigated sample structure. The important finding was that there are quantum-dot like regions where the In content is higher than those of neighboring regions. This is shown in Fig. 8, where the average concentration of In is 20%, while the local distribution of the In content varies between 10% to 32% on a nm scale. In the literature these regions are called nano-clusters or In-rich islands in InGaN [40].

An important question is how the In clustering influences the optical properties of this material. In Fig. 9 [39] local cathodoluminescence spectra monitored at different positions of an InGaN epilayer grown by MBE are shown. The integral luminescence spectrum is seen above on the right hand side. This spectrum is a typical luminescence spectrum of an epilayer with an In content of 8%. The local spectra taken from different spots show the In fluctuation over the sample, which is in agreement with the EDX data. This microstructural behavior is revealed by the cathodoluminescence spectra on the left hand side where light from different local spots (indicated through the open dots in the wavelength image) is monitored. The size of these In rich clusters can be monitored with CL images shown

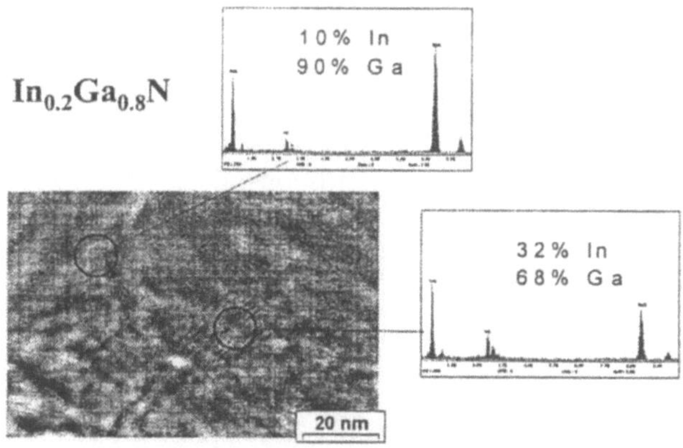

Figure 8 STEM image of an InGaN MQW structure with ring shaped markings indicating areas of the sample from which EDX spectra (upper part and right hand side of the figure) were taken (from [38]).

below, where, at wavelengths of 434.03 nm and 487.00 nm the highest intensities of luminescence are indicated as white spots. These CL images clearly indicate that different local regions of less than 60 nm size emit at different luminescence wavelengths, indicating the varying In content of this region. The histogram (left spectrum above) shows how often a given peak wavelength is observed, corresponding to the number of spots on the sample with a certain In-content. Similar CL investigations of InGaN single quantum wells have been reported by Chichibu et al. [41]. Their CL spectrum mapping revealed an exciton localization at InN rich quantum-disks originating from compositional undulation in the InGaN SQWs. The lateral size of the quantum disks was estimated to be smaller than 60 nm. This assignment was supported by Narukawa et al. [40], claiming the existence of self-formed InGaN quantum dots.

In the following we will discuss the influence of this clustering effect on the optical gain in InGaN heterostructures and quantum wells. Degushi et al. [42] reported on stimulated emission and gain investigations in InGaN MQWs. Two different samples #1 and #2 with different compositional In fluctuations were investigated by electroluminescence (EL) and electroabsorption techniques and the comparison is displayed in Fig. 10 [42]. The compositional In fluctuation in sample #1 is weak and considerably strong in #2 which can be directly proved by the EA and EL signals. The arrows indicate the free-exciton resonance energy obtained from an analysis of the EA spectra. A superlinear increase of light intensity emitted from the sample edge as a function of pump power density I_{exc} for a stripe length L of 500 μm is seen in both spectra. The main difference between the samples is that for the homogenous sample #1 the stimulated emission peak 3.150 eV is close to the free-exciton resonance energy, while in the inhomogeneous sample #2 the stimulated emission coincides with EL (3.069 eV), indicating the luminescence in the lower

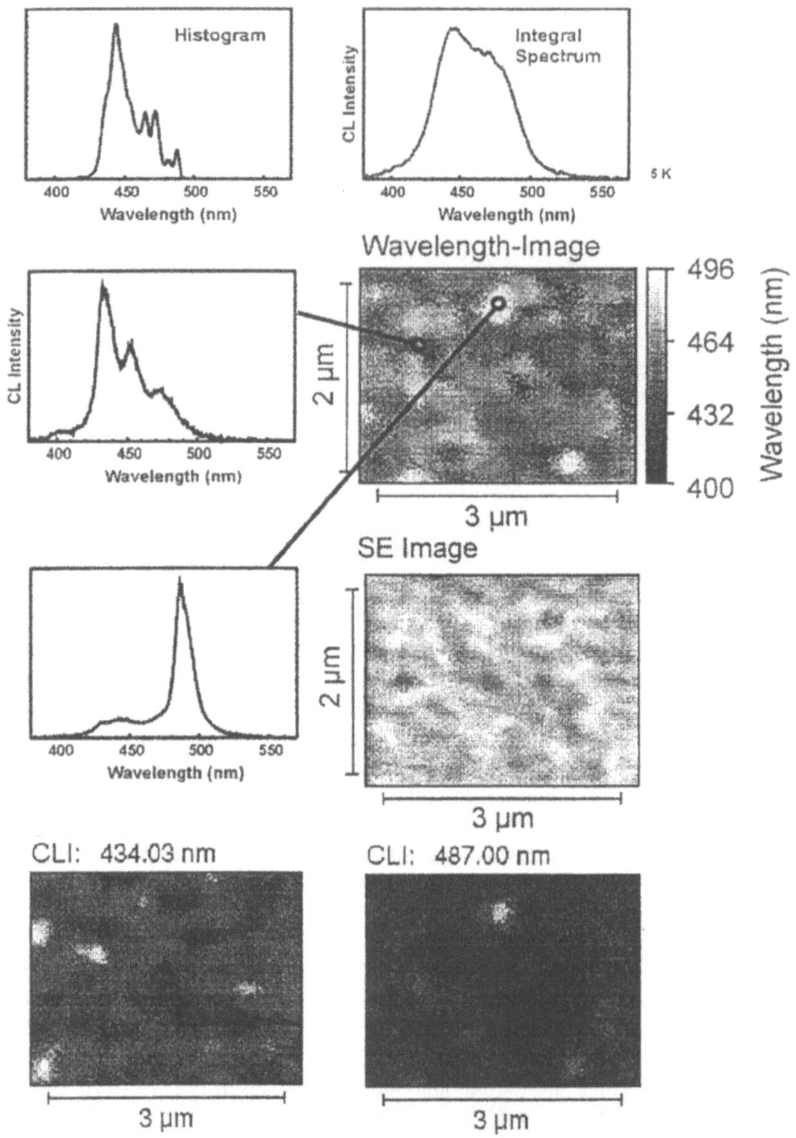

Figure 9 Cathodoluminescence (CL) spectra at different positions of an InGaN layer (arrows), (top left) histogram of the observed wavelength, (top right) integral spectrum, (middle) wavelength image, the brighter the spots the higher is the emitted wavelength, (bottom) CLI spectra at fixed emission wavelengths, brighter spots indicate higher emission intensity (from [39]).

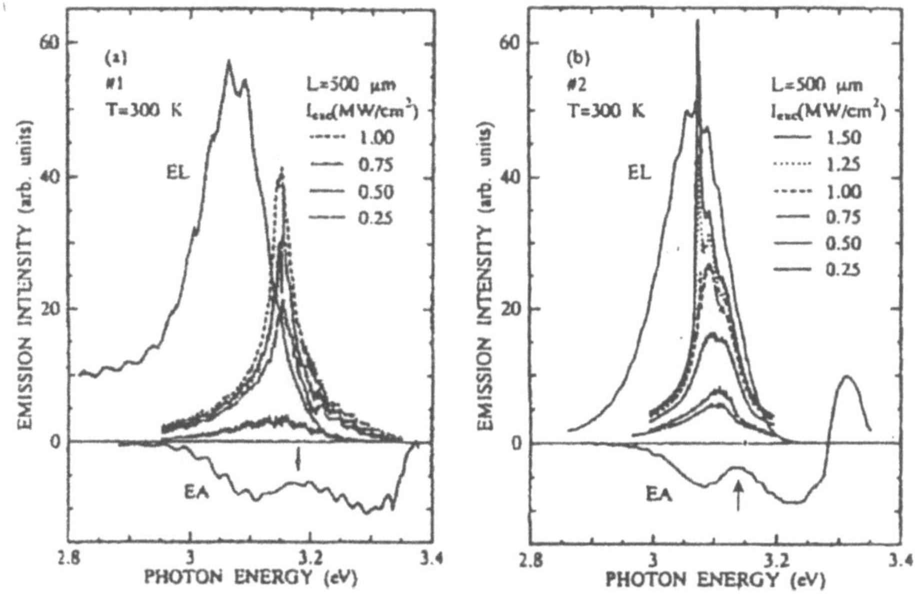

Figure 10 EL spectra of an InGaN MQW at low excitation densities (broad structure), and at various high-excitation densities (smaller structures), the EA spectra are shown below (arrows mark the enrgy of the exciton transition)(from [42])

energy tail of the density of states. The stimulated emission spectra of #2 peaks instantaneously and the FWHM of the line is extremely narrow. The corresponding gain spectra from Degushi et al. [42] elucidating the contributed lasing processes are seen in Fig. 11. The figures show calculated gain as function of photon energy. The optical gain $g(E)$ in the spectrum #1 on the left is relatively broad and can be explained by the electron-hole plasma model. The gain spectra of sample #2 (right hand side) cannot be explained in terms of an electron-hole plasma. Here, an relatively sharp gain peak at 3.07 eV grows up for excitations densities larger than 1 MW/cm^2, while simultaneously dips appear at 3.15 eV corresponding to the averaged free-exciton resonance energy at low enough carrier concentration limits. In case B, when the injected carriers density becomes high the spatial potential undulation leads to regions in which the population inversion is caused by local excitonic states. In summary, Degushi et al. [42] demonstrate that mechanisms producing the optical gain will change according to the degree of compositional fluctuations in InGaN MQW systems.

Recently, Hangleiter et al. [43] reported on ultra-sensitive optical absorption spectroscopy as well as optical excitation spectroscopy under strong pumping conditions on InGaN/GaN double heterostructures and quantum wells. Their absorption spectra show a pronounced excitonic structure and the optical gain is seen close to the low-energy tail of the excitonic absorption. Additionally, they demonstrated that the gain spectra as a function of the excitation density indicate a continuous transition between an excitonic behavior at low carrier density and high carrier densities, where a positive optical gain is obtained. Increasing the excitation densities they observed a typical behavior of an EHP

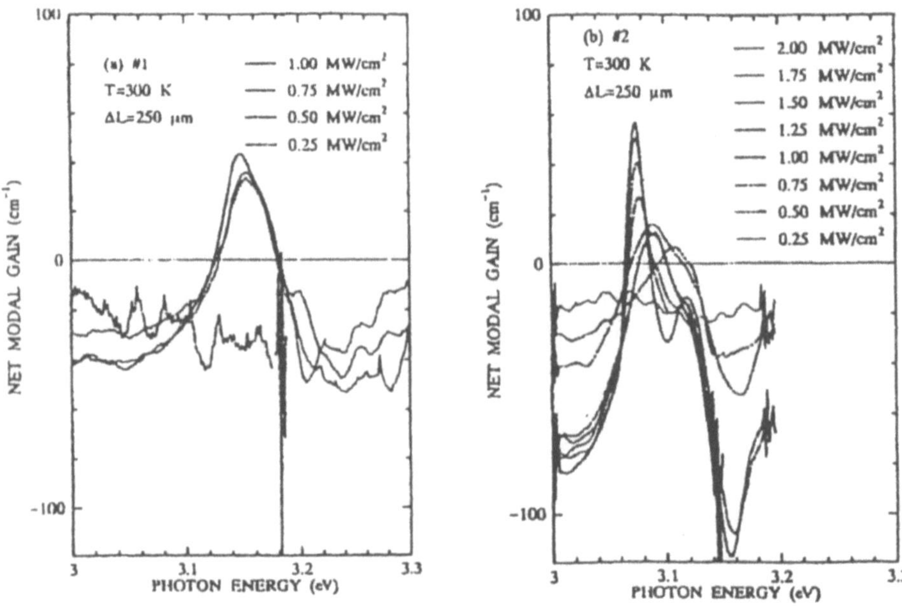

Figure 11 Gain spectra for various excitation densities (a) weak In fluctuations (homogenous) and (b) strong In fluctuations (inhomogenous)(from [42])

where the onset of the gain is shifted to lower energies. They claim that there is no evidence for any localization effect playing an important role for the optical gain in their InGaN/GaN double heterostructures and quantum wells.

5 Piezoelectric Field and Quantum-Confined Stark Effect

Heterostructures and quantum well structures based on group-III nitrides are successfully used as LEDs in the blue, green and yellow spectral range, exhibiting high efficiencies. In contrast laser diodes are only available in the UV and blue spectral range. The difficulty to extend the lasing wavelength was explained by a red-shift of the emission with respect to the absorption in these QW systems [44]. The origin of this red-shift is still a subject of discussion. Beside localization effects from the inhomogenously distributed In clusters piezoelectric fields due to mismatch or induced strain are discussed [44]. Group-III nitrides have large piezoelectric constants along the (0001) orientation [45], and therefore a piezoelectric field in this direction is induced. It is known that (0001)-oriented thin InGaN epilayers coherently grown on GaN are under biaxial stress. The strong piezoelectric fields due to the strain lead to a quantum-confined Stark effect (QCSE) in (0001)-oriented InGaN QWs [46]. This strongly influences the optical properties and lasing mechanisms in this material. Takeushi et al. [46] studied the influence of piezoelectric fields on luminescence properties on InGaN strained quantum well. Fig. 12 from [46] represents low temperature PL spectra of 10 periods of $Ga_{0.87}In_{0.13}N$ (43 Å) $Ga_{0.97}In_{0.03}N$ (43 Å) strained QWs and those of a 400 Å $Ga_{0.87}In_{0.13}N$ strained single layer as a function of the excitation density. Increasing the excitation density a blue shift of the PL is seen. This

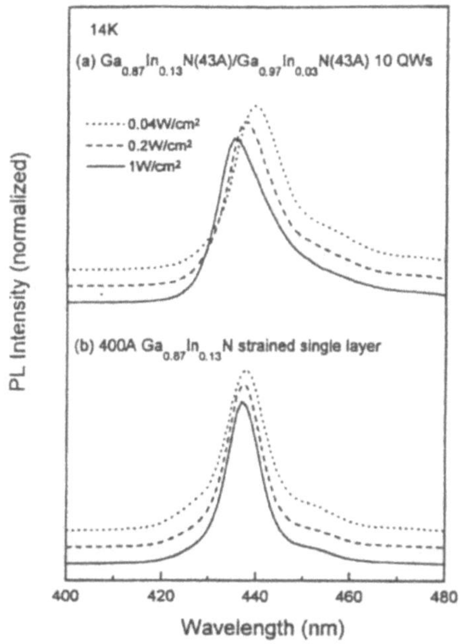

Figure 12
(a) PL spectra of InGaN MQW for various excitation densities (b) PL spectra of starined InGaN layer for various excitation densities (from [46])

behavior was explained in the following way [46]. The transition energy of GaN strained QW's is smaller than that of InGaN unstrained QW's, since the band alignment is tilted by piezoelectric fields. This is called the quantum-confined Stark effect (QCSE). Irradiating the sample with light the piezoelectric fields in the InGaN strained quantum well layer were screened by photogenerated carriers, thus weakening the QCSE. Increasing the excitation density further weakens the QCSE and increases the transition energy which leads to the observed blue shift.

In the 400 Å InGaN single layer a very small shift in PL was seen, even though the InGaN single layer was fully strained. Applying an electric field in this thick layer no QCSE is observed due to the inadequate confinement of carriers. In this case the authors observe the Franz-Keldysh effect which causes a smaller blue shift of recombination energy as compared with the QCSE.

Additionally, Takeushi et al. [46] investigated the influence of piezoelectric fields as a function of the well width seen in Fig. 13. Here, the RT-PL peak energy of $Ga_{0.87}In_{0.13}N$ strained MQW as a function of the well width is shown. The calculation results are also shown for comparison. The solid square represents the measured PL peak energy under high excitation with an N_2-laser, and open squares represent those under low excitation density. The solid line indicates the calculated transition energies assuming that the piezoelectric field of 1.08 MV/cm is induced in the $Ga_{0.87}In_{0.13}N$ strained layer. Transition energies without electric field are shown as dashed line. Again one sees the QCSE weakened by photogenerated carriers in the high-density excitation case in the QW structure. Iyechika et al. [47] investigated electric field modulated PL of InGaN SQW and compared it with PL spectra as a function of the excitation intensity. They could demonstrate that the PL spectra of an InGaN strained quantum well exhibit peak shifts and sharpening with a higher excitation density and that an increasing external electric field causes the same effect on PL spectra. All these results clearly indicate that there exists an internal electric

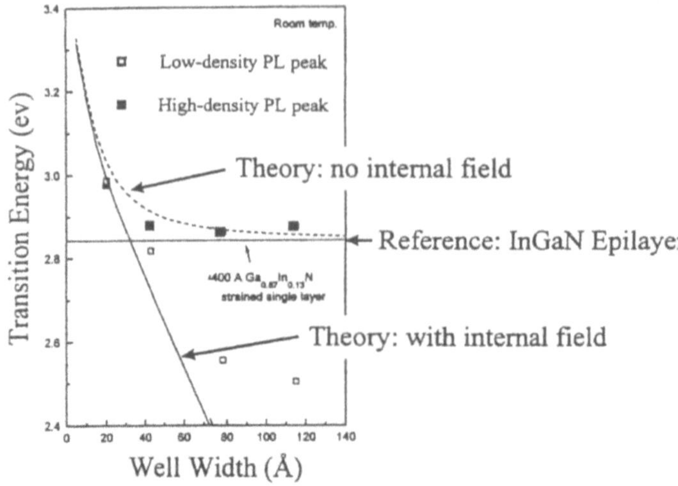

Figure 13 Transition energy versus well width of an MQW. Open squares: low intensity PL; filled squares: high-excitation PL; dashed line: theory with no internal fields; full line: theory with internal fields (from [46]).

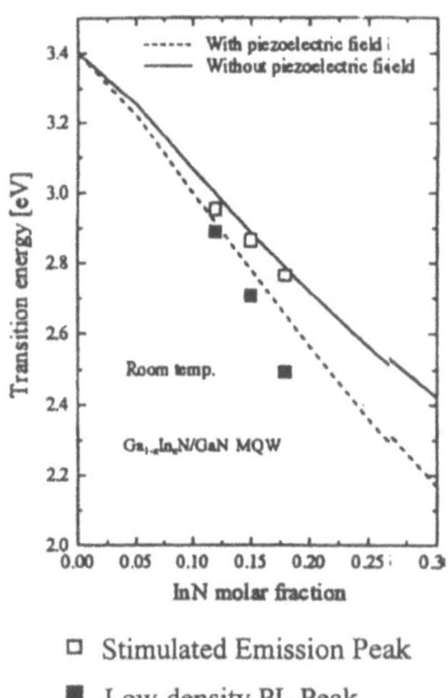

Figure 14
Transition energy vs. InN mole fraction. Full squares: low density PL; open circles: high density PL; dashed line: theory with piezoelectric fields; full line: theory without piezoelectric fields; (from [48]).

field in the QW generated by the piezoelectric field effect (QCSE) and that this effect strongly influences the properties of nitride-based heterostructure devices.

The strong piezoelectric field in strained GaN-based heterostructures has important consequences for the RT stimulated emission. The energy position of the emission is not only a function of the In content in the active layer, but a function of the well thickness and piezoelectric field as well. Sakai et al. [48] also investigated this dependence in an InGaN QW. In Fig. 14 the compositional dependence of the transition energy is shown. The open squares represent the peak energies of the stimulated emission under high excitation density, while the energy position of the low excitation luminescence is given by solid squares. The stimulated emission appears strongly blue shifted with respect to the spontaneous emission. This blue shift increases with increasing In content. The lasing emission of 450 nm for 18% In content is the longest wavelength achieved in nitride-based semiconductors. The broken line in Fig. 14 is the calculated transition energy including piezoelectric fields while the solid line corresponds to a calculation of the transition energies without electric field. In the case of low excitation the calculated transition energies with piezoelectric fields agree with the experimental results. In the high-excitation case the piezoelectric field is again screened by free charge carriers as explained above. Furthermore, the threshold density to achieve lasing strongly increases with higher In content. All this clearly indicates the detrimental influence of the piezoelectric fields on the efficiency of stimulated emission.

Piezoelectric fields are a general problem in ternary group III-nitride QWs. As shown by Im et al. [44] piezoelectric fields due to mismatch-induced strain lead to a Stark shift of the emission in AlGaN QW's as well. The authors demonstrated the strong reduction of the oscillator strength in group-III nitride based QW's, which is again detrimental to the lasing properties.

6 Conclusion

The state-of-the-art laser diode structures by Nakamura et al. [22] are quantum well structures with a low In content, resulting in a low In fluctuation, small piezoelectric fields and, therefore, low laser threshold current densities. The lasing wavelength in InGaN based devices is a function not only of In content but also of the well thickness and piezoelectric fields. As a consequence it is essential for the fabrication of green lasers to achieve a reduction of the piezoelectric fields in QW structures by decreasing the internal strain at higher In incorporation. This holds in the case of achieving UV lasers based on AlGaN QW structures as well.

Acknowledgments

We are indebted to U. von Gfug, A. Kaschner, T. Lüttgert, and I. Broser for their helpful discussion and support. It is a pleasure to thank J. Christen and F. Bertram for their cooperation in the spatially resolved measurements. The authors are very grateful to several growth teams for providing samples T. Detchprohm and K. Hiramatsu (Mie University, Mie), H. Amano and I. Akasaki (Meijo University, Nagoya), D. Ass and K. Lischka (Gesamthochschule Paderborn), O. Ambacher and M. Stutzmann (Walter-Schottky Institut, München). Financial support by the Deutsche Forschungsgemeinschaft and Volkswagenstiftung is gratefully acknowledged. The authors thank S. Fischer and B. K. Meyer (Universität Giessen) for many fruitful discussions.

Bibliography

[1] S. Nakamura, M. Senoh, S. Nagahama, N. Iwasa, T. Yamada, T. Matsushita, H. Kiyoku, Y. Sugimoto, Jpn. J. Appl. Phys. **35**, L 74 (1996)

[2] R. Dingle, K.L. Shaklee, R.F. Lehny, R B. Zetterstrom, Appl. Phys. Lett. **19**, 5 (1971)

[3] R. Dai, W. Zhuang, K. Bohnert, C. Klingshirn, Z. Phys. B.-Condensed Matter **46**, 189 (1982)

[4] H. Amano, N. Sawaki, I. Akasaki, Y. Toyoda, Appl. Phys. Lett. **48**, 353 (1986)

[5] M. A. Khan, R. A. Sgogman, J. M. van Hove, S. Krishnankutty, R. M. Kolbas, Appl.Phys.Lett. **56**, 1257 (1990)

[6] K. Itoh, T. Kawamoto, H. Amano, K. Hiramatsu, I. Akasaki, Jpn. J. Appl. Phys. **30**, 1924 (1991)

[7] M.A. Khan, D.T. Olsen, J.M. van Hove, and JN. Kuznia, Appl. Phys. Lett. **58**, 1515 (1991)

[8] H. Amano, T. Asahi, I. Akasaki, J. Lum. **48+49**, 889 (1991)

[9] H. Amano,T. Tanaka, Y. Kunii, K. Kato, S.T. Kim, I. Akasaki, Appl. Phys. Lett. **64**, 1377 (1994)

[10] K. Kim, W. R. L. Lambrecht, B. Segall, Phys. Rev. **B50**, 1502 (1994)

[11] A. S. Zubrilov, V. I. Nikovlaev, V. A. Dmitriev, K. G. Irvine, J. A. Edmond, C.H. Carter Jr.,

[12] T. Tanaka, K. Ushida, A. Watanabe, S. Minagawa, Electronics Lett. **32**, 35 (1996)

[13] T.J. Schmidt, X.H. Yang, W. Shan, J.J. Song, A. Salvator, W. Kim, O. Aktas, A. Botchkarev, H. Morkoc, Appl. Phys. Lett. **68**, 1820 (1996)

[14] G. Frankowsky, F. Steuber, V. Härle, F. Scholz, A. Hangleiter, Appl. Phys. Lett. **68**, 3746 (1996)

[15] D. Wiesmann, I. Brener, L. Pfeiffer, M.A. Khan, C.J. Sun, Appl. Phys. Lett. **69**, 3384 (1996)

[16] X. P. Zhang, P. Kung, D. Walker, A. Saxler,M. Razeghi, J. Appl. Phys. **80**, 6544 (1996)

[17] I. K. Shmagin, J. F. Muth, S. Krishnankutty, R. M. Kolbas, S. Keller, U. K. Mishra, S. P. DenBaars, Mat. Res. Symp. Proc., vol.449, 1209 (1997)

[18] R. Hofmann, V. Wagner, H-P. Gauggel, F. Adler, P. Ernst, A. Sohmer, H. Bolay, F. Scholz, H. Schweizer, Mat. Res. Symp. Proc., vol.449, 1197 (1997)

[19] L. Eckey, J. Holst, A. Hoffmann, I. Broser, H. Amano, I. Akasaki, T. Detchprohm, K. Hiramatsu, J. Lum. **72–74**, 59 (1997)

[20] I. Akasaki, H. Amano, S. Sota, H. Sakai, T. Tanaka, M. Koike, Jpn. J. Appl. Phys. **34**, L 1517 (1995)

[21] S. Nakamura, M. Senoh, S. Nagahama, N. Iwasa, T. Yamada, T. Matsushita, H. Kiyoku, Y. Sugimoto, T. Kozaki, H Umemoto, M. Sano, K. Chocho, Proc. ICNS'97, Tokushima, Japan, ed. K. Hiramatsu, 444, special issue of Journal of Crystal Growth **189/190**, 820 (1998)

[22] S. Nakamura, M. Senoh, S. Nagahama, N. Iwasa, T. Yamada, T. Matsushita, H. Kiyoku, Y. Sugimoto, T. Kozaki, Appl. Phys. Lett. **72**, 211 (1997)

[23] M. Sugawara, Jpn. J. Appl. Phys. **35**, 124 (1996)

[24] W. W. Chow, Appl. Phys. Lett. **66**, 3000 (1995)

[25] A. Hoffmann, L. Eckey, Mater. Sci. Forum **264–268**, 1283 (1998)

[26] J. Holst, L. Eckey, A. Hoffmann, I. Broser, B. Schöttker, D. J. As, D. Schikora, K. Lischka, Appl.Phys. Lett. **72**, 1439 (1998)

[27] L. Eckey, J. Holst, A. Hoffmann, I. Broser, T. Detchprohm, K. Hiramatsu, MRS Internet J. of Nitride Semiconductor Research **2**, 1 (1997)

[28] J. Holst, L. Eckey, A. Hoffmann, I. Broser, H. Amano, I. Akasaki, MRS Internet J. Nitride Semicond. Res. **2**, 25 (1997).

[29] R. Cingolani, M. Ferrara, M. Lugarà, Solid State Communications **60**, 705 (1986) 1515

[30] Y. Pokrovskii, phys. stat. sol **82**, 385 (1972)

[31] G. Lasher, F. Stern, Phys. Rev. B **133**, A553 (1964)

[32] H. Saito, E. Göbel, Phys. Rev. B **31**, 2360 (1985)

[33] K. Bohnert, phys. stat. sol. **98**, 175 (1980)

[34] L. Eckey, V. Kutzer, J. Holst A. Hoffmann, I. Broser, O. Ambacher, M. Stutzmann, H. Amano, I. Akasaki, Materials Research Soc. Symposium Proc. (1997), Vol. 486, 237 (1997)

[35] S. Stride and H. Morkoc, J. Vac. Sci. Technol. B **10**, 1237 (1992)

[36] Ihsiu Ho, G. B. Stringfellow, Appl. Phys. Lett. **69**, 2701 (1996)

[37] A. Koukitsu, N. Takahashi, T. Taki, and H. Seki. Jpn. J. Appl. Phys.,Part 2, L673 (1996)

[38] P. Veit, J. Christen, F. Bertam to be published

[39] H. Selke, P.L. Ryder, T. Böttcher, S. Einfeldt, V. Kirchner, F. Bertram, T. Riemann, J. Christen, to be published in Apl. Phys. Lett.

[40] Y. Narukawa, Y. Kawakami, M. Funato, S. Fujita, S. Fujita, S. Nakamura, App.Phys. Lett. **70**, 981 (1997)

[41] S. Chichibu, K. Wada, S. Nakamura, Appl. Phys. Lett. **71**, 2346 (1997)

[42] T. Degushi, A. Shikanai, T. Sota, S. Chichibu, S. Nakamura, Proc. ICNS'97, Tokushima, Japan, ed. K. Hiramatsu, 466 (1997)

[43] A. Hangleiter, S. Heppel, J. Off, A. Sohmer, F. Scholz, Materials Science Forum Vol. 264–268, 1287 (1998)

[44] J. S. Im, H. Kollmer, J. Off, A. Sohmer, F. Scholz, A. Hangleiter, Phys. Rev. B **57**, R9435 (1998)

[45] F. Bernardi, V. Fiorentini, D. Vanderbilt, Phys. Rev. B **56**, R 10024 (1997)

[46] T. Takeuchi, S. Sota, M. Katsuragawa, M. Komori, H.Takeuchi, H. Amano, I. Akasaki, Jpn. J. Appl. Phys. **36**, L 382 (1997)

[47] Y. Iyechika, Y. Ono, T. Takada, Materials Science Forum Vol. **264–268**, 1307 (1998)

[48] H. Sakai, T. Takeuchi, S. Sota, M. Katsuragawa, M. Komori, H. Amano, I. Akasaki Proc. ICNS'97, Tokushima, Japan, ed. K. Hiramatsu, 456, special issue of Journal of Crystal Growth **189/190**, 831 (1997)

Growth and Characterization of II-VI Semiconductor Lasers

M. Behringer, H. Wenisch, M. Fehrer, V. Großmann, A. Isemann, M. Klude, H. Heinke, K. Ohkawa, and D. Hommel

Fachbereich Physik, Universität Bremen,
Kufsteiner Str. NW1, D-28359 Bremen, Germany

Abstract: Key components and operating characteristics of II-VI laser diodes have been investigated. The voltage drop at the contact to p-ZnSe could be reduced from above 10 V to below 2 V at high current densities by a multilayer contact with ZnSe and ZnTe layers with variing thickness ratio. By the replacement of MgZnSSe with a ZnSe/MgZnSSe superlattice as material for the cladding layers in the laser diode, doping levels could be increased by a factor of 2 and with the use of ZnSe substrates instead of GaAs, high brightness light emitting devices and laser diodes with long lifetimes could be manufactured.

The II-VI compound semiconductors have been promising materials for optoelectronic devices for decades, but doping limits for n-type as well as p-type layers remained the main obstacle to the realization of electrically pumped devices. Only after the realization of sufficiently effective and stable n- and p-type doping levels in 1991 [1,2], using $ZnCl_2$ and a RF plasma source for doping with atomic nitrogen, the barrier for long lifetime and high brightness light emitting and laser diodes was thought to be overcome and it seemed only a question of time, when room temperature (RT) CW operation would be established. Indeed quick progress had been made. First electrically pumped stimulated emission at 77 K was already reported in 1991 [3,4] and in 1993 CW-operation at RT [5] for a short time could be demonstrated. While the first lasers contained ZnSSe cladding layers [3], their replacement by MgZnSSe [6] offered better optical and electrical confinement and allowed to manufacture devices for CW-operation at room temperature. Further improvements in lifetime could be achieved by the introduction of a GaAs buffer layer and a thorough investigation of the interfaces. The contacts to p-ZnSe also caused difficulties which had to be overcome and the investigation of degradation in the quantum well (QW) [7] and the resulting improvements in the devices finally resulted in lifetimes of 100 h reported already in 1994 [8]. Up to now 400 h could be reached in the SONY laboratories [9] while lifetimes of devices made at universities settle much lower at up to 37 s [10]. In this article, we want to describe some problems in the realization of stable laser devices and possible solutions.

1 Preparation and Design

All investigated samples were grown in a two chamber molecular beam epitaxy (MBE) system EPI 930/95 for III-V and II-VI growth. The III-V chamber is equipped with cells

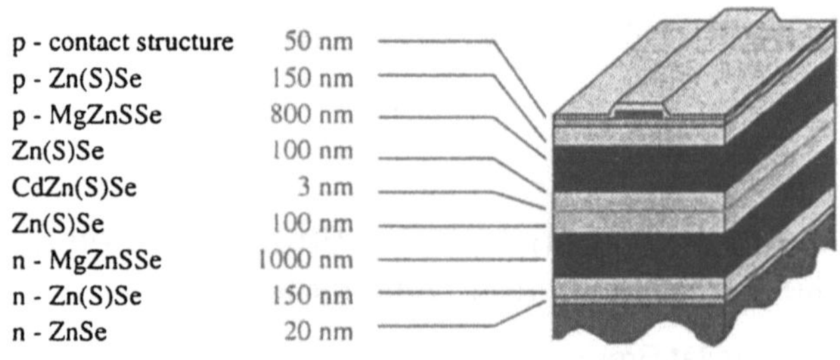

Figure 1
Layer sequence of a typical laser structure. The diode is a double heterostructure, separate confinement single Quantum-Well LD.

for elemental Ga, As and Si for n-type doping. The II-VI chamber allows the evaporation of elemental Zn, Se, Cd, Mg, S and Te, as well as $ZnCl_2$ for n-type doping. A RF-plasma source is used for p-type doping with atomic nitrogen. Both chambers are interconnected by an ultra high vacuum (UHV) transfer system. The heteroepitaxial samples were grown on 2 inch n-type GaAs (001) wafers, the homoepitaxial structures on doped and undoped ZnSe substrates from the Institut für Kristallzüchtung in Berlin, Germany and Sumitomo Electric, Japan. All samples were double heterostructures, separate confinement light emitting (LED) and laser diodes (LD). The structure is shown in Fig. 1. The GaAs substrate was thermally deoxidized in the III-V chamber and a 300 nm GaAs:Si buffer ($N_D - N_A = 1 \times 10^{18}$ cm^{-3}) was deposited prior to II-VI growth. The laser structure consists of ZnSe and ZnSSe buffer layers, MgZnSSe cladding layers, a waveguide (WG) of ZnSSe and a QW made out of CdZnSSe or CdZnSe. The p-type side above the waveguide consists of MgZnSSe cladding layers, a ZnSSe spacer and contact layers out of ZnSe, ZnSeTe and ZnTe. The metal contact is formed by Pd/Au which was annealed after deposition. The free electron concentrations were 2×10^{18} cm^{-3}, 1×10^{18} cm^{-3} and 4×10^{17} cm^{-3} for ZnSe, ZnSSe and MgZnSSe, the free acceptor concentrations were 1×10^{18} cm^{-3}, 2×10^{17} cm^{-3} and 6×10^{16} cm^{-3} for the top contact, the ZnSSe layers and the MgZnSSe cladding layers, respectively [11,12]. In the case of the homoepitaxial structures, the ZnSSe waveguide and buffer layers were replaced by ZnSe. The quaternary cladding layers, and in the heteroepitaxial case the ZnSSe layers were intended to be lattice matched to the corresponding substrate. The MgZnSSe layers had an intended band gap of 2.95 eV at 10 K and the Cd content in the QW was about 35 %. The thicknesses of the individual layers are indicated in Fig. 1. The LDs were contacted with metal stripes with a width variing between 10 μm and 30 μm and the LED had contacts with an area 1 mm $\times$ 1 mm. The devices were mounted to a copper plate with silver paste with the epi-side up. They were contacted at the top side using contact needles. The homoepitaxial devices were contacted from the top only. All electrical measurements were performed at RT. Details concerning the high-resolution X-ray diffraction (HRXRD) can be found in [13,14].

2 Operating Characteristics

LDs grown in Bremen could successfully and repeatedly be operated in CW-mode at RT. Typical threshold current densities were in the range from 350 A/cm^2 to 600 A/cm^2 for resonators without high reflectivity coatings and typical threshold voltages were between 15 V and 20 V. The spectra shown in Fig. 2 were taken just below, at and just above

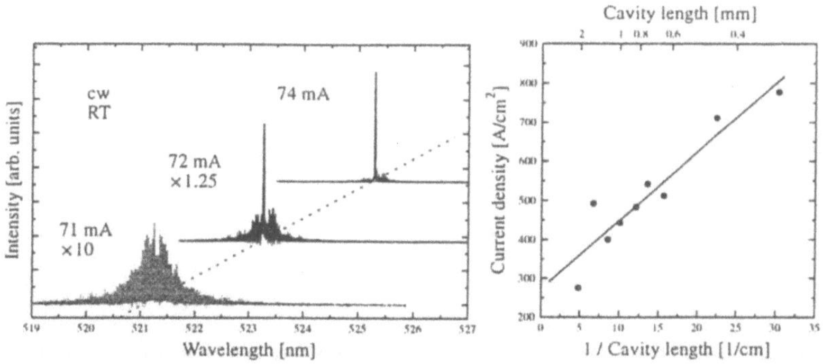

Figure 2 Spectra of stimulated emission in CW-mode at room temperature (left). Resonator length dependent threshold current density (right).

threshold. The lowest curve, taken at a current of 71 mA, still shows a multimode emission with a full width at half maximum (FWHM) of the total emission of about 1 nm. The individual modes are clearly separated and the stimulated emission outdoes the spontaneous emission by a factor of 15. In contrast, the spectrum taken at 72 mA is dominated by a single mode whose intensity is by a factor of 3.5 larger compared to the next intense lasing mode. When the current is further increased to 74 mA this intensity ratio is further increased to 15:1. Therefore, the emission is almost in a single mode already at a driving current 5% above lasing threshold. Simultaneously the maximum intensity is increased by a factor of 10 by a current increase of only 5%.

From *L-I* curves, which were taken in pulsed operation, threshold current densities and external quantum efficiencies could be measured. For the determination of the absorption coefficient and of the nominal threshold current density, these measurements were taken in dependence on the resonator length of the laser devices [15]. The dependence of the threshold current density on the cavity length is shown in the right part of Fig. 2. The investigated samples had uncoated facets with reflectivities of about 21%. It is clearly visible, that the threshold current density for short resonators is as high as 800 A/cm^2, for longer cavities of about 2 mm this is much lower with only about 300 A/cm^2. This corresponds to a stronger compensation of the losses at the mirrors by the longer resonator. By evaluating the resonator length dependent quantum efficiencies, we could derive an absorption coefficient of about $\alpha = 8.8$ cm^{-1} in these layers and a nominal threshold current density of approximately 10×10^4 Acm$^{-2}\mu m^{-1}$. The value for α is higher than for optical pumped structures [16] which could indicate that there is considerable absorption in either the top contact or in the GaAs substrate. That would mean that the cladding layers are still too thin or the refractive index in these layers is too high. The value for the nominal current density is higher by a factor of 3 to 4 in comparison to GaAs LDs which corresponds fairly well with the differences for the effective masses in AlGaAs/GaAs and CdZnSe/ZnSe. Due to the higher masses, the density of states in the bands is higher in the II-VI system and therefore a higher current is necessary to achieve a comparable shift in the quasi Fermi level and to reach the density for transparency [16].

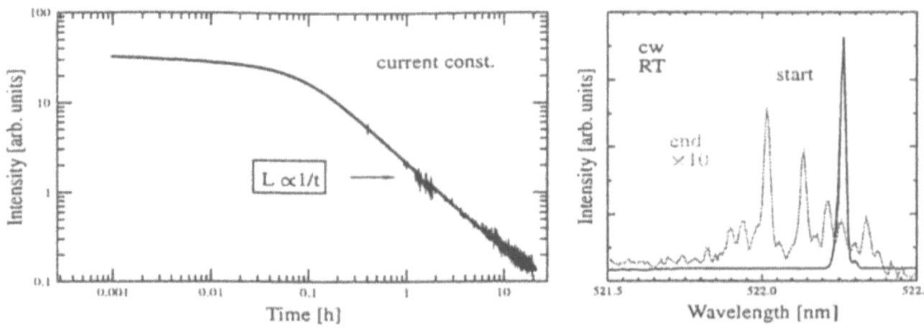

Figure 3 Lifetime measurements (left) and spectra before and after degradation (right).

The overall required current densities are with 500 A/cm^2 comparably low and long lifetimes can be expected. In pulsed operation with a duty cycle of 1% lifetimes in excess of 2 hours and with a duty cycle of 0.1% of more than 15 hours could be reached. Degradation was investigated in CW-mode at current densities of 5 A/cm^2 much below threshold. In Fig. 3 the light output intensity over time at a constant current density is depicted. For very short times, the decrease is exponential, but for very long times, the intensity dwindles only with $1/t$. This slow decrease indicates, that degradation proceeds via the diffusion of point defects [17]. By diffusion of vacancies and interstitials into the quantum well, mid band gap states are increasingly produced in this region and recombination by non-radiative processes takes place and advances. By this, phonons are generated which again induce vibrations in the lattice and thus favour the further diffusion of point defects by increasing their mobilities. Therefore the defect density within the QW will increase proportional to $\sqrt{t}$ and the intensity of the optical emission will decrease as already mentioned proportional to $1/t$. In the right part of Fig. 3 the spectra of a laserdiode are depicted, which was operated in CW-mode at RT. The lifetime of this diode was about 16 s. While one spectrum was taken 1s after turn on, the other spectrum was taken shortly before the end of its lifetime after about 16 s. Both spectra clearly indicate resonator modes of the stimulated emission. But there are distinct differences. The spectrum taken at the beginning shows a clear monomode emission with high intensity in contrast to the degrading diode whose emission spectrum is much broader and by more than one order of magnitude lower in intensity. A similar behaviour could be observed when operated in pulsed mode after more than 20 hours with a duty cycle of 0.1 %. To maintain a constant light output, the necessary current was increased and the FWHM of the emission augmented.

Comparing the emission in pulsed and in CW-mode, one notices, that the maximum of the emission in CW-mode is red-shifted by about 2 nm. This indicates a temperature increase in the QW, probably caused by heat generated at the top p-contact. Looking at the contact surface in Fig. 4, it is clearly visible, that before electrical operation of the laser, the 10 μm wide contact stripe shows hardly any defects or unevenness. After degradation the metal is lifted away partly from the semiconductor material or disappeared totally, presumably due to high temperature. The threshold current density of this structure was about 700 A/cm^2 at the contact and the operating voltage was between 15 V and 20 V. Since most of this voltage probably drops at the contact, a lot of electrical energy is converted into heat in this part of the device resulting in a massive temperature rise and a destruction of the laser. Measurements of the photoluminescence properties of devices, which were degraded in the described manner, showed no acceptor related features in the

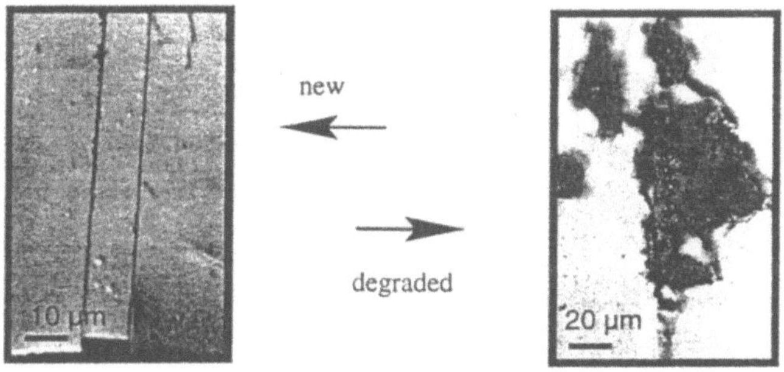

Figure 4 Laser structure top contact before operation (left) and after degradation (right).

spectra any longer [18] which is presumably caused by heat generation at the contact, as well. This degradation probably has to be seen independently of the QW, although the heat generation naturally fastens the degradation in the active region. Therefore we further investigated the contacts to p-ZnSe, as they seem to be the limiting factor for our devices at the present stage.

3 Contacts and Technology

The principal problem of the realization of ohmic contacts to p-ZnSe is the very low lying valence band of p-type doped ZnSe. Since there is no metal with a comparable or higher work function, a barrier is formed at the metal-semiconductor interface resulting in a high resistance for the transport of holes across the interface. For Au on ZnSe, the height of this barrier is approximately 1.2 eV [19]. Carrier transport at the interface is governed by thermionic emission and quantum mechanical tunneling. The exact value of the contact resistance depends on the height as well as on the thickness of the barrier. As the maximal achievable free hole concentration in ZnSe is very low with only 2×10^{18} cm^{-3} [20] several attempts to either lower the barrier or to increase the doping level have been reported in the past. For lowering the barrier, contacts using HgSe [21] or ternary alloys and multilayer contacts with ZnSe/ZnTe [22–24] or ZnSe/BeTe [25] were investigated. ZnSe/ZnTe and ZnSe/BeTe multilayers additionally increase the carrier concentration and therefore further reduce the contact resistance. For ZnTe, however, this advantage is only reached with the accumulation of additional strain and the generation of defects due to the large lattice mismatch of about 7 % between ZnSe and ZnTe. Notwithstanding, we have investigated ZnSe/ZnTe multilayer contacts. The thickness ratio between the ZnSe and the ZnTe layers was varied in a parabolic profile [24] with decreasing layer thickness of the ZnSe layers and increasing thickness of the ZnTe layers towards the top of the structure. An I-V curve of such a contact is shown in Fig. 5, curve (a). The measurement was taken on a ZnSSe pn-diode, therefore the voltage drop is not purely due to the p-contact, but the band gap of the semiconductor and the loss at the contact to the GaAs-substrate have to be considered as well. The remaining voltage drop at the p-contact is about 2 V which is nevertheless still too high for long lifetime LDs. Further investigations are in progress. Alternatively, we have investigated contacts, where the ZnTe concentration was further reduced. Details will be published elsewhere. The I-V curve of such a contact is shown in Fig. 5 , curve (b). This curve was measured on a pn-diode and showed an improvement

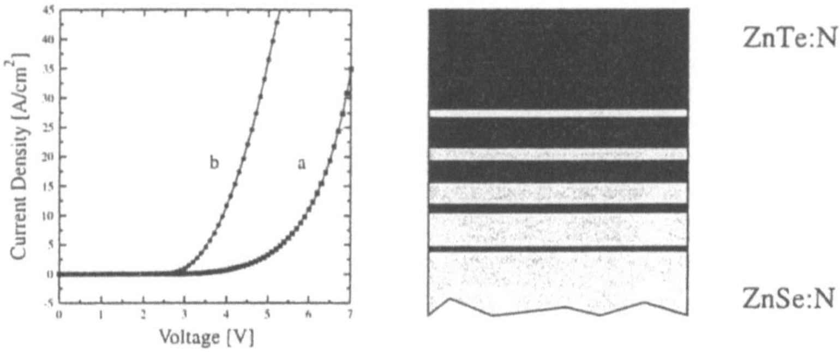

Figure 5 I-V curves through ZnSSe pn-diodes and schematic. The changes are due to improvements at the p-side contact for curve (b).

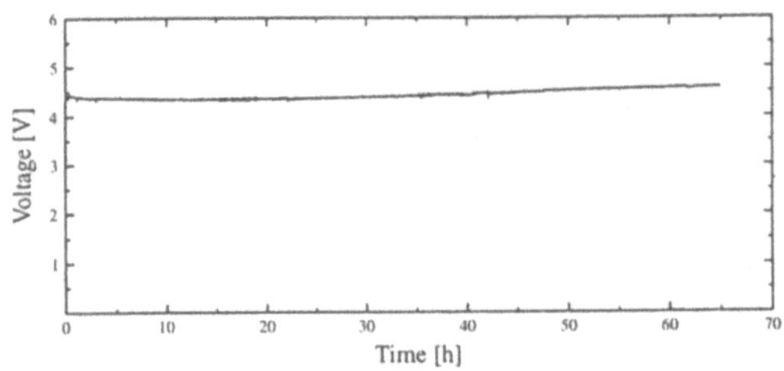

Figure 6 Longtime stability test of a contact structure. The voltage drop across a ZnSSe pn-diode versus time remains almost constant.

in comparison to the conventional design described above. Further investigations of both kinds of contacts will be necessary.
For large contact resistivities, the contact gets very hot and can degrade with time, presumably due to a further reduction of carrier density. Then the resistance is increased and the degradation of the LD becomes even faster. Therefore, the long time behaviour is of great importance. We have investigated structures at current densities of 20 A/cm^2 for more than 60 hours. Although the current density is lower than the density which is needed for stimulated emission, also at these densities, we were able to compare the long-time performances of the different contact designs. The result is shown in Fig. 6. An increase of the voltage drop at the whole device of less than 0.2 V could be detected. This demonstrates, that the contact structures were improved such, that they are no longer the reason for device degradation. Therefore, the investigation of other parts of the device, like the QW could be continued.

A problem of ZnTe based contacts are technological difficulties during the preparation. It was shown that, when using wet chemical etchants for structuring the devices, the etch rates lateral (r_l) and vertical (r_v) towards the layer differ greatly (r_l / r_v= 50), resulting in a tendency to under-etch covered parts. Fig. 7 shows the step of a 50 μm wide laser stripe, which was etched in potassium dichromate $H_2SO_4 : K_2Cr_2O_7$ for one minute. One can distinguish the 350 nm thick metal contact layer and an approx. 5.5 μm wide area, which was originally covered by metal and photoresist. During processing, the contact strip was under-etched on both sides and therefore this part was standing loose. At lift off, the free standing metal part was taken away, and the width of the contact stripe was reduced by about 5 μm on each side.

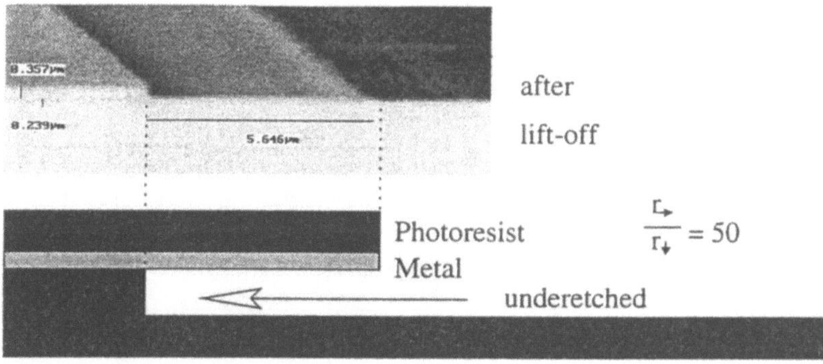

Figure 7 Under-etching of ZnTe contact structures.

As all investigated wet chemical etchants caused the same difficulties, dry chemical etching or physical etching will be necessary.

4 Superlattices Instead of Alloys

Lifetime limiting factors are low carrier concentrations in general and in particular the carrier concentration in the p-type doped barrier material MgZnSSe. Designing LDs, it is desirable to increase the bandgap in these layers as high as possible to effectively confine the optical field [26] and therefore reduce the current density needed. However, the dopability is reduced with increasing band gap [27] and therefore electrical losses augment. To achieve high carrier concentrations simultaneously with sufficiently high bandgaps, we have tried to replace the quaternary alloys MgZnSSe with superlattices (SL) [28]. Instead of 700 nm thick barriers, superlattices with 300 periods were grown. Each period consists of about 1.1 nm ZnSe and 1.4 nm MgZnSSe with an intended bandgap in the quaternary layers of approximately 3.2 eV. Investigations of the dopability indicate, that the maximal achievable hole concentration is by a factor of about 2 higher than in alloys with comparable bandgap. LDs were grown which were identical to the above described structures with the only exception that the alloy cladding layers were replaced by superlattice claddings on both sides of the waveguide. In addition, the alternatively designed contact described in the preceeding section was used [29]. In Fig. 8 a double axis $\omega/2\theta$-scan of the 004-reflection in HRXRD is depicted. Several characteristics are remarkable. The main peak of the SL only appears as a shoulder at the substrate peak at $\omega = 32.8°$, indicating an excellent lattice match to the GaAs substrate. Clearly visible are the first order satellites on both sides of the main peak at angular distances of about $\pm 1.9°$. The existence of the satellites proofs that, despite the small layer thicknesses in the superlattice of little more than 1 nm, no alloy is formed but a SL. The FWHM of the superlattice satellites is $100''$. This small value points to a high structural quality and a good homogeneity of the SL period during growth. This particular structure had ZnSe as active region and emitted light of a wavelength of 465 nm.

The emission of a similar structure with SL as cladding layers but a CdZnSSe quantum well also showed electrically pumped stimulated emission below and above lasing threshold. In Fig. 9 a spectrum at pulsed injection is depicted. Clearly distinguishable is the

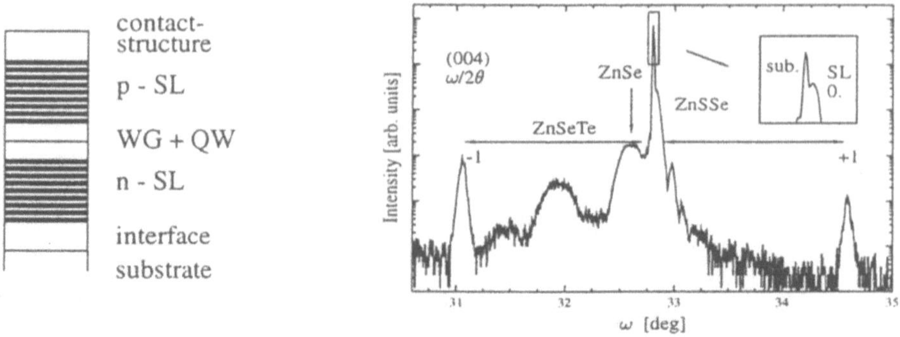

Figure 8 Schematic view and 004 ω/2θ X-ray measurement of a SL-cladding LD.

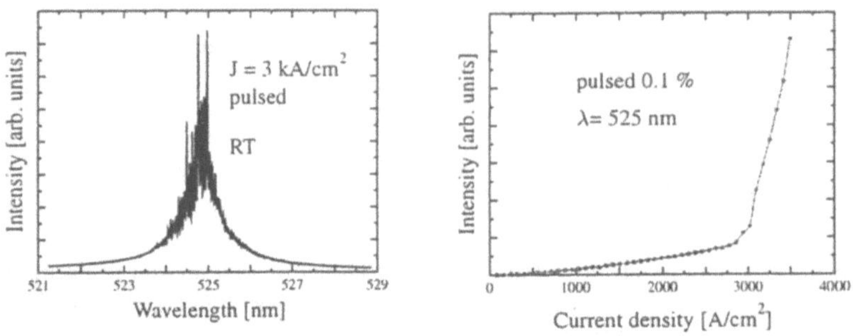

Figure 9 Electroluminescence in a LD with SL-claddings. Left: mode-spectrum. Right: *L-I* curve indicating a threshold current density of 3 kA/cm^2.

mode structure from stimulated emission and the background arising from spontaneous recombination. In the *L-I* curve in the right part of the figure, the difference between the slowly rising intensity in LED mode and the very steeply rising part of the curve at stimulated emission above the threshold can clearly be observed. The threshold current density of this structure was extremely high with approx. 3 kA/cm^2. Reasons for this could be low bandgaps in the cladding layers, resulting in a poor carrier and optical confinement or a broadening of the current path in the superlattices. As the effective masses lateral and normal to the superlattice layers are different, the current path could be much broader in the well than in the contact and therefore the actual current density lower. This assumption is further supported by the observation of the background of spontaneous emission in the far field. This background was very bright with almost the same wavelength as the stimulated emission. Therefore the current density at the contact might be reduced when the current path is further confined with plasma etching technology, for instance. This possibility is currently investigated. In this context it is in any case remarkable, that this high current density can be injected via the presented contact design. Thus, it can be expected, that the contacts are no longer the lifetime limiting factor for LDs shown here.

The reasons for degradation are point defects and extended defects [30], like stacking faults which originate at the substrate/epi-layer interface already during the beginning of the growth. Therefore choice and preparation of the substrate require special care.

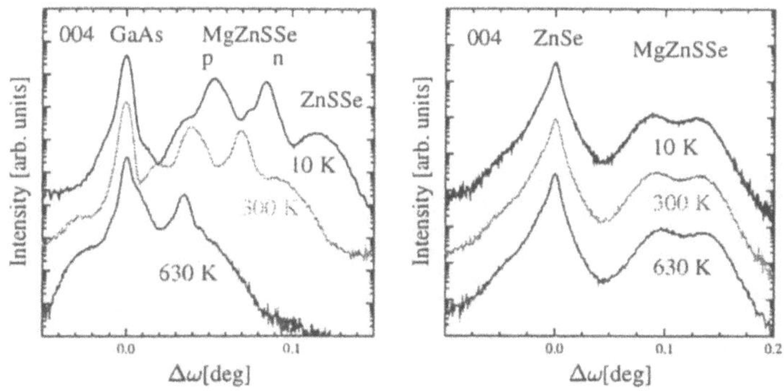

Figure 10 Some 004 ω/2θ-scans of a hetero- (left) and homoepitaxially (right) grown LD at different temperatures.

5 Homo- and Heteroepitaxy

5.1 Thermally Induced Strain

ZnSe and ZnSe based alloys are generally grown on GaAs substrates due to the low lattice mismatch of about 0.27% and the excellent quality of the available substrates. But both material systems have different thermal expansion coefficients. The relative difference between the values for GaAs and ZnSe amounts to approximately 23%. This causes thermally induced strain in the epitaxial layers when the sample is cooled from growth to room temperature. A layer, that was lattice-matched to the substrate during growth, will be mismatched at higher or lower temperatures (up to −0.05% at room temperature). This thermally induced strain can be reduced by the generation of additional dislocations in the layer. When the thermally induced strain is not sufficient to induce dislocations, the lateral lattice mismatch is maintained constant over the whole temperature range and the thermally induced strain becomes evident as the vertical lattice mismatch changes with temperature. A minimization of thermally induced strain is offered by homoepitaxy because the thermal expansion coefficients of substrate and layer naturally are almost identical. The investigated hetero- and homoepitaxially grown laser structures show indeed a constant lateral lattice mismatch over the whole temperature range. The thermally induced strain is therefore completely described by the vertical lattice mismatch. The differences are examplarily demonstrated Fig. 10. The cladding layers of the heteroepitaxially grown structure, indicated by MgZnSSe, clearly show a shift in position relative to the GaAs substrate peak at variing temperatures. The p-doped cladding layer, labeled "p" [14], was well lattice matched at growth temperature of about 280 °C, while at RT the lattice mismatch is as high as −0.045%. In contrast, the homoepitaxially grown structure maintains its lattice mismatch over the whole temperature range. Small variations in the lattice mismatch can be explained by the different expansion coefficients of the quaternary layers in comparison to the ZnSe substrate. In homoepitaxy, this difference amounts to only 10 % of the value in the case of heteroepitaxy and therefore is of minor importance.

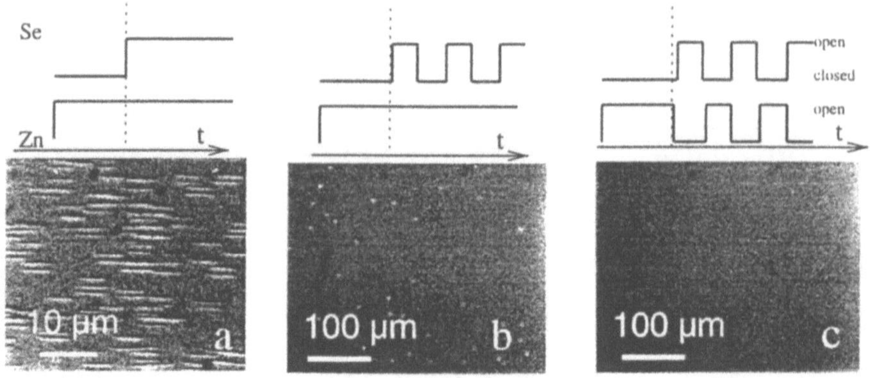

Figure 11 Etch pits on laserstructres started in MBE (a), MEE-MBE (b) and MEE (c) mode.

5.2 Heteroepitaxy on GaAs

In spite of the advantages of the growth on ZnSe substrates in comparison with the growth on GaAs substrates, the layers on GaAs are mostly of high crystalline quality and exhibit defect densities below $1 \times 10^4\ \mathrm{cm}^{-2}$. We were able to achieve that by careful control of not only the GaAs surface, and in this context of the amount of the arsenic coverage on the surface. Also the flux ratios between Zn and Se at II-VI growth start are of distinctive importance. Additionally we have investigated the effect of the mode of growth initiation on the layer quality, by comparing three different approaches. After a Zn irradiation of the GaAs surface in all three cases, we just added the Se flux to initiate growth in the first case, labeled MBE. For the second approach, labeled MEE (=migration enhanced epitaxy), Zn and Se cells were opened alternatively only, so that never both elements were present at the growing surface at the same time. During the third kind of growth start, labeled MBE-MEE, the Zn shutter was always open, while the Se cell was opened for 2 s and then closed for 4 s. The second and third method were repeated for 60 cycles in the described manner, before continuing in normal MBE mode. All three approaches are sketched in Fig. 11. The pictures below show etch pits of structures started in these manners. It is quiet obvious that a start in the MBE mode gives layers with very high defect densities, while layers grown under MEE starting conditions exhibit a drastically lower defect density. Structure started in normal MBE mode typically displayed a defect density of about $5 \times 10^6\ \mathrm{cm}^{-2}$. Little better were structures with MBE-MEE with a defect density of about $5 \times 10^5\ \mathrm{cm}^{-2}$. Only the MEE samples had a defect density of reproducibly less than $1 \times 10^4\ \mathrm{cm}^{-2}$. This defect density allows to manufacture laser structures with no defect within an area of the metal contact and therefore within the current path in the active region. Therefore degradation should be distinctly slowed down and longer lifetime diodes should be possible when growth is initiated in the MEE mode.

5.3 Homoepitaxy

Another approach for the reduction of point defects is homoepitaxy. An advantage besides the much smaller thermal mismatch between substrate and layer (compare Fig. 10)

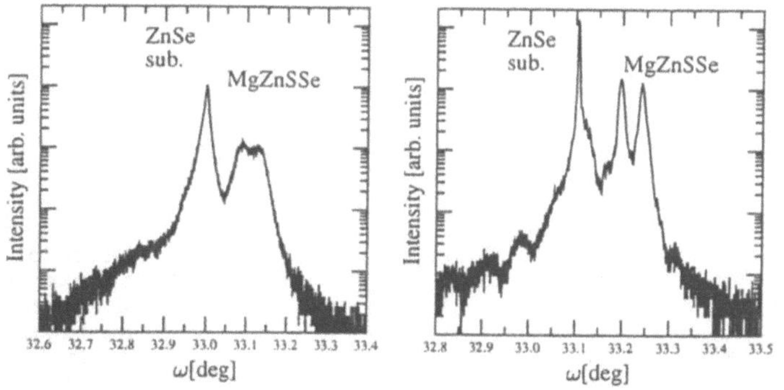

Figure 12 004 $\omega/2\theta$ scans of a homoepitaxial grown laser diode.

is the avoidance of the III-V/II-VI interface and its electrical and structural complications. In addition, it is possible to grow the whole structure in a single growth chamber. On the other hand, homoepitaxy is not yet well investigated [31,32] although first LDs have been demonstrated [33,34]. The reason for this is that good crystals with etch pit densities below 1×10^4 cm^{-2} are difficult to grow [35] and samples are normally small (10mm × 10mm) and expensive. Good substrates are an essential precondition for successful homoepitaxy and, especially for devices, doped substrates [36] will be needed. Unfortunately, those are commercially not available.

Before growth, the surface region, which was damaged during mechanical polishing and oxygen impurities have to be removed from the surface. To remove oxygen, a plasma source (MD 21) from Oxford Applied Research, which is normally used for p-type doping with nitrogen, was run with hydrogen. Recently it became known, that hydrogen radicals can clean ZnSe surfaces [37,38]. The optimization was performed using X-ray photoelectron spectroscopy (XPS) and reflectance anisotropy spectroscopy (RAS). ZnO_x is the problem, while SeO_x can simply be removed with wet chemical etching or thermal etching in the UHV system [39].

A very profound difference could be discovered in the optical, electrical and structural properties between layers, grown with different substrate preparations. In Fig. 12 $\omega/2\theta$ scans of two laser structures are depicted. As the substrates were undoped, a 3 μm thick ZnSe:Cl layer was grown first. The left curve was already shown in Fig. 10. The substrate peak as well as the layer peaks are broadened with FWHM of 46″ and 150″, respectively. The poor layer quality can be explained by the fact, that even at very high temperatures with a Se-flux stabilization, ZnO_x can not be removed completely from the surface. Instead, it has to be assumed, that zinc droplets form at the surface and consequently influence the proceeding growth [31]. A 3 × 1 reconstruction, which can be observed under these circumstances, turns into the known 2 × 1 reconstruction during the growth of the first 5 nm. A similar thickness for a transition layer is reported for a non-optimal growth start in heteroepitaxy of ZnSe on GaAs. The surface morphology of such layers is rough and only optical pumped lasing could be observed for such structures at RT. A different situation arises for growth after optimized substrate preparation. In this case, the substrate was mounted without indium and was cleaned with hydrogen plasma. In addition growth was initiated using MEE instead of MBE and the lattice mismatch was 0.11% to 0.17%. Under these conditions, a $c(2 \times 2)$ reconstruction was established before growth and right from the beginning of growth the surface stayed 2-dimensional and reconstructed. In the

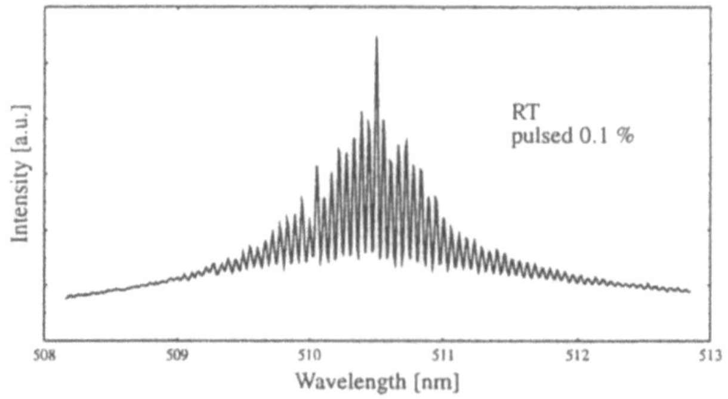

Figure 13 Stimulated emission under electrical injection of a homoepitaxail laser diode.

right curve of Fig. 12, distinct *Pendellösung* fringes can be observed, indicating smooth interfaces, like they could not be observed for quaternary layers grown without plasma cleaning. The X-ray curve for the homoepitaxial sample is comparable with good structures grown on GaAs substrates. The FWHM of the substrate is 13″, and 32″ and 33″ of the quaternary layers. The whole structure is pseudomorphic.

Under current injection, stimulated emission at RT could be observed from these structures [40]. Fig. 13 shows a spectrum which was taken under pulsed injection with a pulse width of 0.3 μs and a duty cycle of 0.1 % above the threshold current density of about 900 to 1000 A/cm^2. The wavelength of emission was approximately 520 nm with a FWHM of about 15 nm in LED mode and 510 nm with a FWHM of only 1 nm in LD mode. The lifetime of this diode in pulsed mode under constant output power was about 1 hour.

6 Conclusion

In conlcusion, II-VI laser diodes have been grown on GaAs and ZnSe substrates and high structural quality could be reached. CW-operation at RT could be achieved routinely and operating characteristics were investigated. Degradation as well as approaches to increase the device lifetime were studied. Contacts to p-ZnSe as well as p-type doping in ZnSe/MgZnSSe superlattices have been studied in detail and destinct improvements could be reported. Finally the growth start on GaAs has been investigated and the perspectives which can be offered by homoepitaxy have been presented. This resulted in the succesful growth of RT-lasing devices. It was shown, that there are still many problems that have to be solved before the question of reliability of II-VI laser diodes can be answered, conclusively.

Bibliography

[1] K. Ohkawa and T. Mitsuyu, J. Appl. Phys. **70**, 439 (1991).

[2] R. M. Park, M. B. Troffer, C. M. Rouleau, J. M. DePuydt, and M. A. Haase, Appl. Phys. Lett. **57**, 2127 (1990).

[3] M. A. Haase, J. Qiu, J. M. DePuydt, and H. Cheng, Appl. Phys. Lett. **59**, 1272 (1991).

[4] H. Jeon, J. Ding, W. Patterson, A. V. Nurmikko, W. Xie, D. C. Grillo, M. Kobayashi, and R. L. Gunshor, Appl. Phys. Lett. **59**, 3619 (1991).

[5] N. Nakayama, S. Itoh, H. Okuyama, M. Ozawa, T. Ohata, K. Nakano, M. Ikeda, A. Ishibashi, and Y. Mori, Electron. Lett. **29**, 2194 (1993).

[6] J. M. Gaines, R. R. Drenten, K. W. Haberern, T. Marshall, P. Mensz, and J. Petruzello, Appl. Phys. Lett. **62**, 2462 (1993).

[7] S. Guha, J. M. DePuydt, M. A. Haase, J. Qiu, and H. Cheng, Appl. Phys. Lett. **63**, 3107 (1993).

[8] S. Taniguchi, T. Hino, S. Itoh, K. Nakano, N. Nakayama, A. Ishibashi, and M. Ikeda, Electron. Lett. **32**, 552 (1996).

[9] E. Kato, H. Noguchi, M. Nagai, H. Okuyama, S. Kijima, and A. Ishibashi, Electron. Lett. **34**, 282 (1998).

[10] D. C. Grillo, M. D. Ringle, G. C. Hua, J. Han, R. L. Gunshor, M. Hovinen, and A. V. Nurmikko, J. Vac. Sci. Technol. B **13**, 720 (1995).

[11] M. Behringer, K. Ohkawa, V. Grossmann, K. Leonardi, M. Fehrer, D. Hommel, M. Kuttler, M. Strassburg, and D. Bimberg, J. Cryst. Growth **184/185**, 580 (1998).

[12] K. Ohkawa, M. Behringer, H. Wenisch, M. Fehrer, B. Jobst, D. Hommel, M. Kuttler, M. Strassburg, D. Bimberg, G. Bacher, D. Tönnies, and A. Forchel, Phys. Stat. Sol. (b) **202**, 683 (1997).

[13] V. Grossmann, M. Behringer, H. Heinke, and D. Hommel, J. Cryst. Growth **184/185**, 100 (1998).

[14] H. Heinke, V. Grossmann, M. Behringer, H. Wenisch, and D. Hommel, J. Cryst. Growth **184/185**, 587 (1998).

[15] H. C. Casey and M. Panish, *Heterostructure Lasers Part A: Fundamental Principles*, Academic Press, Inc., 1978.

[16] K. Kondo, M. Ukita, H. Yoshida, Y. Kishita, H. Okuyama, S. Ito, T. Ohata, K. Nakano, and A. Ishibashi, J. Appl. Phys. **76**, 2621 (1994).

[17] S.-L. Chuang, A. Ishibashi, S. Kijima, N. Nakayama, M. Ukita, and S. Taniguchi, IEEE J. Quant. Electr. **33**, 970 (1997).

[18] M. Vehse, to be published (1998).

[19] Y. Koide, K. Hashimoto, H. Ishikawa, K. Tsukui, T. Oku, M. Murakami, N. Teraguchi, Y. Tomomura, and A. Suzuki, J. Cryst. Growth **159**, 709 (1996).

[20] J. Qiu, J. M. DePuydt, H. Cheng, and M. A. Haase, Appl. Phys. Lett. **59**, 2992 (1991).

[21] S. Einfeldt, M. Behringer, J. Nürnberger, H. Heinke, T. Behr, C. R. Becker, D. Hommel, and G. Landwehr, Phys. Stat. Sol. (b) **187**, 439 (1995).

[22] Y. Fan, J. Han, L. He, J. Saraie, R. L. Gunshor, M. Hagerott, H. Jeon, A. V. Nurmikko, G. C. Hua, and N. Otsuka, Appl. Phys. Lett. **61**, 3160 (1992).

[23] F. Hiei, M. Ikeda, M. Ozawa, T. Miyajima, A. Ishibashi, and K. Akimoto, Electron. Lett. **29**, 878 (1993).

[24] G. L. Yang, H. Luo, L. Lewandowski, and J. K. Furdyna, Phys. Stat. Sol. (b) **187**, 435 (1995).

[25] A. Waag, F. Fischer, H. J. Lugauer, T. Litz, J. Laubender, U. Lunz, U. Zehnder, W. Ossau, T. Gerhardt, M. Möller, and G. Landwehr, J. Appl. Phys. **80**, 792 (1996).

[26] S. Itoh, N. Nakayama, S. Matsumoto, M. Nagai, K. Nakano, M. Ozawa, H. Okuyama, S. Tomiya, T. Ohata, M. Ikeda, A. Ishibashi, and Y. Mori, Jap. J. Appl. Phys. **33**, L938 (1994).

[27] H. Okuyama, Y. Kishita, T. Miyajima, A. Ishibashi, and K. Akimoto, Appl. Phys. Lett. **64**, 904 (1994).

[28] A. Ishibashi, II-VI blue-green light emitters, in *II-VI Compounds 1995*, pp 555–565, 1995.

[29] M. Behringer, to be published (1998).

[30] S. Guha, J. M. DePuydt, J. Qiu, G. E. Hofler, M. A. Haase, B. J. Wu, and H. Cheng, Appl. Phys. Lett. **63**, 3023 (1993).

[31] K. Ohkawa, A. Ueno, and T. Mitsuyu, J. Cryst. Growth **117**, 375 (1992).

[32] H. Wenisch, T. Behr, J. Kreissl, K. Schüll, D. Siche, H. Hartmann, and D. Hommel, J. Cryst. Growth **174**, 751 (1997).

[33] Z. Yu, C. Boney, W.C. Hughes, W.H. Rowland Jr., J.W. Cook Jr., J.F. Schetzina, G. Cantwell, and W.C. Harsch, Electron. Lett. **31**, 1341 (1995).

[34] A. Ohki, T. Ohno, T. Matsuoka, and Y. Ichimura, Electron. Lett. **33**, 990 (1997).

[35] D. Siche, H. Hartmann, K. Böttcher, and E. Krause, Phys. Stat. Sol. (b) **194**, 101 (1996).

[36] H. Wenisch, M. Fehrer, K. Ohkawa, D. Hommel, M. Prokesch, U. Rinas, and H. Hartmann, J. Appl. Phys. **82**, 4690 (1997).

[37] M.H. Jeon, L.C. Calhoun, B.P. Gila, M.H. Ludwig, and R.M. Park, Appl. Phys. Lett. **69**, 2107 (1996).

[38] W.C. Hughes, C. Boney, M.A.L. Johnson, J.W. Cook Jr., and J. Schetzina, J. Cryst. Growth **175/176**, 546 (1997).

[39] Y.-D. Kim, S.L. Cooper, M.V. Klein, and B.T. Jonker, Appl. Phys. Lett. **62**, 2387 (1993).

[40] H. Wenisch, K. Ohkawa, A. Isemann, M. Fehrer, and D. Hommel, Elec. Lett. **34**, 891 (1998).

Many-Body Theory for II-VI Semiconductor Laser Media

K. Henneberger, H. Güldner, G. Manzke, Q.Y. Peng,
M.F. Pereira Jr.[1]

Fachbereich Physik, Universität Rostock, D-18051 Rostock, Germany
[1] Departamento de Fisica, Universidad de Los Andes,
AA 4976 Santafé de Bogotá, D.C.-Colombia

Abstract: We study the interplay between resonant cavity, carrier-quantum confinement and many-body effects by means of a Keldysh Green's function approach which treats the relevant quasi-particles (photons, carriers and plasmons) within a fully quantum-mechanical footing. In order to consistently describe the gain lineshape and spectral position of lasing, it is demonstrated that both Coulomb effects within mean-field contributions and higher order many-body effects as screening, diagonal and off-diagonal dephasing and renormalization of the interband energy have to be taken into account. Strong compensation between diagonal and off-diagonal contributions is found. Numerical results for II-VI heterostructures under quasi-equilibrium conditions and for temperatures ranging from 100 K to 300 K are compared with recent experimental findings. Solving the kinetic equation of photons transient spectral and temporal oscillations experimentally observed in the laser emission are shown to result from memory effects.

1 Introduction

Blue-green light-emitting laser diodes have attracted considerable interest recently, among other reasons due to the short operation wavelengths which allow for massive data storage in optical disks [1]. However, there is no consensus on which microscopic mechanism is responsible for lasing in these structures in the different operation temperature ranges. Due to the large excitonic binding energy, inelastic exciton-exciton scattering in bulk ZnSe and ZnSSe [2,3] and in ZnCdSe/ZnSe MQWs [4], exciton-LO-phonon scattering in ZnCdSe/ZnSe MQWs [5], and a Fermi-like many-body interaction between localized excitons in ZnCdSe/ZnSSe MQWs [6–8] have been suggested in the last few years. Localized tail states as well as biexciton decay in ZnSe/ZnSSe superlattices [9] and ZnCdSe/ZnSe quantum well structures have been considered [10]. In contrast to that, Cingolani et al. [11] have supported electron-hole (e-h) plasma recombination as the dominant mechanism for lasing even at low temperatures. At temperatures above 80 K this mechanism is also favoured by Klingshirn et al. [9] for ZnSe/ZnSSe superlattices. Recently, results from a novel pump-and-probe photoluminescence excitation experiment [12] support e-h plasma related lasing even for low temperatures, and further experimental results under optical pumping and calculations for laser samples containing ZnCdSe/ZnSSe quantum wells (QW) question exciton-related gain mechanisms [13].

A considerable amount of literature exists on laser theory in general and particularly on semiconductor lasers being based on Langevin theory to model the relevant fluctuation and damping processes [14–16]. Only recently a microscopic, rigorous quantum theory for semiconductor lasers has been developed which can connect and relate to each other the correlations of both the light and matter subsystems. The subsystems are integrated with the context of a global interacting many-body system on an equal quantum mechanical footing [17,18]. This theoretical concept, which is based on nonequilibrium Green's functions (GF) [19], relates the output power of a highly excited semiconductor in a unique way to the relevant correlation functions, namely, the spectrally resolved cavity loss and the photon density. The semiconductor plays the dual role of gain medium and resonant cavity.

For steady state lasing well below full gain-loss compensation, the e-h system can be considered to be in thermal quasi-equilibrium and its recombination rate is obtained from the optical gain via the Kubo-Martin-Schwinger (KMS) relation. Then, every physical quantity is given in terms of the gain, or, using the Kramers-Kronig transformation, in terms of the optical susceptibility.

In Section 2 for this situation we demonstrate the influence of the relevant many-body effects beyond Hartree-Fock (HF), namely, dynamical screening, diagonal and off-diagonal dephasing and energy renormalization, as well as excitonic features, on the optical susceptibility. We solve the semiconductor Bloch equations (SBE) for bulk II-VI compounds and densities ranging from well below up to above the Mott transition until optical gain is achieved. Besides room temperature where lasers are intended to operate, our investigations are restricted to temperatures high enough so that both biexcitonic correlations and disorder-induced localization can be ruled out.

In Section 3 an alternative approach to the SBE, namely, the Bethe-Salpeter equation (BSE) for the polarization function is used to investigate the optical gain and emission line shape of II-VI quantum wells. Quantum confinement and valence band coupling are considered by diagonalizing the corresponding Luttinger Hamiltonian. The Photoluminescence Excitation Spectroscopy (PLES) and Two-Beam Photoluminescence Excitation (TBPLE) spectra calculated in this way are in good agreement with experimental data.

In Section 4 the restrictions to steady state in the excitation and to quasi-equilibrium for the carriers are removed. Transient spectral and temporal oscillations experimentally observed in the laser emission are shown to result from memory effects in the quantum kinetics of photons.

2 Many-Body Effects in the Gain Spectra of II-VI Bulk Semiconductors

A peculiarity of the II-VI semiconductors in comparison to III-V materials is their rather high excitonic binding energy even in the bulk case. This reflects strong Coulomb correlations, which have to be taken into account not only considering the excitonic features at low excitation but also in the case of high excitation, particularly in the gain region, where the many-body effects become dominant. In this section we will discuss the relevant many-body effects in the frame of linear response theory assuming the carrier subsystem to be in quasi-equilibrium and show how they influence energy shift and line shape of the absorption/gain spectra.

The linear optical response can be determined solving the semiconductor Bloch equations (SBE) for the susceptibility

$$\chi_k(\omega) = \chi_k^0(\omega) + \chi_k^0(\omega) \sum_q \left\{ v_{k-q} + \frac{\Theta_{k,q}(\omega)}{N_k} \right\} \chi_q(\omega). \tag{2.1}$$

Here $\chi_k^0(\omega)$ is the susceptibility in Random Phase Approximation (RPA)

$$\chi_k^0(\omega) = \frac{N_k}{k^2 - \omega - \Delta e_k^{\mathrm{HF}} - \Sigma_k^r(\omega)} \tag{2.2}$$

including the Coulomb-Hartree Fock (HF) renormalized carrier energies $\Delta e_k^{\mathrm{HF}} = \sum_{a,q} v_{k-q} f_q^a$. $f_k^{\mathrm{e}}/f_k^{\mathrm{h}}$ are the carrier distributions of electrons/holes and $N_k = 1 - f_k^{\mathrm{e}} - f_k^{\mathrm{h}}$ describes the Pauli blocking. The retarded self-energy Σ_k^r comprises the sum over the diagonal (with respect to the band indices) contributions of electrons and holes. The self-energy is generated due to screening of the Coulomb interaction $\Sigma_k^r = \sum_q \Theta_{q,k} = \Delta e_k^{\mathrm{sc}} + i\Gamma_k$, where its real part results in an additional renormalization of the interband energy Δe_k^{sc} and the imaginary part in the dephasing rate Γ_k.

The complex matrix $\Theta_{k,q}$ describes screening by coupling the carrier system to the collective plasma excitations

$$\Theta_{k,q}(\omega) = -\sum_{a \neq b} \int \frac{d\omega'}{2\pi} \frac{\left[1 - f_k^a\right] V_{k-q}^{>}(\omega') - f_k^a V_{k-q}^{<}(\omega')}{\omega - \varepsilon_k^a - \varepsilon_q^b - \omega' + i\delta} = \Delta V_{k,q}^{\mathrm{eff}} - i\Gamma_{k,q}, \tag{2.3}$$

represented by the plasmon GF $V^{\gtrless}$. The real part of the interaction matrix $\Delta V_{k,q}^{\mathrm{eff}}$ acts as a renormalization of the Coulomb interaction due to screening, whereas the imaginary part is the so-called off-diagonal dephasing $\Gamma_{k,q}$ [20]. The complex interaction matrix can be derived considering the scattering terms in the kinetic equation for the coherent polarization p_k. The diagonal terms arise from collisions between carriers only, whereas the off-diagonal terms are generated by collisions between carriers and the interband polarization [21]. The plasmon GF $V^{<}$ is related to the retarded screened potential $V_q^r(\omega)$ via KMS-relation

$$V_q^{<}(\omega) = b(\omega)\, \mathrm{Im}\, V_q^r(\omega), \tag{2.4}$$

where $b(\omega)$ is a Bose function. The dielectric function $\varepsilon^L(q,\omega)$ generating the dynamically screened potential $V_q^r(\omega) = v_q/\varepsilon^L(q,\omega)$ is used in RPA considering the influence of the damping of carrier states. A detailed discussion of the role of dynamical screening was given elsewhere [22].

The role of the different many-body corrections in (2.1) is analysed under gain conditions ($T = 77$ K and $n = 5 \cdot 10^{18}$ cm^{-3}) in Fig. 1. The solid line corresponds to our approach. Neglecting the effective interaction $\Delta V_{k,q}^{\mathrm{eff}}$ the gain is increased (long-dashed line), indicating, that the Coulomb interaction is reduced by $\Delta V_{k,q}^{\mathrm{eff}}$. On the other hand it is shifted by an amount of Δe_k^{sc} to higher energies, when this (diagonal) contribution to the energy shift is dropped (dash-dotted line). A compensation of the dephasing rate Γ_k with the off-diagonal dephasing $\Gamma_{k,q}$ was shown to be necessary to give the correct excitonic line shape [28]. Neglecting the off-diagonal part the dephasing is overestimated and the gain becomes much weaker (dotted line in Fig. 1).

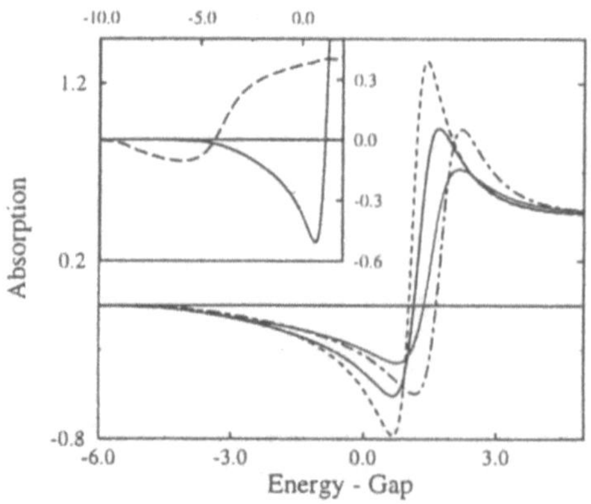

Figure 1
Influence of different many-body corrections in the absorption (ZnSe, $T = 77$ K, $N = 5.0 \times 10^{18}$ cm^{-3}). Complete approach (2.1): solid line; without effective interaction $\Delta V^{\text{eff}}_{k,q}$: dashed line; without correlation energy shift Δe^{sc}_k: dash-dotted line; and without off-diagonal dephasing $\Gamma_{k,q}$: dotted line. Inset: comparison of the complete approach (solid line) with simple static screening (dashed line, for explanation see text).

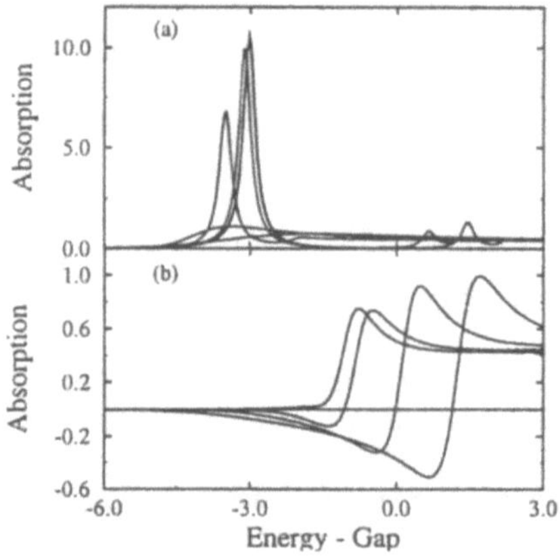

Figure 2
Transition from absorption to gain for ZnSe at 77 K and different carrier densities $N = 1.0 \times 10^{14}$, 1.0×10^{16}, 1.0×10^{17}, 2.5×10^{17}, 5.0×10^{17} cm^{-3} in (a) from top to bottom and $N = 5.0 \times 10^{17}$, 1.0×10^{18}, 2.5×10^{18}, 5.0×10^{18} cm^{-3} in (b) from left to right.

The inclusion of many-body effects by a retarded self-energy is standard in the semiconductor theory. However, in earlier treatments the sum of the HF- and the screening contribution $\Delta e^{\text{HF}}_k + \Delta e^{\text{sc}}_k$ was approximated by a Coulomb-hole and a statically screened HF-contribution [23], and the Coulomb potential in (2.1) was replaced by a statically screened one, approximating the effect of Coulomb + effective interaction $\Delta V^{\text{eff}}_{k,q}$ in (2.1). This treatment leads to an artificial red-shift both of the exciton [24] and the gain [25] as demonstrated in the inset of Fig. 1 [26,27].

In [20] was shown, that an artificial absorption below the band gap is avoided by the consideration of the off-diagonal dephasing. This problem does not appear with our interaction matrix (2.3), which is different from other approaches [20,26,29,30] with respect to the energy conservation relation in the denominator of (2.3) according to [31]. Our approach results in dephasing rates, which are localized for smaller and vanish for higher wave numbers. This is the reason for a correct gain behaviour below the band gap for the RPA result. An alternative way to solve this problem is demonstrated in the next Section.

In Fig. 2 the imaginary part of the susceptibility is compared for different carrier densities at a temperature of 77 K in order to demonstrate the transition from absorption to gain with increasing excitation. At lower densities the 1s- and 2s-exciton states are clearly resolved. Increasing excitation a weak red shift of the exciton appears. The sensitivity of this shift was investigated in [24,32,33] both theoretically and experimentally, where we could show, that our approach is able to describe the experimentally observed crossover of the shift from red to blue with decreasing temperature. With a further increase of the excitation the exciton becomes broadened and bleached above the Mott transition at a density of about $2.5 \cdot 10^{17}$ cm^{-3} (dash-dotted line in Fig. 2a). Finally, gain appears (Fig. 2b) in the vicinity of the 1s-exciton peak, however, after the Mott transition is finished.

As a characteristic feature of the II-VI semiconductors a remaining Coulomb enhancement gives rise to the bump near the band edge. As mentioned in the introduction the nature of the gain in the II-VI materials has been discussed controversially. Our results show a strongly Coulomb correlated e-h plasma to be responsible for the gain. This supports corresponding statements in [12,13,34] for II-VI quantum wells. In comparison to earlier treatments our statement is additionally founded, because our microscopic approach explains the excitonic features in detail, e.g., the shift of the exciton and the carrier induced dephasing [24,32]. Generally, from the experiments it is well known that the exciton undergoes only a weak shift with increasing excitation. This arises in our approach by a nearly complete compensation of the energy shifts in $\chi^0(\omega)$ with the two interaction contributions (Coulomb + effective) in (2.1).

3 Gain and Emission Line Shape in II-VI Quantum Wells

In this section we apply the general theory to the quantum well (QW) case at quasi-equilibrium. The quantum confinement and valence band coupling require the diagonalization of the corresponding Luttinger Hamiltonian. We consider uncoupled (parabolic) conduction and coupled (non-parabolic) valence bands. Each conduction subband is described by the $m_J = \pm 1/2$ projection of the electron spin σ, and a quantization index $j = c_1, c_2, \ldots$, as well as the in-plane quasi-momentum $\vec{k}$. We keep the notation as simple as possible and condense the quantum numbers in a single label, $n = \{\sigma_J\}$. The coupled valence bands have each a heavy $m_J = \pm 3/2$ and a light-hole component $m_J = \pm 1/2$. The band labels are $n_1, n_2, n_3 \ldots$. The block diagonalization label p reduces to a spin label σ in the absence of band-coupling.

Following the prescription in [35] for a homogeneous approximation of the excited medium, we use the QW Hamiltonian basis set of eigenstates [36] to expand the optical susceptibility

$$\chi(\omega) = \sum_{k,n_1,n_2} \Pi_{n_1,n_2}(k)\, \chi_{n_1 n_2}(k,\omega), \tag{3.5}$$

where $\Pi_{n_1,n_2}(k)$ are the matrix elements of the velocity operator and n_i includes all quantum numbers of this set.

For each transition in (3.5) we obtain a Bethe-Salpeter equation (BSE)

$$\chi_{n_1 n_2}(k,\omega) = \chi^0_{n_1 n_2}(k,\omega) + \chi^0_{n_1 n_2}(k,\omega) \sum_{k'} V_{n_1,n_2}(k-k')\chi_{n_1 n_2}(k',\omega). \tag{3.6}$$

For fixed n_1, n_2, parabolic dephasing relation, and the bulk screened potential, this equation corresponds to (2.1). The imaginary part of the RPA term, $\chi_0(k,\omega)$ driving the Bethe-Salpeter equation is obtained in contrast to (2.2) by

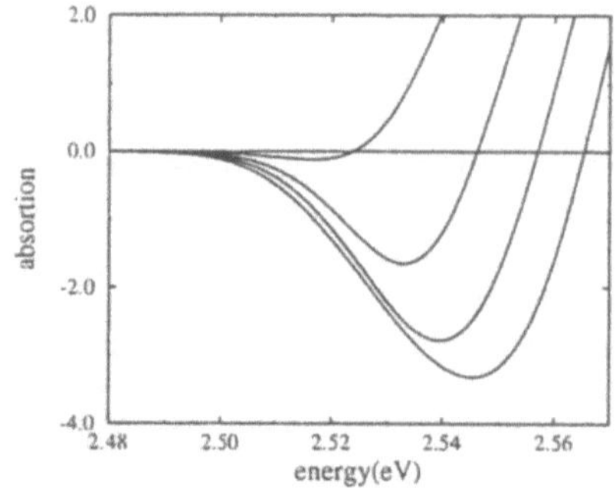

Figure 3
Gain spectra for a 100 Å $Zn_{0.8}Cd_{0.2}Se/ZnSe$ QW at a temperature of $T = 100$ K. From bottom to top the carrier densities are $N = 0.5, 1.0, 1.5, 2 \times 10^{18}$ cm^{-3}.

$$\mathrm{Im}\chi^0_{n_1n_2}(k,\omega) = \int \frac{d\omega'}{2\pi} \frac{2\Gamma_{n_1} 2\Gamma_{n_2} \left[1 - f^{\mathrm{e}}_{n_1}(\omega') - f^{\mathrm{h}}_{n_2}(\omega - \omega')\right]}{\left[(\omega' - e_{n_1})^2 + \Gamma^2_{n_1}\right]\left[(\omega - \omega' - e_{n_2})^2 + \Gamma^2_{n_2}\right]}. \tag{3.7}$$

The real part follows by Kramers-Kronig transformation. The renormalized energies e_n and the dephasing rates Γ_n correspond to those in (2.2), $f^{\mathrm{e,h}}_n(\omega) = 1/\{1 + \exp[\beta(\hbar\omega - \mu^{\mathrm{e,h}}_n)]\}$, is the Fermi distribution. Using the gain $g(\omega) = -8\pi\omega^2/c^2\,\mathrm{Im}\chi(\omega)$ from (3.5) the recombination rate $R(\omega)$ can be determined via the KMS relation

$$R(\omega) = b(\omega)g(\omega). \tag{3.8}$$

Fig. 3 depicts gain spectra for a 100 Å $Zn_{0.8}Cd_{0.2}Se/ZnSe$ QW at a temperature of $T = 100$ K and carrier densities, from bottom to top given by $N = 0.5, 1.0, 1.5$ and 2×10^{18} cm^{-3}. We include in the computation the influence of inhomogeneous broadening, due to fluctuations of the alloy concentration x. A Gaussian distribution around a central value $x_0 = 0.8$ with a dispersion $x = 0.014$ is used, namely, $g(\omega) = \sum g(\omega,x)P(x)/\sum P(x)$, $P(x) = \exp\{-[(x - x_0)/xw]^2\}$. The transitions considered in the optical spectra are e1-hh1 and e2-hh2, since all others have been shown to yield negligible contributions to the optical spectra.

The gain spectra evolve smoothly on the low-energy side to zero and do not change sign into an unphysical absorption [27,37,38]. Indeed, the extended treatment for the RPA-term (3.7) effects concerning this sensitive point in the same way as the approach in the last section considering the off-diagonal dephasing or dephasing rates, which are localized for small wave numbers. However, there is still one further fundamental problem which can be solved with the extended RPA-term (3.7). If the Pauli-blocking factor is written under the quasi-particle approximation, i.e., with the Fermi functions evaluated at the dispersion relations, $f^{\mathrm{e}}(e^{\mathrm{e}}(k))$, $f^{\mathrm{h}}(e^{\mathrm{h}}(k))$, the susceptibility does not satisfy the KMS condition (3.8). This condition must be fulfilled at quasi-equilibrium. Another consequence of its violation is the development of unphysical negative luminescence on the high energy side of the spectrum. None of these artifacts appear in our numerics [27]. Furthermore, our approach automatically guarantees that the switch from absorption to gain occurs exactly at the chemical potential. Consequently, the technique proposed in [39] to extract the gain spectra from spontaneous emission data can be used safely without the discrepancies between "experimental" and "computed" chemical potential differences.

Fig. 4 illustrates the capability of the theory to model Photoluminescence Excitation Spectroscopy (PLES). Not all features of PLES can be reproduced in the model used

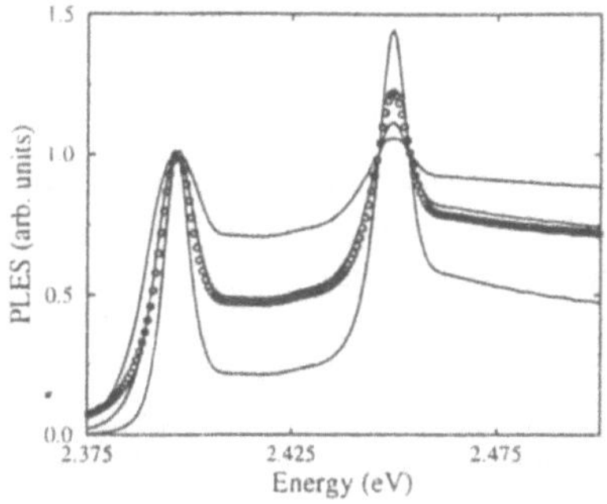

Figure 4
Photoluminescence excitation spectroscopy spectra for a 100 Å $Zn_{0.8}Cd_{0.2}Se/ZnSe$ QW. From bottom to top, the solid curves correspond to excitation intensities of 1 ns pulses given by: $I_{ex} = 8.3, 400, 800\ kW/cm^2$. The spectra are normalized respectively by the ratios 1, 0.002, and 0.0014. The linear absorption spectrum is shown for comparison as circles.

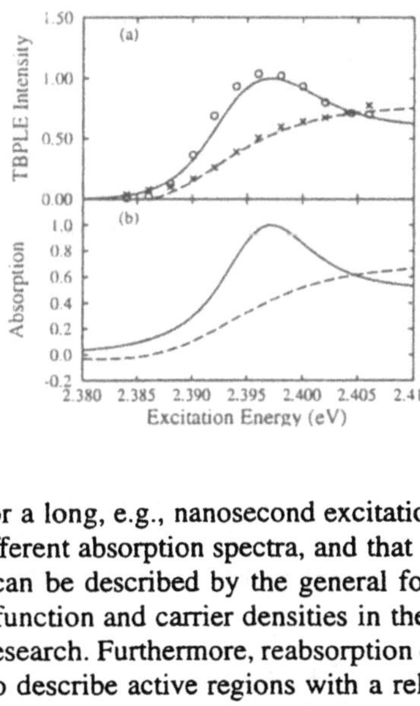

Figure 5
Two-beam photoluminescence excitation spectroscopy spectra for a 100 Å QW at 10 K. The solid and dashed curves are respectively for the pump intensities $I = 0$ and $40\ kW/cm^2$. Symbols: experimental data from [12]. (b) Corresponding absorption spectra.

here. For a long, e.g., nanosecond excitation pulse, the leading edge and the tail experience different absorption spectra, and that should alter the spectrum significantly. These effects can be described by the general formalism solving the equations for the polarization function and carrier densities in the time domain, and this will be the subject of future research. Furthermore, reabsorption effects were not included. The theory is better suited to describe active regions with a relatively small number of quantum wells, uniformly excited. However, the basic features of PLES are consistently reproduced by the numerics presented here. As the excitation intensity and thus the carrier density increases, so does the luminescence yield. The peak to valley contrast is reduced as seen in typical experiments [40]. So, we obtain a connection between absorption (symbols) and PLES, and how they are qualitatively obtained from a microscopic calculation.

Fig. 5a is for the Two-Beam Photoluminescence Excitation Spectroscopy experiment (TBPLE) introduced in [12]. The results are presented for a 100 Å $Zn_{0.8}Cd_{0.2}Se/ZnSe$ QW. The numerical method follows directly the experimental technique. The linear absorption $\alpha(\omega, N_{pu} = 0)$ is computed and from that, assuming that all the light is converted into carriers, the pump-generated density N_{pu} can be estimated. The pump energy is fixed at $\omega_{pu} = 2.48$ eV and the luminescence, $L(N_{pu})$, due to N_{pu} only, is computed at the fixed energy $\omega_{det} = 2.38$ eV. The pump-generated carriers give rise to a nonlinear absorption spectrum $\alpha(\omega, N_{pu})$. So, when the probe beam excites the sample, it is absorbed according to $\alpha(\omega, N_{pu})$ and generates additional carriers, $N(\omega, N_{pu})$. The luminescence due to both beams is detected at ω_{det}, and the constant pump contribution is subtracted, $L(\omega, N(\omega, N_{pu}) + N_{pu}) - L(N_{pu})$. The probe beam intensity is fixed at $92\ kW/cm^2$. For a comparison with the data from [12] (symbols), the TBPLE is normalized to the value at 2.404 eV. The circles and crosses are, respectively, for the experimental intensities $I_{\rm pu} = 0$ and 40 kW/cm^2. The solid and dashed lines are the corresponding computed curves with a plasma density $T_{\rm pl} = 100$ K.

The corresponding absorption spectra also with the normalization at 2.404 eV are shown in Fig. 5b. The absorption and TBPLE are qualitatively related, and it is clear that the excitonic peak disappears for carrier densities required for inversion, i.e., $I_{\rm pu} \geq$ 40 kW/cm^2 and consequently lasing to occur. The gain appears in our calculations near the spectral position where lasing has been measured, which supports our microscopic model for the lasing mechanism. The actual state of the system can not be described by free-carrier or pure excitonic models, but rather by a strongly Coulomb-interacting e-h plasma. Note that the co-existence of an "exciton-like" kink and gain depends on the plasma temperature and thus on the excitation conditions and on the scattering mechanisms. A systematic comparison with experiments over a wide temperature range and different samples further confirms the quality of our model [34,41,42]. Moreover, the detailed study shows that thinner quantum wells should, under ideal conditions, give rise to an enhancement in the Coulomb effects, since the stronger carrier confinement increases the effective Coulomb interaction. However, thinner samples are usually of worse quality and more inhomogeneously broadened due to alloy concentration and well width fluctuation. The inhomogeneous broadening can actually mask the increased Coulomb-related features in the optical spectra [41,42].

4 Laser Dynamics

In order to investigate the temporal evolution of the laser output after pulse excitation we have to go beyond the assumption of quasi-equilibrium in the carrier system. We solve a system of coupled equations for both photon and carrier distributions, which lead to a kinetic equation for the photon distribution $n(\omega,t)$ [43]. Omitting the arguments t and ω, we obtain

$$(\partial_\omega q_1^2)\partial_t n - (\partial_t q_1^2)\partial_\omega n = R\hat{D} + gn - P_{\rm out}, \tag{4.9}$$

where q_1 denotes the real part of the photon dispersion $q^2 = \omega^2/c^2(1+4\pi\chi) = (q_1 + iq_2)^2$. On the RHS we have the spectral function of photons

$$\hat{D}(\omega,t) = g(\omega,t)S(\omega,t) + \hat{D}_0(\omega,t), \tag{4.10}$$

which can exactly be decomposed into medium-induced ($g \cdot S$) and vacuum-induced ($\hat{D}_0$) photon states [17,18], and the spectral output of the cavity,

$$P_{\rm out}(\omega,t) = \kappa(\omega,t)n(\omega,t), \tag{4.11}$$

with the cavity loss function

$$\kappa(\omega,t) = \frac{\hat{D}_0(\omega,t)}{S(\omega,t)}. \tag{4.12}$$

For the numerical treatment we use the generation rate in RPA

$$R(\omega,t) = \frac{4\pi}{c}|j_{cv}|^2 \sum_k \frac{2\Gamma_k(t)}{[\omega - \varepsilon_k(t)]^2 + \Gamma_k^2(t)} f_k^e(t) f_k^h(t), \tag{4.13}$$

where the dephasing and the interband energy correspond to those in (2.2).

In contrast to the treatment in the previous sections the nonequilibrium distributions are obtained by consistently solving a Boltzmann equation in which enter elastic carrier-carrier scattering via a screened Coulomb potential, recombination of e-h pairs connected with light emission, the thermalization of the e-h plasma due to interaction with the lattice, described by a bath of equilibrium LO-phonons, and an incoherent optical pump.

The solution of (4.9) is carried out on its characteristics ω_t, which obey

$$\frac{d\omega_t}{dt} = -\frac{\partial_t q_1^2(\omega,t)}{\partial_\omega q_1^2(\omega,t)}, \tag{4.14}$$

and the general solution turns out starting at $t = 0$ with a nonexcited medium, where $n = 0$, as

$$n = n_{\mathrm{ad}} - \int_0^t d\tau \frac{dn_{\mathrm{ad}}}{d\tau} exp\left[- \int_\tau^t d\tau' \frac{\kappa - g}{\partial_\omega (q_1^2)} \right], \tag{4.15}$$

with the quantities depending now on ω_t and t. The integral expression on the RHS describes the retardation of the photon distribution in comparison to the corresponding adiabatically following quantity $n_{\mathrm{ad}}(\omega,t) = R(\omega,t)S(\omega,t)$. This retardation depends on the evolution of n_{ad}, and mainly on the exponential memory term, in which the cavity net loss $\kappa - g$ enters.

We first apply this laser theory to a numerical computation of the transient emission in order to compare the adiabatically following photon distribution with that resulting from the kinetic equation. We use Gaussian pulses of variable duration but all with the same total energy, in resonance with the fundamental band gap. In Fig. 6 the dynamics of the photon distribution at a fixed resonator mode is shown within a time scale compatible with the excitation-pulse duration. In the beginning of the excitation process e-h pairs near the band edge are created, giving rise to a steep emission onset. Simultaneously the gap shrinkage and the Coulomb interaction lead to an occupation of higher k-states in the bands which reduces the optical gain, and hence, the emission intensity. The curves in Fig. 6 show that the deviation between the two approaches for n increases both with the temporal gradient of the adiabatic quantity as well as with the resonator length. The dynamically computed photon distribution generally shows a smoother switching behaviour. Obviously, the validity of the adiabatic description is restricted to short cavities and to temporal changes of the photon kinetics on a time scale of few picoseconds, or, for highly excited long cavities, even nanoseconds.

Second, we carried out a numerical description of oscillations in the emission output of few millimeters long ZnSe-slabs, which have been recently measured [44]. In cavities of such length, the Fabry-Perot mode structure in the photon distribution would be extremely dense and sensitive to changes of the resonator length, which doubtlessly occur by the variation of the lattice temperature. Therefore, we use in these computations an envelope function for both spectral function and photon distribution. The oscillations in the emission (Fig. 7), which appear above a certain carrier density come up by the interplay of the retardation in the photon kinetics and the thermal relaxation of the carriers. The adiabatic photon distribution changes simultaneously with the gain. The memory term in

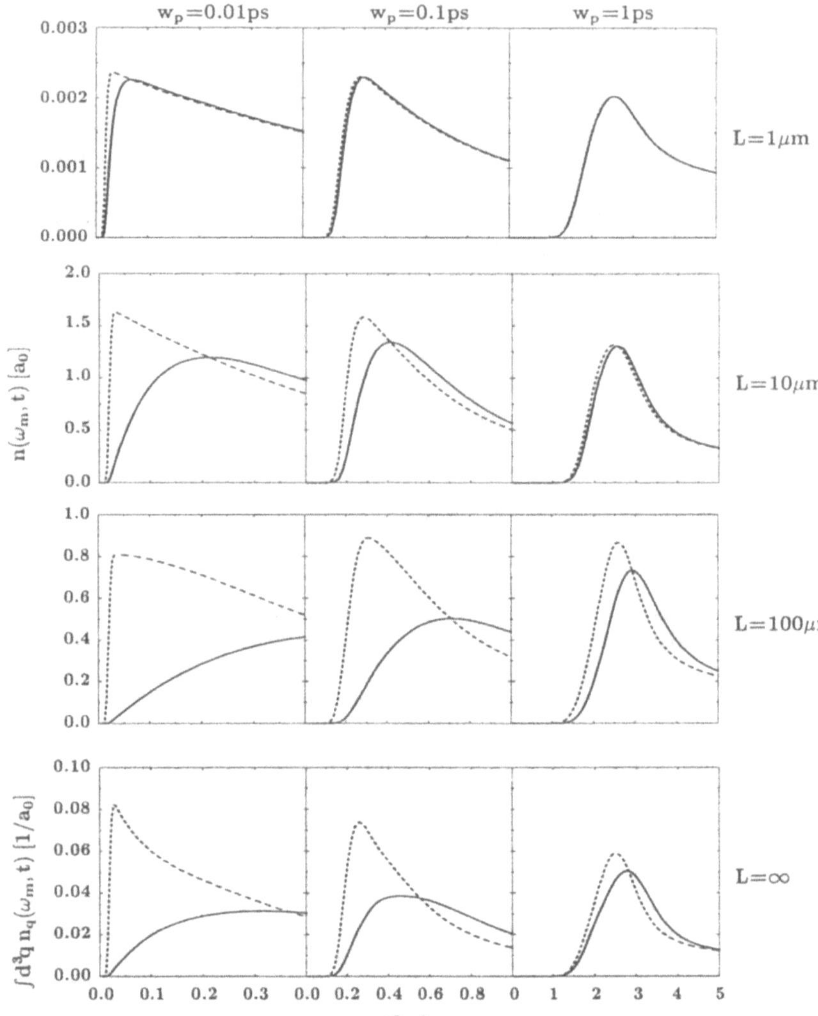

Figure 6 Photon distribution on a fixed laser mode adiabatically following (dashed) and from kinetic equation (solid) for resonators of different length L [μm] = 1, 10, 100, ∞ excited with Gaussian pulses of halfwidths w_p [ps] = 0.01, 0.1, 1 and pulse maxima at 0.02, 0.2, 2 ps.

(4.15) causes a delay in the evolution of the output. At a certain level of carrier density, this delay leads to a diminution of carriers near the band gap being such strong that it cannot be immediately compensated by the thermal relaxation. The temperature of the e-h plasma rises so that the emission is switched off for about 5–10 ps. Further relaxation towards lower temperatures causes a reoccupation of near-band-gap states and leads to a second peak. Moreover, for stronger excitation, repetitions of this process occur, leading to multiple oscillations in the output.

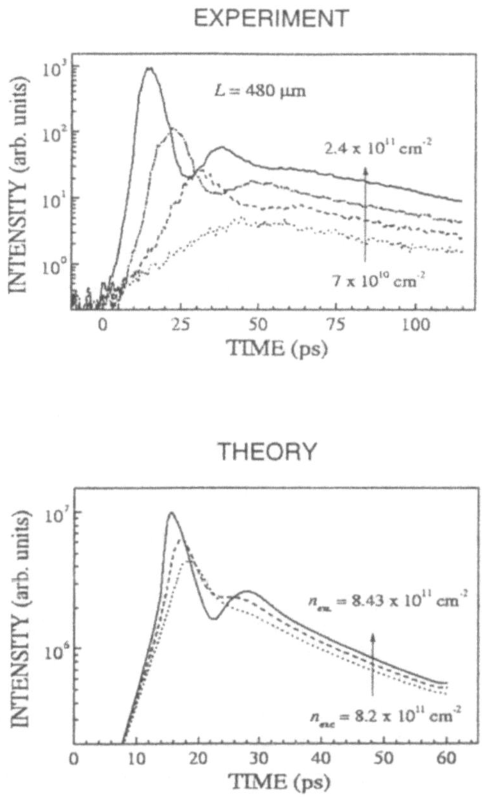

Figure 7
Experiment [44]: Typical transients of stimulated emission at four different excitation intensities (4.8, 5.5, 7.6, 16 MW/cm^2) for the 3 nm ZnCdSe quantum well sample. Length of the excitation stripe: 480 μm. Theory: Temporal evolution of the total emission intensity for three excitation intensities.

5 Conclusion

In summary, the presented first principles theory of light emission and absorption in semiconductors provides a technique to consistently study the various many-body correlation and nonequilibrium effects on equal footing. This concerns resonant cavity effects, excitonic correlations, diagonal and off-diagonal dephasing and band-gap renormalization, dynamical screening, band coupling, and quantum confinement. Quantitative agreement with recent experimental findings was demonstrated to be achievable for the exciton shift, gain lineshape, Photoluminescence Excitation Spectroscopy and transient oscillations in the laser output.

This work has been supported by the Deutsche Forschungsgemeinschaft (DFG), "Schwerpunktprogramm: II-VI Halbleiter, Optoelektronik im blauen Spektralbereich". We gratefully acknowledge the fruitful collaboration and stimulating discussions with the groups of J. Gutowski, D. Hommel (Bremen) and S.W. Koch (Marburg).

Bibliography

[1] T. Ohata, S. Itoh, N. Nakayama, S. Matsumoto, K. Nakano, M. Ozawa, H. Okuyama, S. Tomiya, M. Ikeda, and A. Ishibashi, phys. stat. sol. (b) **187**, 279 (1995).

[2] J.Y. Jen, T. Tsutsumi, I. Souma, Y. Oka, and H. Fujiyasu, Jpn. J. Appl. Phys. **32**, L1542 (1993).

[3] H. Kalt, J.H. Collet, L.S. Dang, J. Cibert, S.D. Baranovskii, R. Saleh, M. Umlauff, K.P. Geyzers, M. Heuken, and C. Klingshirn, Physica B **191**, 90 (1993).

[4] X.H. Yang, J.M. Hays, W. Shan, J.J. Song, and E. Cantwell, Appl. Phys. Lett. **62**, 1071 (1993).

[5] Y. Kawakami, I. Hauksson, H. Stewart, J. Simpson, I. Galbraith, K.A. Prior, and B.C. Cavenett, Phys. Rev. B **48**, 11994 (1994).

[6] J. Ding, H. Jeon, T. Ishihara, M. Hagerott, A.V. Nurmikko, H. Luo, N. Samarth, and J. Furdyna, Phys. Rev. Lett. **69**, 1707 (1992).

[7] J. Ding, M. Hagerott, T. Ishihara, H. Jeon, and A.V. Nurmikko, Phys. Rev. B **47**, 10528 (1993).

[8] J. Ding, M. Hagerott, P. Kelkar, A.V. Nurmikko, D.C. Grillo, L. He, J. Han, and R.L. Gunshor, Phys. Rev. B **50**, 5787 (1994).

[9] C. Klingshirn, H. Kalt, M. Umlauff, W. Petri, F.A. Majumder, S.V. Bogdanov, W. Langbein, M. Gruen, M. Hetterich, K.P. Geyzers, M. Heuken, A. Naumov, H. Stanzl, and W. Gebhardt, J. Cryst. Growth **138**, 786 (1994).

[10] F. Kreller, M. Lowisch, J. Puls, and F. Henneberger, II-VI Workshop Linz (Austria), 1994; F. Kreller, M. Lowisch, J. Puls, V. Rossin, A. Schuelzgen, J. Griesche, N. Hoffmann, M. Rabe, and F. Henneberger in: Proc. 22nd Internat. Conf. Physics of Semiconductors, Ed. D. Loockwood, World Scientific Publ. Co., Singapore 1995 (pp. 1352 to 1355).

[11] R. Cingolani, R. Rinaldi, L. Calcagnile, P. Prete, P. Sciacovelli, L. Tapfer, L. Vanzetti, G. Mula, F. Bassani, L. Sorba, and A. Franciosi, Phys. Rev. B **49**, 16769 (1994).

[12] A. Diessel, W. Ebeling, J. Gutowski, B. Jobst, K. Schuell, D. Hommel, and K. Henneberger, Phys. Rev. B **52**, 4736 (1995).

[13] A. Diessel, W. Ebeling, J. Gutowski, B. Jobst, K. Schuell, D. Hommel, M.F. Pereira, Jr., H. Güldner, and K. Henneberger, J. Cryst. Growth **159**, 661 (1996).

[14] H. Haken, *Laser Theory*, Handbuch der Physik Vol. XXV/2c, Springer-Verlag, Berlin 1970.

[15] M. Sargent, M.O. Scully, and W.E. Lamb, *Laser Physics*, Addison-Wesley, Reading, MA 1977.

[16] M. Lax, *Fluctuation and Coherence Phenomena in Classical and Quantum Physics*, pp. 269-478 in *Brandeis University Summer Institute in Theoretical Physics*, Gordon and Breach, New York 1966.

[17] K. Henneberger and S.W. Koch, in: *Microscopic Theory of Semiconductors: Quantum Kinetics, Confinement and Lasers*, World Scientific Publ. Co., Singapore 1995.

[18] K. Henneberger and S.W. Koch, Phys. Rev. Lett. **76**, 1820 (1996).

[19] K. Henneberger and H. Haug, Phys. Rev. B **38**, 9759 (1988).

[20] S. Hughes, A. Knorr, S.W. Koch, R. Binder, R. Indik, and J.V. Moloney, Solid State Commun. **100**, 555 (1996).

[21] R. Binder and S.W. Koch, Prog. Quant. Electron. **19**, 307 (1995).

[22] G. Manzke, U. Moldzio, and K. Henneberger, phys. stat. sol. (b) **202**, 961 (1997).

[23] C. Ell, R. Blank, S. Benner, and H. Haug, J. Opt. Soc. Am. B **6**, 2006 (1989).

[24] G. Manzke, Q.Y. Peng, U. Moldzio, and K. Henneberger, phys. stat. sol. (b) **206**, 37 (1998).

[25] Q.Y. Peng, G. Manzke, U. Moldzio and K. Henneberger, phys. stat. sol. (b) **206**, 419 (1998).

[26] A. Girndt, F. Jahnke, A. Knorr, S.W. Koch, and W.W. Chow, phys. stat. sol. (b) **202**, 725 (1997).

[27] M.F. Pereira, Jr., and K. Henneberger, Phys. Rev. B, accepted.

[28] F. Jahnke, M. Kira, S.W. Koch, G. Khitrova, E.K. Lindmark, T.R. Nelson, Jr. , D.V. Wick, J.D. Berger, O. Lyngnes, H.M. Gibbs, and K. Tai, Phys. Rev. Lett. **77**, 5257 (1996).

[29] A. Leitenstorfer, T. Elsässer, F. Rossi, T. Kuhn, W. Klein, G. Boehm, G. Traenkle, and G. Weimann, Phys. Rev. B **53**, 9876 (1996).

[30] S. Haas, F. Rossi, and T. Kuhn, Phys. Rev. B **53**, 12855 (1996).

[31] D.B. Tran Thoai and H. Haug, Z. Physik **B91**, 199 (1993).

[32] G. Manzke, Q.Y. Peng, K. Henneberger, U. Neukirch, K. Hauke, K. Wundke, J. Gutowski, and D. Hommel, IQEC'98, San Francisco, 1998.

[33] G. Manzke, Q.Y. Peng, K. Henneberger, U. Neukirch, K. Hauke, K. Wundke, J. Gutowski, D. Hommel, Phys. Rev. Lett. **80**, 4943 (1998).

[34] P. Michler, A. Diessel, W. Ebeling, J. Gutowski, B. Jobst, D. Hommel, K. Schuell, J. Nuernberger, G. Landwehr, M.F. Pereira, Jr., and K. Henneberger, Proc. Intern. Symp. Blue Laser and Light Emitting Diodes, Eds. A. Yoshikawa, K. Kishino, M. Kobayashi, and T. Yasuda, Ohmsha Ltd., Tokyo 1996, paper We-07 (pp. 147 to 150).

[35] M.F. Pereira, Jr., and K. Henneberger, Phys. Rev. B **53**, 16485 (1996).

[36] M.F. Pereira, Jr., S.W. Koch, and W.W. Chow, J. Opt. Soc. Am. B **10**, 765 (1993).

[37] M.F. Pereira, Jr., and K. Henneberger, phys. stat. sol. (b) **202**, 751 (1997).

[38] M.F. Pereira, Jr., and K. Henneberger, phys. stat. sol. (b) **206**, 477 (1998).

[39] C.H. Henry, R.A. Logan, and F.R. Merrit, J. Appl. Phys. **51**, 3042 (1980).

[40] H. Grempel, A. Diessel, W. Ebeling, J. Gutowski, K. Schuell, B. Jobst, D. Hommel, M.F. Pereira, Jr., and K. Henneberger, phys. stat. sol. (b) **194**, 199 (1996).

[41] P. Michler, M. Vehse, J. Gutowski, M. Behringer, D. Hommel, M.F. Pereira, Jr., and K. Henneberger, J. Cryst. Growth **184/185**, 575 (1998).

[42] P. Michler, M. Vehse, J. Gutowski, M. Behringer, D. Hommel, M.F. Pereira, Jr., and K. Henneberger, Phys. Rev. B **58**, 2055 (1998-II).

[43] H. Güldner, K. Henneberger, phys. stat. sol. (b) **206**, 413 (1998).

[44] P. Michler, U. Neukirch, K. Wundke, J. Gutowski, M. Behringer, D. Hommel, H. Güldner, and K. Henneberger, phys. stat. sol. (b) **206**, 399 (1998).

SiC: Polar Properties and their Influence on Technology and Devices

R. Helbig and F. Engelbrecht

Institut für Angewandte Physik, Universität Erlangen-Nürnberg
Staudtstr. 7, D-91058 Erlangen, Germany

Abstract: Silicon Carbide (SiC) is one of the tetrahedrally coordinated semiconductor materials with a wide band gap E_g. Due to the strong asymmetric valence charge density distribution along the 'Si-C' bond, nearly the complete bond charge is closely located at the carbon atom. Therefore, SiC is piezoelectric and shows a strong 'Reststrahl' absorption which allows the investigation of optical transitions due to impurities (vibrational and electronic transitions) only in a narrow band in the infrared spectral region. The polar character of the 'Si-C' bond also strongly influences the physico-chemical properties of SiC which are of fundamental importance for the successful development of process technologies for to electronic devices. The state of the art and new developments in the fields crystal growth, oxidation and reactive ion etching of SiC are reported. In addition to the polar properties of the SiC, also the anisotropic properties have to be taken into account for the design and the processing of electronic devices.

1 Introduction

The wide band gap semiconductor materials have driven new and important developments in the field of semiconductor devices in the last years, some of them already with remarkable commercial success. Shuji Nakamura has reported the GaN-story in his plenary talk at this meeting [1]. Here we will focus our interest on the wide band gap material silicon carbide. Silicon carbide is well known for the existence of a large number of different crystal modifications. The most common of these polytypes (6H, 4H, 3C, 15R) are indirect semiconductors with only very special applications in optoelectronics as UV-photodetectors [2]. The fields of applications of silicon carbide devices are power electronics, high frequency electronics and high temperature electronics [3] due to their large band gaps, high critical electric field strengths, high thermal conductivities and high saturation velocities of carriers. The excellent stability of silicon carbide at high temperatures and against different chemical environments has led to the 'formula' that silicon carbide is good for the electronics 'in harsh environments'. Silicon carbide wafers are available up to a diameter of two inches and the material can be further improved by epitaxial layers (n- or p-type) grown by CVD processes. The increasing interest in silicon carbide is documented in the proceedings of the biennially international conferences on silicon carbide [4–10] and other publications [3,11–17].

However, there are still several technological difficulties for the realization of practical electronic devices made of SiC. Here, we will specially focus on physical aspects of silicon carbide and its technology which are due to the polar character of the Si-C-bond.

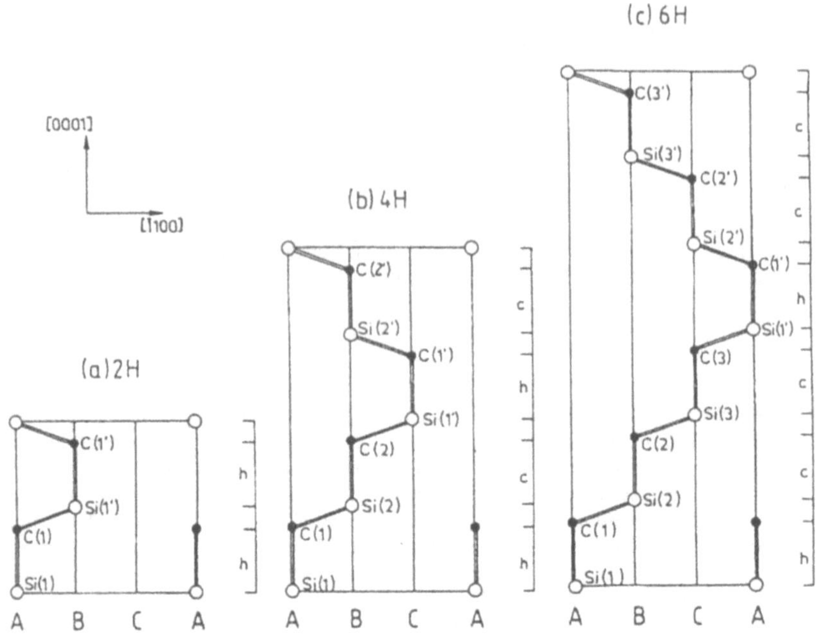

Figure 1 The zigzag chain structure of the different SiC polytypes. The Si (open circles) and C (filled circles) atoms are located in the (11$\bar{2}$0) plane for the hexagonal polytypes (after [34]).

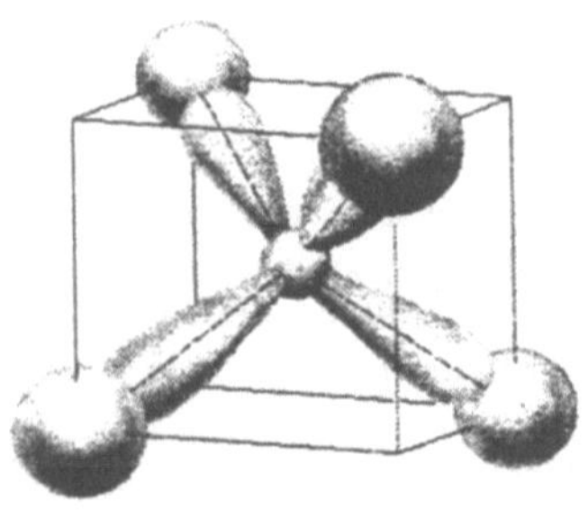

Figure 2
Directed bonding orbitals for an asymmetric tetrahedral sp^3-hybrid (after [39]).

2 Bonds

As most of the commercially used semiconductor materials, silicon carbide belongs to the family of tetrahedrally coordinated compounds. Silicon carbide is the only stochiometric binary compound formed from the elements of the IVth group of the periodic system of elements (PSE). In spite of the fact that we have no conclusive explanation for the existence of the different polytypes, the crystallographic structure of polytypes can easily described by the different zigzag sequences of the Si-C double layers (Fig. 1). The valence charge density distribution of the Si-C bond is very asymmetric. From the more artist point of view this is shown in Fig. 2. For comparison between different semiconductor compounds their valence electron charge density distribution is shown in Fig. 3. It can be seen that the 'bonding charge' between the bonding partners is located symmetrically

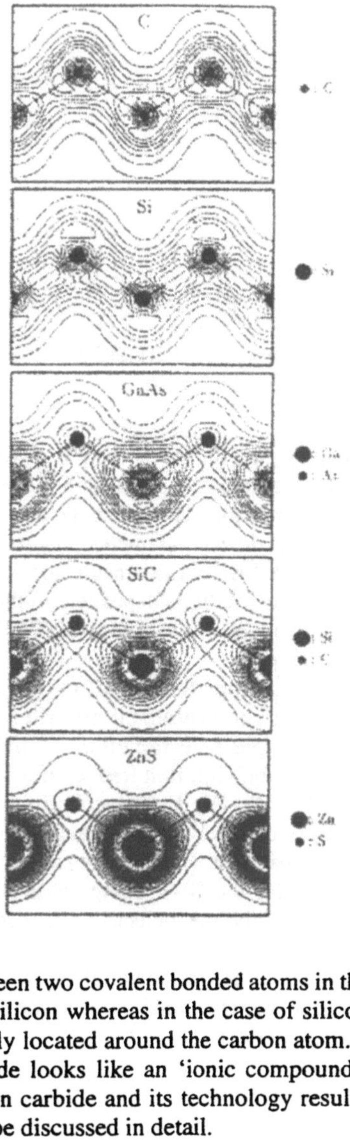

Figure 3
The total valence electron charge density distribution in [001]-[110] plane for C, Si, GaAs, SiC and ZnS. The increments of the contours are 2.5 *e* per unit cell. The lowest contour has this value in each case. The respective bond length are used as length unit in each panel (after [62]).

between two covalent bonded atoms in the case of the elemental semiconductors diamond and silicon whereas in the case of silicon carbide the electron valence charge density is mainly located around the carbon atom. In the map of the valence charge density silicon carbide looks like an 'ionic compound'. In the following sections some properties of silicon carbide and its technology resulting from the polar character of the 'Si-C' bond will be discussed in detail.

3 Lattice Properties

3.1 Piezoelectricity

The ionic character of tetrahedrally coordinated compound semiconductors with zinc-blende or wurtzite structure leads to a piezoelectric nature of the crystal, which gives rise to the piezoelectric effect. If one looks at the sequence of the different Si-C double layers

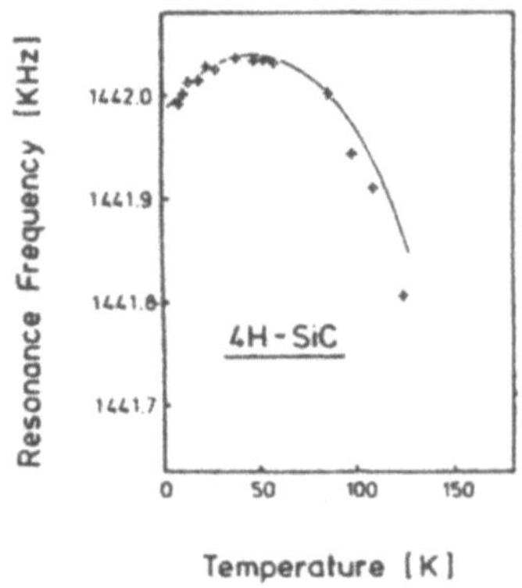

Figure 4
Resonance frequency of a 6H–SiC disc at different temperatures (after [18]).

in the c-direction (Fig. 1) it is easy to imagine that an electric dipole moment is generated by an uniaxial pressure along the c-axis. All silicon carbide polytypes have no inversion symmetry and are as GaAs and the other III-V compounds too. This piezoelectric polarization can result in additional charges and hence change device parameters due to stress in the material. Stress effects can occur in devices due to different expansion coefficients of the materials employed. It can be generated from dielectric films deposited on the surface of a polar semiconductor like SiC or from metallizations of the gate (MESFET) and the ohmic contacts which are close to the active device region.

The difficulty in the investigation of the piezoelectric properties of SiC is due to the finite electrical conductivity of the samples. Free carriers compensate the electric field of the piezoelectric effect. Therefore, we demonstrated the piezoelectric effect of different SiC-polytypes at low temperatures by measurements of the vibration frequencies of resonators of different shape at different temperatures (Fig. 4) [18]. However the numerical value of the electromechanical coupling constant is not precisely known, because not all of the elastic constants of SiC polytypes under investigation have been determined. The sign of the piezoelectric effect of SiC is the same as for the III-V compounds [13] as shown by Rottner [19], which is opposite to the sign of the piezoelectric effect for the II-VI compounds [20].

The piezoelectric properties of SiC make it possible to use the material for actuators and to expand surface acoustic wave (SAW) filters to higher frequencies due to the higher stiffness of SiC in comparison to other piezoelectric materials.

3.2 Lattice Vibrations

The asymmetry of valence charge density between the carbon and silicon leads to a large dipole moment for the totally asymmetric branch of the optical lattice vibrations – the optical lattice vibration with the highest energy. The other branches of optical lattice vibrations can be described in the concept of the small zone scheme [21,22] folding back different branches of the dispersion relation along c-direction of the zincblende polytype to get the dispersion relation of the other polytypes. This leads to small values of dipole moments for the other optical branches of lattice vibrations. The optical lattice vibrations, together with the occurrence of the various combination modes due to anharmonic effects, determine the optical properties of SiC in the infrared part of the spectrum [23]. It is therefore difficult to identify electronic transitions in the infrared part of the spectrum,

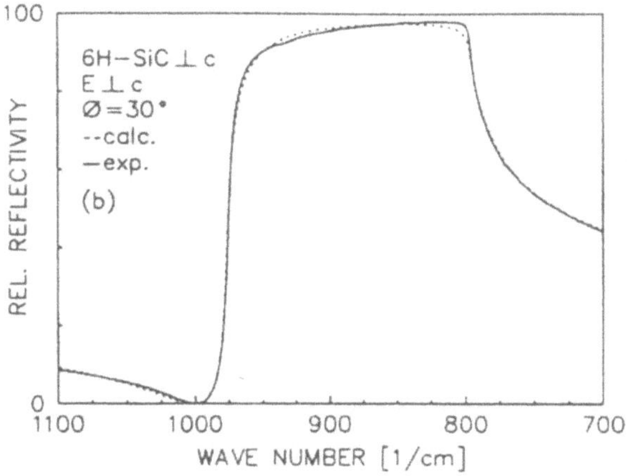

Figure 5
Reflectivity spectrum of 6H-SiC in the energy range 700 cm^{-1} – 1100 cm^{-1} (ordinary ray: $E \perp c$; angle of incidence $\phi = 30°$). The surface of the 6H-SiC wafer is oriented perpendicular towards the *c*-axis (after [27]).

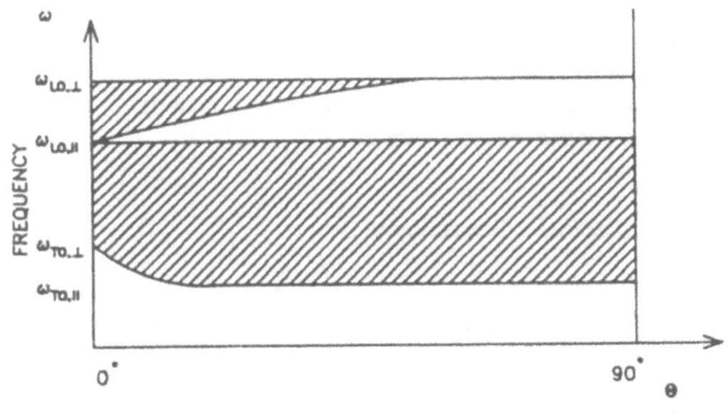

Figure 6
Energy of the main phonon modes ω_{TO} and ω_{LO} in 6H-SiC as a function of the angle θ between the optical axis and the propagation direction (directional dispersion). Two Reststrahl bands can be observed for the extraordinary ray ($E \parallel c$). The hatched parts in figure show the area of high reflectivity (the figure is not to scale, after [27]).

in order to identify electronic transitions between bound states of shallow impurities and localized vibrational modes (LVM) [23–26].

In Figure 5 we show the results of the measurements of the reflectivity in the 'Reststrahlenband' of 6H–SiC together with the directional dispersion of the strongest lattice oscillators (Fig. 6) which is due to the crystal anisotropy of 6H–SiC [27]. The corresponding frequencies and oscillator strengths of the lattice oscillators in 6H-SiC for the description of the optical properties of SiC in the infrared are given in [27].

3.3 Thermal Conductivity

The high energies of the lattice vibrations are the responsible for the excellent thermal conductivity of silicon carbide [28]. Though the thermal conductivity should not be discussed here in detail, it is pointed out that it is the base for applications of SiC in bipolar power devices [29]. In bipolar devices the high band gap energy ($E_g \approx 3$ eV) generates high thermal losses in forward direction, which can be only dissipated by the excellent thermal conductivity of SiC.

4 Electronic Properties

4.1 Bands

The energy band structure of SiC together with the asymmetric charge density distribution between the carbon and the silicon constituents was calculated by several groups [30–38]. To some extend the fine structure of the uppermost valence bands in the non-cubic polytypes can be interpreted in terms of the polar character of the Si-C bond [39]. For the polytype 6H-SiC the energy values of the spin-orbit splitting Δ_{SO} and the crystal field splitting Δ_{CR} were determined trom the low temperature photoluminescence emission spectrum and the hydrostatic shift of the energies of bound excitons [40,41]. The numerical value for the spin–orbit splitting energy of the uppermost valence bands of Δ_{SO} =7.7 meV for 6H-SiC determined by this method is much closer to the numerical value of atomic carbon (5.9 meV) than that of atomic silicon (44.1 meV) [42]. This demonstrates clearly the polar character of the 'Si-C' bond.

The polaron coupling – another feature of the polar character of a material – is not yet investigated in detail and we will not deal further with this problem here.

4.2 Saturation Velocity and Avalanche Breakdown

Due to the energy of the optical phonons the electron saturation velocity v_s in 6H–SiC is very high $v_s \approx 2 \cdot 10^7$cm/s [43,44]. This high values of v_s together with the high electrical breakdown field ($E_c > 2.0$ MV/cm in 6H-SiC) which is caused by the high energy band gap and the high phonon energies are the base of most of the electronic SiC devices. Both properties – band gap and phonon energy – are determined by the asymmetric (polar) Si-C bond.

5 Defects

It is a difficult task in solid state physics to describe and determine the energy levels of defects in a crystalline solid [45]. In this section we want to point out some hand waving arguments – which are probably not strictly true – for the defect chemistry of SiC on the basis of the asymmetric (polar) Si-C bond. For the intrinsic defects one can expect from a simple point of view that a Si-constituent cannot occupy a carbon place and vice versa for energetical reasons. However, numerical results of Wang et al. [46] and Deak et al. [47] show that the antisite defect C_{Si} is energetically the most probable one. An argument for the interpretation of this result is the simple consideration that in the configuration of the antisite defect the tetrahedrally coordinated Si-constituents and C-constituents are the 'smallest silicon and diamond crystals' which are very stable configurations.

Similar arguments can be applied to the incorporation of simple foreign defects, e. g. the incorporation of group-V donors such as nitrogen, on a carbon site and the incorporation of group-III elements on a silicon site. Though it seems that this expected behaviour is observed in most cases [48], there is also some evidence that the incorporation of N on a Si-site has to be taken into account [49].

6 SiC Device Technology

6.1 Crystal Growth

If a SiC single crystal with surfaces oriented perpendicular to the c-axis is prepared, two different configurations are possible: the (0001) face occupied with Si atoms (Si-face) and the $(000\bar{1})$ face occupied with C-atoms (C-face). In the crystal growth process by sublimation the behaviour of these two faces is quite different. It was observed in more than 300 crystal growth experiments that for 6H–SiC seed crystals the growth on a Si-face leads to a 6H–SiC crystal whereas the growth on a C-face leads to a 4H-SiC crystal (on axis orientation, growth temperature 2300 °C) [50]. However, it should be mentioned that there may exist some exceptions (for different growth temperatures) [51,52] and it is not fully clarified if the growing conditions were really identical.

6.2 Oxidation

As the natural thermal oxide of SiC, SiO_2 can be grown on the surface of the former by conventional techniques. The interface SiC/SiO_2 has – in many respects – similar properties as the well-studied system Si/SiO_2. However, the growth of high quality oxides on SiC with a low density of interface, and fixed oxide charge states combined with high dielectric strength, has been proven to be difficult due to the participation of carbon during the oxidation process [53,54].

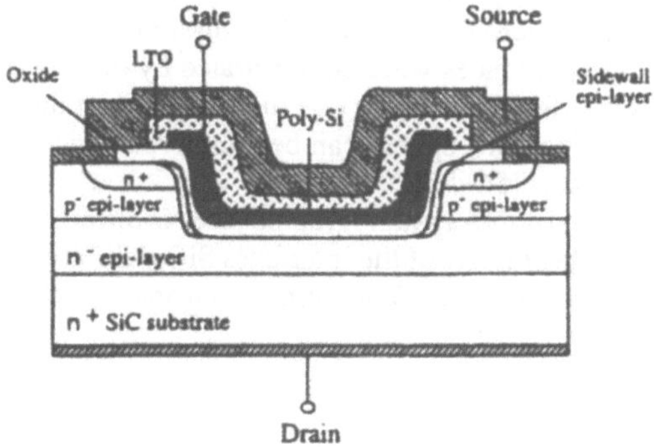

Figure 7
Scheme of a vertical U-groove MOSFET one of the most promising unipolar SiC device structures (after [54]).

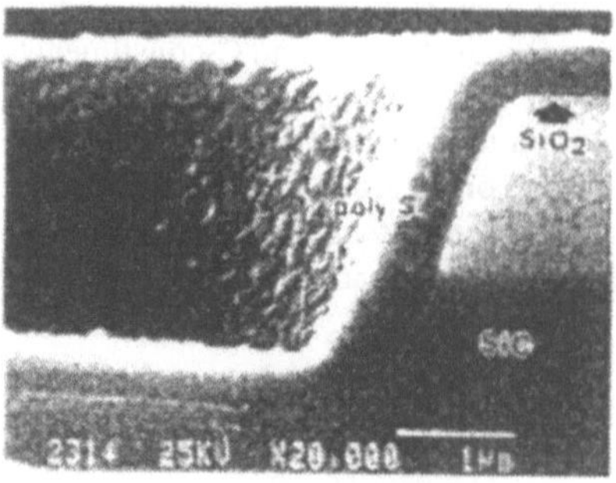

Figure 8
SEM micrographs of the cross-sectional view of the SiC trench MOS diode cell fabricated on (0001) Si-face 6H-SiC wafer (U-groove MOSFET, after [54]).

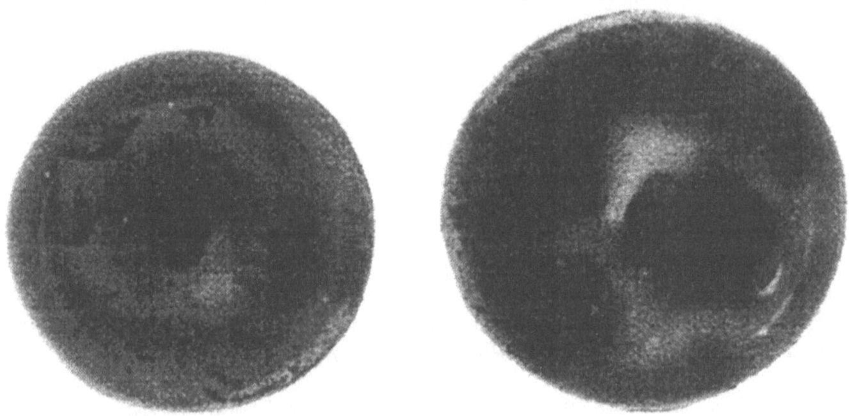

Figure 9 (a) Optical micrograph of an oxidized 6H-SiC sphere with the star-shaped (0001) Si face. Contrast or colours, respectively are due to interference effects with different oxide thickness. (b) Optical micrograph of an oxidation 6H-SiC sphere with the $(000\bar{1})$ C face. Contrast or colours, respectively are due to interference effects with different oxide thickness.

A very promising SiC device structure is the U-groove MOSFET. The scheme of this device is shown in Fig. 7 and an example for its realization is given in Fig. 8. Due to the anisotropic-polar nature of the SiC crystal structure the preparation and the gate oxide growth of these U-groove structures is difficult. It is expected that the oxidation rates of the different involved crystallographic faces are different. The dissimilar oxidation rates for the different oriented crystallographic faces were demonstrated by the oxidation of a 6H-SiC sphere. In Fig. 9, various interference colours due to differences in the thickness of the oxide surface layer of an oxidized SiC sphere can bee observed which are due to the orientation dependent oxidation rates [54–56].

But not only the chemical behaviour of the crystallographic faces is different, the oxide and interface defect related electrical properties of the interfaces SiO_2/SiC are also different [57], and have to be taken into account when fabricating electronic devices based on MOS-structures.

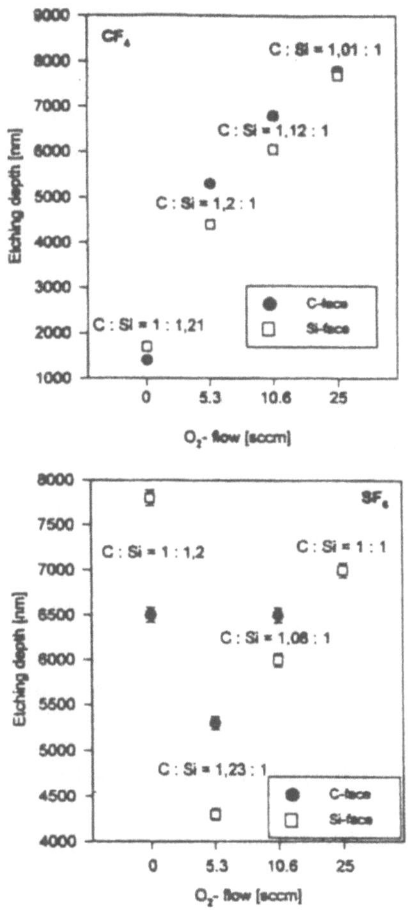

Figure 10
Etch depths of the Si- and the C-face of 6H-SiC by RIE in $CF_4/O_2/N_2$ and $SF_6/O_2/N_2$ plasma. (a, top) Anisotropic etching behavior in $CF_4/O_2/N_2$ plasma. In the absence of O_2 the Si face etches faster than the C face: the situation reverses for O_2 presence. For O_2 gas flows higher than 25 sccm both faces are etched equally. (b, bottom) Anisotropic etching behaviour in $SF_6/O_2/N_2$ plasma. In the absence of O_2 the Si face is etched faster than the C face; the situation reverses for O_2 presence at the costs of reduced etching rates. For O_2 gas flows higher than 25 sccm both sides are etched equally (after [60]).

6.3 Reactive Ion Etching (RIE)

Due to the high chemical stability of SiC there are no chemical methods to prepare lateral structures in SiC besides wet etching at elevated temperatures in alkaline solutions [58] or with electrolytic etching techniques [59]. However, these technologies are difficult to control. Therefore the only practicable process technology for etching of SiC is reactive ion etching. In the past several years, reactive ion etching of SiC polytypes has been investigated in fluorinated gases, usually mixed with oxygen and occasionally with other additives or in a mixture of fluorinated gases [60,61]. For the case of reactive ion etching we can demonstrate the anisotropy of the etching process by measuring the etching rates for the Si-face and the C-face, as shown in Fig. 10.

Acknowledgements

This work was supported by Deutsche Forschungsgemeinschaft (DFG), Bayerische Forschungsstiftung and BMBF through SFB 292, FOROPTO and the program SiC electronic respectively. Many discussions with our colleagues at the Institut für Angewandte Physik of the Universität Erlangen–Nürnberg are gratefully acknowledged.

Bibliography

[1] S. Nakamura, in *62. Physikertagung Regensburg 1998*, edited by DPG (Physik Verlag, Weinheim, 1998).

[2] J. Edmond *et al.*, phys. stat. sol. (a) **162**, 481 (1997).

[3] M. Ruff, H. Mitlehner, and R. Helbig, IEEE Trans. Electron Devices **41**, 1040 (1994).

[4] *Proc. of the First Intern. Conf. on Amorphous and Crystalline Silicon Carbide and Related Materials, Washington D. C.* , Vol. 34 of *Springer Proceedings in Physics*, edited by G. L. Harris and C. Y.-W. Yang (Springer–Verlag, Berlin, 1987).

[5] *Proc. of the 2nd Intern. Conf. on Amorphous and Crystalline Silicon Carbide II: Recent Developments, Santa Clara*, Vol. 43 of *Springer Proceedings in Physics*, edited by M. M. Rahman, C. Y.-W. Yang, and G. L. Harris (Springer–Verlag, Berlin, 1988).

[6] *Proc. of the 3rd Intern. Conf. on Amorphous and Crystalline Silicon Carbide III and other Group IV–IV Materials, Washington D. C.* , Vol. 56 of *Springer Proceedings in Physics*, edited by G. L. Harris, M. G. Spencer, and C. Y.-W. Yang (Springer–Verlag, Berlin, 1990).

[7] *Proc. of the 4th Intern. Conf. on Amorphous and Crystalline Silicon Carbide IV, Santa Clara*, Vol. 71 of *Springer Proceedings in Physics*, edited by C. Y. Yang, M. M. Rahman, and G. L. Harris (Springer–Verlag, Berlin, 1991).

[8] *Proc. of the 5th Intern. Conf. on Silicon Carbide and Related Materials, Washington, DC*, Vol. 137 of *Institute of Physics Conference Series*, edited by M. G. Spencer *et al.* (Institute of Physics Publishing, Bristol, 1993).

[9] *Proc. of the 6th Intern. Conf. on Silicon Carbide and Related Materials, Kyoto*, Vol. 142 of *Institute of Physics Conference Series*, edited by S. Nakashima, H. Matsumani, S. Yoshida, and H. Harima (Institute of Physics Publishing, Bristol, 1995).

[10] *Proc. of the 7th Intern. Conf. on Silicon Carbide, III-Nitrides and Related Materials, Stockholm*, Vol. 264-268 of *Materials Science Forum*, edited by G. Pensl, H. Morkoc, B. Monemar, and E. Janzen (Trans Tech Publications Ltd, Uetikon-Zuerich, 1997).

[11] W. J. Choyke, in *The Physics and Chemistry of Carbides, Nitrides and Borides, NATO Advanced Study Institute, Series B Physics*, edited by R. Freer (Kluwer, Dordrecht, 1990), p. 563.

[12] G. Pensl and R. Helbig, in *Festkörperprobleme*, edited by U. Rössler (Fried. Vieweg & Sohn, Braunschweig, 1990), Vol. 30, p. 133.

[13] *Numerical Data and Functional Relationships in Science and Technology (Landolt–Börnstein New Series)*, edited by O. Madelung (Springer, Berlin, 1982), Vol. 17a.

[14] *Properties of Silicon Carbide*, Vol. 13 of *EMIS Datareviews Series*, edited by G. L. Harris (INSPEC, London, 1995).

[15] *Diamond, SiC and Nitride Wide Bandgap Semiconductors (San Francisco)*, Vol. 339 of *Mat. Res. Soc. Symp. Proc.*, edited by C. H. Carter, G. Gildenblatt, S. Nakamura, and R. J. Nemanich (Materials Research Society, Pittsburgh, 1994).

[16] *Fundamental Questions and Applications of SiC (Part II)*, Vol. 162 of *phys. stat. sol. (a)*, edited by W. J. Choyke, H. Matsunami, and G. Pensl (Akademie Verlag, Berlin, 1997), .

[17] *Fundamental Questions and Applications of SiC (Part I)*, Vol. 202 of *phys. stat. sol. (b)*, edited by W. J. Choyke, H. Matsunami, and G. Pensl (Akademie Verlag, Berlin, 1997).

[18] S. Karmann, R. Helbig, and R. A. Stein, J. Appl. Phys. **66**, 3922 (1989).

[19] K. Rottner, Diplomarbeit, 1991.

[20] *Numerical Data and Functional Relationships in Science and Technology (Landolt–Börnstein New Series)*, edited by O. Madelung (Springer, Berlin, 1982), Vol. 17b.

[21] D. W. Feldman, J. H. Parker, W. J. Choyke, and L. Patrick, Phys. Rev. **170**, 698 (1968).

[22] D. W. Feldman, J. H. Parker, W. J. Choyke, and L. Patrick, Phys. Rev. **173**, 787 (1968).

[23] F. Engelbrecht and R. Helbig, in *Diamond, SiC and Nitride Wide Bandgap Semiconductors (San Francisco)*, Vol. 339 of *Mat. Res. Soc. Symp. Proc.*, edited by C. H. Carter, G. Gildenblatt, S. Nakamura, and R. J. Nemanich (Materials Research Society, Pittsburgh, 1994), p. 529.

[24] W. Suttrop *et al.*, J. Appl. Phys. **72**, 3708 (1992).

[25] F. Engelbrecht. S. Huant, and R. Helbig, Phys. Rev. B **52**, 11008 (1995).

[26] W. Götz *et al.*, J. Appl. Phys. **73**, 3332 (1993).

[27] F. Engelbrecht and R. Helbig, Phys. Rev. B **48**, 15698 (1993).

[28] S. G. Müller *et al.*, in *Proc. of the 7th Intern. Conf. on Silicon Carbide, III-Nitrides and Related Materials, Stockholm*, Vol. 264-268 of *Materials Science Forum*, edited by G. Pensl, H. Morkoc, B. Monemar, and E. Janzen (Trans Tech Publications Ltd, Uetikon-Zuerich, 1998), p. 623.

[29] J. Mottok, priv. Commun.

[30] P. Srichaikul, A. B. Chen, and W. J. Choyke, in *Amorphous and Crystalline Silicon Carbide IV, Santa Clara*, Vol. 71 of *Springer Proceedings in Physics*, edited by C. Y. Yang, M. M. Rahman, and G. L. Harris (Springer–Verlag, Berlin, 1991), p. 170.

[31] M. Rohlfing, P. Krüger, and J. Pollmann, Phys. Rev. B **48**, 17791 (1993).

[32] W. H. Backes, P. A. Bobbert, and W. van Haeringen, Phys. Rev. B **49**, 7564 (1994).

[33] K. Karch *et al.*, in *Proc. 22nd Intern. Conf. Phys. Semicond.*, edited by D. J. Lockwood (World Scientific, Singapore, 1994), p. 401.

[34] P. Käckell, B. Wenzien, and F. Bechstedt, Phys. Rev. B **50**, 10761 (1994).

[35] W. R. L. Lambrecht, in *Diamond, SiC and Nitride Wide Bandgap Semiconductors*, Vol. 339 of *Mat. Res. Soc. Symp. Proc.*, edited by C. H. Carter, G. Gildenblatt, S. Nakamura, and R. J. Nemanich (Materials Research Society, Pittsburgh, 1994), p. 565.

[36] W. R. L. Lambrecht, S. Limpijumnong, and B. Segall, in *International Conference on Silicon Carbide and related Materials, Kyoto*, Vol. 142 of *Inst. Phys. Conf. Ser.*, edited by S. Y. n. H. H. S. Nakashima, H. Matsunami (IOP Publishing Ltd, Bristol, 1995), p. 263.

[37] W. R. L. Lambrecht, S. Limpijumnong, S. N. Rashkeev, and B. Segall, phys. stat. sol. (b) **202**, 5 (1997).

[38] G. Wellenhofer and U. Rössler, phys. stat. sol. (b) **202**, 107 (1997).

[39] J. C. Phillips, *Bonds and Bands in Semiconductors* (Academic Press, New York, 1973).

[40] F. Engelbrecht *et al.*, in *Proc. of the 7th Intern. Conf. on High Pressure Semicond. Phys., Schwäbisch Gmünd*, Vol. 198 of *Phys. Stat. Sol. (b)*, edited by K. Syassen, R. A. Stradling, and A. R. Goni (Akademie Verlag, Berlin, 1996), p. 81.

[41] F. Engelbrecht *et al.*, Phys. Rev. B **56**, 7348 (1997).

[42] F. Herman, C. D. Kuglin, K. F. Cuff, and R. L. Kortum, Phys. Rev. Lett. **11**, 541 (1963).

[43] W. v. Münch and E. Pettenpaul, J. Appl. Phys. **48**, 4823 (1977).

[44] I. A. Khan and J. J. A. Cooper, in *Proc. of the 7th Intern. Conf. on Silicon Carbide, III-Nitrides and Related Materials, Stockholm*, Vol. 264-268 of *Materials Science Forum*, edited by G. Pensl, H. Morkoc, B. Monemar, and E. Janzen (Trans Tech Publications Ltd, Uetikon-Zuerich, 1997), p. 509.

[45] M. Altarelli and F. Bassani, in *Handbook on Semiconductors*, edited by T. S. Moss (North–Holland Publishing Company, Amsterdam, 1982), Vol. 1, p. 269.

[46] C. Wang, J. Bernholc, and R. F. Davis, Phys. Rev. B **38**, 12752 (1988).

[47] P. Deak *et al.*, in *Proc. of the 7th Intern. Conf. on Silicon Carbide, III-Nitrides and Related Materials, Stockholm*, Vol. 264-268 of *Materials Science Forum*, edited by G. Pensl, H. Morkoc, B. Monemar, and E. Janzen (Trans Tech Publications Ltd, Uetikon-Zuerich, 1998), p. 279.

[48] S. Greulich-Weber, phys. stat. sol. (a) **162**, 95 (1997).

[49] T. Stiasny and R. Helbig, phys. stat. sol. (a) **162**, 239 (1997).

[50] R. A. Stein, P. Lanig, and S. Leibenzeder, Mater. Sci. Eng. B **11**, 69 (1992).

[51] Y. A. Vodakov *et al.*, Silicon Carbide Monocrystal Growth (Int. Patent No. WO 97/27350).

[52] V. D. Heydemann, N. Schulze, D. L. Barrett, and G. Pensl, Appl. Phys. Lett. **69**, 3728 (1996).

[53] J. A. Cooper, phys. stat. sol. (a) **162**, 305 (1997).

[54] S. Onda, R. Kumar, and K. Hara, phys. stat. sol. (a) **162**, 369 (1997).

[55] K. Christiansen and R. Helbig, J. Appl. Phys. **79**, 3276 (1996).

[56] K. Ueno, phys. stat. sol. (a) **162**, 299 (1997).

[57] V. V. Afanasev, M. Bassler, G. Pensl, and M. Schulz, phys. stat. sol. (a) **162**, 321 (1997).

[58] J. W. Faust and H. M. Liaw, in *Silicon Carbide - 1973*, edited by R. C. Marshall, J. W. Faust, and C. E. Ryan (Univ. of South Carolina Press, Colombia S. C., 1974), p. 657.

[59] S. Rysy, H. Sadowski, and R. Helbig, to be published.

[60] R. Wolf and R. Helbig, J. Electrochem. Soc. **143**, 1037 (1996).

[61] P. H. Yih, V. Saxena, and A. J. Steckl, phys. stat. sol. (b) **202**, 605 (1997).

[62] M. Sabisch, P. Krüger, and J. Pollmann, Phys. Rev. B **51**, 13367 (1995).

Growth Mechanisms of SiC on α-SiC(0001) Prepared by Solid-Source Molecular Beam Epitaxy

Andreas Fissel, Bernd Schröter and Wolfgang Richter

Friedrich-Schiller-Universität Jena, Institut für Festkörperphysik,
Max-Wien-Platz 1, D-07743 Jena, Germany

Abstract: Epitaxial growth of SiC on hexagonal (or α)-SiC(0001) has been performed between 900 and 1 350 °C by means of solid-source molecular beam epitaxy. Our results demonstrate that similar SiC growth mechanisms act in all vapour phase epitaxial techniques. They also show that the control of the Si/C ratio and the supersaturation (S) is essential for the growth mode and the kind of polytype grown. Low temperature ($T < 1\,200$ °C) deposition on on-axis SiC substrates results in the growth of 3C-SiC, which is significantly improved by an alternating supply of Si and C. On vicinal substrates, a step flow growth mode has been realized at T down to 1 050 °C. In experiments performed at $T > 1\,200$ °C under near equilibrium conditions, different growth modes and conditions stabilizing the growth of certain polytypes have been found. With a step decrease of S, a step-flow growth mode of both 4H- and 6H-SiC occurs and, depending on the step configuration, we also propose a 1D nucleation at steps edges. At even lower S, homoepitaxial growth of 4H-SiC via 2D nucleation occurs on C-stabilized surfaces.

1 Introduction

SiC is a wide-band-gap semiconductor material of great technological interest for devices operating at high temperatures, high power, high frequency and in harsh environments [1,2]. The optical and electrical properties have been reviewed extensively for example by Choyke and Pensl [3,4]. One of the most interesting properties of SiC is the occurrence of different structures (polytypes) with different physical properties. These structures can be described by different stacking of the Si-C double layers perpendicular to the direction of the closed-packed planes, i.e., the cubic (111) or the hexagonal (0001). The different stacking sequences in SiC results in different physical properties. The only cubic polytype 3C, for example, has a band gap of about 2.3 eV, whereas 4H, one of the hexagonal polytypes, has a gap of about 3.2 eV. Therefore, it may be possible to built devices from heterostructures consisting of one semiconducting material with confined electrons in a two-dimensional (2D) gas [5] only by a different stacking of atomic SiC layers in [0001] direction. In this context, thin epitaxial films of SiC on SiC(0001) substrates with definite layer structure are of interest. Moreover, modulation doped structures may also be of interest for future new applications.

The molecular beam epitaxy (MBE) is an attractive method to prepare such structures because of the controlled deposition process within an atomic layer range, a lower deposition temperature and the clean growth conditions. Another benefit, which also originates from the ultra high vacuum (UHV), is the possibility of the in situ evaluation of the growth process by means of reflection high-energy electron diffraction (RHEED). Up to now, however, MBE technique has been used only to a small extent for the epitaxial growth of SiC [6–10].

In general, both bulk growth methods and epitaxial growth techniques are based on a vapour phase deposition process. Therefore, the principle growth mechanisms of SiC should be comparable in both epitaxy and bulk growth.

The processes of SiC growth have been investigated by a number of authors [11–13]. Most of the work concerning the epitaxial growth of SiC has been focused on the chemical vapour deposition technique (CVD) [14–17]. In this field, the epitaxial growth of SiC has experienced a large progress and the SiC epitaxial layers have reached a quality suitable for device application [see for example the special issue of phys. stat. sol.(b) **202** (1997)]. The breakthrough has been realized by performing homoepitaxial SiC growth on vicinal (off-oriented) SiC(0001) surfaces by a so called "step-controlled" epitaxy process [18,19]. In step-controlled epitaxy, the grown layer inherits the stacking order of the substrate steps on the vicinal surface through a step-flow growth mode. The critical growth conditions for the transition from the step-flow growth mode to the growth by 2D nucleation have recently been determined in dependence of the growth rate and the off-angle for the CVD-growth on 6H-SiC(0001) at temperatures between 1 200 and 1 600 °C by Kimoto and Matsunami [20].

The growth of heteropolytypic structures, however, demands definite nucleation conditions. At first, therefore, conditions have to be established not only to realize a perfect layer-by-layer growth in the epitaxial process, but also conditions stabilizing the growth of a certain polytype.

Many papers have dealt with both thermodynamic and kinetic origins of polytype growth of SiC [12,21], but a conclusive theory is still missing. Most of these theories are based on the influence of impurities, the Si/C-ratio and the temperature. It is believed that 3C-SiC is the stable structure in the nucleation stage. Consequently, nucleation far from the equilibrium, where a rearrangement of the growing film is not possible, results only in 3C-SiC growth [22]. The occurence of 3C-SiC can also be explained in the framework of nucleation theory considering that 3C-SiC has the lowest surface energy [12]. At low temperature or high supersaturation, a nucleus consisting of only few atoms is already stable. In this case, practically all atoms are surface atoms and the total potential energy of a nucleus is given by the surface energy.

At high temperatures or low supersaturation, different polytypes have been grown at certain conditions by means of sublimation. An important factor affecting the crystal polytype is the seed polarity. Because of the polarity in the [0001] direction, the SiC surface is covered by a outermost atomic layer of Si (Si-face) or C (C-face). On the (0001) Si-face always the 6H-SiC polytype has been grown, whereas on the (000-1) C-face under certain conditions 4H-SiC occurs [23,24]. This was found regardless of the modification of the substrate. Stein at al. [23], therefore, suggested that the different surface energies of the C-face and the Si-face have a marked influence on the kind of polytype growing on this surface. Maltsev et al. [25] suggest that the occurrence of the 4H-SiC on the C-face is due to the planar character of the bond between the two carbon atoms (sp^2-hybridization), which breaks the original symmetry of the seed. Furthermore, from the thermodynamic point of view, 4H should always occur at more C-rich conditions [12,26]. Moreover, 4H is preferentially grown in the presence of selected impurities like Sn,Ge,Pb in the vapour phase [26]. Only in one case 6H-SiC was grown on the Si-face of 4H-SiC by means of sublimation growth (Modified Lely Method) [27] at temperatures between 2 000 and 2 300 °C and high growth rates in the range of 2.5 mm/h. Under the presence of supersaturated Si vapour, leading sometimes to the occurrence of the Si liquid phase, always 3C-SiC has been grown despite of the high temperature of about 2 000 °C. Often a pure Si thin buffer layer was found within the 3C-6H transition area. In general, on the SiC(0001) Si-face, there is a greater tendency for the occurrence of 3C-SiC than on the (0001) C-face [27].

We report about results of different growth studies of SiC on α-SiC(0001), necessary to realize multi-heterostructures of different polytypes, such as 4H/3C/ 4H. Experiments has been performed between 900 and 1 350 °C by means of solid-source MBE. In this context, we demonstrate the specific role of surface superstructures occurring on SiC(0001) and, moreover, of an alternating supply of Si and C for the stabilization of the low-temperature epitaxial 3C-SiC growth via 2D nucleation. Growth kinetics has been investigated by means of RHEED-intensity measurements. The mechanisms of formation and growth of different polytypes and, moreover, of observed growth features have been investigated and discussed in the framework of growth parameters, such as temperature, supersaturation and Si/C ratio.

2 Experimental Conditions

The SiC films were grown between 900 and 1 350 °C in a RIBER-EVA 32 MBE system. The source materials of high-purity Si and high-purity C were evaporated separately by means of electron beam guns and controlled by a mass-spectrometer based flux meter. More detailed experimental conditions are decribed elsewhere [28,29]. As substrates were used: on-axis and 3–8.5° off-axis SiC(0001) wafers from Cree Research Inc. and 3C-SiC films CVD-grown on Si(100). The MBE growth mode, surface morphology and structure were investigated by in situ reflection high-energy electron diffraction (RHEED), by transmission electron microscopy (TEM), ex situ atomic force microscopy (AFM) and scanning electron microscopy (SEM) investigations. The grown polytype has been identified by means of electron channelling patterns [30]. Ex situ preparation conditions are to be found in [31]. In situ, prior the epitaxial growth, in many cases surfaces has been prepared by in situ sublimation etching in a Si flux to remove surface imperfections. Whereas after ex situ preparation no ordered step structure were found, a well developed step structure has been obtained after a 2 hours in situ etching, with steps typically 2–6 monolayer (ML) in height.

3 Results

3.1 Low Temperature Epitaxial Growth of 3C-SiC

At low temperatures, deposition takes place far from the equilibrium, necessary for the formation of the 3C-SiC polytype. Therefore, the control of the Si/C ratio and the surface mobility of adatoms play a crucial role in the growth of perfect films.

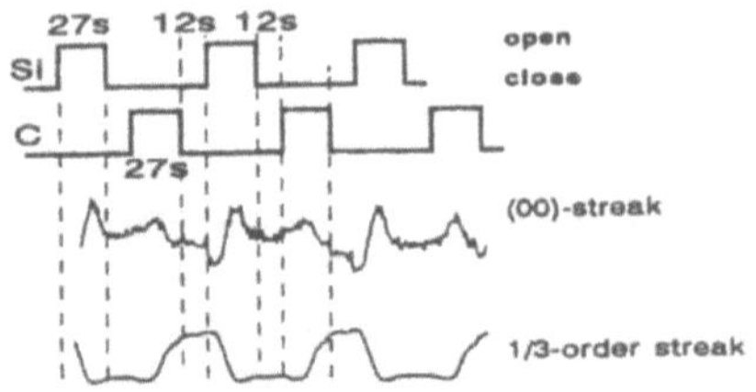

Figure 1
Sequences of source supply (upper) part and the corresponding RHEED intensity oscillation (lower part) of the (00)-spot and the 1/3-order streak during MBE of SiC on 6H-SiC(0001)

Recently, we already demonstrated the significant role of Si-stabilized superstructures for the low-temperature epitaxial process [28,29,32–35]. Annealing the surface under the presence of a slight Si flux changed the initially (1×1) SiC surface structure with an increasing amount of Si to a Si-rich (3×3)-superstructure. This superstructure is formed by an ordered 11/9 [36] to 13/9 ML [37,38] thick Si adlayer on top of the SiC surface. This is supported by recent investigations of the Stranski-Krastanov growth of Si on 6H-SiC(0001) [39]. The (3×3) superstructure is stable up to temperatures of about 1 030 °C in a Si flux in the range of 10^{13} to 10^{14} atoms/cm^{2}s. Increasing the Si flux above these values results in a clustering of Si. An optimum growth, therefore, is achieved by using an excess Si in the supply Si/C ratio slightly lower than the limit determined for non-clustering conditions.

SiC growth at $T < 1\,050$ °C has been investigated by means of RHEED intensity measurements [29,32]. The epitaxial growth was found to take place in two steps. After exposing the surface with both Si and C, at first a Si determined superstructure is formed during the deposition of the first 1.5 ML. Reaction between C and Si was assumed to be initially suppressed because of the high surface energy of SiC and the low heat of formation of SiC. In agreement with a theoretical prediction [38], the formation of Si-rich surfaces seems to be energetically favoured over C-rich surfaces. Thereby, the (3×3) superstructure is the lowest energy configuration and the C-adsorption becomes very easy on this Si-rich surface. Moreover, the relaxation behaviour of the (3×3) superstructure can lead to an enhanced diffusion of adatoms which promotes a layer-by-layer growth, as suggested by Starke at al.[37]. This is supported by our growth experiments [28,29,32,34,35] performed at $T > 900$ °C. The structure of films grown below 1 000 °C, however, was found to consist of different polytypic stacking sequences [29,32,34]. In some films, an abrupt change in the microstructure of the film occured after a certain thickness. This abrupt change was associated with a Si enrichment at the surface as revealed by RHEED [29]. In this case mostly pure 3C-SiC was found. From this findings we concluded that the occurence of different stacking sequences in the low temperature deposition results from fluctuations in the Si/C ratio at the surface occuring easyly under the dynamic conditions of growth.

Stable layer-by-layer growth of 3C-SiC has been realized by an atomic level controlled surface-stabilised growth using Si-rich surface superstructures [33–35]. By an alternating supply of C and Si, the superstructures can be switched between two different phases. The atomic level control of the growth was realized by recording the RHEED intensities of both the (00)-spot and an 1/3-order streak resulting from the superstructures on the SiC surfaces. Exposing, for example, the $(\sqrt{3} \times \sqrt{3})$R30° structure with about 1 ML silicon the structure transforms to a more Si-rich (1×1) and recurs during deposition of one ML carbon, as illustrated in Fig. 1. Between each deposition step the sample was annealed for a certain time (up to an almost constant RHEED intensity) to improve the adatom ordering at the surface. A typical TEM micrograph from a film grown by an alternating

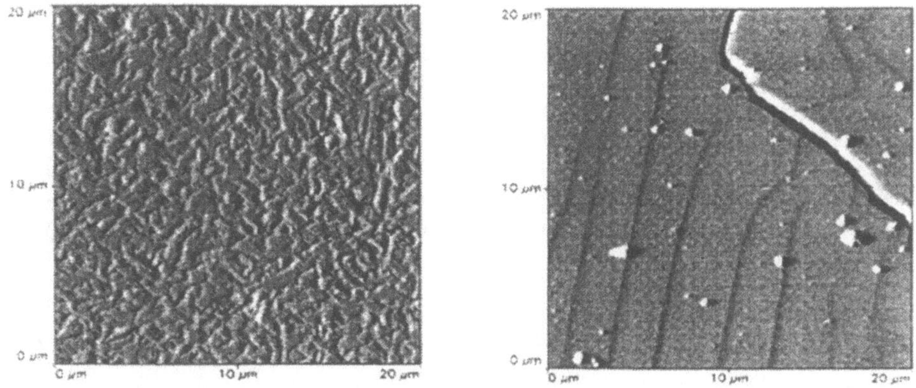

Figure 2 Cross-sectional HRTEM image of 3C-SiC on 6H-SiC(0001) grown by an alternate supply at 950 °C

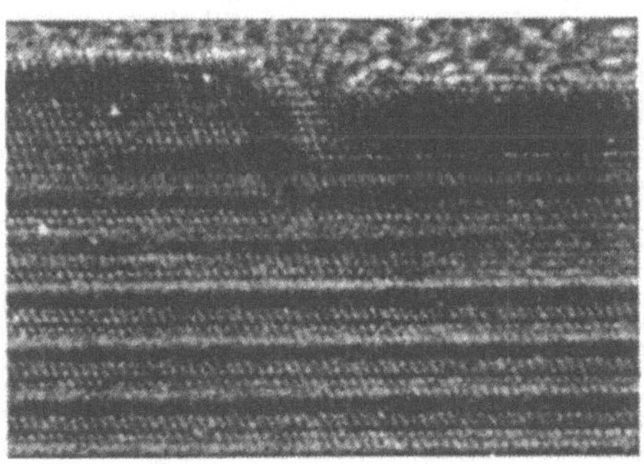

Figure 3
AFM images of 3C-SiC films grown on 6H-SiC(0001) at 1 150 °C with (a) continously supply of Si and C and (b) an alternating supply of C

supply on the (3 × 3) surface of an on-axis 6H-SiC(0001) at 950 °C is presented in Fig. 2. The film was 3C-SiC with only some defects, mostly double-position boundaries (DPB). DPB result from the nucleation and growth of 3C in different stacking sequences, such as ABC and ACB, on adjacent terraces [40]. The size, however, of the domains surrounded by the DPB is strongly dependent on the adatom mobility. This is demonstrated in experiments performed at fixed temperature (1 150 °C) and Si flux on on-axis 6H-SiC, but, with an alternating supply of carbon. The carbon shutter has been switched corresponding to deposit one monolayer in each cycle. Between each deposition step a short 5 s growth interruption was performed. In this way, the epitaxial film quality was significantly improved. As seen in Fig. 3, the sizes of the DPB domains increases up to some hundreds microns, a value which is comparable to thick SiC films grown by CVD at much higher temperatures [41]. Furthermore, many steps can be seen within the DPB indicating an step flow growth mode. XRD-measurements of the film demonstrate the good crystalline quality of the grown film. No differences in the FWHM in the $\theta - 2\theta$ scan (<0.004°) was found between the substrate and the film. Furthermore, the lattice constant c of the film in the growth direction was found to be smaller ($dc/c = 8.7 \times 10^{-4}$) in comparison to 6H. This agrees with results recently published for thick CVD-grown films [40,42]. In contrast to the conclusion given by Chien [42], it demonstrates that the 3C-SiC is fully strained and not relaxed, as we would expected for a pseudomorphic grown film. No difference in the in-plane lattice constant a between the substrate and the film was obtained.

For films containing much higher DPB densities a high degree of relaxation was found, in agreement with recently published results [43].

In summary, the heteroepitaxial growth of 3C-SiC on off-axis α-SiC has to be performed at low temperatures ($T < 1\,000$ °C) by an alternating supply of Si and C controlled to an atomic level using Si-rich superstructures. Those superstructures on the SiC(0001) surface also promote the layer-by-layer growth. At higher temperatures, where the Si desorption is already significant, the 3C-SiC growth can be performed on on-axis α-SiC substrates. In this case, the film quality was found to be significantly improved by an alternating supply of C.

3.2 Investigations of the MBE Growth Kinetics of SiC on 3C-SiC(100) by RHEED

RHEED intensity is very sensitive to morphological changes on the surface, such as changes of nucleation density [44]. Therefore, investigation of RHEED intensities especially of the specular reflected beam can give information concerning the nucleation process or adatom mobility. However, typical RHEED intensity oscillations corresponding to a layer-by-layer growth via 2D nucleation, found for example for the MBE growth of Si on Si(111) [45], were not detected for the growth of SiC on SiC(0001), in contrast to growth experiments performed on 3C-SiC(100) [35,46]. Here, the oscillations occurring in the initial stage of growth were found to correspond to a SiC monolayer growth in each oscillation period. This indicates an accurate layer-by-layer growth via 2D nucleation of SiC on 3C-SiC(100). In our studies we have investigated the initial stage of growth by analysing the change of RHEED specular beam intensity just after the deposition has been started (<0.3 ML) using a simple kinematic scattering theory [46,47]. In this stage, the epitaxial growth at lower temperature is mainly determined by nucleation and the RHEED intensity change is due to the scattering of electrons at edges of formed nuclei. This change becomes maximal if the nucleation density is saturated. Figure 4 shows an Arrhenius plot of the maximum RHEED intensity change, which corresponds to the saturation nucleus density. The plot clearly shows a transition to a stronger temperature dependence at $T > 930$°C. Such a transition has been found in MBE growth experiments of Si on Si(111), too, however, at a much lower temperature of about 500°C [46]. The slope for $T < 930$ °C corresponds to an activation energy of 0.77 ± 0.03 eV, whereas at

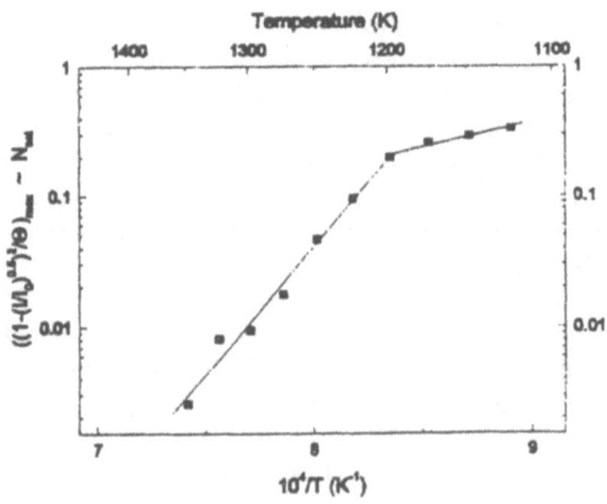

Figure 4
Arrhenius plot of the maximum RHEED intensity change corresponding to the nucleus saturation density

$T > 930$ °C the slope yields a value of $E = 3.9 \pm 0.2$ eV. These activation energies can be explained within the framework of classical rate equation based nucleation theories [48]. The general relationship between the saturation nucleus density N_s, the temperature and the growth rate R can be written as

$$N_s \propto R^{i^*/(i^*+2)} \exp\left(\frac{E_{i^*} + i^* E_D}{(i^*+2)kT}\right),$$

where E_{i^*} is the standard free energy of formation of the smallest stable nucleus, which is around $E_b(i^* - 1)$, where E_b is the bonding energy, E_D the activation energy of surface diffusion and $(i^* + 1)$ the number of atoms in the smallest stable nucleus. Owing to high supersaturation, the smallest stable nucleus can be considered as made of a dimer (i.e. $i^* = 1$). In this case, the nucleation energy is only dependent on the activation energy of surface diffusion. This is consistent with a model recently proposed by Mo et al. [49] for the MBE growth of Si on Si(100). Considering the obtained activation energy for $T < 930$ °C, we get an activation energy for the surface diffusion of $E_D = 2.3 \pm 0.1$ eV, which we assume to be the carbon surface diffusion barrier. From the energy at $T > 930$ °C, we deduce a transition to a value of $i^* = 6$–7 considering a value of $E_b = 3.3$ eV, which is about half the evaporation energy of SiC. The high-temperature value is comparable to those obtained for CVD by Kimoto and Matsunami [50] for the temperature-dependent nucleation density on 6H-SiC(0001) between 1 470 and 1 870 K. Investigation of the dependence of nucleus density from the deposition rate [50], however, yielded an exponent of $p = 1.5$ for the relationship $N \propto R^p$, which is inconsistent with rate equation based theories, where in the case of 2D nuclei, the exponent should be $i^*/(i^* + 2)$, a value always smaller than unity. This may be due to the high hydrogen concentration present on the surface in CVD effecting the growth process, as demonstrated for the CVD-growth of Si on Si(111) [51].

The absence of RHEED intensity oscillations in our investigations demonstrates that the nucleation process on SiC(0001) is different from a monolayer like growth. That means that on top of initially formed nuclei immediately an additional nucleus will be formed and so on. The interpretation as multilayer nucleation process is supported by a TEM picture of a 3C-SiC film grown on a nominal on-axis 6H-SiC presented in Fig. 5. In the vicinity of this picture there occurs a incoherent twin boundary (DPB). On the left side of this boundary, there is a terrace structure with steps 3- and 6-bilayer in height. Because there were no steps on the substrate before, this terrace structure results from the nucleation process and implies that 3 bilayer high nuclei may preferentially formed. This may result from the slight different surface energy of the A, B and C terminated terraces [52].

3.3 Determination of Growth Conditions at High Temperatures

At temperatures above 1 100 °C the graphitization is a serious problem in the MBE growth which can only be prevented if a certain excess of Si over SiC is present in the gas phase. At first, therefore, the Si-flux stabilizing the surface against graphitization was determined from the formation of the C-determined $(\sqrt{3} \times \sqrt{3})$R30°-superstructure detected by the occurence of superstructure spots in RHEED diffraction pattern as a function of temperature. The corresponding vapour pressures of Si have been calculated by the known Hertz-Knudsen relationship

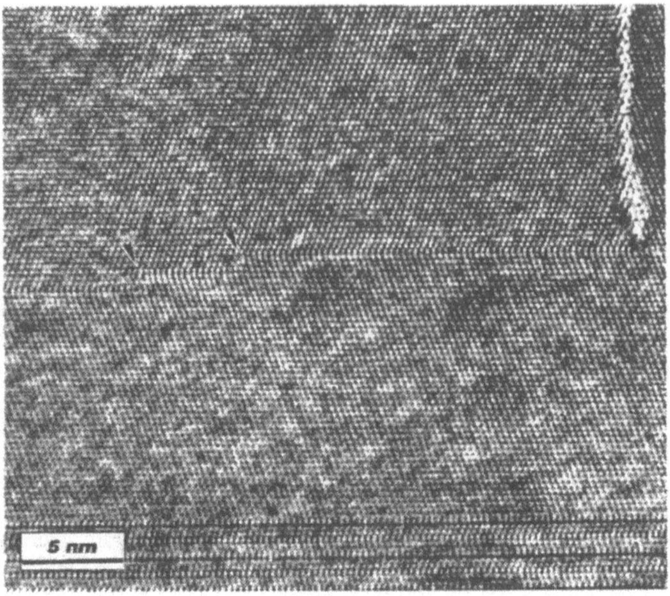

Figure 5 Cross-sectional HRTEM image of a 3C film on 6H-SiC(0001) grown at 1 050 °C and growth rate of 50 nm/h

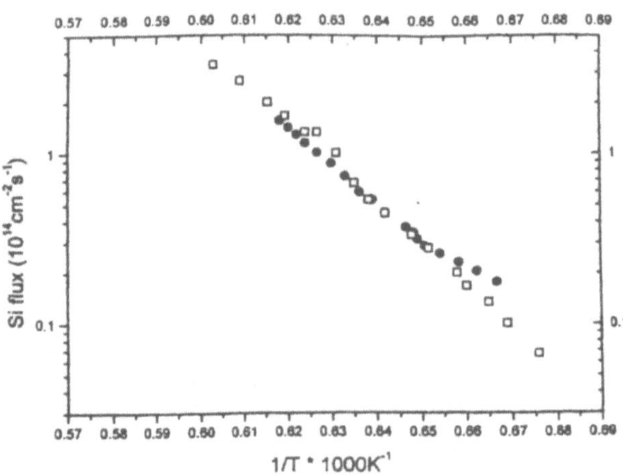

Figure 6 Equilibrium Si flux to stabilize the SiC(0001) surface against graphitization

$$F = p_v/(2\pi mkT)^{1/2},$$

where F is the flux, p_v the vapour pressure of the component, m the molecular mass, k the Boltzmann constant and T the substrate temperature. The results of these investigations are given in Fig. 6. The slope in the Arrhenius plot corresponds to an activation energy of 4.5 ± 0.3 eV. At the adsorption-desorption equilibrium, the vapour over SiC mainly consists of the species Si, Si_2C and SiC_2 [54] in accordance to the dissoziation reactions:

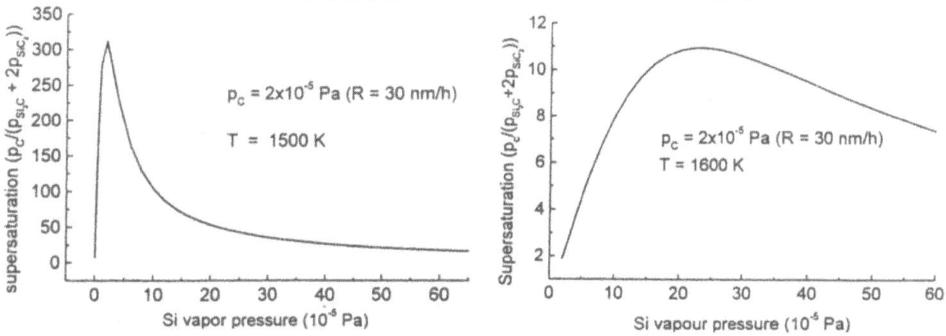

Figure 7 Supersaturation as a function of Si flux

$$\begin{aligned} SiC(s) &\longrightarrow Si(g) + C(s) \\ 2SiC(s) &\longrightarrow SiC_2(g) + Si(g)\ , \\ SiC(s) + Si(g) &\longrightarrow Si_2C(g) \end{aligned}$$

where g and s refer to the gas and solid phase, respectively. Moreover, the vapour phase composition above the SiC surface can be shifted between the SiC-C equilibrium and the SiC-Si one [12,53]. Comparing these results with our data obtained for the Si equilibrium flux we can find agreement to values given by Drowart et al. [54] and Karpov [53] for the so called SiC-C equilibrium. This is consistent with the fact that in our investigations the Si vapour phase is in equilibrium with the C-determined superstructure formed on the SiC surface. The partial pressures of SiC_2 and Si_2C in the SiC-C system are not only determined by temperature, but also by the given Si vapour pressure over the surface [55] in accordance with the reactions given above. That means that the supersaturation S, which is determined by the relation for the minority component carbon,

$$S = \frac{p_C}{p_{Si_2C} + 2p_{SiC_2}},$$

becomes a strong function of the Si overpressure in the high temperature range of our investigations and, consequently, S can not controlled independently concerning Si and C, as shown in Fig. 7. A strong change in S is seen from these curves by varying the Si flux. At a certain Si flux slightly above the equilibrium value a maximum in S occurs. This effect becomes very important if the equilibrium conditions change during the growth for example by effects like a changing surface morphology. A further increase in Si vapour pressure results in a drop of supersaturation down to an almost constant level. Under these conditions a stable growth via step-flow should be possible on terraces with lengths short enough to prevent a SiC nucleation. However, the excess Si flux should be restricted to a value low enough to prevent the formation of Si clusters or droplets at defects or at step edges giving rise to 3C-SiC growth by a subsequent carbonization process. Moreover, we have also to consider that an increase in excess Si-flux leads to a higher adatom density on the surface decreasing the surface diffusion length.

3.4 Epitaxial Growth of SiC on 6H- and 4H-SiC(0001) at Near Equilibrium Conditions

At lower S (< 100) always a step flow growth mode has been realized down to temperatures of 1 050 °C on off-axis substrates of both 6H-SiC and 4H-SiC(0001) cut 3.5 to 8.5° towards the [1120] direction, respectively.

Different growth features and growth modes (Fig. 8) were obtained for the high-temperature ($T = 1\,300$ °C) growth on on-axis substrates (terrace length above 1 μm) in dependence of S, which was varyied by changing the Si flux. At a maximum S (about 50) 2D nucleation of 3C-SiC on terraces frequently occur. Decreasing S by a factor of 2, we have frequently observed triangular shaped inclusions of 3C. Mostly, these triangles are connected with defects. A formation mechanism for such triangular defects in 4H epitaxial films has been proposed by Zhou et al. [56], based on the interaction of partial dislocations with surface steps. However, this model cannot explain the occurrence of these defects also in 6H epilayers, because it is restricted to 4H. Another more likely explanation is based on investigations performed on 6H-SiC films grown on 6H(0001) given by Mokhov et al [26], who found large (up to 600 μm) stacking faults within the epilayers. A mechanism for the preferential occurence of 3C on top of this stacking fault was recently proposed by Konstantinov [17].

Decreasing S again by a factor of about 2, no 3C inclusions were found within the films. However, sawtooth like structures were found at the step edges on larger terraces extending across the terraces. In many cases, nearly equal distances between the singular teeth have been observed.

For that reason, we propose a new growth mechanism based on an 1D dimensional homogeneous nucleation [57,58] at step edges. At high temperatures the adatoms have a high mobility and can reach rapidly existing steps. If there exists a high dangling bond density at the step there is a high probability to incorporate the adatoms directly into the step edge give rise to an step-flow growth mode. Along close-packed step-edges, however, there is a lower probability to incorporate the adatoms and diffusion along the edges is enhanced. Because of the longer lifetime of adatoms at these edges the adatom concentration increases rapidly and can reach the value necessary to form a nucleus at the step edge. This value is in fact lower than the value necessary to form 2D nuclei on the terraces. The formed nuclei grow then rapidly in size across the terrace especially in [1120] direction, where the growth rate is 20% faster than in [1100] direction [59]. The fastest part of the nucleus than can reach the next step edges. This may be also an additional mechanism for step bunching and gives a explanation for the enhanced step bunching under C-rich conditions in CVD [60].

At even lower supersaturation (<6), which corresponds to C-rich conditions represented by the permanent presence of the C-rich superstructure, the film surface of 4H-SiC was found to relatively rough and, moreover, no step structures were visible. The films grown under these conditions, however, were of 4H polytype. This demonstrates that under this conditions the films are grown via nucleation. The vanishing of the step structure can be attributed to a partly sublimation etch process, because of the easier removement of atoms from kink positions and, moreover, to the nucleation process itself. The occurence of 2D nucleation at this low supersaturation is a surprising fact and can only originated from a reduced mobility of adatoms on top of the C-rich surface and, moreover, by the presence of many nucleation sites due to the superstructure. Large vacancies have been reported to form the C-rich superstructure [61], which can act as such nucleation centres.

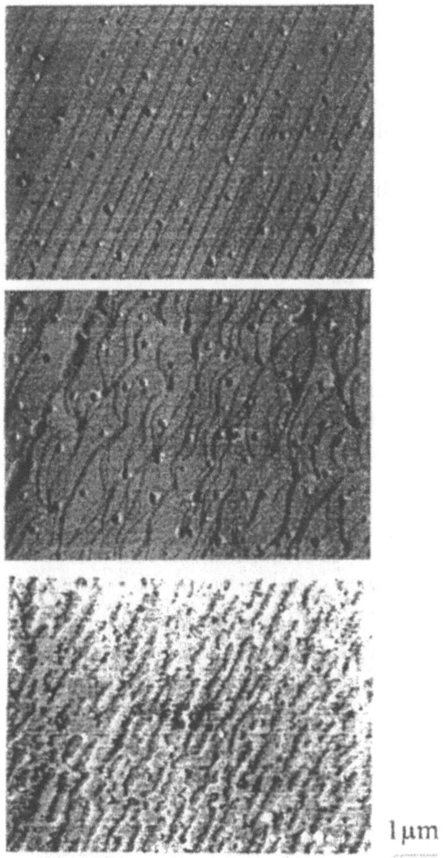

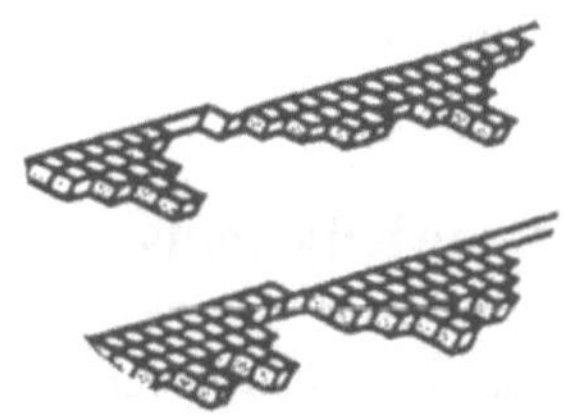

Figure 8
SEM images of different growth features obtained for the epitaxial growth of 4H on 4H-SiC(0001) at 1 300 °C at a supersaturation of (a) $S = 25$, (b) $S = 10$, (c) $S < 10$ and (d) model of the nucleation process at step edges

The occurence of 4H in the nucleation stage on 4H-SiC at low supersaturation can be explained by means of a reversed Ostwald's step rule. After this rule the formation of the metastable phase can take place at first if its nucleation probability is initially higher. Taking such a process into account for the formation of 4H at low S we have to consider that the polytype structure exerts an influence on the partial pressure of vapour components and, consequently, on S and the Si/C ratio at the surface. Moreover, we have also to consider a mechano-chemical activation due to misfit stresses. 3C represents the phase with the lower surface energy in comparison to 4H. The ratio for the (0001) plane is about $\sigma_{3C}/\sigma_{4H} = 1767/1800$ [62]. On the other hand, there exists a misfit between the lattice constant of both phases, which is in the range of $(a_{3C} - a_{4H})/a = 10^{-3}$ [63]. This misfit leads to an increase of saturation vapour pressure above the cubic phase due to the strain energy which is in the range of some meV. At low S, the strain effect on the vapor pressure will overcompensate the influence of the surface energy. We have estimated that at $S < 6$ the occurence of 4H-SiC in the nucleation stage can be expected.

4 Conclusion

Epitaxial growth of SiC on α-SiC(0001) in the temperature range between 900 and 1 350 °C has been performed by means of solid-source MBE.

Low temperature deposition, necessary for 3C-SiC growth, has been investigated. Nearly perfect epitaxial growth has only been achieved by an alternating supply of Si and C. The layers were found to be fully strained on 6H-SiC(0001) up to thicknesses $> 1\ \mu$m. Si-rich superstructures promote the layer-by-layer growth. The nucleation process was found to be different from monolayer like growth.

On vicinal substrates, a step-controlled growth on both 4H and 6H-SiC has been realized at temperature down to 1 050 °C.

At near equilibrium conditions, 1D nucleation at step edges of 4H on 4H-SiC and 6H on 6H-SiC has been proposed as a new growth mechanism. At very low supersaturation and C-stabilized conditions, 2D nucleation of 4H on 4H-SiC has been found. This may open the way to realize multi-heterostructures of different polytypes.

Acknowledgments

The authors acknowledge the support of this research by the Sonderforschungsbereich 196 (project A03) of the Deutsche Forschungsgemeinschaft. We wish to thank Ms U. Kaiser for the TEM investigations, Mr. J. Schulz for the AFM measurements and Mr. J. Kräusslich for performing the XRD measurements.

Bibliography

[1] M.Ruff, H.Hitlehner, R.Helbig, IEEE Transaction of Electronic Devices **41**, 1040 (1994).

[2] R.F.Davis, G.Kelner, M.Shur, J.W.Palmour, J.A.Edmond, Proc. IEEE **79**, 677 (1991).

[3] W.J.Choyke, *The Physics and Chemistry of Carbides, Nitrides and Borides*, 563-587, (Kluwer Academic Publishers, 1990).

[4] G.Pensl and W.Choyke, Physica B **185**, 264 (1991).

[5] F.Bechstedt and P.Käckell, Phys. Rev. Lett. **75**, 2180 (1995).

[6] S.Kaneda, Y.Sakamoto, T.Hirara, T.Tanaka, J. Cryst. Growth **81**, 536 (1995).

[7] T.Yoshinobu, H.Mitsu, I.Izumikawa, T.Fuyuki, H.Matsunami, Appl. Phys. Lett. **60**, 824 (1995).

[8] L.B.Rowland, R.S.Kern, S.Tanaka, R.F.Davis, J. Mater. Res. **8**, 2753 (1993).

[9] S.Tanaka, R.S.Scott, R.F.Davis, Appl. Phys. Lett. **65**, 2851 (1994).

[10] R.S.Kern, K.Järrendahl, S.Tanaka, R.F.Davis, phys. stat. sol. (b) **202**, 379 (1997).

[11] W.F.Knippenberg, Philips Res. Repts **18**, 161 (1963).

[12] Yu.M.Tairov and V.F.Tsvetkov, Prog. Cryst. Growth Character. **7**, 111 (1983).

[13] E.N.Morkhov, I.L.Shulpina, A.S.Tregubova, Yu.A.Vodakov, Cryst. Res. Technol. **16**, 879 (1981).

[14] W.V.Muench and I.Pfaffeneder, Thin Solid Films **31**, 39 (1976).

[15] R.F.Davis, J. Cryst. Growth **137**,161 (1994).

[16] H.Matsunami, Physica B **185**, 65 (1993).

[17] A.O.Konstantinov, C.Hallin, B.Pecz, O.Kordina, E.Janzen, J. Cryst. Growth **178**, 495 (1997).

[18] K.Shibahara, N.Kuroda, S.Nishino, H.Matsunami, Jap. J. Appl. Phys. **26**, L1815 (1987).

[19] H.S.Kong, J.T.Glass, R.F.Davis, J. Appl. Phys. **64**, 2672 (1988).

[20] T.Kimoto and H.Matsunami, J. Appl. Phys. **75**, 850 (1994).

[21] T.F.Page, *The Physics and Chemistry of Carbides; Nitrides and Borides*, pp. 197–214 (Kluwer Academic Publishers, Netherlands,1990).

[22] V.Heine, C.Cheng, R.J.Needs, J. Am. Ceram. Soc. **74**, 2630 (1991).

[23] R.A.Stein, P.Lanig, S.Leibenzeder, Mater. Sci. Eng. **B11**, 69 (1992).

[24] V.D.Heydemann, N.Schultze, D.L.Barrett, G.Pensl, Appl. Phys. Lett. **69**, 3728 (1996).

[25] A.A.Maltsev, A.Yu.Maksimov, N.K.Yushin, Inst. Phys. Conf. Ser. **142**, 41 (1996).

[26] E.N.Mokhov, A.D.Roenkov, Y.A.Vodakov, G.V.Saparin, S.K.Obyden, Inst. Phys. Conf. Ser. **142**, 245 (1996).

[27] K.Chourou, M.Anikin, V.Lauer, G.Guillot, J.M.Bluet, J.Camassel, O.Chaix, R.Madar, Mater. Sci. Forum **264–268**, 17 (1998).

[28] A.Fissel, B.Schröter, W.Richter, Appl. Phys. Lett. **66**, 3182 (1995).

[29] A.Fissel, U.Kaiser, E.Ducke, B.Schröter, W.Richter, J. Cryst. Growth **154**, 72 (1995).

[30] B.Schröter, M.Kreuzberg, A.Fissel, K.Pfennighaus, W.Richter, Mater. Sci. Forum **264–268**, 355 (1998).

[31] E.Ducke, R.Kriegel, A.Fissel, U.Kaiser, B.Schröter, P.Müller, W.Richter, Inst. Phys. Conf. Ser. **142**, 609 (1996).

[32] A.Fissel, U.Kaiser, K.Pfennighaus, E.Ducke, B.Schröter, W.Richter, Inst. Phys. Conf. Ser. **142**, 121 (1996).

[33] A.Fissel, U.Kaiser, K.Pfennighaus, B.Schröter, W.Richter, Appl. Phys. Lett. **68**, 1204 (1996).

[34] A.Fissel, K.Pfennighaus, U.Kaiser, B.Schröter, W.Richter, Mater. Sci. Eng. **B46**, 324 (1997).

[35] A.Fissel, K.Pfennighaus, U.Kaiser, B.Schröter, W.Richter, Diamond and Rel. Materials **6**, 1316 (1997).

[36] M.A.Kulakov, G.Henn, B.Bullemer, Surf. Sci. **346**, 49 (1996).

[37] U.Starke, J.Schadt, J.Bernhardt, M.Franke, K.Reuter, H.Wendler, K.Heinz, J.Furthmüller, P.Käckel, F.Bechstedt, Phys. Rev. Lett. **80**, 758 (1998).

[38] J.Furthmüller, P.Käckell, F.Bechstedt, A.Fissel, K.Pfennighaus, B.Schröter, W.Richter, Mater. Sci. Forum **264–268**, 339 (1998).

[39] A.Fissel, K.Pfennighaus, W.Richter, Appl. Phys. Lett. **71**, 2981 (1997).

[40] H.S.Kong, B.L.Jiang, J.T.Glass, G.A.Rozgonyi, J. Appl. Phys. **63**, 2645 (1988).

[41] W.S.Yoo, N.Hamaguchi, J.M.Carulli, N.I.Buchan, M.A.Tischeler, F.-R.Chien, A.R.Nutt, Inst. Phys. Conf. Ser. **137**, 259 (1994).

[42] F.R.Chien, S.R.Nutt, W.S.Yoo, J. Appl. Phys. **77**, 3138 (1995).

[43] A.N.Andreev, M.P.Schelgov, A.S.Tregubova, *Ext. Abstract of the Intern Conf. on SiC and Rel. Mater., Stockholm, Sweden*, 217 (1997).

[44] T.Shitara, D.D.Vvedensky, M.R.Wilby, J.Zhang, J.H.Neave, B.A.Joyce, Phys. Rev. **B46**, 6815 (1992).

[45] T. Sakamoto, N.J. Kawai, T. Nakagawa, K. Ohta, T. Kojima, Surf. Sci. **174**, 651 (1986).

[46] A.Fissel, M.Oehme, K.Pfennighaus, W.Richter, Surf. Sci. **383**, 370 (1997).

[47] P.Poelsma, A.F.Becker, G.Rosenfeld, R.Kunkel, N.Nagel, L.K.Verheij, G.Comsa, Surf. Sci. **272**, 269 (1992).

[48] J.A.Venables, Surf. Sci. **299/300**, 798 (1994).

[49] Y.W. Mo, J. Kleiner, M.B. Webb, M.G. Lagally, Phys. Rev. Lett. **66**, 1998 (1991).

[50] T. Kimoto and H. Matsunami J. Appl. Phys. **78**, 3132 (1995).

[51] L.Andersohn, Th.Berke, U.Köhler, B.Voigtländer, J. Vac. Sci. Technol. **A14**, 312 (1996).

[52] F.R.Chien, S.R.Nutt, W.S.Yoo, Mat. Res. Soc. Symp. Proc. **339**, 399 (1995).

[53] S.Yu.Karpov, Yu.N.Makarov, M.S.Ramm, phys. stat. sol. (b) **202**, 201 (1997).

[54] J.Drowart, G.De Maria, M.G.Inghram, J. Chem. Phys. **29**, 1015 (1958).

[55] R.C.Glass, D.Henshall, V.F.Tsvetkov, C.H.Carter, phys. stat. sol. (b) **202**, 149 (1997).

[56] L.Zhou, P.Pirouz, J.A.Powell, Mater. Sci. Forum **264–268**, 417 (1998).

[57] F.C.Frank, J. Cryst. Growth **22**, 233 (1974).

[58] J.Zhang and G.H.Nancollas, J. Cryst. Growth **106**, 181 (1990).

[59] N.Nordell, S.Karlsson, A.O.Konstatinov, Ext. Abstract of the Intern Conf. on SiC and Rel. Mater., Stockholm, Sweden, p.144 (1997).

[60] T.Kimoto, A.Itoh, H.Matsunami, J. Appl. Phys. **81**, 3494 (1997).

[61] L.Li, Y.Hasegawa, T.Sakurai, J. Vac. Sci. Technol. **B15**, 1307 (1997).

[62] R.Yakimova, M.Syväjärvi, E.Janzen, Mater. Sci. Forum **264–268**, 159 (1998).

[63] H.Romanus, G.Teichert, L.Spiess, Mater. Sci. Forum **264–268**, 437 (1998).

$Si_{1-x-y}Ge_xC_y$ Alloys: Growth and Properties of a New Semiconducting Material

H. Jörg Osten

Institute for Semiconductor Physics,
Walter-Korsing-Str. 2, D-15230 Frankfurt (Oder), Germany

Abstract: The growth and properties of $Si_{1-y}C_y$ and $Si_{1-x-y}Ge_xC_y$ alloys pseudomorphically strained on Si(001) is reviewed. Although the bulk solubility of carbon in silicon is small, epitaxial layers which more than 1 at.% C can be fabricated by molecular beam epitaxy and different chemical vapor deposition techniques. One of the most crucial questions is the relation between substitutional and interstitial carbon incorporation, which has a large impact on the electrical and optical properties of these layers. Substitutionally incorporated C atoms allow strain manipulation, including the growth of an inversely strained $Si_{1-x-y}Ge_xC_y$ layer. Local ordering effects due to atomic size differences and the growth on reconstructed surfaces, the mechanical and structural properties, and the influence of C atoms on band structure and charge carrier properties will be discussed. We show how lower carbon concentrations can influence dopant diffusion, without affecting strain and band alignment. Finally, we present possible applications of this new semiconducting material.

1 Introduction

Layered systems based on group-IV elements and grown heteroepitaxial on Si substrates extend the material variety for new device concepts in silicon microelectronic technology. $Si_{1-x}Ge_x$ layers on Si(001) are the most investigated heteroepitaxial group IV system [1]. The $Si_{1-x}Ge_x$ on Si(001) system exhibits some severe limitations, like the existence of a critical thickness for perfect pseudomorphic growth that depends on the amount of germanium – equivalent to the introduced strain. Above that thickness strain-relief occurs by plastic flow, i.e. by injection and propagation of misfit dislocations. In addition, metastable SiGe layers – thicker than the critical thickness and grown at low temperatures – tend to relax during post-growth thermal treatments, typical for device processing. The thickness has to be kept always below that critical thickness – free tuning of crystallographic and therefore electronic structure is not possible. The main band offset between Si and strained SiGe is located in the valence band, that is, this system is much better suited for hole-channel than for electron-channel devices.

Recently, it was shown that $Si_{1-y}C_y$ and $Si_{1-x-y}Ge_xC_y$ layers can be grown pseudomorphically on Si(001), using molecular beam epitaxy (MBE) or different chemical vapor deposition (CVD) techniques [2]. This new material might overcome some of the limitations of SiGe on Si(001). In this paper we review the current understanding of growth of $Si_{1-y}C_y$ and $Si_{1-x-y}Ge_xC_y$ layers grown pseudomorphically on Si(001). Further, we present important mechanical and electrical material properties. We discuss SiGe:C layers with lower carbon concentration ($< 10^{20}$ cm^{-3}), that can have a significant influence on dopant diffusion without affecting strain and band alignment. Finally, we will propose possible device applications for this new semiconducting material.

2 Growth of Pseudomorphic $Si_{1-y}C_y$ and $Si_{1-x-y}Ge_xC_y$ Alloys on Si(001)

According to the binary Si-C phase diagram, stoichiometric SiC is the only stable compound. Carbon has a low bulk solubility in Si (3×10^{17} atoms/cm^3 at the melting point [3]). Any alloy with a substitutional C concentration above the bulk solubility limit is thermodynamically metastable. However, it is not the equilibrium bulk solubility that is important during growth but rather a "surface solubility". The increased solubility results from two factors [4]. First, the presence of a surface breaks the bulk symmetry and therefore creates sites for adatoms, that are energetically inequivalent to bulk sites. The stress associated with the atomic size mismatch between C and Si is partially relieved and the energy of the substitutional C decreases smoothly as it approaches the surface. Second, there is a stress field near the surface, associated with the atomic reconstruction there [5]. The coupling of the carbon stress to this spatially varying surface stress substantially lowers the carbon energy on certain sites. For a (2×1) reconstructed silicon(001) surface there are two inequivalent sites. One site lies beneath the surface dimer, and the other site lies between the dimers. As discussed by Kelires and Tersoff [5], the sites beneath the dimers are under compressive stress. Thus they are favorable sites for a smaller atom such as C. The adatoms are rapidly buried on subsurface sites, which serve like a sink, and thus immobilized. The high carbon density near the surface is frozen in as the crystal grows, permitting the growth of a highly supersaturated solid-solution, e.g. high carbon concentrations in silicon or silicon/germanium. Recent calculations for carbon near the Si(001) surface suggest an enhancement of 10^4 over equilibrium bulk solubility [4].

The lowest-energy position of carbon in Si is found to be the substitutional site. The observed activation energy for the diffusion of substitutional carbon (3.04 eV) is quite large [3]. Both effects are responsible for the existence of these metastable alloys. Carbon precipitation is a rare process that has been reported to occur only in the presence of a supersaturation of silicon-self-interstitials or oxygen [6]. Once deposited as a random alloy at low temperatures, such carbon-containing layers appear to be able to withstand annealing at significantly higher temperatures [7,8]. Stable solutions of C in Si or SiGe have already been grown with concentrations in the percentage range.

One of the critical challenges for further development and applications of strained C containing heterostructures is the ability to incorporate the carbon atoms solely in substitutional positions, which has a large impact on electrical and optical properties of these layers. For example, there is a variety of interstitial Si-C complexes with energy levels within the $Si_{1-y}C_y$ gap, which can drastically influence transport properties or luminescence behavior [3]. Molecular beam epitaxial growth of C containing alloys pseudomorphically strained on the (2×1) reconstructed Si(001) has been investigated as a function of growth conditions [9]. The interstitial to substitutional carbon ratio is strongly influenced by the growth conditions, such as temperature and the Si deposition rate. Figure 1 shows substitutional carbon concentrations obtained from X-ray rocking curves for a series of samples grown at different temperatures with the same Si growth rate (0.3 Å/s)

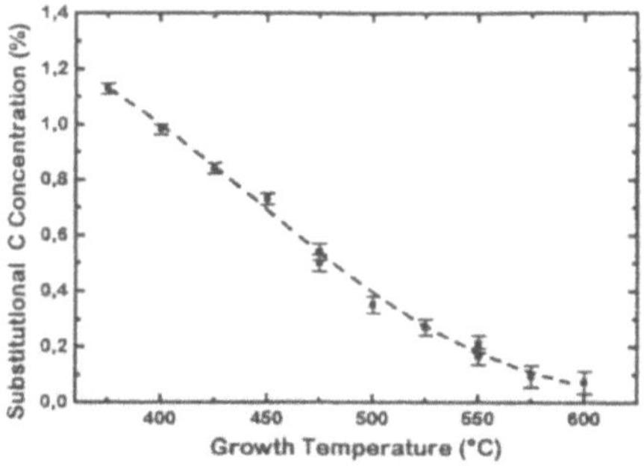

Figure 1
Substitutional carbon concentration obtained from X-ray rocking curves as a function of growth temperature. All samples were grown with a silicon rate of 0.3 Å/s and identical C flux. The drawn error bars show only the accuracy of the X-ray measurements and not possible fluctuations in the carbon flux. The dashed line represents the best fit according to a first order kinetic approach.

and carbon flux. There is a significant decrease of the substitutional carbon concentration with increasing growth temperature.

If the temperature is high enough for Si epitaxy, we can assume that all arriving C adatoms are rapidly incorporated in positions beneath the Si dimers. When another Si adatom arrives above a surface position filled with a C atom, (i) it migrates farther on the surface, (ii) it binds to the carbon ("freezing in" of the substitutional C), or (iii) it kicks the C atom out of a lattice position and forms a Si-C interstitial complex on the surface. The incorporation behavior can be well described by first order kinetics. The dashed line in Fig. 1 represents a fit based on a first order kinetic approach [9]. Both reduction in growth temperature and increase of the overall growth rate leads to an increase of the substitutional to interstitial carbon ratio. However, these can also cause some deterioration in crystal quality. Strategies to change the surface kinetics seem to be necessary to shift the window for a high substitutional to interstitial carbon ratio to process conditions suitable for good epitaxial quality. It is known that the presence of so-called surfactant atoms on the surface during growth can drastically alter the growth kinetics. Antimony was successfully applied as a surfactant to increase the quality of $Si_{1-y}C_y$ and $Si_{1-x-y}Ge_xC_y$ layers [10]. In another paper, atomic hydrogen was supplied during MBE growth using an rf source [11]. The layers were grown at a hydrogen pressure above 10^{-5} mbar. Epitaxial layers could be grown under such conditions. Comparing X-ray rocking curves for layers grown under identical conditions (the same Si and C fluxes, and the same growth temperature) with and without hydrogen, the authors found a significant enhancement of the amount of substitutionally incorporated carbon atoms in hydrogen-mediated growth compared to commonly used MBE growth.

3 Mechanical Properties of Carbon Containing Alloys

Adding a constituent with a covalent radius much smaller than that of silicon to a layer containing the larger Ge (r_{Si} = 1.17 Å, r_{Ge} = 1.22 Å, but r_C = 0.77 Å) opens the possibility to manipulate the strain. It was shown that it is possible to form an "inversely distorted" SiGe cell, i.e. a cell that is tetragonally distorted due to tensile stress instead of the usual compressive stress by adding some carbon [12]. Using simple linear extrapolations between the different lattice constants and the appropriate relative concentration (Vegard's law for a ternary system (Fig. 2)), we can estimate the effective lattice constant in the $Si_{1-x-y}Ge_xC_y$ system

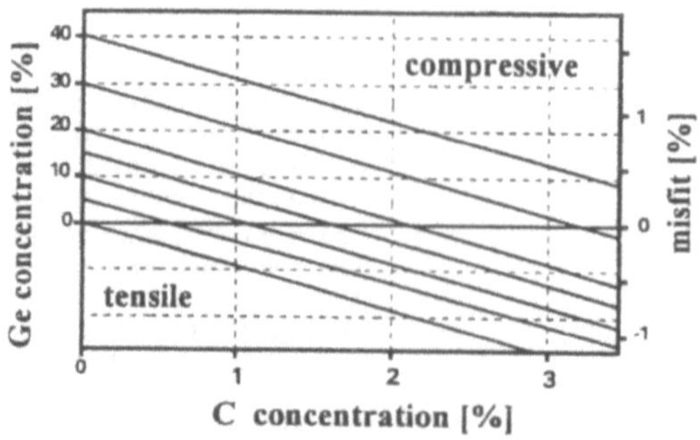

Figure 2
Estimated misfit reduction of $Si_{1-x-y}Ge_xC_y$ layers on Si(001) caused by the incorporation of carbon (ternary Vegard's law).

$$a_{SiGeC} = (1-x-y) \cdot a_{Si} + x \cdot a_{SiGe} + y \cdot a_{SiC}. \tag{3.1}$$

Experience has shown that carbon concentrations above 5 at.% start to destroy the epitaxial relation between the layer and the substrate [13]. In a global picture, a $Si_{1-y}C_y$ alloy with 5% carbon is strain-equivalent to a $Si_{1-x}Ge_x$ layer with approximately 50% Ge (they have the same absolute lattice mismatch).

An interesting feature is the condition for zero misfit. The perpendicular lattice constant of such a layer equals that of cubic silicon, opening the possibility for the growth of thick alloy layers on Si substrates. We can estimate that approximately 1 at.% of carbon is needed to reduce the lattice mismatch between silicon and a $Si_{0.9}Ge_{0.1}C_y$ layer to zero. From X-ray rocking curves of $Si_{1-x-y}Ge_xC_y$ layers grown on Si(001), we found that for a Ge/C ratio of about ten the lattice constant in growth direction is indistinguishable from that of the substrate, indicating the absence of macroscopic strain. The microscopic structure of such a layer has been investigated in detail in [14]. The Si-C bonds are stretched by about 7% with respects to the bond length in cubic silicon carbide, whereas the C atoms mainly strain the Si-Si bond in their neighborhood, with more distant Si-Si bonds less affected. Although the average lattice constant of the strain-compensated $Si_{1-x-y}Ge_xC_y$ layer is the same as that of pure silicon, the alloy contains large internal (microscopic) strain. Calculations support the picture of only partial relaxation of the Si-Si and Ge-Ge bonds towards a common bond length [14]. The Si-C and Si-Ge bonds in strain-compensated $Si_{1-x-y}Ge_xC_y$ are strongly stretched. The calculated 7% stretching of the Si-C in strain-compensated $Si_{1-x-y}Ge_xC_y$ corresponds to 35% relaxation towards the average bond length of the alloy (predicted by Vegard's law).

By combining theoretical and experimental results for local SiC phonon modes in dilute $Si_{1-y}C_y$ alloys, information concerning the short-range order was obtained [15]. Calculations using an anharmonic Keating model predict satellite peaks near the vibrational frequency of an isolated C impurity, associated with second-, third- etc., nearest-neighbor C-C pairs. By comparing theoretical spectra with those obtained by Raman and infrared absorption spectroscopy, it was concluded that the probability for a third-nearest-neighbor coordination of the C atoms is considerably above that for a purely random alloy. This confirms earlier work predicting an attractive third-nearest-neighbor interaction of the C impurities due to elastic interactions [16].

The temperature stability of strained $Si_{1-y}C_y$ layer was investigated during and after post growth annealings [7,8,17–19]. Despite the tensile strain in a 100 nm thick layer and the high carbon supersaturation, the samples were stable up to 800 °C [8]. Beyond this temperature range, the substitutional carbon content decreased exponentially during isothermal annealing. This effect can be explained by precipitation of carbon and/or nucleation and diffusion limited growth of SiC nanocrystals. Although the number of possible nucleation centers for dislocation generation was quite high, we could not find any relaxation by the formation of extended misfit dislocation networks [8]. We conclude that contrary to the mechanism of strain relief in $Si_{1-x}Ge_x$, in comparably strained $Si_{1-y}C_y$ epilayers, the main high temperature process is precipitation. From the temperature/time behavior of the measured changes in lattice spacing, we were able to extract an activation energy for that process of 3.2 eV ,in good agreement with the energy for bulk diffusion of substitutional carbon [3]. Similar results were obtained for the thermal stability of compressive strained $Si_{1-x-y}Ge_xC_y$ alloys [17,18]. Recently, the growth of a relaxed, only one micron thick, stepwise graded buffer based on a combination of $Si_{1-x}Ge_x$ and $Si_{1-x-y}Ge_xC_y$ was demonstrated [20]. This buffer concept relies on the retardation of dislocation glide in $Si_{1-x-y}Ge_xC_y$ relative to strain equivalent $Si_{1-x}Ge_x$ on silicon. The homogeneous $Si_{1-x}Ge_x$ layer with $x = 30\%$ on top of the buffer structure is $(73 \pm 5)\%$ relaxed. For the nonoptimized buffer growth the authors found a threading dislocation density below 10^5 cm^{-2}. A stepped $Si_{1-x}Ge_x$ buffer with the identical thickness and strain profile grown with the same temperature ramp yield a threading dislocation density above 10^7 cm^{-2}. This indicates that the addition of carbon is a promising way for new relaxed buffer concepts with low threading dislocation densities.

4 Electrical Properties of $Si_{1-y}C_y$ and $Si_{1-x-y}Ge_xC_y$ Layers

The current knowledge about optical and electrical properties of tensile strained $Si_{1-y}C_y$ or strain-free $Si_{1-x-y}Ge_xC_y$ layers on Si(001) is still limited. Assuming an average band structure for ternary $Si_{1-x-y}Ge_xC_y$ alloys, Soref [21] has suggested an empirical interpolation technique for the band gap. His technique results in an increase in the fundamental gap of $Si_{1-y}C_y$ layers with increasing y. Alternatively, Demkov and Sankey [22] have modeled $Si_{1-y}C_y$ alloys with supercells with random occupation of the lattice sites according to the alloy composition. They found that the fundamental gap is reduced when a few percent of carbon is added to the silicon lattice. Recently, measurements of the dielectric function and critical points in $Si_{1-y}C_y$ and $Si_{1-x-y}Ge_xC_y$ strained alloys were published using spectroscopic ellipsometry and electroreflectance [23–25]. From these experimental results for the optical transitions, we would state that the virtual crystal approximation is not able to describe the behavior of all optical transitions, i.e. the band structure, for $Si_{1-y}C_y$ and $Si_{1-x-y}Ge_xC_y$ alloys correctly. Also the carbon effects on the band structure cannot be reduced to changes in strain only. In order to describe adequately the observed energy shifts for pseudomorphic carbon-containing layers, we have to consider at least strain-induced effects and effects due to alloying [24]. First investigations of the direct transition and first electrical measurements [26] indicate that the carbon effect on band structure is smaller than on the crystallographic structure. Therefore, a completely strain-compensated (cubic) $Si_{1-x-y}Ge_xC_y$ layer still shows significant "Ge effects" (smaller gap than Si, valence band offset to Si) [25]. It is possible to grow pseudomorphically strained $Si_{1-y}C_y$ layers on Si that are under tensile stress. In such strained

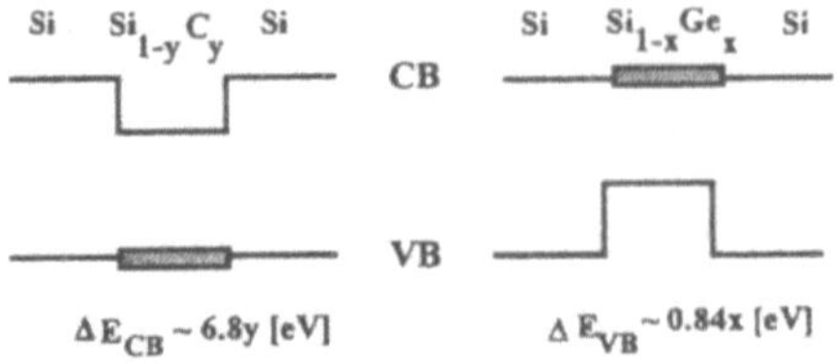

Figure 3
Predicted bandstructure for tensile strain $Si_{1-y}C_y$ layers on Si in comparison to compressive strained $Si_{1-x}Ge_x$ layers. The given numbers are estimated from recent PL measurements [27].

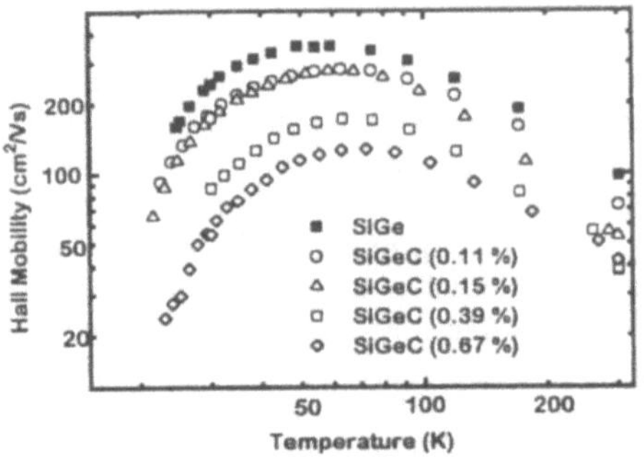

Figure 4
Hall mobility of holes as a function of temperature for heavily B-doped strained $Si_{1-x}Ge_x$ and $Si_{1-x-y}Ge_xC_y$ layers with 18% germanium on Si(001).

layers the gap will be determined by the Δ_2-minimum of the conduction band. Recent photoluminescence measurements on tensile strained $Si_{1-y}C_y$ layers yield a reduction in the fundamental gap, with the main offset occurring in the conduction band [27]. A prediction for the bandstructure of strained $Si_{1-y}C_y$ layers based on these results is shown in Fig. 3 in comparison to compressively strained $Si_{1-x}Ge_x$. The absence of a significant valence band offset caused by carbon incorporation was also shown by Kim et al. [28]. These authors found that carefully performed *in situ* XPS measurements allow the evaluation of valence band offsets in strained heteroepitaxial systems on Si(001). The results obtained for $Si_{0.75}Ge_{0.25}$ alloy layer agrees very well with the known values, indicating the reliability of the used method. For $Si_{0.977}C_{0.023}$ alloy layer tensile strained on Si(001), we could not find any significant valence band offset to Si. The observed band gap narrowing is dominated by the conduction band offset. Ternary strained $Si_{0.727}Ge_{0.25}C_{0.023}$ alloy layer shows the same valence band offset to Si as an appropriate binary $Si_{0.75}Ge_{0.25}$ alloy layer, that is, the valence band offset is dominated by Ge effects.

In [29] we have investigated the temperature dependence of charge carrier densities and Hall mobilities in tensile strained $Si_{1-y}C_y$ and in compressively strained $Si_{1-x-y}Ge_xC_y$ layers. In both cases, the measured charge carrier densities at room temperature are not affected substantially by the addition of a small concentration of carbon ($< 1\%$) under identical growth conditions and dopant fluxes. The measured Hall mobilities monotonically decrease with increasing carbon content for electrons in $Si_{1-y}C_y$, and for holes in $Si_{1-x-y}Ge_xC_y$, respectively (Fig. 4). From analyses of the temperature dependencies of the charge carrier densities and of the Hall mobilities, we conclude that electrically active defects are formed with the addition of carbon. The results indicate that those defects should be connected to carbon/Si-interstitials or other C-related complexes. Brunner et al. [30] compared Hall mobilities of holes in modulation doped $Si_{1-x}Ge_x$ and $Si_{1-x-y}Ge_xC_y$ layers. The found an increase in mobility for layers with 49% Ge, especially at low temperature. This gain in mobility can be attributed to an increased critical thickness for strain relaxation, and thus to a lower mobility degradation in the $Si_{1-x-y}Ge_xC_y$ due to dislocation formation compared to the more distorted $Si_{1-x}Ge_x$ layer.

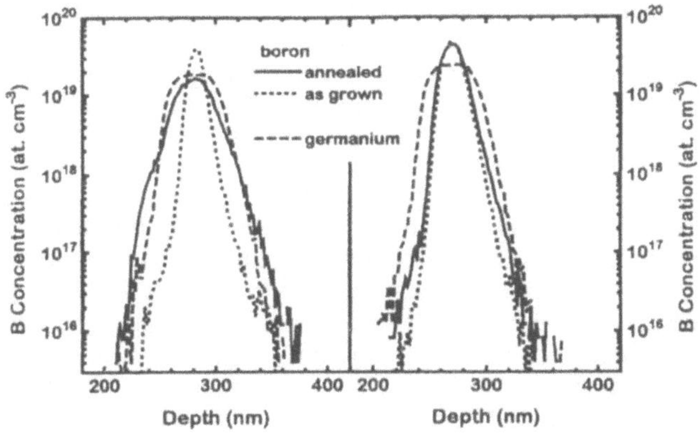

Figure 5 SIMS profiles of boron before and after the formation of a phosphorous implanted poly-Si emitter annealed at 950 °C for 30 s for a buried, box-shaped 30 nm thick $Si_{0.8}Ge_{0.2}$ layer without (left) and with 10^{20} at. cm^{-3} carbon background (right).

5 The Influence of Low Carbon Concentrations

So far we discussed only layer properties with carbon concentration in the percentage range. Such concentrations allow to manipulate lattice and band structure. It has been shown, that carbon concentrations below 10^{20} cm^{-3} can significantly suppress boron diffusion during post-epitaxial processing steps [31] without affecting strain and band alignment. Strong enhancement in boron diffusivity in $Si/Si_{1-x}Ge_x/Si$ heterobipolar transistors (HBTs) has been reported for emitter contact implantation into single crystalline Si and rapid thermal annealing [32]. This enhancement is presumably associated with the injection of excess point defects from the implant-damaged region into the B-doped $Si_{1-x}Ge_x$. A strong influence on boron diffusion has also been found when using P-doped poly-Si emitters [33]. To evaluate the potential of carbon for suppressing boron outdiffusion we choose such an emitter with a heavy annealing (950 °C/30 s). Figure 5 shows SIMS boron profiles for buried, 30 nm thick, box-shaped SiGe structures with and without carbon background (within the SiGe layer) measured in the as-grown state and after processing. The layered structure without carbon exhibits strong boron outdiffusion. A background of substitutionally incorporated carbon atoms ($< 10^{20}$ at. cm^{-3}) region stops the boron outdiffusion nearly completely. Within the accuracy of SIMS we could not find any differences in the boron profiles after a phosphorous emitter formation plus annealing (950 °C/30 s) in comparison to the as-grown state (Fig. 5). For a deeper understanding of this effect we have to look at carbon impurities in Si in concentrations below the solid bulk solubility range, that have been widely investigated over the last years [3]. The lowest energy position of carbon in bulk Si is the substitutional site (C_S). During epitaxial growth of SiGe:C layers, carbon atoms will mainly be incorporated on substitutional sites. Besides interstitial carbon incorporation during growth [9], C_i can also be formed by a replacement mechanism with silicon self-interstitial atoms (I) [6]

$$I + C_S \rightarrow C_i. \qquad (5.2)$$

This reaction is mainly associated with the injection of excess interstitials from post-growth implantation damage and annealing. Interaction of substitutional carbons with vacancies is very unlikely, because the C dangling-bond energy is larger than that of Si, which would imply a repulsive interaction [34]. Group-III acceptors also act as traps for the interstitials I, and undergo their own displacement mechanism [35], giving, e.g.,

$$B_S + I \rightarrow B_i. \tag{5.3}$$

Boron interstitials are mobile already near room temperature [36]. Reaction (5.3) is the physical reason for the observed strong boron diffusion during interstitial-creating post-epitaxial processing steps [37]. Thus, the suppression of boron outdiffusion is governed by the competition between the boron and the C_S atoms for capture of silicon self-interstitials. Boron interstitials can also complex with C_S to give B_i-C_S at $E_C - 290$ meV, or form pairs of B_i-B_S atoms at $E_V + 300$ meV [38]. At temperatures above 200 °C the following reaction has been observed [39]:

$$C_i + B_S \rightarrow B_i - C_S. \tag{5.4}$$

All the different defect formations discussed so far should act as scattering or recombination centers, and thus influence charge carrier properties. In a recent publication [40], we described some effects of carbon on the electrical properties of highly boron-doped SiGe:C layers for C concentration of around 10^{20} cm^{-3}. We found no significant influence of carbon on charge carrier concentration and hole mobility. However, carbon-related defects can significantly influence carrier lifetimes. Specifically, the generation lifetime can be reduced by more than an order of magnitude compared to analogous SiGe layers. The measured lifetimes of several hundreds nanoseconds are sufficiently high for device applications. We already demonstrated cutoff and maximum oscillation frequencies of more than 70 GHz and delays per stage down to 15 ps for ring oscillators with integrated SiGe:C heterobipolar transistors (HBTs) [41]. We were also able to obatin almost ideal base current characteristics for SiGe:C HBTs, even with carbon in the depleted regions. The measured performances for SiGe:C HBTs demonstrate that these transistors should be suitable for high frequency circuit applications.

6 Summary and Outlook

We reviewed some important mechanical and electrical material properties of $Si_{1-y}C_y$ and $Si_{1-x-y}Ge_xC_y$ layers grown pseudomorphically on Si(001). This new material might overcome some of the limitations of strained $Si_{1-x}Ge_x$ and open new field for device applications of heteroepitaxial Si-based systems. In detail, we found the following advantages for C containing materials:

- Increase thickness, stability, Ge content of $Si_{1-x}Ge_x$
  (for p-channel FET, npn HBT).
- Use strained $Si_{1-y}C_y$ on Si instead of Si on relaxed buffer
  (for n-channel FET, pnp HBT).
- Design new buffer concepts with $Si_{1-x-y}Ge_xC_y$
  (as virtual substrates for hetero FETs).

- Tensile/compressive strained superlattices on Si(001) for optics.
- Increase performance and process margins for HBTs.

Acknowledgment

The author would like to thank B. Dietrich, E. Bugiel, G.G. Fischer, P. Gaworzewski, J. Griesche, M. Kim, W. Kissinger, D. Knoll, D. Krüger, G. Lippert, M. Methfessel, K. Pressel, H. Rücker, R. Sorge, H.J. Thieme, M. Weidner, and P. Zaumseil, for their contributions to the SiGeC research reviewed in this report.

Bibliography

[1] S.C. Jain, *Germanium-Silicon Strained Layers and Heterostructures* (Academic Press, Boston, 1994).

[2] S.C. Jain, H.J. Osten, B. Dietrich, and H. Rücker, Semicond. Sci. Technol. **10**, 1289 (1995).

[3] G. Davies, and R.C. Newman, *Carbon in Monocrystalline Silicon*, in Handbook of Semiconductors Vol. 3, ed. by T.S. Moss, Elsevier Science B. V.

[4] J. Tersoff, Phys. Rev. Lett. **74**, 5080 (1995).

[5] P.C. Kelires, and J. Tersoff, Phys. Rev. Lett. **63**, 1164 (1989).

[6] W.J. Taylor, T.Y. Tan, and U. Goesele, Appl. Phys. Lett. **62**, 3336 (1993).

[7] M.S. Goorsky, S.S. Iyer, K. Eberl, F.K. LeGouech, J. Angilello, and F. Cardonne, Appl. Phys. Lett. **60**, 2758 (1992).

[8] G.G. Fischer, P. Zaumseil, E. Bugiel, and H.J. Osten, J. Appl. Phys. **77**, 1934 (1995).

[9] H.J. Osten, M. Kim, K. Pressel, and P. Zaumseil, J. Appl. Phys. **80**, 6711 (1996).

[10] E.T. Croke, A.T. Hunter, P.O. Petterson, C.C. Ahn, and T.C. McGill, Thin Solid Films **294**, 105 (1997).

[11] G. Lippert, P. Zaumseil, H.J. Osten, and M. Kim, J. Crystal Growth **175/176**, 476 (1997).

[12] H.J. Osten, E. Bugiel, and P. Zaumseil, Appl. Phys. Lett. **64**, 3440 (1994).

[13] A.R. Powell, K. Eberl, B.A. Ek, and S.S. Iyer, J. Crystal Growth **127**, 425 (1993).

[14] B. Dietrich, H.J. Osten, Rücker, H., M. Methfessel, and P. Zaumseil, Phys. Rev. B **49**, 17185 (1994).

[15] Rücker, H., M. Methfessel, B. Dietrich, K. Pressel, and H.J. Osten, Phys. Rev. B **53**, 1302 (1996).

[16] Rücker, H., M. Methfessel, E. Bugiel, and H.J. Osten, Phys. Rev. Lett. **72**, 3578 (1994).

[17] P. Warren, J. Mi, F. Overney, and M. Dutoit, J. Cryst. Growth **157**, 414 (1995).

[18] P. Zaumseil, G.G. Fischer, K. Brunner, and K. Eberl, J. Appl. Phys. **81**, 6134 (1997).

[19] H.J. Osten, D. Endisch, E. Bugiel, B. Dietrich, G.G. Fischer, M. Kim, D. Krüger, and P. Zaumseil, Semicond. Sci. Technol. **11**, 1678 (1996).

[20] H.J. Osten, and E. Bugiel, Appl. Phys. Lett **70**, 2813 (1997).

[21] R.A. Soref, J. Appl.Phys. **70**, 2470 (1991).

[22] A.A. Demkov, and O.F. Sankey, Phys.Rev. B **48**, 2207 (1993).

[23] W. Kissinger, M. Weidner, H.J. Osten, and M. Eichler, Appl. Phys. Lett. **65**, 3356 (1994).

[24] S. Zollner, J. Appl. Phys. **78**, 5209 (1995).

[25] W. Kissinger, H.J. Osten, M. Weidner, and M. Eichler, J. Appl. Phys. **79**, 3016 (1996).

[26] K. Rim, S. Takagi, J.J. Welser, J.L. Hoyt, and J.F. Gibson, Mater. Res. Soc. Symp. Proc. Vol. **379**, 327 (1995).

[27] K. Brunner, K. Eberl, and W. Winter, Phys. Rev. Lett. **76**, 303 (1996).

[28] M. Kim, and H.J. Osten, Appl. Phys. Lett. **70**, 2702 (1997).

[29] H.J. Osten, and P. Gaworzewski, J. Appl. Phys. **82**, 4977 (1997).

[30] K. Brunner, W. Winter, K. Eberl, N.Y. Jin-Phillipp, and F. Phillipp, J. Crystal Growths **175/176**, 451 (1997).

[31] P.A. Stolk, H.-J. Gossmann, D.J. Eaglesham, and J.M. Poate, Materials Science and Engineering B **36**, 275 (1996).

[32] T. Ghani, J.L. Hoyt, A.M. McCarthy, and J.F. Gibbons, J. Electron. Materials **24**, 999 (1995).

[33] D. Knoll, B. Heinemann, K.E. Ehwald, P. Schley, W. Rœpke, D. Bolze, J. Schlote, F. Herzel, and G. Fischer, Proc. 25th ESSDERC, pp.627–630 (1995).

[34] J. Tersoff, Phys. Rev. Lett. **64**, 1757 (1990).

[35] G.D. Watkins, in *Radiation Effects in Semiconductors*,Ed. M. Hulin, Dunod, Paris, p. 97 (1965).

[36] P.M. Fahey, P.B. Griffin, and J.D. Plummer, Rev. Mod. Phys. **61**, 289 (1989).

[37] P.J. Drevinsky, C.E. Caefer, S.P. Tobin, J.C. Mikkelsen, and L.C. Kimerling, Mater. Res. Soc. Symp. Proc. Vol. **104**, 167 (1988).

[38] T.T. Fang, W.T.C. Fang, P.B. Griffin, and J.D. Plummer, Appl. Phys. Lett. **68**, 791 (1996).

[39] L.C. Kimerling, M.T. Asam, J.L. Benton, P.J. Drevinsky, and C.E. Caefer, Material Sci. Forum **38–41**, 141 (1989).

[40] H.J. Osten, G. Lippert, P. Gaworzewski, and R. Sorge, Appl. Phys. Lett. **71**, 1522 (1997).

[41] H.J. Osten, G. Lippert, D. Knoll, R. Barth, B. Heinemann, H. Rücker, and P. Schley, IEDM Techn. Digest, pp. 803 (1997).

Correlations Between Microstructure and Electronic Properties in Amorphous Carbon Based Materials

T. Frauenheim, T. Köhler, G. Jungnickel, P. Sitch, P. Blaudeck

Institut für Physik, Theoretische Physik III, Technische Universität, D-09107 Chemnitz, Germany, E-mail: frauenheim@physik.tu-chemnitz.de

Abstract: We describe the present status of molecular-dynamics modelling of amorphous carbon materials and the atomic level insights, particulary for the electronic structure, that such investigations have provided. The main factors influencing chemical bonding will be discussed, namely the atomic density, the annealing time for structural relaxation and the incorporated hydrogen fraction. By comparing the π-orbital characteristics, such as *HOMO-LUMO* gaps and defect states, we define the conditions for high quality amorphous carbon materials that would be promising for both shallow p- and n-type doping. We investigate doping by substitution of nitrogen and boron into existing high-density tetrahedrally bonded amorphous carbon (ta-C) models. Additionally, we present first results on N-defects in ta-C which naturally develope during simulated annealing of a liquid carbon-nitrogen mixture at varying N concentration between 3 and 12%.

1 Introduction

The outstanding chemical, mechanical and electronic properties of tetrahedrally bonded amorphous carbon materials suggest possible applications in electronic devices operating under harsh environmental conditions. To realize this goal, experimental groups have put considerable effort in fabricating thin carbon films of a high electronic quality, i.e. with large band gaps, steep band edges and reasonably low defect concentrations. Such materials promise to remove some of the problems currently associated with the technological application of diamond. Despite the world-wide success in growing CVD films, diamond has not yet been successfully shallow n-type doped. The most obvious candidate, substitutional N, produces a donor level in natural diamond 1.7 eV below E_{cond}, too deep in the gap for practical doping purposes. This has been attributed to a strong off-site relaxation of the N atom in a {111}-bonding direction away from one of its four nearest C-neighbours, caused by the electrostatic repulsion of a lone pair of electrons at N and a predominantly p-orbital dangling bond on the carbon [1]. The situation in CVD diamond is still being investigated. The CVD-films can be doped p-type by substitutional boron, which yields a shallow acceptor with a level 0.37 eV above the valence band top [2]. However, since ionization of this level requires a temperature at which the electron mobility is greatly reduced, the superficial excellence in electronic operation is somehow lost. The behaviour of N in CVD grown films is apparently contrary to that in natural diamond, almost all N atoms preferring the substitutional tetrahedral onsite incorporation [3]. We have explained this to be due to a Fermi level effect: in CVD diamond, E_{f} is pinned below the donor state associated with the onsite N electron by deep lying defect bands stemming from bulk-C defects or π^* surface-like states in grain boundaries. Charge is transferred from the N environment to these defect states. Hence, there is now no "excess" charge close to N, and no off-site N motion [4,5]. Consequently, n-type doping is compensated by donating these charges into C-π^*-states.

The emergence at the turn of the decade of highly tetrahedral amorphous carbons (ta-C) [6] offerred an alternative way of solving the doping problem. These materials, combine the physical properties of diamond with a reduced band gap and thus the prospect of both n- and p-doping[7,8]. The outstanding physical properties of ta-C arise from the predominance (up to 80%) of sp^3 bonding, whereas the reduced bandgap is a consequence of the existence of small π-bonded clusters and sp^2 pairs in the material. Electronically, these introduce π–π^* shoulders into the σ–σ^* band gap, thus giving a band gap ranging from 2.2 to 3.0 eV. Whether doping has actually been experimentally achieved is, however, controversial. Veerasamy et al. [9] claimed success using phosphorus and nitrogen, where they observed increased conductivity of up to five orders of magnitude and a switch from intrinsic p-type to n-type. However, Ronning et al. [10] attributed this change in conductivity not to extended "doping" states but to variable range hopping via localized states at the Fermi energy. Consequently, a successful shallow n-dopant is still lacking. In ta-C, a potential acceptor or donor level is in competition with bonding and anti-bonding states associated with π orbital overlap between 3-fold coordinated carbon atoms which form π bonded clusters. These states generate broad bands within the σ–σ^* gap of the covalent network, the tails of which overlap with the acceptor and donor levels. The injected hole or electron then migrates to an occupied high energy π or a low energy unoccupied π^* level, respectively. As a result, the π cluster undergoes substantial relaxation and the Fermi level is pinned at the associated, partially filled, localized deep state. Electrical conduction is then only possible via variable range hopping [5,11].

Since there are still ongoing efforts to further improve the electronic quality of amorphous carbon films for doping purposes by influencing the microstructure of the films during deposition, we will in the following provide a molecular level insight into the chemical bonding nature of amorphous carbon based materials. The main influencing factors, the density, annealing time for structural relaxation, and incorporated hydrogen fraction, will be discussed. Reviewing the present status of theoretical amorphous carbon modelling and doping, we will finally define the bonding constraints that would successfully allow accomodation of both boron and nitrogen at reasonable solubilities as shallow substitutional dopants.

2 Theoretical Modelling

All amorphous carbon models have been generated using molecular-dynamics (MD) annealing simulations, cooling an equilibrated liquid of a given composition, under constant volume conditions down to room temperature. By using Γ-point Brillouin-zone sampling for total energy calculations in sufficiently large 3-dimensional periodic supercells (containing 128 C-atoms plus other type atoms), interatomic forces are obtained within the density-functional based tight-binding (DF-TB) scheme [14]. This method has been shown to be transferable to systems of various type and scale and successful in modeling heteroatom interactions with H, B, and N as well [5,12,13]. For investigating the doping capabilities of amorphous carbon we substituted B or N atoms into various sites in the amorphous matrix. We subsequently annealed the structure at 1000 K for 2 ps followed by a final conjugate gradient relaxation. Additionally, we performed first simulations in which the N-defects in ta-C are naturally incorporated by simulated cooling of a liquid carbon-nitrogen mixture for a N concentration varying between 3 and 12%.

3 Microstructure and π-Bonding Related Electronic Properties

For pure a-C, the changes in chemical bonding versus mass density have been studied systematically by empirical and *ab initio* based TB-schemes [15–19]. A few structures at selected densities have been published using Car-Parrinello 1st principle MD and *ab initio* multicenter TB-MD studies: a 54-atom structure at 2.0 g/cm^3 [20] and 64-atom supercells of 3.0 g/cm^3 [8,21]. Further systematic analysis of other factors influencing the structural properties, such as simulation regime (e.g. constant pressure), annealing time [22,23], hydrogen [24,25] or other heteroatom incorporation [13] and finally different starting precursor situations [26], is still a very active research field. A reliable statistical proof of particular conclusions, such as the significance of 3- and 4-fold carbon rings [21] –thereby avoiding uncertainties due to the small supercells – is, however, missing. In the following, we shall address some of this questions and summarize the results of recent simulations performed by our group.

3.1 Variation of Density and Annealing Time

In analyzing the final metastable a-C modifications generated under identical annealing conditions at different mass densities, we find clear trends for a pronounced π-clustering of undercoordinated sp and sp^2 sites at any density, stabilizing the model structures by lowering the strain and increasing the cohesive energies.

At lower densities (up to 2.4 g/cm^3), the π-clusters are dominant, (high sp- and sp^2-fractions, Fig. 1) percolating through the system. However, at and above 3.0 g/cm^3, they decrease in size and become dispersely embedded into a strained sp^3-matrix. The structural, chemical bonding and electronic gap data of the high-density modifications are summarized in Table 1. The constant atom number in the supercells allows for a comparison of the cohesive energies for different densities. We found 3.0 g/cm^3 to be a "magic" density at which energetically most stable a-C modifications are formed. Figure 2 shows

Table 1 Structure, chemical bonding and band gap data of high-density ta-C models

Data	ta-$C_{3.0}^{rapid}$	ta-$C_{3.0}^{ext}$	ta-C:$H_{3.0}^{rapid}$	a-$dia_{3.52}^{rapid}$	a-$dia_{3.52}^{ext}$
$C3$ (%)	36	31	32	12	8
$C4$ (%)	64	69	68	88	92
k_{CC}	3.64	3.69	3.25	3.88	3.92
$N_{rings}^{(4...8)}$	6,36,54, 20,32	3,43,58, 12,39	4,17,29, 7,25	5,41,79, 33,67	1, 46, 98, 18, 67
$E_{\pi-\pi^*}$ (eV)	2.88	2.07	3.30	5.59	3.22
$N_{defects}$	4	4	3	8	2
$N_{(size)}^{\pi-cl}$	$16_2,4_3,1_6$	$9_2,1_4,1_6,1_8$	$11_2,1_4,1_{11}$	$1_2,3_3$	5_2

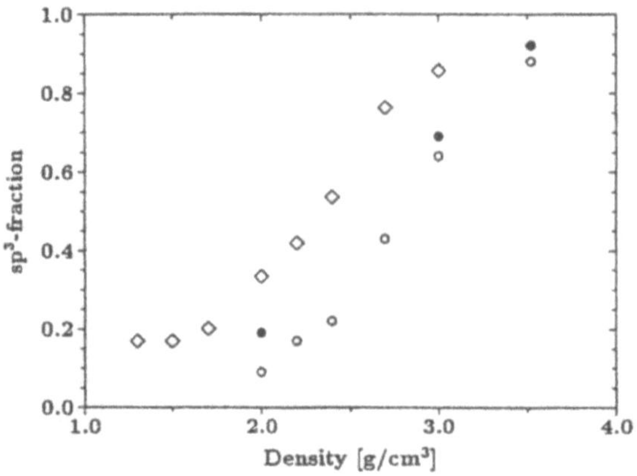

Figure 1
Carbon sp^3-content versus mass density, rapid and extended cooled a-C models (○,•), a-C:H (◇). Note, the variable H-content of a-C:H models: 1.30–2.20 g/cm^3 (35 at%), 2.40–2.70 g/cm^3 (30 at%), 3.00 g/cm^3 (25 at%).

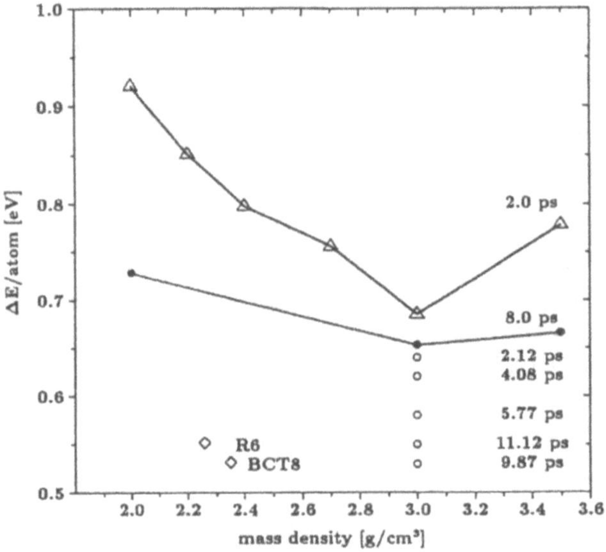

Figure 2
Cohesive energy per atom versus mass density, refering to the binding energy per atom of diamond, rapid and extended cooled a-C models (△,•), optimized annealing by parallel processing a-C models (○), novel crystalline sp^2-bonded carbon forms (◇). The total relaxation time is denoted by the numbers.

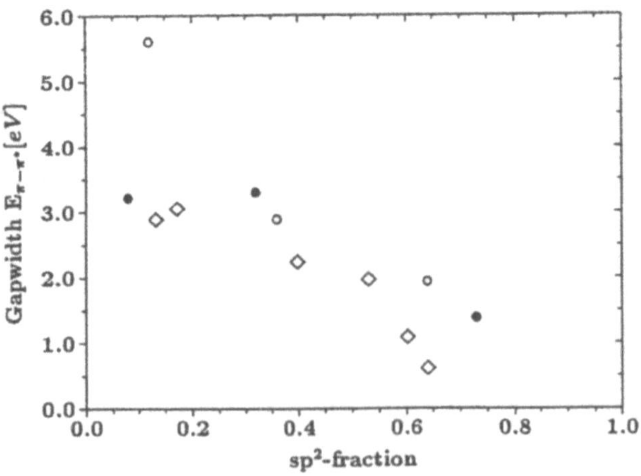

Figure 3
Formation of π–π^* - gap versus sp^2-content, notation as in FIG. 2.

the calculated cohesive energy differences per atom of a representative number of a-C structures versus density, including data of rapid quench (2 ps), extended stochastically cooling (8 ps) [22] and optimized annealing by parallel processing at different time scale (2–11 ps) [23]. When extending the cooling for longer periods, there is a clear tendency to form energetically more stable structures and to reduce the energy differences at the various densities.

This "flatness" in the energy curve can be explained as the result of the energetic degeneracy of sp^2 and sp^3 C-phases at equilibrium conditions. In particular, the lowest energy structures at 3.0 and 3.5 g/cm^3 become almost isoenergetic with annealing time.

3.2 Low-Density Amorphous Carbon, a-C

The models at 2.0 g/cm^3 show that the extension of the cooling time strongly favours an increase in the network connectivity, due to a lowering of the sp-fraction and an increase in the sp^2, sp^3-fractions from (27,9,64) to (8,19,73)%, respectively (Fig. 1). The percolating π-clusters simultaneously change from a composition of more chain-like segments and a few (1,9,5,8) 4,5,6,7-fold rings with alternating single-, double, and triple bonds (C-C coordination number $k_{CC} = 2.80$) to largely enhanced ring agglomerations (2,17,28,11) 4- to 7-fold rings, including strained aromatic subunits ($k_{CC} = 3.11$). Differences in the medium-range order give rise to some peak shifts and changes in the theoretically calculated interference functions and radial distributions [18]. From comparison with neutron diffraction data [27], this suggests that the extended stochastically cooled model is a more realistic representative of rf-sputtered a-C films.

Electronically, the low-density a-C modifications are of lesser interest, due to the occurance of broad π-bands reflecting the bonding in highly defective extended clusters. There is a consequent reduction of the size of the band gap, which is found to be smaller than 2 eV after removal of unbonded defect states.

3.3 High-Density Amorphous Carbon, ta-C

Many efforts have been made to increase the sp^3 C-fraction by growing so-called highly tetrahedrally bonded materials, ta-C, using ion beam techniques [28] and vacuum arc deposition [29,30] and hydrogenated ta-C:H, using a monoenergetic C_2H_2 rf-plasma beam source [31]. Though the structures are formed under nonequilibrium conditions, the actual existence of the high density ta-C(:H) films confirms their theoretically predicted high stability. The characteristic properties of the films in relation to the density are strongly dependent on the impact energy of deposited carbon atoms. The sp^3 content, compressive stress, hardness and optical gap are all found to pass through a maximum correlated to the highest density achievable. In the optimum energy range of about 90-100 eV per C-atom, densities of 2.9 to 3.3 g/cm^3 (in some cases even 3.5 g/cm^3) [32] corresponding to sp^3 fractions of 60 to $\geq 90\%$ are reported. In particular, the opening of wide optical gaps, 2–3 eV, at reasonably small defect concentrations, $\approx 10^{-17}$ cm^{-3}, have been considered to be very promising for further doping investigations.

The formation of very high density ta-C films has been attributed to subplantation processes of energetic carbon species beneath the amorphous surface. In the subsurface compressive stress is accumulated by interstitial formation of impact atoms and/or into the substrate displaced surface near atoms, which results in an overall density increment [33,34]. As shown recently, a considerable fraction of atoms are bonded in metastable configurations, which relax into the desired sp^3-rich ta-C phases [35].

A number of recent papers [8,17–19] have related the observed electronic properties of ta-C to the embedding of small π-bonded clusters (preferably pairs) within a strained sp^3-matrix. The balance between the π-bonding cohesive energy gain, defect formation and the overall reduction of compressive stress is found to be crucial in determining the band gap properties and the related electronic quality for doping.

We consider now the annealing time influences on structure formation. The more extended regimes at 3.0 g/cm^3 yield a slightly increased sp^3-fraction, from 64 to 69% (Fig. 1). The π-clusters simultaneously increase in size (Table 1). As a consequence, the band gap shrinks from 2.88 to 2.07 eV and the π, π^* bands are considerably broadened. This causes any doping level to overlap with the π-bands and to be compensated by defect stabilization or emptying/filling of π and π^*-states, respectively.

At diamond density, the highest sp^3-fractions of 88 and 92% are obtained. The defect density in the rapidly cooled structure is extremely high, indicating a non optimal evolution of π-clustering. However, in the more extended annealed modification, we find, except for two isolated sp^2-defect sites, an almost ideal structure for doping. All other undercoordinated atoms exclusively arrange in pairs, i.e. the smallest possible defect-free π-clusters. Thus narrow π and π^* bands with steep band edges are formed.

3.4 Incorporation of Hydrogen at Different Densities

The addition of hydrogen in amorphous carbon deposition causes the microstructure to change considerably. The results of a DF-MD [36] preliminary study using very small supercells (64 C atoms plus H) have recently been confirmed by considering systems twice as large [37]. As can be seen in Figure 1 the sp^3-fraction at densities ≤ 2.4 g/cm^3 is strongly enhanced by a factor of ≈ 2 at H-concentrations of 35%. The hydrogen is well distributed throughout the system. As Table 2 shows, the fraction of H bonded as sp^3C-H increases with density. While the numbers of H bonded as sp^3C-H_2 does not show a systematic density dependence, the number of sp^2C-H groups is simultaneously strongly reduced with density, the maximum number of the latter sites being observed at 1.7 g/cm^3. In this particular structure, 2 large graphitic regions with pronounced aromaticity, consisting of 6 and 4 fused almost planar 6-fold rings, respectively are observed. The occurance of extended sp^2 clusters has been also reported by other groups in H-free low-density a-C modifications [15,20]. In contrast to their structures, in the presence of hydrogen, a large fraction of sp^2 cluster edge sites become saturated by hydrogen. The extended graphitic regions are interconnected by chain-like hydrocarbon segments, representing a large portion of the amorphous matrix.

Hence, the introduction of H into a-C reduces the network connectivity and removes sp-bonded sites. While the total number of small carbon rings is reduced, too, there are increasing tendencies in forming larger subunits (graphitic-like inclusions). At 2.0 to 2.4 g/cm^3, a few buckled ring clusters and chain-like sp^2-segments are partially saturated by hydrogen and become interconnected by mono- or dihydride sp^3-carbon. Similar tendencies are reported by Iarlori et al., who considered a 2.6 g/cm^3 64-C-atom supercell

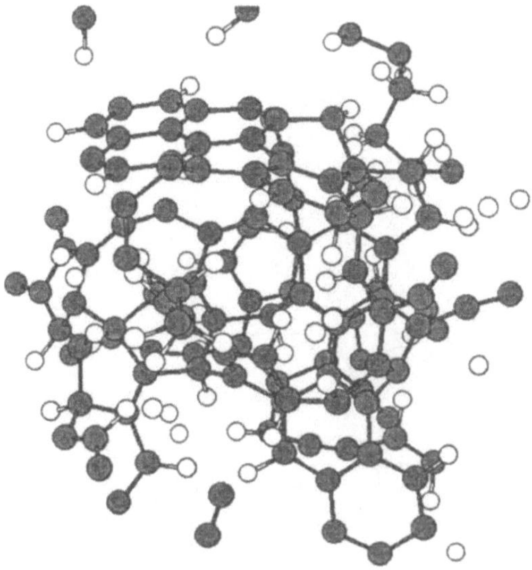

Figure 4
Structural snapshot of the 1.70 g/cm^3 a-C:H model.

Table 2 Statistical data for the occurance of hydrogen in CH_x groups (%) in a-C:H models at various densities ρ.

ρ (g/cm^3)	x_H (%)	sp^2-CH	sp^3-CH	sp^2-CH_2	sp^3-CH_2	sp^3-CH_3
1.30	35	26.1	11.6	5.8	23.2	4.3
1.50	35	27.5	13.0	5.8	11.6	21.7
1.70	35	37.7	21.7	2.9	14.5	0.0
2.00	35	26.1	28.9	2.9	23.2	4.3
2.20	35	15.9	30.4	2.9	28.9	13.0
2.40	30	12.7	50.9	0.0	25.5	0.0
2.70	30	5.4	67.0	0.0	21.8	5.4
3.00	25	4.6	86.0	0.0	9.3	0.0

including 16 at.% hydrogen [24]. The simulated structures are in reasonable agreement with recent neutron scattering data, reported at similar mass densities [24,38]. The electronic gap values obtained at comparable densities are very similar to those in the H-free systems. At 2.7 and 3.0 g/cm^3, the sp^3-C-fraction for slightly reduced H-concentrations (30 and 25%) reaches almost 80 and 90% [37] (Fig. 1). The corresponding structures may be seen as hydrogenated ta-C analogues, which open similar large band gaps at reasonably low defect numbers as in ta-C. Thin film materials of this type have been successfully grown by monoenergetic C_2H_2 plasma beam techniques [31]. However, due to the hydrogen incorporation, the clustering of the different carbon hybrids changes considerably. While in the most stable ta-C structures small even-membered π-clusters are randomly embedded in an otherwise strained sp^3-matrix, there is a tendency for a two phase separation of the C-hybrids in ta-C:H [25]. We find relatively large sp^3-clusters (diamond-like subunits), now becoming embedded in an sp^2-enriched sp^2-sp^3-matrix, in which again the formation of small π-clusters yields a maximum stress reduction from the network favoring strong π-bonds and maximum values of the band gap around 3.0 eV (compare Figure 3).

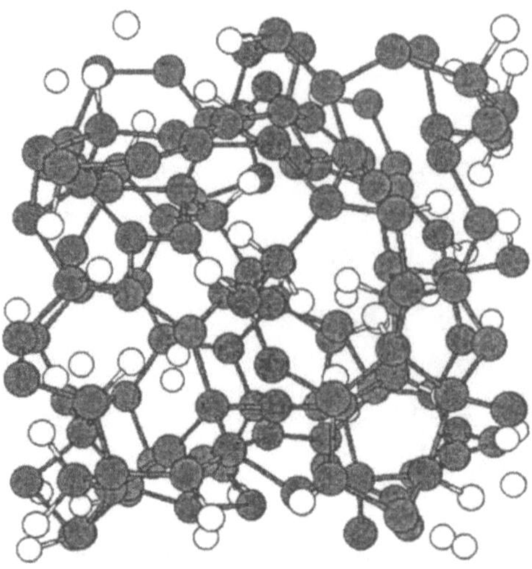

Figure 5
Structural snapshot of the 2.40 g/cm^3 a-C:H model.

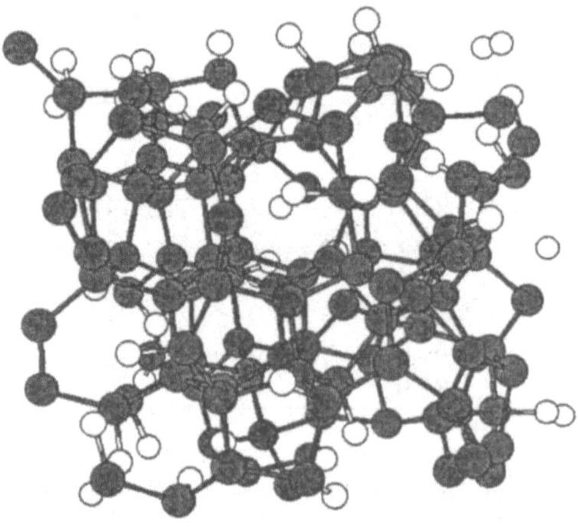

Figure 6
Structural snapshot of the 3.00 g/cm^3 ta-C:H model.

In the following, we investigate the doping capabilities of the most suitable ta-C modifications with respect to electronic applications as discussed in Section 3.3. We have been chosen a structure without defects that has steep band edges, narrow π- and π*-bands and an overall *HOMO-LUMO*-splitting of abaut 3.7 eV.

4 The p- and n-Type Doping Capabilities of ta-C Using B and N

First, we describe results for substitution of B and N into various sites of the amorphous matrix. All structures after substitution have been subsequently annealed at 1 000 K for 2 ps and finally relaxed using a conjugate gradient technique.

4.1 Nitrogen Substitution

The most stable site for the N atom is a stretched sp^3 site. The long bond subsequently breaks, leaving the N atom 3-fold coordinated with slightly distorted σ bonds of lengths 1.57, 1.53 and 1.48 Å and a lone pair of electrons, as indicated by a Mulliken charge analysis. This also shows that there is a movement of charge from the newly created 3-fold coordinated nearest-neighbour C atom to a π cluster elsewhere in the amorphous matrix. This causes considerable rearrangement of the atoms in this π cluster, where the extra charge is accomodated in anti–bonding levels, and results in a partially filled defect state pinning the Fermi energy 0.63 eV above the π valence band top, being localised there. In addition, a relatively undistorted sp^3 site was found to be metastable with an energy of incorporation for the N atom being 0.54 eV higher than that of the minimum energy cluster mentioned above. The N atom remains 4–fold coordinated and fully sp^3–bonded with back bond lengths 1.46, 1.54, 1.47 and 1.42 Å. To facilitate the stabilization of this configuration, we also observe a migration of charge from the N atom to remote π clusters, where the resultant partially filled defect level appearing 0.68 eV above the π valence band top is localised. Substitution into the π bonded site is found to be unstable, the N atom spontaneously breaks the π bond and transforms into the 3–fold σ bonded plus lone pair arrangement described above.

4.2 Boron Substitution

For all lattice positions tested, including stressed and strained 4-fold coordinated sp^3 and 3-fold coordinated sp^2 sites, we observe an overwhelming preponderance for the B atom to modify its local environment so that it sits in a fully 4-fold coordinated sp^3 environment. When substituted into a sp^3 bonded environment, B forms C-B bonds of lengths 1.62, 1.60, 1.66 and 1.63 Å. Electronically, the *HOMO* is a partially occupied defect state 0.32 eV above the valence band edge, which is strongly localised in remote π clusters. The physical reasoning for this is that the B level lies in the π shoulder of states comprising the valence band edge in ta–C and is filled by electrons migrating from these π clusters. Thus, an intrinsic *p*-type behaviour of a ta-C-film by incorporation of boron is going to be enhanced. An sp^2 site is energetically unfavoured – the B atom spontaneously rearranges to form 4 nearest neighbour bonds. A strongly localized defect is then present on one of its undercoordinated nearest neighbours.

These results show that even in zero-defect structures with narrow π and π^*-bands, the doping compensation cannot be altered, neither by substituting B and N, nor other types of atoms.

However, the doping studies described so far and similarly all what has been reported by other groups during the years, are not real doping investigations. In real thin film deposition the doping atoms are either implanted by high-energy ion beams or being more relevant for a-C materials they are incorporated during growth. That is why we next present first results on N defects in tetrahedral amorphous carbon, ta-C, that are naturally incorporated by simulated annealing of a liquid carbon-nitrogen mixture at varying N concentration.

Table 3 Hybridization of carbon atoms: C-hybrid fractions

N-content (%)	sp^2 (%)	sp^3 (%)	sp^{1+x} (%)	sp^{2+x} (%)
0.0	20.3	69.5	0.8	8.6
3.0	29.7	58.6	1.6	9.4
5.9	25.8	61.7	0.8	11.7
7.2	32.8	53.9	3.1	8.6
8.6	36.7	51.6	3.1	8.6
9.9	33.6	49.2	4.7	12.5
11.1	46.1	40.6	3.1	9.4

4.3 N Incorporation upon Annealing

We have generated seven ta-C:N models at 3.0 g/cm^3 density with N-fractions between 3 and 11 at.%. We started out of random atom distributions representing a given stoichiometry (supercell including 128·C plus x·N atoms) and density. These cells at fixed volume in a first run of 0.5 ps (500 MD-steps) are equilibrated at 8000 K. Following this initialization the low-energy amorphous structures are obtained by applying simulated annealing/cooling to the cells within 2 ps to 300 K. After additional 1 ps equilibration at room-temperature no more changes in chemical bonding occur and the structures for the final characterization are relaxed via conjugate gradient techniques.

The physical property most relevant to our investigation is that of the hybridization of the atoms. The hybrid formation at the C-sites determines the gap properties while that at the N-sites controls the doping capability. In our studies we follow Stephan et al. [7] and determine the sp^x-hybrid fractions of incorporated C and N atoms. For the different models all values are summarized in Tables 3 and 4. From Table 3 we clearly see that the inclusion of nitrogen up to about 11% yields a considerable reduction in the C-sp^3-content from 70 to 40%. As a consequence the π-bonding in the network is enhanced, the π- and π^*-bands become broadened and the *HOMO-LUMO*-gaps are shrinking. This tendency naturally limits the doping quality of related structures which more favourably should have large gaps, steep edges and narrow bands.

Nevertheless, the doping should be related to the incorporation of nitrogen. Here we find at the different concentrations (Table 4) various types of bonding and related hybrid formation. Two of them are electronically saturated non-doping lone-pair configurations: sp^{2+x} (3-fold σ-bonded), sp^{1+x} (2-fold σ + π-bonded), while the remaining two: sp^3 (4-fold σ-bonded) and sp^2 (3-fold 2-σ- + π-bonded) are doping-like. Since both types due to almost isoenergetic formation energies occur at similar probabilities the doping efficiency is established to be low, which is confirmed by many experiments. However, regarding the doping-like N-sites, they do not contribute shallow donor electrons in the classical sense. For both types of sites the donor levels are in competition with anti-bonding states associated with π orbital overlap between 3-fold coordinated carbon atoms which form π bonded clusters. The associated states generate broad bands within the $\sigma - \sigma^*$ gap of the covalent network. The injected electron from the N atom migrates to an unoccupied π^* level. As a result, the π cluster undergoes substantial relaxation and the Fermi level is shifted towards n-type behaviour pinned at the associated, partially filled π^*-state. Since the small π-clusters usually are screened within an sp^3 matrix conductivity may only become possible through variable range hopping, which has been observed experimentally [10].

Table 4 Hybridization of nitrogen atoms: N-hybrid fractions.

N-content (%)	sp^2 (%)	sp^3 (%)	sp^{1+x} (%)	sp^{2+x} (%)
0.0	50.0	0.0	25.0	25.0
3.0	50.0	25.0	25.0	0.0
5.9	30.0	20.0	10.0	40.0
7.2	25.0	16.7	33.3	25.0
8.6	36.7	51.6	3.1	8.6
9.9	64.3	14.3	7.1	14.3
11.1	62.5	12.5	12.5	12.5

Summarizing these studies: Nitrogen incorporation during deposition (simulated-annealing molecular-dynamics) of high-density ta-C materials is enhancing the π-bonding formation of the network simultaneously yielding a reduction of the C-sp^3 fraction. This limits the desirable doping properties. The N atoms at almost equal portions prefer to stay in both nondoping and doping-like sites. While the nondoping N's form electronically saturated lone-pairs, the doping-like N's either as fourfold coordinated sp^3 atoms or as threefold coordinated sp^2 atoms (with two σ- and one π-bond) yield donor-states competing (overlapping) with empty π^*-states of the small π-clusters in the network. The N-injected electron migrates to such an unoccupied energetically lower-lying π^*-state slightly shifting the Fermi-level towards n-type behaviour. Since this is a charge donation into "localized" carbon π^*-states that are lower in energy than the actual donor level would be, the presence of classical shallow n-type doping of N in ta-C can not be confirmed. As described by Ronning and Hofsäss, the increase of electrical conduction by several orders of magnitude rather has to be explained by variable range hopping between small localized π-clusters embedded within an insulating strained sp^3-matrix.

5 Conclusions

Considering the atomistic amorphous models, the high-density ta-C and ta-C:H modifications provide the most promising electronic properties for doping, namely large band gaps and low defect concentrations. However, the broad distribution of π- and π^*-states still present strongly counteracts doping. Any shallow acceptor or donor level for B and N will overlap with these states and consequently will either trap or transfer an electron from the π/π^*-band, thus merely shift the Fermi-energy to the corresponding band edges.

We therefore develop principles that make both shallow *p*- and *n*-type doping by B and N in a carbon based material most likely and approach this problem from a different point of view. Instead of searching for a possible new element or atom complex that would act as a shallow donor in known carbon allotropes we have been looking for a suitable crystalline host for shallow N- and B-doping. As the result we recently postulated novel crystalline sp^2-bonded, wide band gap forms of carbon, which by the isolation of strong localized π-bonds provide constraints that accomodate both boron and nitrogen as shallow substitutional dopants [39].

Acknowledgements

We gratefully acknowledge the support from the Deutsche Forschungsgemeinschaft and the trinational DACH cooperation.

Bibliography

[1] S. A. Kajihara, A. Antonelli, J. Bernholc, R. Car, Phys. Rev. Lett. **66**, 2010 (1991).

[2] R. Locher, C. Wild, N. Herres, D. Behr, P. Koidl, Appl. Phys. Lett. **65**, 35 (1994).

[3] R. Samlenski, C. Haug, R. Brenn, C. Wild, R. Locher, P. Koidl, Diamond and Related Materials **6**, 947 (1996).

[4] C. E. Nebel, J. Münz, M. Stutzmann, R. Zachai, H. Güttler, Phys. Rev.B **55**, 1 (1997).

[5] P. Sitch, Th. Köhler, G. Jungnickel, D. Porezag, Th. Frauenheim, Solid State Commun. **100**, 549 (1996).

[6] D. R. Mckenzie, D. A. Muller and B. A. Pailthorpe, Phys. Rev. Lett. **67**, 773 (1991).

[7] U. Stephan, Th. Frauenheim, P. Blaudeck and G. Jungnickel, Phys. Rev. B **50**, 1489 (1994).

[8] D. A. Drabold, P. D. Fedders and P. Stumm, Phys. Rev. B **49**, 16415 (1994).

[9] V. S. Veerasamy, G. A. J. Amaratunga, C. A. Davies, A. E. Timbs, W. I.Milne, D. R. Mckenzie, J. Phys. Condens. Matter **5**, L169 (1993).

[10] C. Ronning, U. Griesmeier, M. Gross, H. C. Hofsäss, R. G. Downing and G. P. Lamaze, Diamond and Related Materials **4**, 666 (1995).

[11] C. Ronning, U. Griesmeier, M. Gross, H. C. Hofsäss, R. G. Downing, G. P. Lamaze, Diamond and Related Materials **4**, 666 (1996).

[12] J. Widany, F. Weich, Th. Köhler, D. Porezag, Th. Frauenheim, Diamond and Related Materials **5**, 1031 (1996).

[13] F. Weich, J. Widany, Th. Frauenheim, Phys. Rev. Lett. **78**, 3326 (1996).

[14] D. Porezag, Th. Frauenheim, Th. Köhler and G. Seifert, Phys. Rev. B **51**, 12947 (1995).

[15] C. Z. Wang, K. M. Ho, Phys. Rev. Lett. **70**, 611 (1993).

[16] C. Z. Wang, K. M. Ho, C. T. Chan, Phys. Rev. Lett. **71**, 1185 (1993).

[17] Th. Frauenheim, P. Blaudeck, U. Stephan, G. Jungnickel, Phys. Rev.B **48**, 4823 (1993).

[18] U. Stephan, Th. Frauenheim, P. Blaudeck, G. Jungnickel, Phys. Rev.B **49**, 1489 (1994).

[19] Th. Frauenheim, G. Jungnickel, Th. Köhler, U. Stephan, Journ. Non-Cryst. Mat. **182** (1995) 186.

[20] G. Galli, R. M. Martin, R. Car, M. Parrinello, Phys. Rev.B **42**, 7470 (1990).

[21] N. A. Marks, D. R. McKenzie, B. A. Pailthorpe, M. Bernasconi, M. Parinello, Phys. Rev. Lett. **76**, 768 (1996).

[22] Th. Köhler, G. Jungnickel, Th. Frauenheim, Phys. Rev.B **52**, 11837 (1995).

[23] P. Blaudeck, K. Hoffmann, G. Jungnickel, Th. Köhler, Th. Frauenheim, submitted to Computers in Physics, July 1997.

[24] S. Iarlori, G. Galli, O. Martini, Phys. Rev.B **49**, 7060 (1994).

[25] Th. Frauenheim, G. Jungnickel, U. Stephan, P. Blaudeck, S. Deutschmann, M. Weiler, S. Sattel, K. Jung, H. Ehrhardt, Phys. Rev.B **50**, 7940 (1994).

[26] A. P. Sutton, P. D. Godwin, A. P. Horsfield, MRS Bulletin **21**, 42 (1996).

[27] F. Li, J. Lannin, Phys. Rev. Lett. **65**, 1905 (1991).

[28] D. R. McKenzie, G. B. Smith, Z. Q. Liu, Phys. Rev.B **37**, 8875 (1988).

[29] V. S. Veerasamy, G Amaratunga, W. I. Milne, P. Hewitt, P. J. Fallon, D. R. McKenzie, C. A. Davis, Diamond and Related Materials **2**, 782 (1993).

[30] G. Jungnickel, M. Kühn, S. Deutschmann, F. Richter, U. Stephan, P. Blaudeck, Th. Frauenheim, Diamond and Related Materials **3**, 1056 (1994).

[31] M. Weiler, S. Sattel, K. Jung, H. Ehrhardt, V. S. Veerasamy, Diamond and Related Materials **3**, 608 (1994).

[32] Y. Lifshitz, S. R. Kasi, J. W. Rabalais, W. Eckstein, Phys. Rev.B **41**, 10468 (1990).

[33] C. A. Davis, Thin Solid Films **226**, 30 (1993).

[34] J. Robertson, Diamond and Related Materials **2**, 984 (1993).

[35] S. Uhlmann, Th. Frauenheim, S. Lifschitz, Phys. Rev. Lett. **81**, 641 (1998).

[36] Th. Frauenheim, P. Blaudeck, U. Stephan, G. Jungnickel, Phys. Rev.B **48**, 4823 (1993).

[37] Th. Köhler, G. Jungnickel, Th. Frauenheim, to be published.

[38] G. Jungnickel, Th. Frauenheim, D. Porezag, P. Blaudeck, U. Stephan and R. Newport, Phys. Rev.B **50**, 6709 (1994).

[39] G. Jungnickel, P. Sitch, Th. Frauenheim, B. Eggen, M. Heggie, C. Latham, C. Cousins, Phys. Rev. B **57**, R661 (1998).

Characterization of CVD-Diamond Layers: Electronic States

Rolf Sauer

Abteilung Halbleiterphysik, Universität Ulm, D-89069 Ulm, Germany

Abstract: Electronic states in CVD-diamond layers are discussed with emphasis on luminescence studies. Included are data on characteristic selected defects, intrinsic band-edge luminescence and excitons localized at boron acceptors introduced into the layers by doping during layer deposition or by ion implantation. The problem of shallow donors is also addressed, and in particular promising recent electrical and optical data on phosphorus-doped diamond layers are presented.

1 Introduction

Two "natural" donor and acceptor impurities exist in diamond, nitrogen and boron, respectively. Single nitrogen atoms on substitutional lattice sites possess a donor level $\approx$ 1.7 eV below the conduction band [1,2] (Fig. 1). This large ionization energy provides only very low electrical n-type conductivity at room temperature. However, high-temperature diamond p-n junctions made from a B-doped homoepitaxial layer on N-doped substrate have been successfully demonstrated [3]. Aggregated nitrogen produces a second, much deeper donor level 4.0 eV below the conduction band [1,2]. Boron occupies substitutional sites forming an acceptor level with a hole activation energy of 370 meV [1,2]. The level is shallow enough to produce p-type conductivity at room temperature that can be used for electronic devices, as e.g. Schottky diodes and field effect transistors [4]. For such applications it is important to note that increasing doping levels above $\approx 10^{16}$ cm^{-3} reduce the

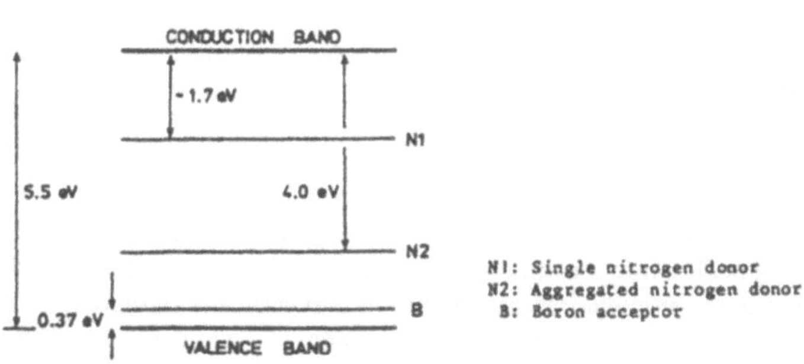

Figure 1 Nitrogen donor and boron acceptor levels in diamond (after [1]).

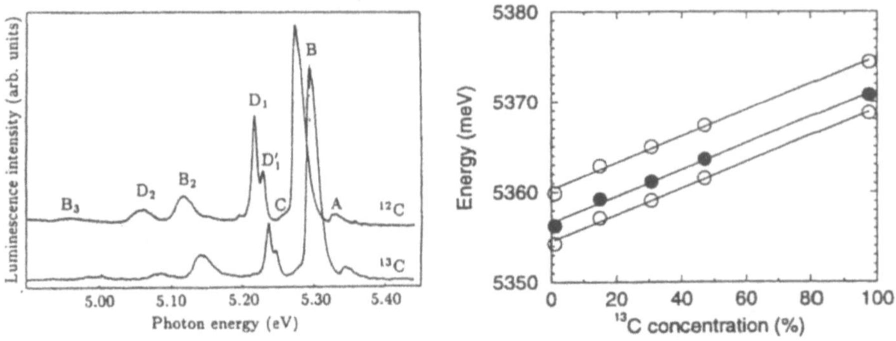

Figure 2 Left: Shift of near-band-edge electronic states from ^{12}C to ^{13}C diamond. B intrinsic free exciton (FE) transition, D_1/D'_1 recombination lines from boron bound excitons (BE) both with the participation of wave-vector-conserving TO-phonons [11]. Right: Linear dependence of bound exciton multiplet lines on the isotopic composition [13].

effective ionization energy monotonically (see below) thus enforcing the role of boron as a useful acceptor.

A large number of defects is known in diamond [5]. There are nearly hundred paramagnetic centers whose electronic properties (Landé g-factors, hyperfine interaction constants, triplet splitting parameters) have been studied [6,7]. Many of them have not yet been positively identified. In the present article, we will focus on optically active defects and electronic states. Numerous optical defect spectra have been reported in natural bulk diamonds and synthetic high pressure-high temperature (HPHT) diamonds [5,8,9]. Most of the sharp-line luminescence or absorption spectra are due to point defect complexes created after irradiation or ion implantation incorporating vacancies, carbon interstitials, nitrogen atoms or the implanted elements, as e.g. transition metals [8,9]. Currently, the possibility to produce HPHT diamonds with controlled $^{12}C/^{13}C$ isotopic composition is used to investigate isotope shifts in the optical spectra; various $^{12}C/^{13}C$ shifts in no-phonon (NP) transitions and $^{12}C/^{13}C$ plus $^{14}N/^{15}N$ shifts in local vibrational mode (LVM) satellites have been reported promoting the identification of the optical defects [10]. Such studies include the variation of the optical band-gap E_{gx} with isotopic composition from nearly pure ^{12}C crystals to almost pure ^{13}C which is reflected in intrinsic free exciton and bound exciton radiative transitions [11–13] (Fig. 2).

2 CVD-Diamond Layers

2.1 Optical Defects

Diamond layers grown by microwave-assisted CVD or by the hot filament technique exhibit many of the optical defect spectra emerging in bulk diamonds as mentioned above [14]. One spectrum is of particular interest because it is specifically observed in diamond layers grown on silicon substrates or in a CVD equipment using silica in the reactor. The spectrum, with its NP-transition at 1.681 eV, was originally discovered in CVD-grown homoepitaxial films [15] and then also observed in Si-implanted natural diamonds [16]. From its intensity depending quadratically on the Si ion dose it was ascribed to a complex

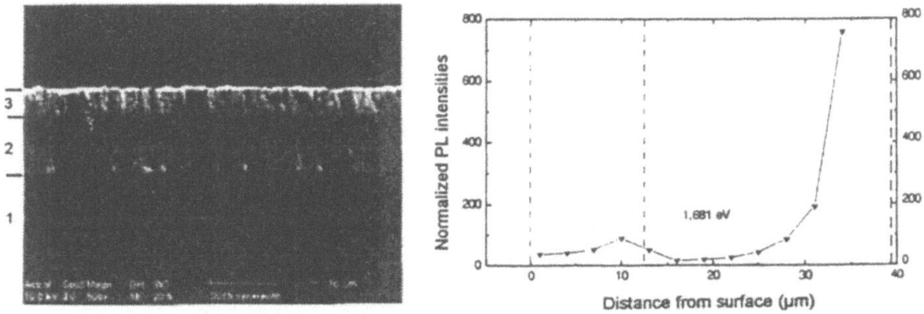

Figure 3 Spatial distribution of the 1.681 eV optical defect. Left: Cross-sectional micrograph of a CVD-diamond layer grown on Si substrate; 1: Si substrate, 2: undoped 27 μm thick diamond layer, 3: boron-doped 12 μm thick diamond layer. Right: Luminescence intensity of the 1.681 eV line as a function of the distance from the surface (0 $\mu m \triangleq$ surface, 39μm $\triangleq$ Si substrate) [19].

incorporating two Si atoms. Many groups have later observed the spectrum in CVD layers and confirmed its relationship to Si by the circumstantial evidence [17,18]. Fig. 3 shows an experiment [19] where a CVD-diamond layer was grown on silicon, nominally undoped for 27 μm layer thickness followed by a 12 μm boron-doped layer. Micro-Raman and micro-PL measurements were performed on a cross-sectional area of the sample. Graphitic components in the layer (characterized by the so-called D- and G-bands in the Raman spectrum) are strong close to the substrate but disappear monotonically with increasing distance from the substrate indicating improved sp^3-bonded crystal structure. The intensity of the 1.681 eV PL line, normalized to the strength of the diamond Raman phonon line at 1332 cm^{-1} or taken absolutely, exhibits similar behavior (Fig. 3). It was argued that Si can be introduced into the growing diamond layer due to plasma etching either from the silica reactor or from the Si substrate. The latter process ceases once the diamond crystallites coalesce and form a closed layer explaining the strong reduction of the defect concentrations in the first few micrometers of diamond layer growth. Direct proof of Si participation and evidence for one Si atom per optical center originates from spectroscopically highly resolved luminescence. Usually the 1.681 eV transition is found in the CVD layers at very large linewidths probably due to large inhomogeneous stress in the grains of the polycrystalline layer. The linewidths were significantly reduced in free-standing plates of polycrystalline CVD-diamond [20]. Here, the line was resolved into a doublet (Fig. 4). Further line components were seen in a Si-doped homoepitaxial diamond film where the stress is even lower [21]. It was shown that the basic luminescence pattern is a quartet originating in electronic transitions between split initial and final states. The splitting was traced back to built-in overpressure in the defect sized too big for the small and rigid cage of the diamond lattice [21]. Finally, up to 12 sharp line components were observed in a Si-doped HPHT-diamond (Fig. 4) [22]. A convincing interpretation of this complex spectrum could be given ascribing the dominant line quartet to ^{28}Si, superimposed by identical but shifted quartets from the ^{29}Si and ^{30}Si isotopes with relative intensities proportional to the natural abundances of the three Si isotopes. This showed unambiguously the presence of one Si atom per optical center [22].

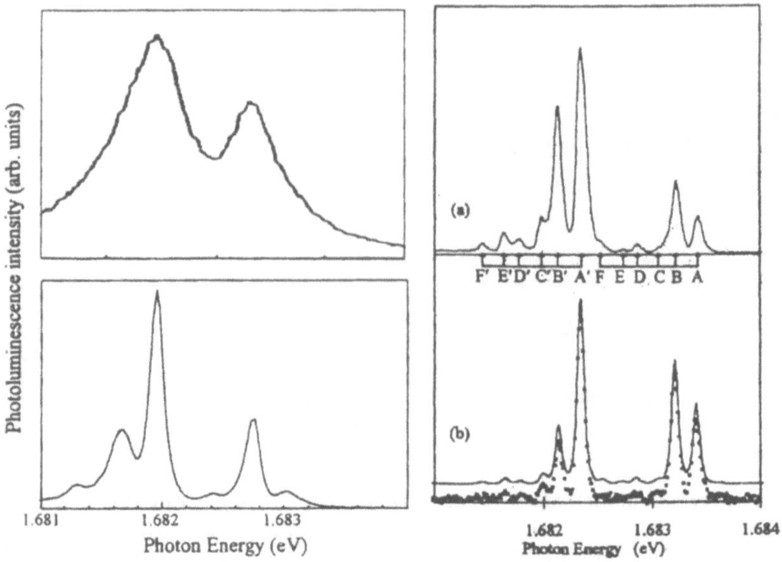

Figure 4 Fine structure of the 1.681 eV line. Left top: Doublet structure in a free standing CVD film [20]; left bottom: Highly resolved doublet with additional line components in a Si-doped homoepitaxial diamond film [21]; Right (a): 12-component-spectrum in a Si-doped HPHT diamond, (b): synthesis of the experimental spectrum by superposition of three identical quartet patterns shifted from each other for ^{28}Si, ^{29}Si, and ^{30}Si with relative strengths according to the natural abundances of the isotopes [22].

2.2 Intrinsic Band-Edge Luminescence

When impurities and defects capture excess electrons, holes, or free excitons (FE) the lifetime of these species is reduced and intrinsic band-edge luminescence is suppressed. The observation of FE luminescence, in diamond preferentially as a wave-vector conserving TO-phonon-assisted transition at $\approx$ 5.29 eV (cf. Fig. 2), is therefore an indication of low impurity and defect concentrations. In the past years, routinely grown CVD-diamond layers have reached high enough optical quality for FE emission to be visible, first reports dating to 1990 [23]. Fig. 5 shows a comparison between sample morphology and intrinsic band-edge luminescence [24]. The CVD-diamond film grown on Si forms a highly textured, oriented closed layer in the center of the sample. Towards the edges, crystallites of cubo-octahedral structure are partially coalesced but the film is no longer completely closed, and finally, single crystallites of cubo-octahedral morphology are grown at the edges. These isolated crystallites show FE-luminescence which disappears when the closed film develops. Since the crystallite sizes are comparable in the three cases the loss of FE-luminescence is probably not due to enhanced surface recombination but must be associated with the stress in the closed film and with higher defect concentrations not visible in the SEM-micrographs of Fig. 5. However, this situation is not symptomatic for CVD-layers: Some closed layers, with fiber texture and oriented crystallites as in Fig. 5c, show fairly intense FE-luminescence whereas other films grown in the same apparatus under the same nominal growth conditions fail. Evidently, present optical qualities of CVD-diamond layers are a critical issue and not yet completely under control.

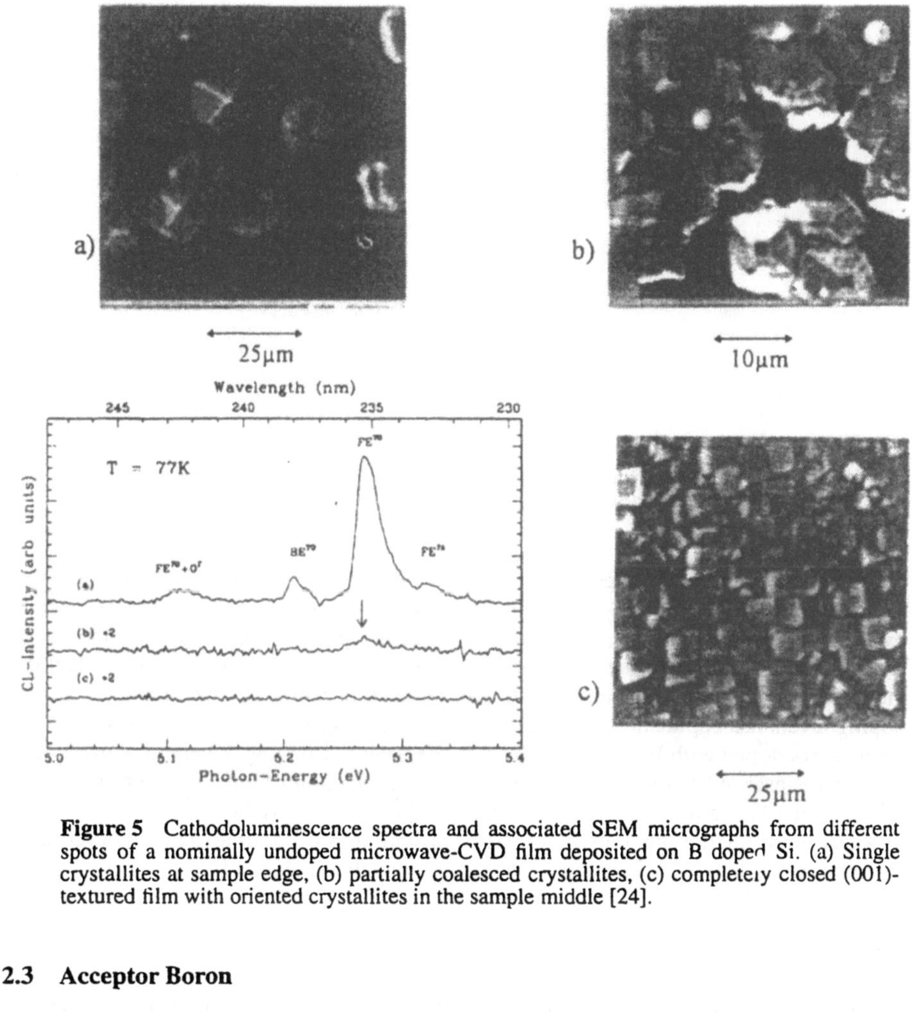

Figure 5 Cathodoluminescence spectra and associated SEM micrographs from different spots of a nominally undoped microwave-CVD film deposited on B doped Si. (a) Single crystallites at sample edge, (b) partially coalesced crystallites, (c) completely closed (001)-textured film with oriented crystallites in the sample middle [24].

2.3 Acceptor Boron

The acceptor causing p-type conductivity of semiconducting IIb-diamonds was previously believed to be aluminium. Experiments on laboratory-grown semiconducting HPHT diamonds when combined with other data on resistivity and activation energy for conduction indicated that the dominant semiconducting properties of both natural and synthetic diamond are due to one acceptor, boron [25,26]. CVD-diamond layers can be doped with boron by adding appropriate boron compounds, such as diboran, to the gas mixture [27,28]. For low boron doping levels, up to $\approx 10^{16}$ cm^{-3}, resistivity and Hall measurements yield the ionization energy $E_a = 0.37$ eV as known from natural and synthetic bulk diamonds. For higher effective acceptor concentrations, the measured activation energy decreases until it becomes zero at around 10^{20} cm^{-3} doping concentration (Fig.6) [28]. The data in Fig. 6 corroborate earlier results showing more scatter of the data points [29]. The experimental data can be fairly well described by the expression $E_a = 0.37$ eV - $\alpha n_{dot}^{1/3}$ eVcm where n_{dot} is the effective doping concentration and $\alpha = 1.646 e^2/4\pi\varepsilon\varepsilon_0 \cdot (4\pi/3)^{1/3} = 6.7 \times 10^{-8}$ eVcm with $\varepsilon = 5.7$ [28]. The expression

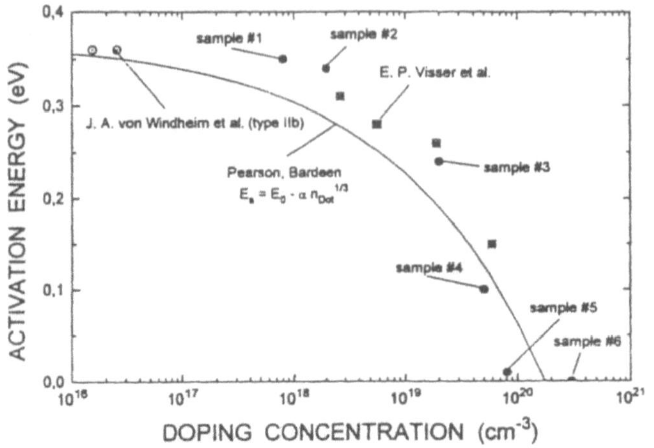

Figure 6 Thermal activation energy of boron vs effective doping concentration (Full line: Model according to Pearson and Bardeen [30].) [after [28]]

goes back to a model developed by Pearson and Bardeen for p-doped silicon [30] where the screening of the impurity Coulomb potential by the free holes is taken into account by three terms, the potential energy of the attraction between the holes and the acceptor ion, the self-energy of the holes, and a correlation energy resulting from the fact that the mobile carriers tend to keep apart. The reduced acceptor activation energy at high doping levels has application potential, e.g. for pulse-doped MESFET structures [31,32]. Thin layers doped with B to $\approx 10^{19}$ cm^{-3} in the peak ("δ-doping") were obtained during microwave-assisted homoepitaxial CVD-growth of diamond films by inserting a boron rod into the plasma for ≈ 5 s [32]. The δ-doped layer served as the p-conductive channel of a diamond junction field effect transistor providing 50–100 times larger source-drain currents in the output I-V-characteristic than previously obtained [4].

2.4 Boron-Bound Excitons

Mobile free excitons can be localized at shallow acceptor and donor impurities forming bound excitons. Bound excitons (BE) are known for a great variety of semiconductors from optical studies of their characteristic luminescence spectra [33]. In diamond, boron acceptors bind excitons with a localization energy of $E_{\mathrm{loc}} \approx 55$ meV, and their radiative recombination spectra have been studied as early as 1965 [34]. Similar to the situation with FEs, CVD-diamond layers need to have high enough optical quality (long-range crystallographic order in relatively large grains with sufficiently low defect concentrations) for the observation of BE recombination spectra. Such qualities were achieved no sooner than 1990 [23] . Polycrystalline CVD-diamond films doped by adding boron compounds to the gas mixture in quantitative amounts showed corresponding intensities of the BE luminescence spectrum (Fig. 7) [35]. This is firm direct optical evidence for boron being the impurity which localizes the excitons. It is not the purpose of the present article to discuss subtle features of the exciton spectra and the excitonic structure; this has been done in other papers [34], lately in investigating fine structure of the excitons observed in almost strain-free bulk HPHT-diamonds [12,13]. Here, we take the observation of the BE spectrum as a sensor for the presence of the shallow impurity boron in view of potential

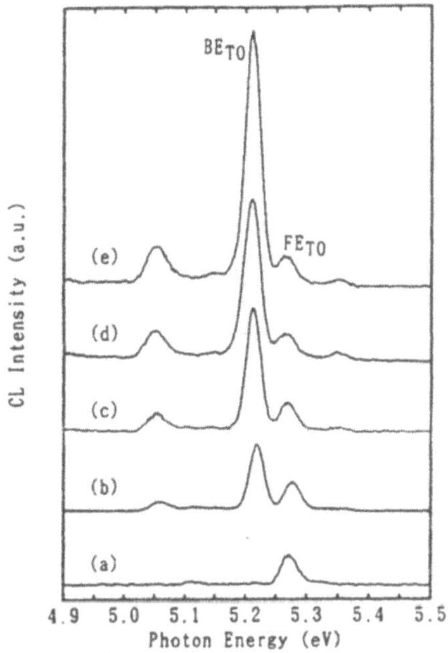

Figure 7
Near-band-edge cathodoluminescence spectra at 80 K (FE: free exciton, BE: boron bound exciton) of a polycrystalline 4 μm thick CVD-film for increasing boron/carbon ratios during deposition: (a) 0 ppm, (b) 50 ppm, (c) 100 ppm, (d) 250 ppm, (e) 500 ppm, and (f) 1000 ppm [35].

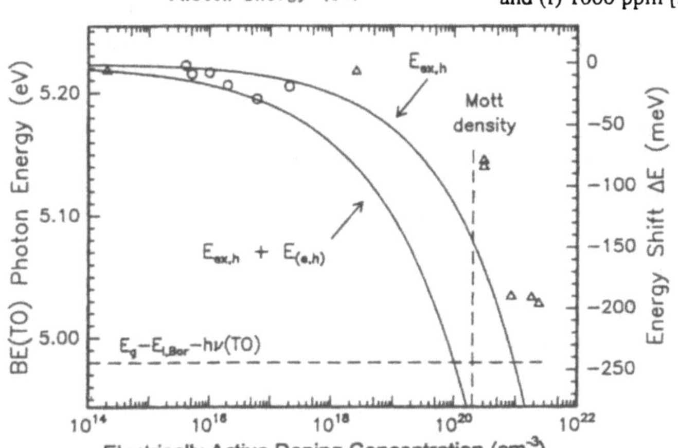

Figure 8 Shift of the boron bound exciton (BE) emission energy vs electrically active doping concentration. Triangles and circles are experimental, full lines are theoretical calculations taking into account hole-hole exchange ($E_{ex,h}$) alone or together with electron-hole correlation ($E_{e,h}$) [24].

shallow donors to be discussed below. The possibility to dope CVD-diamond relatively freely with boron in a wide concentration range has enabled studies of the BE spectra as a function of the doping level or the electrically active carrier concentration, respectively. It is found (Fig. 8) that the BE^{TO} line (the dominant TO-phonon assisted bound exciton transition) shifts to lower energy, starting from the low concentration value at ≈ 5.22 eV and approaching nearly 5 eV for $\approx 10^{21}$ cm^{-3} carrier concentration [36]. This reduction of the BE recombination energy resembles the reduction in activation energy of the neutral boron acceptor (Fig. 6) but is much more difficult to explain for the present four-particle complex. Basically, interactions of the three charged particles, the two holes and the electron, and the acceptor ion in the BE complex are responsible for the energy reduction. For a quantitative explanation, a theory was employed where all these interactions are included [37]. The theory originally formulated for n-type Si or Ge was applied to the case of p-type diamond. Two leading energy terms emerge which are plotted in Fig. 8 versus the carrier concentration p. These terms are the hole-hole exchange energy $E_{\mathrm{ex,h}}$ scaling as $p^{1/3}$ and the electron-hole interaction energy $E_{\mathrm{e,h}}$ scaling as $p^{1/4}$. Comparison of the experimental data and the theoretical terms show (Fig. 8) that the exchange energy alone describes the experimental situation fairly well. It is unclear why the electron-hole energy is not needed given the fact that there is no fit parameter in the theory to adjust its strength. The similar behavior of the bound exciton and neutral acceptor energies versus carrier concentration raises the question whether both effects have the same physics in common; this is not evident from the simple Pearson-Bardeen model.

Except for doping diamond with boron in the HPHT-synthesis [38] or during CVD growth by adding B-containing compounds to the gas mixture [39,28] boron can be incorporated as an acceptor by ion implantation and subsequent proper annealing [40]. It is an interesting question in itself but also in view of the potential n-type doping of diamond whether implantation and annealing techniques leave the lattice sufficiently undamaged so that bound excitons with their more extended wavefunction in comparison to neutral acceptors can form and decay radiatively. This question has been positively answered by implantation and luminescence experiments on natural type IIa diamonds [41]. When $^{11}B^+$ is implanted into bulk diamond crystals at 350 keV energy at doses of some 10^{11} cm^{-3} and higher, the BE spectrum emerges after annealing at $T_a \geq 800$ °C. The luminescence intensities increase monotonically up to $T_a \approx 1200$ °C (Fig. 9) [41]. The presence of the BE spectrum and its spectral position – unique for doping levels below $\approx 10^{16}$ cm^{-3} – show unambiguously that B has been incorporated on acceptor-active substitutional lattice sites. Control experiments with $^{12}C^+$, $^{31}P^+$, and $^{75}As^+$ ion implantation exhibited only luminescence features due to radiation damage but no ion-specific features. When for B implantation the annealing temperature is kept constant at 1 200 °C, the BE luminescence becomes stronger for increasing ion doses from 2×10^{11} cm^{-3} to 2×10^{13} cm^{-3} (Fig. 9); beyond this dose, the lattice is too heavily damaged for excitons to form [41].

2.5 Shallow Donors

There is no "natural" shallow donor in diamond. Intuitively, one would expect the light atoms P out of the group V elements and Li out of the group I elements to act as substitutional or interstitial donors, respectively, similar to the other elementary semiconductors Si and Ge. Previous attempts to dope diamond with P were unsuccessful [26]. Theoretically, first-principles supercell calculations using the plane wave pseudopotential method predict that Li and Na act as shallow donors on interstitial lattice sites and P as a shallow donor on substitutional sites [42]. The binding energies were determined to be in the range of a few hundred millielectronvolts, $\approx$ 100 meV for Li, $\approx$ 300 meV for Na, and $\approx$ 200 meV for P. However, independent first-principles calculations employing a cluster approached based on the local density approximation yield a much higher value, 1 090 meV,

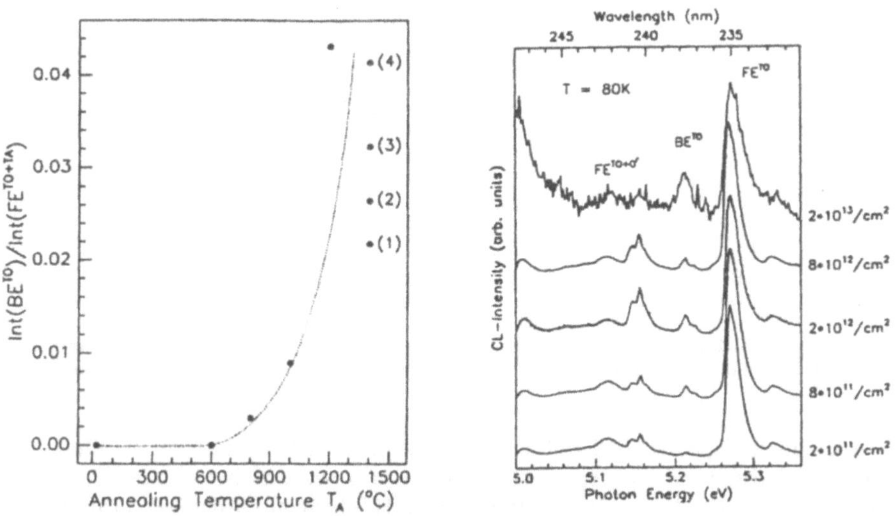

Figure 9 Left: Dependence of boron BE luminescence intensity on annealing temperature in natural IIa diamonds implanted with $^{11}B^+$ ions at a dose of 3×10^{12} cm^{-2} at 350 keV. Right: Increase of boron BE^{TO} luminescence normalized to the intrinsic free exciton (FE^{TO}) radiation for increasing boron ion-doses from 2×10^{11} cm^{-2} to 2×10^{13} cm^{-3} at 350 keV implantation energy and 1 200 °C annealing temperature [41].

for P [43]. Li is estimated to be a fast diffusor and therefore unstable even at room temperature for longer storage times [42]. In all cases, extremely low solid solubility values are obtained theoretically. Therefore, the probability that these elements are incorporated into the lattice under quasi-equilibrium conditions such as diffusion or during the growth is considered very low.

An alternative way to dope semiconductors under non-equilibrium conditions is ion implantation and annealing. In diamond, this technique is basically problematic since diamond is the metastable phase of carbon. When dopants are introduced by ion-implantation the lattice is damaged. This question was already addressed briefly above in the context of boron implantation. Damage-induced states and graphitic bonds can give rise to electrical conductivity. This conductivity may persist at high temperatures due to incomplete annealing of the damage but does not represent the desired conductivity due to impurity-to-band activation of a chemical dopant. A survey of the state of ion implantation is given in [44]. Recent experiments focus on Li, Na, and P as prospective dopants. Li ion implantation of natural type IIa diamonds [45] and synthetic diamonds [46] followed by annealing led to n-type conductivity with largely varying thermal activation energies in the range from ≈ 80 meV to ≈ 270 meV. In the first case, control implantations with $^{12}C^+$ ions showed no remaining conductivity after annealing. From this result it was concluded that the conductivity is due to the presence of Li and not to implantation-related damage [45]. However, the temperature dependence followed rather $\sigma \sim \exp(T/T_0))^{1/4}$ typical for variable-range hopping than $\sigma \sim \exp(\Delta E/kT)$ as for donor-to-conduction band activation. Experiments with Na implantation were found to be similarly inconclusive [45]. Homoepitaxial diamond CVD-films doped with Li showed a thermal activation energy of ΔE = 160 meV in a limited temperature range but the authors felt the conductivity mechanism had further to be investigated [28].

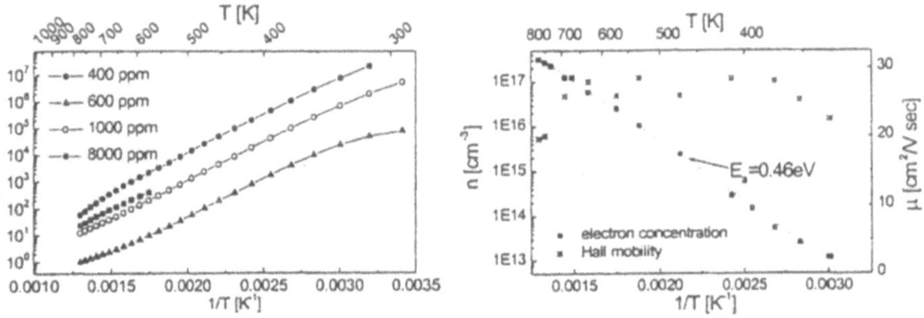

Figure 10 Left: Resistivity vs temperature of n-conducting phosphorus doped homoepitaxial diamond films. Right: Electron concentration n and mobility μ versus temperature [50].

Attempts to dope diamond with P include synthesis of hot filament CVD-films on Si substrates [47,48] and homoepitaxial growth by microwave-assisted CVD [49,50], with P_2O_5 or PH_3 as the doping source, growth of HPHT-diamonds using P as a catalyst [51] and P ion implantation of CVD-films [53] and of natural IIa diamonds [54,55]. The samples were not in all cases electrically characterized. Electrical data for homoepitaxially grown layers are shown in Fig. 10 [50]. Resistivity-vs-temperature curves in an Arrhenius plot are essentially straight lines for all P concentrations with an activation energy close to 0.57 eV. After correction for the electron mobilities the corresponding plots of the electron concentration yield activation energies of 0.46 eV (Fig. 10). The authors conclude that this energy is related with a donor 0.46 eV below the conduction band [49,50].

Cathodoluminescence studies were performed on selected P-doped samples from various groups including samples as described in [48] through [53] and [55]. Some samples showed BE-luminescence due to B though B was no intentionally doped. (B contamination of the growth chamber is often found and due to previous p-doping experiments.) A few of the P-doped samples exhibited a new BE-line at ≈ 5.17 eV (Fig. 11). The line originates from a shallow bound exciton with a localization energy of $E_{loc} \approx 100$ meV which, when thermally dissociated from the binding complex, enhances the free exciton luminescence [57]. As these samples are intentionally P-doped, the impurity binding the exciton is likely to be a P-related shallow center. Unambiguous evidence for a P-related donor comes from discrete donor-acceptor pair (DAP) spectra in the band-edge region. DAP broad bands and resolved DAP line spectra had been reported previously in the energy ranges of 1.4 ... 2.0 eV [58] and 2.2 ... 2.7 eV [59], respectively. Here, the novel DAP spectra are close to the band-edge and are seen in samples which were grown and P-doped by different methods in different groups. Four samples out of a large group show almost identical spectra typical in appearance for discrete DAP transitions (Fig. 12). Following the classical case of line-resolved DAP recombination spectra in GaP [60] or the later examples in Si [61] theoretical spectra were synthesized using the standard expressions for line positions as a function of the DA interspacing R_{AD} (see Fig. 12) [62] and the distribution of sites in the lattice (see [60–62]). The synthesized spectra were energy-shifted for optimal agreement with the measured spectra (Fig. 12). This is equivalent to assigning experimental line positions to "shell numbers" or to R_{AD} values. The shallow acceptor involved in the DAP spectra is very likely B since the samples under discussion exhibit boron BE luminescence. Taking consequentially $E_A = 370$ meV, a value of $E_D = (650 \pm 50)$ meV is obtained for the donor involved [62]. This value is consistent

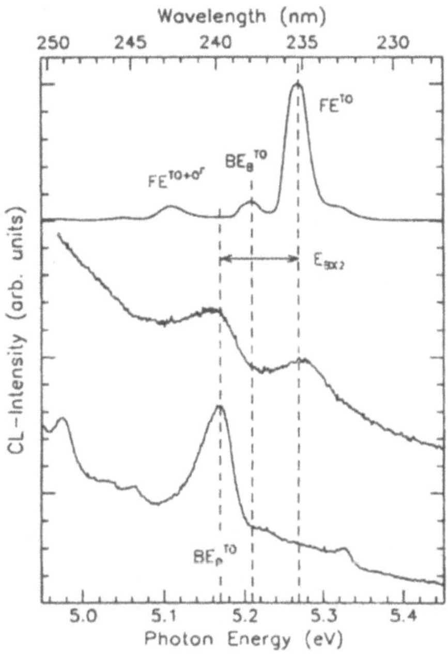

Figure 11
Lower traces: Near-band-edge cathodoluminescence spectra at 80 K showing a new excitonic peak ($\approx$ 5.17 eV) in phosphorus doped CVD diamond films grown by the hot-filament technique on Si:B substrate. Bottom: higher P-doping [P] = 70 ppm. Middle: lower P-doping. Upper trace: Cathodoluminescence spectrum of a nominally undoped MW-CVD diamond grown on Si:B, the boron BE^{TO} line originates from boron contamination of the growth chamber [56].

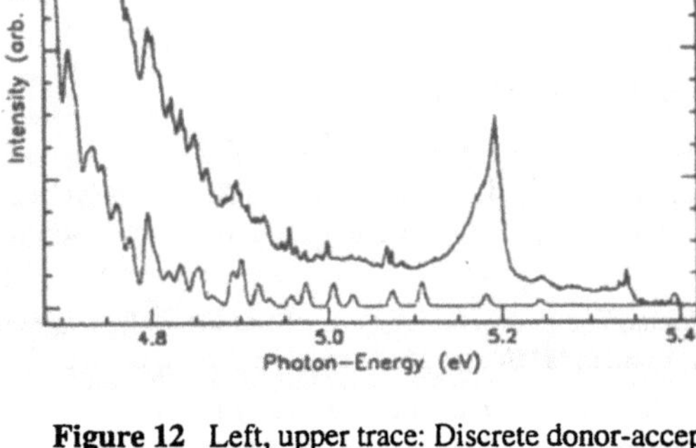

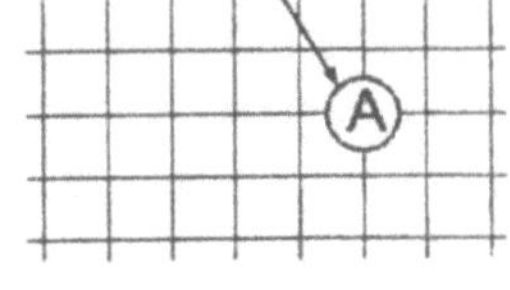

Figure 12 Left, upper trace: Discrete donor-acceptor pair (DAP) line spectrum in a P-doped (and B contaminated) CVD diamond film grown at NIRIM. Lower trace: Synthesized spectrum using standard DAP transition theory (see text). All lines in the theoretical spectrum are Gaussian broadened by 10 meV, the spectrum was shifted in energy for coincidence with characteristic line groups in the experimental spectrum. Right: Spatial dependence of DAP recombination energy $h\nu$ by DA-Coulomb interaction ($\sim R_{AD}^{-1}$) and electron-hole polarization $J(R_{AD})$ [60,61].

with the thermal activation energy $\Delta E = 0.46$ eV from the resistivity and Hall measurements [49,50] to the extent that the true donor ionization energy E_D can vary between ΔE and $2\Delta E$ depending on the degree of compensation taken in evaluating the electrical data.

In Si, shallow impurity ionization energies E_i and excitonic localization energies E_{loc} are related by an empirical law $E_{loc} \approx 0.1 E_i$, the so-called Haynes' rule [63]. When this rule is assumed to be also valid in diamond, one could correlate the boron data, $E_i(B) = 370$ meV and $E_{loc}(B) = 55$ meV with the ionization energy of a shallow donor with $E_i(\text{donor}) = E_D = 650$ meV to yield $E_{loc}(\text{donor}) = 55\ \text{meV} \times 650/370 \approx 96$ meV in close agreement with the exciton localization energy (≈ 100 meV) found experimentally in the P-doped sample (Fig. 11). The combined and consistent results of resistivity and Hall measurements and of the luminescence experiments on the same B- and P-doped samples give hope that P was in fact introduced into bulk diamond and CVD-layers as a shallow donor with an activation energy of around 650 meV.

Acknowledgements

The author is grateful to H. Sternschulte and K. Thonke (Abteilung Halbleiterphysik, Universität Ulm) for helpful discussions. The financial support by the Deutsche Forschungsgemeinschaft (Contract Sa 520) carried out under the auspices of the trinational D-A-CH cooperation of Germany, Austria, and Switzerland on the "Synthesis of Superhard Materials" is gratefully acknowledged.

Bibliography

[1] A.T. Collins, Mat. Res. Soc. Symp. Proc. Vol. **162**, 3 (1990) and further references therein.

[2] *Properties and Growth of Diamond*, edited by G. Davies, EMIS Datareviews Series No. 9, INSPEC, the Institution of Electrical Engineers, London (1994).

[3] T.H. Borst, S. Strobel, and O. Weis, Appl. Phys. Lett. **67**, 2651 (1995).

[4] see, e.g. G.S. Gildenblatt, S.A. Grot, and A. Badzian, Proc. IEEE **79**, 647 (1991).

[5] see the survey of J. Walker, Rep. Prog. Phys. **42**, 1608 (1979).

[6] J.H.N. Loubser and J.A. van Wyk, Rep. Prog. Phys. **41**, 1201 (1978).

[7] C.A.J. Ammerlaan, Landolt-Börnstein, *Numerical Data and Functional Relationships in Science and Technology*, New Series III **22b**, edited by O. Madelung and M. Schulz (Springer, 1990), p. 117.

[8] C.D. Clark, A.T. Collins, and G.S. Woods, *The Properties of Natural and Synthetic Diamond*, edited by J.E. Field, Academic Press, London (1992), p. 35.

[9] A comprehensive listing is given in A.M. Zaitsev, *Optical Properties of Diamond*, in *Handbook of Industrial Diamonds and Diamond Films*, edited by M.A. Prelas, G. Popovici, and L. Bigelow, Marcel Dekker (1998), p. 227.

[10] G. Davies, *Properties and Growth of Diamond*, edited by G. Davies, EMIS Datareviews Series No. 9, INSPEC, the Institution of Electrical Engineers (1994), p. 248.

[11] A.T. Collins, S.C. Lawson, G. Davies, and H. Kanda, Phys. Rev. Lett. **65**, 891 (1990).

[12] H. Sternschulte, S. Wahl, K. Thonke, R. Sauer, T. Ruf, M. Cardona, and T.R. Anthony, Mater. Sci. Forum **258–263**, 757 (1997).

[13] T. Ruf, M. Cardona, H. Sternschulte, S. Wahl, K. Thonke, R. Sauer, P. Pavone, and T.R. Anthony, Solid State Commun. **105**, 311 (1998).

[14] see, e.g. L.H. Robins, L.P. Cook, E.N. Farabough, and A. Feldman, Phys. Rev. **B39**, 13367 (1989).

[15] V.S. Vavilov, A.A. Gippius, A.M. Zaitsev, B.V. Deryaguin, B.V. Spitsyn, and E.A. Aleksenko, Fiz. Tekh. Poluprovodn. **14**, 1811 (1980) [Sov. Phys. Semicond. **14**, 1078 (1980)].

[16] A.M. Zaitsev, V.S. Vavilov, and A.A. Gippius, Kratkie Soobsh. Fiz. **10**, 20 (1981) [Sov. Phys. Leb. Inst. Rep. **10**, 15 (1981)].

[17] Y. Saito, S. Matsuda, and S. Nogita, J. Mater. Sci. Lett. **5**, 565 (1986).

[18] L. Bergman, B.R. Stoner, K.F. Turner, J.T. Glass, and R.J. Nemanich, J. Appl. Phys. **73**, 3951 (1993).

[19] D. Rehm, Diploma thesis, Abteilung Halbleiterphysik (Universität Ulm), in cooperation with Daimler-Benz Research Center (Ulm, 1995).

[20] A.T. Collins, L. Allers, C.J.H. Wort, and G.A. Scarsbrook, Diam. Rel. Mater. **3**, 932 (1994).

[21] H. Sternschulte, K. Thonke, R. Sauer, P.C. Münzinger, and P. Michler, Phys. Rev. B **50**, 14554 (1994).

[22] C.D. Clark, H. Kanda, I. Kiflawi, and G. Sittas, Phys. Rev. **B51**, 16681 (1995).

[23] H. Kawarada, Y. Yakata, and A. Hiraki, Appl. Phys. Lett. **57**, 1889 (1990).

[24] H. Sternschulte, J. Horseling, T. Albrecht, K. Thonke, and R. Sauer, Diamond Rel. Mater. **5**, 585 (1996).

[25] A.T. Collins and A.W.S. Williams, J. Phys. C: Solid State Phys. **4**, 1789 (1971).

[26] R.M. Chrenko, Phys. Rev. B **7**, 4560 (1973).

[27] J. Mort, D. Kuhman, M. Machonkin, M. Morgan, F. Jansen, K. Okamura, Y.M. LeGrice, and R.J. Nemanich, Appl. Phys. Lett. **55**, 1121 (1989).

[28] T.H. Borst and O. Weis, Diam. Rel. Mater. **4**, 948, (1995).

[29] J.C. Bourgoin, J. Krynicki, and B. Blanchard, phys. status solidi (a) **52**, 293 (1979).

[30] G.L. Pearson and J. Bardeen, Phys. Rev. **75**, 865 (1949).

[31] H. Shiomi, Y. Nishibayashi, N. Toda, and S. Shikata, IEEE Electron Device Lett. **16**, 36 (1995).

[32] A. Vescan, D. Gluche, W. Ebert, and E. Kohn, IEEE Electron Device Lett. **18**, 222 (1997).

[33] *Excitons*, edited by E.I. Rashba and M.D. Sturge, North-Holland, Amsterdam (1982).

[34] R.J. Dean, E.C. Lightowlers, and D.R. Wight, Phys. Rev. **140**, A352 (1965).

[35] H. Kawarada, Y. Yakata, H. Matsuyama, T. Sogi, and A. Hiraki, *Proc. 2nd Internat. Symp. Diamond Materials*, edited by A.J. Purdes, The Electrochemical Society, Pennington (1991), Vol. 91, p. 420; H. Kawarada and A. Yamaguchi, Diamond Rel. Mater. **2**, 100 (1993).

[36] H. Sternschulte, T. Albrecht, K. Thonke, and R. Sauer, *The Physics of Semiconductors*, edited by M. Scheffler and R. Zimmermann, World Scientific, Singapore (1996), p. 169.

[37] G.D. Mahan, J. Appl. Phys. **51**, 2634 (1980).

[38] C.M. Huggins and P. Cannon, Nature **194**, 829 (1962).

[39] K. Okano, H. Naruki, Y. Akiba, T. Kurosu, M. Iida, and Y. Hirose, Jpn. J. Appl. Phys. **27**, L173 (1988).

[40] G. Braunstein and R. Kalish, J. Appl Phys. **54**, 2106 (1983).

[41] H. Sternschulte, S. Wahl, K. Thonke, R. Sauer, M. Dalmer, C. Ronning, and H. Hofsäss, Appl. Phys. Lett. **71**, 2668 (1997).

[42] S.A. Kajihara, A. Antonelli, and J. Bernholc, Mat. Res. Soc. Symp. Proc. Vol. **162**, 315 (1990).

[43] K. Jackson, M.R. Pederson, and J.G. Harrison, Phys. Rev. **B41**, 12641 (1990).

[44] J.F. Prins, *The Properties of Natural and Synthetic Diamond*, edited by J.E. Field, Academic Press, London (1992), p. 301.

[45] S. Prawer, C. Uzan-Saguy, G. Braunstein, and R. Kalish, Appl. Phys. Lett. **63**, 2502 (1993).

[46] R. Job, M. Werner, A. Demisenko, A. Zaitsev, and W.R. Fahrner, Diam. Rel. Mater. **5**, 757 (1996).

[47] K. Okano, H. Kiyota, T. Iwasaki, Y. Nakamura, Y. Akiba, T. Kurosu, M. Iida, and T. Nakamura, Appl. Phys. **A51**, 344 (1990).

[48] S. Bohr, R. Haubner, and B. Lux, Diamond Rel. Mater. **4**, 133 (1995).

[49] S. Koizumi, M. Kamo, Y. Sato, S. Mita, A. Sawabe, A. Reznik, C. Uzan-Saguy, and R. Kalish, Diamond Rel. Mater. **7**, 540 (1998).

[50] R. Kalish, A. Reznik, C. Uzan-Saguy, S. Koizumi, and M. Kamo, *Advanced Materials '98*, The 5th NIRIM Internat. Symp. Advanced Materials (ISAM'98), Tsukuba, Ibaraki, Japan, edited by M. Kamo et al., National Institute for Research in Inorganic Materials (1998), p. 105.

[51] M. Akaishi, H. Kanda, and S. Yamaoka, Science **259**, 1592 (1993).

[52] S. Koizumi, M. Kamo, I. Sakaguchi, Y. Sato, S. Mita, A. Sawabe, A. Reznik, C. Uzan-Saguy, and R. Kalish, *Advanced Materials '98*, The 5th NIRIM Internat. Symp. Advanced Materials (ISAM'98), Tsukuba, Ibaraki, Japan, edited by M. Kamo et al., National Institute for Research in Inorganic Materials (1998), p. 109.

[53] CVD-films grown by P. Koidl et al. at FHG-Institut für Angewandte Festkörperforschung (Freiburg, Germany) ion-implanted at room temperature and annealed at 1200 °C by M. Schreck et al. at Institut für Experimentalphysik (Universität Augsburg, Germany), unpublished.

[54] G.S. Sandhu, C.T. Kao, M.L. Swanson, and W.K. Chu, Mat. Res. Soc. Symp. Proc. Vol. **162**, 321 (1990).

[55] R. Kalish et al. (1998), private communication.

[56] H. Sternschulte, Ph.D. thesis (1998).

[57] H. Sternschulte, T. Albrecht, K. Thonke, R. Sauer, M. Grießer, and M. Grasserbauer, Mat. Res. Soc. Symp. Proc. Vol. **423**, 693 (1996).

[58] J.A. Freitas,Jr., P.B. Klein, and A.T. Collins, Appl. Phys. Lett. **64**, 2136 (1994).

[59] B. Dischler, W. Rothemund, C. Wild, R. Locher, H. Biebl, and P. Koidl, Phys. Rev. B **49**, 1685 (1994).

[60] D.G. Thomas, M. Gershenzon, and F.A. Trumbore, Phys. Rev. **133**, A269 (1964).

[61] U.O. Ziemelis and R. R. Parsons, Can. J. Phys. **59**, 784 (1981). W. Schmid, U. Nieper, and J. Weber, Solid State Commun. **45**, 1007 (1983).

[62] H. Sternschulte, K. Thonke, and R. Sauer, *Advanced Materials '98*, The 5th NIRIM Internat. Symp. Advanced Materials (ISAM'98), Tsukuba, Ibaraki, Japan, edited by M. Kamo et al., National Institute for Research in Inorganic Materials (1998), p. 113.

[63] R.J. Haynes, Phys. Rev. Lett. **4**, 361 (1960).

Electron Ground States in a Few-Electron Quantum Dot.

T.H. Oosterkamp[1], W.G. van der Wiel[1], L.P. Kouwenhoven[1], D.G. Austing[2], T. Honda[2], S. Tarucha[2,3]

[1] Department of Applied Physics and DIMES, Delft University of Technology, P.O. Box 5046, 2600 GA Delft, The Netherlands
[2] NTT Basic Research Laboratories, 3-1, Morinosoto Wakamiya, Atsugi-shi, Kanagawa 243-01, Japan
[3] Tokyo University, 7-22-1 Roppongi, Minato-ku, Tokyo 106, Japan

Abstract: We review the electronic ground states in few-electron quantum dots studied by measuring Coulomb oscillations in the linear transport regime. The number of electrons in our vertical quantum dots are changed one-by-one from zero up to about a hundred by means of the gate voltage. We identify the quantum numbers of the states by measuring the magnetic field dependence. We find that transitions in the ground states originate from crossings between single-particle states and Hund's rule. Most features can be explained with an extended constant interaction (CI) model. This model describes the electron-electron interactions by a charging energy plus an exchange energy which are both independent of magnetic field.

1 Introduction

Semiconductor quantum dots can be regarded as artificial atoms, since their electronic properties resemble the ionization energy and discrete excitation spectrum of atoms [1]. Quantum dots can be fabricated between source and drain contacts so that the atom-like properties can be probed in current-voltage (I-V) measurements. Additionally, with a gate electrode nearby, one can vary the exact number of electrons, N, on the quantum dot by changing the gate voltage, V_g. When an electron is added, the total charge on the dot changes by the elementary charge, e. The associated energy change, known as the addition energy, is a combination of the single-electron charging energy and the change in single-particle energy. Charging effects and discrete single-particle states have been studied in a variety of quantum dot systems, defined not only in semiconductors but also in metal grains and molecules [2].

Quantum dot devices usually contain some disorder due to impurities or because the shape is irregular [2]. Clean quantum dots, in the form of regular disks, have only recently been fabricated in a semiconductor heterostructure [3]. The circular symmetry of two dimensional (2D) disks gives rise to a 2D shell structure in the addition energies, analogous to the 3D shell structure in atomic ionization energies [4]. The 3D spherically symmetric potential around atoms yields the shells 1s, 2s, 2p, 3s, 3p, ... The ionization energy has a large maximum for atomic numbers 2, 10, 18, ..., corresponding to complete filling of a shell. Up to atomic number 23 these shells are filled sequentially with electrons (i.e.

mixing between shells starts at atomic number 24). Within a shell, Hund's rule determines whether a spin-down or a spin-up electron is added [4].

Our vertical quantum dots have the shape of a disk with a diameter roughly ten times its thickness [3]. We find that their lateral confinement potential has good cylindrical symmetry with a rather soft boundary profile, which can be approximated by a harmonic potential. The symmetry of such a 2D harmonic potential leads to complete filling of 2D shells by 2, 6, 12, ... electrons. This sequence of electron numbers are the 'magic numbers' for a 2D harmonic dot. In this paper we review the atomic-like properties observed in single-electron transport. We find unusually large addition energies when the electron number coincides with a magic number. Furthermore, we can identify the quantum numbers of the single-particle states by studying their magnetic field dependence. At sufficiently small magnetic fields ($B < 0.4$ T) we see that spin filling obeys Hund's rule. At higher magnetic fields we observe that spin-degenerate states are filled by spin-up electrons alternating with spin-down electrons (i.e. an even-odd filling of the single-particle states).

The measured addition energies can be understood qualitatively in terms of an extended constant interaction (CI) model. In the CI model the direct Coulomb energy, E_c, between any two electrons on the dot is assumed to be independent of B. The energy $E_c = e^2/C$ can be thought of as the charging energy of a single-electron on a capacitor C. To account for effects like Hund's rule we extend the CI model with a constant exchange energy E_{ex}. This extended CI model qualitatively explains the evolution of the ground state energies with B.

2 Device Parameters and Experimental Set up

Our vertical quantum dot is a miniaturized resonant tunneling diode [5]. The pillar in Fig. 1a is etched from a semiconductor double-barrier heterostructure (DBH) and a metal gate electrode is deposited around it. Electrons are confined on the dot in all three dimensions. The surface potential together with the gate potential confines the electrons in the lateral x- and y-directions while the DBH provides the confinement in the growth z-direction.

The DBH consists of an undoped 12.0 nm $In_{0.05}Ga_{0.95}As$ well and undoped $Al_{0.22}Ga_{0.78}As$ barriers of thickness 9.0 and 7.5 nm (the thinner one is closest to the substrate). The source and drain contacts are made from Si doped n-GaAs which are separated by 3.0 nm undoped GaAs spacer layers from the barriers. The concentration of the Si dopants increases stepwise, on both sides starting with $0.75 \cdot 10^{17}$ cm^{-2} at 3 nm away from the barriers and increasing to $2.0 \cdot 10^{18}$ cm^{-2} at 400 nm from the barriers. An electron moving through the conduction band experiences a varying electrostatic potential landscape due to the changing band gap in the growth direction. Fig. 1b shows the calculated self-consistent energy profile in the growth direction for the large unpatterned DBH [5,6].

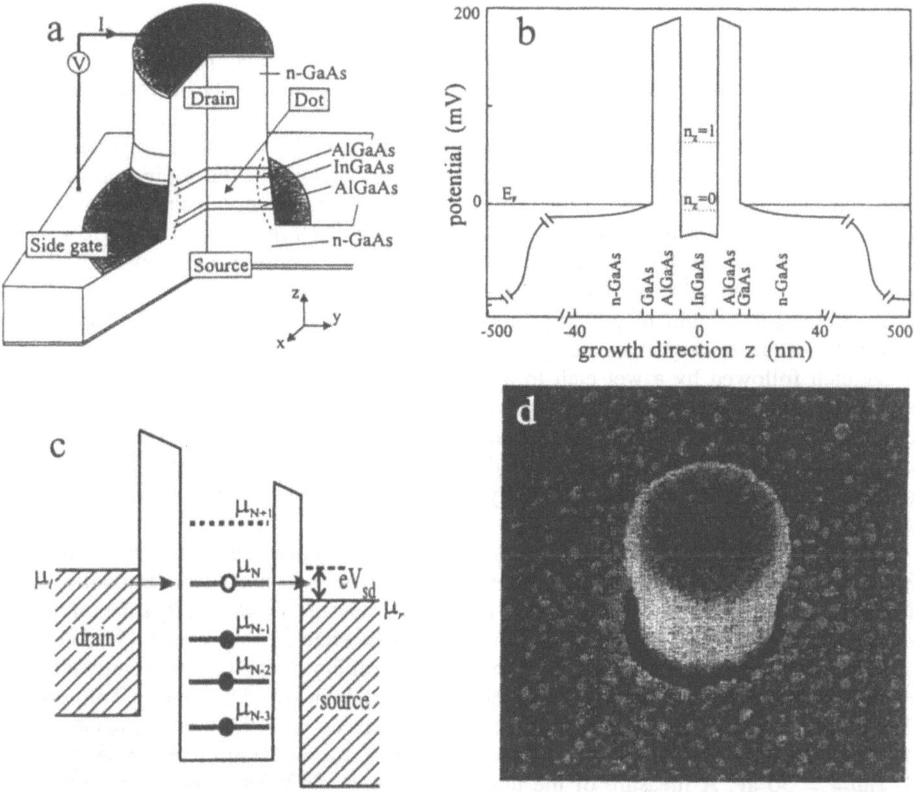

Figure 1 (a) Schematic diagram of the device. The dot is located between the two heterostructure barriers. (b) Self-consistent calculation [6] of the energy diagram of the unpatterned double barrier heterostructure from which the device is fabricated. (c) Schematic energy diagram of the electron states in the dot. Shaded areas represent the states in the leads which are continuously filled up to the Fermi levels. A voltage, V_{sd}, applied between the source and the drain shifts one Fermi level relative to the other. Current may flow when the electro-chemical potential of the dot for a certain number of electrons (solid lines) lies between the Fermi levels of the leads. (d) SEM photo of a circular quantum dot. The diameter of the pillar is $D = 0.5\ \mu$m.

A key ingredient of this material system is the inclusion of 5% indium in the $In_{0.05}Ga_{0.95}As$ well. This lowers the conduction band bottom in the well to 32 meV below the Fermi level of the n-GaAs contacts. From the confinement potential in the z-direction and the effective mass m^* in $In_{0.05}Ga_{0.95}As$ it follows that the lowest quantum state in the well is at an energy $E_z = 26$ meV above the conduction band bottom. This is still 6.1 meV below the Fermi level of the contacts. To reach equilibrium the well is filled with electrons until the Fermi level of the dot is as close as possible to the Fermi levels in the contacts. Our vertical quantum dot system thus contains electrons without applying external voltages. This is important since it allows to study the linear transport regime; i.e. the current in response to a very small source-drain voltage, V_{sd}.

Previously, GaAs without indium was used as the well material [7,8]. The lowest state in the well is then above the Fermi level of the contacts. In this case, it is necessary to apply a large V_{sd} to force electrons into the well and to obtain a current flow. Subsequently, all experiments then have to be performed far out of equilibrium, which limits the energy resolution. From these materials two-terminal [7] as well as three-terminal devices [8] have been fabricated. These early three-terminal devices could only be studied in the non-linear regime. Other techniques, in which the barriers are doped with Si, have been successful to accumulate some electrons in the well [9,10]. This has the disadvantage that it introduces strong potential fluctuations in the dot. By optimizing the doping profile and using capacitance spectroscopy (which allows one barrier to be very thick) Ashoori et al. [11] succeeded in measuring the linear response of few-electron quantum dots. However, these capacitance-dots did not show signatures of circular symmetry. Our new dots allow, for the first time, to study both the linear and the non-linear response.

To put the gate around the pillar we first define a metal circle with geometrical diameter D which will serve later as the top contact. The metal circle is then used as a mask for a dry etch followed by a wet etch to a point just below the region of the DBH. This etch combination gives the pillar a somewhat smaller diameter than the top such that evaporation from above deposits the metal gate on the substrate and against the lower part of the pillar [12]. We find that when the diameter is around $D \sim 0.5\ \mu$m we can vary the number of electrons on the dot, N, one-by-one from 0 to more than 80, by increasing the gate voltage, V_g.

The electron density in the well follows from self-consistent calculations and is equal to $n_e = 1.67 \cdot 10^{15}\ \mathrm{m}^{-2}$, which agrees with $n_e = 1.7 \cdot 10^{15}\ \mathrm{m}^{-2}$ deduced from Shubnikov-de Haas oscillations measured on large area ($D = 50\ \mu$m) devices. From density times area we estimate to have about 10 electrons in the dot when the effective diameter $d_{\mathrm{eff}} = 100$ nm.

The geometry of our samples leads to strong screening of the Coulomb interactions inside the dot by the metallic regions in the source and drain which are only 10 nm away. This is illustrated when we compare the self-capacitance with the total capacitance of the dot. The self-capacitance of a disk with diameter $d_{\mathrm{eff}} = 100$ nm is given by $C_{\mathrm{self}} = 4\varepsilon_r\varepsilon_0 d_{\mathrm{eff}} = 50$ aF. A measure of the unscreened Coulomb interactions is the charging energy $e^2/C_{\mathrm{self}} = 4$ meV. The real capacitance is much better approximated by two parallel capacitors between dot and source and drain leads, $C = C_s + C_d = 2\varepsilon_r\varepsilon_0\pi d_{\mathrm{eff}}^2/(4d) \sim 200$ aF. Here $d \sim 10$ nm is the distance between dot and leads. So, a measure of the screened Coulomb interactions follows from $e^2/C = 1$ meV; a value four times smaller than in the unscreened case. Electrons in the dot will thus have a short-range interaction. The strength of this interaction is further weakened by the finite thickness of the disk.

An important parameter for our device is the ratio of the charging energy to the confinement energy. To estimate the confinement energy we assume a harmonic potential with an oscillator frequency ω_0 and note that for $N \sim 10$ electrons the third shell is partially filled. From equating the energy at the classical turning point to the energy of the third shell ($\frac{1}{2}m^*\omega_0^2(d_{\mathrm{eff}}/2)^2 = 3\hbar\omega_0$) we get $\hbar\omega_0 = 3$ meV. This implies that the separation of the single-particle states is of order or even larger than the charging energy. This puts vertical dots in a very different regime compared to lateral quantum dots where the separation between single-particle states is typically 5 to 10 times smaller than the charging energy [2].

The well can in good approximation be regarded as a 2D system. The second quantum state due to confinement in the growth direction is 63 meV above the Fermi level. This energy is much larger than the lateral confinement energy. As we already mentioned, the finite thickness of the well has important consequences for the electron-electron interactions since the thickness of the well (12 nm) is comparable to the screening length due to the near vicinity of the leads (10 nm).

The devices are mounted in a dilution refrigerator for measurements at 50 mK. Due to external noise and interference the effective electron temperature is typically 100–200 mK. We measure the current, I, flowing vertically through the dot in response to a dc voltage, V_{sd}, applied between the source and drain contacts. The gate voltage V_g can typically be varied from -2.5 to $+0.7$ V. Beyond these values leakage occurs through the Schottky barrier between the gate and source. We can use a magnetic field B up to 16 T directed parallel to the current.

3 Addition Energies and Shell Filling

Figure 2a shows the current as a function of V_g. Clear Coulomb oscillations are observed with each peak corresponding to a change of exactly one electron on the dot. As we show below, the first peak corresponds to the transition from 0 to 1 electrons on the dot. This means that we know the absolute number of electrons on the dot; i.e. $N = 1$ between the 1st and 2nd peaks; $N = 2$ between the 2nd and 3rd peaks, etc. For $N \lesssim 20$, the spacings between peaks fluctuate. This increasing 'irregularity' for smaller N has previously been reported for dots containing a few electrons [7–11]. In this study, however, we find that the 'irregularity' in peak spacings is in fact systematic with respect to N.

In Fig. 2b the measured size of the Coulomb gap (i.e. the source-drain voltage at which current starts to flow) is plotted as a function of gate voltage. The dark regions indicate zero current due to Coulomb blockade, while in the light regions the Coulomb gap is

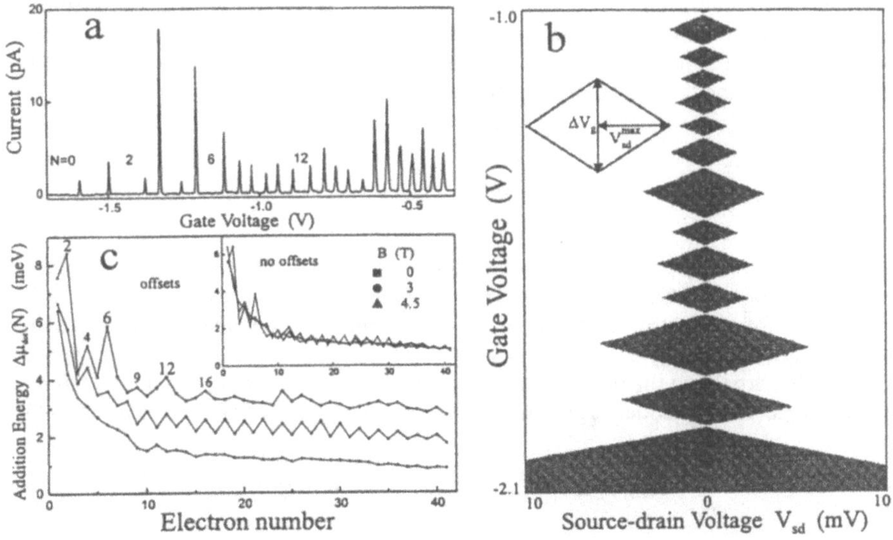

Figure 2 (a) Coulomb oscillations in the current vs. gate voltage at $B = 0$ T observed for a $D = 0.5$ μm dot. V_{sd} = 150 μV. (b) Measured size of the Coulomb gap versus gate voltage. In the dark diamond shaped areas the device is in Coulomb blockade. In the light regions current flows. The sizes of the diamonds directly correspond to the peak spacings in (a). (c) Addition energy $\Delta\mu(N)$ vs. electron number at different magnetic fields for a $D = 0.5$ μm dot. The curves for $B = 0$ and 3 T have been given offsets of 2 and 1 meV, respectively for clarity. The inset shows the same curves without offsets.

overcome and current is flowing through the dot. The vertical $V_{sd} = 0$ axis corresponds directly to the data in Fig. 2a. Note, for instance, that the peak spacings along the $V_{sd} = 0$ axis are the same as the peak spacings for the corresponding N in Fig. 2a. The width of the zero-current region, labeled as $N = 0$, keeps increasing when we make V_g more negative. This implies that here the dot is indeed empty which allows us to assign absolute electron numbers to the blockaded regions.

As we discuss in Section 4, the peak spacing is a measure for the amount of energy necessary to add an electron to the dot. Figure 2c shows the addition energy as a function of N for three different magnetic fields. As N is decreased, the addition energy generally becomes larger due to the increase of the Coulomb interaction as the effective dot size is reduced. We find that the addition energy at $B = 0$ T is unusually large for $N = 2$, 6 and 12. As we discuss below, these are the expected magic numbers for a 2D harmonic dot with circular symmetry.

In about twenty devices with D between 0.4 and 0.54 μm we find that the addition energy is unusually large for $N = 2$ and 6. A large addition energy for $N = 12$ is observed in about ten devices. We also observe relatively large addition energies for $N = 4$, 9 and sometimes also 16. We note that traces as in Fig. 2a reproduce in detail (including peak spacings and heights, but not the precise gate voltages) after we cycle the device to room temperature. In the remaining part of this paper we focus on one particular $D = 0.5$ μm device. All the main features, however, have been reproduced in several other devices.

The observation of the magic numbers at $B = 0$ implies that the 2D lateral potential is parabolic within the resolution of our experiments. Note that for any 2D potential with circular symmetry there remains a twofold orbital degeneracy associated with positive and negative angular momentum states (except for the lowest energy state which has zero angular momentum). This makes the magic numbers 2 and 6 very robust. The assumption of a parabolic potential should be valid when N is small [3,7]. For larger N, the potential will be flattened in the center due to screening effects. This could be the reason why we do not observe the third shell in all of the devices.

4 Constant Interaction Model

In this section we introduce the constant interaction CI model that describes the electronic states of the dot. The CI model is based on two important assumptions. First, the Coulomb interactions of an electron on the dot with its environment and with other electrons on the dot are parametrized by a constant capacitance C. Second, the discrete energy spectrum of a single particle on the dot, which we discuss in the next section, is not affected by the interactions. The CI model approximates the total energy $U(N)$ of an N electron dot by

$$U(N) = \frac{\left[e(N-N_0) - C_g V_g\right]^2}{2C} + \sum_N E_{n,l}(B) \tag{4.1}$$

where N_0 is the number of electrons on the dot at zero gate voltage. The term $C_g V_g$ is a continuous variable and represents the charge that is induced on the dot by the gate voltage V_g through the capacitance C_g. The total capacitance between the dot and the source, drain and gate is $C = C_s + C_d + C_g$. The last term is a sum over the occupied states $E_{n,l}(B)$ which are the solutions to the single-particle Schrödinger equation described in Section 5. Note that only the single-particle states depend on magnetic field.

The electro-chemical potential of the dot is defined as $\mu_{\rm dot}(N) \equiv U(N) - U(N-1)$. Electrons can flow from left to right when $\mu_{\rm dot}$ is between the electro-chemical potentials, $\mu_{\rm left}$ and $\mu_{\rm right}$, of the leads: $\mu_{\rm left} > \mu_{\rm dot}(N) > \mu_{\rm right}$. From (4.1) we get the electro-chemical potential of the dot

$$\mu_{\rm dot}(N) = (N - N_0 - \frac{1}{2})E_{\rm c} - e\frac{C_{\rm g}}{C}V_{\rm g} + E_N \tag{4.2}$$

The addition energy is given by

$$\Delta\mu(N) \equiv \mu_{\rm dot}(N+1) - \mu_{\rm dot}(N) = E_{\rm c} + E_{N+1} - E_N \tag{4.3}$$

with E_N the topmost filled single-particle state for an N electron dot.

The electro-chemical potential is changed linearly by the gate voltage with a proportionality factor $\alpha = eC_{\rm g}/C$ (see 4.2). This α factor also relates the peak spacing to the addition energy. We define the spacing between the $N^{\rm th}$ Coulomb peak at $V_{\rm g}^N$ and the next peak at $V_{\rm g}^{N+1}$ as $\Delta V_{\rm g}(N) = V_{\rm g}^{N+1} - V_{\rm g}^N$. The addition energy follows from $\Delta\mu(N) = \alpha\Delta V_{\rm g}(N)$.

The inset to Fig. 2b illustrates how to determine the α factor from the slopes of the sides of the Coulomb diamonds: $\alpha = \left|eV_{\rm sd}^{\rm max}/\Delta V_{\rm g}\right|$. Since the gate voltage changes the dot area, the α factor changes with N. We therefore determine α for each N. We find that, in the $D = 0.5\ \mu$m dot, α varies from 57 to 42 meV/V for $N = 1$ to $N = 6$, and then gradually decreases to 33 meV/V as N approaches 20. On close inspection we can see that the sides of the diamonds for small N are a little curved indicating that α changes somewhat even inside a diamond. These notions on addition energy and peak spacings are valid for any kind of confining potential. In the next section, we discuss the single-particle states in a specific potential.

5 Single Particle States of a 2D Harmonic Oscillator

For the simplest explanation of the 'magic numbers' we ignore, for the moment, Coulomb interactions between the electrons on the dot. The familiar spectrum of a one-dimensional harmonic oscillator $E_n = (n + 1/2)\hbar\omega$ becomes $E_{n,l} = (2n + |l| + 1)\,\hbar\omega_0$ in two dimensions. Here n (= 0, 1, 2,$\cdots$) is the radial quantum number, l (= 0, $\pm$1, $\pm$ 2,$\cdots$) is the angular momentum quantum number of the oscillator and ω_0 is the oscillator frequency.

The electronic states are expected to be significantly modified by a magnetic field, B, applied perpendicular to the plane of the dot. The eigenenergies $E_{n,l}$ as a function of B can be solved analytically for a 2D parabolic confining potential $V(r) = m^*\omega_0^2 r^2/2$ [13]

$$E_{n,l}(B) = (2n + |l| + 1)\,\hbar\sqrt{\frac{1}{4}\omega_{\rm c}^2 + \omega_0^2} - \frac{1}{2}l\hbar\omega_{\rm c} \tag{5.4}$$

where $\hbar\omega_0$ is the electrostatic confinement energy, $\hbar\omega_{\rm c} = \hbar eB/m^*$ is the cyclotron energy and $m^* = 0.067m_e$ is the effective mass in GaAs ($\hbar\omega_{\rm c} = 1.7$ meV at 1 T). Since GaAs has a small g-factor ($g = -0.44$ and $g\mu_B B = 25\ \mu$eV at 1 T) we can neglect the Zeeman energy. This makes each orbital state two-fold spin-degenerate.

In Fig. 3a we plot $E_{n,l}$ vs. B for $\hbar\omega_0$= 3 meV. It shows that the orbital degeneracies at $B = 0$ are lifted in the presence of a B-field. As B is initially increased from 0 T, a single-particle state with a positive or negative l shifts to lower or higher energy, respectively.

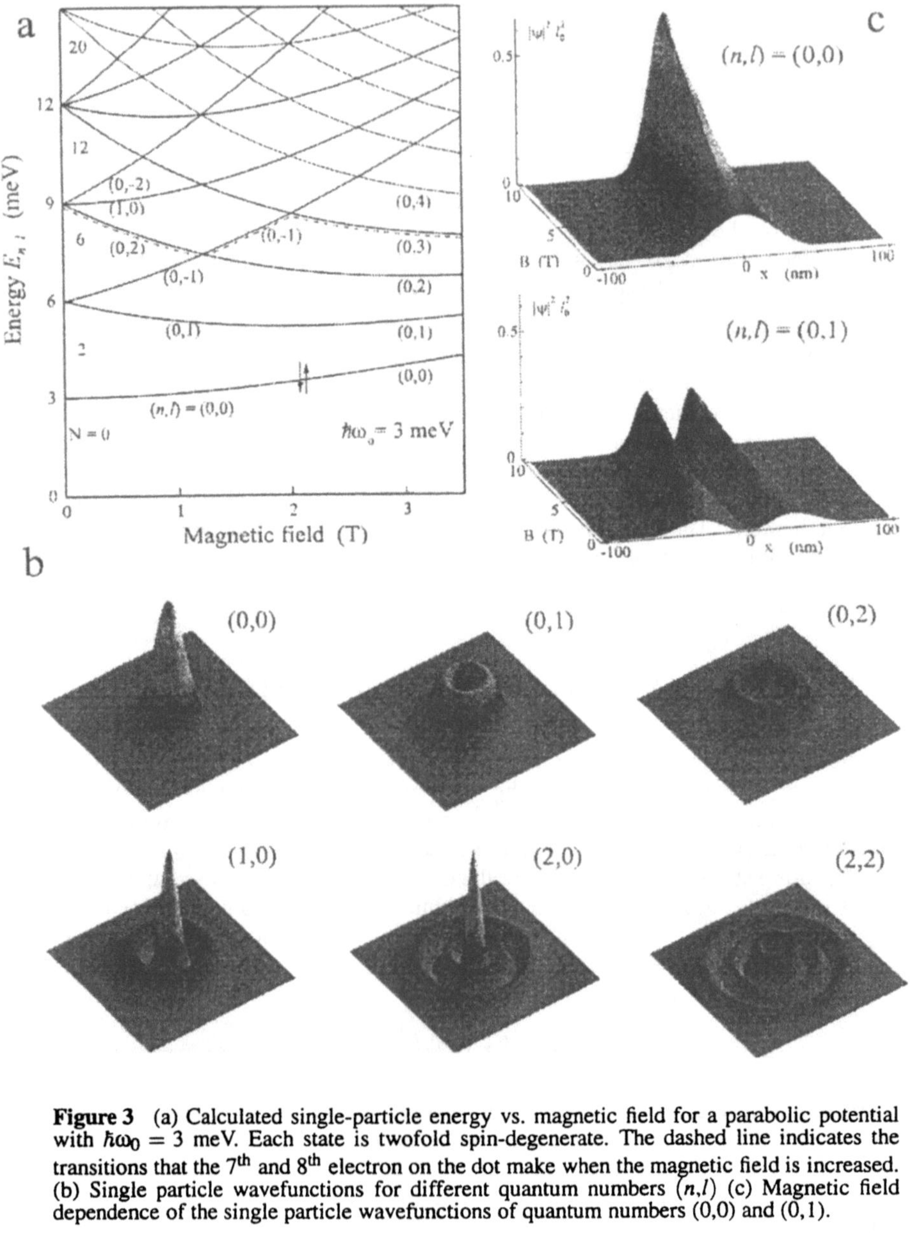

Figure 3 (a) Calculated single-particle energy vs. magnetic field for a parabolic potential with $\hbar\omega_0$ = 3 meV. Each state is twofold spin-degenerate. The dashed line indicates the transitions that the 7th and 8th electron on the dot make when the magnetic field is increased. (b) Single particle wavefunctions for different quantum numbers (n,l) (c) Magnetic field dependence of the single particle wavefunctions of quantum numbers (0,0) and (0,1).

At $B = 0$ the lowest energy state has $(n,l) = (0,0)$ and is twofold spin degenerate. The next state has a double orbital degeneracy $E_{0,1} = E_{0,-1}$. We denote these degenerate states the second shell. Including spin degeneracy this shell can contain up to 4 electrons, so it will be filled when there are 6 electrons on the dot. In the third shell states with quantum numbers (1,0), (0,2) and (0,−2) are degenerate. With spin this shell may contain up to 6 electrons, leading to the magic number 12. Note that the degeneracy of the (1,0) state with the (0,2) and (0,−2) states is lifted if the potential has a non-parabolic component [15].

When we put 7 non-interacting electrons in the dot and increase B, the 7th electron will make several transitions into different states. We have indicated this in Fig. 3a with a dashed line. The transitions are from (0,2) to (0,−1) at 1.3 T and then to (0,3) at 2 T. Similar transitions are also seen for other N. After the last crossing the electrons occupy states forming the lowest orbital Landau Level which is characterized by the quantum numbers $(0,l)$ with $l \geq 0$. Together with spin we can label the last crossing as filling factor 2; this in analogy to the quantum Hall effect in a 2D electron gas. In contrast to the bulk 2D case, the confinement lifts the degeneracy in this Landau level. The calculated separations between single-particle states at, for example, $B = 3$ T, is still quite large (between 1 and 1.5 meV in Fig. 3a). If we would speak of magic numbers at $B = 3$ T the sequence would simply be 2, 4, 6, 8, etc.

In the CI model a Coulomb charging energy is added to the non-interacting single-particle states of (5.4) to take into account the interaction. The addition spectrum of a quantum dot then follows from the electro-chemical potential $\mu_{\text{dot}}(N)$ (4.2) where the topmost filled state E_N in Fig. 3a is added to the charging energy. The addition spectrum versus B is shown in Fig. 4b. Note that spin-degenerate states are now separated by E_c, so each line appears double. The magic numbers 2, 6, 12, 20, etc. are still visible at $B = 0$ as enhanced energy separations which are equal to $E_c + \hbar\omega_0$. An even-odd parity effect is seen at $B = 5$ T where the energy separations for $N =$ even are larger than those for $N =$ odd.

For some applications it is helpful to know the wavefunctions belonging to the eigenenergies of (5.4). They are given by

$$\psi_{n,l}(r,\phi) = \frac{1}{\sqrt{2\pi}} e^{il\phi} \frac{1}{l_B} \sqrt{\frac{n!}{(n+|l|)!}} e^{-r^2/4l_B^2} (r/\sqrt{2}l_B)^{|l|} L_n^{|l|}(r^2/2l_B^2) \tag{5.5}$$

where $l_B = (\hbar/m^*\Omega)^{1/2}$ with $\Omega = \sqrt{\omega_c^2/4 + \omega_0^2}$ is the characteristic length and $L_n^{|l|}$ are generalized Laguerre polynomials. The square of the wavefunction $|\psi_{n,l}(r,\phi)|^2$ is plotted in Fig. 3b for different quantum numbers (n,l). Note that two wavefunctions with quantum numbers $(n, \pm l)$ only differ in the phase factor $e^{\pm il\phi}$. The number of nodes of the wavefunction going out from the center is given by the radial quantum number n. If the angular momentum quantum number is not zero there is an additional node at $r = 0$. The larger $|l|$ the wider the node around $r = 0$. When a magnetic field is applied, the characteristic length l_B set by the oscillator frequency Ω becomes smaller and the extent of the wavefunction shrinks as is shown in Fig. 3c for the square of the wavefunctions with quantum numbers (0,0) and (0,1). This turns out to be an important effect that leads to deviations from the CI model when B is changed on a relatively large scale (e.g. 10 T). For example, two electrons in the (0,0) state are pushed closer together when B is increased which enhances their Coulomb interaction. In this case, E_c will not be completely independent of B.

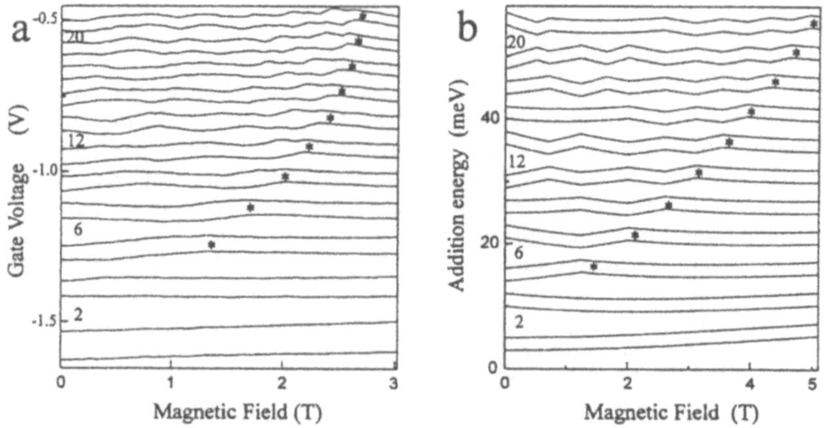

Figure 4 (a) Gate voltage positions of the current oscillations vs. magnetic field for a dot with $D = 0.5\ \mu$m. (b) Calculated electro-chemical potential $\mu_{dot}(N)$ using the constant interaction (CI) model (4.1), with $E_c = 2$ meV and $\hbar\omega_0 = 3$ meV.

6 Magnetic Field Dependence of Ground States

Figure 4a shows the measured B-field dependence of the positions of the current peaks. It is constructed from I-V_g curves for B increasing from 0 to 3.5 T in steps of 0.05 T. The evolution of the first 22 current peaks are plotted. The positions of the first three peaks depend monotonously on B, whereas the other peaks oscillate up and down a number of times. The number of 'wiggles' increases with N. Each N = odd peak has a neighbour for $(N+1)$ = even that wiggles in-phase. This pairing implies that the Nth and $(N+1)^{th}$ electrons occupy the same single-particle state with opposite spin. We see this pairing up to $N = 40$.

Around 3 T the peaks have stopped wiggling and now evolve smoothly with B. Close inspection shows that the peak spacing alternates between 'large' for even N and 'small' for odd N. This is particularly obvious when we convert peak spacing to addition energies. Then a clear even-odd parity is seen in the addition energy versus N at 3 T (see the middle trace of Fig. 2c). The amplitude of the even-odd oscillations is a good measure of the separation between the single-particle states at 3 T. We observe a slowly decreasing amplitude from ~ 1 to ~ 0.5 meV for N increasing to 40, suggesting that the confining energy decreases with increasing gate voltage. This trend is also seen in the B dependence of the last transition (i.e. filling factor 2 which is indicated by $*$'s). We find that for larger N the transition into the lowest Landau level occurs at a B-value lower than calculated (note the different field scales of figures 4a and b). The evolution of the $*$'s saturate at ~ 3 T for $N \sim 50$ which means that for these larger electron numbers increasing V_g increases the area and N such that the electron density stays constant. Note that at 4.5 T the addition energy versus N in Fig. 2c becomes smoother, suggesting that the dot becomes more and more spin polarized as B is increased.

A more detailed comparison can be made from Fig. 5a, which focuses on the range in B from -5 to 5 T for N = 4 to 7. For comparison Fig. 5b shows the calculated addition spectrum. It is clear that the 5th and 6th peaks form a pair. At 1.3 T the evolution of the 6th peak has a maximum whereas the 7th peak has a minimum. This corresponds to the crossing of the energy curves of the $(0,-1)$ and $(0,2)$ quantum states at 1.3 T in Fig. 3a.

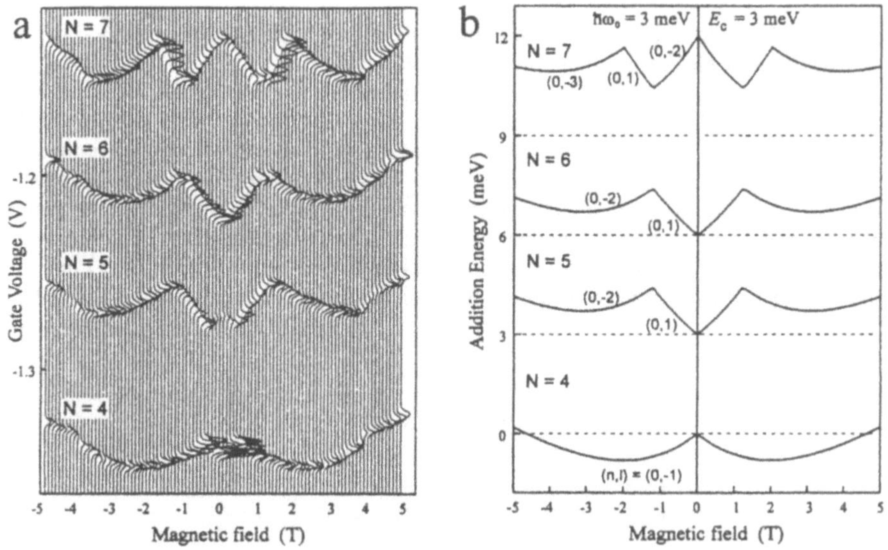

Figure 5 (a) Evolution of the 4th, 5th, 6th and 7th current peaks with B-field from -5 to 5 T observed for the $D = 0.5\ \mu$m dot. The original data consist of current vs. gate voltage traces for different magnetic fields which are offset. (b) Calculated electro-chemical potential $\mu_{\text{dot}}(N)$ using the CI model, with $E_c = 3$ meV and $\hbar\omega_0 = 3$ meV.

The effective dot diameter when there are 6 or 7 electrons on the dot is derived from the wavefunctions ($d \sim 5l_B$ for (0,2), see Fig. 3b). From the estimated confinement potential $\hbar\omega_0 = 3$ meV it follows that the diameter is about $d_{\text{eff}} \sim 100$ nm.

A closer look at Fig. 5a shows that peaks 4 and 5 are not exact replicas. In particular around $B = 0$, peak 4 has a sharp cusp down while peak 5 has a cusp up in the range $-0.4 < B < 0.4$ T. We discuss in the next section that this deviation from the CI model results from the exchange interaction between electrons with parallel spins in the second shell.

7 Hund's Rule and Exchange Energy

We now focus on the evolution of the peak positions near $B = 0$ T and show that deviations from the CI model are related to Hund's rule. Figure 6a shows the B-field dependence of the 3rd, 4th, 5th and 6th current peaks for a B-field up to 2 T. The pairing of the 3rd and 4th peaks and the 5th and 6th peaks *above* 0.4 T is clearly seen. However, we find that *below* 0.4 T the 3rd and 5th peaks are paired, and the 4th and 6th peaks are paired. Note that this pairing is also seen in the peak heights. The evolution as a pair of the 3rd and 5th peaks for $B < 0.4$ T is continued by the 3rd and 4th peaks for $B > 0.4$ T. Similarly, the evolution as a pair of the 4th and 6th peaks for $B < 0.4$ T is continued by the 5th and 6th peaks for $B > 0.4$ T. From these observations, we conclude that the 4th electron undergoes an angular momentum transition from $l = -1$ to $l = 1$ at 0.4 T, whereas the 5th electron undergoes an angular momentum transition from $l = 1$ to $l = -1$. These 'extra' transitions can be understood in terms of Hund's rule, which states that a degenerate shell of states will be filled by electrons with parallel spins [4].

To explain the above transitions we need to extend the CI model. If we leave out the contributions from the gate voltage one can write for the total energy $U(N) = \frac{1}{2}N(N-1)E_c + \sum E_{n,l} - E_{exch}(\sigma)$. The last term allows to take into account a reduction of the total energy due to the exchange interaction between electrons with parallel spins. For simplicity we assume that only electrons in quantum states with identical radial quantum number and opposite angular momentum $(n, \pm l)$ have an appreciable exchange interactions and ignore all other contributions. Let us first write out explicitly $U(N)$ and $\mu(N)$ for N from 1 to 6 assuming that spin is always minimized (i.e. total spin $S = \frac{1}{2}$ or 0):

$$\begin{aligned}
U(1) &= E_{0,0} \\
U(2) &= E_c + 2E_{0,0} \\
U(3) &= 3E_c + 2E_{0,0} + E_{0,1} \\
U(4) &= 6E_c + 2E_{0,0} + 2E_{0,1} \\
U(5) &= 10E_c + 2E_{0,0} + 2E_{0,1} + E_{0,-1} - E_{exch} \\
U(6) &= 15E_c + 2E_{0,0} + 2E_{0,1} + 2E_{0,-1} - 2E_{exch}
\end{aligned}$$

and

$$\begin{aligned}
\mu(1) &= U(1) - 0 = E_{0,0} \\
\mu(2) &= U(2) - U(1) = E_c + E_{0,0} \\
\mu(3) &= U(3) - U(2) = 2E_c + E_{0,1} \\
\mu(4) &= U(4) - U(3) = 3E_c + E_{0,1} \\
\mu(5) &= U(5) - U(4) = 4E_c + E_{0,-1} - E_{exch} \\
\mu(6) &= U(6) - U(5) = 5E_c + E_{0,-1} - E_{exch}
\end{aligned}$$

For $N = 4$ we can also put the 3rd and 4th electron with parallel spins (i.e. $S = 1$) in the separate single-particle states $E_{0,1}$ and $E_{0,-1}$. In this case we get: $U^*(4) = 6E_c + 2E(0,0) + E(0,1) + E(0,-1) - E_{exch}$. The ground state for $N = 4$ has $S = 1$ if $U^*(4) < U(4)$, or $E_{0,-1} - E_{0,1} = \hbar\omega_c < E_{exch}$. A transition from $S = 1$ to $S = 0$ occurs for larger B-fields. The experimentally observed transition field $B = 0.4$ T yields $E_{exch} = 0.7$ meV.

Below $B = 0.4$ T not only the electro-chemical potential of $N = 4$ but also of $N = 5$ is affected by this extra transition due to E_{exch}:

$$\begin{aligned}
\mu^*(4) &= U^*(4) - U(3) = 3E_c + E_{0,-1} - E_{exch} \\
\mu^*(5) &= U(5) - U^*(4) = 4E_c + E_{0,1}
\end{aligned}$$

Note that $\mu^*(4)$ follows the B-dependence of $E_{0,-1}$ which implies that the 4th peak should pair with the 6th peak. Similarly, the 5th peak should pair with the 3rd peak for $B < 0.4$ T.

We find a remarkable agreement between what we measure in Fig. 6a and the calculation in Fig. 6b if we assume $E_c = 3$ meV and $E_{exch} = 0.7$ meV. In Fig. 6b we include quantum numbers (n,l) to identify the angular momentum transitions, and pictorial diagrams to illustrate the spin configurations.

At $B = 0$ the ground-state addition energies $\Delta\mu(N) \equiv \mu(N+1) - \mu(N)$ become:

$$\begin{aligned}
\Delta\mu(1) &= E_c \\
\Delta\mu(2) &= E_c + E_{0,1} - E_{0,0} \\
\Delta\mu(3) &= E_c - E_{exch} \\
\Delta\mu(4) &= E_c + E_{exch} \\
\Delta\mu(5) &= E_c - E_{exch}
\end{aligned}$$

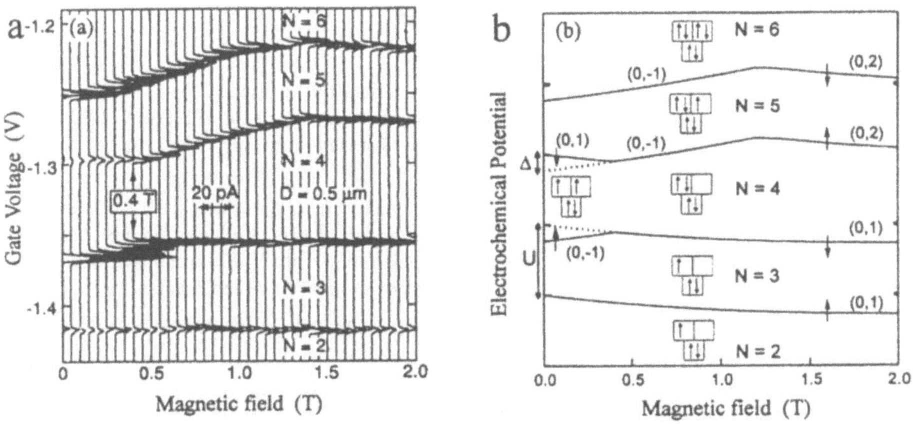

Figure 6 (a) Evolution of the 3rd, 4th, 5th and 6th current peaks with B-field from 0 to 2 T. The original data consist of current vs. gate voltage traces for different magnetic fields which are offset. (b) Calculated electrochemical potential vs. magnetic field for the model including a constant exchange energy as described in the text. Parameters are $E_c = 3$ meV, $\hbar\omega_0 = 3$ meV and $E_{exch} = 0.7$ meV.

While the peak spacing for $N = 2$ is enhanced due to the separation in single-particle energies, the spacing for $N = 4$ is expected to be larger than the spacings for $N = 3$ and 5 by twice the exchange energy $2E_{exch}$ (= 1.4 meV). These enhancements are indeed observed in the addition curve for $B = 0$ in Fig. 2b. Similar bookkeeping as above also explains the enhancements of $N = 9$ and 16 which correspond to a spin-polarized half-filled third shell (S=3/2) and fourth shell (S=2), respectively. The above simple example shows that inside a shell the symmetry can be broken and degeneracies can be lifted due to interactions. This spin-polarized filling and symmetry-breaking is completely analogous to Hund's rule in atomic physics. Although our bookkeeping method is very simple it explains the data well. Self-consistent calculations of several different approaches have confirmed this model of constant charging and exchange energies [14,15].

In conclusion, we use single-electron tunneling spectroscopy to probe electronic states of a few-electron vertical quantum dot-atom. At zero magnetic field the addition energy reveals a shell structure associated with a 2D harmonic potential. As a function of magnetic field, current peaks evolve in pairs, arising from the antiparallel filling of spin-degenerate states. Close to zero magnetic field, however, this pairing behavior is altered to favour the filling of states with parallel spins in line with Hund's rule. The results are in good agreement with an extended constant interaction model.

Acknowledgments

We thank H. Akera, R.J. van der Hage, T. Fujisawa, J. Janssen, Y. Kervennic, J.E. Mooij, S.K. Nair, H. Tamura, Y. Tokura, and T. Uesugi for help and discussions. The work was supported by the Dutch Foundation of Fundamental Research on Matter (FOM) and L.P.K. by the Royal Netherlands Academy of Arts and Sciences (KNAW).

Bibliography

[1] See for reviews in popular magazines: M. Reed, Scientific American **268**, 118 (1993); M.A. Kastner, Physics Today **46**, 24 (1993); R.C. Ashoori, Nature **379**, 413 (1996); C.J.P.M. Harmans, Physics World **5**, 50 (March 1992); L.P. Kouwenhoven and C.M. Marcus, Physics World **11**, (June 1998).

[2] See for a review on quantum dots: L.P. Kouwenhoven, C.M. Marcus, P.L. McEuen, S. Tarucha, R.M. Westervelt and N.S. Wingreen, *Electron transport in quantum dots*, in *Mesoscopic Electron Transport*, edited by L.L. Sohn, G. Schön and L.P. Kouwenhoven, June 1996 (Kluwer, Series E 345, 1997). see also: http://vortex.tn.tudelft.nl/~leo/papers/.

[3] S. Tarucha, D.G. Austing, T. Honda, R.J. van der Hage and L.P. Kouwenhoven, Phys. Rev. Lett. **77**, 3613 (1996).

[4] M. Alonso and E.J. Finn, *Quantum and Statistical Physics* (Addison-Wesley, 1968); L. I. Schiff, *Quantum Mechanics* (MacGraw-Hill, 1949).

[5] H. Mizuto and T. Tanoue, *The Physics and Applications of Resonant Tunneling Diodes*, (Cambridge, 1996).

[6] Y. Tokura (private communications).

[7] B. Su, V.J. Goldman and J.E. Cunningham, Phys. Rev. B **46**, 7644 (1992); M. Tewordt, L. Martín-Moreno, V.J. Law, M.J. Kelly, R. Newbury, M. Pepper, D.A. Ritchie, J.E.F. Frost and G.A.C. Jones, Phys. Rev. B **46**, 3948 (1992); T. Schmidt, M. Tewordt, R.H. Blick, R.J. Haug, D. Pfannkuche, K. von Klitzing, A. Förster and H. Lüth, Phys. Rev. B **51**, 5570 (1995).

[8] M.W. Dellow, P.H. Beton, C.J.G.M. Langerak, T.J. Foster, P.C. Main, L.Eaves, M. Henini, S.P. Beaumont and C.D.W. Wilkinson, Phys. Rev. Lett. **68**, 1754 (1992); P. Gueret, N. Blanc, R. Germann and H. Rothuizen, Phys. Rev. Lett. **68**, 1896 (1992).

[9] S. Tarucha, D.G. Austing and T. Honda, Superlattices and Microstructures **18**, 121 (1995).

[10] R.C. Ashoori, H.L. Stormer, J.S. Weiner, L.N. Pfeiffer, S.J. Pearton, K.W. Baldwin and K.W. West, Phys. Rev. Lett. **68**, 3088 (1992).

[11] R.C. Ashoori, H.L. Stormer, J.S. Weiner, L.N. Pfeiffer, K.W. Baldwin and K.W. West, Phys. Rev. Lett. **71**, 613 (1993).

[12] D.G. Austing, T. Honda, Y. Tokura and S. Tarucha, Jap. J. Appl. Phys. **34**, 1320 (1995); D. G. Austing, T. Honda and S. Tarucha, Semicond. Sci. Technol. **11**, 388 (1996).

[13] C. G. Darwin, Proc. Cambridge Philos. Soc. **27**, 86 (1930); V. Fock, Z. Phys. **47**, 446 (1928).

[14] M. Macucci, K. Hess and G.J. Iafrate, Phys. Rev. B **48**, 17354 (1993); A. Natori, M. Fujito, H. Yasunaga, Superlattices and Microstructures **22**, 65 (1997); A. Angelucci and A. Tagliacozzo, Phys. Rev. B **56**, R7088 (1997); M. Koskinen, M. Manninen, S.M. Reimann, Phys. Rev. Lett. **79**, 1389 (1997); I.H. Lee, V. Rao, R.M. Martin, J.P. Leburton, Phys. Rev. B **57**, 9035 (1998); M. Rontani, F. Rossi, F. Manghi and E. Molinari, Appl. Phys. Lett. **72**, 957 (1998).

[15] V. Halonen, P. Hyvönen, P. Pietiläinen and Tapash Chakraborty, Phys. Rev. B **53**, 6971 (1996).

Atomically Precise, Coupled Quantum Dots Fabricated by Cleaved Edge Overgrowth

W. Wegscheider, G. Schedelbeck, M. Bichler and G. Abstreiter

Walter Schottky Institut, Technische Universität München,
Am Coulombwall, D-85748 Garching, Germany

Abstract: Recent progress in the fabrication of quantum dots by molecular beam epitaxy along three directions in space is reviewed. The optical properties of different sample structures consisting of individual quantum dots, pairs of coupled dots as well as of linear arrays of dots are studied by microscopic photoluminescence spectroscopy. The high degree of control over shape, composition and position of the $7 \times 7 \times 7$ nm^3 size GaAs quantum dots, which form at the intesection of three orthogonal quantum wells, allows a detailed investigation of the influence of coupling between almost identical zero-dimensional objects. In contrast to the inhomogeneously broadened quantum well and quantum wire signals originating from the complex twofold cleaved edge overgrowth structure, the photoluminescence spectrum of an individual quantum dot exhibits a single sharp line (full width at half maximum < 70 μeV) almost free of background signal. Microscopic photoluminescence excitation spectroscopy directly reveals the discreteness of the energy levels of the zero-dimensional structures and justifies the denomination "artificial atoms" for the quantum dots. It is further demonstrated that an "artifical molecule", characterized by the existence of bonding and antibonding states, can be assembled from two of such "artificial atoms". The coupling strength between the "artificial atoms" is adjusted by the "interatomic" distance and is reflected in the energetic separation of the bonding and antibonding levels and the linewidths of the corresponding interband transitions.

1 Introduction

The optical absorption and emission spectra of atoms exhibit sharp lines which originate from the discrete energy level scheme of electrons within the different shells. In contrast, as a result of the periodic potential, electrons in crystalline solids reside within bands which are separated by forbidden energy gaps. Consequently, the optical properties of semiconductors are usually dominated by wide absorption bands, resulting from the smooth and continuous density of states profiles. Semiconductor quantum wires (QWRs) and dots (QDs) with quantum confinement of charge carriers to one (1D) or zero dimensions (0D), respectively, have attracted extensive research activities recently, as they possess unique properties which arise due to sharp peaks in their density of states. While charge carriers in 1D systems can still move freely along the QWR axis and, thus, still have a continiuous energy level spectrum, the complete confinement present in 0D systems results in discrete δ-function like density of states peaks. The successful realization of different types of QDs in semiconductors over the last five years [1–18] led to the discovery of effects like a "shell-like" like energy staircase for single-electron charging [18] or extremely narrow homogeneous linewidths for optical transistions [9] which resemble in many ways the characteristic features of atoms. As a result of this analogy such 0D objects are often referred to as "artificial atoms". The growing interest in QD systems is, however, not only of fundamental nature, but is strongly driven by applications such as semiconductor laser devices [19], where discrete energy level schemes should be advantageous in many respects [21].

Several highly sophisticated techniques have been implemented to fabricate these "artificial atoms" from molecular beam epitaxially prepared layers which often serve as a starting material since quantum confinement along the growth direction can be readily achieved. Lateral patterning of two-dimensional (2D) quantum well (QW) layers, however, severely affects the optical quality of the structures and typical photoluminescence (PL) linewidths of single QDs prepared in this way exceed 0.5 meV [1–3]. In addition, substantial fluctuations in composition and size from one QD to another cannot be avoided. The realization of QDs confining excitons to within about 10 nm, the equivalent of the hydrogen-like Bohr orbit for excitons in GaAs, whose emission is governed by narrow "atom-like" lines was up to now limited to natural formation techniques. These rely on monolayer thickness fluctuations in narrow QWs [6–9] or on a thermodynamically driven instability during strained layer epitaxy which leads to spontaneous island formation [10–13]. These natural and self-assembled QDs do not suffer from crystal imperfections or free surfaces which are inevitably accompanied by a high number of nonradiative recombination centers. However, as a consequence of the inherent randomness in the formation process highly local spectroscopic techniques are essential in order to extract the properties of individual objects from an ensemble of such QDs, which differ in size and shape and whose position is not under precise experimental control. The long-standing question of whether the analogy between QDs and artificial atoms can be taken even further – that is, whether QDs act like coupled atoms in a molecule [20] – had therefore to be answered in a different kind of experiment.

Molecular beam epitaxial (MBE) growth on the cleavage face of a previously prepared GaAs/AlGaAs multilayer, a technique pioneered by Pfeiffer et al. [22], has attracted increasing attention because of its capability for the fabrication of high-quality QWR structures suitable for optical [23–27] and transport investigations [28,29]. The high perfection of these structures led to the demonstration of enhancement of exciton binding [24,25] and lasing from excitons in 1D [24]. As predicted in [30] cleaved edge overgrowth (CEO) can be also employed to produce QDs, in which charge carriers are confined in all three dimensions due to atomically precise QW potentials, and that are, therefore, highly uniform in size and shape. Here we report on the optical properties of different structures consisting of individual QDs, pairs of coupled QDs as well as of linear arrays of QDs which were obtained by twofold application of the CEO method [31,32]. The complex sample designs make explicit use of atomic control of both size and position of twofold CEO QDs. The latter seem to be ideally suited to perform fundamental studies on a novel class of well isolated or controllably coupled 0D objects.

This article is organized as follows. After the introduction the origin for quantum mechanical carrier confinement to 1D and 0D at the right-angle intersection of two and three QWs, respectively, is outlined in Section 2. In addition, the technical implementation of this concept for the fabrication of atomically precise QDs by twofold application of the CEO method is presented together with a description of the two QD sample structures investigated experimentally. The final Section is devoted to the optical properties of these two sample structures; a superlattice in form of a linear array of QDs as well as an isolated QD and a series of QD pairs with varying coupling strength.

2 T-shaped Quantum Well Intersections

The precise control of layer thicknesses and layer compositions currently achievable by MBE allows the design of very sophisticated band-gap engineered multilayer structures with applications ranging from light emitters and detectors based on tailored transitions between quantum confined states to high electron mobility transistors (HEMTs) where Coulomb scattering is suppressed because of spatial separation of the charge carriers from the ionized donors. The possibilty to overgrow a heterostructure by another set of layers just adds another degree of freedom in the design parameters. From the vast variety of structures which can be prepared by CEO [33] we concentrate here on QWR and QD systems suitable for optical interband transitions.

2.1 Carrier Confinement to One and Zero Dimensions

Figure 1 illustrates the physics of QWR formation at the T-shaped intersection of two QWs [24,34]. The figure shows the cross-section of such an intersection together with contours of constant probability for electrons. The origin of the quantum mechanical bound state is the relaxation of QW confinement at the intersection. While a classical particle would be unbound for the given T-shaped potential, the "expansion" of the wavefunction into the larger available volume at the junction results in a smaller kinetic contribution to the total energy of electrons and holes. Consequently, motion of the 1D carriers is limited to the line defined by the intersecting planes of the two QWs. In contrast to other QWR fabrication techniques which introduce additional confinement of the carriers in a previously prepared 2D system, for example by partial removal of a QW layer and subsequent overgrowth with barrier material, the QWR states in this structure are energetically located below the ground state QW transitions. The existence of such confinement relaxation 1D states was proposed already more than ten years ago by Störmer [35] and Chang et al. [36]. The theoretical treatment of single-particle quantum confined states at the intersection of two QWs requires solving of the 2D Schrödinger equation for the T-shaped potential. Although simplified calculations [36,37] based on factorization of the structure potential, thus, reducing the problem to an effective one-dimensional Schrödinger equation, give physical insight and lead to analytical results, they are not accurate enough for this purpose. In particular, Coulomb correlation between an electron and a hole bound to such a QW intesection, i.e. the formation of a 1D exciton, demands for more advanced numerical schemes which have been carried out using a simple variational procedure [38]

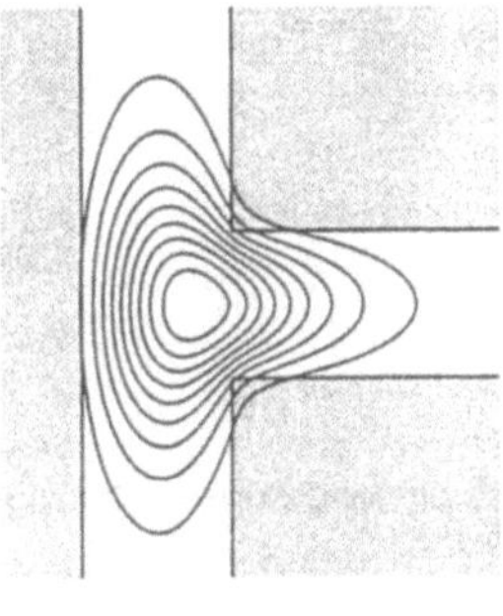

Figure 1
T-shaped intersection of two quantum wells in cross section. The contours are lines of constant probability ($|\psi|^2 = 0.1, 0.2, \ldots, 0.9$) for electrons confined in the T-shaped potential.

or in the framework of the complete six-dimensional Hamiltonian [39]. In good agreement with the experimental results [24], these calculations indicate that for the T-shaped intersection of two 7 nm wide QWs, a structure which is also relevant for the experimentally realized QD systems described below, the 1D excitonic state is about 13 meV deep with respect to the 2D excitons.

Similar to the formation of a 1D state at the T-shaped intersection of two QWs further "expansion" of the electron and hole wavefunctions into a third QW due to confinement relaxation leads to a lowering of the corresponding 0D ground state energy. Carriers which can move freely in two directions within the QWs and along one direction in the QWRs are completely localized in such QDs. The corresponding lowering of the 0D excitonic states relative to the 1D QWR states has been theoretically modelled recently [40] and amounts to about 6 meV for the intersection of three 7 nm wide QWs.

2.2 Technical realization and sample structures

The CEO technique has been described in detail elsewhere [22,33]. Figure 2 summarizes the three step growth procedure for the fabrication of T-shaped QWRs and QDs. The resulting sample structure after growth of QW layers along the three orthogonal directions is shown in Fig. 3a. The two different QD structures investigated here are sketched in Fig. 3b and c. The QD superlattice structure of Fig. 3b consists of a 22-period GaAs/$Al_{0.35}Ga_{0.65}As$ multiple QW (MQW_1) structure with well and barrier thicknesses of 7 and 30 nm, respectively, which was two times *in situ* cleaved along the [110] and [$\bar{1}$10] cleavage planes and subsequently overgrown with two 7 nm wide GaAs single QWs (QW_2, QW_3) and $Al_{0.35}Ga_{0.65}As$ barriers. As can be seen in Fig. 3a three types of QWRs (QWR_{12}, QWR_{13}, QWR_{23}), resulting from intersections of MQW_1 with QW_2, MQW_1 with QW_3 and QW_2 with QW_3 form. At a length of about 800 nm 22 equally spaced and nominally identical QDs, each of which is constituted by the junction of three QWs, are located along QWR_{23}. In order to study the transition from an isolated QD ("artificial atom") to a system of two QDs with adjustable coupling strength ("artificial molecule") the QD structure of Fig. 3c has been prepared. The layer sequence of the first growth step consists in this case of a single QW followed by three pairs of QWs. The barriers separating adjacent QWs and also QDs are d = 15 nm, 30 nm and 60 nm. The spatial distance between the QDs adjusts the coupling strength of the zero-dimensional QD levels. The

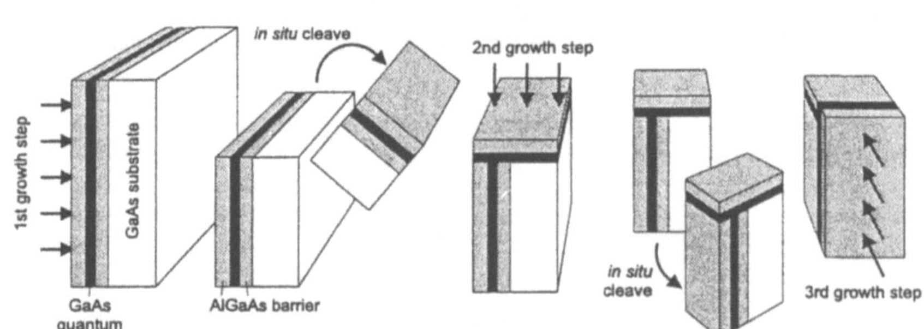

Figure 2 The concept of molecular beam epitaxial growth on the cleavage faces of a previously prepared multilayer structure (cleaved edge overgrowth).

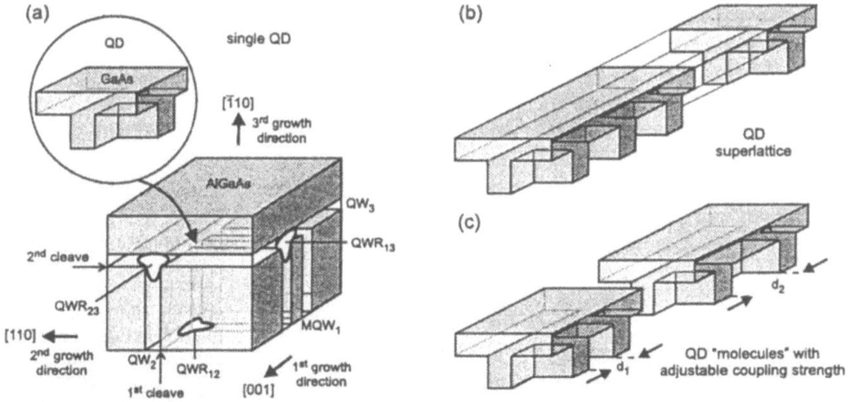

Figure 3 (a) Schematic illustration of the quantum dot (QD) structure (not to scale) obtained after three growth steps separated by two *in situ* cleaves. The junction of three quantum wells and wires, at which a QD forms, is shown in the magnified part of the figure. The T-shaped contours are lines of constant probability for electrons confined in the quantum wires. The two sample structures investigated here consist of a QD superlattice (b) and a series of three coupled QDs with different coupling strengths as well as an individual QD (c).

center-to-center distance from the single QD to the first and the following coupled QDs was chosen to be 800 nm. Further details of the two sample structures are given in [31,32].

3 Optical Properties of Isolated and Coupled Quantum Dot Systems

The CEO QD samples were investigated using microphotoluminescence (μPL) and μPL excitation (μPLE) spectroscopy combining high spatial and spectral resolution. Excitation was performed by means of a tunable dye laser which was focused through a microscope objective lens (numerical aperture NA = 0.75) onto the sample. The power density was kept below 50 W/cm^2. Under such conditions less than one exciton at a time is expected to be in the QDs. The samples were mounted on the cold-finger of a He cryostat and studied at nominally 5 K. The emitted light was collected by the same objective lens and directed to a confocal imaging system, which defined a nearly diffraction limited detection range of about 800 nm. Finally, the PL signal was dispersed using a triple grating Raman spectrometer (spectral resolution $\approx$ 40 μeV) and detected with a liquid nitrogen cooled charge-coupled device camera. We would like to point out that in spite of diffraction limitation, spectroscopy of single QDs is possible using conventional optical microscopy since the CEO method allows the fabrication of QD structures with adjustable QD separations larger than the spatial resolution of the instrument. This is the case for the sample, designed to monitor subsequent phases in the formation of artificial molecules out of artificial atoms, in which all relevant QD objects, i.e. the single QD and the three coupled QD systems, can be probed individually without the use of non-diffraction limited probes like near-field optical spectroscopy [7,9].

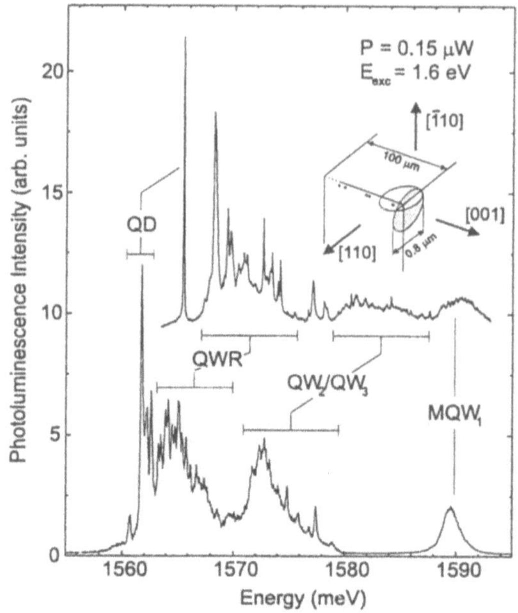

Figure 4
μPL spectra of the samples containing the 22-period QD superlattice (lower curve) and the individual and coupled QDs (upper curve). The spectra were obtained by excitation and detection from the (001) (lower curve) and ($\bar{1}10$) sample surfaces (upper curve), respectively, as indicated by the shaded circles in the inset (not to scale) which also schematically depicts the positions of the QDs (black dots). The two spectra are vertically offset against each other for clarity.

Figure 4 shows typical μPL spectra of the two samples recorded from the (001) and ($\bar{1}10$) surfaces as indicated in the inset. The spectrum of the superlattice QD sample (lower curve) was taken at a position in the vicinity of the QDs. As a result of the excitation geometry given in the inset of Fig. 4 all 22 QDs are excited simultaneously and contribute to the signal in the spectral range of 1560–1563 meV. It should be noted at this point that due to the large excitation area, which is unavoidable in diffraction limited spectroscopy, and exciton diffusion over a length of about 0.5 μm [41], the carriers, most of which are generated in the QW regions, recombine in the quantum dots, wires and wells. The relatively strong response from the QDs in spite of their extremely small volume results from efficient collection of excitons from the adjacent QWR and QW regions. Comparison with spectra taken at positions excluding excitation of the dots or wires identifies the PL response observed at 1589 meV and in the range of 1570–1580 meV as emission of MQW_1 and the two wells QW_2/QW_3. The PL energies of QW_2 and QW_3 of 1574 meV and 1576 meV determined independently in this way almost coincide, but are, as a result of the large heavy hole mass for $\langle 110\rangle$ type confinement directions, well below the MQW_1 response. Emission of the three types of QWRs present in the QD structure is detected around 1565 meV. Again, the PL maxima of QWR_{12} and QWR_{13}, determined by moving the detection window away from the QD region coincide within about 1 meV. With the high spatial resolution used for these experiments we find that the QWR PL and partly that of QW_2 and QW_3 starts to decompose into sharp lines as a result of exciton localization. This effect becomes even more pronounced in the μPL spectrum of the second sample which has been taken at a position where only one QD can contribute to the signal (upper curve). Consequently, the number of localized 1D states within the excitation and collection volume is also reduced substantially and individual sharp lines start to emerge from the broad QWR background. The QD response, however, is in this case characterized by a single sharp line at an energy of 1565.9 meV, which indicates that the corresponding

0D exciton ground state is ≈ 5 meV deep with respect to the 1D QWR exciton states centered around an energy of 1571 meV. This value agrees well with the theoretical prediction for our QD structure of 6 meV [40]. Comparison of the two spectra in Fig. 4 also reveals that all PL transitions associated with the QDs, QWRs and QW_2/QW_3 appear at lower energies in the QD superlattice sample. This is a result of slightly larger widths of QW_2 and QW_3 in this sample. The unambigous assignment of the peaks to the different low-dimensional structures in our samples follows from a detailed three-dimensional μPL mapping which is presented in extracts below.

3.1 Quantum Dot Superlattice

Figure 5 shows μPL spectra taken at different positions from the ($\bar{1}10$) surface of the QD superlattice sample together with grey scale intensity plots (pixel size 200×200 nm^2) at selected PL energies. For the position chosen in the upper panel of this figure, marked by a white cross in the grey scale image, emission from MQW_1 and QWR_{13} can be detected. The grey scale intensity plot recorded with the detection system kept fixed at the MQW_1 PL line, indicated by an arrow in the spectrum, clearly reveals quenching of the multiple quantum well emission in the vicinity of the QDs due to exciton diffusion into wires and dots. The fact that the MQW_1 PL signal is not accordingly suppressed at positions away from the QDs but still on top of QWR_{13} is a result of the finite excitation and collection depth of our probe which exceeds the exciton diffusion length. In accordance with the detection position used for the spectrum displayed in the middle panel of Fig. 5, only signals

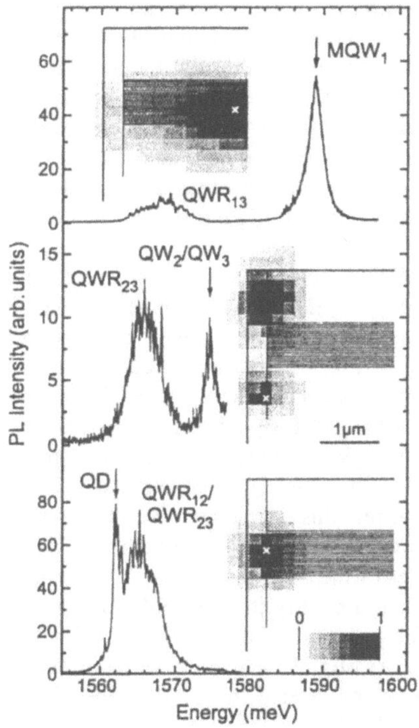

Figure 5
μPL spectra (E_{exc} = 1.6 eV, P = 0.15 μW) and grey scale intensity plots recorded from the ($\bar{1}10$) surface of the QD superlattice sample.The positions on the sample where the spectra were taken are indicated by the white crosses in the grey scale images which are superimposed on a schematic view of the sample. The arrows define the spectral position of the corresponding grey scale plot.

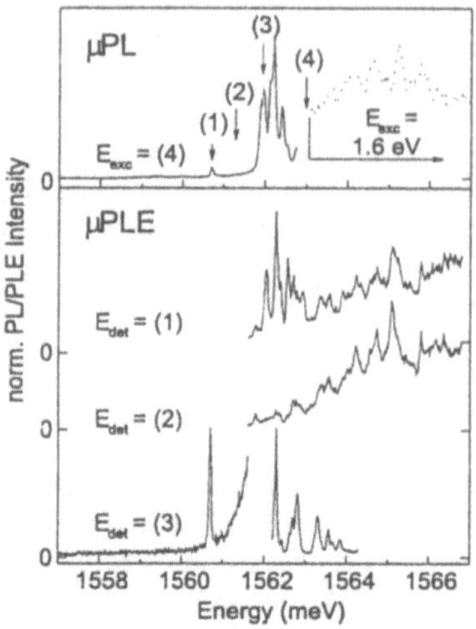

Figure 6
Microscopic PL and PLE spectra (P = 0.4 μW) taken from the ($\bar{1}10$) surface of the QD superlattice sample at a position centered on top of the quantum dot array. The arrows mark the energetic positions used for excitation (μPL) or detection (μPLE).

originating from recombination within the two single QWs (QW_2, QW_3) and the single wire (QWR_{23}) which is formed by these two wells can be detected. Again, bleaching of the QW emission in the vicinity of the multiple quantum wires and dots is observed, as can be seen in the corresponding grey scale plot. Exclusively at the intersection of the wells and wires, i.e. on top of the QD array, the characteristic sharp peaks, interpreted before as PL originating from the QDs, emerge (lower panel of Fig. 5). Although the relative intensities of these peaks have changed with respect to the spectrum shown in Fig. 4 (lower PL curve), their energetic positions exactly coincide. The intensity decrease of the QD emission is in all directions limited by the experimental resolution, as demonstrated by the grey scale image taken at one of the sharp resonances. Similar intensity distributions are recorded for the other sharp peaks. They are, however, along the extension of the QD array slightly offset against each other [31]. This implies that the different QD resonances can be attributed to exciton recombination in individual spatially separated QDs of the array. The energy variations might be explained by slight differences in the widths of the participating QWs at the dot positions.

Near-resonant μPL and μPL excitation (μPLE) spectra of this sample are collected in Fig. 6. When the excitation is tuned to the energy labeled (4) in the upper panel of Fig. 6, the PL signal of the QDs is much clearer resolved and almost free of background. The μPLE spectrum displayed in the lower panel of Fig. 6 with the detection at the energy labeled (1), the lowest QD level, exhibits sharp resonances exactly at the QD PL energies. For energies above 1563 meV the lineshape of the μPLE signal coincides with the QWR PL finestructure (dotted line). If the detection energy is slightly detuned with respect to the QD levels, for instance to the energy labeled (2), the sharp QD resonances are completely suppressed. In contrast, the QWR μPLE signal remains unchanged. For the lowest μPLE spectrum displayed in Fig. 6 the detection was moved to the energy labeled (3). Again, sharp resonances appear exactly at the energies of the other dot levels. Surprisingly, even for excitation energies smaller than the detection energy it is possible to observe such a sharp resonance at the lowest dot level. The pronounced excitation of individual QDs under resonant pumping of QD states which are slightly above or below in energy and localized at spatially separated QDs has to be interpreted in terms of coupling of the 0D states. Such a coupling seems to be possible although the QDs are separated by 30 nm wide barriers, because the effective barrier height, represented by QWR_{23} is only $\approx$ 5 meV. This has been confirmed by a one-dimensional Kronig-Penney type superlattice calculation, that resulted in a ground state bandwidth of 1.7 meV. Assuming a temperature slightly higher than the nominal 5 K, the thermal energy of about 0.7 meV is comparable to the energy difference from the lowest to the higher QD levels. The observed anti-Stokes luminescence is then explained by thermally activated tunneling of excitons from one QD to another.

3.2 Transition from an Artifcial Atom to a Molecule

Deeper insight into the effects of quantum mechanical coupling between 0D exciton states in two QDs can be gained by a detailed analysis of the sample which contains a single QD as well as a series of pairs of QDs with differently adjusted coupling strength. Figure 7 shows the spectral image of a linescan along QWR_{23} in this sample on an enlarged energy scale. The image is dominated by the single line, which is also visible in the upper PL spectrum of Fig. 4, and which originates from excitons recombining within the single QD. Moving the excitation and detection window to the coupled QDs, a series of double peaks appears. The spatial positions of all peaks exactly coincide with the designed positions of our individual and coupled QDs as can be seen by the integrated PL intensity plots of Fig. 8, in which the nominal QD positions are marked by arrows. The spatial

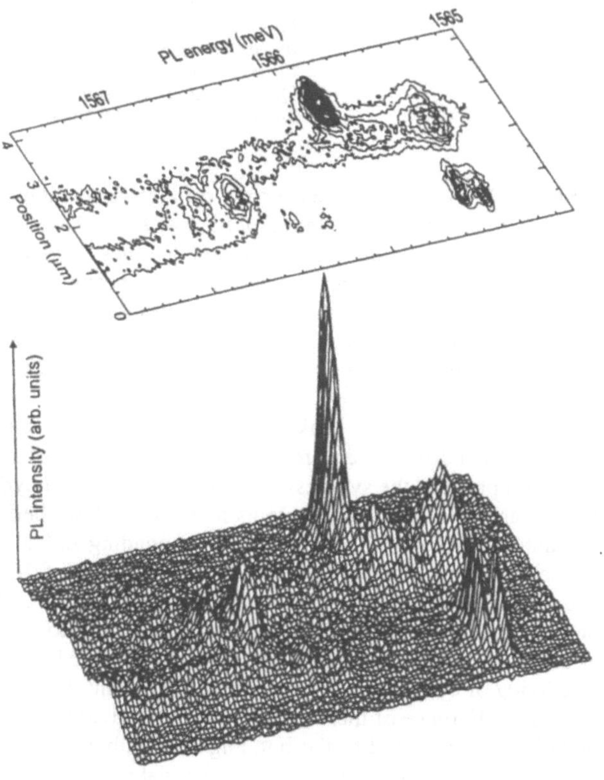

Figure 7
Spectral image resulting from a high-resolution μ-PL linescan (E_{exc} = 1588 meV) along QWR_{23}, which connects the single QD with the three different coupled QDs, taken from the $(\bar{1}10)$ sample surface (lower panel). The contour lines in the upper panel of this figure represent regions of constant PL intensity of 10%, 15%, 20%,..., 95% of the maximal intensity.

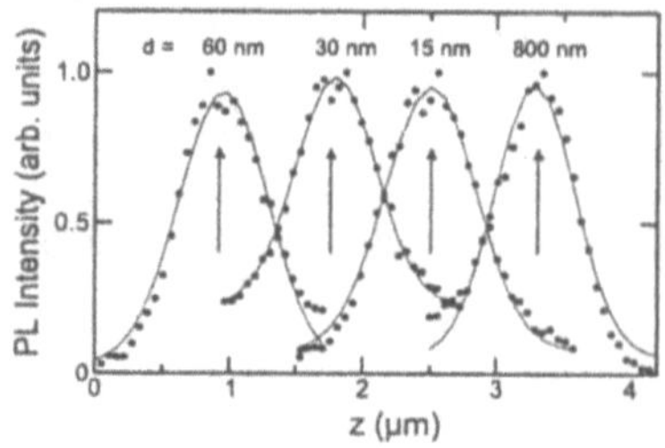

Figure 8
Normalized PL intensities integrated over the spectral regions of maxima in Fig. 7, centered at about 1565.42 meV (d = 60 nm), 1566.63 meV (d = 30 nm), 1565.70 meV (d = 15 nm) and 1565.89 meV (d = 800 nm, single QD) and plotted as a function of the position along the linescan. The nominal positions of the different QD structures are indicated by the arrows in this plot. The solid curves are just guides to the eye.

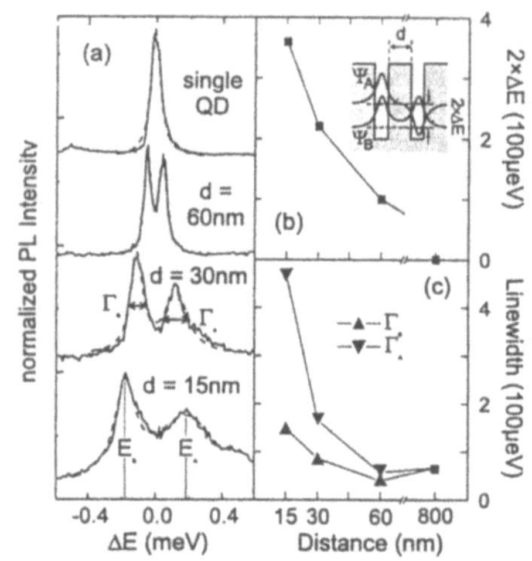

Figure 9
(a) Selected normalized μPL spectra from the linescan of Fig. 7 of the single and the coupled QDs (at the positions marked by the arrows), plotted on a relative energy scale and offset vertically against each other for clarity, (b) energetic splitting between the bonding and antibonding levels as indicated in the inset and (c) linewidths of the two levels as a function of the QD distance d. The solid lines in (b) and (c) are a guide for the eye.

width of these peaks is comparable to that observed for the different transitions in the QD superlattice sample and is limited by the spatial resolution of the μPL setup. This clear correlation of the sharp lines with position directly proofs that they originate from excitonic recombination in the CEO QDs. The energetic spread of the peaks in the range of 1 to 2 meV can be well described by monolayer thickness fluctuations in the overgrown QWs. However, the wavefunctions of the coupled QDs probes only a very short part of the constituting QWs with respect to the length scale of monolayer fluctuations and coupling between almost identical objects can be studied.

The appearance of two energetically distinct peaks from coupled QDs, as shown in Fig. 9a, is clear evidence that their excitonic levels are coupled. This scenario of an exciton bound to two QDs can be compared to the synthesis of a positively charged hydrogen molecule H_2^+ from an electron and two protons, a systems that can be described in the framework of elementary quantum mechanics by the formation of bonding and antibonding states. The influence of coupling on the excitonic levels, on the order of 100 μeV, is, however, small compared to the binding energy of an exciton confined to a CEO QD of more than 10 meV. Therefore, the exciton is considered as one particle whose internal structure is not affected by coupling. In analogy to the two coupled electronic levels of the H_2^+ molecule, the wavefunction ψ of our system of two coupled excitonic 0D levels can be approximated by the sum and difference of the wavefunctions of individual 0D states. These wavefunctions, ψ_B and ψ_A, correspond to the bonding and antibonding states as visualized in the schematic drawing of the one-dimensional potential profile for excitons in the coupled system of Fig. 9b. From first-order perturbation theory, the energy levels of the coupled system can be written as $E_{B/A} = E_0 \mp \Delta E$, where E_0 is the energy level of an individual QD and ΔE is a matrix element that describes the energetic splitting $2 \times \Delta E$ caused by coupling. Following this argument, we assign the two energetically distinct PL lines of each coupled QD to emission from bonding and antibonding energy levels, which are separated by $2 \times \Delta E$. The observed systematic dependence of the energetic splitting on the distance between the coupled dots (Fig. 9b) confirms our interpretation in terms of bonding and antibonding states. The enhancement of the energetic splitting with decreasing distance between the coupled QDs directly reflects the increasing coupling strength of the 0D levels.

In agreement with results obtained on naturally formed QDs [9] the spectra shown in Fig. 9a are well fitted with Lorentzian functions (dashed curves). This implies that the PL linewidths corresponding to the bonding and antibonding states are life-time broadened. These linewidths increase systematically with increasing coupling strength (Fig. 9c). The broadening is especially pronounced for emission from the antibonding level. This decrease of the exciton lifetime for the antibonding state with increasing level separation might be due to acoustical phonon scattering from the upper into the lower state [42]. In such a simple two-level picture one would expect a cubic dependence of the linewidth of the upper level on the energy level separation, which is also observed experimentally.

Acknowledgements

The cleaved edge overgrowth technique used in this work was adopted from L. N. Pfeiffer and K. W. West. One of the authors (W.W.) wishes to thank them for their excellent tutoring in this field which was provided during his postdoctoral stay at Bell Laboratories. Helpful discussions with A. Zrenner and programming support for data analysis by C. Obermüller are also gratefully acknowledged. The project is financially supported by the Deutsche Forschungsgemeinschaft in the framework of SFB 348 and the Bundesministerium für Bildung und Forschung via contract 01 BM 630/1.

Bibliography

[1] K. Brunner, U. Bockelmann, G. Abstreiter, M. Walther, G. Böhm, G. Tränkle and G. Weimann, Phys. Rev. Lett. **69**, 3216 (1992).

[2] Y. Nagamune, H. Watabe, M. Nishioka, and Y. Arakawa, Appl. Phys. Lett. **67**, 3257 (1995).

[3] M. Bayer, A. Schmidt, A. Forchel, F. Faller, T. L. Reinecke, P. A. Knipp, A. A. Dremin, and V. D. Kulakovskii, Phys. Rev. Lett. **74**, 3439 (1995).

[4] N. C. van der Vaart, S. F. Goodijn, Y. V. Nazarov, C. J. P. M. Harmans, J. E. Mooij, L. W. Moolenkamp, and C. T. Foxon, Phys. Rev. Lett. **74**, 4702 (1995).

[5] F. R. Waugh, M. J. Berry, D. J. Mar, R. M. Westervelt, K. L. Campman, and A. C. Gossard, Phys. Rev. Lett. **75**, 705 (1995).

[6] A. Zrenner, L. V. Butov, M. Hagn, G. Abstreiter, G. Böhm, and G. Weimann, Phys. Rev. Lett. **72**, 3382 (1994).

[7] H. F. Hess, E. Betzig, T. D. Harris, L. N. Pfeiffer, and K. W. West, Science **264**, 1740 (1994).

[8] K. Brunner, G. Abstreiter, G. Böhm, G. Tränkle, and G. Weimann, Appl. Phys. Lett. **64**, 3320 (1994); Phys. Rev. Lett. **73**, 1138 (1994).

[9] D. Gammon, E. S. Snow, B. V. Shanabrook, D. S. Kratzer, and D. Park, Phys. Rev. Lett. **76**, 3005 (1996); Science **273**, 87 (1996).

[10] J.-Y- Marzin, J.-M. Gérard, A. Izraël, D. Barrier, and G. Bastard, Phys. Rev. Lett. **73**, 716 (1994).

[11] R. Leon, P. M. Petroff, D. Leonard, and S. Safard, Science **267**, 1966 (1995).

[12] M. Grundmann, J. Christen, N. N. Ledentsov, J. Böhrer, D. Bimberg, S. S. Ruvimov, P. Werner, U. Richter, U. Gösele, J. Heydenreich, V. M. Ustinov, A. Yu. Egorov, A. E. Zhukov, P. S. Kopév, and Zh. I. Alferov, Phys. Rev. Lett. **74**, 4043 (1995).

[13] A. Kurtenbach, K. Eberl, and T. Shitara, Appl. Phys. Lett. **66**, 361 (1995).

[14] M. A. Kastner, Phys. Today **46**, 24 (1993).

[15] O. Klein, C. de Chamon, D. Tang, D. M. Abusch-Magder, U. Meirav, X.-G. Wen, M. A. Kastner, and S. J. Wind, Phys. Rev. Lett. **74**, 785 (1995).

[16] R. C. Ashoori, H. L. Stormer, J. S. Weiner, L. N. Pfeiffer, K. W. Baldwin, and K. W. West, Phys. Rev. Lett. **71**, 613 (1993).

[17] R. C. Ashoori, Nature **379**, 413 (1996).

[18] S. Tarucha, D. G. Austing, T. Honda, R. J. van der Hage, and L. P. Kouwenhoven, Phys. Rev. Lett. **77**, 3613 (1996).

[19] N. Kirstaedter, N. N. Ledentsov, M. Grundmann, D. Bimberg, V. M. Ustinov, S. S. Ruvimov, M. V. Maximov, P. S. Kopév, Zh. I. Alferov, U. Richter, P. Werner, U. Gösele, J. Heydenreich, Electron. Lett. **30**, 1416 (1994).

[20] L. Kouwenhoven, Science **268**, 1440 (1995).

[21] Y. Arakawa and A. Yariv, IEEE J. Quantum Electron. **22**, 1887 (1986).

[22] L. Pfeiffer, K. W. West, H. L. Störmer, J. P. Eisenstein, K. W. Baldwin, D. Gershoni, and J. Spector, Appl. Phys. Lett. **56**, 1697 (1990).

[23] A. R. Goñi, L. N. Pfeiffer, K. W. West, A. Pinczuk, H. U. Baranger and H. L. Stormer, Appl. Phys. Lett., **61**, 1956 (1992).

[24] W. Wegscheider, L. N. Pfeiffer, M. M. Dignam, A. Pinczuk, K. W. West, S. L. McCall, and R. Hull, Phys. Rev. Lett. **71**, 4071 (1993).

[25] T. Someya, H. Akiyama and H. Sakaki, Phys. Rev. Lett. **74**, 3664 (1995); **76**, 2965 (1996).

[26] H. Gislason, W. Langbein and J. M. Hvam, Appl. Phys. Lett. **69**, 3248 (1996).

[27] J. Hasen, L. N. Pfeiffer, A. Pinczuk, S. He, K. W. West, and B. S. Dennis, Nature **390**, 54 (1997).

[28] W. Wegscheider, W. Kang, L. N. Pfeiffer, K. W. West, H. L. Stormer, und K. W. Baldwin, Solid-State Electron. **37**, 547 (1994).

[29] A. Yacoby, H. L. Stormer, N. S. Wingreen, L. N. Pfeiffer, K. W. Baldwin, and K. W. West, Phys. Rev. Lett. **77**, 4612 (1996).

[30] W. Wegscheider, L. N. Pfeiffer, and K. W. West, in "Festkörperprobleme/Advances in Solid State Physics", Vol. 35, ed. by R. Helbig (Vieweg, Wiesbaden, 1996), p. 155.

[31] W. Wegscheider, G. Schedelbeck, G. Abstreiter, M. Rother, and M. Bichler, Phys. Rev. Lett. **79**, 1917 (1997).

[32] G. Schedelbeck, W. Wegscheider, M. Bichler, and G. Abstreiter, Science **278**, 1792 (1997).

[33] L. Pfeiffer, H. L. Störmer, K. W. Baldwin, K. W. West, A. R. Goñi, A. Pinczuk, R. C. Ashoori, M. M. Dignam, and W. Wegscheider, J. Cryst. Growth **127**, 849 (1993).

[34] W. Wegscheider, L. N. Pfeiffer, K. W. West, A. A. Kiselev, M. Hagn, and R. E. Leibenguth, Proc. of the NATO Advanced Study School on "Optical spectroscopy of low dimensional semiconductor" (Ankara, Sept. 9 – 20, 1996), G. Abstreiter, ed. by A. Aydinli, end J. P. Leburton (Kluwer Academic, Dordrecht, 1997), p. 127.

[35] H. L. Störmer, unpublished.

[36] Y.-C. Chang,L. L. Chang, and L. Esaki, Appl. Phys. Lett. **47**, 1324 (1985).

[37] G. Bastard, and J. Y. Marzin, Solid State Commun. **91**, 39 (1994).

[38] A. A. Kiselev, and U. Rössler, Semicond. Sci. Technol. **11**, 203 (1996).

[39] S. Glutsch, F. Bechstedt, W. Wegscheider, and G. Schedelbeck, Phys. Rev. B **56**, 4108 (1997); physica status solidi (a) **164**, 405 (1997).

[40] M. Grundmann and D. Bimberg, Phys. Rev. B **55**, 4054 (1997).

[41] R. D. Grober, T. D. Harris, J. K. Trautman, E. Betzig, W. Wegscheider, L. Pfeiffer, and K. West, Appl. Phys. Lett. **64**, 1421 (1993).

[42] P. Platzman, personal communication.

Raman Spectroscopy of Quantum Dots

Christian Schüller

Institut für Angewandte Physik und Zentrum für Mikrostrukturforschung, Universität Hamburg, Jungiusstraße 11, D-20355 Hamburg, Germany

Abstract: We review recent resonant Raman experiments on GaAs-AlGaAs quantum dots. We were able to observe simultaneously single-particle-like electronic excitations (SPE) and collective spin-density and charge-density excitations (SDE and CDE). This allows us to determine the influence of many-particle interactions in the quantum dots. Moreover, by a detailed analysis of wave-vector and magnetic-field dependent measurements, we can identify transitions between different electronic shells of the quasiatomic systems which predominantly contribute to the observed low-energy SDE's and resemble an atomic fine structure.

1 Introduction

The first experiments of inelastic light scattering by collective plasma oscillations (plasmons) of free electrons, coupled to LO phonons, in *n*-type bulk GaAs were performed by Mooradian and Wright in 1968 [1]. Later Mooradian also observed under resonant excitation, i.e., the laser frequency is in the vicinity of the optical $E_0 + \Delta$ energy gap of the semiconductor, excitations which – at that time – were interpreted as single-particle excitations [2]. According to the experimental findings, Hamilton and McWhorter deduced in their theoretical work that this single-particle scattering, which results from so called spin-density fluctuations, can be observed in zincblende-type semiconductors in depolarized scattering geometry, i.e., the polarization directions of incoming and scattered light are perpendicular to each other. Scattering by plasmons due to charge-density fluctuations is visible in parallel polarization configuration (polarized geometry) [3]. In contrast to light scattering by optical phonons, the electronic Raman signals are strongly dependent on resonance enhancement effects at optical energy gaps. In many cases only these enhancement effects, which occur if the laser frequency is in the vicinity of such energy gaps, allow for the observation of electronic excitations. In 1978, E. Burstein proposed that, due to these resonance enhancements, light scattering should be sensitive enough to observe electronic excitations of quasi-two-dimensional (2D) electron gases with densities as low as 10^{11} cm^{-2} [4]. Such 2D electron systems can be realized today in a nearly perfect way in modulation-doped GaAs-AlGaAs heterostructures or quantum wells, grown by molecular-beam epitaxy (MBE). Soon after this proposal, the first observations of 2D intersubband excitations were reported by A. Pinczuk [5] and G. Abstreiter [6] in their pioneering works. In the following decade a wealth of experiments on 2D electron systems followed which demonstrated the versatility of the resonant light scattering technique [7]. But through all the years it was commonly accepted that the electronic excitations, which can be observed by inelastic light scattering, fall into two main categories:

Spin-density excitations (SDE) which were interpreted as single-particle excitations because exchange-correlation effects were assumed to be small (observed in depolarized geometry) and charge-density excitations (CDE, plasmons) which can be observed in polarized geometry [7]. The latter are depolarization shifted with respect to the corresponding single-particle energy spacings due to Coulomb interactions. Very surprisingly, in contradiction to this long lasting assumption, Pinczuk et al. demonstrated in another pioneering work in 1989 that in high-mobility quantum well samples, *additionally* to the intersubband SDE and CDE, excitations can be observed with energies in between those of the SDE and CDE and which occur in *both* polarization configurations [8]. These excitations showed all features which one expects from pure single-particle excitations. This completely changed the point of view and from there on also SDE's were regarded as *collective* excitations of the electron gas, whereas the excitations which show no polarization selection rules were interpreted as *single-particle* excitations (SPE). Theoretical considerations by state of the art calculations in the local-density [9,10] or time-dependent Hartree Fock approximation [10] confirmed the experimentally observed *collective* shift of the SDE with repect to the SPE concerning the energetic positions. Nevertheless, there was so far no explanation for the occurence itself of single-particle-like excitations in the Raman spectra. The calculations of Raman spectra, which for simplicity were made almost throughout for *nonresonant* conditions, exhibited no single-particle peak because these excitations are naturally screened by the interaction [10]. More recent experimental investigations showed strong evidence that the *resonant* scattering process, which in the experiments is practically always present, plays a crucial role in the occurence of quasi unscreened SPE's in low-dimensional electron systems [11]. So far there was no theoretical publication at all available where all these effects – many-particle interactions and resonant scattering, i.e., inclusion of the valence band – were properly taken into account. Very recently, a theoretical work appeared which demonstrated that in a Luttinger liquid *resonant* scattering leads to different polarization selection rules [12].

The development of sophisticated structurizing techniques in the past decade allowed to reduce the dimensionality further and produce quasi-one-dimensional (1D) quantum wires and quasi-zero-dimensional (0D) quantum dots, starting from 2D systems. The quantum dots can be regarded as some kind of artificial atoms. In 1989 the first Raman experiments on electronic excitations in quantum wires were reported [13,14]. Since then a number of papers appeared about, e.g., many-particle interactions in those systems [11,15,16] and investigations with applied external magnetic field [17–19]. In the past few years also first Raman experiments on 0D quantum dots have been reported [11,20–22] In this article we review our recent Raman investigations of deep-etched modulation-doped GaAs-AlGaAs quantum dots.

2 Experimental Setup and Sample Characteristics

The experiments were performed using a tunable Ti:sapphire laser. For detection a triple Raman spectrometer and a charge-coupled-device camera was used. For measurements without magnetic field, the samples were mounted in a closed-cycle cryostat at $T = 12$ K. To apply an external magnetic field, an optical split-coil magnet was used where the samples were immersed in superfluid helium at $T = 2$ K. The wave-vector transfer q was varied by adjusting different directions of the incident and scattered beam with respect to the sample normal. In the closed-cycle cryostat conventional backscattering geometry

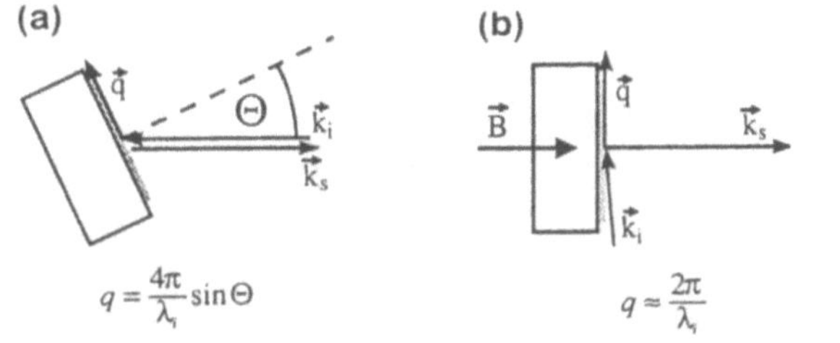

Figure 1
Sketch of the scattering geometry which was used in the experiments (a) without and (b) with magnetic field. $\vec{k}_i$ ($\vec{k}_s$) is the wave vector of the incident (scattered) light. $\vec{q}$ is the wave-vector transfer in lateral direction.

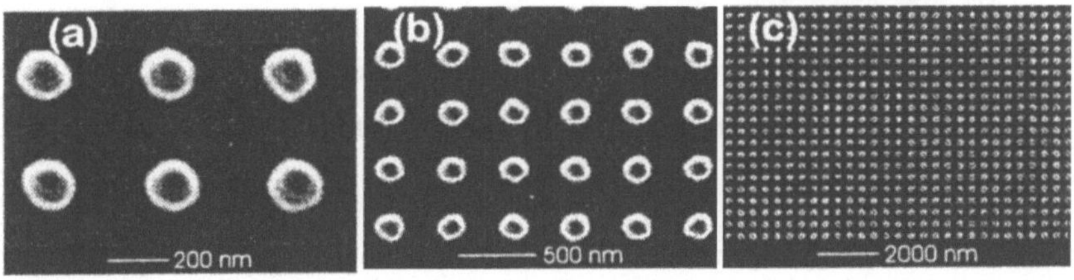

Figure 2 Scanning electron micrographs of deep-etched quantum dots with different magnification.

was used, as displayed in Fig. 1a. In this configuration, the wave-vectior transfer q in lateral direction is given by the expression which is written in Fig. 1a. For the experiments in the split-coil magnet, a glancing incidence of the laser beam onto the sample from a side window was used while the sample normal was oriented parallel to the magnetic-field direction. The detection direction in this configuration was also parallel to the sample normal (Fig. 1b). The corresponding expression for q is given in Fig. 1 (b).

The quantum-dot samples were prepared starting from one-sided modulation-doped GaAs-$Al_{0.3}Ga_{0.7}As$ single quantum wells. The low-temperature carrier densities and mobilities of the unstructured samples were in the range of $(7-8) \times 10^{11}$ cm^{-2} and 7×10^5 cm^{2}V^{-1} s^{-1}, respectively. Arrays of quantum dots have been defined by holographic lithography using the 458 nm line of an Argon laser. The pattern was transferred into the sample by a reactive-ion-deep-etching process, i.e., etching through the 2D electron gas of the quantum well. The lateral geometrical dot diameters and periods are in the range of $(170-250)$ nm and $(500-800)$ nm, respectively. For Raman experiments, i.e., under illumination with laser light, the lateral *electronic* diameters are typically $(30-40)$ nm smaller than the geometrical widths, due to side-wall depletion [23].

Figure 2 shows scanning electron micrographs of such deep-etched structures. By simple geometrical considerations, taking into account the 2D carrier densities and geometrical size of the dots, the number of electrons per dot can be estimated to be on the order of 100 in our samples.

Figure 3a shows a sketch of the cross section of a deep-etched quantum dot with circular symmetry. The external lateral potential, which acts in x and y directions on a test electron within the dot, is determined by the homogeneously distributed ionized remote donors in the AlGaAs barrier and negative surface charges. By analytical considerations [24], this external lateral potential can be estimated to be to a good approximation parabolic. This implies that the eigenenergies are [25]

$$E_{nm} = \hbar\Omega_0 \left(2n + |m| + 1\right), \tag{2.1}$$

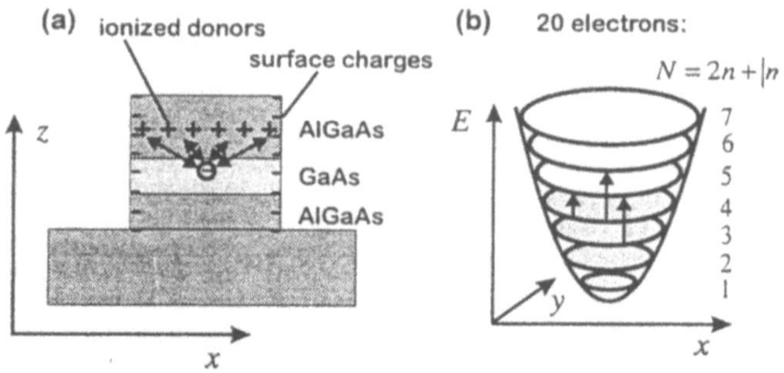

Figure 3 Schematic cross section of a deep-etched quantum dot. (b) Sketch of the two-dimensional parabolic potential in a quantum dot. The grey shaded circles indicate levels which are occupied with electrons. n is the radial and m the angular momentum quantum number.

which are the energy eigenvalues of a two-dimensional harmonic oscillator. Here $n = 0,1,2,...$ is the radial and $m = 0,\pm 1,\pm 2,...$ the azimuthal or angular momentum quantum number of the circularly symmetric structure, and Ω_0 is the quantization energy. The degeneracy of the energy levels, including spin degeneracy, is $2N$, where N is defined as $N = 2n + |m| + 1$. In a more atomistic language this means that electronic shells of the quasiatomic system are completely filled at *magic* electron numbers 2,6,12,20,....

By the crude assumption that the effective lateral potential is not changed by filling in a small number of electrons in the quantum dot, this results in, e.g., four occupied 0D levels for 20 free electrons within the dot. This situation is sketched in Fig. 3b. In reality, the screening behavior of the electrons results in a flattening of the effective *self-consistent* potential in the center of the dot, which means that the effective potential is not parabolic. In the extreme limit of a large number of electrons in the dot, the potential is more square-well-like. Neglecting for a moment many-particle interactions, the electronic elementary excitations, which can be observed in inelastic light scattering experiments, are transitions of single electrons from occupied to unoccupied states as sketched by vertical arrows in Fig. 3 (b). These transitions can in principle have changes in the lateral quantum number $\Delta N = 1,2,3,...$. Naturally, as in the case of quantum wells and wires, the Coulomb interaction will lead to the formation of *collective* excitations, CDE's and SDE's, where, normally, CDE's have larger and SDE's have smaller energies than pure single-particle transitions. To calculate these many-particle excitations, one has to sum over all possible single-particle transitions including the interaction. Recently, we were able to observe all types of electronic elementary excitations – CDE's, SDE's, and SPE's – simultaneously in one and the same quantum-dot sample [11].

3 Electronic Raman Scattering in Quantum Dots

3.1 Scattering Mechanisms

The inelastic light scattering by electronic excitations in semiconductor microstructures with relatively small numbers of electrons can be observed only under specific interband resonance conditions. That means that in the scattering process valence band states are involved as intermediate states. Nevertheless, most of the theoretical work concerning

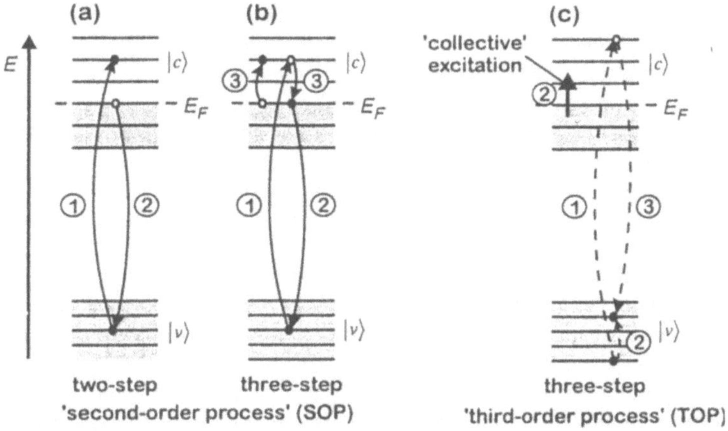

Figure 4 Schematic picture of the (a) two-step and (b) three-step second-order scattering mechanisms, and (c) the three-step third-order excitonic scattering mechanism for inelastic light scattering by electronic excitations in quantum dots.

the dynamic Raman response in microstructures is performed for nonresonant conditions, neglecting the valence band structure. We summarize here the dominant *resonant* scattering processes, which are fragmentarily present in the literature, and which we find in our experiments on quantum wells, wires and dots to be the dominant ones.

The coupling of the radiation with the electron system is maintained by replacing the momentum $\vec{p}$ of the electron by $\vec{p}+e\vec{A}$ in the Hamiltonian H_0 of the unperturbed system. $\vec{A}$ is the vector potential of the electromagnetic field.

$$\begin{aligned} H &= \frac{1}{2m}\sum_i\left[\left(\vec{p}_i+e\vec{A}(\vec{r}_i)\right)^2+U(\vec{r}_i)\right]+\hat{V}_{e-e}+\hat{V}_{e-ph}= \\ &= H_0+\frac{1}{2m}\sum_i\left[\left(\vec{A}(\vec{r}_i)\right)^2+\vec{p}_i\vec{A}(\vec{r}_i)+\vec{A}(\vec{r}_i)\vec{p}_i\right] \end{aligned} \tag{3.2}$$

$\hat{V}_{e-e}$ is the Coulomb interaction, $\hat{V}_{e-ph}$ the electron-phonon interaction, and U includes the lattice-periodic potential as well as all types of external potentials. For simplicity, the spin-orbit coupling is neglected in (3.2). In a perturbation theory, the last three terms in the second line of (3.2) can be treated as a perturbation. Here the $\vec{A}^2$ terms give contributions to the light scattering cross section in *first* order, the $\vec{p}\vec{A}$ terms in *second* order perturbation theory. The $\vec{p}\vec{A}$ terms together with, either the Coulomb interaction $\hat{U}_{e-e}$ of a photoexcited exciton with the Fermi sea, or the electron-phonon interaction $\hat{U}_{e-ph}$ of a photoexcited exciton with the phonon bath, give contributions in *third* order perturbation theory. $\hat{U}_{e-ph}$ leads to the scattering by phonons which shall not be discussed here. The second- (SOP) and third-order light scattering processes (TOP) appear to be the dominant ones in electronic Raman scattering on microstructures because they exhibit a resonant behavior [26]. In Fig. 4, transitions which contribute to the second (Fig. 4a,b) and third-order (Fig. 4c) processes in quantum dots are sketched. Valence-band levels are marked by $|v\rangle$, and conduction-band levels by $|c\rangle$.

In the first step of the two-step SOP, an electron is excited from a valence-band level to a conduction-band state. In the second step, an electron, either with the same or with opposite spin as the photo-excited electron, recombines with the hole in the valence band. Thus, as a net effect, either a nonspinflip or spinflip single-particle excitation has been created in the conduction band. In the third step of the three-step SOP, another SPE is created by Coulomb interaction. This represents the screening of the SPE by other SPE's in the system. This screening also leads to the formation of collective CDE's and SDE's by direct and exchange Coulomb coupling. For nonresonant conditions, the SPE's are screened by the interaction. This is formally implemented by the cancellation of the two-step SOP contribution by the three-step SOP in the scattering cross section. We find in our experiments for all our samples that under conditions of extreme resonance, which means that the laser frequency is in the vicinity of the fundamental bandgap of the underlying 2D structure, quasi unscreened SPE's [29] occur and they show stronger resonance enhancements than the collective excitations [11]. This means that the above mentioned cancellation of terms does not hold in the case of extreme resonance and thus single-particle-like excitations can be observed. A quite similar behavior was previously reported for intra-band excitations in *n*-type GaAs bulk samples [30]. For laser frequencies well above the bandgap, an excitonic third-order scattering process, as for the case of quantum dots displayed in Fig. 4c, has appeared to be dominant (for 2D excitations, see [27,28]). In this scattering mechanism, which is assisted by the Coulomb interaction, in the first step the incident photon creates an exciton with a hole in the valence band and an electron in a higher conduction-band level. This exciton is scattered by direct and exchange Coulomb interaction with the Fermi sea into another state, where for example, as drawn in Fig. 4c, the hole is scattered into another state. By this scattering process, a collective excitation (SDE or CDE) is created in the conduction-band system. In the third step, the scattered exciton recombines. This scattering process exhibits very sharp and intense resonance profiles in a plot of the scattered intensity versus laser frequency [27]. We find that this TOP is the dominant scattering mechanism for excitation of collective SDE's and CDE's in all our low-dimensional samples (see below).

In the following we want to elucidate these scattering processes in some more detail. The scattering cross section is given by [3]

$$\frac{d^2\sigma}{d\Omega d\omega} = \frac{\omega_S}{\omega_I}\frac{e^4}{c^4 m^4} S(\omega)\,, \tag{3.3}$$

where the structure factor $S(\omega)$ is defined as

$$S(\omega) = \sum_F |\langle F|\hat{V}_{\text{eff}}|I\rangle|^2\,\delta(E_F - E_I - \omega)\,. \tag{3.4}$$

ω_I (ω_S) is the frequency of the incident (scattered) photon, and $\omega = \omega_I - \omega_S$ the energy transfer. $\hat{V}_{eff}$ is the effective operator, which describes the transition of the system from the many-particle initial state $|I\rangle$ with energy E_I to the final state $|F\rangle$ with energy E_F. Using the second-quantization technique, this effective operator of the light scattering $\hat{V}_{\text{eff}}$ can in a many-particle system be expanded in terms of creation ($\hat{c}^\dagger$) and annihilation ($\hat{c}$) operators of single-electron states [28]

$$\hat{V}_{\text{eff}} = \sum_{\alpha,\beta} \gamma_{\alpha\beta}\hat{c}^\dagger_\beta \hat{c}_\alpha\,. \tag{3.5}$$

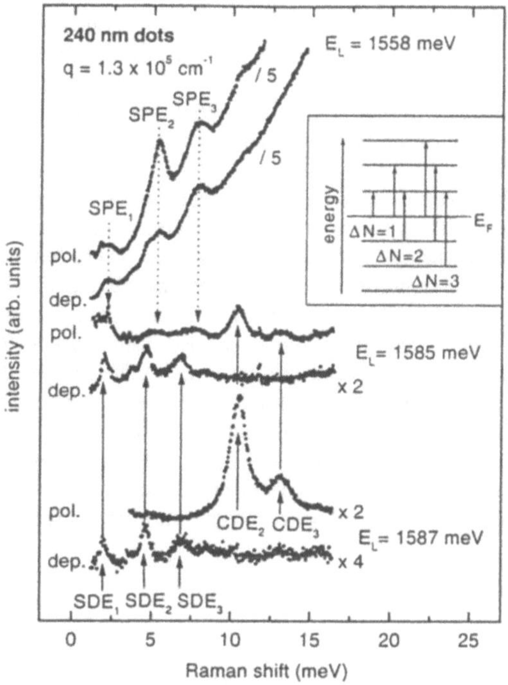

Figure 5
Depolarized and polarized Raman spectra of electronic excitations in a quantum-dot sample for different laser frequencies E_L. The background in the upper spectra is due to hot luminescence.

Here each α and β represents a set of quantum numbers, (n,m,σ), for, respectively, the radial, azimuthal, and spin quantum numbers of the corresponding state. If we treat for a moment the excitons as simple electron-hole pairs without interaction, which is for a quantitative analysis certainly not correct, the scattering amplitudes $\gamma_{\alpha\beta}$ can approximately be written as [31,28]

$$\gamma_{\alpha\beta} \sim \langle\alpha|e^{i\vec{q}\vec{r}}|\beta\rangle\vec{e}_I\vec{e}_S + \frac{1}{m}\sum_{\beta'}\frac{\langle\alpha|\vec{p}\vec{A}_S|\beta'\rangle\langle\beta'|\vec{p}\vec{A}_I|\beta\rangle}{E_\beta - E_{\beta'} + \hbar\omega_I} + \\ + \sum_{\nu\nu'}\frac{\langle\beta|\vec{p}\vec{A}_S|\nu'\rangle\langle\nu'|\hat{U}_{e-e}|\nu\rangle\langle|\vec{p}\vec{A}_I|\alpha\rangle}{\left(E_\beta - E_{\nu'} - \hbar\omega_S\right)\left(E_\alpha - E_\nu - \hbar\omega_I\right)}, \quad (3.6)$$

where in the last two terms we have written only the strongest resonant term. The first term in (3.6) represents the nonresonant contributions from the $\vec{A}^2$ terms in first order perturbation theory. Here the scalar product of the polarization vectors of the incident ($\vec{e}_I$) and scattered ($\vec{e}_S$) photon shows that this nonresonant term yields only scattering by plasmons (CDE's). Spinflip processes, which are necessary for the excitation of SDE's, are only possible in the resonant second and third term. The second term describes second-order processes as schematically shown in Fig. 4a, and the third term represents third-order resonant scattering as displayed in Fig. 4c.

Figure 5 shows Raman spectra of electronic excitations in quantum dots with 240 nm geometrical diameter. The spectra in the lower part were recorded at laser energies well above the effective bandgap. In the lowest spectrum at $E_L = 1587$ meV, three peaks can be observed which can be identified as SDE's due to polarization selection rules. Correspondingly, the peaks in the polarized spectrum can be identified as CDE's. We find that especially the CDE's exhibit a very sharp ($FWHM \approx 3.5$ meV) resonance profile in this range of laser frequencies which leads us to the interpretation that this scattering by *collective* excitations is caused by the third-order TOP as discussed before.

If the laser frequency is slightly lowered (spectra at $E_L = 1585$ meV in Fig. 5), additional broad features appear, especially in the polarized spectrum, which then slowly evolve into intense peaks, by further lowering E_L towards the bandgap. These features are marked in Fig. 5 by dotted arrows. Under conditions of extreme resonance (e.g. $E_L = 1558$ meV in Fig. 5), these peaks dominate the spectra and are present in both polarization configurations. We therefore interpret them within the above described framework as SPE's, which are caused by a SOP under conditions of extreme resonance. The inset in Fig. 5 shows a sketch of the discrete level scheme in a quantum dot. The vertical arrows mark some possible single-particle transitions. Here ΔN is the change in the lateral quantum number as defined above. The index i of the labels SPE_i in Fig. 5 gives the corresponding ΔN of the observed SPE's. From the experimental spectra in Fig. 5 we can deduce for the corresponding collective SDE's and CDE's directly the energy renormalizations due to many-particle interactions.

3.2 Many-Particle Interactions

We have measured large series of spectra for many different laser frequencies and wavevector transfer q. From these spectra – some exemplary ones are plotted in Fig. 5 in section 3.1 – we can deduce the energy shifts of the collective excitations with respect to the SPE's. This gives us the possibility to determine the relative strengths of the direct and exchange Coulomb interactions in the quasiatomic system [11]. Figure 6 displays the energetic positions of the observed electronic excitations versus q.

The first important fact that we can extract from Fig. 6 is that the energies do not depend on q. This is a clear indication for a 0D behavior. For isolated quantum dots, q is

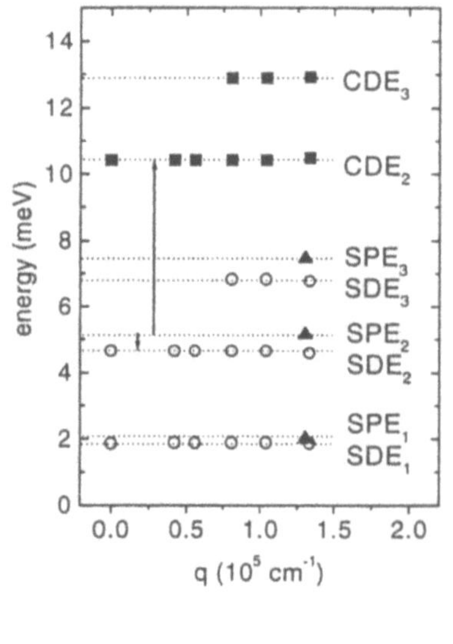

Figure 6
Energies of electronic excitations in a 240 nm quantum-dot sample versus transferred lateral wave vector q. From Ref. [11]

not a good quantum number because the system has no translation symmetry in the x-y-plane. We will discuss the meaning and consequences of the wave-vector transfer q in our experiments on quantum dots in detail below in Section 4. From Fig. 6 we can also see that the SDE's are slightly shifted to lower energies, and the CDE's are shifted to higher energies with respect to the corresponding SPE's. For clarity, vertical arrows indicate the energy renormalizations of the SDE_2 and CDE_2 with respect to the SPE_2. By comparing the energy renormalizations, we find the ratio $(E_{SPE_i} - E_{SDE_i})/(E_{CDE_i} - E_{SPE_i}) \approx 0.1$. This ratio seems to increase with increasing i.

4 Quasiatomic Fine Structure in Quantum Dot Atoms

So far, the observed excitations have been interpreted in terms of a simple level scheme, where the individual levels are characterized by a lateral quantum number N. But what we are looking at here, is some kind of an artificial atom, which, in analogy to real 3D atoms, is characterized by electronic shells and distinct quasiatomic orbitals. In order to highlight this internal structure, which should be inherently present in our spectra of elementary electronic excitations, we have done careful further studies of the wave-vector and especially also magnetic-field dependence of the observed excitations [22].

Figure 7 shows depolarized (Fig. 7a) and polarized (Fig. 7b) spectra for different wave-vector transfers q, which were recorded at a laser energy where dominantly collective excitations are excited due to the TOP. For large q, the SDE's and CDE's, as discussed in the previous sections, are visible. In the spectra with the smallest q, only one peak can be observed for each scattering geometry (the small peak at 15 meV in the polarized spectrum is due to hot luminescence). This confirms our interpretation that these peaks are the SDE_2 and CDE_2: The Raman process is a two-photon process, which has as a consequence that in symmetric systems excitations with even parity are Raman allowed. This is in contrast to direct absorption, which is a one-photon process. Therefore, we conclude that in our axially symmetric dots the SDE_2 and CDE_2, which have even parity, are Raman-allowed excitations. By increasing the wave-vector transfer q, the symmetries of the excitations are broken and also the Raman-forbidden excitations with odd parity appear. A remarkable fact, which we can not explain at the moment, is that this effect is much stronger for the SDE's than for the CDE's. Furthermore, an additional peak occurs at the low-energy side of the SDE_2 for large q. This peak can be resonantly enhanced by slightly changing the laser frequency (spectrum with an asterisk in Fig. 7a). We emphasize that the SDE_1, SDE_2 and the additional peak have essentially the same linewidth (≈ 0.5 meV) whereas the SDE_3 has more than twice the linewidth, which means that it probably represents a superposition of two or more excitations, which is not resolved. Furthermore, we can see in Fig. 7a that the additional peak has almost exactly twice the energy of the SDE_1. In order to learn something about the internal structure of these excitations in the quasiatomic system, we analyze our observations in a simple model: As we have learned from the investigations in Section 3.2, exchange-correlation effects are small in our structures so that we can neglect for a moment many-particle interactions and treat the SDE's as simple single-particle transitions. As a first approach, we also start with a two-dimensional parabolic potential for the effective lateral potential of our quantum dot. The eigenenergies of this two-dimensional harmonic oscillator were given in (2.1). The corresponding energy spectrum is displayed in Fig. 8a. Each filled square in Fig. 8 represents an allowed energy state, which can be occupied by two electrons due to

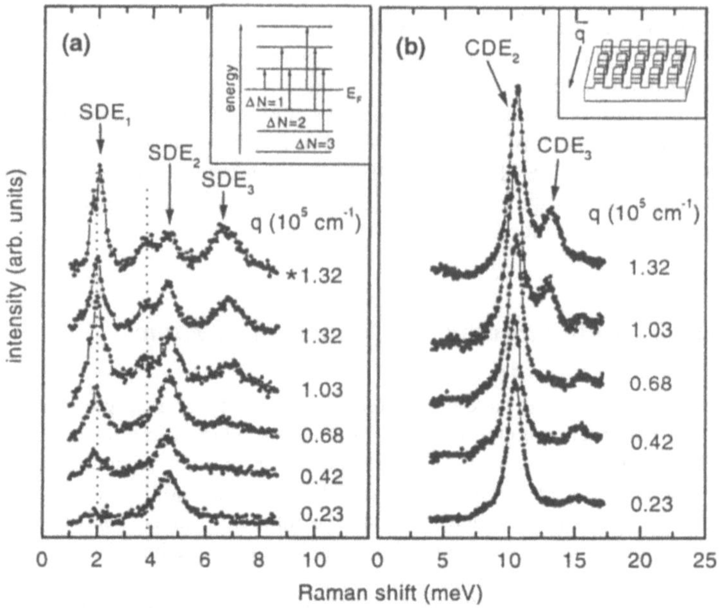

Figure 7 (a) Depolarized and (b) polarized Raman spectra of a quantum-dot sample with 240 nm dot diameter for different wave-vector transfer q. The depolarized spectra in (a) were recorded at a laser energy $E_L = 1600.7$ meV, the spectrum marked with (*) at $E_L = 1601.3$ meV. The insets sketch (a) those single-particle transitions which predominantly contribute to the observed excitations, and (b) the direction of wave-vector transfer. From Ref. [21]

spin degeneracy. Here we have considered, for simplicity, 42 electrons per dot, which makes no qualitative difference in our interpretation. The solid lines connect points which belong to the same radial quantum number n, where $n = 0,1,2,...$ rises from the bottom to the top. All points which can be connected by a horizontal line in Fig. 8a belong to the same lateral quantum number $N = (2n + |m| + 1)$ and represent the discrete levels, which were schematically shown in the inset of Fig. 5. We have marked some states with $'s'$, $'p'$, etc., to indicate the character of the quasiatomic orbitals as, e.g., quasi-s or quasi-p orbitals. Actually, these orbitals have not the same symmetry as orbitals of real atoms, therefore we call them *quasi*. In a classical picture, the electrons with finite angular momentum quantum number m move on circular orbits within the dot, where the radius R is proportional to the square root of $|m|$ as shown in the left part of Fig. 9. With increasing radial quantum number n, the set of orbits increases in energy as can be seen in the right part of Fig. 9. The circular movement of the electrons represents a current which produces a magnetic moment $\vec{m}$, perpendicular to the plane in which the current flows. This magnetic moment can have opposite directions depending on the sign of the quantum number m. If we now apply a magnetic field $\vec{B}$ perpendicular to the x-y-plane, $\vec{B}$ can be oriented either parallel or antiparallel to $\vec{m}$. This has as a consequence that states with $\vec{m}$ parallel to $\vec{B}$ decrease in energy, while states with $\vec{m}$ antiparallel to $\vec{B}$ increase in energy. This is shown in Fig. 8b, where a finite magnetic field has been considered. In general, the energy spectrum in a parabolic quantum dot in dependence on a perpendicular magnetic field is given by

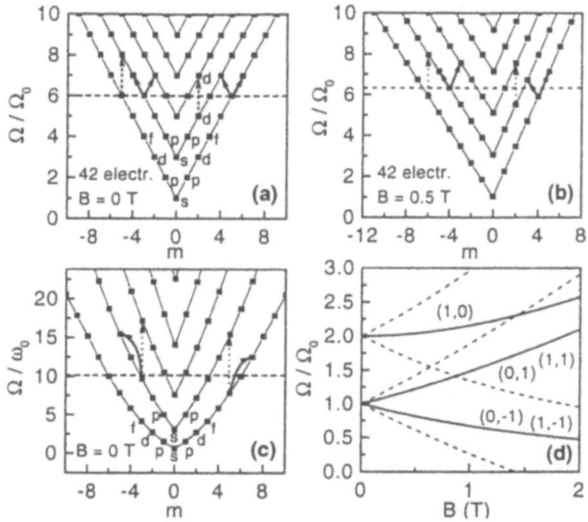

Figure 8 Calculated energies for (a) a quantum dot with parabolic effective potential without and (b) with external magnetic field. (c) calculated energies in a dot with cylindrical symmetry and hard walls. (d) single-particle transition energies for a parabolic quantum dot with external magnetic field. The solid lines mark the experimentally observed transitions, the dashed lines give the forbidden next higher transitions. For the two dashed lines starting at $\Omega/\Omega_0 = 1$ the corresponding transitions are $(-1, \pm 3)$ and $(2, \pm 3)$, for the dashed lines starting at $\Omega/\Omega_0 = 2$, $(0, \pm 2)$ and $(2, \pm 2)$. In a hard-wall potential, the $(0, \pm 2)$ and $(2, \pm 2)$ are shifted to lower energies, which results in a fine structure. Ω_0 is the quantization energy of the parabolic potential, and $\omega_0 = \hbar/(2m^*)(\pi/a)^2$. a is the dot diameter in the case of the hard-wall potential. [From Ref. 21]

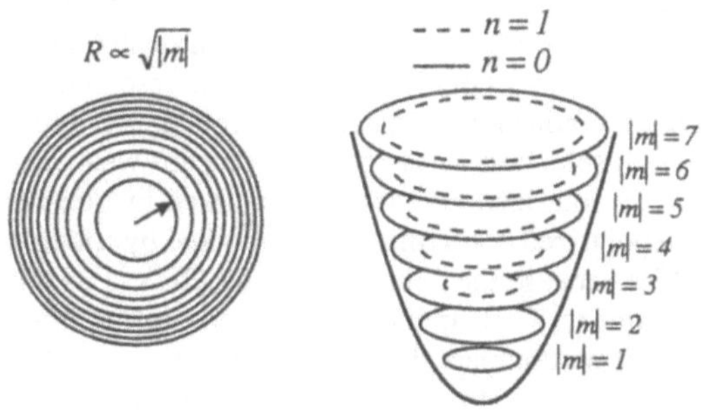

Figure 9
Sketch of the classical electron orbits in a quantum dot with parabolic lateral potential.

$$E_{mn}(B) = (2n + |m| + 1)\hbar\sqrt{\Omega_0^2 + \left(\frac{\omega_c}{2}\right)^2} + m\frac{\omega_c}{2}, \tag{4.7}$$

where $\omega_c = eB/m^*$ is the cyclotron resonance frequency [25]. Within the model described so far, the SDE_l consists of transitions from the highest occupied to the first unoccupied level in Fig. 7a, i.e., $\Delta N = 1$. In principle several such transitions are possible with different changes in radial and azimuthal quantum numbers. In Fig. 7a only some transitions with $(\Delta n = 0, \Delta m = \pm 1)$ and $(\Delta n = 1, \Delta m = \pm 1)$ are indicated by solid arrows. We claim that only these transitions predominantly contribute to the observed SDE_l. This can be

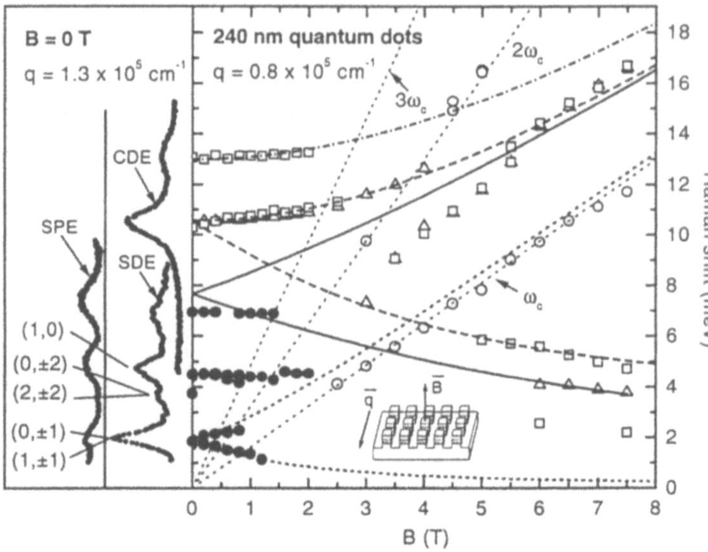

Figure 10 Magnetic field dispersions of the observed modes in a quantum dot sample. In the left panel spectra of SDE's and CDE's for $B = 0$ are displayed. The spectrum of SPE's was recorded at a laser energy $E_L = 1561$ meV under conditions of extreme resonance. The spectra of SDE's and CDE's were taken at $E_L = 1587$ meV. The numbers in the brackets, $(\Delta n, \Delta m)$, give the changes in radial (Δn) and azimuthal (Δm) quantum numbers for the observed transitions. [From Ref. 21]

deduced from magnetic-field dependent measurements: Figure 10 displays the magnetic-field dependence of the SDE's (solid symbols) and the CDE's (open symbols). In the left panel of Fig. 10, characteristic spectra of SDE's and CDE's, at $B = 0$ and large wave-vector transfer q, are displayed for illustration. For comparison we have also included a spectrum, which is dominated by SPE's under conditions of extreme resonance, where in this spectrum we have substracted the hot-luminescence background. We find that in a magnetic field the usual polarization selection rules for plasmons are weakened. Excitations which are most prominent in polarized geometry are marked by squares, and excitations which are dominant in depolarized configuration are marked by triangles. We want to concentrate here on the SDE's because the behavior of the CDE's is well known from far-infrared transmission experiments [32]. In Fig. 10 we can see that the SDE_1 mode splits with finite magnetic field. If we compare this splitting with the expected splitting of the transitions $(\Delta n = 0, \Delta m = \pm 1)$ and $(\Delta n = 1, \Delta m = \pm 1)$, calculated with (4.7) (Fig. 8 (d)), this gives the correct result. Also the dotted lines starting at $E = 2$ meV and $B = 0$ in Fig. 10 are calculated with (4.7) for the above given $(\Delta n, \Delta m)$. Transitions with $\Delta m \geq 2$ would result in larger splittings (see, e.g., dashed lines in Fig. 8d), which are experimentally not observed. Within this parabolic model, the splitting between the two branches with $\Delta n, |\Delta m| \leq 1$ is exactly ω_c. The slight deviation of the upper branch from the theoretical curve can be due to a nonparabolicity of the effective potential or to the remaining collective character of the SDE's. The collective nature of the SDE's furthermore manifests in the vanishing of the excitations at relatively small B. This is an effect of Landau damping: At finite B, all degeneracies are lifted and a wealth of single-particle lines cross the energies of the SDE's so that they get Landau damped. For illustration, the forbidden next higher transitions are displayed in Fig. 8d (dashed lines).

Also for the SDE_2 in principle several transitions are possible as can be seen in Fig. 8a. The experiments show neither a splitting nor a shift with magnetic field. Among all possible transitions, only transitions with ($\Delta n = 1, \Delta m = 0$), as indicated by dotted arrows in Fig. 8a, show such characteristics (see also Fig. 8d). Transitions with $|\Delta m| \geq 2$ would exhibit splittings $\geq 2\omega_c$ (see also dotted lines in Fig. 8d). We note here that due to the scattering geometry, which we use in the magnet cryostat (see Fig. 1b), we are restricted to a wave-vector transfer $q \approx 0.8 \times 10^5$ cm^{-1}. At this relatively small q we can not follow all peaks, especially the SPE's and the additional peak in the spectrum of SDE's, in a magnetic field. Nevertheless, from the energetic position of the additional peak, which lies below the position of the SDE_2, we can deduce that it consists of transitions with $\Delta n = 0,2$ and $\Delta m = \pm 2$, although these transitions are in a parabolic potential energetically degenerate with the transitions ($\Delta n = 1, \Delta m = 0$) (dashed arrows in Fig. 8a): From geometrical considerations we can estimate about 200 electrons per dot in our structures. At this relatively large number of electrons screening is important and the real effective potential would certainly deviate from a parabolic shape. As an extreme limit we have calculated the eigenenergies for a square-well potential with cylindrical symmetry. The real dot potential should be in between these limits. The corresponding energy spectrum is shown in Fig. 8c. Indeed, from Fig. 8c we can see that transitions with $|\Delta m| = 2$ (solid arrows) have lower energies than transitions with ($\Delta n = 1, \Delta m = 0$) (dotted arrows). Furthermore, transitions with $|\Delta m| = 2$ have almost exactly twice the energy of transitions with $|\Delta m| = 1$, if we move along lines with constant n in Fig. 8d. This is observed in experiment as can be seen by the two vertical dotted lines in Fig. 7a.

5 Conclusions

We have observed all types of electronic elementary excitations – SDE's, SPE's, and CDE's – in one and the same quantum-dot sample. This allows us to estimate the relative strengths of direct and exchange Coulomb interactions. Compared to the energy renormalizations of the CDE's, we find that SDE's are shifted only by about 10% due to exchange-correlation effects. The dependence of the scattered intensities on laser frequency shows that the collective excitations are created dominantly by a third-order excitonic scattering process, and the SPE's occur due to a second-order process under conditions of extreme resonance. Furthermore, investigations with applied magnetic field and varying wave-vector transfer q allowed us to identify transitions between different electronic shells of the quasiatomic systems which predominantly contribute to the observed low-energy SDE's. We find that these excitations are formed by transitions with $\Delta n, |\Delta m| \leq 1$, whereas transitions with $\Delta n, |\Delta m| \leq 2$ can only be observed at relatively large q and form a fine structure.

Acknowledgements

The experiments reviewed here were only possible due to an excellent cooperation with many colleagues, as listed in the references. I like to thank Detlef Heitmann, Gernot Biese, Christoph Steinebach, Edzard Ulrichs, Katharina Keller, Lucia Rolf, Silke Bargstädt-Franke, Christian Heyn, Vidar Gudmundsson, and Karl Eberl. This work was supported by the Deutsche Forschungsgemeinschaft via projects He1938/6 and He1938/7 and the Graduiertenkolleg 'Physik nanostrukturierter Festkörper'.

Bibliography

[1] A. Mooradian and G. B. Wright, Phys. Rev. Lett. **16**, 999 (1966).

[2] A. Mooradian, Phys. Rev. Lett. **20**, 1102 (1968).

[3] D. Hamilton and A. L. McWhorter in: *Light Scattering Spectra of Solids*, ed. G. B. Wright (Springer, New York, 1969), p. 309.

[4] E. Burstein, A. Pinczuk, and S. Buchner in: *Physics of Semiconductors*, ed. B. L. H. Wilson (The Institute of Physics, London, 1979), p. 1231.

[5] A. Pinczuk, H. L. Störmer, R. Dingle, J. M. Worlock, W. Wiegmann, and A. C. Gossard, Solid State Commun. **32**, 1001 (1979).

[6] G. Abstreiter and K. Ploog, Phys. Rev. Lett. **42**, 1308 (1979).

[7] For an overview see: A. Pinczuk and G. Abstreiter in: *Light Scattering in Solids V*, Topics in Applied Physics Vol. 66, eds. M. Cardona and G. Güntherodt (Springer, Berlin, 1988) p. 153.

[8] A. Pinczuk, S. Schmitt-Rink, G. Danan, J. P. Valladares, L. N. Pfeiffer, and K. W. West, Phys. Rev. Lett. **63**, 1633 (1989).

[9] T. Ando and S. Katayama, J. Phys. Soc. Jpn. **54**, 1615, (1985).

[10] M. S.-C. Luo, S. L. Chuang, S. Schmitt-Rink, and A. Pinczuk, Phys. Rev. B **48**, 11086 (1993).

[11] C. Schüller, G. Biese, K. Keller, C. Steinebach, D. Heitmann, P. Grambow, and K. Eberl, Phys. Rev. B **54**, R17304 (1996).

[12] M. Sassetti and B. Kramer, Phys. Rev. Lett. **80**, 1485 (1998).

[13] J. S. Weiner, G. Danan, A. Pinczuk, J. Valladares, L. N. Pfeiffer, and K. W. West, Phys. Rev. Lett. **63**, 1641 (1989).

[14] T. Egeler, G. Abstreiter, G. Weimann, T. Demel, D. Heitmann, P. Grambow, and W. Schlapp, Phys. Rev. Lett. **65**, 1804 (1990).

[15] A. R. Goñi, A. Pinczuk, J. S. Weiner, J. S. Calleja, B. S. Dennis, L. N. Pfeiffer, and K. W. West, Phys. Rev. Lett. **67**, 3298 (1991).

[16] A. Schmeller, A. R. Goñi, A. Pinczuk, J. S. Weiner, J. S. Calleja, B. S. Dennis, L. N. Pfeiffer, and K. W. West, Phys. Rev. B **49**, 14778 (1994).

[17] A. R. Goñi, A. Pinczuk, J. S. Weiner, B. S. Dennis, L. N. Pfeiffer, and K. W. West, Phys. Rev. Lett. **67**, 1151 (1993).

[18] C. Steinebach, R. Krahne, G. Biese, C. Schüller, D. Heitmann, and K. Eberl, Phys. Rev. B **54**, R14281 (1996).

[19] E. Ulrichs, G. Biese, C. Steinebach, C. Schüller, D. Heitmann, and K. Eberl, Phys. Rev. B **56**, R (1997).

[20] R. Strenz, U. Bockelmann, F. Hirler, G. Abstreiter, G. Böhm, and G. Weimann, Phys. Rev. Lett. **73**, 3022 (1994).

[21] D. J. Lockwood, P. Hawrylak, P. D. Wang, C. M. Sotomayor Torres, A. Pinczuk, and B. S. Dennis, Phys. Rev. Lett. **77**, 354 (1996).

[22] C. Schüller, K. Keller, G. Biese E. Ulrichs, L. Rolf, C. Steinebach, D. Heitmann, and K. Eberl, Phys. Rev. Lett. **80**, 2673 (1998).

[23] T. Demel, D. Heitmann, P. Grambow, and K. Ploog, *Phys. Rev. Lett.* **64**, 788 (1990), F. Perez, B. Jusserand, S. Zanier, Y. Guldner, and B. Etienne, Physica E (to be published).

[24] U. Merkt, Physica B **189**, 165 (1993).

[25] V. Fock, Z. Phys. **47**, 446 (1928).

[26] For scattering by 2D intra- and intersubband excitations see, e.g., for the SOP Refs. [4,7], for the TOP Refs. [27,28].

[27] G. Danan, A. Pinczuk, J. P. Valladares, L. N. Pfeiffer, K. W. West, and C. W. Tu, Phys. Rev. B **39**, 5512 (1989).

[28] A. O. Govorov, J. Phys. Condens. Matter **9**, 4681 (1997).

[29] Within the approximations applied so far in the described theory, these excitations are *single-particle* excitations. Nevertheless, we believe that in a many-particle system all excitations must be *many-particle* excitations. Thus, a more rigorous theoretical treatment should prove that for these excitations the collective effects may be small but that they are still present.

[30] A. Pinczuk, L. Brillson, and E. Burstein, Phys. Rev. Lett. **27**, 317 (1971).

[31] F. A. Blum, Phys. Rev. B **1**, 1125 (1970).

[32] see D. Heitmann and J. P. Kotthaus in *Physics Today*, June 1993, p. 56, and references therein.

Charged Excitons in InAs Quantum Dots

R.J. Warburton[1], C. Bödefeld[1], C.S. Dürr[1], K. Karrai[1], J.P. Kotthaus[1], G. Medeiros-Ribeiro[2], and P.M. Petroff[2]

[1] Sektion Physik der Ludwig-Maximilians-Universität, Geschwister-Scholl-Platz 1, 80539 München, Germany
[2] Center for Quantized Electronic Structures (QUEST) and Materials Department, University of California, Santa Barbara, California 93106

Abstract: We report studies of charged excitons in InAs quantum dots. The self-assembled dots are embedded in a field effect device so that the dots can be loaded with electrons by applying a bias between a metal gate and back contact. The dots are homogeneous enough for an ensemble of dots to show Coulomb blockade in the vertical transport. We have measured the optical activity of excitons by detecting minute changes in the transmission of light through the sample. We show how Coulomb blockade in the transport is related to Pauli blocking in the absorption. Furthermore, the higher excitations exhibit substantial shifts on occupation of the electron ground state. This is a consequence of carrier-carrier interactions.

1 Introduction

Quantum dots are small regions in a semiconductor crystal which confine carriers in all three directions. With lithographic techniques, it is possible to create highly tunable systems which exhibit single-electron tunneling in the transport properties [1,2]. These dots are typically hundreds of nm in size which means that the quantization energies are small. Although quantum mechanical effects can be seen, the physics is dominated by classical charging effects. Recently, it has been possible to measure single electron transport on smaller devices by defining and contacting a small mesa in a vertical tunneling structure [3]. In this case, the quantization energy is comparable to the charging energy. The addition energies in the transport show strong quantum mechanical effects which can be explained by considering the quantization in the dot's confinement potential and the direct and exchange Coulomb interactions [4]. It would be of great interest to study also excitons in such a charge-tunable system, measuring how the excitonic properties depend on electron occupation. Generally speaking, a variety of splittings and changes in oscillator strengths are expected in the optical properties of charged dots [5]. With the advent of near-field techniques it is perhaps possible to measure the excitonic response of a single charge-tunable dot. However, these experiments are difficult, and it is more straightforward to measure the response of an ensemble of dots. In this case, the challenge is to manufacture dots which are uniform enough so that the electron occupancy is the same in each dot.

Self-assembled dots have much to offer this field. First, they are much smaller than lithographically defined dots, leading to much larger quantization energies. Secondly, they are extremely homogeneous with dot to dot fluctuations of only $\sim 10\%$ in the lateral size [6]. Such dots can be located in a metal-insulator-semiconductor structure which enables the dots to be loaded sequentially with electrons [7,8]. We report here a summary of experiments [9] on these charge-tunable self-assembled quantum dots by measuring the absorption of light in the near infrared.

2 Samples and Experimental Details

The wafer was grown using the Stranski-Krastanow growth mode, as described elsewhere [6]. The InAs dots are about 20 nm in diameter with a height of about 7 nm. The dots are separated by 25 nm of intrinsic GaAs from a back contact, a highly doped layer of GaAs. This layer acts as a tunneling barrier. A blocking barrier, a GaAs/AlAs superlattice, prevents tunneling to the surface. We evaporate a semi-transparent gate onto the sample surface. The electron occupancy of the dots, N, can be controlled by applying a voltage, V_g, between gate and back contact. The capacitance, C, between gate and back contact can be used to determine N [7,8,10,11]. Whenever tunneling can take place between the back contact and the dots a rise in the device's capacitance occurs. The $C - V_g$ profile for the sample used here is shown in Fig. 1. The doublet structure around $V_g \sim -0.75$ V corresponds to tunneling of electrons into the first dot level, s. The splitting arises from Coulomb repulsion: the loading of the second electron into a given dot is hindered by the presence of the first. The important point is that although we measure almost 10^8 dots, the inhomogeneous broadening is small enough for such Coulomb effects to be observed. For $V_g < -1.2$ V, all the dots are empty; for $V_g = -0.75$ V, we estimate that at least 80% of the dots are singly occupied; and for $V_g = -0.4$ V to all intents and purposes all the dots are doubly occupied. On further increasing V_g to ~ 0 V, tunneling into the second electron state, the p state, takes place. In this case, we cannot resolve the individual charging peaks. However, at $V_g \sim 0.2$ V each dot is completely filled with six electrons. Further increases in V_g simply cause the wetting layer to be occupied.

We measure the transmission of the device at wavelengths around 1 μm with a Fourier transform spectrometer. The change in transmission at resonance is very small, $\sim 0.01\%$. In order to detect such small changes one must achieve a very high signal:noise ratio, and one must be able to distinguish dot absorption from artefacts and drifts in the spectral response of the setup. We achieved very good signal:noise by using a Ge p-i-n diode mounted behind the sample (both at 4.2 K) and a commercial current amplifier to detect the light. This combination gave very good results, limited in fact for our $\sim \mu$W light intensities only by the dynamic range of the amplifier. In order to remove drifts and the spectral response of the setup, we recorded for each gate voltage of interest a reference spectrum at a large and positive gate voltage where the dots are all completely filled with electrons and the interband transitions blocked.

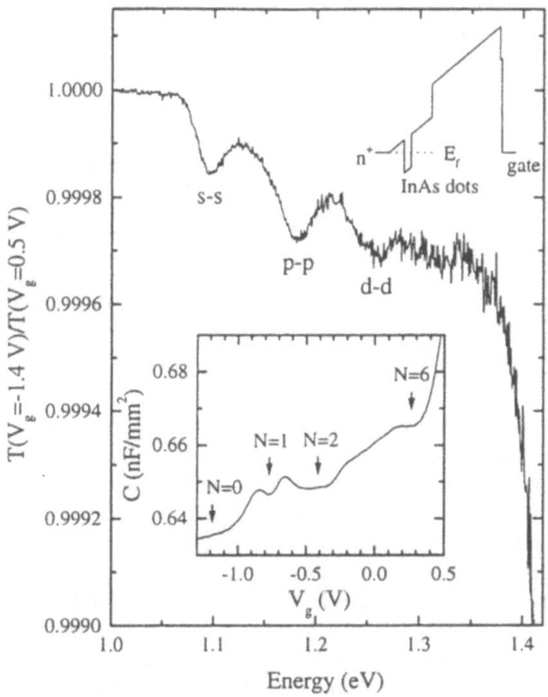

Figure 1
The inset shows the conduction band profile of the device in the growth direction with Fermi energy, E_f. The capacitance, C, of the device at 4.2 K is shown against gate voltage, V_g. Charging of the electron levels in the dots corresponds to peaks in the capacitance enabling the electron occupancy per dot, N, to be determined. The main figure is the transmission spectrum at $V_g = -1.4$ V ratioed against the spectrum at 0.5 V. Taken from [9].

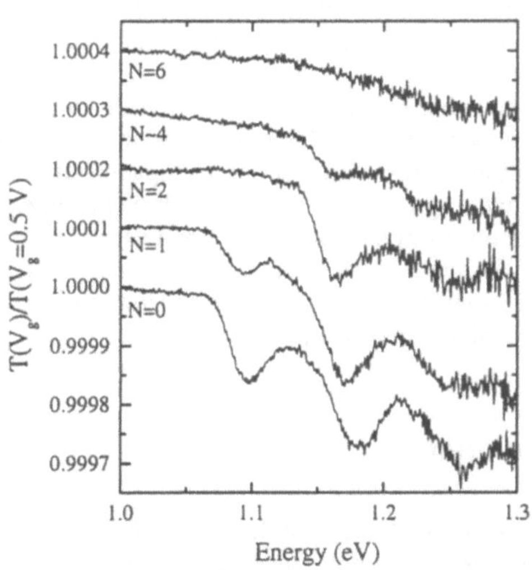

Figure 2
Interband transmission spectra of the dot ensemble taken at $V_g = -1.2, -0.75, -0.4, 0.0, 0.2$ V all referenced to the spectrum at 0.5 V. The curves are labelled with the electron occupation, $N = 0, 1, 2, 4, 6$ respectively, and are offset from 1 for clarity. Taken from [9].

3 Transmission Spectroscopy of the Charge-Tunable Dots

Fig. 1 shows the transmission spectrum for $V_g = -1.4$ V, i.e. for electrically neutral dots. There are three transitions between $\sim$ 1.1 and 1.3 eV which correspond to the s–s, p–p, and d–d excitons. s–s refers to an excitonic transition between the hole s and electron s

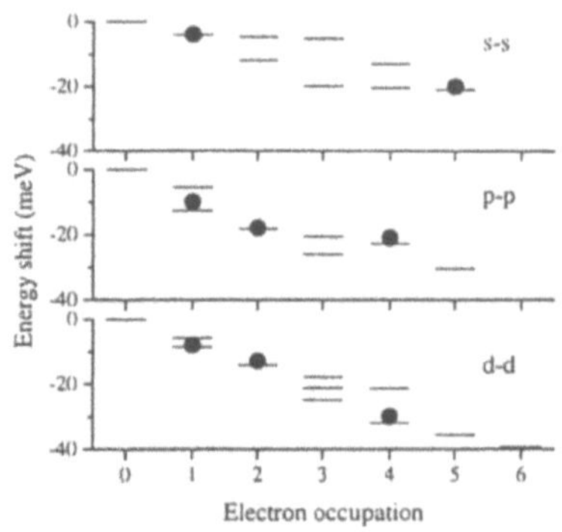

Figure 3
The shift in transition energy is plotted against electron occupation for the three transitions, s–s, p–p, and d–d. The symbols represent the results of the experiments, the sold lines the results of the perturbation theory calculations. For s–s, the absorption is blocked for $N > 2$; we plot in this regime the photoluminescence energy, where N is the number of electrons after recombination. For d–d, only the calculated shifts for $|m| = 2$ are plotted.

states, and etc.. At higher energy a much stronger absorption process sets in, most probably from absorption in the wetting layer. The behaviour at higher V_g is shown in Fig. 2. V_g's were chosen where the electron occupancy per dot, N, is well defined: $N = 0,1,2$ and 6. $N = 4$ is also plotted but it should be borne in mind that there is a distribution of N here. The s–s transition loses half its intensity at $N = 1$, and at $N = 2$ disappears altogether. This is a simple consequence of the Pauli exclusion principle. The s–s transition is clearly forbidden once the s level is fully occupied at $N = 2$. For $N = 1$, only one of two possible light polarisations can be absorbed. Similarly, the p–p transition disappears once the p level is fully occupied at $N = 6$. The d–d transition should persist at $N = 6$ yet it is hard to say from Fig. 2 if this is the case or not. It could be that the d–d transition becomes broad at large N, merging into the signal from the wetting layer. However, it could also be the case that the d electron level is no longer bound at $N = 6$ in which case the d–d transition will also disappear with the p–p.

A closer examination of Fig. 2 shows that the peaks not only reduce in amplitude with increasing N but also exhibit N-dependent shifts. This information is summarised in Fig. 3, a plot of the shifts in transition energy with electron occupation. It might be thought that the shifts arise from the changing vertical electric field through some sort of Stark effect. However, we can rule this out from the data at large and negative V_g where we change the electric field but not N. In this regime, we cannot detect any shifts in the transition energies. We therefore argue that the energy shifts arise from an interaction between the exciton and the electrostatically stored electrons.

4 Perturbation Theory Calculations

In order to account for the energy shifts in Fig. 3, we treat the Coulomb interactions between the electrons and between the electrons and holes as perturbations to the confinement potential. The justification is that the quantization energy in the conduction band is $\hbar\omega_e = 49$ meV, considerably larger than the ~ 20 meV Coulomb energy between two s electrons. The former has been measured by far infrared spectroscopy on these samples [10]; the latter by capacitance spectroscopy [11]. The in-plane confinement potentials are very close to parabolic. This can be judged from the absence of strong electron-electron interactions in the far infrared spectra at large N [10] (no energy shifts at all would be observed for a perfectly parabolic potential [12]) and from calculations of the dots' structure [13]. Furthermore, the vertical quantization energy is almost an order of magnitude larger than the lateral quantization energy. We therefore make the approximation that the dots' confinement potential is two-dimensional. The spirit of the calculations is outlined with an example below.

When the p–p transition is excited in a dot occupied with two electrons, the electron in the electron-hole pair is repelled by the existing two electrons. This increases the energy of the exciton by $2E^{\mathrm{c}}_{\mathrm{sp}} - E^{\mathrm{x}}_{\mathrm{sp}}$, where $E^{\mathrm{c}}_{\mathrm{sp}}$ and $E^{\mathrm{x}}_{\mathrm{sp}}$ are the direct and exchange interactions, respectively, between an s and a p electron. The exchange term arises because the total three electrons in the dot are indistinguishable and must be described by an anti-symmetrised wave function. On the other hand, the hole in the valence band p state is attracted by the two existing s electrons. This decreases the energy of the exciton by a term $2E^{\mathrm{eh}}_{\mathrm{sp}}$. The hole is considered distinguishable from the electrons and so, at least to a first approximation, electron-hole exchange is unimportant. The net shift in energy is then $2E^{\mathrm{c}}_{\mathrm{sp}} - E^{\mathrm{x}}_{\mathrm{sp}} - 2E^{\mathrm{eh}}_{\mathrm{sp}}$. For each transition we have performed a similar analysis.

The energies $E^{\mathrm{c}}_{\mathrm{sp}}$ etc. are matrix elements of the Coulomb interaction with the single particle states, for example

$$E^{\mathrm{c}}_{\mathrm{sp}} = \frac{e^2}{4\pi\varepsilon_0\varepsilon_{\mathrm{r}}} \int \frac{|\psi_{\mathrm{s}}(\vec{r}_1)|^2 |\psi_{\mathrm{p}}(\vec{r}_2)|^2}{|\vec{r}_1 - \vec{r}_2|} \, \mathrm{d}\vec{r}_1 \mathrm{d}\vec{r}_2,$$

and are evaluated with the two-dimensional harmonic oscillator wave functions. The integrals can all be performed analytically, with results depending on the effective lengths for the electrons (l_{e}) and holes (l_{h}):

$$l_{\mathrm{e}} = \sqrt{\frac{\hbar}{m^*_{\mathrm{e}}\omega_{\mathrm{e}}}}$$

with, obviously, a similar equation for the holes. $\hbar\omega_{\mathrm{e}}$, $\hbar\omega_{\mathrm{h}}$, m^*_{e}, m^*_{h} are the electron quantization energy, hole quantization energy, electron effective mass, and hole effective mass, respectively. $\hbar\omega_{\mathrm{e}} = 49$ meV is known from intraband spectroscopy [10]. Both intraband spectroscopy [10] and capacitance [11] have determined m^*_{e}; there is a small discrepancy between the two results but this is largely irrelevant for our purposes and we have taken an average value, $0.07m_0$, which is also the GaAs mass. We then have $l_{\mathrm{e}} = 4.7$ nm. We determine the hole parameters from the energy splittings between the s–s, p–p and d–d excitons for neutral dots. These are dominated by $\hbar\omega_{\mathrm{e}} + \hbar\omega_{\mathrm{h}}$, giving an estimate of $\hbar\omega_{\mathrm{h}} = 25$ meV. However, the splitting between d–d and p–p is smaller than that between p–p and s–s because the exciton binding energy decreases with energy. This allows us also to estimate l_{h} to be 3.5 nm. Reassuringly, we find a hole mass $m^*_{\mathrm{h}} = 0.25m_0$, close to the bulk heavy hole mass of GaAs as expected.

With these parameters we have evaluated the energy shifts as shown by the horizontal lines in Fig. 3. Note that the $N = 0$ energies have been used to determine the dot parameters and so it is only the energy shifts which can be usefully compared to the calculations. The agreement is very good. The p–p transition moves down in energy by some 18 ± 2 meV at $N = 2$ and the agreement with the calculations is here excellent. For $N = 1$, there is a splitting in the calculation because the stored s electron and the excited p electron can have either parallel or antiparallel spins. The splitting is ~ 7.3 meV, the exchange energy $E^{\mathrm{x}}_{\mathrm{sp}}$. We do not resolve it in the experiment, presumably because it is too small compared to the p–p linewidth (~ 40 meV). At higher N, we inevitably integrate over a distribution in N, but nevertheless the experiments appear to show that further large decreases in the p–p exciton energy do not occur. The calculations show approximately this behaviour if one performs a weighted average over several N.

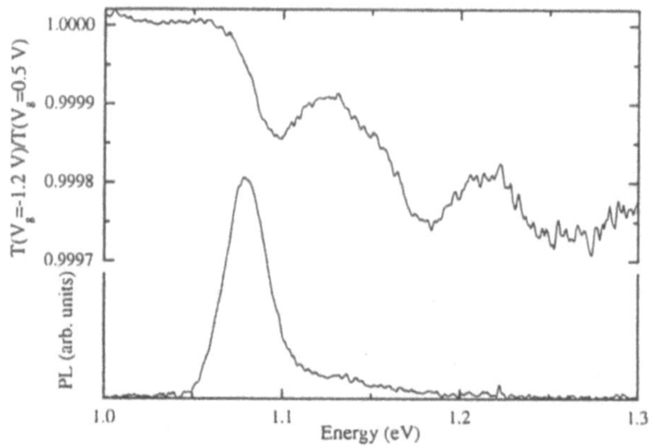

Figure 4 Photoluminescence excited with a pump power of ~ 8 W/cm^2 of 514 nm radiation, and transmission, taken for empty dots, $V_g = -1.2$ V. The laser light causes the dots to fill with electrons, resulting in the strong redshift with respect to the transmission minimum.

The d–d exciton also shifts to lower energies on occupation, but by smaller amounts. The theory is complicated by the presence of two transitions with L_z quantum number $m = 0$ and $m = \pm 2$ (only the transitions with $|m| = 2$ are plotted in Fig. 3), but at the level of the experimental resolution, agreement with the calculation is achieved.

The s–s exciton is shifted by -4 ± 2 meV on going from $N = 0$ to $N = 1$. This redshift implies that the X^- is stable in this system, i.e. given two electrons and one hole it is energetically advantageous to confine all three particles in the same dot. This property can be very simply related to the fact that $l_e > l_h$ [14]. At higher N, the s–s transition obviously disappears in the transmission experiment, but can be followed by photoluminescence experiments. Experiments on similar samples have shown that the transition shifts monotonically to the red on increasing N [15,16]. We have extended these photoluminescence experiments to high pump powers; Fig. 4 shows a spectrum taken with ~ 8 W/cm^2 of 514 nm laser light. We excite here a sizeable number of electron-hole pairs in the GaAs barrier material. The electrons are either trapped in the dots or swept away to the back contact; some of the holes however are trapped between the dots and surface by the blocking barrier. This positive charge tends to screen out the in-built electric field, giving approximately flat band conditions, filling the dots with electrons. We can verify that this indeed takes place from transmission experiments: at $V_g = -1.2$ V for instance, a spectrum taken with laser light on the sample mimics very closely one taken with a large and positive V_g. Fig. 4 shows how the photoluminescence is strongly redshifted with respect to the transmission. This comes about from the negative charge induced in the dots in the conditions generated in the photoluminescence experiment. Assuming that the dots are filled with six electrons, we calculate a redshift of ~ 20 meV, in excellent agreement with the measured 20 ± 5 meV.

5 Simulations of the Transmission

At voltages intermediate to those at which $N = 0$, $N = 1$ and $N = 2$ are well defined, we observe splittings and distortions in the lineshape (Fig. 5). For instance, the p–p line can be seen to split into two at $V_g = -0.84$ V. We estimate from the capacitance that at this V_g half the dots are empty and the other half are occupied with one electron. The occupied dots have a down-shifted resonance, causing the line to split. A similar effect occurs at

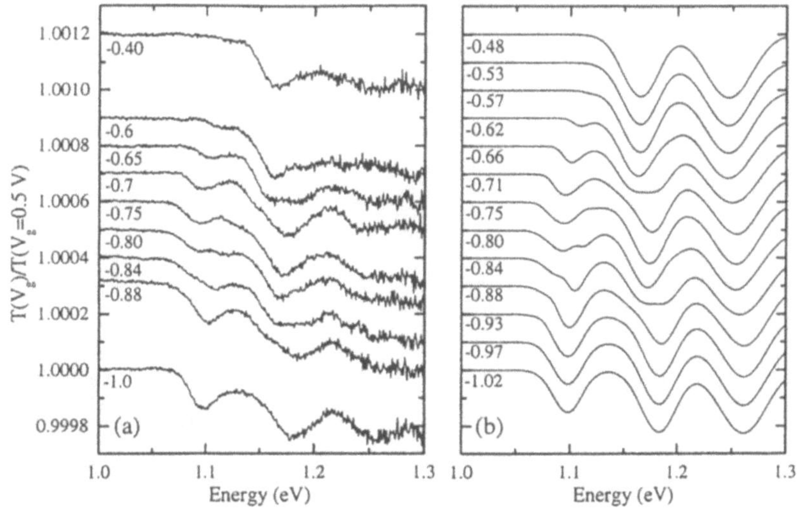

Figure 5 (a) Transmission spectra taken at various V_g for which the dots have N between 0 and 2. (b) Simulated spectra. The curves are labelled with V_g and are offset from 1 for clarity.

$V_g = -0.65$ V where half the dots are singly and half doubly occupied. Analogous effects for the d–d transition are not resolvable because of the larger linewidth. The s–s transition is more complicated because population of a particular dot inevitably implies a reduction in its oscillator strength. At $V_g = -0.84$ V the unoccupied dots, which are typically those with a large s state energy and hence larger than average interband energy, absorb more strongly than the occupied dots with a small s state energy. This skews the line to the high energy side. At $V_g \geq -0.65$ V, only a small number of high energy dots can contribute to the absorption which pushes the line to higher energy and, also, reduces the linewidth slightly. In between these two voltages, rather complex lineshapes are observed.

We have simulated the transmission by modelling the inhomogeneous broadening. The dominant broadening arises from fluctuations in the vertical direction, giving rise to a Gaussian distribution (FWHM Γ_0) in the dots' s state energy, E_0, with respect to the Fermi energy. These fluctuations give a contribution of $(1+\alpha)\Gamma_0$ to the broadening of the excitonic transitions, where $\alpha\Gamma_0$ comes from the holes. There is also a distribution in the lateral size. We assume that the dots with a particular E_0 have a Gaussian distribution (FWHM Γ_1) in the confinement energy $\hbar\omega_e + \hbar\omega_h$, where Γ_1 is independent of E_0. For empty dots, the interband transitions then have widths $\sqrt{(1+\alpha)^2\Gamma_0^2+\Gamma_1^2}$, $\sqrt{(1+\alpha)^2\Gamma_0^2+(2\Gamma_1)^2}$, and $\sqrt{(1+\alpha)^2\Gamma_0^2+(3\Gamma_1)^2}$ for s–s, p–p, and d–d respectively. The contribution from Γ_1 enters according to the contribution of $\hbar\omega_e + \hbar\omega_h$ in the energy. For occupied dots, we divide the distribution of E_0 into three groups corresponding to $N = 0$, 1 and 2. For each group, we calculate the transmission including Pauli blocking and the N-dependent shifts. We related the change in transmission to the dipole matrix element with a standard approach [17].

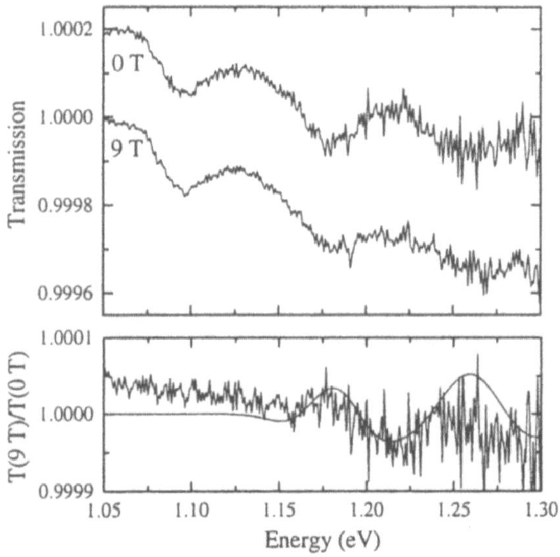

Figure 6
The upper plot shows the transmission at $V_g = -1.2$ V (empty dots) for magnetic fields 0 and 9 T. The curves are offset for clarity. The lower plot shows the ratio of the 9 T and 0 T traces. The result of the simulations is also shown.

Results are plotted in Fig. 5. We took $\Gamma_0 = 18$ meV directly from the capacitance, and made a rough fit to determine $\Gamma_1 = 13.3$ meV and $\alpha \simeq 0.6$. The results reproduce the measured spectra remarkably well. The remaining discrepancy is that the measured transmissions are somewhat smaller than the simulated transmissions at high energies. This probably comes from the tail of the wetting layer absorption.

6 Measurements in Magnetic Field

We have applied a magnetic field along the growth direction and measured the transmission spectrum for empty dots. Shown in Fig. 6 is the transmission for both 0 and 9 T, and also the result of dividing the 9 T trace with the 0 T trace. It can be seen that the magnetic field has a very weak effect on the optical response. The magnetic field lifts the degeneracy with respect to L_z at $B = 0$. For example, the p–p transition should split into two, with a splitting $e\hbar B(1/m_e^* + 1/m_h^*)$. This amounts to 21 meV at 10 T. In our case, this splitting is smaller than $\hbar\omega_e + \hbar\omega_h$ and also smaller than the linewidth of the p–p transition. Nevertheless, a weak effect can be detected. In Fig. 6, we have also included the effect of the magnetic field in the simulations. The agreement with the experiments suggests that the p–p transition does split into two, but either a smaller inhomogeneous broadening or much higher magnetic fields are required to study this in detail.

7 Conclusions and Outlook

We have presented a study of charged excitons in InAs self-assembled quantum dots. As the dots are filled with electrons, interband transitions are blocked, exactly as expected from the Pauli exclusion principle. The Coulomb interactions between the electrons and the hole lead to substantial shifts in the exciton energies. These shifts can be understood with a simple model in which the Coulomb interactions between the carriers are treated by first order perturbation theory. The accuracy of the measurement is limited at present by the inhomogeneous broadening of the transitions. However, the homogeneous width of excitonic transitions in quantum dots can be as small as 100 μeV [18,19] which is a motivating factor for future single dot experiments.

Acknowledgments

We would like to thank A. O. Govorov for some very helpful comments and suggestions. The work was funded by the Deutsche Forschungsgemeinschaft (SFB 348), by QUEST, a National Science and Technology Center (grant DMR 20007), and by the BMBF through a Max Planck Research Award. J.P.K. would like to thank the staff at QUEST for their hospitality.

Bibliography

[1] P. L. McEuen, E. B. Foxman, U. Meirav, M. A. Kastner, Y. Meir, N. S. Wingreen, and S. J. Wind, Phys. Rev. Lett. **66**, 1926 (1991).

[2] R. C. Ashoori, H. L. Stormer, J. S. Weiner, L. N. Pfeiffer, S. J. Pearton, K. W. Baldwin, and K. W. West, Phys. Rev. Lett. **71**, 613 (1993).

[3] S. Tarucha, D. G. Austing, T. Honda, R. J. van der Hage, and L. P. Kouwenhoven, Phys. Rev. Lett. **77**, 3613 (1996).

[4] M. Eto, Jpn. J. Appl. Phys. **36**, 3924 (1997).

[5] A. Barenco and M. A. Dupertuis, Phys. Rev. B **52**, 2766 (1995).

[6] D. Leonard, K. Pond, and P. M. Petroff, Phys. Rev. B **50**, 11 687 (1994).

[7] H. Drexler, D. Leonard, W. Hansen, J. P. Kotthaus, and P. M. Petroff, Phys. Rev. Lett. **73**, 2252 (1994).

[8] G. Medeiros-Ribeiro, F. G. Pikus, P. M. Petroff, and A. L. Efros, Phys. Rev. B **55**, 1568 (1997).

[9] R. J. Warburton, C. S. Dürr, K. Karrai, J. P. Kotthaus, G. Medeiros-Ribeiro, and P. M. Petroff, Phys. Rev. Lett. **79**, 5282 (1997).

[10] M. Fricke, A. Lorke, J. P. Kotthaus, G. Medeiros-Ribeiro, and P. M. Petroff, Europhys. Lett. **36**, 197 (1996).

[11] B. T. Miller, W. Hansen, S. Manus, R. J. Luyken, A. Lorke, J. P. Kotthaus, S. Huant, G. Medeiros-Ribeiro, and P. M. Petroff, Phys. Rev. B. **56**, 6764 (1997).

[12] L. Brey, N. F. Johnson, and B. I. Halperin, Phys. Rev. B **40**, 10647 (1989).

[13] A. Wojs, P. Hawrylak, S. Fafard, and L. Jacak, Phys. Rev. B **54**, 5604 (1996).

[14] Ph. Lelong and G. Bastard, Solid State Commun. **98**, 819 (1996).

[15] K. H. Schmidt, G. Medeiros-Ribeiro, M. Oestreich, P. M. Petroff, and G. H. Döhler, Phys. Rev. B **54**, 11346 (1996).

[16] K. H. Schmidt, U. Kunze, G. Medeiros-Ribeiro, J. M. Garcia, P. Wellman, and P. M. Petroff, in *Proceedings of Modulated Semiconductor Systems, Santa Barbara, July 1997*, and to be published in Physica E.

[17] R. S. Knox, *Theory of Excitons*, Solid State Physics, Suppl. 5 (Academic Press, New York, 1963).

[18] J. -Y. Marzin, J. -M. Gérard, A. Izraël, D. Barrier, and G. Bastard, Phys. Rev. Lett. **73**, 716 (1994).

[19] M. Grundmann *et al.*, Phys. Rev. Lett. **74**, 4043 (1995).

Charge Fluctuations, Chaotic Trajectories, and the Coulomb Blockade

H. Buhmann, S. Möller, and L.W. Molenkamp

2. Physikalisches Institut, RWTH-Aachen,
Templergraben 55, D-52056 Aachen, Germany

Abstract: Two different experimental methods have been used to determine the charging energy U^* of a quantum dot as a function of the coupling to the external leads. A powerlaw scaling behaviour depending on the number of opening leads was observed for the weak coupling regime in both types of experiments, in agreement with theories based on the Tomonaga-Luttinger formalism. For stronger coupling ($t > 0.5$), in asymetric quantum dots with only one opening lead, U^* appears to reach a constant value. This latter behaviour can be explained by taking into account chaotic motion of the electrons in the quantum dot.

1 Introduction

During the last few years there has been a considerable interest in the transport properties of quantum dots [1]. Quantum dots are small metallic islands where the charge carriers are confined in all three dimensions. Transport through these small structures is possible via the coupling to external leads by tunnel barriers. At low temperatur.. the transport properties of such an island are determined the electrostatic energy which is given by its capacity and charge. In order to add an extra charge to the dot, an excess energy is necessary to account for an increase of the free energy F of the system when the electron number changes from n to $n+1$. At zero temperature the free energy is equal to the electrostatic energy for a dot containing n electrons and the sum over all occupied energy level of the dot: $F(n,T=0) = U_{\text{elec.stat.}} + \sum_{p=1}^{n} E_p$. If this necessary excess energy exceeds the thermal energy ($k_B T$) transport through the dot is suppressed. This effect is called Coulomb blockade. Transmittance can be reachieved by adjusting the electrostatic potential of the dot, which can be done conveniently by using external gate electrodes. By varying the gate voltage V_g it is possible to adjust the electro-chemical potential of the dot, $\mu(n) := F(n+1) - F(n)$, in such a way that it becomes equal to the Fermi energy E_F of the external leads. At these voltages transport is again possible. This situation is illustrated in Fig. 1. The conductance results in a periodic series of sharp conductance peaks, known as Coulomb oscillations, which are smeared out for finite temperatures. The electro-chemical potential oscillates in a sawtooth manner between $E_F + (e^2/C + \Delta E)$ and $E_F - (e^2/C + \Delta E)$ with increasing electron number on the dot. Neglecting the energy level-splitting ΔE of the dot, this leads within this classical picture to a charging energy of $U = e^2/2C$. Coulomb oscillations can only be observed when the electrons are tightly confined to the quantum dot. The conductance G of the barriers connecting the dot to external leads has to be below $2e^2/h$, i.e. in the tunnelling regime [2,3]. When the

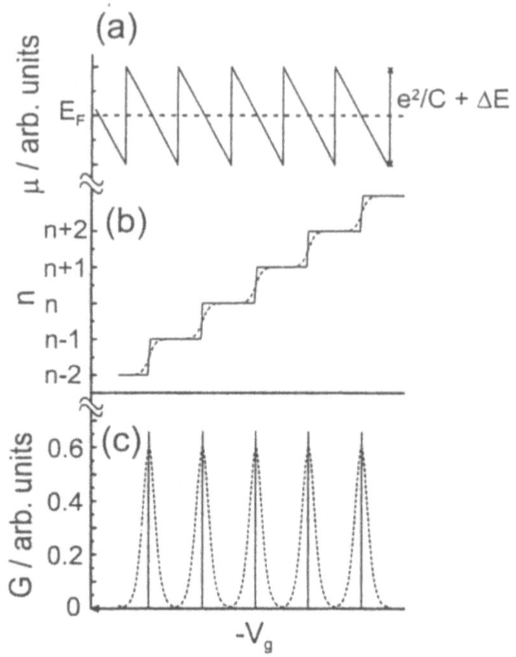

Figure 1
(a) electro-chemical potential μ, (b) number of electrons, and (c) Coulomb oscillations G of a quantum dot at zero and finite temperature (solid and dashed line, respectively).

barrier conductance approaches $2e^2/h$, quantum fluctuations occur in the charge on the dot, which reduce the effect of the Coulomb blockade and therefore reduce the charging energy with respect to its classical value.

This phenomenon has recently attracted considerable attention from theorists, since it proved possible to describe this behaviour using scaling- and Luttinger-liquid theory based on a Tomonaga-Luttinger formalism [4–9]. Flensberg showed[6,8] that the charging energy associated with the addition of an electron to the dot should scale with the conductance of the barriers, according to

$$U^* = U(1-t)^{N_c}, \tag{1.1}$$

where $U \equiv e^2/2C$ is the bare charging energy of the dot in the tunnel regime, C its self-capacitance, U^* the effective (or "renormalized") charging energy observed for finite barrier conductance, N_c the total number of leads with a the transmission probability t varied in a scaling experiment.

Unfortunately, conductance measurements [2,3] are not well suited to test (1.1) because of the occurence of complicating cotunnelling processes [10,11] when the barrier conductance approaches $2e^2/h$. In this article we describe two different experimental methods that were designed to verify the validity of (1.1). In both cases we use fully adjustable gate-defined quantum-dot devices in a (Al,Ga)As modulation doped heterostructure. First, we discuss a double dot structure in an electrometer configuration [12] where the oscillating electrostatic potential of one dot results in a sawthooth modulation of the Coulomb-blockade oscillation of the other. Depth and line shape of these modulations are directly related to the effective charging energy U^* of the first dot. Second, we present thermopower measurements of a single quantum dot. The amplitude of the sawtoothlike thermopower oscillations [13,17] depends on the ratio $k_B T/U^*$. At fixed temperatures these oscillations provide information about U^* as a function of the barrier height.

2 Experiments

2.1 Samples

The devices used in this study are fabricated from GaAs-(Al,Ga)As modulation doped heterostructures. The two-dimensional electron gas (2DEG) with mobility $\mu \approx 2 \times 10^6$ cm^{2}V^{-1}s^{-1} ($\mu \approx 1 \times 10^6$ cm^{2}V^{-1}s^{-1}) and sheet carrier concentration $n_s \approx 1.8 \times 10^{11}$ cm^{-2} ($n_s \approx 3.4 \times 10^{11}$ cm^{-2}) for double-dot and single-dot structure respectively. The nanostructures are defined electrostatically using a split-gate approach. A schematic layout of the devices is shown in Fig. 2. The diameter of the dots is about 1 μm in Fig. 2a and about 0.8 μm in Fig. 2b. All experiments are performed at temperatures below 50 mK. The electrical characteristics of the devices are measured using standard low-frequency lock-in techniques.

2.2 Electrometer Experiment

The electrometer experiment (cf. Fig. 2a) is performed by monitoring the conductance G of dot 1, measured between ohmic contacts 1 and 4, as a function of the voltage on gate II, V_{gII}. Note, this experiment is the semiconductor analogue of the metallic two-dot electrometer device described by Lafarge et al. [12]. The advantage of the semiconductor device is that it allows for a continuous voltage controlled adjustement of the barrier transparancy.

In our experiment, barriers AB, BE, and EF are all adjusted in the tunneling regime (G_{AB},G_{EF},$G_{BE} \approx 0.26 \times e^2/h$), so that dot 1 is always well defined. The transmission probability of the barriers BC and EC is adjusted to value between $0.26 \times e^2/h$ and $0.65 \times e^2/h$ ($t = 0.13$ and $t = 0.325$) for each sweep of gate II. In Fig. 3 (a) the conductance G of dot 1 is shown as a function of V_{gII} (only one Coulomb-blockade oscillation of dot 1 is shown here). The traces for different values of t are displayed with a constant offset.

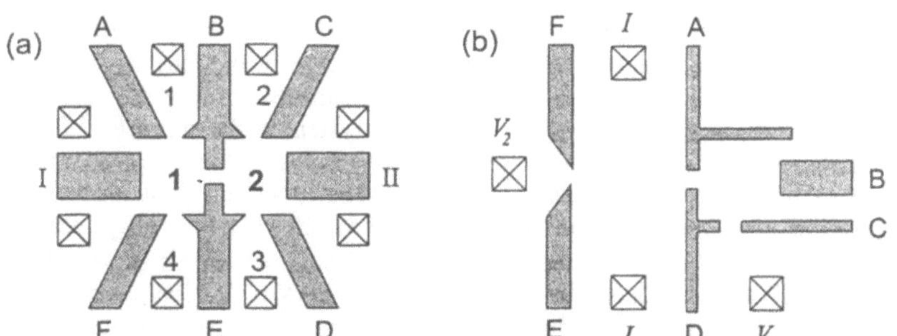

Figure 2 Schematic Sample structure: (a) double dot and (b) single dot with heating channel. The hatched areas denote gates and the crossed squares ohmic contacts.

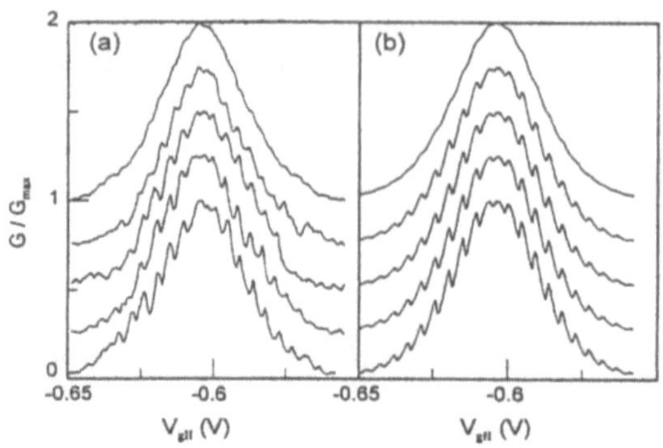

Figure 3
(a) Traces of G versus V_{gII} in an electrometer experiment, where the conductance of barriers BC and DE is varied. (a) Experimental data, where from top to bottom G_{BC},$G_{DE} \approx 0.65, 0.375, 0.215, 0.07$ and $0.026 \times e^2/h$. In all traces G_{AB},G_{BE},$G_{EF} \approx 0.026 \times e^2/h$. (b) The results of calculations using the theoretical approach of Ref. [6,8] with U^* as fit parameter.

Due to the relatively large distance between gate II and dot 1, the effect of V_{gII} on dot 2 is much larger than on dot 1 and therefore a large number of Coulomb-blockade oscillations occurs on dot 2 while the number of electrons changes only by one on dot 1. With open leads ($t = 0.325$) dot 2 is not expected to exhibit strong modulations in the electrostatic potential due the Coulomb-blockade effect. Therefore, the measured oscillation on dot 1 is smooth. On closing the leads to dot 2, the oscillations of dot 1 show the gradual appearance of a sawtooth pattern on the originally smooth Coulomb-blockade oscillations. This modulation results from sawtooth oscillations in the electrostatic potential of dot 2 [14,12], due to the Coulomb-blockade oscillations that V_{gII} induces in that dot. The electrostatic potential of dot 2 adds up to the total potential that is sensed by dot 1, resulting in the sawtooth dips in the G vs. V_{gII} trace at each gate voltage where the number of electrons changes in dot 2. Note, that the period of the sawtooth feature is unaffected by the changes in G_{BC} and G_{DE}. The sawtooth period is determined solely by the classical electrostatics.

It was demonstrated in [14] that the depth and lineshape of the sawtooth oscillations are determined by the charging energy of dot 2. Thus dot 1 acts as an electrometer that measures the changes in the potential of dot 2. Hence, the data of Fig. 3 can be used to deduct the change of the charging energy of dot 2 as a function of the conductance of barriers BC and DE. The experiment shows that for increasing conductances G_{BC} and G_{DE} the sawtooth feature is much less pronounced. These observations can be attributed to the scaling of the charging energy U_2^* as a function of the conductance of barriers BC and DE.

The curves in Fig. 3b are the results of a theoretical modeling of these experiments using a theory that incorporates the influence of quantum fluctuations according to (1.1). Full details of this theory are given elsewhere [14]. Briefly, the modeling starts with an expression for the electrostatics of the double-dot system in the absence of fluctuations [15], which allows to extract the relevant elements of the capacitance matrix of the system from a fit to the data in the bottom traces of Fig. 3a. The curves in Fig. 3b are then obtained by fitting the experiment with an optimum value for U_2^*. Fig. 4 displays the obtained values of U_2^* versus $(1-t)^2$, where t is the transmission of barriers BC and DE. Note, the number of varied leads is two for this experiments, $N_c = 2$. The observed behavior yields strong support for the validity of (1.1) in describing in the scaling behavior of the quantum-dot charging energy.

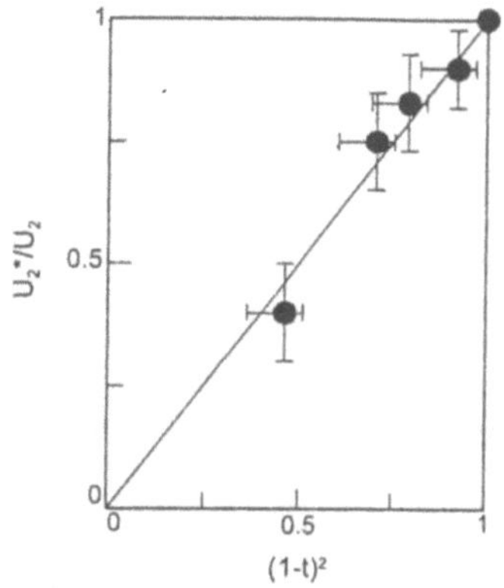

Figure 4
Plot of the ratio U_2^*/U_2, determined from the modeled curves in the left panel of (a) vs. $(1-T)^2$, where T is the transmission of the barriers.

2.3 Thermopower Measurement

The sample structure for this experiment is shown in Fig. 2b. The device consists of a quantum dot defined by gates labeled A-D with two leads (AD and CD) and dimensions 700 nm × 800 nm. The electrochemical potential of the dot can be varied by applying an external voltage on gate B. Additionally, the quantum dot is coupled on one side to a 2 μm wide and 20 μm long electron channel (gates: A, D, E and F) by one quantum pointcontact QPC_{dot}^{AD} (lead: AD). Opposite to QPC_{dot}^{AD} is another quantum pointcontact QPC_{ref} (gates: E and F) in the channel boundary, which is used as a reference for thermopower-measurement purposes (see below).

Passing a heating current throught the channel creates a temperature difference across the dot. For low heating currents I the temperature increase ΔT of the electron system in the channel is proportional to I^2. The thermopower of the dot S_{dot} can be determined by measuring the voltage difference between ohmic contacts V_1 and V_2 [16]. The measured thermovoltage contains also the thermopower of the reference pointcontact S_{ref} according to

$$V_{Th} := V_1 - V_2 = (S_{ref} - S_{dot})\Delta T . \tag{2.2}$$

In order to obtain the thermopower of the dot as a function of the barrier transparency the thermopower of QPC_{ref} is adjusted to a constant minimum value at a fixed temperature difference ΔT. Thus variations in V_{Th} reflect directly changes of S_{dot}. In addition, the transmission probability of QPC_{dot}^{CD} was measured as function of gate CD and corrected for the experimentally determined electrostatic crosstalk of the other active gates.

The measured thermovoltage of the dot is displayed in Fig. 5 as a function of the gate voltage V_B for different transmission probabililies t of the QPC_{dot}^{CD}. The observed thermopower shows the expected sawtooth oscillation [13,17]. The amplitude decreases with increasing conductance of pointcontact QPC_{dot}^{CD}, however, oscillation are still visible for $t > 0.82$. Using the theoretical approach of [17] it is possible deduce the effective charging energy U^* of the dot from the amplitude of these oscillations. Fig. 6 shows the calculated dependence of the thermopower amplitude as a function of the ratio of the electron temperature and charging energy. The solid line is an exponential fit-function to the discrete values of the theoretical approach which was used to determine the effective

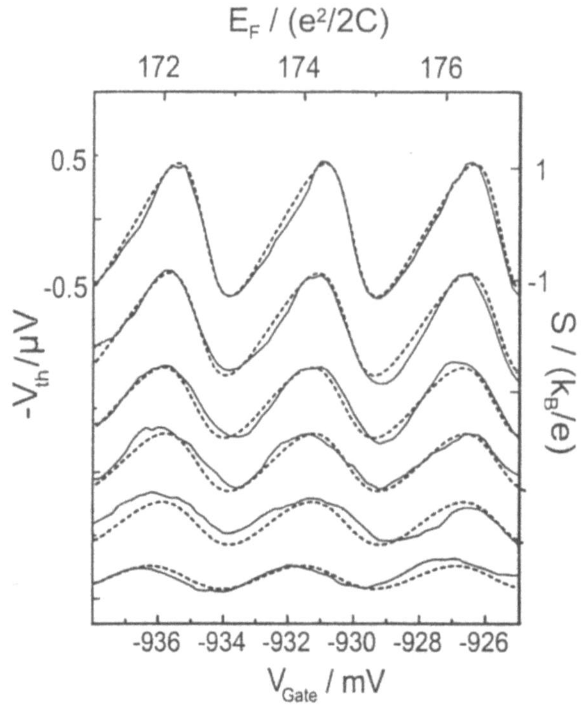

Figure 5
Measured thermovoltage oscillations for different transmission probabilities of the QPC^{CD}_{dot}. From top to bottom: $t = 0.06$, 0.19, 0.29, 0.38, 0.43, and 0.82 (solid lines) and calculated thermopower for $k_BT/U^* = 0.22$, 0.25, 0.3, 0.33, 0.37, and 0.45 (dashed lines) (from top to bottom).

charging energy for the experimental data. The dashed lines in Fig. 5 show the calculated thermopower for different values of k_BT/U^* indicated in the figure caption. The normalized effective charging energy U^*/U deduced from these fits is plotted against the reflection probability $1-t$ of QPC^{CD}_{dot} in Fig. 7. The bare charging energy U of 150μeV was determined from thermal activation studies of the Coulomb-blockade oscillations in the Coulomb-bockade regime ($t \leq 0.05$). Note that in the scaling experiment only one lead to the dot is opened and therefore the exponent of (1.1) is equal to one. For values $0.5 < (1-t) < 1$ a linear scaling is found, in agreement with (1.1). A striking observation is that for higher values of t the validity of (1.1) seems broken. A breakdown of the scaling theory for large values of t is unexpected because the mapping of the quantum-dot system to the behaviour of a one-dimensional lead should have produced the best agreement exactly in this regime [6,8]. However, the experimentally found effective charging energy appears to approach a constant value $U^*/U \approx 0.45$.

The deviation of the scaling behaviour from the predicted dependence can be understood by taking into account random electron scattering by the boundaries of the quantum dot. Aleiner and Glazman [19] conjectured that for chaotic scattering a residual oscillatory dependence of the transport properties of the quantum dot is maintained even when one of the pointcontacts leading to the dot carries a fully transmitted mode, as long as the other pointcontacts are still in the tunnelling regime. In terms of renormalized charging energy this can be expressed by making the following substitution in (1.1) (for $N_c = 1$)

$$(1-t) \rightarrow \frac{\Delta E}{U} \ln^2\left(\frac{U}{\Delta E}\right) . \tag{2.3}$$

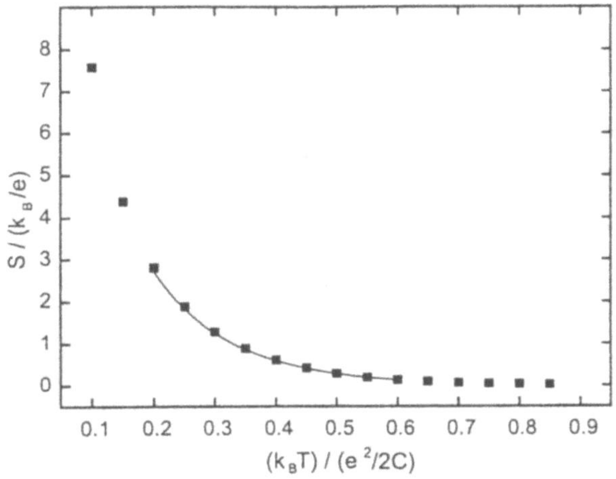

Figure 6
Calculated dependence of the thermopower amplidude on temperature and charging energy [17]. The solid line is an exponential fit, which is used to extract the charging energy from the experimental data in Fig. 5.

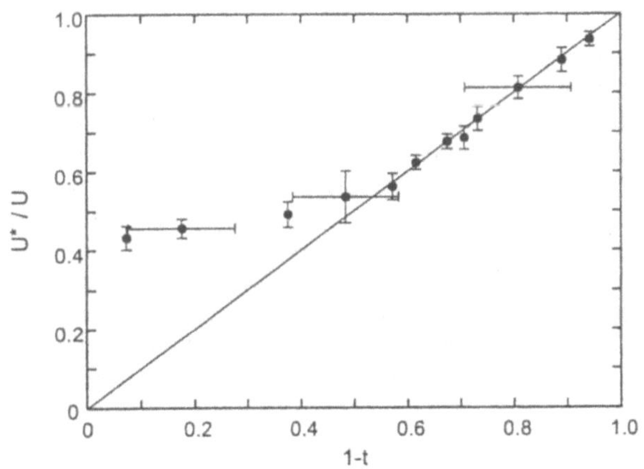

Figure 7
Renormalized charging energy versus $1-t$. The solid line represents the linear behaviour expected form Eq. 1.1 ($N_c = 1$).

Here, ΔE is the mean spacing of the electronic energy levels in the dot. Estimating ΔE from the density of states of a two-dimensional electron system with the size of the actual quantum dot leads to $\Delta E \approx 23\,\mu$eV, which gives an effective charging energy for $t = 1$ of the order of

$$U^*(t = 1) = 0.54\,U\,. \tag{2.4}$$

The extrapolation of the data in Fig. 5 gives

$$U^*(t = 1) \approx (0.45)\,U\,. \tag{2.5}$$

The remarkable agreement between these two values seems indeed to indicate the exsistance of a residual Coulomb blockade at a fully transmitted channel, in agreement with the theory [19].

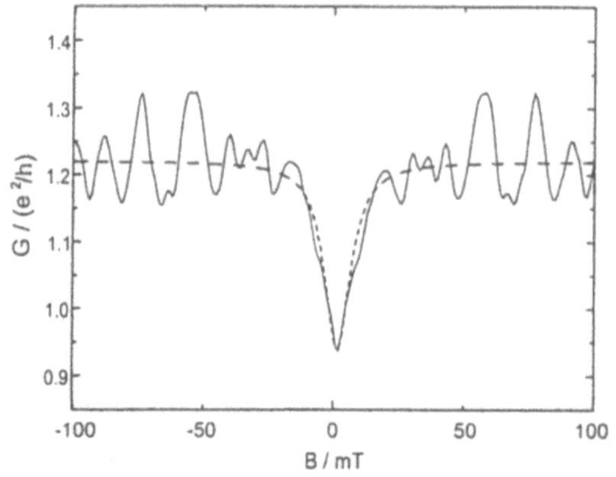

Figure 8
Averaged conductance G vs. magnetic field for 30 individual bias gate-voltage V_B (solid line) and fit of the weak localisation peak around $B = 0\,\mathrm{T}$ according to (3.6) (dashed line).

3 Chaotic Trajectories

One prerequisite for our interpretation of the results of the previous section is that the electron motion in the dot is chaotic. In order to validate this assumption the transport characteristics of the dot have to be investigated separately in the ballistic regime. An electron entering through one lead the dot may eventually leave the same way if one assumes a random scattering inside the dot. If the average dwell time τ_d of an electron in the dot exceeds the ergodic time τ_{erg}, the time each trajectory explores the entire phase space, the motion is chaotic. A description of a chaotic system is possible in the framework of the Random Matrix Theory (RMT) [20]. In the framework of the RMT the static of transport properties of an open chaotic system are evaluated by calculating the scattering matrix of the system which contains random elements. This theory gives the following expression for the averaged magneto conductance of a chaotic quantum dot in the ballistic regime:

$$\langle G(B)\rangle = \frac{\delta G}{1+(\Phi/\Phi_c)^2}, \qquad \text{with } \Phi_c = \frac{h}{e}\left(\frac{\tau_{erg}}{\tau_d}\right)^{1/2} \tag{3.6}$$

Measuring the conductance of the dot with open leads as a function of an external magnetic field for a large number of different dot configurations, i.e., a large number of different gate voltages V_B, it is possible to obtain the statistical ensemble to verify the chaotic behavior of the electron motion in the dot. Fig. 8 displays the averaged conductance of the dot as a function of a magnetic field over 30 individual measurements. Similar experiments have been reported on chaotic dots in [21]. The observed dip around $B = 0$ T is due to the weak localisation effect. The ration τ_d/τ_{erg} is obtained by fitting the lineshape according to (3.6) which yields a value of $\tau_d/\tau_{erg} \approx 5.3$. This result implies that the motion of the electron in the dot is chaotic and the conjecture for the observed deviation from the predicted scaling behaviour of the effective charging energy seems to be appropriate. Further evidence from thermopower measurements in the ballistic regime will be reported elsewhere.

4 Conclusion

We have presented two experimental methodes to determine the scaling behaviour of the charging energy of a quntum dot with the opening of the tunnel barriers, $0 < t < 1$. For transmission probabilities $0 < t < 0.5$ the scaling follows a power-law behaviour according to theoretical predictions. However, using thermopower measurements it was possible to detect oscillations for barrier transmission probabilities far above $t = 0.5$. The measured behaviour seems to contradict the expected vanishing of the Coulomb blockade for $t \to 1$. The observation of a constant effective charging energy for high transmission of the barriers can be explained by taking into account the classical chaotic trajectories of the electrons inside the dot.

Acknowledgments

We would like to thank S.F. Godijn for participating in the thermopower measurements, K. Flensberg, C.M. Marcus and L.I. Glazman for fruitful discussions. Part of this work was supported by the DFG, SFB-341.

Bibliography

[1] For a review see: H. v. Houten, C.W.J. Beenakker, and A.A.M. Staring in *Single Charge Tunnelling*, edited by H. Grabert and M. Devoret, NATO ASI Series B, Vol.294 (Plenum, New York 1992).

[2] L. P. Kouwenhoven, N. C. van der Vaart, A. T. Johnson, W. Kool, C. J. P. Harmans, J. G. Williamson, A. A. M. Staring, and C. T. Foxon, Z. Phys. B **85**, 367 (1991).

[3] E. B. Foxman, P. L. McEuen, U. Meirav, N.S. Wingreen, Y. Meir, P. A. Belk, N. R. Belk, M. A. Kastner, and S. J. Wind, Phys. Rev. B **46**, 10020 (1993).

[4] K. A. Matveev, Zh. Eksp. Teor. Fiz. **99**, 1598 (1991) [Sov. Phys. JETP **72**, 892 (1991)].

[5] C. L. Kane and M. P. A. Fisher, Phys. Rev. Lett. **68**, 1220 (1992); K. A. Matveev and L. I. Glazman, Phys. Rev. Lett. **70**, 990 (1993).

[6] K. Flensberg, Phys. Rev. B **48**, 11156 (1993).

[7] K. A. Matveev, Physical Review B **51**, 1743 (1995).

[8] K. Flensberg, Physica B **203**, 432 (1994).

[9] G. Falci, J. Heins G. Schön, and G.T. Zimanyi, Physica (Amsterdam) **203B**, 409 (1994); H. Schoeller and G. Schön, Physica (Amsterdam) **203B**, 423 (1994).

[10] D. V. Averin and Yu. V. Nazarov, Phys. Rev. Lett. **65**, 2446 (1990). D. C. Glattli, C. Pasquier, U. Meirav, F. I. B. Williams, Y. Jin, and B. Etienne, Z. Phys. B **85**, 375 (1991).

[11] M. Kemerink and L. W. Molenkamp, Appl. Phys. Lett. **65**, 1012 (1994).

[12] P. Lafarge. H. Pothier, E. R. Williams, D. Esteve, C. Urbina, and M. Devoret, Z. Phys. **B 85**, 327 (1991).

[13] A.A.M. Staring, L.W. Molenkamp, B.W. Alphenaar, H. van Houten, O.J.A. Buyk, M.A.A. Mabesoone, C.W.J. Beenakker, and C.T. Foxon, Europhys. Lett. **22**, 57 (1993).

[14] L.W. Molenkamp, K. Flensberg, and M. Kemerink, Phys. Rev. Lett. **75**, 4282 (1995).

[15] I.M. Ruzin, V. Chandrasekhar, E.I. Levin and L.I. Glazman, Phys. Rev. **B 45**, 13469 (1992).

[16] L.W. Molenkamp, H. van Houten, C.W.J. Beenakker, R. Eppenga, and C.T. Foxon, Phys. Rev. Lett. **65**, 1052 (1990).

[17] C.W.J. Beenakker, and A.A.M. Staring, Phys. Rev. B **46**, 9667 (1992).

[18] K.A. Matveev, Phys. Rev. B **51**, 1743 (1995).

[19] I.L. Aleiner, and L.I. Glazman, preprint cond-mat/9710195 (1997), and L.I. Glazman, private communication.

[20] For a review see: C.W.J. Beenakker, cond-mat/9612179, to be published in Rev. Mod. Phys.

[21] C.M. Marcus, A.J. Rimberg, R.M. Westervelt, P.F. Hopkins, and A.C. Gossard, Phys. Rev. Lett. **69**, 506 (1992).

Diode Lasers Based on Quantum Dots

M. Grundmann[1], F. Heinrichsdorff[1], N.N. Ledentsov[1,2], D. Bimberg[1], Zh.I. Alferov[2]

[1] Institut für Festkörperphysik, Technische Universität Berlin, Hardenbergstraße 36, D-10623 Berlin, Germany
[2] A.F. Ioffe Physico-Technical Institute, Politekhnicheskaya 26, 194021 St. Petersburg, Russia

Abstract: Semiconductor diode lasers are presented whose active medium consists of quantum dots (QDs). Laser operation is based on zero-dimensionally localized carriers. High density arrays of uniform QDs are fabricated using epitaxy in the Stranski-Krastanow growth mode. The strong localization of carriers in the QDs leads to non-thermal population statistics directly impacting the gain and threshold. The excitonic nature of recombination leads to a linear gain-current relation. Edge emitting and vertical cavity surface emitting lasers have been fabricated with some of their properties already exceeding the performance of conventional semiconductor diode lasers based on quantum wells.

1 Introduction

A new generation of photonic devices emerges from the incorporation of quantum dots (QDs) into the active layer. The zero-dimensional (OD) carrier confinement and strong carrier localization lead to profound modifications of the electronic structure, density of states, and carrier statistics as compared to quantum wells. Significant improvements were proposed for injection lasers using quantum dots (QDs) as active media [1,2] already more than a decade ago.

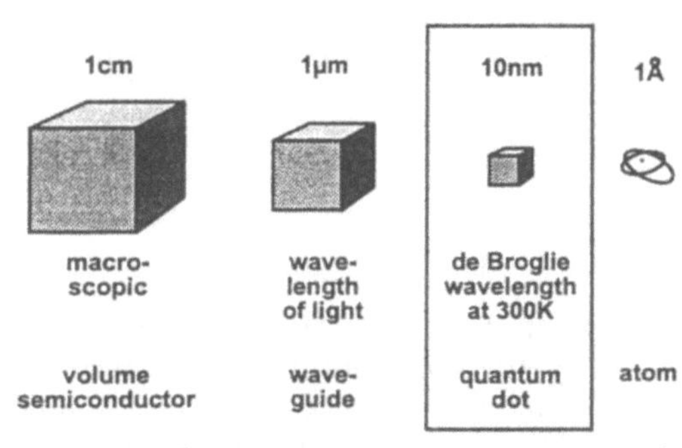

Figure 1
Size scale of quantum dots.

Quantum dots provide the ultimate limit of size quantization in solids (Fig. 1) and result in the strongest possible modification of electronic properties as compared to quantum wells and wires. For laser applications, δ-function like density of states and perfect overlap of electron and hole wavefunctions in a QD result in decreased threshold current density, higher temperature stability of threshold current, higher material and differential gain. On the other hand, until recently, there was a lack of fruitful ideas to fabricate QDs in a way compatible to modern device technology. Furthermore fundamental limitations towards efficient relaxation of carriers in QDs were expected [3]. The limits are overcome now by using Stranski-Krastanow (SK) growth. Simultaneously the possibilities for band gap engineering in strained heterostructures, initially discussed in [4], are extended, e.g. by emission at 1.3 μm or longer wavelengths of InAs QDs on GaAs substrate.

In this paper results are presented for structures in the InGaAs/AlGaAs model material system. SK growth is, however, a general phenomenon observed in a wide class of heterostructures, such as other III-V compounds, Ge/Si, group-III nitrides or II-VI compounds.

2 Growth

A layer of a material having a lattice constant different from that of the substrate, after some critical thickness is reached, can spontaneously transform to an array of three-dimensional islands [5–8]. More recently it was shown, that there can exist a range of deposition parameters, where the islands are small ($\approx$ 10 nm), have a similar size and shape and form dense arrays [9,10]. Due to the strain-induced renormalization of the surface energy of the facets, an array of equisized and equishaped 3D islands can represent a stable state of the system [11,12]. This is directly confirmed for the most extensively studied MBE growth in the InAs/GaAs materials system by the reversibility of the islanding - surface planarization process [13]. Stable arrays of 3D InAs islands have been also fabricated using MOCVD growth [14,15]. When the surface density of QDs is high, interaction of the islands via the substrate makes also their lateral ordering favorable [11,12]. If strained InAs islands are covered with a thin GaAs layer, the islands in a second sheet are formed over the dots in the first sheet (Fig. 2), resulting in a three-dimensional ordered array of QDs either isolated or strongly vertically coupled [16–18]. The size and the shape of InAs islands can be changed by changing the deposition mode. Bright luminescence (300 K) from InGaAlAs QDs in (Al,Ga)As or (In,Ga)P matrices can be tuned in the range 0.7 – 1.39 μm. Modification of the QD PL energy is also possible via post-growth annealing [19,20].

3 Electronic Spectrum and Gain

Experimentally observed energy levels for electrons and holes agree with theoretical calculations based on QD geometry derived by HREM images [21]. The remarkable property of a QD is a narrow (FWHM<0.1 meV) atom-like luminescence line [22,23] which does not show broadening with temperature [23] and directly manifests the formation of an electronic QD.

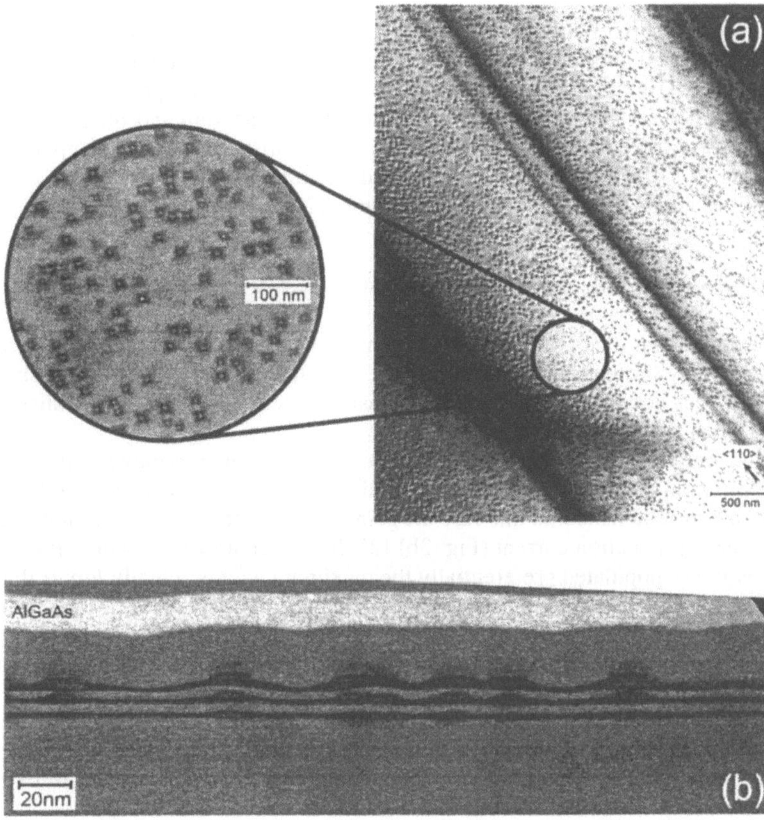

Figure 2 Plan view and cross section TEM images of a 3-fold stack of InAs/GaAs quantum dots grown by MOCVD. The white band in the cross section TEM is an AlGaAs layer [15].

3.1 Carrier Statistics

The carrier statistics in QD ensembles is profoundly modified compared to higher dimensional structures. In double heterostructure and quantum well lasers carriers are injected into a plasma with quasi Fermi levels for electrons and holes. Recombination is bimolecular. Transparency is obtained when the difference of the quasi Fermi levels is equal the band gap.

In a quantum dot laser the injected carriers are captured into laterally uncoupled QDs and form excitons. The carriers do not leave the dot (unless they evaporate at sufficiently high temperatures, see Section 3.2) before they suffer spontaneous recombination or stimulated emission induced by photons in the laser cavity. This represents an extreme *non-equilibrium* situation (NTC, non thermal equilibrium) since dots with different ground state energies have the same population; a Fermi function does not exist. Such NTC situation was first experimentally observed in [24].

The transparency condition in a QD with a spin-degenerate level for electrons and holes is realized when it captures one exciton. Then, the probability to emit or to absorb light is equal [25,26] (in the strong confinement limit). Positive gain is provided by biexcitons; one should also consider charged excitons which provide gain and bleach ground state absorption. As there is no transport between QDs (particularly at low temperatures) the population of QDs with electrons, holes and excitons is defined solely by the capture probabilities and is not a function of temperature [27].

3.2 Temperature Dependence of Threshold

The gain spectrum of a QD ensemble directly reflects the level population and the inhomogeneously broadened density of states. The spectrum has a Gaussian shape. The maximum does not exhibit a shift with increasing injection (see Fig. 3a). Transparency is reached when *all* dots reach transparency.

At high temperatures, thermal evaporation of carriers from dots may result in a predominant population of deeper QDs and a Fermi distribution of carriers over the QD ensemble is eventually reestablished. In this case the gain maximum shifts to higher photon energies with increasing injection current (Fig. 3b) [25,26]. Since now QDs with a lower ground state energy are populated preferentially the maximum gain is actually *larger* than in the NTC case for a given carrier density. This is visualized in Fig. 4 where the maximum gain as a function of carrier density is compared for NTC and ensembles in thermal equilibrium with the lattice at different temperatures. Therefore the smaller threshold is expected for the thermal ensemble. Since the establishment of a thermal ensemble proceeds with increasing temperature, a corresponding *decrease* of threshold is the consequence. An increase of threshold with temperature will result eventually since excited states or barrier levels become populated and represent competing recombination channels or carrier loss mechanisms.

3.3 Gain-Current Relation

The gain-carrier density relation in a QD ensemble is known to be linear [28]. The experimentally observable gain-current relation, however, is not trivial. The formula for the recombination rate in bulk material $R \sim np$, where n and p represent the average electron and hole densities, can obviously not be applied for QD ensembles. Instead rate equations for the micro-states have to be solved [27]. As a consequence of the excitonic nature of the recombination process, the gain-current relation is *linear* before saturation (Fig. 5a). Differences arise between simultaneous (excitonic) and uncorrelated capture of electrons and holes. In the first case only neutral dots exist. In the second case charged dots are also present; the extra carrier does not lead to a recombination current but to a bleaching of absorption and thus to gain at no cost of injection current. Accordingly the lower threshold (solid line in Fig. 5a) follows.

In Fig. 5b the dependence of threshold on the area coverage with quantum dots is shown. Generally the higher the number of quantum dots the larger the expected threshold because more dots have to be inverted. However, if the coverage drops below a certain value ζ_{min} the maximum achievable gain does not suffice to overcome the losses and thus lasing becomes impossible ($j_{th} \rightarrow \infty$). In this case lasing might occur on an excited quantum dot state or even in the wetting layer or barrier [25].

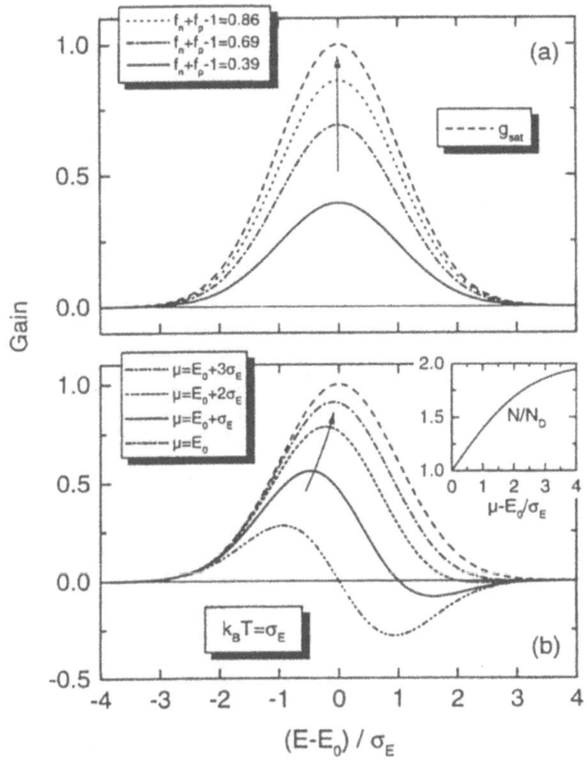

Figure 3
Gain spectra (in units of maximum saturated gain) for a quantum dot array. A two-fold degenerate ground state is assumed. (a) for vanishing thermal escape from the dots (NTC, non-equilibrium population) and different inversion levels. (b) Gain for dot population in thermal equilibrium (Fermi statistics) as a function of the chemical potential μ for $k_B T = \sigma_E$. Inset depicts the carrier number as a function of chemical potential.

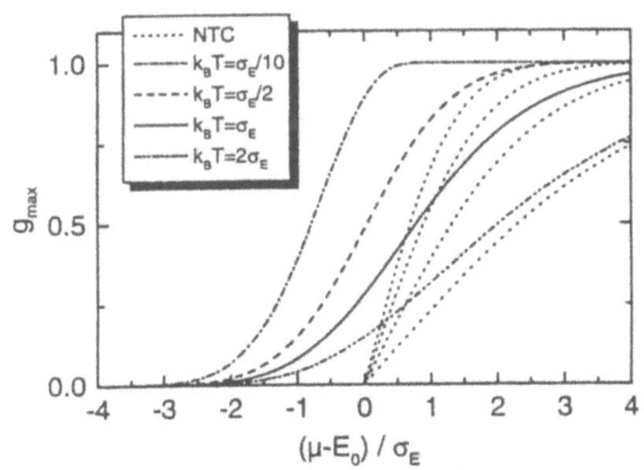

Figure 4
Maximum gain (in units of maximum saturated gain) for a quantum dot array assuming a two-fold degenerate ground state. Fermi distributions for various temperatures (σ_E is the inhomogeneous broadening of the ensemble) are compared to non-equilibrium (NTC) population of dots (dashed lines) for the same carrier number for a given value of the chemical potential μ.

4 Capture and Relaxation of Carriers in a Quantum Dot

Carrier capture is very fast (< 1 ps) in properly designed structures with a high concentration of QDs [30]. The carrier relaxation to the InAs/GaAs QD ground state was found to be about 40 ps for excitation in the GaAs matrix and 25 – 40 ps for direct excitation in QD excited states [30,31]. In the latter case the lifetime of the exciton in the QD excited state and the population time of the QD ground state coincide [30]. Lower values are measured

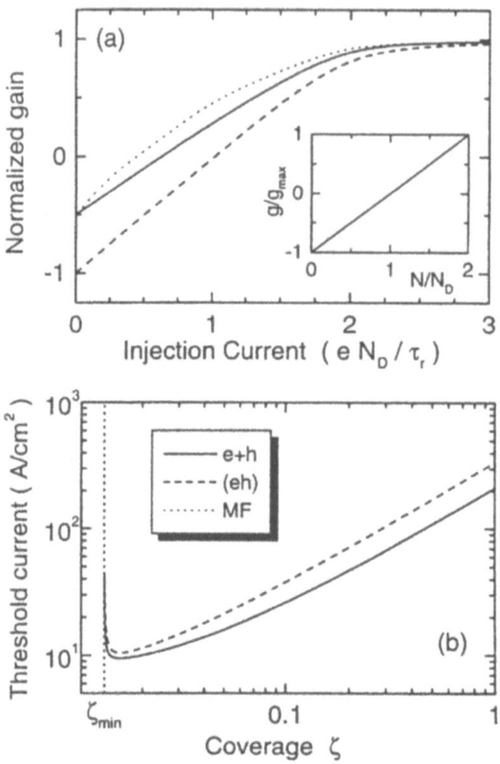

Figure 5
(a) Gain of a quantum dot ensemble as a function of injection current for separate (e+h, solid line) and simultaneous (X, dashed line) capture of electron and holes. For comparison the result from mean field theory (MF, dotted line) is shown. The gain is given in units of the maximum gain. A barrier (wetting layer) which provides an additional recombination channel is included and leads to the gradual saturation at currents $j > 1.5\, e\, N_D/\tau$, N_D being the dot number. We have assumed identical recombination time constants τ_r in the barrier and dot and a capture time into the dot of $\tau_0 = \tau_r/100$ for e, h, and X. The inset shows the relation bet- ween gain and carrier density N (in units of N_D) in the dot ensem- ble, being identical and linear for all models. (b) Threshold current for both capture models and MF theory as a function of coverage for a typical dot ensemble and a total loss of $\alpha_{tot} = 10\,\text{cm}^{-1}$ as a function of area coverage ζ.

when the energy separation between the QD ground and excited state fits to integer number of InAs QD LO phonons [30] whose energy coincides with those calculated in [32]. These results are in agreement with the PLE data also showing enhanced relaxation at multiple LO energies [33]. The modeling of transient experiments requires rate equations for the microstates, whereas a conventional description using average level populations is not adequate, in particular in the case of fast inter-level relaxation and saturation effects [34].

5 Realization of Quantum Dot Lasers

InGaAs/GaAs QD lasers operating under photo-pumped excitation both at low and at room temperature have been fabricated already in 1993 [35]. The first report on injection lasing already revealed the most important advantages of QD lasers [36]. At low observation temperatures very high characteristic temperatures (> 400 K) were observed. At high temperatures, however, evaporation of excitons from dots resulted in a decrease of QD gain. Electroluminescence from QDs in edge geometry is significantly depolarized [13] due to hole quantization in all three dimensions. Lasing energy resonant to the

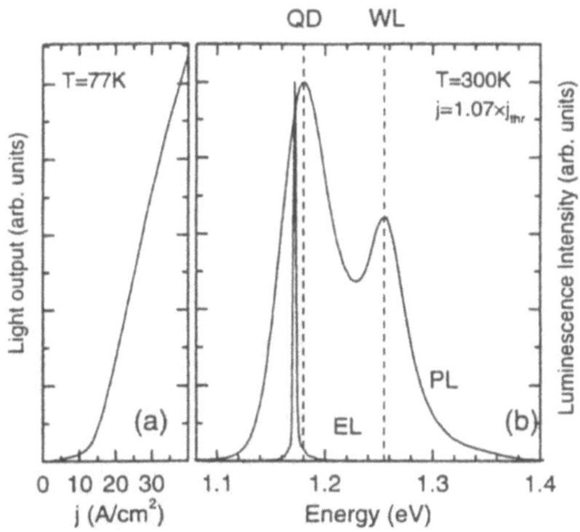

Figure 6
(a) Optical power vs. injection current density at 77 K (threshold 12.7 Acm^{-2}) for MOCVD-grown InAs/GaAs/AlGaAs QD laser. (b) Laser emission at room temperature starts mono- mode on the lower energy side of the QD ground state visible in the photoluminescence spectrum (3× InAs/GaAs/AlGaAs QDs).

ground state (GS) of QDs and low threshold current densities have been demonstrated at low temperatures. At higher temperatures, evaporation of carriers from InGaAs QDs into a GaAs matrix resulted in an increase in j_{th}, and at 300 K the lasing energy comes closer to the wetting layer energy. To overcome this problem, vertically electronically coupled QDs have been applied reducing j_{th} both at low and at high temperatures down to 90 Acm^{-2} (300 K, $N = 10$) and resulting in a GS QD lasing up to room temperature. Ultrahigh material gain and suppression of gain saturation have been demonstrated [24]. In Fig. 6 results are shown for MOCVD grown QD lasers.

Further improvements were obtained by putting coupled QDs in an AlGaAs or InGaP matrix, providing better localization of carriers in QDs. The transparency current density j_{tr} was further reduced down to 60 A/cm^{-2} [18] and room temperature cw lasing with 1.5 W has been realized for a structure with 10 stacks of InGaAs/AlGaAs QDs [37]. High internal (> 96%) and differential (70%) efficiencies at 300 K are also obtained for InGaAs/AlGaAs lasers based on vertically coupled QDs [38].

In Fig. 7 the threshold current of a laser with a three-fold stack of InGaAs/GaAs quantum dots is shown as a function of temperature. As discussed in Section 3.2 j_{th} decreases first, i.e. $T_0 < 0$. After reaching a minimum around 140 K the threshold then increases for $T > 160$ K, probably due to thermal loss of carriers into the barrier. In the entire temperature range the lasing energy (Fig. 7a) stays close to the position of the ground state transition found in photoluminescence.

6 Basic Advantages of Quantum Dot Lasers

6.1 Edge Emitting Lasers

Low transparency currents and complete temperature insensitivity of the laser parameters to temperature for appropriate QD and barrier parameters provide basic advantages of QD lasers. For currently achieved 10% QD relative size dispersion the theoretical threshold current density equals to only 10 Acm^{-2} and much lower values can be obtained in structures with improved QD homogeneity [25,40]. Values of j_{th} of about 12 – 16 Acm^{-2} are

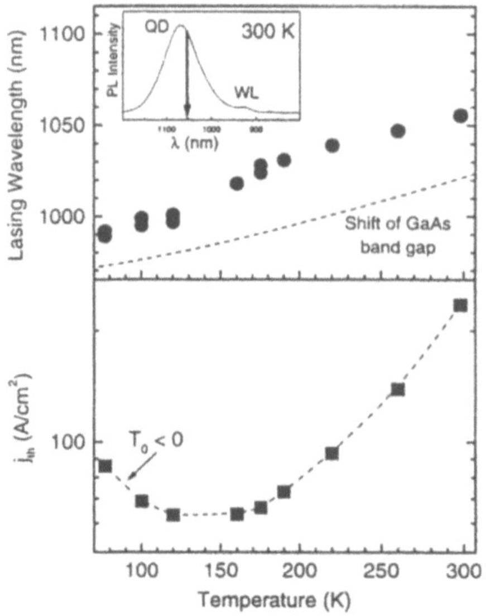

Figure 7
Temperature dependence of QD laser properties for MOCVD grown device incorporating a three-fold stack of InGaAs/GaAs QDs. Cavity length was 1.5 mm, contact stripe width 24 μm. (a) Wavelength of laser emission compared to band gap shift of bulk GaAs. Inset compares room temperature PL spectrum and lasing emission. (b) Threshold current density.

measured for QD lasers in a temperature range up to 180 K [15]. Characteristic temperatures $T_0 > 350$ K have been measured up to 30 °C [41]. Saturation material gain as high as 150000 cm^{-1} is reported in [42,43] as compared to QW values of about 3000 cm^{-1}. This observation agrees fairly well with calculated values [25]. Maximum differential gain at threshold reported for QD lasers approaches 10^{12} cm^2 and exceeds the QW laser values by about three orders of magnitude [24,42].

6.2 Dynamics

Any absorption or gain peak causes a modulation of the refractive index near the resonance energy in agreement with Kramers-Kronig transformation. Thus, the photon wavelength in the crystal can change during the excitation pulse, described by the linewidth enhancement factor α. Highly asymmetric absorption and gain profiles in QWs cause the α-factor to be close to 2. For symmetric QD gain (or absorption spectrum), the change in the refractive index near the gain maximum is zero. In real QDs, due to the finite contribution of the QD excited states to the total gain α is expected to be about 0.5 [44].

Direct observation of relaxation oscillations reveals presently cut-off frequencies close to 10 GHz (Fig. 8) [44–46], limited by gain compression [44].

6.3 Surface-Emitting Lasers

First vertical cavity surface-emitting lasers (VCSEL) operating via the QD GS have been reported for photo-pumped excitation at low temperatures [47]. Injection VCSELs based on stacked InGaAs/AlGaAs QDs were realized later [48]. The lasing at 300 K occurred via excited states and j_{th} was about 5000 Acm^{-2}. Using oxide defined 7 μm mesa and a single sheet of InGaAlAs/GaAs QD structure, injection lasing via the QD GS at about 1000 Acm^{-2} (300 K) has been realized with a ≈ 0.5 mA total threshold current [49]. Using the novel concept of electronically-coupled InGaAs/GaAs QDs and oxide-defined

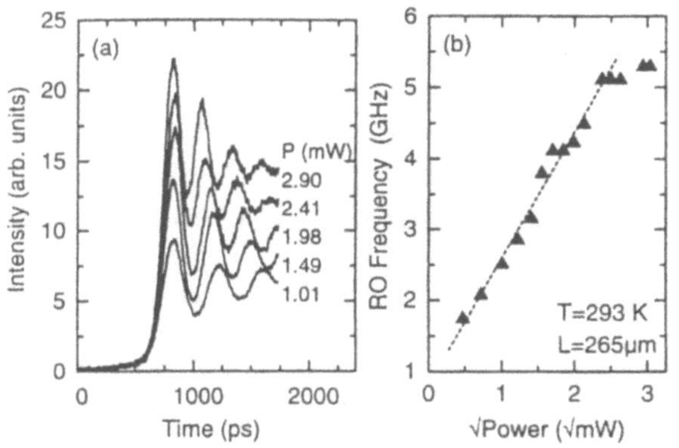

Figure 8
(a) Streak camera traces of relaxation oscillations (RO) of MOCVD grown quantum dot laser with a three-fold stack of InAs/GaAs quantum dots at different power levels. (b) RO frequency versus the square root of the output power per facet, dashed line is best fit of linear region with slope 1.8 $GHz\,mW^{-1/2}$. After [45].

10 μm mesas, cw lasing on the QD ground state with a total current of 180 μA has been recently manifested (180 Acm^{-2}, 300 K) [50]. Maximum efficiencies amount to 16 %. Total currents of 68 μA are measured for 1 μm apertures.

7 Conclusion

It took about seven years since the first realization of photo-pumped lasing in quantum wells [51] until quantum well lasers were created with parameters improved with respect to double heterostructure lasers [52]. The recent progress in the area of injection lasers based on self-organized quantum dots lets us hope that quantum dot lasers with dramatically improved characteristics (low threshold, high temperature stability, low chirp, extended wavelength range) as compared to recent quantum well lasers will be fabricated soon, taking advantage of the fundamentally different electronic structure, gain, carrier statistics, and the large extension of emission wavelengths on standardized material systems.

Acknowledgement

We are grateful for the contributions of many colleagues, in particular M.-H. Mao, N. Kirstaedter, A. Krost, M. Maximov, O. Schmidt, A.O. Kosogov, P. Werner, V.M. Ustinov, and P.S. Kop'ev. Parts of this work have been supported by Deutsche Forschungsgemeinschaft in the framework of Sfb 296, Volkswagenstiftung, and INTAS. N.N.L. is particularly grateful to the Alexander von Humboldt-Stiftung.

Bibliography

[1] Y. Arakawa, H. Sakaki, Appl. Phys. Lett, **40**, 939 (1982).

[2] M. Asada, Y. Miyamoto, Y. Suematsu, IEEE J. Quantum Electron. **QE-22**, 1915 (1986).

[3] H. Benisty, C.M. Sotomayor-Torrès, C. Weisbuch, Phys. Rev. B **44**, 10945 (1991).

[4] G.C. Osbourn, J. Appl. Phys. **53**, 1586 (1982)

[5] L. Goldstein, F. Glas, J.Y. Marzin, M.N. Charasse, and G. Le Roux, Appl. Phys. Lett. **47**, 1099 (1985)

[6] J.M. Moison, F. Houzay, F. Barthe, L. Leprince, E. André, O. Vatel, Appl. Phys. Lett **64**, 196 (1994).

[7] D. Leonard, M. Krishnamurthy, C.M. Reaves, S.P. DenBaars, P.M. Petroff, Appl. Phys. Lett. **63**, 3203 (1993).

[8] N.N. Ledentsov, M. Grundmann, N. Kirstaedter, J. Christen, R. Heitz, J. Böhrer, F. Heinrichsdorff, D. Bimberg, S. S. Ruvimov, P. Werner, U. Richter, U. Gösele, J. Heydenreich, V. M. Ustinov, A. Yu. Egorov, M. V. Maximov, P. S. Kop'ev, Zh. I. Alferov, Proc. 23rd Int. Conf. on the Physics of Semiconductors, Vancouver, vol. 3, 1855 (World Scientific, Singapore, 1995),.

[9] M. Grundmann, N.N. Ledentsov, J.Christen, J. Böhrer, D. Bimberg, S.S. Ruvimov, P. Werner, U. Richter, U. Gösele, J. Heydenreich, V.M. Ustinov, A.Yu. Egorov, A.E. Zhukov, P.S. Kop'ev, Zh.I. Alferov, phys. stat. sol (b) **188**. 249 (1995).

[10] M. Grundmann, Festkörperprobleme (Advances in Solid State Physics) **35**, 123 (1996).

[11] V.A. Shchukin, N.N. Ledentsov, P.S. Kop'ev, and D. Bimberg, Phys. Rev. Lett. **75**, 2968 (1995).

[12] V.A. Shchukin, A.I. Borovkov, N.N. Ledentsov, P.S. Kop'ev, M. Grundmann, D. Bimberg, Phys. Low-Dim. Struct. **12**, 43 (1995).

[13] N.N. Ledentsov, M. Grundmann, N. Kirstaedter, O. Schmidt, R. Heitz, J. Böhrer, D. Bimberg, V.M. Ustinov, V.A. Shchukin, A.Yu. Egorov, A.E. Zhukov, S. Zaitsev, P.S. Kop'ev, Zh.I. Alferov, S.S. Ruvimov, P. Werner, U. Gösele, J. Heydenreich, Solid State Electronics **40**, 785 (1996).

[14] F. Heinrichsdorff, M. Grundmann, A. Krost, D. Bimberg, A. Kosogov, P. Werner, Appl. Phys. Lett. **68**, 3284 (1996).

[15] F. Heinrichsdorff, M.-H. Mao, N. Kirstaedter, A. Krost, D. Bimberg, A. O. Kosogov, P. Werner, Appl. Phys. Lett. **71**, 22 (1997).

[16] Q. Xie, A. Madhukar, P. Chen, N. Kobayashi, Phys. Rev. Lett **75**, 2542 (1995).

[17] G.S. Solomon, J.A. Trezza, A.F. Marshall, J.S. Harris, Phys. Rev. Lett. **76**, 952 (1996).

[18] N.N. Ledentsov, V.A. Shchukin, M. Grundmann, N. Kirstaedter, J. Böhrer, O. Schmidt, D. Bimberg, V. M. Ustinov, A. Yu. Egorov, A. E. Zhukov, P. S. Kop'ev, S. V. Zaitsev, N. Yu. Gordeev, Zh. I. Alferov, A. I. Borovkov, A. O. Kosogov, S. S. Ruvimov, P. Werner, U. Gösele, and J. Heydenreich, Phys. Rev. B **54**, 8743 (1996).

[19] A.O. Kosogov, P. Werner, U. Gösele, N.N. Ledentsov, D. Bimberg, V.M. Ustinov, A.Yu. Egorov, A.E. Zhukov, P.S. Kop'ev, Zh.I. Alferov, Appl. Phys. Lett. **69**, 3072 (1996).

[20] F. Heinrichsdorff, M. Grundmann, *unpublished*

[21] M. Grundmann, N.N. Ledentsov, O. Stier, D. Bimberg, V.M. Ustinov, P.S. Kop'ev, Zh.I. Alferov, Appl. Phys. Lett. **68**, 979 (1996).

[22] J.-Y. Marzin, J.M. Gerard, A. Izraël, D. Barrier, G. Bastard, Phys. Rev. Lett. **73**, 716 (1994).

[23] M. Grundmann, J. Christen, N.N. Ledentsov, J. Böhrer, D. Bimberg, S.S. Ruvimov, P. Werner, U. Richter, U. Gösele, J. Heydenreich, V.M. Ustinov, A.Yu. Egorov, A.E. Zhukov, P.S. Kop'ev, Zh.I. Alferov, Phys. Rev. Lett. **74**, 4043 (1995).

[24] O.G. Schmidt, N. Kirstaedter N.N. Ledentsov, M.-H. Mao, D. Bimberg, V.M. Ustinov, A.Yu. Egorov, A.E. Zhukov, M.V. Maximov, P.S. Kop'ev, Zh.I. Alferov, Electronics Lett. **32**, 1302 (1996).

[25] M. Grundmann and D. Bimberg, Jpn. J. Appl. Phys. **36**, 4181 (1997).

[26] M. Grundmann and D. Bimberg, phys. stat. sol. (a) **164**, 279 (1997).

[27] M. Grundmann and D Bimberg, Phys. Rev. B **55**, 9740 (1997).

[28] K.J. Vahala, IEEE J. Quantum Electron. **QE-24**, 523 (1988).

[29] N.N. Ledentsov, M.V. Maximov, P.S. Kop'ev, V.M. Ustinov, M.V. Belousov, B.Ya. Meltser, S.V. Ivanov, V.A. Shchukin, Zh.I. Alferov, M. Grundmann, D. Bimberg, S.S. Ruvimov, W. Richter, P. Werner, U. Gösele, J. Heydenreich, P.D. Wang, C.M. Sotomayor Torres, Microelectronics Journal **26**, 871 (1995).

[30] R. heitz, M. Veit, N. N. Ledentsov, A. Hoffmann, D. Bimberg, V. M. Ustinov, P. S. Kop'ev, Zh. I. Alferov, Phys. Rev. B **56**, 10435 (1997).

[31] R. Heitz, A. Kalburge, Q. Xie, M. Grundmann, P. Chen, A. Hoffmann, A. Madhukar, D. Bimberg, Phys. Rev. B, April 1998.

[32] M. Grundmann, O. Stier, D. Bimberg, Phys. Rev. B **52**, 11969 (1995).

[33] R.Heitz, M.Grundmann, N.N. Ledentsov, L. Eckey, M. Veit, D. Bimberg, V.M. Ustinov, A.Yu. Egorov, A.E. Zhukov, P.S. Kop'ev, Zh.I. Alferov, Appl. Phys. Lett. **68**, 361 (1996).

[34] M. Grundmann, R. Heitz, D. Bimberg, J.H.H. Sandmann, J. Feldmann, phys. stat. sol. (b) **203**, 121 (1997).

[35] N.N. Ledentsov, V.M. Ustinov, A. Yu. Egorov, A. E. Zhukov, M. V. Maximov, I. G. Tabatadze, P. S. Kop'ev, Semiconductors **28**, 832 (1994).

[36] N. Kirstaedter, N. N. Ledentsov, M. Grundmann, D. Bimberg, V. M. Ustinov, S. S. Ruvimov, M. V. Maximov, P. S. Kop'ev, Zh. I. Alferov, U. Richter, P. Werner, U. Gösele, J. Heydenreich, Electron. Lett. **30**, 1416 (1994).

[37] Yu.M. Shernyakov, A.Yu. Egorov, A.E. Zhukov, A.V. Zaitsev, A.R. Kovsh, I.L. Krestnikov, A.V. Lunev, N.N. Ledentsov, M.V. Maximov, A.V. Sakharov, V.M. Ustinov, Zhao Zhen, P.S. Kop'ev, Zh.I. Alferov, D. Bimberg, Tech. Phys. Lett. **23**, 51 (1997).

[38] M.V. Maximov, *unpublished*

[39] R. Mirin, A. Gossard, J. Bowers, Electronics Lett. **32**, 1732 (1996).

[40] L.V. Asryan, R.A. Suris, Semicond. Sci. Technol. **11**, 1 (1996).

[41] M. V. Maximov, N. Yu. Gordeev, S. V. Zaitsev, P. S. Kop'ev, I. V. Kochnev, N. N. Ledentsov, A. V. Lunev, S.S. Ruvimov, A.V. Sakharov, A.F. Tsatsul'nikov, Yu. M. Shernyakov, Zh. I. Alferov, D. Bimberg, Semiconductors **31**, 124 (1997).

[42] N. Kirstaedter, O.G. Schmidt, N.N. Ledentsov, D. Bimberg, V.M. Ustinov, A.Yu. Egorov, A.E. Zhukov, M.V. Maximov, P.S. Kop'ev, Zh.I. Alferov, Appl. Phys. Lett. **69**, 1226 (1996).

[43] K. Kamath, P. Bhattacharya, T. Sosnowski, T. Norris, J. Phillips, Electron. Lett. **32**, 1374 (1996).

[44] D. Bimberg, N. Kirstaedter, N.N. Ledentsov, Zh.I. Alferov, P.S. Kop'ev, V.M. Ustinov, IEEE J. Selected Topics in Quantum Electronics **3**, 1 (1997).

[45] M.-H. Mao, F. Heinrichsdorff, A. Krost, D. Bimberg, Electron. Lett. **33**, 1641 (1997).

[46] K. Kamath, J. Phillips, H. Jiang, J. Singh, P. Bhattacharya, Appl. Phys. Lett. **70**, 2952 (1997).

[47] R. Schur, F. Sogawa, M. Nishioka, S. Ishida, Y. Arakawa, Jpn. J. Appl. Phys. **35**, L357 (1997).

[48] H. Saito, K. Nishi, I. Ogura, S. Sugou, and Y. Sugimoto, Appl. Phys. Lett. **69**, 3140 (1996).

[49] D.L. Huffaker, O. Baklenov, L.A. Graham, B.G. Streetman, and D.G. Deppe, Appl. Phys. Lett. **70**, 2356 (1997).

[50] J.A. Lott, N.N. Ledentsov, V.M. Ustinov, A.Yu. Egorov, A.E. Zhukov, P.S. Kop'ev, Zh.I. Alferov, D. Bimberg, Electron. Lett., **33**, 1150 (1997).

[51] J.P. van der Ziel, R. Dingle, R.C. Miller, W. Wiegmann, and W.A. Nordland Jr., Appl. Phys. Lett. 26, 463 (1975).

[52] W.T. Tsang, Appl. Phys. Lett. **40**, 217 (1982).

Sonoluminescence: When Bubbles Glow

Detlef Lohse[1] and Sascha Hilgenfeldt[2]

[1] Theoretical Physics, Ludwig-Maximilians University Munich, Theresienstr. 37, D-80333 Munich, Germany
[2] Division of Engineering and Applied Sciences, Harvard University, 29 Oxford Street, Cambridge, MA 02138, USA

Abstract: Sound driven gas bubbles in water can emit light. This phenomenon is called sonoluminescence (SL). Two types of questions arise: 1. What is the origin of the light? and 2. When does SL occur? Whereas the first question is still not completely answered, the second question could be resolved within the hydrodynamical/chemical approach towards sonoluminescence we elaborated in the last three years. For SL to occur, the bubble collapse has to be violent enough to ensure energy transfer from the fluid to the gas in the bubble. Moreover, three kinds of instabilities have to be considered: (i) The bubble has to be shape stable. (ii) Diffusive stability distinguishes between unstable and stable SL. (iii) Chemical stability, i.e., molecular gases dissociate, react to water soluble gases and only inert gases remain within the bubble. Our results resolve various experimental paradoxes and quantitatively account for many measurements.

1 The Phenomenon

In 1990, Felipe Gaitan discovered that a single gas bubble levitated in a strong acoustic field can emit bursts of light so intense as to be observable by the naked eye [1]. This phenomenon, later experimentally explored in detail [2], is called *single bubble sonoluminescence* (SBSL). The energy focusing quality of the system is amazing: Typical acoustical energies employed are 10^{-12} eV/molecule, whereas light photon emission implies energies of ~ 1 eV/molecule. Therefore, the energy is focused by 12 orders of magnitude!

The experimental set up is so simple that sets for home construction are meanwhile commercially available for a couple of hundred dollars [3]. Degassed fluid (normally water) is filled into a glass container several centimeters high. The container is used as an acoustic resonator [4] by driving the fluid through two transducers mounted at opposite sides. As frequency a *resonance frequency* of the cell is chosen, $\omega/2\pi = 20$ kHz – 50 kHz (corresponding to a wavelength in the centimeter range and a period $T = 2\pi/\omega = 20\,\mu$s – 50 μs). Therefore, though the forcing of the piezos is small, a large acoustic pressure can build up in the cell. A standing wave $P(t) = P_a \cos\omega t$ is generated; amplitudes $P_a \sim 1.1$ – 1.5 atm are required for SBSL. Now a single gas bubble either directly injected or created through local boiling of the water can be trapped in the pressure antinode at the center of the cell. The bubble is excited to strong oscillations. In a certain range of the high dimensional space of experimental parameters the bubble emits a short, intense light pulse once per cycle.

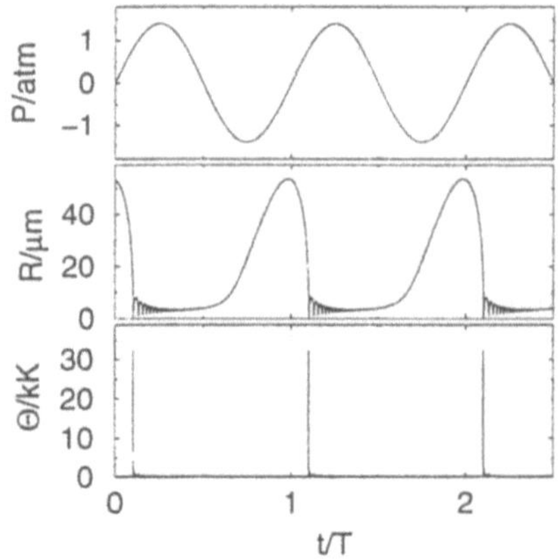

Figure 1
Forcing pressure $P(t)$, bubble radius $R(t)$, and the bubble's interior temperature resulting from the RP-SL approach. The forcing pressure is chosen to be $P_a = 1.4$ atm, the bubble's ambient radius $R_0 = 4\ \mu$m. All material constants are chosen for an argon bubble in water.

2 Bubble Dynamics

To describe the bubble dynamics, we use an approximation which has proved reliable in the context of cavitation dynamics. The idea is to consider the bubble as a perfectly spherical cavity, with the pressure inside the bubble having no spatial variations. The temporal variation of the pressure follows from a polytropic process equation. In this approach the full dynamics is reduced to the *Rayleigh-Plesset (RP) ODE* [5,6],

$$R\ddot{R} + \frac{3}{2}\dot{R}^2 = \frac{1}{\rho_w}\left(p(R,t) - P(t) - P_0\right) + \frac{R}{\rho_w c_w}\frac{d}{dt}p(R,t) - 4\nu_w\frac{\dot{R}}{R} - \frac{2\sigma}{\rho_w R} \tag{2.1}$$

Here, $T = 2\pi/\omega = 37\ \mu$s is the driving period, ρ_w,c_w, and ν_w are the density, speed of sound, and viscosity of water, and σ is the surface tension at the water/gas interface. The pressure $p(R,t)$ and the temperature $\Theta(R,t)$ of the gas inside the bubble are both assumed to be spatially homogeneous; their dependence on each other is modeled by a van der Waals like equation of state [7].

The RP-SL bubble approach allows for very long time calculations of the bubble dynamics, but it completely ignores the dynamics inside the bubble producing the light. Clearly, considering the bubble's interior in a more sophisticated way [8,9] will modify the dynamics of $R(t)$, so will the consideration of heat exchange between the bubble and its surrounding [10,11]. However, our results are robust towards the resulting changes.

A typical example for the dynamics resulting from the RP-SL bubble approach is shown in Figure 1. When the external pressure $P(t) + P_0$ is negative, the bubble expands to ten times its ambient radius R_0 (i.e., the bubble radius at ambient normal conditions of $P_0 =$ 1 atm and 293 K) and beyond. When the pressure increases again, the bubble collapses down to nearly its hard core radius and then oscillates with its eigenfrequency before the cycle repeats. The collapse is so violent and fast that the gas inside the bubble can be heated up to several tens of thousands of Kelvin.

While the experimentally found bubble dynamics seems to be well described by the RP-equation, two questions arise in connection with the SBSL experiment: 1. What is the origin of the light? and 2. When does SBSL occur, i.e., what is the parameter space of SBSL? The second question can be answered within the RP-SL bubble approach and we will do so in Section 4. The first question however still is not completely solved.

The current experimental status and theoretical model considerations concerning the first question will be discussed in the next Section.

3 Origin of the Light

3.1 Experimental Information

The spectra of single SL bubbles are continuous [2], in contrast to those of multi bubble sonoluminescence (MBSL) [12], a phenomenon already discovered in the 30's [13] and later analyzed in detail by Kuttruff [14], in which a bubbly fluid is strongly sonicated (with pressure amplitudes considerably larger than in SBSL) and irregular light pulses can be detected. The MBSL spectra show distinct lines. The differences between the MBSL and the SBSL spectra are not completely understood. A possible interpretation is that the bubble collapse in MBSL is less spherical because of the irregular driving sound field and bubble-bubble interactions. Therefore, the gas inside the bubbles heats up to lower temperatures and less light is emitted.

For most gas bubbles, the SBSL spectrum presumably has its peak in the ultraviolet. It is not, however, directly detectable as water absorbs light below 200 nm wavelength very effectively. Only for xenon bubbles a clear peak is visible at optical wavelengths (around 250 nm [2]). If the visible tail of a typical SBSL spectrum for air bubbles is fitted with the black body radiation formula, temperatures of 25 000 K and beyond seem appropriate.

Another essential experimental information is the width of the light pulse. Up to recently 50 ps was considered to be an upper bound [2], giving rise to various speculations on the origin of the light. It was the ingenious experiments by Gompf, Eisenmenger, et al. [15] using time correlated single photon counting which corrected this former estimate. Meanwhile several further experiments [16,17] have confirmed Gompf et al.'s result [15] that the width of the light pulse is between 60 and 300 ps, depending on the adjustable experimental parameters. Moreover, Gompf, Eisenmenger, et al. [17] found that the light pulse is asymmetric and that this asymmetry increases with increasing forcing pressure. As an example Figure 2 shows the time resolved light pulse, as measured with a streak camera.

Gompf, Eisenmenger et al. [15,17] also obtained the remarkable result that the light pulse in the UV spectral regime is as long as in the red spectral regime.

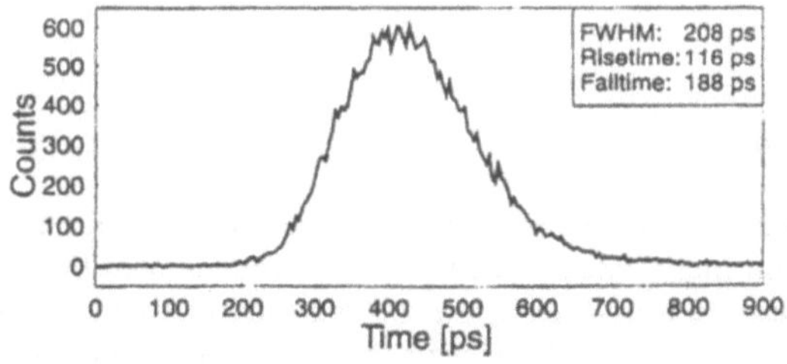

Figure 2
Streak camera measurement of the SBSL light pulse for a forcing pressure of $P_a = 1.3$ atm; the water has an O_2 concentration of 2.1 mg/l and a temperature of 8 °C. The pulse shape is asymmetric with a risetime (10% – 90%) of 116 ps and a falltime (90% – 10%) of 188 ps. The Figure is reproduced from [17] with kind permission by the authors.

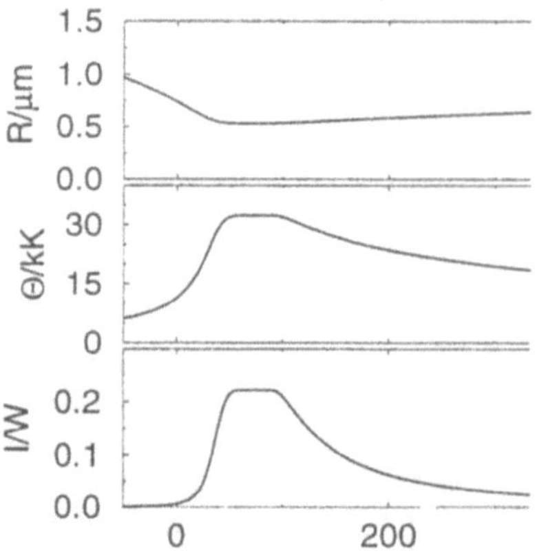

Figure 3
$R(t)$, $\Theta(t)$, and $I(t)$ at the bubble collapse within the RP-SL bubble approach, assuming an adiabatic collapse and black body radiation as assumed light emitting process. $P_a = 1.4$ atm and $R_0 = 4\ \mu$m.

3.2 Theory

Both the very high temperatures as 25 000 K and beyond and the short width of the light pulse (60 ps – 300 ps) and the asymmetry of the light pulse can in principle be accounted for within the RP-SL bubble approach, with the additional simple assumption of black body radiation as light emitting process. In Fig. 3 we show $R(t)$, $\Theta(t)$, and the light intensity $I(t)$ for the example of $P_a = 1.4$ atm and $R_0 = 4\ \mu$m. The asymmetry seen in the pulse also increases with increasing forcing pressure, as in experiment [17].

However, what can not (yet) be accounted for with this simple approach is the uniform pulse width in the red and in the UV spectral domain [15,17]. Only for very large temperatures $k_B\Theta(t) \gg h\nu$ one would expect the length of a light pulse to become independent of its frequency ν, but then one would also expect a higher light intensity than what is seen in the experiments.

Perhaps this paradox may be resolved by considering the spatial inhomogeneities in the bubble and/or the opacity of the gas [9].

4 Phase Space of SBSL

We now come to the second question: When does SBSL occur, i.e. what determines the range of experimental parameters in which SBSL can be observed? This region turns out to be so small that SBSL could remain undiscovered for six decades after light emission from MBSL bubbles was found.

4.1 Experimental information

First, we discuss the main experimental parameters which can be varied. As already stated above, the driving frequency is normally fixed at the resonance frequency of the cell. The adjustable parameters are the driving pressure P_a, the gas pressure overhead p_∞ (corresponding to the gas concentration in the liquid), the water temperature Θ_w, and the chemical mixture of the gas; the quantity often varied in experiment to change the gas mixture

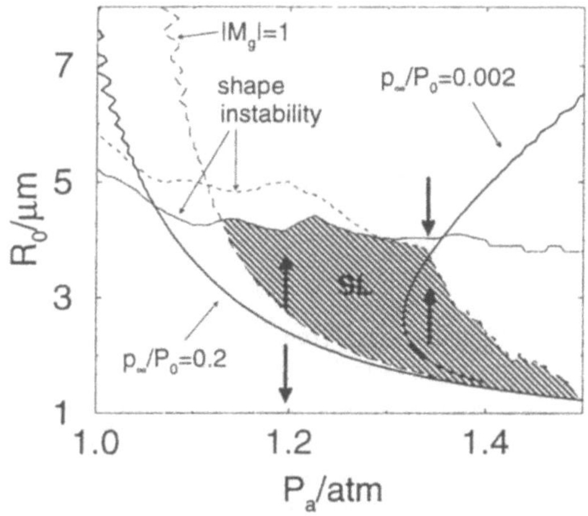

Figure 4
Stability conditions for a bubble in P_a–R_0 parameter space. Bubbles above the line $|M_g| = 1$ fulfill the energy focusing condition. Bubbles below the shape instability lines are stable towards non-spherical surface oscillations. Bubbles on the thick lines are in diffusive equilibrium for $p_\infty/P_0 = 0.2, 0.002$, respectively. The arrows indicate diffusive growth or shrinking.

is the argon percentage ξ_l (or the percentage of another inert gas). SBSL is found only for large forcing pressure $P_a \sim 1.1 - 1.5$ atm and small gas pressure overhead p_∞ (both compared to the ambient pressure P_0).

The ambient radius R_0 of the bubble is not an adjustable parameter but the system chooses R_0 dynamically through exchange of gas with the liquid by diffusion and/or chemical reaction (see Section 4.2 below). It can be measured by Mie scattering techniques [18].

We report the results on argon bubbles first. Two distinct phases of SBSL have been proposed on account of experimental observations [2]: (i) *Unstable SBSL* has been observed in the pressure overhead range $p_\infty/P_0 = 6\% - 26\%$. This phase is characterized by an increase of the relative phase of light emission with respect to the driving on the slow (diffusive) timescale ~ 1 s, followed by a rapid breakdown and another subsequent increase. The light intensity itself behaves in the same way and the bubble is reported to be dancing or jiggling [1,2]. This state of SBSL is also unstable in the sense that often all of a sudden the bubble disappears, i.e., it cannot be trapped in the acoustic field for a very long time. (ii) *Stable SBSL* occurs in argon bubbles at very low gas pressure $p_\infty/P_0 \sim 0.4\%$. The diffusively stable state is characterized by the constancy of the relative phase of light emission over *billions* of cycles, i.e., days (!). The same is true for the light intensity. This remarkable stability may even be considered more surprising than the emission of light pulses.

For air bubbles the same two phases exist, however, for gas concentrations about two orders of magnitude larger than for argon bubbles: Stable SBSL is observed for $p_\infty/P_0 \sim 10 - 40\%$ [1,2,19], unstable SBSL for even higher gas concentrations. A further difference between Ar and air bubbles is that the transition to SBSL via increasing P_a is marked by a sudden breakdown of the radius of air bubbles, whereas for Ar bubbles the transition is smooth.

Also the dependence on the water temperature is dramatic: While at 30 °C typically 10^5 photons are emitted per light pulse, this number increases by more than a factor of 100 at 2 °C [2]. Note, by the way, that this number is still small as compared to the about 10^{10} gas particles in the bubble. Typically only 0.1% of the total energy contained in the compressed bubble are emitted by the light. Most of the kinetic energy of the collapse is lost through sound radiation.

4.2 Theory

All of the above outlined findings can be accounted for within the RP-SL bubble approach. In particular, the necessary conditions for SBSL to occur can be derived within that approach. We will do so in the following Subsections. We illustrate the conditions in Fig. 4, which shows a portion of the P_a–R_0 parameter space.

Shape Stability and Energy Focusing

An essential condition for SBSL is sufficient *energy transfer* from the fluid to the gas bubble and focusing of the energy. We take as criterion for energy transfer and the resulting heating of the gas that the Mach number of the bubble wall (with respect to the speed of sound in the gas) has to be larger than one ($|M_g| = |\dot{R}|/c_{\text{gas}} \gtrsim 1$). The criterion is motivated by the fact that the typical velocity of gas particles is $\sim c_{\text{gas}}$ and if the bubble wall velocity does not have that speed, efficient energy transfer is not possible. From RP simulation, it is easy to compute a corresponding line in P_a–R_0 parameter space (dashed in Fig. 4).

To be able to oscillate for a large number of driving cycles, the bubble should be (spherically) *shape stable* to guarantee sufficient focusing power and to allow for its existence at all. The idea of the RP approximation can be extended to incorporate deviations from spherical shape [20–22,7], so that the onset of *shape oscillations* can also be calculated within the RP-SL bubble approach. Different types of shape stabilities exist [22,7]. The bubble can survive those shape instabilities which act on longer timescales, pinching off microbubbles. This pinch off causes the aforementioned break in the relative phase of light emission in the unstable SBSL state [2] and the recoil of the remaining bubble is the origin of the observed "jiggling" or "dancing" of the bubble. In effect, the spherical instabilities limit the ambient radius of a bubble to $R_0 \lesssim 5\ \mu$m (solid and dashed line in Fig. 4, calculated for room temperature). Together with the energy focusing criterion, this defines a small region of sonoluminescing bubbles in P_a–R_0 parameter space.

Diffusive Stability

Diffusive processes can also be understood within the RP approach to the SBSL bubble [7,23–25]. At first sight this is surprising because there is no diffusive timescale in the RP equation. For a qualitative understanding why this works nevertheless look at the bubble radius $R(t)$ (Fig. 1, middle), resulting from the RP equation. For large $R(t)$ the pressure inside the bubble will be low and gas diffuses into the bubble (rectified diffusion). For small $R(t)$, on the other hand, the bubble will shrink because of the enormous pressure inside [26]. This concept was made quantitative by Fyrillas and Szeri [25] and Löfstedt et al. [24]. The main idea is a separation of the slow diffusive timescale from all timescales in the RP equation. We therefore call this approach the *adiabatic* approximation of diffusion.

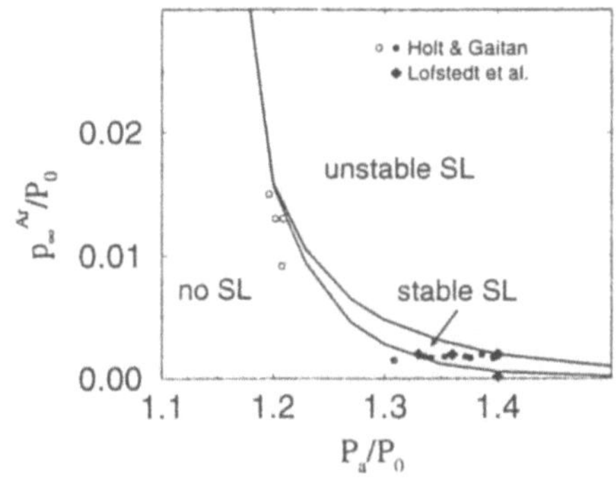

Figure 5
SBSL phase diagram in the p_∞^{Ar}/P_0 versus P_a/P_0 parameter space. Stable SBSL is only possible in a very small window of argon partial pressure. The experimental data points included for comparison refer to observed stable SBSL (filled symbols) or stable non-SL bubbles (open symbols) from refs. [24] (diamonds) and [19] (circles) and show good agreement with the theory. Note that only those data can be included for which all of P_a, p_∞, and ξ_l are known from experiment.

These calculations yield the thick lines in Fig. 4, which give the location of diffusive equilibria in P_a–R_0 parameter space. Also, the stability of the equilibria can be immediately read off from these curves: points with positive slope are stable, those with negative slope unstable. For relatively large partial gas pressure p_∞/P_0 (e.g. 20%) the bubble behaves as described above: if the radius is below the equilibrium line, it shrinks and dissolves, whereas above the line it grows by rectified diffusion, until shape instabilities cause mass loss and limit the growth. These diffusive equilibria are *unstable*, but for high enough P_a and moderate pinch-offs at the shape instability line, *unstable SBSL* can be maintained, with the light phase and brightness changing abruptly at every pinch-off.

For very small p_∞/P_0 (e.g. 0.2%) the situation is entirely different. The curve of diffusive equilibria bends back, displaying a branch of positive slope and thus *stable* equilibria. If this branch intersects the shaded region of SBSL bubbles, *stable SBSL* becomes possible, as the light emitting bubbles have a defined R_0, which does not change by diffusion. These bubbles oscillate in precisely the same way in every cycle, keeping fixed phase and intensity, again in agreement with experiment.

Our main result is the phase diagram [7,27] Fig. 5 in the $P_a - p_\infty^{Ar}$ phase space. It has been obtained through evaluation of diffusive stability curves for many overhead pressures p_∞^{Ar}/P_0. We find that for SBSL stability only the *partial pressure of argon* (or any other inert gas) $p_\infty^{Ar} = \xi_l p_\infty$ is relevant as molecular gases dissociate [27,28] at the high temperatures achieved in the bubble and the dissociation products dissolve in water. This idea will be discussed in greater detail in the next Subsection. For given forcing pressure amplitude P_a and partial argon pressure overhead p_∞^{Ar} we predict with the phase diagram Figure 5 whether the bubble is in the diffusively unstable SBSL state, the diffusively stable SBSL state, or in a non-sonoluminescing state. The phase diagram is in remarkable agreement with the available experimental data. We stress that our theory does *not* contain any fit parameter.

Chemical Stability

Apart from shape instabilities and diffusive instabilities also *chemical instabilities* (i.e., dissociation and reactions) have to be considered [27–29]. The high central bubble temperatures of several 10^4 K are more than sufficient to dissociate any molecular gas constituent, including N_2 and O_2 in the case of an air-filled bubble. The formed radicals immediately react with water molecules to a variety of products, e.g., HNO_2 and H_2O_2.

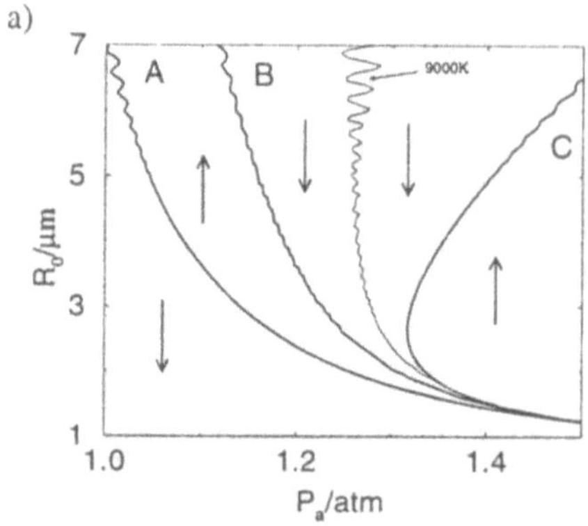

Figure 6 (a) Phase diagram for air at $p_\infty/P_0 = 0.20$ in the R_0 - P_a space. The arrows indicate whether the ambient radius grows or shrinks at this parameter value. Curve A denotes the equilibrium for an air bubble, on curve C the bubble only contains argon. The intermediate curve B necessarily exists because of the topology of the diagram, and represents an additional stable equilibrium. The thin curve shows when the nitrogen dissociation threshold ~ 9000 K is reached. (b) Experimental measurement of the same phase diagram by Holt and Gaitan [19]. Bubbles in shaded areas are shape stable.

These substances are extremely soluble in water and are readily absorbed by the surrounding liquid, leaving only the inert gas component of air, i.e. argon, inside the bubble.

To model this mechanism within the RP-SL bubble approach, we have to assume an equation of state for the gas in the bubble and an Arrhenius type reaction law for molecular gases [27]. This approach allows us to account for the very discrepant SBSL stability regimes (difference of a factor of 100 in the pressure overhead p_∞, see above) for air and argon. Because of this discrepancy, Löfstedt et al. had hypothesized [24] an "as yet unidentified mass ejection mechanism" which we have revealed as chemical here: As molecular gases dissociate at the high temperatures achieved in the bubble at collapse, only the partial argon pressure p_∞^{Ar} is relevant for stability. Air contains one percent argon, so we immediately understand the missing factor of 100. To put it in a nutshell, sonoluminescing "air" bubbles are in fact argon bubbles!

The chemical dissociation mechanism introduces new features into the P_a–R_0 phase diagram of Fig. 4. Consider an air bubble at a partial pressure of $p_\infty/P_0 = 20\%$. For low P_a, the (non-SL) bubble does not heat sufficiently to initiate chemical reactions. Thus, p_∞ is the relevant quantity for the diffusive stability – in this case, the bubble is unstable (curve A in Fig. 6a). For high P_a, the bubble contains only argon; thus, the relevant quantity is now the argon partial pressure $p_\infty^{Ar}/P_0 \approx 0.2\%$ – which allows for stable equilibria (curve C). Because of the topology of the diagram Fig. 6a, another *stable* equilibrium line must necessarily exist in between. Indeed, such a curve (B) is found in our computations. These equilibria describe bubbles for which the mass loss by chemical reactions is exactly compensated by the growth through rectified diffusion; they do not, in general, show sonoluminescence. Holt and Gaitan [19] have measured the phase diagram Fig. 6b experimentally using air, and have identified both stable SBSL equilibria at the expected location for pure argon concentrations (curve C) and stable non-SL bubbles for smaller driving pressures (curve B). The agreement with our theory is very good, indicating that the dissociation mechanism is indeed a key process behind sonoluminescence.

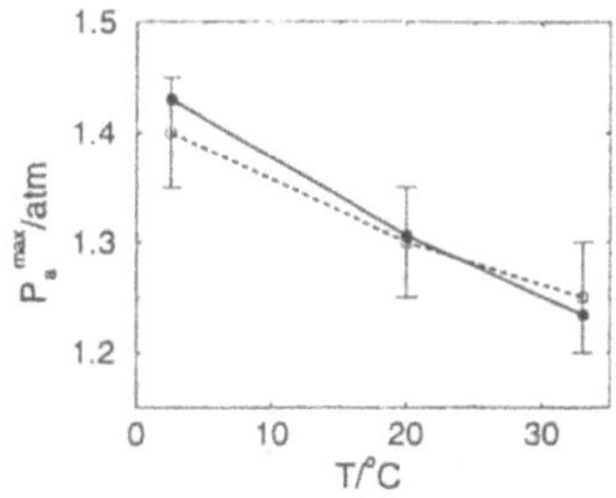

Figure 7
Maximal forcing pressure P_a^{max} before the bubble breaks as a function of water temperature. The open symbols are the experimental values taken from [2], the closed ones the theoretical results from [30].

Dependence of SBSL on the Water Temperature

Finally also the water temperature dependence of SBSL can be accounted for within the hydrodynamical/chemical approach to SBSL [30]. The reason is that the material constants of water depend on its temperature, most importantly, the viscosity, the gas solubility, and the vapor pressure. If these dependences are included in the hydrodynamical approach to SBSL, it turns out that bubbles at low temperature can be forced much stronger before breaking apart – one essential reason for this is the increased viscosity [31] – and therefore emit more light. Figure 7 shows a quantitative comparison between the experimental and theoretical maximal forcing pressure. Again, we stress that there is no fit parameter in our model.

5 Concluding Remarks

The RP-SL bubble approach is a bubble dynamical model which is able to predict and explain many salient features of SBSL. Its success is rather surprising, considering the complexity of the full nonlinear hydrodynamics of the violent collapse of a gas bubble in a liquid, and comparing this to the simplicity of our uniform bubble model. Our picture of "classical bubble dynamics with a twist", namely, including the effects of molecular dissociation and recombination, has so far passed the crucial ordeal for every theory – the comparison of its predictions to experiment, which in this case turns out to be very favorable. The light emission computed from this model (using its most trivial extension, i.e. black body radiation), although not in complete agreement with observation, shows that it is indeed possible to explain pulse time scales below 1 ns purely as a consequence of bubble dynamics. This is contrary to earlier statements and shows that the classical work of Lord Rayleigh and Plesset [6], on which the dynamical models are based, still has its bearing upon today's physical problems.

Bibliography

[1] D. F. Gaitan, Ph.D. thesis, The University of Mississippi, 1990.

[2] B. P. Barber *et al.*, Phys. Rep. **281**, 65 (1997).

[3] R. A. Hiller and B. B. Barber, Scientific American **272**, 78 (1995).

[4] L. A. Crum, J. Acoust. Soc. Am. **57**, 1363 (1975).

[5] C. E. Brennen, *Cavitation and Bubble Dynamics* (Oxford University Press, Oxford, 1995).

[6] L. Rayleigh, Philos. Mag. **34**, 94 (1917); M. Plesset, J. Appl. Mech. **16**, 277 (1949).

[7] S. Hilgenfeldt, D. Lohse, and M. P. Brenner, Phys. Fluids **8**, 2808 (1996).

[8] C. C. Wu and P. H. Roberts, Phys. Rev. Lett. **70**, 3424 (1993); W.Moss, D. Clarke, J. White, and D. Young, Phys. Fluids **6**, 2979 (1994); L. Kondic, J. I. Gersten, and C. Yuan, Phys. Rev. E **52**, 4976 (1995); V. Q. Vuong and A. J. Szeri, Phys. Fluids **8**, 2354 (1996).

[9] W. Moss, D. Clarke, and D. Young, Science **276**, 1398 (1997).

[10] M. Plesset and A. Prosperetti, Ann. Rev. Fluid Mech. **9**, 145 (1977); A. Prosperetti, J. Fluid Mech. **222**, 587 (1991).

[11] A. Prosperetti and Y. Hao, Phil. Trans. Roy. Soc., in press (1998).

[12] T. Matula *et al.*, Phys. Rev. Lett. **75**, 2602 (1995).

[13] H. Frenzel and H. Schultes, Z. Phys. Chem. **27B**, 421 (1934).

[14] H. Kuttruff, Acustica **12**, 230 (1962).

[15] B. Gompf *et al.*, Phys. Rev. Lett. **79**, 1405 (1997).

[16] R. Hiller, S. J. Putterman, and K. R. Weninger, Phys. Rev. Lett. **80**, 1090 (1998).

[17] R. Pecha, B. Gompf, G. Nick, and W. Eisenmenger, Phys. Rev. Lett., submitted (1998).

[18] B. P. Barber *et al.*, Phys. Rev. Lett. **72**, 1380 (1994).

[19] G. Holt and F. Gaitan, Phys. Rev. Lett. **77**, 3791 (1996).

[20] M. Plesset, J. Appl. Phys. **25**, 96 (1954).

[21] A. Prosperetti, Quart. Appl. Math. **34**, 339 (1977).

[22] M. Brenner, D. Lohse, and T. Dupont, Phys. Rev. Lett. **75**, 954 (1995). M. Brenner, T. Dupont, S. Hilgenfeldt, and D. Lohse, Phys. Rev. Lett. **80**, 3668 (1998).

[23] P. Epstein and M. Plesset, J. Chem. Phys. **18**, 1505 (1950); A. Eller, J. Acoust. Soc. Am. **46**, 1246 (1969).

[24] R. Löfstedt, K. Weninger, S. J. Putterman, and B. P. Barber, Phys. Rev. E **51**, 4400 (1995).

[25] M. M. Fyrillas and A. J. Szeri, J. Fluid Mech. **277**, 381 (1994).

[26] A. Eller and H. G. Flynn, J. Acoust. Soc. Am. **37**, 493 (1964).

[27] D. Lohse *et al.*, Phys. Rev. Lett. **78**, 1359 (1997).

[28] M. P. Brenner, S. Hilgenfeldt, and D. Lohse, in *Nonlinear Physics of Complex Systems – Current Status and Future Trends*, edited by J. Parisi, S. C. Müller, and W. Zimmermann (Springer Lecture Notes in Physics, Berlin, 1996), p. 79.

[29] D. Lohse and S. Hilgenfeldt, J. Chem. Phys. **107**, 6986 (1997).

[30] S. Hilgenfeldt, D. Lohse, and W. Moss, Phys. Rev. Lett. **80**, 1332 (1998).

[31] S. Hilgenfeldt, M. P. Brenner, S. Grossmann, and D. Lohse, J. Fluid Mech., J. Fluid Mech. **365**, 171 (1998).

Bloch Oscillations in Semiconductor Superlattices

V.G. Lyssenko[1], M. Sudzius[2], F. Löser, G. Valusis[3], T. Hasche, K. Leo, M.M. Dignam[4], and K. Köhler[5]

Institut für Angewandte Photophysik, Technische Universität Dresden, 01062 Dresden, Germany
[1] permanent address: Institute of Microelectronics Technology, Chernogolovka, 142432 Russia
[2] permanent address: Physics Department, Vilnius University, 2600 Vilnius, Lithuania
[3] permanent address: Semiconductor Physics Institute, 2600 Vilnius, Lithuania
[4] Physics Department, Lakehead University, Thunder Bay, ON, Canada P7B 5E1
[5] Fraunhofer-Institut für Angewandte Festkörperphysik, 79108 Freiburg, Germany

Abstract: More than 60 years ago, Zener has predicted that an electron in a periodic potential subject to a static electric field will perform harmonic oscillations (so-called Bloch oscillations) in real space. We present here results which directly prove the existence of these spatial oscillations: By observing the field-induced shift of the Wannier-Stark ladder, the macroscopic dipole moment associated with the Bloch oscillations of the electron wave packets can be traced. The displacement of the electron wave packet can then be derived as a function of time, with the excitation density as the sole parameter. We discuss the dependence of Bloch oscillation dynamics on the optical excitation conditions. In particular, we show that the motion can be continuously tuned between the harmonic spatial motion envisioned by Zener and a symmetric breathing-mode motion where no macroscopic dipole oscillation is present. The results further show that the dynamics of the wave packets can be significantly influenced by excitonic effects due to the photo-excited holes.

1 Introduction

We are used to consider electronic transport in solids with a static field applied in the picture of drift transport: carriers are treated as quasi-free particles and accelerate till they are subject to a scattering process [1]. It has been shown by Bloch in 1928 [2] that the k-space velocity during this acceleration is linearly proportional to the electric field:

$$\hbar \frac{dk}{dt} = eF. \tag{1.1}$$

Due to scattering, the carrier lose momentum until the acceleration sets in again. The drift velocity of the carriers is then determined by a balance between the momentum and energy gain from the field during ballistic motion and by the momentum and energy changes due to elastic and inelastic scattering processes.

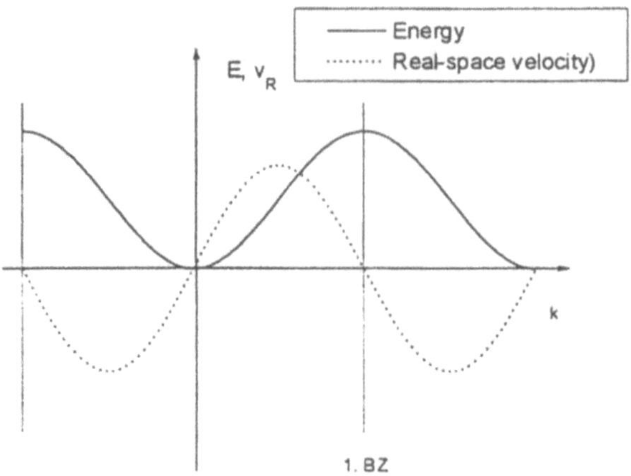

Figure 1
Bloch oscillations in the semi-classical picture: An electron at $k = 0$ starts to move with constant velocity in k-space once the field is turned on. Till the edge of the first Brillouin zone, it gains energy. When it leaves the first Brillouin zone, the energy starts to decrease and the velocity (dashed line) becomes negative. When it reaches the center of the second Brillouin zone, it has returned to its original spatial position (after [4]).

However, the drift transport picture is a specialized case of transport since it requires the carriers to stay close to the lower band edge where they can be described by a parabolic dispersion relation. If the fields are large and the scattering rates low enough, carriers will reach the upper half of the band where the have negative mass and move against the field. If no scattering is present, the crystal momentum will grow linearly with time

$$k(t) = k_0 + \frac{eF}{\hbar}t. \tag{1.2}$$

Due to the periodicity of the band structure, it is immediately obvious that the electron will perform a periodic motion, as first pointed out by Zener [3]. If we assume a harmonic dispersion relation

$$E(k) = \frac{\Delta}{2} - \frac{\Delta}{2}\left(\cos(kd)\right), \tag{1.3}$$

where Δ is the band width, the real-space velocity is given by

$$v_R = \frac{1}{\hbar}\frac{\partial E}{\partial k} = \frac{\Delta d}{2\hbar}\sin(kd) \tag{1.4}$$

i.e. the electron moves forth and back (Fig. 1). It is then easily derived that the spatial position of the electron is harmonically changing with time:

$$z(t) = z_0 - \frac{\Delta}{2eF}\cos(\omega t) \tag{1.5}$$

where

$$\omega = \frac{eFd}{\hbar} \tag{1.6}$$

is the period of the oscillation [3]. Obviously, the total range which is traced by the electron motion [5] is the field-induced localization length

$$L = \frac{\Delta}{eF}. \tag{1.7}$$

In another picture, the oscillation can be regarded as a Bragg reflection of the electron at the lattice once its wavelength reaches half of the lattice period.

The effect just described has never been observed in any bulk solid. For metals, it is straightforward to estimate that for the small electric fields which can be applied, the BO frequencies are many orders of magnitude lower than the scattering rates [4].

A crucial prerequisite for experimental studies of BO was the development of the semiconductor superlattice [6]: by the alternative growth of semiconductor layers with small and large bandgap, a lattice with much larger lattice constant, and therefore much smaller band width, can be designed. In the last few years, BO in superlattices have been addressed by a number of different experimental techniques [7]. One approach have been electric transport studies, which brought some key results for the understanding of high-field transport in narrow bands. For instance, experiments have shown that electrons can reach higher parts of the band [8]. The experiments by Sibille and coworkers [9] have brought a direct observation of negative differential conductivity (NDC). However, it is obvious that classical transport experiments do always integrate over a large number of carriers with random phase. Therefore, BO of individual carriers are masked by the averaged motion of the ensemble.

Another approach which has been recently used is to study the influence of far-infrared (FIR) radiation on the transport in superlattices. It was shown that the $I(V)$-curve of the superlattice can change significantly if FIR is focused onto the sample [10,11]. Recently, the *inverse Bloch oscillator* was observed [12]. In this experiment, the current through the superlattice showed resonances when the frequency of the radiation was tuned through multiples of the BO frequency.

A rather direct approach to observe the oscillations are interband optical experiments. By excitation with a laser, it is possible to generate an ensemble of carriers with a well-defined phase relation. Thus, the oscillating motion lead to a macroscopic dipole moment which can be observed by time-resolved optical detection methods. For instance, transient four-wave mixing [13] (FWM) is very well suited for the detection of the coherent motion since it is only sensitive to coherent signals. The experiments based on pulsed optical excitation were performed by biasing the superlattice first and then creating a wave packet by excitation with a broad laser pulse.

This way of generating a coherent motion has to be distinguished from the Gedankenexperiment of Zener outlined above: In the Gedankenexperiment, the electron is initially described by a wave packet localized around $k = 0$. Once the field is turned on, one can describe it in two pictures: The description given above is based on the extended k-states, i.e. on the eigenstates without applied field (Fig. 2). These eigenstates become time-dependent when the field is turned on. Simply spoken, one has to replace

$$k \longrightarrow k + \frac{eF}{\hbar}t. \tag{1.8}$$

The other description of the wave packet uses the new eigenstates when the field is turned on. These states are the so-called Wannier-Stark states (Fig. 2), which form energetically a ladder [14] with energies

$$E = E_0 + n\Delta E, \tag{1.9}$$

where ΔE is the ladder spacing given by

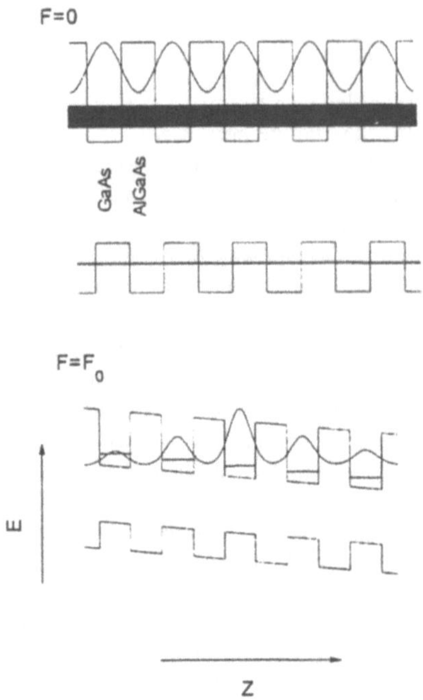

Figure 2
Schematic picture of a Type I semiconductor superlattice without applied field (upper part) and with applied field (lower part). The probabilities amplitudes of the electronic states are also sketched.

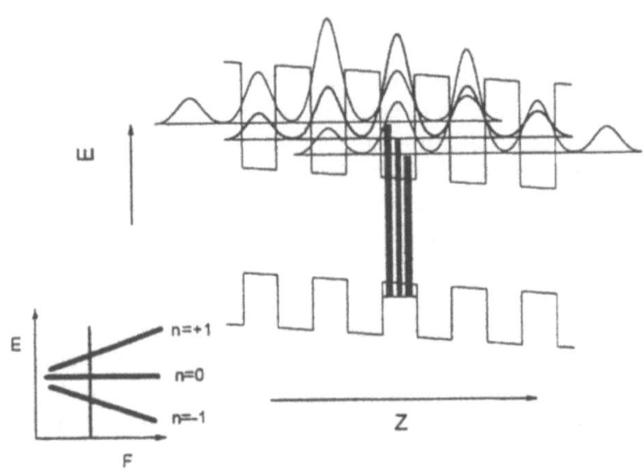

Figure 3
Scheme of the optical transitions in a biased semiconductor superlattice characterized by spatially direct and indirect transitions, leading to a fan chart when the field is swept (inset).

$$\Delta E = eFd, \tag{1.10}$$

Here, n is a integer number and E_0 is the energy of a particular reference state. The Wannier-Stark ladder can be directly observed in optical experiments in superlattices [15,16]. The heavy-holes are rather strongly localized due to their large mass. Thus, the optical transitions between the heavy-hole states and the electrons trace the ladder spectrum as given by (1.9). If the field is swept, a fan chart of the transitions is obtained (Fig. 3).

It is obvious from (1.6) and (1.10) that the temporal periodicity of laser-excited wave packets will be identical to the one in the transport Gedankenexperiment. However, *it is not a priori obvious that the spatial dynamics follow the predictions of Zener:* It has been shown by Ferreira and Bastard that the optically generated wave packets can undergo breathing mode motions which have spatial dynamics very different from transport BO [17].

The two limiting cases are shown in Figs. 4 and 5. These display single-particle calculations of wave packets dynamics (based on the tight-binding approximation as discussed in [17]) for excitation below the center of the WSL (Fig. 4), which leads to a rather well localized wave packet with well pronounced center-of-mass dynamics; and for symmetric excitation (Fig. 5), associated with a breathing-mode motion without any center-of-mass motion. We will show below how both types of motion can be indeed experimentally observed in the optical experiments discussed here. Note, however, that the influence of the Coulomb interaction between electron and hole has strong influence on the dynamics of the wave packets and modify the single particle results presented in Figs. 4 and 5. This will be discussed in detail below.

2 Experimental studies of Bloch Oscillations in Superlattices

In the last years, a number of optical wave packet experiments in superlattices have been performed [18–23]. In these experiments, the wave packets were generated by sub-ps laser excitation of biased superlattices. The first experiments were using four-wave mixing (FWM) [18,19] as detection technique. The first experiment [18] reported a recovery of the FWM signal, with a delay changing with bias field. The data from [19] showed several oscillations of the wave packet with a period tunable over a large range by changing the bias field. Soon after, it was shown [20] that the photoexcited wave packets emit radiation in the THz range. The superradiant character of this radiation was proven recently; the dipole dynamics was inferred indirectly from the emission [24]. After the initial studies, the dynamics of photo-excited wave packets were investigated in detail by further experiments [21–23]. However, these experiments did not prove the original prediction of a harmonic spatial motion of a wave packet made by Zener.

It is interesting to note that BO have been investigated in a totally different system. Using atoms resting in the potential wells of a laser-made optical standing wave, BO were observed in a frequency range about *ten orders of magnitude lower* than in the semiconductor experiments [25]. In this experiment, the composition of the wave packet in k-space was directly measured. The data have beautifully shown the motion of the wave packet through the whole Brillouin zone. Probably, it would have been possible to measure the displacement of the atom wave packets directly since the amplitudes were on the order of μm. However, the spatial motion of a well-localized wave packet is shown indirectly in these experiments by the analysis of the k-distribution. One distinct advantage of this system is the very low dephasing rate compared to experiments in the solid state.

(a)

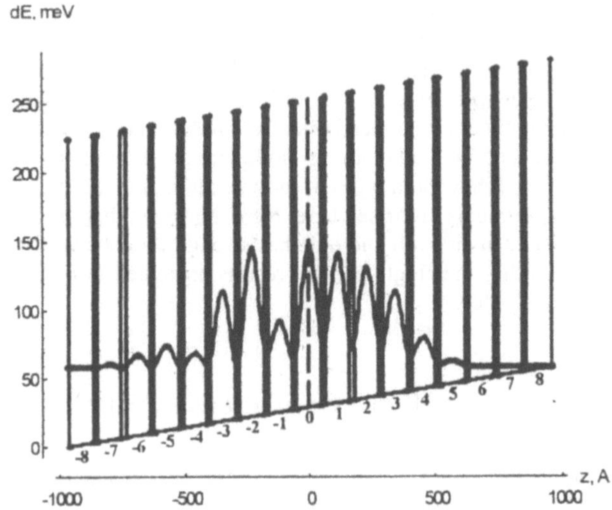

(b)

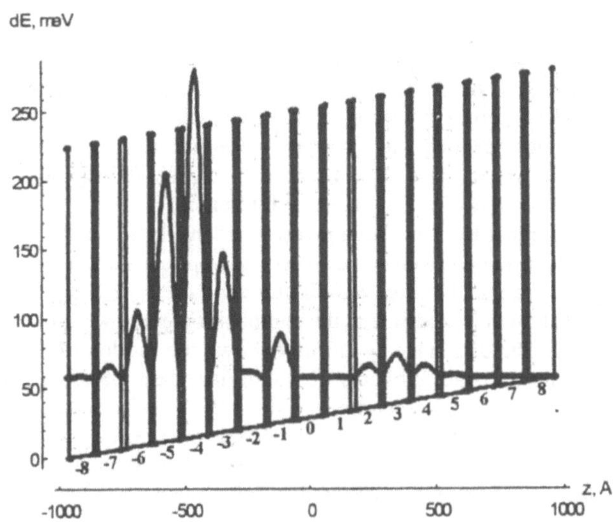

Figure 4 Single-particle calculation of the wave packet dynamics for excitation below the center of the Wannier-Stark ladder. Excited are the $n = -2, -1$, and 0 transitions. (a) wave packet at the time of excitation, (b) after one quarter of a Bloch oscillation. The vertical dashed line marks the center-of-mass of the wave function.

(a)

(b)

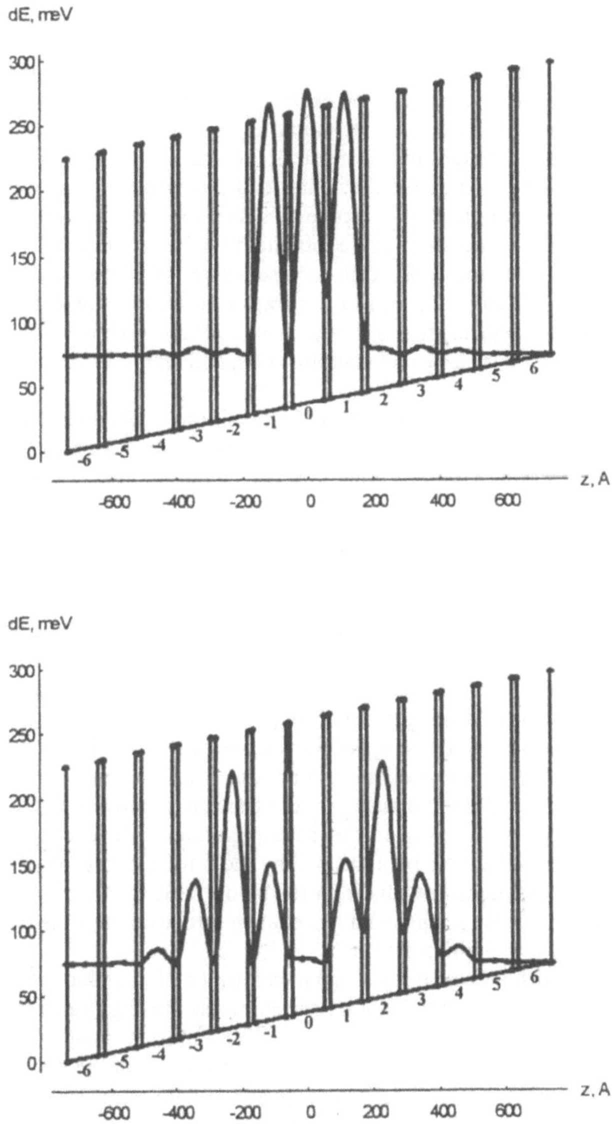

Figure 5 Single-particle calculation of the wave packet dynamics for excitation symmetric to the center of the Wannier-Stark ladder. Excited are the $n = -1$, 0, and +1 transitions. (a) wave packet at the time of excitation, (b) after one quarter of a Bloch oscillation.

3 Real-Space Dynamics of Bloch Wave Packets

3.1 Direct Measurement of the Displacement of the Wave Packet

We have recently introduced a novel experimental technique which is capable of directly resolving the spatial motion of the wave packet [26,27]. The basic idea of the technique is as follows: If the wave packet is created by excitation of electrons and heavy holes, the holes will stay mostly localized due to their much larger mass causing less penetration of the hole wave functions through the SL barriers. The spatial motion of the electrons will thus lead to a dipole moment due to the center-of-mass displacement of the electron wave function. This macroscopic dipole moment is superimposed onto the static bias field in the SL (Fig. 6). The modulation of the total field by the oscillating dipole moment can then be detected by using the WSL itself as a sensitive field sensor: The oscillation of the electric field leads to an oscillation of the transition energies (except for the hh_0 transition, which is field-independent in a first approximation). A characteristic property of these oscillations is that their phase and magnitude depend on the index of the transition (Fig. 6, inset).

A key feature of these oscillations of the WSL transitions is the fact that the energy shift can be related to the relative displacement $z(t)$ of the center-of-mass of the electron and hole wave function. Using Possions equation, one can easily derive that in the static case

$$z(t) = \frac{\varepsilon_0 \varepsilon_r}{ne^2 n_{\text{well}}} \Delta E(t) \tag{3.11}$$

where n_{well} is the carrier density, ε_r is the relative dielectric constant, and ΔE is the energy shift of a Wannier-Stark ladder transition with WSL index n as given in Eq.1.9. When determining $z(t)$, the only parameter needed is the carrier sheet density per well n_{well}, which can, at least in principle, determined with high precision.

It has been recently shown [28] that a dynamic treatment of the modulation of the WSL transitions by the THz field of the oscillating electrons predicts similar motions of the WSL peaks as discussed for the static case, i.e. the transitions move up and down with time as for the limit of $\omega \to 0$. However, there are quantitative deviations from (3.11). These deviations are within the error bars (caused by the limited precision of our density determination) of our experimental results reported below, at least if the $n = 1$ transition is analyzed. However, these deviations need to be taken into account if the displacement is to be determined with very high precision.

When calculating the displacement for a given experimental situation, one has to take into account that the density per well is not uniform, but decreasing from the front to the rear of the SL due to absorption. However, it is straightforward to include this and solve the Poisson equation numerically [26]. Another important point in the derivation of the displacement is the fact that the shifts of the WSL peaks deviate from (1.9) and (1.10) due to excitonic interactions, as discussed below. We thus always use for the analysis the field shift coefficients determined by cw absorption spectroscopy.

In our experiments, we have used spectrally resolved transient FWM to determine the peak shifts. A similar shift could also be observed in pump-probe experiments. However, the FWM experiment suppresses background contributions and yields very well resolved WSL peaks due to the high nonlinearity of the experimental technique.

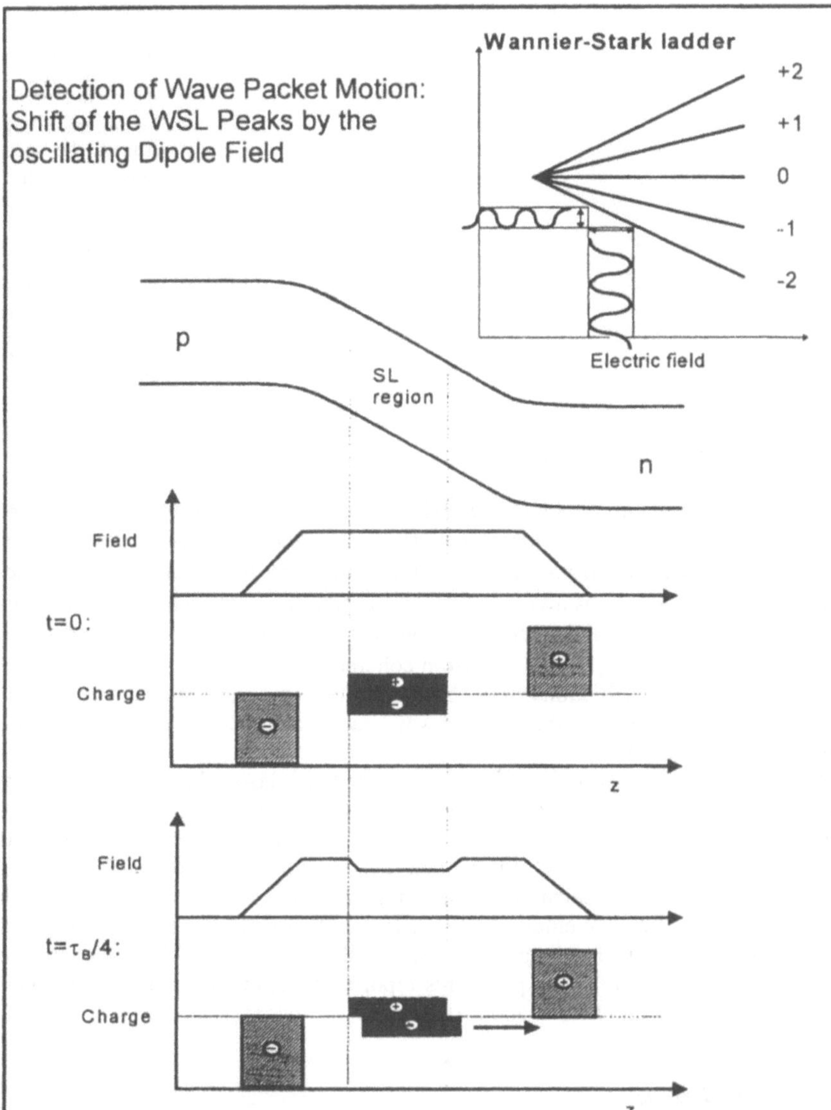

Figure 6 Scheme of the displacement experiment using the Wannier-Stark ladder as field sensor: When both electrons and holes are at the same place, the field is given by the static field created by the space charges of the pn junction. If the electron wave packets move to the right, the field is reduced (bottom), for motion to the other side, the field would be enhanced. The oscillating field associated with the electron oscillation causes a motion up and down the WSL fan chart (upper right corner)

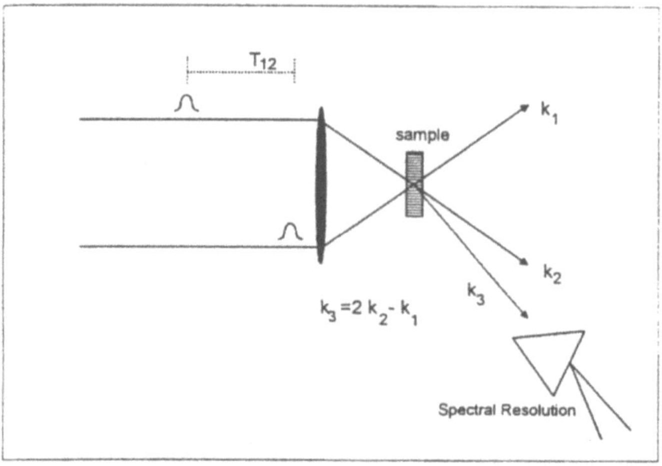

Figure 7
Scheme of the self-diffracted two-beam four-wave mixing experiment: The first pulse in direction $\vec{k}_1$ creates a polarization, the second pulse in direction $\vec{k}_2$ forms a grating. Part of pulse 2 is then diffracted in direction $\vec{k}_3$. In the spectrally resolved variation of the experiment, the diffracted beam is analyzed in a spectrometer.

Experimental Technique and Samples

Most of the experiments which studied BO were using the two-beam transient four-wave mixing technique [13]. In this experiment (Fig. 7), a first optical pulse with wave vector $\vec{k}_1$ creates a wave packet from a superposition of the WSL transitions. A second, delayed pulse in direction $\vec{k}_2$ forms a grating with the oscillating polarization the first pulse has left over. Detected is the light diffracted from this grating into the background-free direction $\vec{k}_3$. In this direction, only signal from coherent polarizations is emitted. In previous experiments, Bloch oscillations were observed in the temporal observation of this signal: the dipole moment between the electron wave packet and the hole will oscillate strongly due to the spatial motion [29].

The experiments reported here are performed in GaAs/$Al_{0.3}Ga_{0.7}As$ superlattices which have been described in detail elsewhere [21]. In particular, we discuss here results which were taken on three SL with 67, 97 and 111 Å well width and 17 Å barrier width. A Kronig-Penney calculation yields miniband (MB) widths of 38, 20, and 12 meV, respectively. The samples are labeled in the following according to their well/barrier width, e.g. 97/17. As laser source, we employ a 120 fs Ti-Sapphire laser. For the spectrally resolved transient FWM, the detected signals are spectrally resolved by a CCD-camera coupled to a spectrometer. In all the experiments reported here, the samples are mounted in a close-cycle cryostat and held at about 10 K.

Experimental Results

Figure 8 shows a three-dimensional plot of the FWM data. In the temporal dynamics, the data clearly show the Bloch oscillations. Along the spectral axis, the data reveal a pronounced modulation due to the occurrence of the different WSL transitions. However, the salient feature of the data are only visible by a careful inspection of the transient spectra: Along with their strong amplitude oscillations as a function of delay time, the WSL peaks show a small oscillating spectral shift as a function of delay time. Figure 9 displays the transient FWM spectra for the 67/17 sample. The spectra shown correspond to delay times of the first part (solid lines) and the second part (dashed lines) of one BO

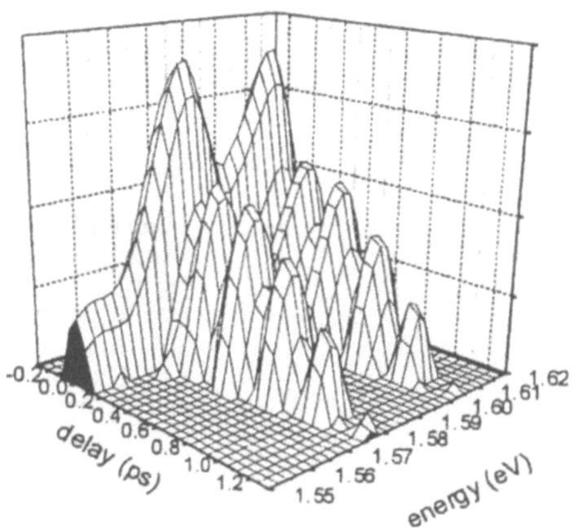

Figure 8
Three-dimensional plot of the low-temperature four-wave mixing signal as a function of the delay time and the voltage across the sample. Flat band is at +0.5 V. The sample is a 97 Å GaAs/17 Å $Al_{0.3}Ga_{0.7}As$ superlattice. (from [36])

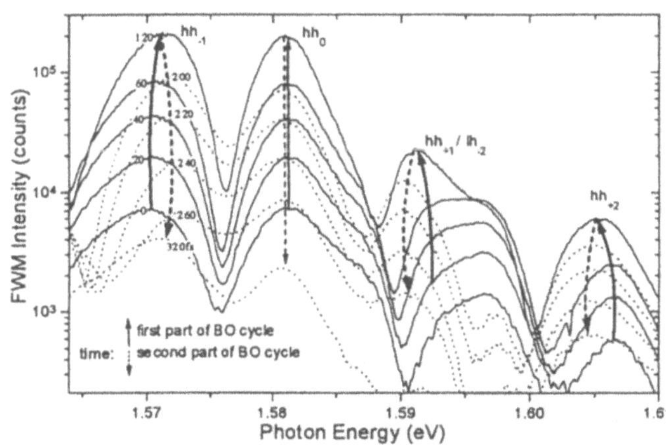

Figure 9 Shift of different transitions of the Wannier-Stark ladder in the 67/17 sample for various delay times during one oscillation cycle. Solid arrows mark the behavior during the first part of the cycle, dashed arrows during the second.

cycle. Visible are the center transition of the WSL (hh_0), the first transition below the center (hh_{-1}), and two transitions above the center. The hh_{-1} transition shows a shift to lower energy during the first part of the BO cycle and shifts back during the second part of the Bloch cycle. The hh_0 center transition virtually stays at constant energy. The hh_{+2} transition shifts first to higher energy and then to lower energies, with an amplitude which is larger than for the hh_{-1} transition (but less than twice the shift, as expected for the static model Dynamic theory confirms the smaller shift [28]). The experimental behavior is thus in excellent agreement with the predictions of the model as discussed above and outlined in Fig. 6. The behavior of the hh_{+1} transition is complicated due to an anticrossing with the lh_{-2} transition. The data in Figure 9 were taken with a broad band laser excitation above the center of the WSL. As will be discussed later, such excitation creates a large number of free carriers which lead to fast interband dephasing. Thus, the Bloch oscillation signal as traced by FWM (where the signal strength is determined by the interband polarization) decays very quickly.

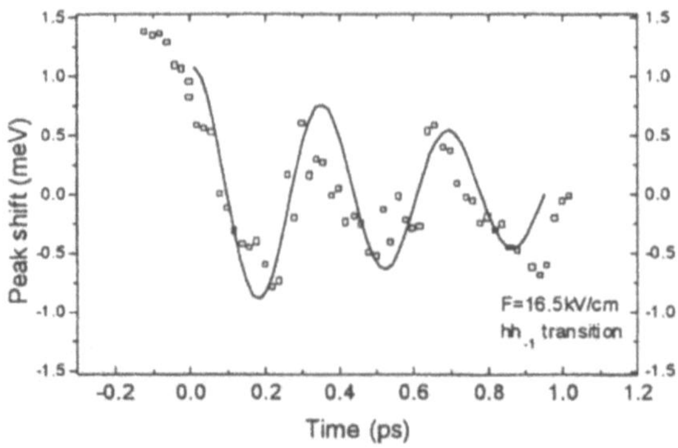

Figure 10 Wannier-Stark ladder peak shift as a function of delay time for a 67 Å/17 Å superlattice. The solid line is a damped harmonic with a damping time of 1.2 ps.

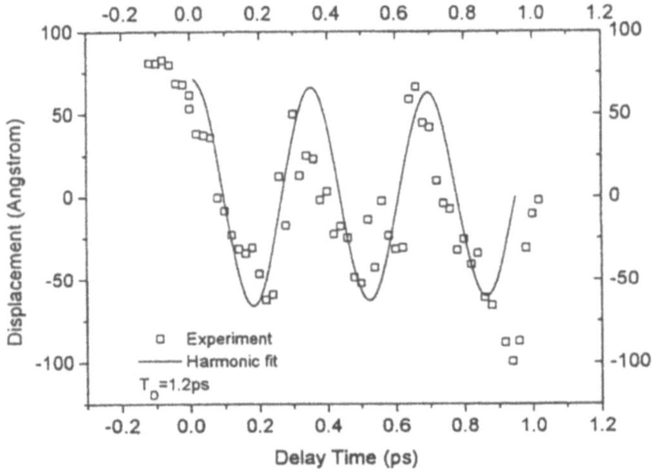

Figure 11 Bloch electron displacement as a function of delay time for a 67 Å/17 Å superlattice.

When determining the detailed temporal shift behavior, we have taken data where the excitation was well below the center of the WSL. The shift of the hh_{-1} WSL transition is displayed in Figure 10. To obtain these data, a rather high excitation density was chosen to make the effect of the field more pronounced. However, we are still in the perturbative limit since the transient field is below 10% of the static bias field in the present experiments. The peak positions plotted in the Fig. 10 are obtained by a fit of a Gaussian to the spectra. The data show an oscillating shift of the peak with delay time, with a decrease of the amplitude with increasing delay time. We interpret this decay of the field shift being due to a decrease of the number of coherent carriers which participate in the BO: Carriers which have lost their coherence do not contribute to the oscillating electric field anymore. A comparison of the dephasing time of the FWM signal with the decrease of the field shift amplitude shows supports this explanation. The solid line is a harmonic function assuming an intraband dephasing time of 1.2 ps, a damping constant we have also obtained for the given experimental conditions in independent measurements [30].

Figure 11 shows the displacement of the wave packet as derived from the peaks shifts

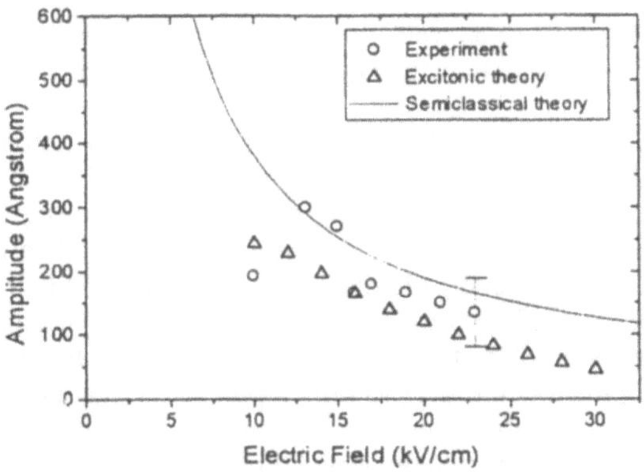

Figure 12
Total oscillation amplitude as a function of the electric field. Circles: Experiment, Line: Semiclassical theory, Triangles: Theory including optical excitation of the wave packet and excitonic interactions.

using the procedure described above. The motion of the wave packet can be described by an oscillation with a total amplitude of approximately 160 Å.

The field detection by shifts of the WSL can also be used to trace the dynamics of *incoherent* carriers: electrons and holes which have lost the coherence are moved by drift transport to the contacts. This will screen the built-in field and lead to a linear shift superimposed onto the harmonic shift of the WSL transitions, which was subtracted in Fig. 10. Assuming that the number of carrier which become incoherently transported away is given by the same 1.2 ps decay as observed for the oscillation, we can roughly derive the average carrier velocity from the field shift. Using a simple analysis, we obtain a drift velocity of the carriers of 3×10^6 cm/s. Due to the uncertainty, mainly in the number of carriers which result from dephased excitons relative to the number of free carriers created by direct photoexcitation, this number is only an order-of magnitude estimate.

Figure 12 displays the dependence of the total Bloch oscillation amplitude on the bias field. The experimental data are the circles; the solid line is the semiclassical limit given by (1.7). Within error, the experimental data agree with the semiclassical theory. In particular, the amplitude is inversely decreasing with electric field as predicted by Zener. For a quantitative discussion of the amplitudes, one has to take into account that the properties of the WSL are strongly modified due to excitonic effects [31]. This leads also to pronounced changes of the BO dynamics, which are discussed in the next section in more detail. The triangles in Fig. 12 show a calculation using the excitonic model described in [32], with the parameters of our experiment. The amplitudes are somewhat lower than the semiclassical limit and in good agreement with the experimental data. The main reason for the lower amplitudes predicted by the excitonic theory (in comparison to the semiclassical results) are not the electron-hole coupling, but the fact that the laser excitation chosen here was rather close to the center of the WSL.

3.2 Control of the Amplitude by Changing Laser Excitation: Tuning Between Breathing Modes and Spatial Oscillations

As already discussed in the introduction, the displacement dynamics and the absolute amplitude of the optically excited BO depend critically on the way the wave packet is composed by the laser excitation. The examples shown in Figs. 4 and 5 are based on single-particle calculations. Since the binding energy of excitons in superlattices is on the same order than the WSL splitting, it can be expected that excitonic effects will play a role in the dynamics of the wave packets as well. We thus need to discuss the excitonic effects in some more detail to understand how they influence the spatial dynamics of the wave packets.

The effects of the excitonic interaction on the optical transitions in superlattices have been investigated in detail by theoretical [31,33] and experimental studies [34]. The three main effects are (i) the WSL fan chart shows a strong asymmetry due to the fact that the transitions below the center (negative n) gain higher oscillator strength compared to the transitions above the center (ii) the energetic distance between the WSL transition is not constant, but varies from transition to transition. In simple terms, this is caused by different excitonic binding energies for the different transitions. (iii) as also shown experimentally [34], the development of the WSL is fully suppressed when the miniband band width is below the excitonic binding energy. In this regime, it is still possible to investigate quantum beats [35], but the behavior expected for BO is not observed.

Dignam, Sipe and Shah have theoretically investigated the dynamics of Bloch wave packets with and without the excitonic interaction [32]. In particular, they have calculated the displacement of the wave packet as a function of laser wavelength and band width. The main results of the theory can be summarized as follows: (i) the upper limit of the amplitude as given by the semiclassical theory (Eq. 1.7) remains largely unchanged and can be nearly reached if the center of the excitation is chosen well below or above the center of the WSL. This corresponds to forming a wave packet localized around $k = 0$ and $k = \pi/d$, respectively. (ii) for excitation close to the center of the WSL, the amplitude of the wave packet reaches a minimum, both in the single-particle and the excitonic picture. When the Coloumb interaction is neglected, one can reach exactly zero amplitude (breathing mode motion), as displayed in Fig. 5. This corresponds to a wave packet distributed over the whole Brillouin zone. When excitons are included, an asymmetry for excitation below and above the center of the WSL develops, mainly since the oscillator strengths are asymmetric due to the excitonic interaction.

We have recently shown [36] that the spatial amplitude of the oscillating wave packets can be tuned between symmetric breathing-mode oscillations and a harmonic spatial motion with the amplitude as predicted by Zener. Bloch-oscillating center-of-mass motion is achieved for excitation well below and above the Wannier-Stark-ladder center whereas breathing mode motion is achieved for symmetric excitation of the WSL.

Figure 13 displays the laser excitation conditions of this experiment. Shown as example are FWM (solid lines) and laser spectra (dashed lines) for excitation below the center of the WSL (a), close to the center (b) and above the center (c). The solid lines are FWM spectra with the hh transitions labeled. Note that we have chosen a field region where the center transition hh_0 is rather weak. As discussed below, this will lead to the observation of higher harmonics in the signals. The relative position of the laser center wave length is defined in units of the WSL splitting, with ω_c as the distance relative to the (experimentally observed) hh_0 transition.

Figure 14 shows the shift of the hh_{-1} peak vs. delay time for various excitation conditions. The time range shown is during the first oscillation, with laser excitation still present. This time range has to be analyzed in our experiment despite the fact that the polarization is still built up by the laser. However, for excitation close to or above the center, the BO are damped out very quickly by fast loss of the interband coherence due to the large number of photo-excited free carriers. This causes fast damping of the FWM signal by interband dephasing, although the intraband coherence (which is the relevant dephasing for the BO) is still present. Lowering the excitation density is not possible since we need a certain minimum density to observe the small shifts of the WSL. For excitation

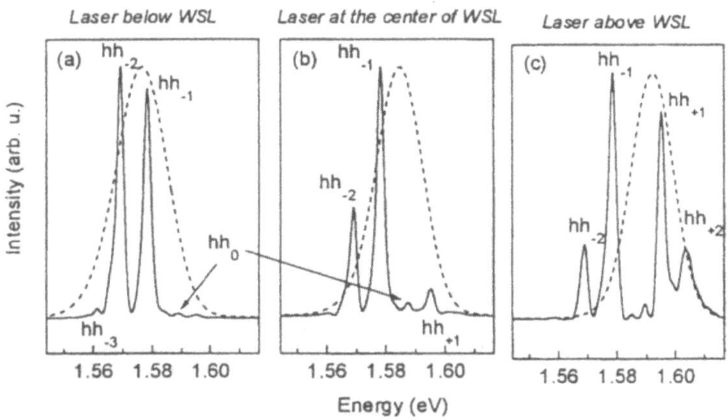

Figure 13 Laser excitation conditions: FWM spectra (solid lines) showing the WSL heavy-hole (*hh*) transitions. Shown are three examples for excitation below the center of the WSL; (a) close to the center (b) and above the center.

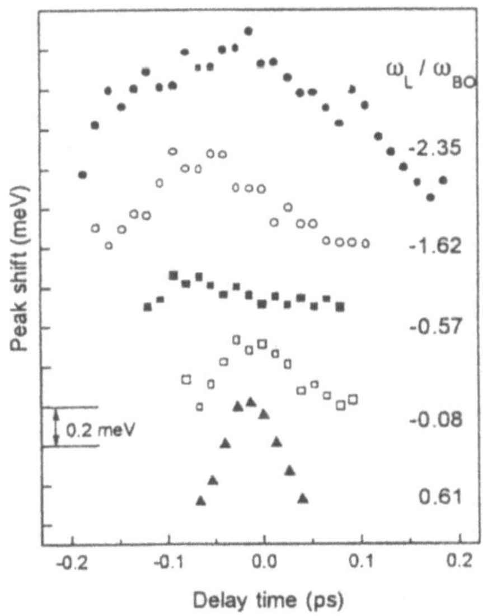

Figure 14
Shift of the hh_{-1} peak vs. delay time for various excitation conditions.

below the WSL, pronounced BO are still observed in the FWM signals [36]. The FWM response shows double BO frequency components due to the suppression of the hh_0 WSL transition in the field range we have investigated.

Figure 15 shows the amplitude (as derived from the FWM peak shift) as a function of the excitation energy (the center of the laser spectrum). The energy is once again given in units of the WSL splitting relative to the experimentally observed energy of the hh_0 transition. The total amplitude of the center-of-mass oscillation of the wave packet is given in units of the WSL period (i.e. 84 Å in this case). The data are taken at a field of 13 kV/cm and an excitation density (in the first well) of 1.2×10^9 cm^{-2}.

The data clearly show that the dipole amplitude is close to zero when the excitation is near the center of the WSL. For excitation well below and above the center of the WSL,

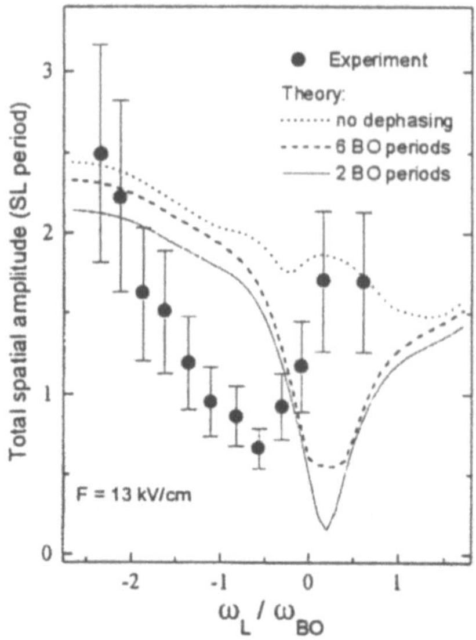

Figure 15
Dots: Experimentally determined amplitude of the Bloch-oscillating wave packet as a function of the excitation energy at a field of 13 kV/cm. Lines: Theoretical calculation of the amplitude for various field, with the thick line corresponding to the field of the experiment. The energy is given in units of the WSL splitting relative to the experimentally observed energy of the hh_0 transition.

the amplitudes reaches several periods of the SL (the semi-classical amplitude as given by (1.7) would be about 3.4 times the SL period and is indicated as the dashed line in Fig. 15).

For comparison to experiment, we have performed theoretical calculations for our specific sample structure using a slightly modified algorithm based on the work of Dignam et al. described above [32]. The comparison of the experiment with theory (full lines in Fig. 15) shows that qualitative agreement between experiment and theory exist concerning the behavior as a function of laser energy: The theoretical curves also show a minimum for excitation close to the center of the WSL. However, this minimum is only as pronounced as in the experiment if dephasing is taken into account (dashed and solid lines in Fig. 15). The theoretical curve neglecting dephasing (dotted line) predicts that the minimum amplitude is still about 1.5 periods of the SL, whereas the experiment reaches nearly zero amplitude. If dephasing is taken into account, the agreement is much better. Given here are curves where the inverse dephasing time is 6 (dashed line) and 2 (solid line) BO periods. The minimum amplitude reaches then values which compare well with the experimental value (with its rather large error). The inclusion of damping also improves the agreement of the energy of the amplitude minimum.

The influence of damping on the amplitude is caused by a rather subtle effect of wave packet physics: Since the energy spacing of the excitonic WSL levels are not equal, different components of the wave packet oscillate with different frequency. The minimum of the amplitude for excitation near the $n = 0$ transition is, in simple terms, caused by the fact that the equal excitation of levels with negative and positive n leads to two components of the wave packet moving in opposite direction and canceling each other. For excitation near the $n = 0$ WSL level, the wave packet is thus initially created with a breathing-mode-type motion. However, due to the different frequencies caused by the exciton-induced distortion of the energy ladder, the motion of the two constituents does not cancel anymore after some delay time, and the overall motion continuously evolves from a breathing-mode motion into an asymmetric motion with considerable total amplitude.

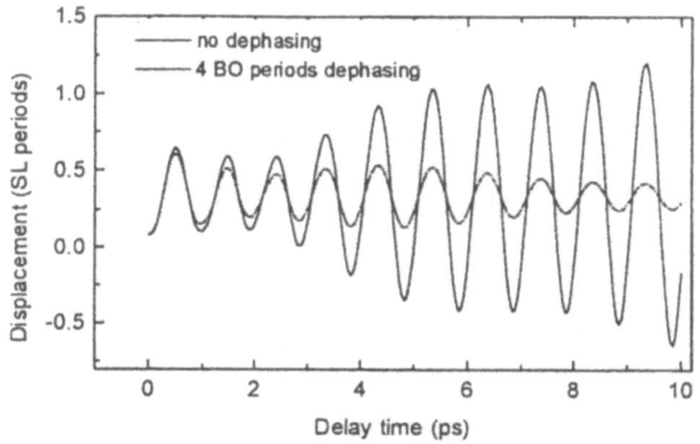

Figure 16 Effect of the dephasing in the time evolution of displacement for symmetric excitation of the wave packet, when excitonic effects are included. Solid line: no damping; dashed line: damping time of 4 BO periods.

In a real system, dephasing reduces the number of coherent carriers quite quickly, so that the amplitude can only be derived for small delay times. The large amplitudes to the out-of-phase motion of the constituents of the wave packet has not developed for these small delay time, and thus the low amplitudes of the initial breathing mode motion is detected. Figure 16 shows a theoretical calculation of the effect of the dephasing in the time evolution of dipole for the wave packet excited with the laser central frequency at the $n = 0$ WSL state. The development of the oscillating dipole amplitude with increasing delay time when dephasing is neglected is clearly visible.

A major difference between experiment and theory is that the latter predicts that the minimum of the amplitude to be reached for excitation *above* the center of the WSL, whereas the former observes minimum amplitude *below* the center. At present, we cannot offer conclusive explanations for this deviation. Possible reasons are the many subtleties of the experimentally observed WSL ladder which up till now are not taken into account: light-hole (*lh*) transitions, anti-crossings of *hh* and *lh* transitions, anticrossings with transitions of higher minibands, optical excitation of free carriers in the exciton continua, etc. In the next section, we show for some selected examples how these effects change the dynamics.

3.3 Influence of Light-Holes and Anticrossings in the Wannier-Stark Ladder on Bloch Oscillation Dynamics

In the previous sections, we have seen several deviations of the results from the predictions of theory, even if excitonic effects were taken into account. For instance, it is visible in Fig. 11 that in the low-field region, the total amplitude deviates from the semiclassical theory and from the theory including excitons. This results from the fact that the WSL in a real superlattice sample is much more complicated than the predictions of the single particle model (1.9) or even the prediction of a theory for the heavy-hole excitons [31].

An illustrative example for the complexity of the optical transitions in a biased superlattice is shown in Fig. 17, which displays linear absorption data of the 111/17 sample as a function of bias voltage and photon energy. Note that this is a state-of-the-art sample and that the complex transitions are not caused by inferior sample quality. The energy range is chosen large enough to observe both the first and the second miniband. For both minibands, the heavy-hole and light-hole exciton transitions are visible in the flat-band

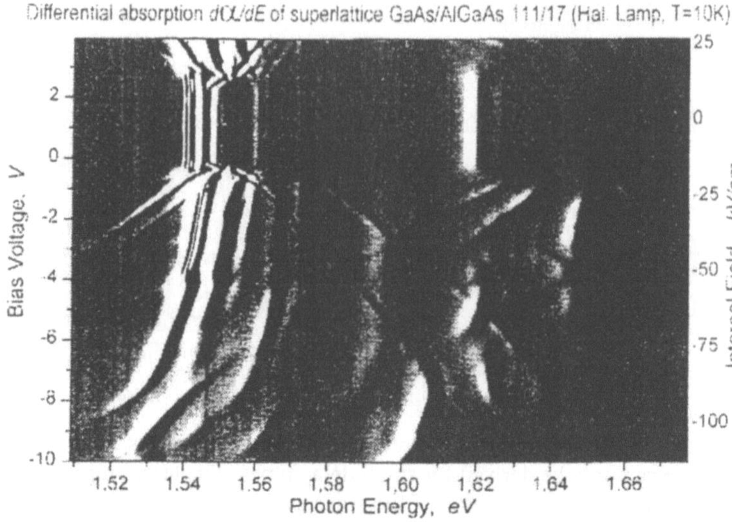

Figure 17 Absorption spectrum of the 111/17 sample taken with a lamp. The signal has been differentiated with the energy as differential variable to show the transitions more pronounced.

region (+2 to 0 V). For increasing reverse bias, the heavy-hole transitions of the first MB are clearly visible. The respective transitions of the second MB are barely recognizable, due the large number of anticrossings they undergo.

Besides the *hh* transitions, a set of *lh* transitions is visible for both minibands. In general, they move in parallel to the *hh* transitions if they have the same index. However, for transitions with different index, anticrossings occur. These will be discussed in detail below. In Fig. 17, the anticrossings between *hh* and *lh* are not well visible since they are most important for the bias region between 0 and −1.5 V.

Since the oscillator strength of the *lh* transitions is only slightly weaker that that of the heavy-hole transitions, they contribute significantly to the dynamics of photo-excited wave packets. Also, due the larger bandwidth, the light holes are not as strongly localized than the electrons and will show spatial dynamics, too.

Anticrossings also occur between the *hh* states of the first and second miniband. A pronounced example is the anticrossing between the first MB hh_0 state and the second MB hh_{-1} state at a voltage of about 7 V and an energy of about 1.53 eV. The strong interaction between these two states is immediately visible in the rather large splitting between the states at resonance.

Besides these two main complications, there are a number of further effects: (i) there is a second fan chart evolving out of the excitons at the upper edge of the first MB. These upper-edge excitons, which have been observed previously [38], are at an energy of about 1.555 eV for flat band conditions (ii) for the hh_0 and lh_0 transitions of the first MB, a red shift due to the quantum-confined-stark effect [39] is clearly visible (iii) at about −4 V bias, there is a kink in all transitions which we attribute to the formation or motion of a field domain. Such effects obviously occur despite the very low light level used for taking these data.

Most of the features described above are not taken into account in the present state of theory, which is restricted to hh excitons [31,33]. However, the field dependence of the *hh* transition energies and oscillator strength is nevertheless matched very well by the calculations based on this theory [31]. We have actually used a comparison between experiment and theory to perform a very accurate field calibration.

For somewhat higher fields, it is possible to have larger field regions where the effects discussed above are small. For lower fields, however, these complicating effects cannot be avoided easily. We believe that lead to the deviations of the total amplitude from theory as visible in Fig. 11. The complicated superposition of many nonequidistant transitions can actually lead to a complete disappearance of the BO due to destructive effects, as recently shown [37].

Anharmonicity due to Anticrossings

As a first example how the different effects in the WSL can be explored in their effect on wave packet dynamics experimentally, we have studied the influence of a *hh-lh* anticrossing. The major influence such an anticrossing will have on our experiments is the fact that the WSL transitions will not linearly depend on the field in the regions of the anticrossing. Figure 18 shows the spectral shift of the transition versus delay time. It is obvious that the behavior is anharmonic. To analyze these data, we have performed cw-spectroscopy of the region and have fitted the transitions with an anticrossing equation. Assuming harmonic spatial oscillations of the wave packet and multiplying this with the nonlinear behavior of the transition as obtained by the described fitting procedure, we are able to replicate the experimental results quite nicely (solid line in Fig. 18). This confirms that the anharmonic behavior is not caused by an anharmonic motion of the wave packet, but from the use of a nonlinear field sensor: a transition which energy function is curved due to the anticrossing. However, a conclusive discussion of this issue, one has to take into account that the dynamic theory of the THz modulation of the WSL can also lead to anharmonic behavior under certain circumstances. This issue is not fully understood yet [28].

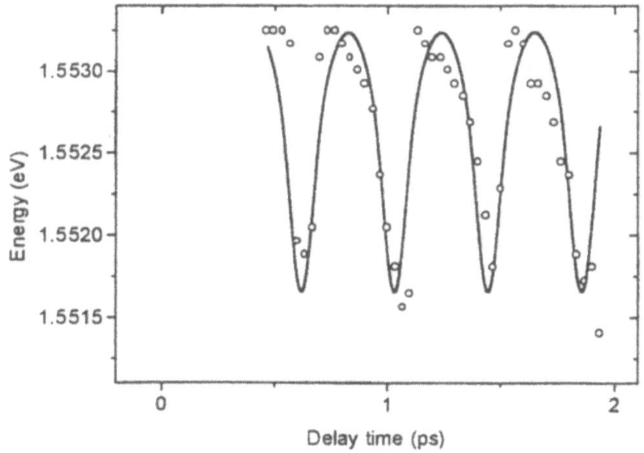

Figure 18 Circles: Spectral shift of the hh_{-1} transition in the 97/17 sample vs. delay time. Line: Results of model as described in the text.

It would also be possible to obtain an anharmonic motion of the wave packet if the SL design would lead to a strongly anharmonic dispersion of the MB. As discussed above, the minibands in the samples investigated here are close to harmonic behavior. Strong anharmonic contributions could be realized if the miniband is close to the upper edge of the barriers: Then, the mass of the electron is large when it is close to the bottom of the band and small when it is close to the upper edge of the band. The experiments with atoms in light lattices mentioned above [25] have beautifully shown the effect of anharmonic bands. In these experiments, the potential height could be controlled in-situ by the intensity of the light beams forming the standing wave potential. The anharmonicity was visible in the real space velocity characteristics of the carriers.

Spectral Shifts due to Polarization Interference between Heavy-Hole and Light-Hole Transitions

The photo-excitation of the WSL with the short laser pulse creates a superposition of polarization associated with the different transitions within the laser spectrum. The BO observed as the beating of the total polarization due to the quantum interference of the various levels is absolutely the same effect like in standard quantum beats experiment; albeit with the specific spatial motion associated with the beating. As we have shown, this spatial motion leads to a macroscopic dipole field which leads to a periodic change of the total field which can be detected in the spectral position of the WSL peaks. However, the presence of both heavy-hole and light-hole transitions leads to additional effects. The polarizations connected with the different *hh* WSL transitions are all linked to a common hole ground state and can therefore be described by quantum beats of transitions with one common ground state and a multitude of excited states. The light-hole transitions, however, are originating from a different ground state. If *hh* and *lh* transitions are simultaneously excited, one has therefore no quantum interference of *hh* and *lh* transitions, but *polarization interference.*

The distinction of quantum and polarization interference has been previously made for other beat phenomena in quantum wells [40]. Both interference phenomena lead to pronounced beating in the far-field response of the samples, although their physical origin is very different. An easy and clear way to distinguish the two phenomena in FWM experiments is the analysis of the FWM spectra [41]. For QB, all signals are in phase, the polarization has common maxima and minima for all frequencies. This is easy to understand: For quantum interference, the polarization disappears completely when the two wave function are out of phase, and therefore no polarization can be present at whatsoever frequency. This behavior is nicely displayed by the signal shown in Fig. 8. For the conditions of this figure, only *hh* transitions play a role and only quantum interference is present. The maxima and minima are thus at the same delay time for all detection energies. The spectral shifts which are important to detect the displacement can thus be analyzed neglecting the temporal dynamics of the peaks which energy is measured.

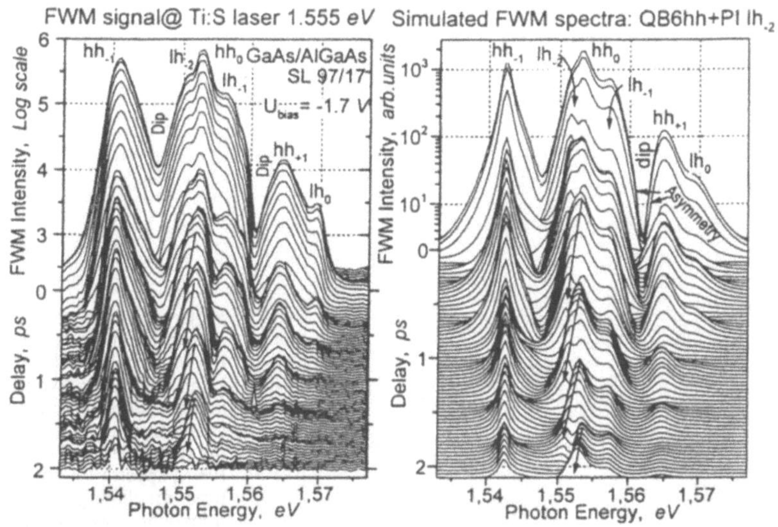

Figure 19 Spectrally resolved FWM plots for the 97/17 sample as a function of delay time. The FWM intensity is shown in a log scale. Left: experimental data. Right: simulation as described in the text.

This is distinctly different if we are in a field region where both heavy-hole and light-hole wave packets are excited: For the polarization interference due to *hh* and *lh* superposition, spectral shifts are expected since the phase of maxima and minima in the spectrally resolved FWM signal are expected to be dependent on the detection energy [41]. The left part of Fig. 19 shows experimental results for the 97/17 superlattice. Note that the FWM signal is plotted in a logarithmic scale. In the field region investigated (-1.7 V bias corresponds to a field of about 10 kV/cm), the lh_{-1} and lh_{-2} transitions are in the same spectral region than the hh_0 transitions. The most prominent feature is the pronounced spectral shift of the hh_0 transition indicated by the arrows. This shift is not related to the shifts caused by the dipole field change, which is very small in the present experiment due to the intentionally chosen low excitation density, but is caused by the periodic shift of the maxima and minima due to the polarization interference of hh and lh transitions. This is also illustrated by the fact that this spectral shift shows no apparent decay over a delay time of 2 ps, although the number of coherent carriers has decreased by an order of magnitude during this delay time.

The right part of Fig. 19 shows the result of a simple theoretical model of the polarization. To calculate this, the contributions of the different transitions were summed up and the total polarization was calculated. As input, we have taken the transition energies and the product of the spectral weight times the laser intensity for the various transition. Thus, the model is parameter-free except for a phenomenological damping time which has to be inserted. Nevertheless, the agreement with the experiment is astonishingly good, even minor features are reproduced well. For instance, the dips between the hh_{-1} and the lh_{-2} transition are clearly visible. What is still missing is a theory which calculates the spatial dynamics of such a complicated superposition of transitions, with inclusion of the Coulomb interaction. Due to the rather large contribution of delocalized *lh* transitions, it can be expected that the total macroscopic dipole moment has a strongly changed spatial

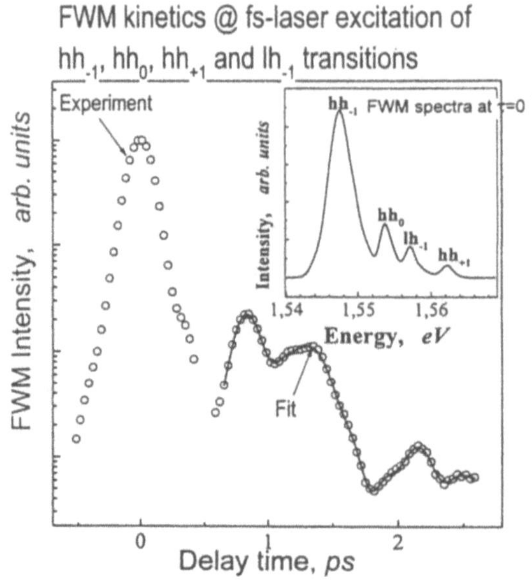

Figure 20
Spectrally integrated transient FWM signal for the 67/17 sample. Circles: Experimental data. Solid line: result of simulation as described in the text. The inset shows the FWM spectra for zero delay time, displaying the transitions which were excited by the laser.

behavior compared to the excitation of *hh* transitions only. A qualitative prediction is that the *lh* wave packets will reduce the overall dipole moment since they should move in the opposite direction. This is indeed the tendency we observe in the low-field region where *lh* excitation becomes important (see Fig. 11). However, a quantitative investigation has not been performed yet.

The simple modeling picture outlined in the previous paragraph is also able to predict the temporal behavior quite nicely. Figure 20 displays the temporal behavior of the time-integrated FWM signal for a case where three hh and one lh transition is excited. For modeling, we were using again only the peak weight and position information available from the absorption spectrum. The complicated beating pattern of the experimentally observed spectrally integrated FWM signal (circles) is reproduced quite well by the simulation (solid line).

4 Future Experiments on Bloch Oscillations

In this last section, we want to briefly outline a few open questions which might be addressed in future experiments. One obvious question is: Is it possible to realize the Bloch oscillation Gedankenexperiment? A first step from the interband optical experiment towards the transport Gedankenexperiment would be an study where the excitation of holes is avoided. This would also greatly simplify the physics of the experiment. The challenge in such experiments is to create Bloch electron wave packets with well-defined relative phase. Up to now, no proposal for the electrical excitation of such a wave packet has been put forward. Clearly, one would need a wave packet which has a spectral width larger than about 10 meV, which requires some injection scheme with a rather short transient.

An intermediate step would be to still use the laser excitation to optically create a well-defined wave packet, but then inject the electrons into the superlattice structure where no photo-excited holes are present. We are presently preparing such an experiment.

Instead of generating the wave packet by a short electrical pulse, a more promising approach to achieve defined phase relations is to implement some feed-back mechanisms. Such an approach has already been proposed for the similar system of coupled double quantum wells by Luryi [42]. He has shown that such a structure with feedback will emit radiation if it is excited with a cw laser.

The real challenge is, however, the realization of an electrically driven emitter: the Bloch oscillator. Only an electrically driven device, ideally monolithically integrated on a single chip, will have a chance for widespread commercial use. To the authors knowledge, no feasible proposals for such a device have been made yet. There is plenty of room for good ideas!

Acknowledgments

We would like to thank Dietmar Klatt and Karl-Heinz Pantke for valuable experimental contributions and Stephan Glutsch for helpful discussions. Working on Bloch oscillations for a couple of years, one of the authors (K.L.) has profited from collaboration and discussions with many people, including D. Dunlap, J. Feldmann, P. Haring, H. Kurz, P. Leisching, G. von Plessen, H. Roskos, and J. Shah. The work reported here was supported by the Deutsche Forschungsgemeinschaft (contracts Le 747/11-1, 11-2 and through the Graduiertenkolleg "Struktur- und Korrelationseffekte in Festkörpersystemen").

More information about Bloch oscillations, including computer simulations, can be found at `http://ppprs1.phy.tu-dresden.de/`.

Bibliography

[1] P. Drude, Ann. Physik **1**, 566; ibid., **3**, 369 (1900). For a lucid discussion of Drude transport, see Ref. [4], Chapter 1.

[2] F. Bloch, Z. Phys. **52**, 555 (1928).

[3] C. Zener, Proc. R. Soc. London Ser. A **145**, 523 (1934).

[4] See, e.g., N.W. Ashcroft and N.D. Mermin, *Solid State Physics*, (Holt-Saunders Int. Ed. 1981), p.225.

[5] It is interesting to note that scattering processes are necessary to have transport occurring in a single band: If no scattering processes were present, the carriers would perfom periodic oscillations in space and time, but would stay on the average at the same spatial position.

[6] L. Esaki and R. Tsu, IBM J. Res. Dev. **61**, 61 (1970).

[7] K. Leo, Semiconductor Science and Technology (Topical Review) **13**, 249-263 (1998). J. Feldman, Festkörperprobleme/Adv. in Sol. St. Phys. **32**, 81 (1992).

[8] L. Esaki, L.L. Chang, W.E. Howard, and V.L. Rideout, Proc. 11 th Conf. Physics of Semiconductors, Warsaw, 1972, p.431.

[9] A. Sibille, J.F. Palmier, H. Wang, and F. Mollot, Phys. Rev. Lett. **64**, 52 (1990); A. Sibille, J.F. Palmier, M. Hadjazi, H. Wang, G. Etemadi, and E. Dutisseuil, Superlattices & Microsctructures **13**, 247 (1993).

[10] A.A. Ignatov, E. Schomburg, K.F. Renk, W. Schatz, J.F. Palmier, and F. Mollot, Ann. Physik **3**, 137 (194).

[11] E. Schomburg, A.A. Ignatov, J. Grenzer, K.F. Renk, D.G. Pavel'ev, Yu. Koschurinov, B. Ja. Melzer, S. Ivanov, S. Schaposchnikov, and P.S. Kop'ev, Appl. Phys. Lett. **68**, 1096 (1996).

[12] K. Unterrainer, B.J. Keay, M.C. Wanke, S.J. Allen, D. Leonard, G. Medeiros-Ribeiro, U. Bhattacharya, and M.J. Rodwell, Phys. Rev. Lett. **76**, 2973 (1996).

[13] T. Yajima and Y. Taira, J. Phys. Soc. Jpn. **47**, 1620 (1979).

[14] G.H. Wannier, Phys. Rev. **117**, 432 (1969). The first prediction for the ladder-like structure was actually given by H.M. James, Phys. Rev. **76**, 1611 (1949).

[15] E. E. Mendez, F. Agullo-Rueda, and J. M. Hong, Phys. Rev. Lett. **60**, 2426 (1988).

[16] P. Voisin, J. Bleuse, C. Bouche, S. Gaillard, C. Alibert, and A. Regreny, Phys. Rev. Lett. **61**, 1639 (1988).

[17] G. Bastard and R. Ferreira, in *Spectroscopy of Semiconductor Microstructures,* Vol. 206 of NATO Advance Study Institute, Series B: Physics, eds. G. Fasol and A. Fasolino (Plenum Press, New York 1989), p. 333.

[18] J. Feldmann, K. Leo, J. Shah, D.A.B. Miller, J.E. Cunningham, S. Schmitt-Rink, T. Meier, G. von Plessen, A. Schulze, and P. Thomas, Phys. Rev. B **46**, 7252 (1992).

[19] K. Leo, P. Haring Bolivar, F. Brüggemann, R. Schwedler, and K. Köhler, Solid State Comm. **84**, 943 (1992).

[20] C. Waschke, H.G. Roskos, R. Schwedler, K. Leo, H. Kurz, and K. Köhler, Phys. Rev. Lett. **70**, 3319 (1993).

[21] P. Leisching, P. Haring Bolivar, W. Beck, Y. Dhaibi, F. Brüggemann, R. Schwedler, H. Kurz, K. Leo, and K. Köhler, Phys. Rev. B **50**, 14389 (1994).

[22] T. Dekorsy, P. Leisching, K. Köhler, and H. Kurz, Phys. Rev. B **50**, 8106 (1994).

[23] G.C. Cho, T. Dekorsy, H.J. Bakker, H. Kurz, A. Kohl, and B. Opitz, Phys. Rev. B **54**, 4420 (1996).

[24] R. Martini, G. Klose, H.G. Roskos, H. Kurz, H.T. Grahn, and R. Hey, Phys. Rev. B 54, R14325 (1996).

[25] M. Ben Dahan, E. Peik, J. Reichel, Y. Castin, and C. Salomon, Phys. Rev. Lett. **76**, 4508 (1996).

[26] V.G. Lyssenko, G. Valusis, F. Löser, T. Hasche, K. Leo, M.M. Dignam, and K. Köhler, Phys. Rev. Lett. **79**, 301 (1997).

[27] V. G. Lyssenko, G. Valusis, F. Löser, T. Hasche, K. Leo, K. Köhler, and M.M. Dignam, Proc. 23th Int. Conf. Phys. Semic., Berlin 1996, Eds. M. Scheffler and R. Zimmermann (*World Scientific, Singapore 1996*), p. 1763.

[28] M.M. Dignam, Bull. Am. Phys. Soc. **43**(1), 784 (1998), and unpublished.

[29] G. von Plessen and P. Thomas, Phys. Rev. B **45**, 9185 (1992).

[30] G. Valusis, V.G. Lyssenko, D. Klatt, K.-H. Pantke, F. Löser, K. Leo, K. Köhler, Proc. 23th Int. Conf. Phys. Semic., Berlin 1996, Eds. M. Scheffler and R. Zimmermann (World Scientific, Singapore 1996), p. 1783.

[31] M. M. Dignam and J. E. Sipe, Phys. Rev. Lett. **64**, 1797 (1990).

[32] M. Dignam, J.E. Sipe, and J. Shah, Phys. Rev. B **49**, 10502 (1994).

[33] M.M. Dignam and J.E. Sipe, PRB.

[34] A.M. Fox, D.A.B. Miller, J.E. Cunningham, W.Y. Jan, C.Y.P. Chao, and S.L. Chuang, Phys. Rev. B **46**, 15365 (1992).

[35] G. Cohen, I. Bar-Joseph, and H. Shtrikman, Phys. Rev. B **50**, 17316 (1994).

[36] M. Sudzius, V.G. Lyssenko, F. Löser, K. Leo, M.M. Dignam, K. Köhler, Phys. Rev. B **57**, 12693 (1998).

[37] F. Löser, M. Sudzius, V.G. Lyssenko, T. Hasche, K. Leo, M.M. Dignam, and K. Köhler, phys. stat. sol. (b) **206**, 315 (1998).

[38] N. Linder, K.H. Schmidt, W. Geißelbrecht, H.T. Grahn, K. Ploog, H. Schneider, and G.H. Döhler, Phys. Rev. B **52**, 17352-17365 (1995).

[39] D.A.B. Miller, D.S. Chemla, T.C. Damen, A.C. Gossard, W. Wiegmann, T.H. Wood, and C.A. Burrus, Phys. Rev. B **32**, 1043 (1985).

[40] K. Leo, J. Shah, E.O. Göbel, T.C. Damen, S. Schmitt-Rink, W. Schäfer, J.F. Müller, K. Köhler, P. Ganser, Phys. Rev. B **44**, 5726 (1991).

[41] T. Tokizaki, A. Nakamura, Y. Ishida, T. Yahima, I. Akai, and T. Karasawa, in *Ultrafast Phenomena VII*, eds. C.B. Harris, E.P. Ippen, G.A. Mourou, and A.H. Zewail, Springer Series in Chemical Physikcs Vol. 53 (Springer Verlag Berlin 1990), p.253.

[42] S. Luryi, IEEE J. Quant. Electr. **27**, 54 (1991).

Optical Properties of Localized Excitons in Nanostructures: Theoretical Aspects

Erich Runge and Roland Zimmermann

Humboldt-Universität zu Berlin, Institut für Physik, Arbeitsgruppe Halbleitertheorie, Hausvogteiplatz 5–7, D-10117 Berlin, Germany

Abstract: A detailed description of excitonic spectra in semiconductor nanostructures needs to take into account roughness-induced disorder. Results are presented for a kinetic equation, which is formulated in terms of disorder eigenstates and includes phonon relaxation and radiative exciton decay. Depending on the excitation scheme, the solutions describe photoluminescence, photoluminescence excitation spectroscopy, or absorption. Their dependence on time, temperature, disorder strength, and spatial resolution can be studied. Spatially resolved spectra turn out to be particularly interesting: Their autocorrelation contains information on the spatial extension of the wave functions, which is otherwise inaccessible.

1 Quantum Wells, Quantum Dots, and Spatially Resolved Spectroscopy

The last years have seen an enormous interest in optical properties of semiconductor quantum wells. A detailed understanding of the spectra proved most useful for the characterization of interface quality. The enterprise to understand, and even more important, to minimize the unavoidable interface-related disorder due to well width fluctuations, interface roughness, and alloying – which on an atomic scale are indistinguishable – is still very much alive [1]. At the same time, the focus of interest has somehow shifted towards even lower dimensional structures, quantum wires and, in particular, quantum dots. The concept of quantum dots thereby covers a wide field ranging, with decreasing confinement strength, from, e.g., chemically created II-VI dots in a glass matrix over etched structures and self-organized quantum dots in the Stranski-Krastanov wetting layer InGaAs/GaAs system to so-called quantum well dots and well width fluctuations [2]. The latter two result from unavoidable variations of composition and interface position and, in consequence, variations of the local energy. Whether this leads to that kind of physics generally associated with the concept of quantum dots, is a matter of degree and depends among others on temperature.

A central issue is whether carriers generated in a particular spatial region are confined to that region or can propagate into other spatial regions. Even in systems thought to contain well defined dots, considerable transfer between dots has been seen. The question of relaxation into potential minima is also of key importance for both dot-based laser devices and quantum-well laser where the suggested lasing mechanism depends on the disorder [3].

In the present work, we discuss the relaxation of excitons created in a quantum well with disordered interfaces into local energy minima. We then look at the consequences for the interpretation of optical measurements. The focus is on spatially resolved spectroscopy, such as micro photoluminescence (μ-PL) and near-field scanning optical microscopy (NSOM). A related technique which allows particularly good resolution but requires large preparation efforts is the use of masks.

2 Excitons in Disordered Quantum Structures

Most GaAs and ZnCdSe quantum wells show an inhomogeneous line width of a few meV, thus much less than the respective exciton binding energies. This suggests, that excitons *as a whole* are subject to local energy variations. This concept has been verified numerically in [4].

Adopting this picture, a single Schrödinger equation for the center of mass motion of the $1s$ exciton state can be derived [5]

$$\left(-\frac{\hbar^2\Delta_{\underline{R}}}{2M_X}+v(\underline{R})\right)\psi_\alpha(\underline{R}) \quad = \quad \mathcal{E}_\alpha\,\psi_\alpha(\underline{R}). \tag{2.1}$$

Here a factorization of the full wave function Ψ in center of mass (COM) $\psi_\alpha(\underline{R})$, relative $\varphi_{1s}(\underline{r})$ and confinement functions $u_e(z_e)$, $u_h(z_h)$ for electron and holes has been assumed. The exciton mass is $M_X = m_e + m_h$ and abbreviations $\beta_{e,h} = M_X/m_{h,e}$ are used throughout. For the case of local band edge fluctuations $\Delta E_{e,h}$ dominating the disorder, the effective potential is given by

$$v(\underline{R}) \quad = \quad \int d^2\underline{R}' \sum_{a=e,h} \beta_a^2\,\varphi_{1s}^2\left(\beta_a(\underline{R}-\underline{R}')\right)\int dz\,u_a^2(z)\,\Delta E_a(\underline{R}',z). \tag{2.2}$$

The average potential fluctuation $\langle v^2(\underline{R})\rangle$ will henceforth be denoted by σ^2. Even if $\Delta E_a(\underline{R},z)$ is correlated only over short distances, the resulting effective potential is rather smooth and correlated over distances given by the exciton Bohr radius. In this case, it is Gaussian distributed, but correlated.

Depending on what quantity one is interested in and thus how many solutions of (2.1) one needs, different numerical methods can be used to solve this Schrödinger equation effectively for many realizations of the random potential. Absorption spectra can be calculated from the time propagation of a single wave function [6]. In contrast, the relaxation kinetics requires all eigenstates in an energy window at the lower band edge, using, e.g., the ARPACK package [7].

We are particularly interested in the optically active (dipole-allowed) states. Assuming the exciton to be of heavy hole ($l_Z = 3/2$) character, the optical weight is proportional to the matrix element [8]

$$M_\alpha^2 \quad = \quad \frac{3}{8\pi Q_{\max}^2}\int d^2\underline{Q}\,\psi_{\alpha\underline{Q}}^2\left(\frac{Q_{\max}}{\sqrt{Q_{\max}^2-Q^2}}+\frac{\sqrt{Q_{\max}^2-Q^2}}{Q_{\max}}\right), \tag{2.3}$$

in terms of the Fourier transformed COM wave function $\psi_{\alpha\underline{Q}}$ and the maximum momentum $Q_{\max} = E_X n/\hbar c$ which couples to photons for given exciton transition energy E_X and index of refraction n.

For strongly localized excitons (2.3) reduces to [9]

$$M_\alpha = \int d^2\underline{R}\, \psi_\alpha(\underline{R}). \tag{2.4}$$

The neglect of the finite photon momentum and thus polariton effects [10], i.e. (2.4) instead of (2.3), leads for the parameter range considered here only to minor quantitative changes. In this approximation, the absorption $\alpha(E)$ coincides with the 'optical density' $D(E)$ (spectral function for $\underline{Q}=0$),

$$\alpha(E) \sim D(E) = \pi \sum_\alpha M_\alpha^2\, \delta(E - \mathcal{E}_\alpha), \tag{2.5}$$

up to a global factor that relates the radiative rates r_α to the M_α^2

$$r_\alpha = \frac{4\, e^2\, p_{cv}^2\, n\, E_X}{3\, \hbar^2\, c^3}\, \varphi_{1s}(0)^2\, \langle u_e|u_h\rangle^2\, M_\alpha^2 \tag{2.6}$$

with the momentum matrix element $p_{cv} = \langle s|\hat{p}|p_z\rangle$. A reduction by scattering into dark exciton states is not included here. The confinement wave function overlap $\langle u_e|u_h\rangle$ typically is close to unity for the strong transitions.

Among the optical active wave functions states of rather different character are found. In the very low-energy tail, the COM states are well localized and approximately Gauss shaped, whereas at higher energies fractal shapes appear [11] (Fig. 1).

The Gaussian shapes can be understood both in the limit of smooth long-range correlated potentials, which locally can be approximated by parabolic minima (Fig. 1), and in the opposite limit of a short-range correlated potential (white noise character) where an 'optimum fluctuation' picture applies [12].

The fractal shapes can be looked at as resonances between plane wave like states localized in different spatial regions with different local wave numbers, as illustrated in Fig. 1.

The arguments given above can be transferred directly to typical quantum *wire* structures. However, it should be noted, that for, e.g., structurally defined quantum *dots* with strong confinement, a rather different pictures has to be used. In that case, the Coulomb correlation is a small correction on top of well defined single-particle states for electron and hole separately. In other words, a separation of relative and center of mass motion is not possible.

It has been argued by Nurmikko and coworkers [3] that for the low-intensity, low-temperature limit the most relevant optical non-linearity is phase space filling of the low-energy minima. This can lead to excitonic lasing and inversion at rather low total densities. If spatially confined biexcitons are bound, a certain exciton population in the low-energy states first induces biexcitonic absorption. But as soon as local minima are filled with biexcitons, further absorption is suppressed and gain is expected. If on the other hand it were energetically preferential that two excitons occupy two separate minima, one would expect gain at even lower intensities.

For structurally defined, strongly confining dots, the concept of two excitons in the same energy minimum is well defined. For self-organized grown dots and quantum well dots, this concept is blurred and details of the disorder wave functions for excitons and biexcitons matter.

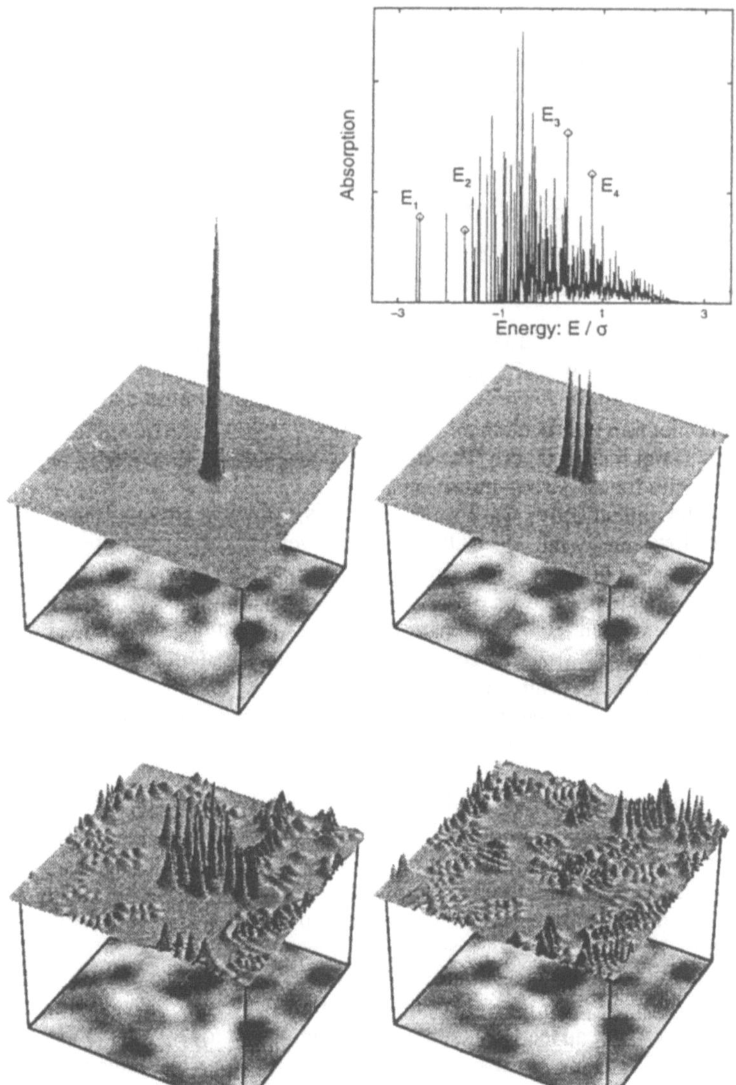

Figure 1 Wave functions of four different eigenvalues, marked in the calculated absorption spectrum by $E_1, \ldots, E_4$ (from upper left to lower right panel). Note that ψ_1 and ψ_2 look like ground state and second excited state of a parabolic potential. The potential distribution is given as gray-scale at the bottom.

3 Level Repulsion in Low-Dimensional Systems

3.1 Spectral Autocorrelations

Spatially resolved techniques such as $\mu - PL$ and NSOM typically resolve at low temperatures the inhomogeneous exciton line into hundreds of extremely narrow lines with μeV width, which is of the order of the radiative life time [13].

We will now address a central claim of this work, namely

- The sharp luminescence lines seen in spatially resolved spectroscopy are not 'just random noise', but fingerprints of individual eigenstates.

- The autocorrelation of line spectra yields information on the quantum character of the states involved !

It is a general feature of all quantum systems that they show level repulsion – *if* states overlap! Simple well known text book examples include (1) the 'avoided crossing' of two levels, (2) the regular and large level spacing of particles in a box, and (3) random matrix theory dating back to Wigner [14].

The latter can be understood in simple terms as an ensemble of a large number of states having random energies and all interacting with a randomly distributed strength. If we ask for the level distance, i.e. the separation of one eigenstate from the one with the closest energy, we can look at these two as an avoided crossing problem and find, that they repel each other. This can be quantified with the famous Wigner-Dyson distributions.

The field of random matrix theory and level repulsion has seen a splendid renaissance in the last years as high-lighted by reviews like [15]. However, the wealth of knowledge in this field can not be directly used for the interpretation of spatially resolved spectra, for the following simple reason: Energy eigenvalues of states that do not overlap, trivially do not repel each other. For quantum wells, quantum wires, and quantum well dots it is not a priori known, to what degree exciton states overlap.

But we can turn the question around and argue: From the level statistics of measured spectra and in particular from its dependence on the detection area $\mathcal{A}_D$, we can deduce how much exciton states overlap or to what degree they are localized in different spatial regions.

In the well understood random matrix limit, where all states interact, the probability $g_2(E)$ to find two states with an energetic distance E,

$$g_2(E) \quad \sim \quad \left\langle \sum_{\alpha \neq \beta} \delta(\mathcal{E}_\alpha - \mathcal{E}_\beta - E) \right\rangle, \tag{3.7}$$

tends to zero for small arguments. Here and later 'small' is meant in the sense of 'small compared to the average level distance Δ_E'. Random matrix theory distinguishes three different universal behaviors depending on the time reversal properties of the Hamiltonian under consideration. For the so-called Gaussian orthogonal ensemble (GOE) of Hamiltonians describable by orthogonal matrices, as valid in the present case, the level correlation function g_2 behaves like

$$g_2(E) \quad \approx \quad |E|/\Delta_E \quad \text{for small } |E|/\Delta_E. \tag{3.8}$$

In the opposite limit of independently distributed levels, no such repulsion exists ($g_2 \approx 1$).

The positions of the very narrow spikes seen in spatially resolved spectra are eigenvalues of the random Schrödinger equation (2.1). They can be measured with high accuracy and their spectral properties should be investigated. The physical quantity that corresponds to g_2 is the *averaged convolution* of a series of measured spectra $I(E)$ with themselves,

$$C_I(E) = \left\langle \int I(E')I(E'-E)\,dE' \right\rangle . \quad (3.9)$$

Here, the central peak at $E = 0$ has to be disregarded. Inserting for I the density of states, $\rho(E) = \sum_\alpha \delta(E - \mathcal{E}_\alpha)$, yields immediately $C_\rho(E) = g_2(E)$.

Level repulsion shows up in the difference between $C_I(E)$ and the *convolution of the average spectrum*

$$C_{\langle I\rangle}(E) = \int \langle I(E')\rangle \, \langle I(E'-E)\rangle \, dE' \; . \quad (3.10)$$

The averages have to be taken over many spectra measured at different spots on the same sample.

3.2 Poor Man's Approach to Level Repulsion in Extended Systems

For the system under consideration here, we are inbetween the limits of uncorrelated levels and random matrix theory-type eigenvalues. The wave functions are extended, but not infinitely, and they are, in general, smaller than the detection area.

Two simple concepts can help us here. However, a definite answer can, at least up to now, only be given by numerical simulations. One way to look at the situation is to argue, that for the superposition of M independent spectra [15]

$$g_2 \approx \frac{M-1}{M} + \frac{1}{M}\frac{|E|}{\Delta_E} \quad \text{for } |E| \ll \Delta_E . \quad (3.11)$$

Thus $C_I(E)$ is reduced at small arguments compared to $C_{\langle I\rangle}$ by a factor (1-1/M).

This leads to the following, simplified but qualitative correct approach: if the measured autocorrelation function shows a relative dip of order $1/M$ one would argue, that M 'independent areas' contribute to the spectrum of the detection area $\mathcal{A}_D$. Thus a wave function area $\approx \mathcal{A}_D/M$ is concluded.

The other helpful concept is to consider instead of the usual correlation function of eigenenergies $g_2(E)$, (3.7), a quantity, that explicitly involves *spatial* information. For a first understanding, we consider a model of two energetically nearest levels $\mathcal{E}_1{}^0 = \Delta$, $\mathcal{E}_2{}^0 = -\Delta$ interacting with each other proportional to their wave function overlap,

$$\langle \psi_1 \mid \psi_2 \rangle \approx \langle \psi_1 \mid \hat{v} \mid \psi_2 \rangle \approx E_m \exp(-R/R_0) \equiv m(R), \quad (3.12)$$

which is parameterized by a strength and an exponential distance dependence. This yields the 2×2 problem

$$\hat{h}_m \approx \begin{pmatrix} -\Delta & m(R) \\ m^*(R) & \Delta \end{pmatrix} . \quad (3.13)$$

Its solutions,

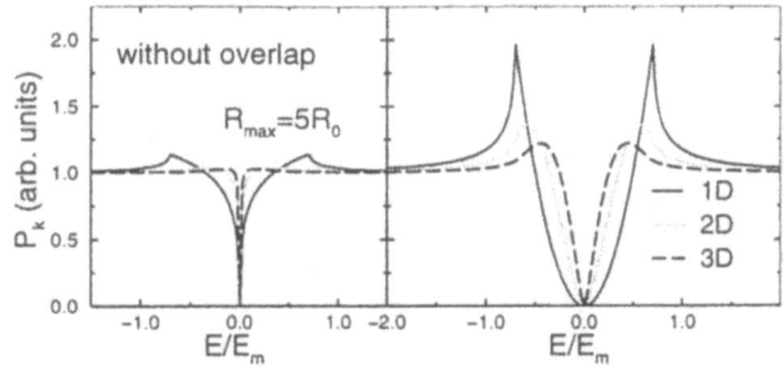

Figure 2 Level repulsion in the 2 × 2 toy model of (3.12)–(3.15): Level repulsion without (left, $k = 0$) and with inclusion of overlap weight (right, $k = 2$)

$$\mathcal{E}_{1,2} \quad = \quad \pm\sqrt{\Delta^2 + |m(R)|^2}, \tag{3.14}$$

are then averaged over distances R in D-dimensional space and over the non-interacting level distance 2Δ. Fig. 2 compares the resulting distribution for the level distance

$$\begin{aligned} P_k(E) \quad &\equiv \quad \langle M^k_{\alpha,\beta}\,\delta(E - \mathcal{E}_\alpha + \mathcal{E}_\beta)\rangle \\ &\approx \quad \int_{-\infty}^{\infty} d\Delta \int_0^{R_{max}} dR\, R^{(D-1)}\, |m(R)|^k\, \delta\left(E - 2\sqrt{\Delta^2 + |m(R)|^2}\right) \end{aligned}$$

without (left panel, $k = 0$) and including (right panel, $k = 2$) the wave function overlap as weight factor.

The left panel shows a slight reminiscence of level repulsion only because integration was cut off at some distance R_{max} corresponding to the detection size. In contrast, a strong level repulsion dip, which depends on dimensionality, is seen in the right panel. No cut-off R_{max} is needed for P_2.

Note, that the details of Fig. 2 depend on the (physically motivated) functional form of $m(R)$. The very narrow drop of P_0 (left panel) to zero at zero level distance reflects the very small but still finite repulsion even for the largest distances.

We found quite general, that whenever eigenvalue correlations appear in formulas for physical relevant quantities, they typically are weighted by matrix elements coupling only those states that overlap in real space. Phonon scattering, which will be discussed below, is an example.

As mentioned before, in lack of a better theory numerical simulations are at present the only way to quantitatively draw conclusions on the wave function extension from level repulsion dips in the correlation function of measured spectra. Nevertheless, it is a fascinating prospective to determine the size of wave functions from spectra such as the inset of Fig. 1 and its experimental counter parts.

3.3 Optical Matrix Elements

Two further aspects complicate a detailed analysis of spatially resolved spectra and their level correlations.

Optical spectra involve besides the energy levels the radiative matrix elements of (2.3) or (2.4) as well. In the conceptually simplest case of absorption, the convolution is

$$C_D(E) \;=\; \left\langle \sum_{\alpha,\beta} \pi^2 M_\alpha^2 M_\beta^2 \, \delta(\mathcal{E}_\alpha - \mathcal{E}_\beta - E) \right\rangle . \tag{3.15}$$

Note that this combination does *not* depend on the overlap of the states ψ_α and ψ_β. On the contrary, one could even expect, that optically active states are *less* correlated than the average, as oscillator strength should be more or less equally distributed in real space. And, one might think that locally only one state could look like the ideal plane wave with $\underline{Q} = 0$. This argument is easily refuted by looking at the eigenstates of a harmonic oscillator potential, where in 2 dimensions every even state has exactly the same optical matrix element, as defined in (2.4). A illustrated by Fig. 1, these are good approximations for some eigenstates. A detailed understanding on the influence of the matrix elements on the correlation properties of absorption spectra is still missing.

As an aside, we mention that the correlation of the absorption spectra (3.15) is related to resonant Rayleigh scattering and to the squared time-dependent polarization. This simplifies its numerical determination [16].

If not absorption, but luminescence is measured, the spectrum, and thus its correlation properties, further depends on the occupation N_α of the individual states,

$$I_{PL}(E) \;=\; \sum_\alpha r_\alpha N_\alpha \delta(E - \mathcal{E}_\alpha). \tag{3.16}$$

We next consider the relaxation process in detail and come back later to level repulsion features.

4 Kinetics of Exciton Relaxation

4.1 Phonon Scattering

The energy relaxation process depends on more material properties than the optical matrix element, which basically reflected the spatial extension of the COM wave functions, only. We focus here on deformation potential scattering involving acoustic phonons ($E = \hbar\omega_q = \hbar u q$), which is thought to dominate the exciton relaxation at least in the III-V materials at low temperatures.

We discuss relaxation in terms of a kinetic equation for the occupation N_α of the disorder eigenstates ψ_α

$$\partial_t N_\alpha \;=\; g_\alpha + \sum_\beta \gamma_{\alpha\leftarrow\beta} N_\beta - \left(r_\alpha + \sum_\beta \gamma_{\beta\leftarrow\alpha} \right) N_\alpha \tag{4.17}$$

with generation rates g_α, radiative recombination r_α, (2.6), and scattering rates [17,18]

$$\gamma_{\beta\leftarrow\alpha} \;=\; \frac{2\pi}{\hbar} \sum_{\vec{q}} \left| t^q_{\alpha\beta} \right|^2 \left[n(\hbar\omega_q)\delta(\mathcal{E}_\beta - \mathcal{E}_\alpha - \hbar\omega_q) \right.$$

$$\left. + (1 + n(\hbar\omega_q))\delta(\mathcal{E}_\beta - \mathcal{E}_\alpha + \hbar\omega_q) \right] \tag{4.18}$$

$$t^q_{\alpha\beta} \;=\; \sqrt{\frac{\hbar\omega_q}{2u^2\rho V}} \int d\vec{r}_e \int d\vec{r}_h \, \Psi^\star_\alpha(\vec{r}_e \vec{r}_h) \left(D_c e^{i\vec{q}\vec{r}_e} - D_v e^{i\vec{q}\vec{r}_h} \right) \Psi_\beta(\vec{r}_e \vec{r}_h)$$

involving deformation potentials $D_{c,v}$, the mass density ρ, sample volume V, and thermal phonon occupation functions $n(\hbar\omega_q)$. In the factorization approximation for the wave function,

$$\Psi(\vec{r}_e\vec{r}_h) \quad = \quad \psi(\underline{R})\ \varphi_{1s}(\underline{r})\ u_e(z_e)\ u_h(z_h), \tag{4.19}$$

this can be rewritten in terms of the Fourier transforms K_e, K_h, and $\chi(\underline{r})$ of the squared confinement wave functions and the squared relative wave function of the exciton, respectively. We quote the result for $D = 2$ dimensions,

$$\gamma_{\beta\leftarrow\alpha} = \frac{\pi}{\hbar^2 u^3 \rho} n(\mathcal{E}_\alpha - \mathcal{E}_\beta)\ (\mathcal{E}_\alpha - \mathcal{E}_\beta)$$
$$\times \int \frac{d^2Q}{(2\pi)^2}\ \frac{q_E}{q_z} \left|(\psi_\alpha\psi_\beta)_{\underline{Q}}\right|^2 |D_c K_e(q_z)\chi(Q/\beta_e) - D_v K_h(q_z)\chi(Q/\beta_h)|^2 \tag{4.20}$$

with abbreviations

$$q_E = \frac{|\mathcal{E}_\alpha - \mathcal{E}_\beta|}{\hbar u}, \qquad q_z = \sqrt{q_E^2 - Q^2}. \tag{4.21}$$

Here $(\psi_\alpha\psi_\beta)_{\underline{Q}}$ stands for the Fourier transform of the *product* of the wave functions. In the approximation, that the phonon momentum is mainly taken over by the confinement wave functions ($Q \to 0$), the wave function dependence reduces to the form given in [18]

$$\gamma_{\alpha\leftarrow\beta} \quad \sim \quad W_0(\mathcal{E}_\alpha - \mathcal{E}_\beta) \int d^2\underline{R}\ \psi_\alpha^2(\underline{R})\ \psi_\beta^2(\underline{R}) \tag{4.22}$$

Note, that this approximation necessarily fails for very small energy transfers, where $(\psi_\alpha\psi_\beta)_{\underline{Q}}$ vanishes due to wave function orthogonality.

We note two important features of (4.20). The first is rather trivial, but nevertheless noteworthy: phonon scattering happens only between states that have a finite overlap in real space ($(\psi_\alpha\psi_\beta)_{\underline{Q}} \neq 0$). It is thus vanishingly small between states localized in different deep and well separated potential minima.

Secondly, note that the total outscattering rate of given energy transfer E

$$\Gamma_{\text{ph}}(E) \quad = \quad \sum_{\alpha\neq\beta} \delta(E - \mathcal{E}_\alpha + \mathcal{E}_\beta)\ \gamma_{\beta\leftarrow\alpha}\ , \tag{4.23}$$

is of the form discussed in the context of (3.15), i.e. a sum over pairs of energies weighted with an overlap-dependent factor. We will come back to that later.

We further want to state explicitly that merely writing down a kinetic equation in the form (4.17) already involves approximations: It is not invariant under a change of basis, e.g., going from eigenstates ψ_α to a basis of plane wave states, as it is often used in the Boltzmann equation community. This comes from the neglect of correlations in the density matrix's off-diagonal elements. However, one can expect the disorder eigenstate basis to be particularly suitable for the description of relaxation processes in the low-energy tail involving rather localized states and small energy transfers.

Equation (4.17) can easily be solved using numerically calculated wave functions and matrix elements. Among other quantities, such as the time dependence of the luminescence, one can calculate the stationary photoluminescence $I_{PL}(E_D,E_E)$ at detection energy E_D and excitation energy E_E. Writing (4.17) in a short hand notation in terms of a phonon scattering matrix γ, diagonal matrices r und Γ containing the radiative rates and the total outscattering rates, respectively, and vectors for generation rates and occupations, this requires a single matrix inversion for each temperature,

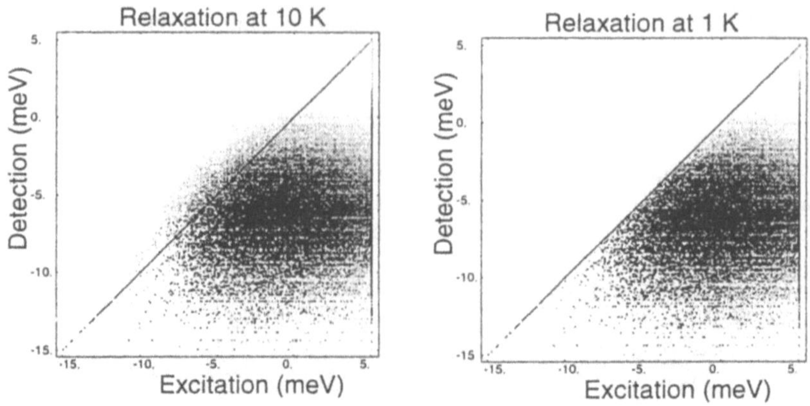

Figure 3 Average relaxation matrix $I_{PL}(E_D,E_E)$ at $T=$ 1 and 10 K for a 5 nm thick $GaAs/Ga_{.30}Al_{.70}As$ quantum well and $\sigma = 8$ meV disorder strength.

$$I_{PL}(E_D,E_E) = \sum_{\alpha\beta} \delta(E_D - \mathcal{E}_\alpha)\, r_\alpha \left(\underline{\underline{r}} + \underline{\underline{\Gamma}} - \underline{\underline{\gamma}}\right)^{-1}_{\alpha\beta} r_\beta\, \delta(E_E - \mathcal{E}_\beta). \qquad (4.24)$$

An illustrative example of $I_{PL}(E_D,E_E)$ for two different temperatures is given as Fig. 3. Integrating over the first argument yields absorption; integrating $I_{PL}(E_D,E_E)$ over the second argument yields luminescence for spectrally broad excitation; each horizontal cut is a photoluminescence excitation (PLE) spectrum; the diagonal presents resonant Rayleigh scattering.

4.2 Non-Monotonous Stokes Shift

Luminescence in inhomogeneously broadened systems is, in general, red-shifted compared to the absorption profile; for quantum wells see, e.g., [19,5], and references therein.

The simplest prediction for this so-called Stokes shift is

$$S = -\frac{\sigma^2}{k_B T}, \qquad (4.25)$$

which is based on multiplication of a Gaussian-shaped, $\exp(-E^2/2\sigma^2)$, inhomogeneous line with a Maxwell distribution. However, this equilibrium argument must fail for temperatures low enough that its estimate (4.25) becomes considerable larger than the line width. A theory for the zero temperature limit has been put forward in [19].

The kinetic equations (4.17), (4.24) allows to study the temperature dependence in detail. Unexpectedly, the calculated Stokes shift turned out to depend non-monotonously on temperature [5]. The red-shift goes along with a narrowing of the luminescence line [18]. Both effects are demonstrated in Fig. 4. A non-monotonous behavior of the luminescence line width for spatially indirect excitons has recently been modelled in [20].

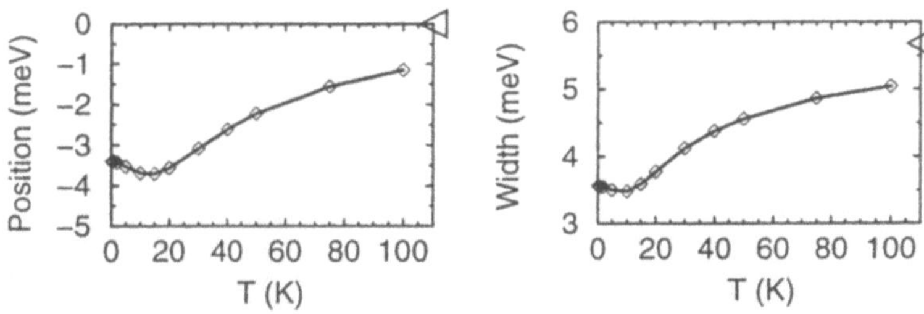

Figure 4 Stokes shift between absorption and luminescence (left) and luminescence line width (right) for different temperatures (diamonds). Simulation parameters as in Fig. 3. Triangles mark the high-temperature limits.

A qualitative understanding of the downward shift with increasing T at low temperatures can easily be developed starting with the picture of [19], which postulates that at $T = 0$ each exciton relaxes into its spatially nearest *local* energy minimum, where it stays for the rest of its radiative life time. At slightly elevated temperatures phonon absorption becomes possible. This allows activation into states with energies higher by about $k_B T$. With some probability, the exciton subsequently relaxes into a much deeper energy minimum.

This poses the question, what the low-temperature behavior of S exactly is. The numerical simulations show a T dependence which is much weaker than the linear behavior predicted by a simple activated hopping picture. Reconsidering the above arguments and converting energy differences in powers of T, one finds, that, again, we need the probability to find a state close in energy *and* with a large overlap. Thus, the low-temperature behavior of S might be a second independent manifestation of level repulsion effects in the optical properties of low-dimensional semiconductor nanostructures.

4.3 Effective Mobility Edge for Exciton Relaxation

Another surprising result based on the kinetic equation (4.17) is the existence of an effective mobility edge for exciton relaxation in narrow quantum wells. Furthermore, we found that photoluminescence excitation spectra agree with absorption only on the high energy side but show an abrupt drop on the low energy side. Thus their peak positions do, in general, not coincide. This can be considered as an additional contribution to the easily measurable shift between luminescence and PLE. These experimental and theoretical findings have been discussed in some detail in [21].

4.4 Level Statistics of Luminescence Spectra

Finally, we come back to the question of quantum mechanical level repulsion in luminescence spectra. Due to the task to calculate *all* eigenstates in the energy range of interest and then to solve the kinetic equation, much more numerical effort is needed to get well converged results; one is restricted to smaller simulation sizes. On the other hand, experiments will also average over a moderate number of spectra, only. The better the spatial resolution, the fewer quantum mechanically independent wave functions contribute, and

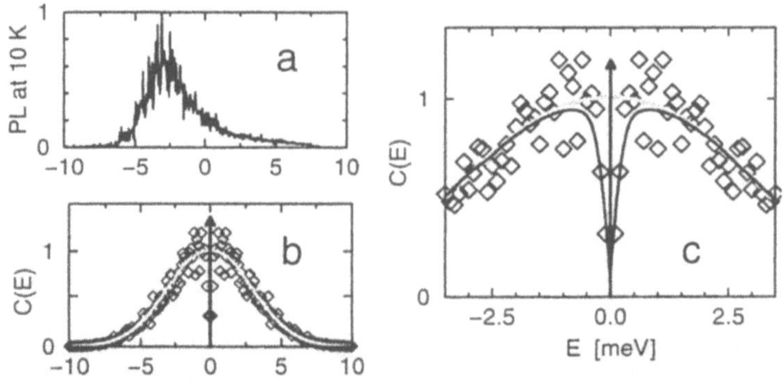

Figure 5 (a) Simulated PL Spectrum as sum of 250 NSOM quantum well spectra, each corresponding to an area 120 nm × 120 nm. Parameters as in Fig. 3. (b) Spectral autocorrelation $C(E)$ for the spectrum a (gray line) and the sum of the 250 individual autocorrelations (diamonds). Only a limited spectral region (-6 meV ... -1 meV) is used for $C(E)$. (c) Enlargement of panel b with GOE-fit (black line) to the data (from Ref. [16]).

the larger is the level correlation dip. Consequently, fewer spectra need to be added to make it visible. Then, the finite width of the central $E = 0$ peak in the convolution, whose wings will tend to fill the level repulsion dip, is less problematic.

Fig. 5 illustrates first of all, that one can expect to find the quantum mechanical level repulsion and thus information on the size of the wave functions even in luminescence spectra. Second, it shows that for typical NSOM resolution about 250 spectra should suffice to exhibit the level repulsion dip.

5 Concluding Remarks

The theoretical understanding of the linear optical properties of disordered quantum wells is rather satisfactory and allows detailed comparison with a wide range of experimental data. The main fit parameter is the strength of the disorder which, of course, varies from sample to sample.

Obvious challenges for the theory come from the exciton-exciton interaction in disordered nanostructures and its consequences for the lasing mechanism. The statistics of level correlations weighted with wave function overlaps deserves a better understanding.

On the experimental side, the verification of the predicted non-monotonous Stokes shift and the level-repulsion dip in spatially resolved spectra are still missing. Details of the very low temperature behavior of the Stokes shift would be very interesting, but probably are difficult to resolve.

Bibliography

[1] *Proc. 23nd ICPS Berlin*, ed. M. Scheffler and R. Zimmermann. (World Scientific Publ. Co., Singapore, 1996).

[2] MRS Bulletin, February 1998: *Semiconductor Quantum Dots*.

[3] J. Ding, H. Leon, T. Ishihara, M. Hagerott, A. V. Nurmikko, H. Luo, N. Samarth, and J. Furdyna, Phys. Rev. Lett **69**, 1707 (1992).

[4] R. Zimmermann, F. Große, and E. Runge, Pure & Appl. Chem. **69**, 1179 (1997).

[5] R. Zimmermann, E. Runge, and F. Große, in [1], p. 511.

[6] S. Glutsch, D. S. Chemla, and F. Bechstedt, Phys. Rev. B **54**, 11592 (1996).

[7] *ARPACK Package* by D. Sorensen, R. Lehoucq, and Chao Yang, available from NetLIB at AT&T and ORNL.

[8] D. S. Citrin, Phys. Rev. B **47**, 3832 (1993).

[9] U. Bockelmann, Phys. Rev. B **48**, 17637 (1993).

[10] L. C. Andreani, F. Bassani, and F. Tassone, in *Optics of Excitons in Confined Systems,* p. 25. (IOP Conference Series Number 123, ed., A. D'Andrea, R. Del Sole, R. Girlanda, and A. Quattropani, Bristol 1992).

[11] M. Schreiber in: K. H. Hoffmann and M. Schreiber, Ed., *Computational Physics*, (Springer, Berlin. 1996), p 147.

[12] Al. L. Efros, C. Wetzel, and. J. M. Worlock, Phys. Rev. B **52**, 8384 (1995).

[13] H. F. Hess, E. Betzig, T. D. Harris, L. N. Pfeiffer, and K. W. West, Science **264**, 1740 (1994).

[14] M. L. Mehta, *Random Matrices*, 2. ed. (Academic Press, San Diego, 1990).

[15] T. Guhr, A. Müller-Groeling, and H. Weidenmüller, *Random Matrix Theories in Quantum Physics: Common Concepts.*, preprint cond-mat/9707301, submitted to Physics Reports.

[16] E. Runge and R. Zimmermann, phys. stat. solidi (b) **206**, 167 (1998).

[17] T. Takagahara, Phys. Rev. B 31, 6552 (1985).

[18] R. Zimmermann and E. Runge, phys. stat. solidi (a), **164**, 511 (1997).

[19] F. Yang, M. Wilkinson, E. J. Austin, and K. P. O'Donnell, Phys. Rev. Lett. **70**, 323 (1993).

[20] S. D. Baranovskii, R. Eichmann, and P. Thomas, phys. stat. solidi (b), **205**, R191 (1998).

[21] U. Jahn, M. Ramsteiner, R. Hey, H. T. Grahn, E. Runge, and R. Zimmermann, Phys. Rev. B **56**, R4387 (1997).

Superradiant Exciton-Light Coupling in Semiconductor Heterostructures – Theory

T. Stroucken[1], S. Haas[1], B. Grote[1], S.W. Koch[1], M. Hübner[2,3], D. Ammerlahn[2], and J. Kuhl[2]

[1] Department of Physics and Materials Sciences Center, Philipps University, Renthof 5, D-35032 Marburg, Germany
[2] Max-Planck-Institut für Festkörperforschung Heisenbergstraße 1, D-70569 Stuttgart, Germany
[3] Present address: Optical Science Center, University of Arizona, Tucson AZ 85721, USA

Abstract: Superradiant coupling of excitons in multiple quantum well Bragg structures is investigated. The superradiance leads to characteristic signatures in the spectral and time-resolved optical response. It is shown, that the superradiant mode gradually vanishes for increasing excitation intensity and for larger static disorder due to excitation induced dephasing or dephasing trough static disorder, respectively.

1 Introduction

Many recent theoretical and experimental investigations on the optical response of femtosecond excited semiconductor multiple quantum well (MQW) structures focus on light propagation effects, i.e. the interaction of excitonic resonances by the exchange of transverse photons[1–11]. Direct radiative decay of an exciton polariton in a bulk semiconductor is inhibited by the requirement of momentum conservation. For a QW exciton with an in-plane momentum $\hbar k_{\|} < \hbar\omega/c$, where $\hbar\omega$ is the excitation energy, a radiative decay channel exists due to the lack of translational invariance in the growth direction of the structure. Values of the radiative lifetime in the range of a few ps have been postulated theoretically[1,2,4,5] and could recently be observed experimentally[6–8]. Since the field which is emitted by any QW in the sample can subsequently interact with other QWs, the short radiative lifetime of the exciton causes an efficient radiative coupling of the QWs, potentially leading to collective effects. The dynamics of the collective excitations depend strongly on the number of QWs N and the interwell separation d. A special situation arises in a MQW Bragg structure, where the interwell spacing d equals an integer multiple of half the exciton wavelength λ in the medium. At low excitation intensities, the dominant coupling mechanism is an stimulated polarization decay due to the reemitted optical field. This type of coupling creates a so called superradiant mode (SM), which ideally is characterized by a radiative decay rate that is N times enhanced in comparison to the single QW (SQW). The remaining $N-1$ "dark" modes then have a vanishing radiative coupling strength. According to theoretical predictions[4], the formation of a superradiant state in a MQW Bragg structure depends critically on the phase matching between optical excitations in different QWs. Therefore, the dynamics of the optical polarization is extremely sensitive to various dephasing mechanisms like scattering at impurities or phonons, interface roughness or excitation-induced dephasing[9,10,12].

In this paper, we summarize our recent theoretical investigations concerning superradiance in MQW Bragg structures. Section 2 gives an outline of the theoretical framework within which we investigate the superradiance phenomenon in such structures. In Section 3, the concept of superradiance, which is based on the linear optical response, and its validity are discussed. In Sections 4 and 5, we investigate the influence of excitation-induced dephasing and dephasing by static disorder on the formation of the SM. >From a theoretical point of view, there are several fundamental differences between excitation induced dephasing and dephasing due to static disorder. Whereas disorder leads to dephasing in the linear regime, where the solutions of the semiconductor Maxwell-Bloch equations (SMBE) can be expressed in terms of stationary exciton-photon modes, an increase of the excitation intensity requires the solution of the full nonlinear SMBE for the nonequilibrium system. Since the concept of superradiance is based on the existence of (quasi-)stationary coupled exciton-photon modes, the formation of a superradiant state is not only affected by dephasing mechanisms but also by optical nonlinearities like phase space filling, Coulomb screening etc.. Static disorder modifies both the homogeneous and the inhomogeneous linewidth, while Coulomb scattering only influences the homogeneous linewidth. >From an experimental point of view, there is another important difference. Whereas effects of excitation-induced dephasing can be controlled by varying the excitation intensity, a certain amount of static disorder is always present in real systems.

Except indicated otherwise, excitonic units $\hbar = m_r = e^2/\varepsilon = 1$ will be used, where m_r is the reduced mass of the electron-hole-pair. The speed of light in the crystal is related to the background dielectric constant ε by $c = c_v/\sqrt{\varepsilon} = \sqrt{\varepsilon}/\alpha$ where c_v is the in vacuum speed of light and α is the fine structure constant.

2 The Semiconductor Maxwell-Bloch Equations

2.1 Maxwell's Equations

In this Section, we concentrate on the wave equation, describing the propagation of a classical optical field interacting with a MQW. Experimentally, the solution of Maxwell's equation

$$\left(\vec{\nabla}^2 - \frac{1}{c^2}\frac{\partial^2}{\partial t^2}\right)\vec{E}(\vec{r},t) = \frac{4\pi}{\varepsilon}\frac{1}{c^2}\frac{\partial^2}{\partial t^2}\vec{P}(\vec{r},t) - 4\pi\vec{\nabla}\vec{\nabla}\cdot\vec{P}(\vec{r},t) \tag{2.1}$$

for the classical radiation field is detected after propagating through the interaction region. >From the measured absorption, transmission or reflection, the optical response of the semiconductor, i.e. the field induced macroscopic polarization density $\vec{P}$, must be deduced. In an inhomogeneous structure like a MQW, the induced intraband contributions to the total polarization density, as well as the divergence of the interband polarization are in general nonvanishing quantities, coupling the different polarization components of the optical field. However, if the semiconductor is initially in its ground state, for excitation with an ultrashort optical pulse with central frequency ω_L and spectral width $\Delta\omega$, we can choose the conditions such that the pulse resonantly excites only the lowest heavy-hole exciton transition. For this situation, the contribution from the light holes to the interaction Hamiltonian can be neglected. Furthermore, if the laser pulse propagates in the growth direction of a MQW, the interband polarization is homogeneous with respect to the in-plane coordinates and the intraband contribution to the macroscopic polarization density vanishes. Note, that for incidence perpendicular to the QW plane, no center of mass motion of the electron-hole pair can be excited. Since the dipole matrix element for the heavy-hole transition has no z−component, the wave equations for the different polarization components are decoupled by momentum conservation within the QW plane. For each circular $\sigma^{\pm}$ polarization component, the wave equation therefore simplifies to

$$\left(\frac{\partial^2}{\partial z^2} - \frac{1}{c^2}\frac{\partial^2}{\partial t^2}\right)E(z,t) = \frac{4\pi}{\varepsilon}\frac{1}{c^2}\frac{\partial^2}{\partial t^2}P(z,t). \tag{2.2}$$

Within the envelope function approximation[13], the z−dependence of the macroscopic interband polarization is determined by the electron and hole confinement wave functions located at the positions z_n of the nth QW

$$P(z,t) = \sum_n P^n(t)\varphi_e^n(z)\varphi_h^n(z) \approx \sum_n P^n(t)\delta(z - z_n). \tag{2.3}$$

Here we neglected the finite thickness L of the QWs in comparison to the wavelength of light, i.e. we have approximated the product of the confinement wave functions by δ functions in Eq. (2.3). Denoting by k_L the wave vector of the light field, and assuming perfect confinement of the electron and hole wavefunctions within the individual wells, the corrections to our equations due to the finite thickness of the QW are of the order $(k_L L)^2$, and can be neglected for appropriately chosen structures.

Formally, Eq. (2.2) can be solved without a detailed knowledge of the polarization by putting

$$E(z,t) = E_n^+(t - z/c) + E_n^-(t + z/c) \tag{2.4}$$

in the nth barrier ($z_{n-1} < z < z_n$) and applying the appropriate boundary conditions at the interfaces[4]. The first set of boundary conditions simply is the requirement of continuity of the tangential components of the electric field at the interfaces. By integrating Eq. (2.2) from $z_n - \delta$ to $z_n + \delta$ and taking the limit $\delta \to 0$ we obtain the second set of boundary conditions. Combining the boundary conditions, we obtain[4]

$$E(z,t) = E_1^+(t - z/c) + E_{N+1}^-(t + z/c) - \frac{2\pi}{\varepsilon}\sum_{m=1}^{N}\frac{1}{c}\frac{\partial}{\partial t}P^m(t - d_{nm}/c) \tag{2.5}$$

where $E_1^+(t - z/c)$ and $E_{N+1}^-(t + z/c)$ is the external applied laser field propagating in the positive and negative z-direction, respectively and $d_{nm} = |z_n - z_m|$.

Eq. (2.5) is still quite general, since no other assumptions besides the dipole selection rules and the planar localization of the microscopic polarization have been made. Due to the QW coupling by the light field, both the transmission and the absorption spectrum in general depend on the number of QW wells as well as on the sample geometry.

2.2 The Semiconductor Bloch Equations

For explicit model calculations of the optical response of MQW structures, we must compute the macroscopic polarization density from a microscopic model. The Hamiltonian, from which the equations of motion are derived, is given by:

$$H = H_0 + H_C + H_I. \tag{2.6}$$

For a MQW within the "Envelope Function Approximation"[13],

$$H_0 = \sum_{n,k} \varepsilon^e_{n,k} c^\dagger_{n,k} c_{n,k} + \sum_{n,k} \varepsilon^h_{n,-k} d^\dagger_{n,-k} d_{n,-k} \tag{2.7}$$

is the free-particle Hamiltonian in the electron-hole picture, where n labels the QW and k all other relevant quantum numbers. $\varepsilon^{e/h}_{n,k}$ is the electron/hole single-particle energy (band structure) in the nth QW, given by the eigenstates of H_0. Restricting our analysis to the lowest confinement subband of a two-band semiconductor, $k = \{\vec{k}_\|, \sigma\}$, combines the in-plane wavevector $\vec{k}_\|$ and the total angular momentum along the axis of growth $\sigma = \pm 3/2, \pm 1/2$ for the heavy holes and the conduction band electrons, respectively. $c^\dagger_{n,k}$ /$c_{n,k}$ and $d^\dagger_{n,-k}$ /$d_{n,-k}$ are the creation/annihilation operators for electrons and holes.

$$\begin{aligned} H_C = &\frac{1}{2} \sum_{n,m,k_1 k_2 k_3} V^{nm}_{k_3} c^\dagger_{n,k_1+k_3} c^\dagger_{m,k_2-k_3} c_{m,k_2} c_{n,k_1} \\ &+ \frac{1}{2} \sum_{n,m,k_1 k_2 k_3} V^{nm}_{k_3} d^\dagger_{n,k_1+k_3} d^\dagger_{m,k_2-k_3} d_{m,k_2} d_{n,k_1} \\ &- \sum_{n,m,k_1 k_2 k_3} V^{nm}_{k_3} c^\dagger_{n,k_1+k_3} d^\dagger_{m,k_2-k_3} d_{m,k_2} c_{n,k_1}. \end{aligned}$$

describes the Coulomb-interaction between carriers in different QWs with the quasi-2-dimensional Coulomb interaction

$$V^{nm}_q = \frac{2\pi}{A} \frac{1}{q_\|} e^{-q_\| d_{nm}} \zeta^{nm}(q_\| L). \tag{2.8}$$

Here A is the normalization area and $\zeta^{nm}(q_\| L)$ accounts for corrections due to the finite width of the QWs[4].

H_I contains the electron-light interaction. Although H_I is diagonal in the QW indices if the electronic overlap between different QWs is sufficiently small, it is non-diagonal in the in-plane wave numbers. The in-plane momentum of the absorbed light excites the center-of-mass motion of the electron-hole-pair. However, on short time-scales, electron-hole pair excitations are not able to move appreciably and in-plane light propagation can be treated within a parametric space dependence, thus neglecting the momentum of the light-field. As a consequence, the total charge is a locally conserved quantity, thus, both the intraband contributions to the polarization density and the polarization charge $\vec{\nabla} \cdot \vec{P}$ vanish. In the z-direction of the MQW, the space dependence of the current is not determined by the exciting light field but by sample geometry. Even if the optical wavelength is much larger than the well width, the symmetry breaking in the growth direction requires a nonlocal description. Thus, if an explicit space dependence in the z-direction is kept but the in-plane momentum of the light field is neglected, which becomes exact for light propagation in the growth direction, the interaction Hamiltonian can be obtained within the dipole approximation as:

$$H_I = - \sum_{\sigma,\sigma',n,\vec{k}_{\parallel}} \vec{d}_{cv}^{\sigma\sigma'} \cdot \vec{E}_n(t)\, c_{n,k}^{\dagger} d_{n,-k}^{\dagger} + h.a., \tag{2.9}$$

where $\vec{d}_{cv}^{\sigma\sigma'}$ is the optical dipole moment between the electron and hole states. It describes the interaction of in the z-direction localized dipolar planes with a classical optical laser field at the position of the plane, which is given by the matrix element $E_n(t) = \langle n|\vec{E}(x,t)|n\rangle$, where $|n\rangle = \varphi_n(z)$ labels the confinement functions in the wells.

In our analysis, we assume excitation with circularly polarized light. Since the optical dipole moment is nonvanishing only for the transitions $|3/2,3/2\rangle \rightarrow |1/2,1/2\rangle$ and $|3/2,-3/2\rangle \rightarrow |1/2,-1/2\rangle$, with $\vec{d}_{vc}^{\sigma\sigma'} = \mu(\vec{e}_x \mp i\vec{e}_y)/\sqrt{2}$ respectively, excitations with left and right circular polarized light are decoupled and can be treated independendly. For this reason, all indices can be omitted.

To calculate the expectation value of the macroscopic interband polarization, the electron and hole distribution functions, f^e and f^h, and the interband transition amplitude p will be used as dynamic variables for the description of the electronic part of the system. These quantities are defined as:

$$f^e_{n,\vec{k}_{\parallel}} = \langle c^{\dagger}_{n,\vec{k}_{\parallel}} c_{n,\vec{k}_{\parallel}}\rangle, \quad f^h_{n,\vec{k}_{\parallel}} = \langle d^{\dagger}_{n,\vec{k}_{\parallel}} d_{n,\vec{k}_{\parallel}}\rangle, \quad p_{n,\vec{k}_{\parallel}} = \langle d_{n,-\vec{k}_{\parallel}} c_{n,\vec{k}_{\parallel}}\rangle. \tag{2.10}$$

The macroscopic interband polarization entering Maxwell's equations is obtained by summing the interband transition amplitude over all different momentum states, $P^n = \mu \sum_{\vec{k}_{\parallel}} p_{n,\vec{k}_{\parallel}}$. The equations of motion can be subsequently obtained from the Heisenberg equations, $d/dt\langle A\rangle = -i\langle [A,H]\rangle$. Seperating the Hartree-Fock terms from the correlation contribution [16–18], we obtain the following equations:

$$\begin{aligned}
\frac{\partial}{\partial t} p_{n,\vec{k}_{\parallel}} =& - i(\omega^e_{n,\vec{k}_{\parallel}} + \omega^h_{n,\vec{k}_{\parallel}}) p_{n,\vec{k}_{\parallel}} \\
& - i\Omega_{n,\vec{k}_{\parallel}} (f^e_{n,\vec{k}_{\parallel}} + f^h_{n,-\vec{k}_{\parallel}} - 1) + \left.\frac{\partial p_{n,\vec{k}_{\parallel}}}{\partial t}\right|_{\text{incoh}} \\
\frac{\partial}{\partial t} f^e_{n,\vec{k}_{\parallel}} =& - i(\Omega^*_{n,\vec{k}_{\parallel}} p_{n,\vec{k}_{\parallel}} - \Omega_{n,\vec{k}_{\parallel}} p^*_{n,\vec{k}_{\parallel}}) + \left.\frac{\partial f^e_{n,\vec{k}_{\parallel}}}{\partial t}\right|_{\text{incoh}} \\
\frac{\partial}{\partial t} f^h_{n,-\vec{k}_{\parallel}} =& - i(\Omega^*_{n,\vec{k}_{\parallel}} p_{n,\vec{k}_{\parallel}} - \Omega_{n,\vec{k}_{\parallel}} p^*_{n,\vec{k}_{\parallel}}) + \left.\frac{\partial f^h_{n,-\vec{k}_{\parallel}}}{\partial t}\right|_{\text{incoh}} .
\end{aligned}$$

Here

$$\Omega_{n,\vec{k}_{\parallel}} = \mu E_n(t) + \sum_{\vec{k}'_{\parallel}} V^{nn}_{\vec{k}'_{\parallel}-\vec{k}_{\parallel}} p_{n,\vec{k}'_{\parallel}}, \tag{2.11}$$

and

$$\omega^{e}_{n,\vec{k}_{\|}} + \omega^{h}_{n,\vec{k}_{\|}} = \varepsilon^{e}_{n,\vec{k}_{\|}} + \varepsilon^{h}_{n,\vec{k}_{\|}} - \sum_{\vec{k}'_{\|}} V^{nn}_{\vec{k}_{\|}-\vec{k}'_{\|}} (f^{e}_{n,\vec{k}'_{\|}} + f^{h}_{n,\vec{k}'_{\|}}), \tag{2.12}$$

are the renormalized Rabi-frequencies and electron and hole energies, respectivelly [14]. Note that due to the local charge conservation, the Coulomb interaction does not couple different QWs although the Coulomb matrix element is nondiagonal in the QW indices. The last terms on the RHS of Eqs. (2.2) refer to correlation contributions and the various scattering processes. These terms are most important for the case of elevated excitation levels. In order to keep the analysis as simple as possible, we treat the scattering terms phenomenologically within a T_2 approximation as long as we restrict the analysis to weak optical fields. For elevated excitation levels, both the screening of the Coulomb interaction and Coulomb scattering become important. We include corrections to the Hartree-Fock approximation in second Born approximation for the polarization and population distributions in the n-th QW[9,19]:

$$\left.\frac{\partial p_{n,\vec{k}_{\|}}}{\partial t}\right|_{\mathrm{incoh}} = -S^{D}_{\vec{k}_{\|},n} + S^{OD}_{\vec{k}_{\|},n} + V^{D}_{\vec{k}_{\|},n} - V^{OD}_{\vec{k}_{\|},n} \tag{2.13}$$

$$\left.\frac{\partial f^{e/h}_{n,\vec{k}_{\|}}}{\partial t}\right|_{\mathrm{incoh}} = s^{f}_{\vec{k}_{\|},n} + s^{D}_{\vec{k}_{\|},n} + s^{OD}_{\vec{k}_{\|},n} + v^{D}_{\vec{k}_{\|},n} - v^{OD}_{\vec{k}_{\|},n}, \tag{2.14}$$

which are second order in the screened Coulomb interaction. The right hand side of Eq. (2.13) contains the dephasing processes due to carrier scattering as well as corrections to the Hartree-Fock self-energies, while the RHS of Eq. (2.14) describes a redistribution of the carrier density due to carrier-carrier collisions as well as scattering processes driven by the coherent interband polarization. For a detailed expression of S,V,s,v see Ref. [19].

Inserting the solution of Maxwell's equations (2.5) in the generalized Rabi-frequency Eq. (2.11) reduces the problem to that of finding the solution of the SBE with a selfconsistently determined radiation field

$$\Omega_{n,\vec{k}_{\|}} = \mu\left(E^{+}_{1}(t - z_n/c) + E^{-}_{N+1}(t + z_n/c)\right) - \frac{2\pi\mu^2}{\varepsilon} \sum_{m=1}^{N} \sum_{\vec{q}_{\|}} \frac{\partial}{\partial t} P^{m}_{\vec{q}_{\|}}(t - d_{nm}/c) + \sum_{\vec{q}_{\|}\neq\vec{k}_{\|}} V^{nn}_{\vec{q}_{\|}-\vec{k}_{\|}} P^{n}_{\vec{q}_{\|}}. \tag{2.15}$$

From this equation, one recognizes that the different QWs are coupled by a dynamical dipole-dipole interaction mediated by the transverse optical field. Due to this dipole-dipole interaction, the dynamics of the electronic polarization not only depends on the total QW number, but via the retarded time arguments also on the spacing between the QWs.

3 Linear Optical Response

In the linear or quasi-stationary regime, Eq. (2.5) can be solved in frequency space by inserting

$$P^n(\omega) = \chi(\omega)\, E(\omega, z_n), \tag{3.16}$$

for the polarization in the n-th QW, where $\chi(\omega)$ is the 2-dimensional QW susceptibility, computed from the SBE. Inserting Eq. (3.16) into Eq. (2.5), one obtains

$$t(\omega) = 1 + \frac{2\pi i\omega}{\varepsilon c} \mathcal{U}^{\dagger} \left(\mathbb{I} - \frac{2\pi i\omega}{\varepsilon c} \chi(\omega)\, \mathcal{D}(\omega) \right)^{-1} \chi(\omega)\, \mathcal{U} \tag{3.17}$$

$$r(\omega) = \frac{2\pi i\omega}{\varepsilon c} \mathcal{U}^{T} \left(\mathbb{I} - \frac{2\pi i\omega}{\varepsilon c} \chi(\omega)\, \mathcal{D}(\omega) \right)^{-1} \chi(\omega)\, \mathcal{U} \tag{3.18}$$

for the MQW transmission and reflection coefficient. Here $\mathbb{I}$ is the $N \times N$ unit matrix, $\mathcal{D}(\omega) = (\exp i\omega d_{nm}/c)_{n,m=1}^{N}$ and $(\mathcal{U})_{n=1}^{N} = \exp i\omega z_n/c$. Apart from the denominator $\left(\mathbb{I} - \frac{2\pi i\omega}{c\varepsilon}\chi(\omega)\,\mathcal{D}(\omega)\right)^{-1}$, Eq. (3.17) looks very similar to the expression relating the transmitted field to the input fields by the linear susceptibility in the absence of propagation effects. The denominator can oe interpreted as field renormalization, accounting for reemission and reabsorption of light. In this regime, the solutions of the SMBE consist of coupled exciton-photon modes, having energies and radiative decay rates that are given by the poles of Eqs. (3.17) and (3.18). The poles depend on the positions z_n of the QWs and the number of coupled wells N. If the susceptibility exhibits a single strong resonance within the spectral range of the exciting laser pulse at $\omega = \omega_0$, we can approximate $\exp(i\omega z_n/c) \approx \exp(i\omega_0 z_n/c)$ and $\mathcal{D}(\omega) \approx \mathcal{D}(\omega_0)$. Within this single exciton-pole approximation, it can be shown that the observable quantities depend only on the phase $\exp(2i\omega_0 d/c)$ accumulated while traveling back- and forward from one QW to another.[4] In a perfect Bragg structure where d is equal to an integer multiple of half the exciton resonance wavelength $\lambda = 2\pi c/\omega_0$ inside the medium, at arrival, the field reemitted from the coherent polarization in a certain QW oscillates in phase with the polarization of all other QW. In these structures, the field induced polarization decay due to reemitted light is the dominating coupling mechanism between excitons in the different wells leading to an N times enhanced radiative dephasing rate. Mathematically, this can be seen from Eqs. (3.17) and (3.18) by diagonalizing the matrix $\mathcal{D}(\omega_0)$. For $z_n = d_1 + (n-1)\lambda/2$, the vector $1/\sqrt{N}((-1)^n)_{n=1}^{N}$ is an eigenvector of $\mathcal{D}(\omega_0)$ with eigenvalue N. This eigenvector corresponds to a superradiant mode. All other eigenvectors have the eigenvalue 0, corresponding to dark or subradiant modes. Only the superradiant mode contributes to the observed signals:

$$t(\omega) = 1 + \frac{2\pi i\omega_0}{\varepsilon c} \frac{N\chi(\omega)}{1 - \frac{2\pi i\omega_0}{\varepsilon c} N\chi(\omega)} = 1 + e^{-2id_1\omega/c} r(\omega). \tag{3.19}$$

>From Eq. (3.19) it can be recognized, that within the exciton-pole approximation, the transmission and reflection of a perfect Bragg structure differs from a SQW only by the replacement $\chi \rightarrow N\chi$, expressing the enhancement of the coupling between the semiconductor and the optical field. To investigate the limitations of the exciton-pole approximation, we insert

$$\chi(\omega) = -\frac{\mu^2|\sum_{\vec{k}}\varphi(\vec{k})|^2}{\omega-\omega_0+i\gamma_0} \tag{3.20}$$

for the linear 2-dimensional 1s exciton, yielding e.g.

$$t(\omega) = 1 - \frac{iN\Gamma}{\omega-\omega_0+i\gamma_0+iN\Gamma}, \tag{3.21}$$

where we defined

$$\Gamma = \frac{2\pi\mu^2\omega_0}{c\varepsilon}|\sum_{\vec{k}}\varphi(\vec{k})|^2 = \frac{4\pi^2\mu^2}{\varepsilon\lambda}|\sum_{\vec{k}}\varphi(\vec{k})|^2 \tag{3.22}$$

as the radiative dephasing rate of a SQW. >From Eq. (3.21) it can be recognized that, within the exciton-pole approximation, the radiative decay rate of a MQW Bragg exciton increases linearly with the QW number. However, as soon as the radiative width approaches the same order of magnitude as the exciton binding energy, radiative coupling of the 1s exciton to the higher excitonic bound states and the continuum states can no longer be neglected and the single exciton-pole approximation fails. Numerical calculations for the transmission $|t|^2$, reflection $|r|^2$ and absorption $1-|t|^2-|r|^2$ spectra of ideally 2D MQW Bragg structures[20] with different QW number are shown in Fig. 1a. They indicate significant deviations from the exciton-pole approximation for $N \gtrsim 20$, where the coupling to the continuum states yields a clearly asymmetric lineshape although the linewidth is still approximately proportional to the QW number. Only in the superlattice limit $N \to \infty$ and the absence of surface effects, the radiative width of a periodic MQW vanishes and stationary polariton states with the dispersion relation

$$\cos qd = \cos\omega d/c + \frac{2\pi\omega}{\varepsilon c}\chi(\omega)\sin\omega d/c \tag{3.23}$$

are recovered, where d is the period of the structure[1–3].

It is worth to note that the formation of the superradiant mode depends extremely sensitively on phase coherence between excitations in different QWs. Small deviations from the Bragg condition lead to a finite coupling of the subradiant modes to the optical field, while simultaneously the coupling strength of the superradiant mode is drastically decreased. This is illustrated in Fig. 1b, where the spectra of a $N = 50$ perfect Bragg structure are compared with the spectra of a structure containing 50 QWs with a spacing of $d = 0.52\lambda$, corresponding to a deviation of 4% of the Bragg condition. While the absorption of the perfect Bragg structure is strongly reduced due to the enhanced radiative decay rate, the slightly detuned structure exhibits very sharp peaks and a rich fine structure that can be attributed to subradiant modes.

4 Intensity Dependence of the Superradiance Phenomenon

In this Section, the influence of excitation-induced dephasing on the superradiant coupling will be investigated. While a systematical investigation of the effects of dephasing by static disorder requires a whole series of samples, the effects of excitation induced-dephasing can be studied systematically in one sample by varying the excitation intensity. For this purpose, the full nonlinear SMBE including the Coulomb scattering rates are

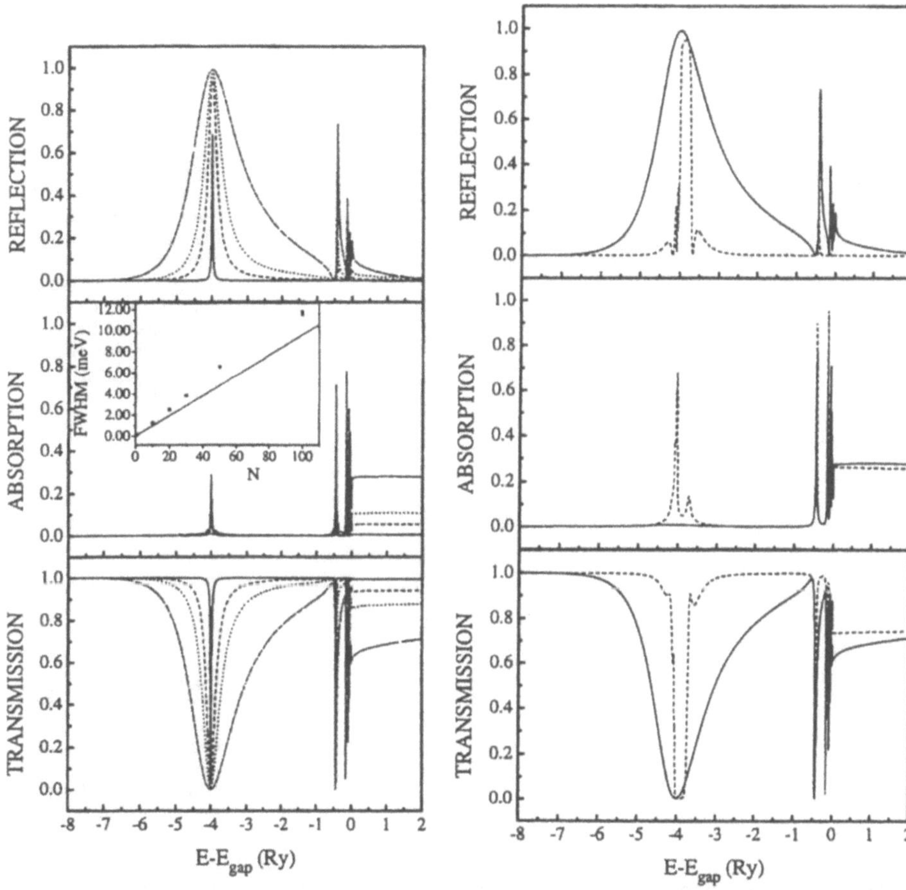

Figure 1 (a) Linear spectra for a $N = 1$ (solid line), $N = 10$ (dashed line), $N = 20$ (dotted line) and a $N = 50$ (dashed-dotted line) Bragg structure. Inset: FWHM of the 1s exciton resonance. (b) Linear spectra for a $N = 50$ Bragg structure (solid line) and a structure with 50 QWs and a period deviating 4% from the Bragg condition (dashed line).

solved numerically using massively paralized numerical schemes[9]. Fig. 2 shows the resulting linear time-resolved transmission of a SQW and a $N = 10$ Bragg structure containing 80 Å GaAs QWs [21] for resonant excitation of the 1s-exciton with a Gaussian pulse of 530 fs FWHM. The signal intensity has been normalized to the maximum intensity transmitted by the SQW that occurs at $t = 0$. Of special importance are those parts of the signal which are seen after the direct transmission of the incident pulse. These components result from reemission of light which is transiently absorbed by the quantum well(s). The signals show a double peak signature as function of time resulting from the destructive interference between the directly transmitted signal and the signal coherently emitted from the quantum wells. For a pulse duration of 530 fs (FWHM) the signal intensity reaches its maximum approximately 1 ps after the arrival of the pulse maximum. At later times the signals reveal a single exponential decay that is dominated by the radiative

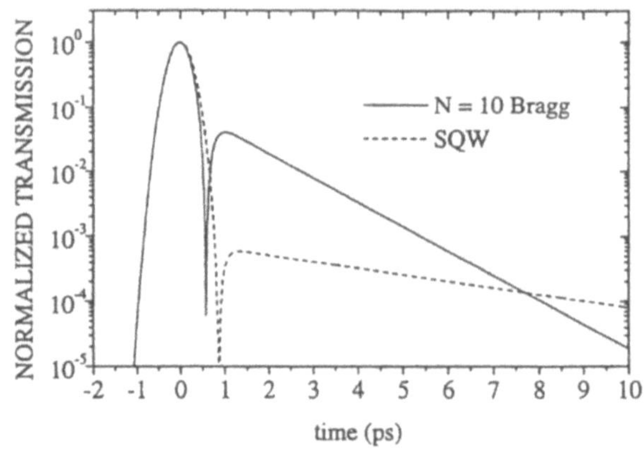

Figure 2
Linear time resolved transmitted signals from a SQW (dashed line) and a $N = 10$ Bragg structure (solid line) if excited resonantly to the heavy-hole transition with a laser pulse of 530 fs FWHM.

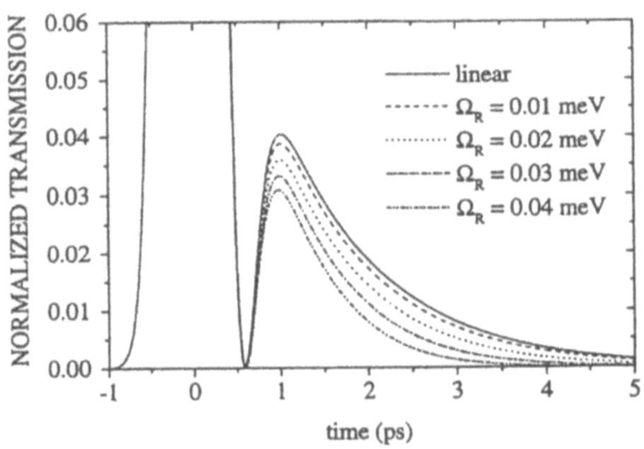

Figure 3
Transmission of laser pulses resonant to the heavy hole transition from a $N = 10$ Bragg structure for different excitation intensities.

decay of the excitation in the QWs. Using $1/\Gamma = 17$ ps and a phenomenological nonradiative contribution $\gamma_0 = 1/50$ ps $^{-1}$ to the dephasing rate, $T_2 = (\gamma_0 + \Gamma)^{-1} = 12.7$ ps for the SQW and $T_2 = (\gamma_0 + N\Gamma_s)^{-1} = 1.64$ ps for the $N = 10$ Bragg sample. Thus the theory predicts a pronounced decrease of the signal decay time for the $N = 10$ sample.

To study the influence of the nonlinear Coulombic effects, we show in Figure 3 the computed normalized transmitted signal of a $N = 10$ MQW Bragg structure for different excitation intensities. The excitation intensity is given in terms of the Rabi energy $\Omega_R = \mu E_0(t = 0)$. Fig. 3 clearly shows that the second peak in the signal, i.e. that part which originates from the reradiation of the MQW structure, is strongly reduced for increasing excitation intensities. To analyze the origin of these intensity dependent effects, we plot in Fig. 4 the ratio of the Bragg and the SQW signals emitted under identical excitation conditions. Apart from modifications for times less than the laser pulse duration, which can be explained by interference of the QW signal and the initial laser pulse, all signal amplitude ratios start at a value of $N^2 = 100$ due to the constructive interference of the signal emitted from different QWs in the Bragg sample. In the linear regime, the ratio decreases with increasing time due to the enhanced radiative decay rate in the Bragg structure. However, at higher excitation levels the decrease of the signal ratio slows down until the total decay rate of the SQW even exceeds that of the Bragg structure such that the signal ratio even increases with increasing time. This surprising result can be

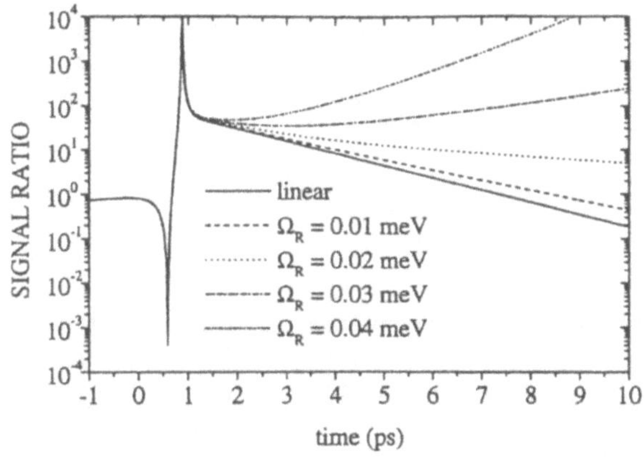

Figure 4 Ratio of the transmitted signals emitted from a $N = 10$ Bragg stucture and a SQW for various excitation intensities.

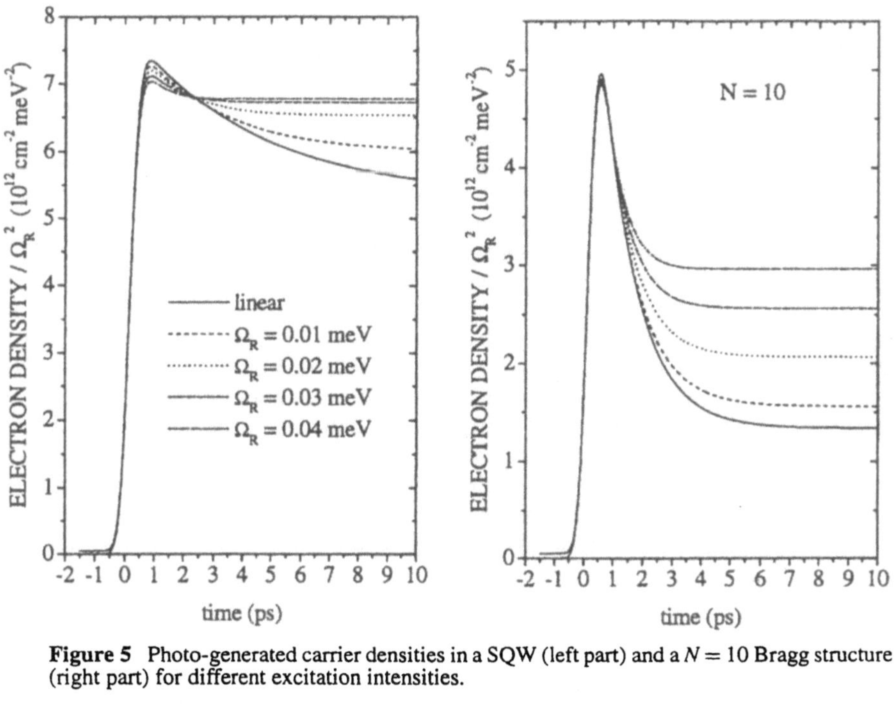

Figure 5 Photo-generated carrier densities in a SQW (left part) and a $N = 10$ Bragg structure (right part) for different excitation intensities.

understood by noting that the photo-generated carrier density depends on the detailed interplay between radiative and non-radiative decay processes. The fast radiative decay in a MQW Bragg sample prevents the coherent polarization to be efficiently scattered into non-radiative states, i.e. being incoherently absorbed. Thus, the carrier density generated in a MQW Bragg structure can become much smaller than the carrier density in a SQW under identical excitation conditions, which is shown in Fig. 5.

Consequently, for moderate excitations, the excitation-induced dephasing in the SQW is sufficiently more pronounced than in the Bragg structure for the same excitation intensities. Hence, the total decay rate of the SQW eventually exceeds that of the Bragg structure. It is worth to note that for increasing excitation intensity, the ratio of the generated carrier density in a Bragg structure and a SQW eventually approaches unity. In this limit, the dynamics of the reflected signal is dominated completely by the Coulomb scattering, while the radiative coupling is of minor importance.

5 Influence of Disorder

In this Section, we investigate the effects of dephasing by static disorder on the formation of a superradiant mode in MQW Bragg structures. We distinguish between well width and barrier thickness fluctuations. Although in a general sample interface roughness and barrier thickness fluctuations occur simultaneously, we treat them seperately in order to investigate the basic features of the two different types of disorder. Neglecting light scattering between different growth islands, we treat an ensemble of statistically independent, slightly different MQW structures. Within each configuration, the thickness of the barriers or the confinement energies within the individual wells are Gaussian distributed, acting as a disorder potential for the electrons and holes. The optical response in the presence of interface roughness is calculated in the classical limit, where the disorder correlation length is much larger than the exciton Bohr radius. In this limit, the effective bandgap varies locally within each QW, while the eigenfunctions are still given by the solution of the Wannier equation. The resulting configuration average of the imaginary part of the linear susceptibility displays the potential distribution of the excitonic resonances [22]. For numerical simplicity, calculations were performed in the ideal 2D limit.

In Fig. 6, we plot the linear absorption spectrum calculated in the vicinity of the 1s-resonance of the hh-exciton for a $N = 10$ Bragg structure. In Fig. 6 a) the individual QWs are treated as perfect while the barrier thickness is Gaussian distributed around an average value $d = \lambda/2$ with a standard deviation σ_d. The absorption spectra in Fig. 6 b) are calculated for a perfectly periodic $N = 10$ Bragg structure where the excitonic resonances within each QW are Gaussian distributed around its mean value assuming a standard deviation σ_E. Without disorder, a broad absorption line is found, which can be attributed to the superradiant mode. In the presence of barrier thickness fluctuations, the disorder induces sharp resonances, in contrast to the more common situation where disorder leads

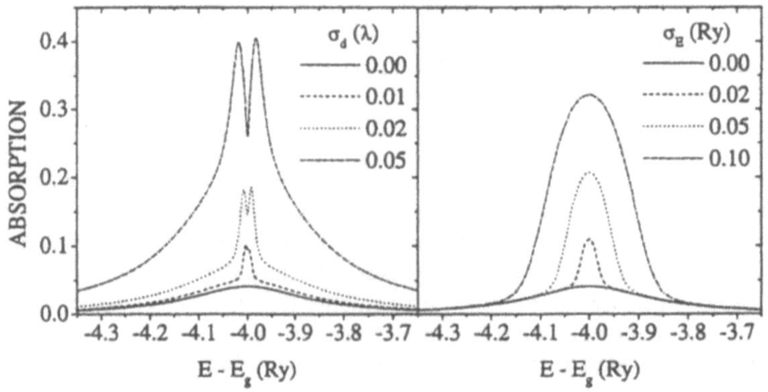

Figure 6 Linear absorption of a $N = 10$ Bragg structure for several values of barrier thickness fluctuations σ_d (a) and several values of the inhomogeneous broadening σ_E (b).

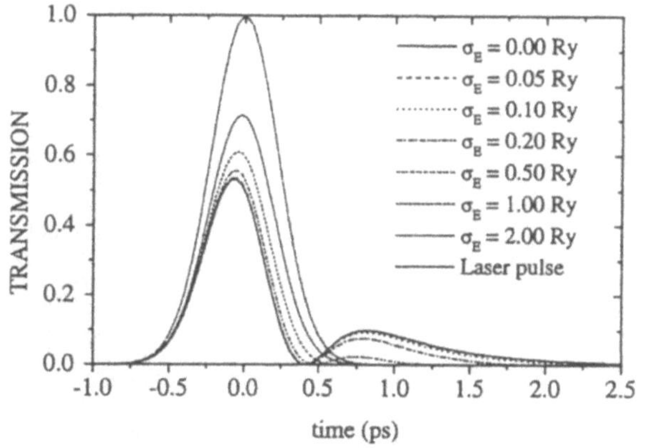

Figure 7
Linear time resolved transmission for $N = 10$ (dashed line) Bragg structure excited resonantly to the heavy-hole exciton transition with a Gaussian laser pulse of 530 fs FWHM for various disorder parameters σ_E.

to broadening of the optical resonances. The sharp structures can be attributed to subradiant modes. Those configurations, that fulfill the Bragg condition exhibit the most strongly decreased absorption rate at the exciton resonance, while those deviating from the Bragg condition display strong but slightly shifted resonances. However, the radiative broadening of these disorder induced modes is small compared to that of the superradiant mode. Therefore, they appear as sharp lines on a spectral background of the superradiant mode. Since the Bragg condition is defined with respect to the average exciton resonance, a coupling to the subradiant modes results from deviations from this resonance position (interface roughness) as well as from fluctuations of the optical barrier thickness. In the case of barrier thickness fluctuations only, the spectral position of the Bragg condition is fixed, leading to a minimum of the absorption at the exciton resonance. Increased disorder leads to a more pronounced splitting of the absorption line (Fig. 6a), which is no longer Lorentzian.

In the presence of interface roughness, the spectral position of the Bragg condition is inhomogeneously distributed. Therefore, the absorption spectrum is inhomogeneously broadened and the minimum in the absorption spectrum cannot be resolved. In Fig. 7, the linear time-resolved transmission is shown for a perfectly periodic $N = 10$ Bragg structure for different values of the inhomogeneous broadening σ_E. It can clearly be recognized from Fig. 6, that the reemitted signal of the MQW structure, is strongly reduced for increasing disorder. To compare these results with those obtained for excitation-induced dephasing, we plot in Fig. 8 the ratio of the SQW and Bragg signals for identical amounts of static disorder within the individual wells. In the absence of disorder, we see again a decreasing signal ratio due to the enhanced radiative decay in the Bragg structure. In the presence of disorder, a variety of modes, each having different radiative decay rates and slightly different resonance positions is excited simultaneously, leading to a complex interference pattern of the signal. For small disorder, the signal emitted from the Bragg structure is initially dominated by the SM, leading to a decreasing signal ratio, while at larger times the signal is dominated by subradiant modes, which cause an increase of the signal ratio. For larger disorder parameters, the radiative widths of the optically active modes become comparable to the radiative width of the SQW mode. Here, the signal ratio oscillates around a constant value, indicating the destruction of superradiance.

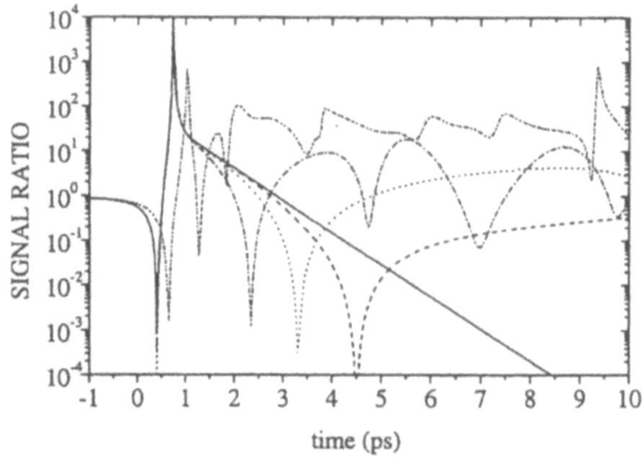

Figure 8
Ratio of the transmitted signals from a SQW and a $N = 10$ Bragg structure for the disorder parameters $\sigma_E = 0$ (solid line), $\sigma_E = 0.05$ Ry (dashed line), $\sigma_E = 0.1$ Ry (dotted line), $\sigma_E = 0.2$ Ry (dashed-dotted line), and $\sigma_E = 1.0$ Ry (dashed-double-dotted line).

6 Conclusions

In conclusion, we have shown that radiative coupling leads to the formation of a SM in multiple quantum well Bragg structures. Efficient radiative coupling requires phase coherence between the excited excitons. Our theory predicts a suppression of the superradiant mode if dephasing processes become important. For excitation with intense laser pulses, the analysis reveals an intricate interplay between excitation-induced dephasing and superradiant decay of the excitation in a MQW Bragg structure. In the presence of interface roughness, the oscillator strength of the SM is partly transferred to the subradiant modes, leading to a comlex interference pattern in the time-resolved signals.

Partial funding of this work through the Deutsche Forschungsgemeinschaft and a grant of CPU time on the T3E at the HLRZ Jülich is gratefully acknowledged.

Bibliography

[1] E.L. Ivchenko, A.I. Nesvizhskii and S. Jorda, Phys. Sol. State **36**, 1156 (1994),E.L. Ivchenko, Sov. Phys. Solid State **33**, 1344 (1991).

[2] L.C. Andreani, Phys. Stat. Sol. (b) **188**, 29 (1995), L. C. Andreani, F. Tassone, and F. Bassani, Solid State Commun. **77**, 641 (1991).

[3] D. S. Citrin, *Phys. Rev. B*, **47**:3832, 1993, D. S. Citrin, *Phys. stat. sol. (b)*, **188**:43, 1995.

[4] T. Stroucken, A. Knorr, P. Thomas, and S.W. Koch, Phys. Rev. B **53**, 2026 (1996).

[5] E. Hanamura, Phys. Rev. B **38**, 1228 (1988).

[6] M. Hübner, J. Kuhl, T. Stroucken, A. Knorr, S. W. Koch, R. Hey, and K. Ploog, Phys. Rev. Lett. **76**, 4199 (1996).

[7] B. Deveaud, F. Clérot, N. Roy, K. Satzke, B. Sermage, and D. S. Katzer, Phys. Rev. Lett. **67**, 2355 (1991).

[8] A. Vinattieri, J. Shah, T. C. Damen, D. S. Kim, L. N. Pfeiffer, M. Z. Maialle, and L. J. Sham, Phys. Rev. B **50**, 10868 (1994).

[9] S. Haas, T. Stroucken, M. Hübner, J. Kuhl, B. Grote, A.Knorr, F. Jahnke, S.W. Koch, R. Hey, K. Ploog, Phys. Rev. B (in press).

[10] B. Grote, T. Stroucken, S. Haas, P. Thomas, and S.W. Koch, Phys. Stat. Sol. (b) **205**, 209 (1998).

[11] M. Hübner, J. Kuhl, S. Haas, T. Stroucken, S.W. Koch, R. Hey, and K. Ploog, Sol. State Comm. **105**, 105 (1998).

[12] H. Wang, Kyle Ferrio, D. G. Steel, Y. Z. Hu, R. Binder, and S. W. Koch, Phys. Rev. Lett. **71**, 1261 (1993).

[13] G. Bastard, *Wave Mechanics applied to Semiconductor Heterostructures* (Les Editions de Physique, France, 1988).

[14] see e.g. H. Haug, S.W. Koch, *Quantum Theory of the Optical and Electronic Properties of Semiconductors* (World Scientific, Singapore, 3rd ed. 1994) and references therein.

[15] T.Stroucken, A. Knorr, C. Anthony, A. Schulze, P. Thomas, S.W. Koch, M. Koch, S.T. Cundiff, J. Feldmann, E.O. Göbel, Phys. Rev. Lett. **74**, 2391 (1995).

[16] M. Lindberg, and S.W. Koch, Phys. Rev. B **38**, 3342 (1988).

[17] S. Schmitt-Rink, D.S. Chemla, and H. Haug, Phys. Rev. B **37**, 941 (1988).

[18] I. Balslev, R. Zimmermann, and A. Stahl, Phys. Rev. B **40**, 4095 (1989).

[19] F. Jahnke, Habilitation thesis, Philipps-Universität Marburg (1996); F. Jahnke, M. Kira, and S. W. Koch, Z. Physik B. **104**, 559 (1997).

[20] All parameters correspond to GaAs: (Δ_{LT}=0.08 meV, ε=12.9, 1 Ry=4.2 meV, E_{gap}=1.52 eV). For our calculations we use a very small homogeneous nonradiative linewidth of $\gamma_0 = 0.013$ meV.

[21] The parameters used for the numerical calculation are $m_e = 0.071\, m_0$, $m_h = 0.25\, m_0$, where m_0 is the free electron mass, $\varepsilon = 12.53$, $\gamma_0^{-1} = 50$ ps, and $\mu = 6.75$ Å.

[22] R. Zimmermann, Phys. stat. sol. (b), **173**, 129 (1992).

Superradiant Exciton/Light Coupling in Semiconductor Heterostructures – Part II: Experiments

J. Kuhl[1], M. Hübner[1], D. Ammerlahn[1], T. Stroucken[2], B. Grote[2], S. Haas[2], S.W. Koch[2], G. Khitrova[3], H.M. Gibbs[3], R. Hey[4], and K. Ploog[4]

[1] Max-Planck-Institut für Festkörperforschung, D-70569 Stuttgart, Germany
[2] Department of Physics and Materials Science Center, Philipps Universität, D-35032 Marburg, Germany
[3] Optical Sciences Center, University of Arizona, Tucson, AZ 85721, USA
[4] Paul-Drude-Institut für Festkörperelektronik, D-10177 Berlin, Germany

Abstract: Time-integrated and time-resolved degenerate four-wave mixing as well as reflection measurements with subpicosecond optical pulses are applied to study the nonlinearity and dynamics of coupled exciton/photon modes in multiple quantum well structures. The photon-mediated interwell exciton/exciton coupling leads to strong changes of the linear and nonlinear optical properties, such as the transmission, reflectivity, absorption, 3rd order susceptibility, homogeneous linewidth, dephasing time and radiative lifetime of the 2D exciton. The strongest effects are observed in Bragg structures with $d = \lambda/2$ where λ is the wavelength of the heavy hole exciton transition. The collective behavior is observed in high-quality samples at low density and low temperature and rapidly disappears if the coherence is destroyed by intrawell excitation-induced dephasing, exciton/phonon scattering or sample disorder.

1 Introduction

Studies of excitonic dephasing in semiconductor quantum wells (QWs) have been the subject of intensive experimental and theoretical research during the last decade. Various dephasing mechanisms such as scattering by defects, interface roughness, alloy fluctuations or other quasi-particles (phonons, excitons, free electrons or holes) could be identified. Generally, it has been assumed that the dephasing mechanisms and times as well as the nonlinearity of the excitonic polarization are similar for single QWs (SQWs) and nominally identical multiple QWs (MQWs) of the same quality if the barrier thickness is large enough to prevent carrier tunneling between neighboring wells and that the dephasing rates of a few ps measured for 2D excitons in III-V semiconductors at low temperature and low excitation density are due to deviations from perfect long range order in the crystal [1]. Surprisingly, radiative recombination processes have not been considered as an efficient polarization decay channel although radiative contributions to the dephasing time T_2 of the 2D exciton of the order of 10 ps or even less have been postulated [2–5] and radiative lifetimes of a few tens of ps have been measured in photoluminescence experiments [6–8]. Very recently, it has been demonstrated, however, that radiative recombination supplies a nonnegligible contribution to optical dephasing in very high-quality

SQWs and that the importance of this process is strongly enhanced for 2D excitons in structures with simultaneous carrier and photon confinement such as QWs in semiconductor microcavities [9,10] and MQW Bragg structures [11] consisting of a large number N of identical QWs with interwell distances d which correspond to multiples of half the wavelength λ of the lowest heavy hole exciton transition.

The interaction of the excitonic polarization with coherent laser pulses propagating through MQW structures results in the formation of coupled exciton/photon modes since the field reemitted by a QW interacts with the excitons in the other wells. For arbitrary spacings of the QWs, the coupling removes the N-fold degeneracy of the excitonic resonances [12]. Here, we will focus onto the features of Bragg samples for which this exciton interwell coupling via dynamic dipole-dipole coupling results in a superradiant polarization decay. The modification of the exciton lifetime caused by phase coherent photons resembles the behavior of QWs embedded in microcavities [9,10] or excitons irradiated by phase-controlled optical pulse trains [13]. The collective response can be calculated from the semiconductor Maxwell Bloch equations (SMBE) as described in the previous article [14]. The model predicts remarkable modifications of the magnitude and dynamics of the linear and nonlinear optical properties and their strong variation with N, d and the propagation direction of the coherent photons with respect to the growth direction. Since the theory has in particular analyzed the role of a mismatch between d and $\lambda/2$ and stochastic sample disorder, a comparison of experimental data with the theoretical predictions is very intriguing.

In this paper, we summarize experimental investigations of coupled exciton/photon modes in two different III-V semiconductor compounds. The paper is organized as follows. Section 2 gives a description of the experimental techniques and samples. Section 3 reports on the observation of superradiant and subradiant modes. Section 4 deals with the role of reflection from the air/surface interface for the radiative recombination rate of coherent excitons. Section 5 demonstrates a real-time transition from a free-polarization decay (FPD) to a photon-echo (PE). Section 6 explores the influence of the mismatch between d and $\lambda/2$ on T_2 and the temporal and spectral shape of the degenerate four-wave mixing (DFWM) signal. Section 7 deals with the influence of the superradiance onto the sample absorption. The final Section 8 presents a brief summary.

2 Experimental Techniques and Samples

The formation of the superradiant mode requires macroscopic phase coherence in the sample. Due to momentum conservation, reemission of radiation by the coherently driven interband polarization is confined to the directions of the transmitted or reflected excitation beam. Thus time-integrated (TI), time-resolved (TR) or spectrally-resolved (SR) DFWM with subpicosecond optical pulses as well as TI and TR linear and nonlinear reflection measurements are appropriate spectroscopic techniques to investigate the nonlinearity and dynamics of radiatively coupled exciton/photon modes in dependence on N, the detuning between d and $\lambda/2$ and the excitation intensity.

The optical pulses for resonant excitation of the excitonic transition were taken from a mode-locked Ti:sapphire laser, providing 100 fs-pulses with an average power of 800 mW at a repitition rate of 76 MHz. In order to prevent simultaneous generation of higher excitonic states and free carriers, the pulses used in most studies were sent through an external pulse shaper which limits their bandwidth to 3.3 meV. Behind the pulse shaper the pulse duration amounts to 700 fs. To minimize effects of biexcitonic contributions and excitation-induced dephasing (EID), cocircularly polarized, low energy (2 pJ, focused to a spot of 50 μm) pulses with wavevectors $\vec{k}_1$ and $\vec{k}_2$ were used for DFWM in two-pulse self-diffraction geometry. The signal diffracted in the direction $2\vec{k}_2 - \vec{k}_1$ has been measured time-integrated by a slow photodiode as a function of the delay τ_{12} between the two excitation pulses. Backward or backward and forward diffraction have been used for samples with absorbing and transparent substrates, respectively. For resolution of the real-time shape, the DFWM signal is cross-correlated with a 100 fs pulse from the Ti:sapphire laser in a 2 mm thick $LiIO_3$ crystal. Its spectral distribution is analyzed by a 0.75 m double monochromator.

We used two different $N = 10$ GaAs/Al_xGa_{1-x}As Bragg samples with 20 nm well width and $x = 0.3$, and two (In,Ga)As/GaAs Bragg samples with 10 and 30 QWs with a width of 8 nm and an In content of 3%. All samples were grown by molecular beam epitaxy on GaAs substrates which are opaque and transparent at the wavelengths of the lowest heavy hole exciton transitions in the GaAs and the (In,Ga)As QWs, respectively. The 94.5 nm thick (Al,Ga)As barriers of the GaAs QW samples yield a QW separation very close to $\lambda/2$ in the material at 8 K. For identification of the exciton/photon-coupling-induced dynamical and spectral features, the measured signals are compared to corresponding curves taken on a high-quality GaAs SQW. The (In,Ga)As samples were intentionally wedge-shaped, so that d varied continuously with sample position thus enabling systematic studies of the coupled exciton/photon modes in dependence on the mismatch between d and $\lambda/2$. The samples were kept at a temperature of 8 K.

3 Dephasing of Super- and Subradiant Modes

In a perfect Bragg structure, the radiative interwell coupling generates one superradiant mode with an N times enhanced oscillator strength whereas the residual $N-1$ coupled modes obtain vanishing oscillator strength, i.e. they are "optically dark". However, in the presence of sample disorder, e.g. inhomogeneous broadening of the exciton transition or stochastic fluctuations of the barrier width, a finite amount of oscillator strength is transferred from the superradiant to the "forbidden" modes [5,11,15]. The exciton/photon coupling implies a reduced radiative decay of these subradiant modes as compared to the SQW exciton. The presence of disorder creates a complex splitting of the exciton resonance since each coupled mode has a slightly different frequency and homogeneous linewidth $\gamma_n = \gamma_0 + \beta_n \gamma_{rad}$ where $n = 1, \ldots, N$ is the mode index, γ_0 and γ_{rad} are the non-radiative and radiative contributions to the dephasing rate of a nominally identical SQW. The factors $0 < \beta_n < N$ are depending on the coupling strength. In a perfect Bragg structure, $\beta_1 = N$ for the superradiant mode and $\beta_i = 0$ for the subradiant modes ($i = 2, \ldots, N$). Because of transfer of oscillator strength from the superradiant to the subradiant modes, β_1 becomes smaller than N whereas the β_i for $i > 1$ assume finite values between 0 and 1 depending on the reduction of interwell exciton coupling caused by disorder.

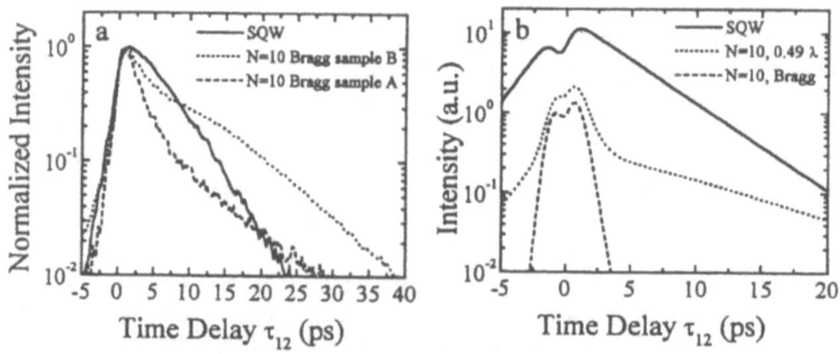

Figure 1 (a) Normalized TI-DFWM signals for two $N = 10$ GaAs Bragg structures (dashed and dotted line) and a SQW (solid line) measured in backward diffraction geometry. (b) Calculated TI-DFWM signals for a GaAs SQW (solid line), a $N = 10$ perfect (dashed line), and a slightly detuned ($d = 0.49\lambda$, dotted line) Bragg structure ($\gamma_0 = 0.084$ meV, $\gamma_{rad} = 0.134$ meV).

Figure 1a depicts TI-DFWM signals measured for the two different $N = 10$ GaAs wells and the corresponding SQW. For comparison, Fig. 1b shows theoretical predictions obtained from the SMBE. For this calculation, we used $T_{rad} = 9.8$ ps for the radiative contribution to phase relaxation of the lowest hh exciton in the GaAs SQW, and considering the high sample quality, we assumed a rather small nonradiative contribution ($T_0 = 50$ ps) to the total dephasing rate $\gamma_{hom} = (T_2)^{-1} = (T_{rad})^{-1} + (T_0)^{-1} = \gamma_{rad} + \gamma_0$.

Radiative coupling leads to biexponential decay of the signal for the MQWs. The signal decay time for the GaAs MQW Bragg sample A increases from $\tau_{dec} = 2.5$ ps for the superradiant mode to 9.5 ps at later times. The signal decay time is equal to $T_2/2$ and $T_2/4$ for homogeneously and inhomogeneously broadened transitions, respectively [16]. Taking into account that the transition of the coupled exciton/photon mode switches from a homogeneous line at the beginning when the signal is dominated by the superradiant mode to an inhomogeneously broadened line at large delays when the signal is due to the surviving subradiant modes (see chapter 4 and [17]), we obtain $T_2^{sup} = 5$ ps and $T_2^{sub} = 38$ ps for the superradiant and subradiant modes, respectively. T_2^{sub} provides a lower limit for T_0. The initial superradiant decay is less pronounced for Bragg sample B but the signal exhibits the same slow decay for the subradiant modes. The weaker superradiance is due to a larger deviation of the barrier width from the Bragg condition. The curve for the SQW exhibits a slight increase of the slope with increasing delay which is typical for the signal in the presence of inhomogeneous broadening (linewidth γ_{inh}) which is comparable to the homogeneous linewidth (γ_{hom}). Analysis of the signal decay yields $T_2 = 9.6$ ps corresponding to $\gamma_{hom} = 0.137$ meV and $\gamma_{inh} \approx 0.15$ meV. From the data, we calculate a measured $T^*_{rad} = 12.8$ ps. This value agrees very well with the radiative contribution to dephasing of 13.3 ps calculated from simple quantum mechanical expressions in the 2D limit. Differences in comparison to the values of [11] are due to the fact that we have now regarded the finite γ_{inh} and taken $T_0 = 38$ ps from the decay of the subradiant modes of the Bragg structure.

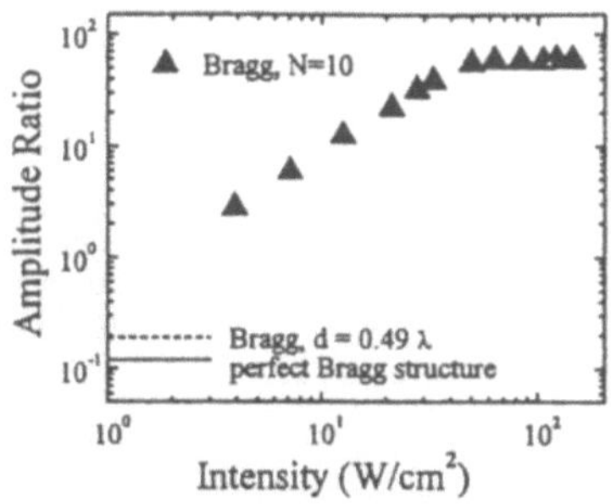

Figure 2
Ratio of the signal amplitudes measured on the $N = 10$ Bragg sample A and the SQW as a function of excitation density. The solid and the dashed line mark theoretical predictions at low excitation for a perfect Bragg structure and a sample with $d = 0.49\lambda$ assuming $T_{rad} = 9.8$ ps and $T_0 = 50$ ps.

The data reveal strong influence of energetic and geometric disorder on the strength of the superradiant coupling in the Bragg samples and the resulting radiative contribution to excitonic dephasing. The mismatch between d and $\lambda/2$ can be intentionally varied if the Bragg structure is rotated with respect to the incident laser beams. Systematic investigations of the signal shape in dependence on the angle α_{ext} between the bisector of the two laser beams and the normal to the sample surface have been reported in [11] and [18].

Superradiant reemission of photons on a subpicosecond time scale should prevent effective excitation of the sample. This conjecture is confirmed by comparing the amplitudes A_{10} and A_1 of the diffracted signals measured on the $N = 10$ GaAs Bragg structure and a SQW reference sample, respectively (Fig. 2). In the absence of radiative interwell coupling, the signal diffracted by a Bragg structure is predicted to be proportional to N^2. At low excitation intensity, we measure, however, a much smaller ratio $R = A_{10}/A_1 = 2$ for sample A. The discrepancy of this value with respect to the theoretical predictions of 0.1 to 0.2 calculated for the perfect sample and a sample with $d = 0.49\lambda$ using $T_0 = 50$ ps and $T_{rad} = 9.8$ ps can be attributed to significant sample disorder. Exciton/photon coupling in MQW structures requires phase coherence between the excitons in the different wells and the exciting radiation field. Thus the characteristic coupling features will rapidly disappear at higher excitation intensities where intrawell excitation-induced dephasing leads to rapid destruction of the phase coherence. Accordingly, R rapidly increases with excitation intensity and saturates at values between 70 and 80 for intensities > 50 W/cm^2.

Because of momentum conservation, the superradiant emission is restricted to the transmission or reflection direction of the incident laser beam. A comprehensive analysis of linear and nonlinear TR reflectivity measurements on Bragg samples has been published in [4,5,19,20].

The results of this section demonstrate that contrary to prevailing assumptions radiative recombination strongly contributes to excitonic dephasing in high-quality GaAs QWs and that scattering by defects plays at most a comparable role. Due to superradiant enhancement, radiative processes are the dominant dephasing mechanism in high-quality Bragg structures. Since formation of the superradiant state necessitates macroscopic phase coherence, the dynamics of the optical polarization are extremely sensitive to competing dephasing mechanisms such as scattering by defects, interface roughness, phonons or EID.

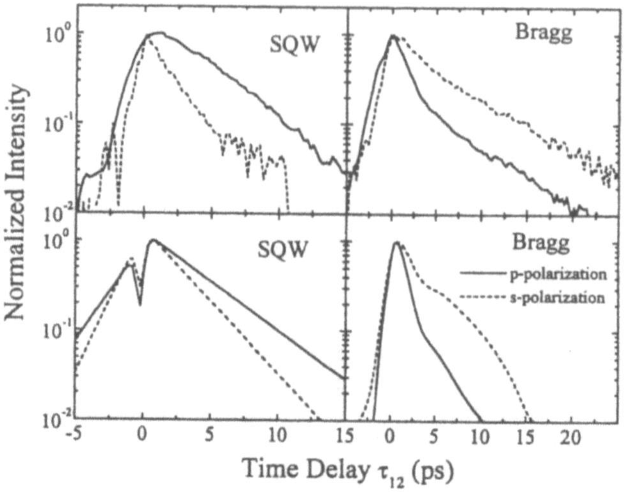

Figure 3 Calculated (a) and measured (b) TI-DFWM signals for Bragg sample B (right) and the SQW (left) at Brewster's angle for laser beams polarized parallel (solid curves) and perpendicular (dashed curves) to the plane of incidence.

4 Influence of the Sample/Air Surface Reflection

The experiments demonstrate that γ_{rad} of the MQWs critically depends on sample geometry because the effective local field is a superposition of the laser field with fields reemitted from the QWs. This observation raises the question concerning the modifications of the radiative lifetime due to the sample/air interface, which reflects part of the backward propagating waves. The electric field reflected at the sample/air interface will increase or decrease the local field at the QW positions depending on the thickness d_l of the cladding layer. As shown in [20], the effective radiative decay rate observed in an experiment is

$$\gamma_{rad}^{*} = \gamma_{rad}\left[1 + \frac{\sqrt{\varepsilon}-1}{\sqrt{\varepsilon}+1}\cos\left(4\pi\frac{d_l}{\lambda}\right)\right],$$

where $\sqrt{\varepsilon}$ is the refractive index of the cladding layer and γ_{rad} the radiative dephasing rate for the bare QW. The corresponding influence of the substrate/air interface at the sample backside is negligible for our GaAs QWs because the substrate is strongly absorbing at the QW exciton transition. It should be noticed that this cavity effect even modifies the radiative lifetime of a single QW. The shortest and longest lifetimes are expected for $d_l = m\lambda/2$ and $(2m+1)\lambda/4$, respectively, whereas for $d_l = (4m \pm 1)\lambda/8$ (m is an integer number) the influence of the surface should be negligible. Using this formula, we calculate a radiative contribution to dephasing $T_{rad} = 20$ ps (equivalent to a radiative lifetime of 10 ps) for a bare GaAs SQW taking into account $d_l = 0.51\lambda$ for our sample.

The influence of surface reflection on dephasing can be studied by comparing DFWM signals generated with two laser beams which are polarized either parallel (p-polarization) or perpendicular (s-polarization) to the plane of incidence on a sample mounted under Brewster's angle ($\alpha_{ext} = 74°$). Experimental curves for the GaAs Bragg sample B (effective $d_l = 0.09\lambda$) and the reference SQW (effective $d_l = 0.56\lambda$) are plotted in the upper parts of Fig. 3. The average excitation intensity was adjusted to 75 and 148 μW for the p- and s-polarization, respectively, to compensate for the difference of the surface reflectivity. The initial superradiant decay on the Bragg sample reveals time constants of 2 ps and 4.5 ps for p- and s-polarisation, respectively. For the SQW, the superradiant decay is accelarated by approximately a factor of 2. The gross features of the polarization dependence are qualitatively well reproduced by the theoretical curves shown in the lower part. Comparison confirms a distinct effect of surface reflectivity on the dephasing dynamics of excitons.

5 From a Free Polarization Decay to a Photon-Echo

Time resolution of the DFWM signal is a powerful tool to differentiate between homogeneous and inhomogeneous broadening of the excitonic transition [17]. For a homogeneously broadened two-level system. the response has a distinctly asymmetric shape and the trailing edge decays with the time constant $T_2/2$. It appears at $t = \tau_{12}$ in real time, i.e. immediately after the interaction of the second pulse with the system, whereas the response of a dominantly inhomogeneously broadened system ($\gamma_{\mathrm{inh}} > 10\gamma_{\mathrm{hom}}$) is a symmetric PE occuring at $t = 2\tau_{12}$. In the intermediate regime, the temporal evolution is more complex. The TI-DFWM curve exhibits a continuous transition from $\tau_{\mathrm{dec}} = T_2/2$ for small delays to $T_2/4$ at large τ_{12} and the TR signal shifts from τ_{12} to $2\tau_{12}$.

Figure 4a shows normalized TR-DFWM signals measured on the $N = 10$ GaAs MWQ A for different delays τ_{12}. α_{ext} was set to zero so that the sample is very close to the Bragg condition. The curves recorded for different τ_{12} reveal remarkable changes from the asymmetric shape of a FPD observed for $\tau_{12} < 2$ ps to a symmetric pulse typical for PE at $\tau_{12} > 5$ ps. Further strong evidence for a transition from a FPD to a PE at $\tau_{12} \approx 2$ ps is provided by the plot of the real-time position of the signal maximum $t_{\max}$ vs. τ_{12} (Fig. 4b). Up to 2 ps, the data are fitted by a line with slope 1 whereas the points beyond 2 ps are well approximated by a line with slope 2. Since the reference GaAs SQW exhibits a FPD for time delays up to 8 ps [21], this phenomemon has to be attributed to the radiative coupling in the MQW Bragg sample.

The transition from FPD to PE is caused by the coupling-induced distribution of γ_{hom}. During the first 2 ps, the signal is dominated by the superradiant mode which has a broad homogeneous linewidth because of the fast radiative decay. Consequently the response shows the behavior of a FPD. After rapid depletion of this mode, the residual signal originates from the subradiant modes for which both the frequency and the linewidth are

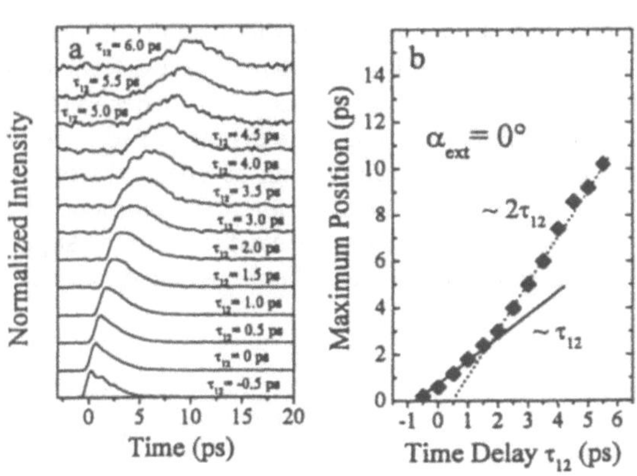

Figure 4
(a) Measured TR-DFWM signals for increasing delay τ_{12} on the $N = 10$ GaAs QW Bragg sample A at $\alpha_{\mathrm{ext}} = 0$. (b) Peak position of the signal vs. τ_{12}.

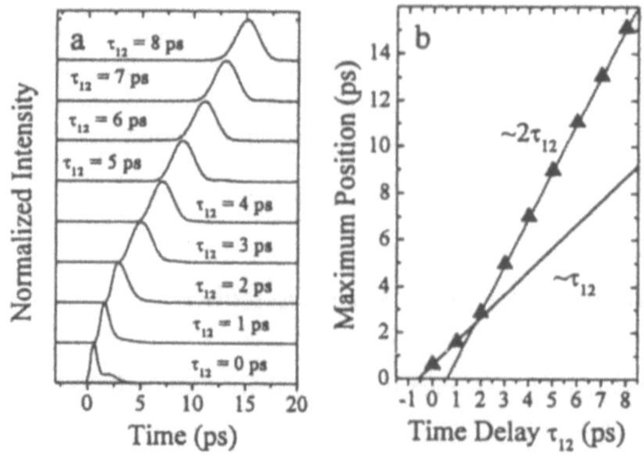

Figure 5
(a) Calculated TR-DFWM in reflection geometry from a $N = 10$ Bragg structure for various time delays τ_{12} and $\alpha_{ext} = 0$. (b) Real time position of the signal maximum as a function of τ_{12}.

inhomogeneously distributed. As simultaneously their γ_{hom} is much smaller, the ratio of $\gamma_{inh}/\gamma_{hom}$ strongly grows thus explaining the observed transition from FPD to PE. Contributions of FPD and PE to the DFWM signal have been previously reported in case of simultaneous excitation of quasi-delocalized and localized excitons [22,23].

Figure 5 depicts theoretical TR-DFWM signals for a $N = 10$ Bragg structure calculated with GaAs material parameters and using $\gamma_0 = 25.2\ \mu$eV, $\gamma_{rad} = 0.108$ meV, and $\gamma_{inh} = 0.84$ meV which is in good agreement with the measured photoluminescence linewidth of approximately 0.7 meV. Comparison with the experimental data of Fig. 4 proves that the theory reproduces all the prominent features of the experimental curves. Based on the fact that the subradiant modes emit a PE, their total dephasing rate and thus γ_{hom} is related to the decay rate of these modes observed in TI-DFWM by $\gamma_{hom} = h/4\pi\tau_{dec} = 36.6\ \mu$eV. Since the radiative decay of subradiant modes is strongly suppressed, γ_{hom} can serve as an upper limit for the nonradiative contribution γ_0 to γ_{hom}. Assuming that γ_0 is the same for subradiant and superradiant modes, we have to conclude that even for the SQW γ_{hom} is dominated by radiative decay.

Careful inspection of the data in Figs. 4 and 5 reveals that the PE does not occur exactly at $t_{max} = 2\tau_{12}$ but is slighty shifted by a constant amount. This effect can by explained by the theoretical model [17] and is due to the fact that the third order polarization in a radiatively coupled sample is driven by a coherent superposition of the external laser field and the propagating polarization wave reradiated by the excitonic oscillators in the residual QWs. Both terms have a different temporal behavior and are not in phase with each other. Interference between the two source terms leads to a small shift of the PE of the order of 1 ps as compared to the PE expected for noninteracting two-level systems.

The transition from a homogeneously to an inhomogeneously broadened state in real time is an intrinsic property of Bragg samples and has to be distinguished from previous observations that the line broadening of a given sample can be changed from inhomogeneous to homogeneous character by choice of the experimental conditions [24,25]. It necessitates homogeneous broadening dominated by radiative decay for the superradiant mode and significant structural disorder which tranfers a finite oscillator strength to the subradiant modes.

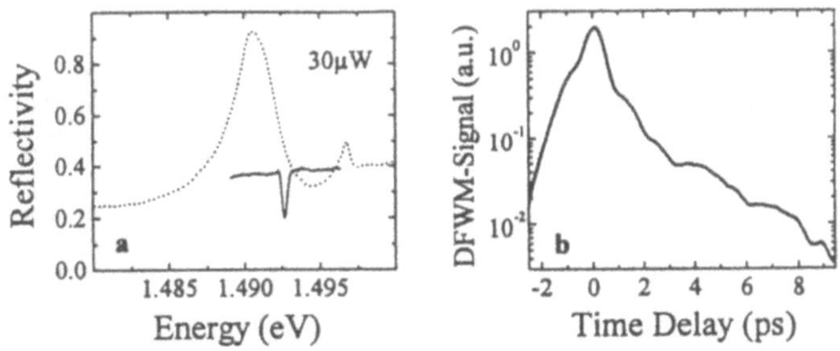

Figure 6 (a) Linear reflection spectra measured on the $N = 30$ (In,Ga)As sample in resonance (dashed line) and for a detuning of $d = 0.40\lambda$ (solid line). (b) TI-DFWM signal measured with 270 fs laser pulses at the Bragg position of the wedge-shaped $N = 30$ (In,Ga)As/GaAs sample in backward diffraction configuration.

6 Exciton/Photon Modes in Mismatched Bragg Samples

To obtain further insight into the role of detunings between d and $\lambda/2$, we have performed a systematic study of the spectral and temporal behavior of the third order nonlinear polarization of the heavy hole exciton transition on wedge-shaped (In,Ga)As/GaAs MQWs where the barrier width could be varied in the range up to 20% with respect to the Bragg resonance [26].

Since GaAs is transparent at the heavy hole exciton transition of the (In,Ga)As wells, reflection and transmission of the exciton line can be analyzed without the need to remove the substrate. Moreover the transparent substrate enables simultaneous studies of forward and backward diffraction DFWM. In the past, both configurations have been expected to provide the same results. Recent theoretical studies [4,5,27] have predicted, however, substantial differences in the temporal evolution and spectral composition of the two signals if $d \neq \lambda/2$.

Linear reflection experiments on the wedge-shaped $N = 30$ (In,Ga)As sample reveal strong variations of the exciton linewidth with sample position in the low density regime which disappear at higher excitation intensities. Figure 6a compares line profiles of the hh exciton measured with very weak 120 fs pulses (spectral full-width-half maximum 14.5 meV) and the spectrometer (resolution 0.1 meV) for the Bragg resonance and a position with $d = 0.40\lambda$ where the radiative coupling is strongly suppressed. Tuning from the Bragg position to off-resonance results in a dramatic narrowing of the reflectivity linewidth from ≈ 3.2 meV to less than 0.4 meV and a concomittant drop of the reflectivity from a value close to 1 to less than 0.3. These values agree fairly well with linewidth data recorded on another piece of the same wafer [28]. For $d = 0.5\lambda$, the reflectivity at the exciton transition is strongly enhanced as compared to the nonresonant background reflectivity because of the superradiant reemission. In contrast, the reflection exhibits a dip at the exciton resonance for the strongly detuned sample position. The large spectral width at the Bragg resonance reflects strong homogeneous line broadening due to superradiant damping of the excitation. The linewidths of 3.2 meV is equivalent to a dephasing time of $T_2 = 0.4$ ps. The TI-DFWM signal (Fig. 6b) measured in the backward diffraction configuration at the Bragg resonance position on the $N = 30$ (In,Ga)As sample with 270 fs pulses (application of still shorter pulses implies remarkable contributions of EID due to simultaneous excitation of free electron/hole pairs) reveals a fast initial decay of the polarization with a time constant of 350 fs which is limited by the pulse duration.

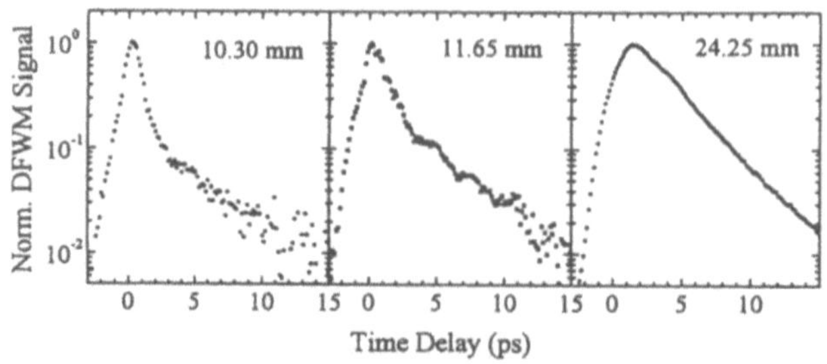

Figure 7 Normalized TI-DFWM signals measured for different positions on the wedge-shaped $N = 30$ (In,Ga)As sample in backward diffraction geometry with 700 fs laser pulses.

Deconvoluting the pulse assuming Gaussian profiles yields an estimate of 220 fs for the signal decay time. The corresponding dephasing time $T_2 = 440$ fs agrees fairly well with the linewidth measured in linear reflection.

Figure 7 shows normalized TI-DFWM traces measured on $N = 30$ (In,Ga)As structure for different sample positions with two laser beams of 1 mW average power and a pulse duration of 700 fs. The variations of the signal shape with sample position are caused by the corresponding changes of the barrier width which is plotted in Fig. 8a. At $x = 10.3$ mm where $d \approx \lambda/2$, the signal exhibits a double exponential decay. The initial peak which decays to less than 10% of the maximum value with a time constant of 0.65 ps is due to of the superradiant mode. The decay time is limited by the laser pulse duration. The real dephasing time of this mode is close to 0.4 ps as demonstrated by the TI-DFWM experiment with 270 fs pulses (Fig. 6b) and the width of the reflection spectrum measured in the frequency domain (Fig. 6a). The subsequent slower decay with a time constant of 4 ps (equivalent to a dephasing time of $T_2 = 16$ ps in case of dominating inhomogeneous broadening) has to be attributed to the subradiant modes. The considerably larger difference in the dephasing times of the two mode types as compared to the GaAs samples is due to the 3 times larger number of coupled wells. The initial decay is remarkably slowed down if the QW separation is tuned to values larger or smaller than $\lambda/2$, whereas the decay of the weakly coupled modes remains almost unchanged. Assuming that the subradiant modes are inhomogeneously broadened, we estimate a value of 80 μeV for the nonradiative contribution γ_0 to the homogeneous linewidth. Comparison of the time domain and frequency domain data yields then an inhomogeneous linewidth of the subradiant modes of approximately 0.3 meV.

Comparison of DFWM signals measured in backward and forward diffraction geometry shows significant differences if d deviates from $\lambda/2$ by only a few percent. These differences are particularly pronounced in the spectra of the diffracted beams. Normalized spectra of the TI-DFWM signal (Fig. 9a) measured with 700 fs for $\tau_{12} = 0.66$ ps at various sample positions exhibit remarkable changes of the spectral shape, width and peak positon as a function of the detuning between d and $\lambda/2$ and very pronounced differences for the signal spectra diffracted in the forward and backward direction. These characteristic features are well reproduced by the calculated spectral profiles in Fig. 9b. For the perfect Bragg sample and a laser pulse delay which is comparable to or smaller than the pulse duration, the theory predicts a very broad symmetric spectrum with a pronounced dip at the top. This dip is due to partial overlap of the two pulses. In this situation, the DFWM signal consists of two contributions which are not in phase and interfere destructively. In spectra recorded for larger pulse delays (negligible overlap of the two fields), the

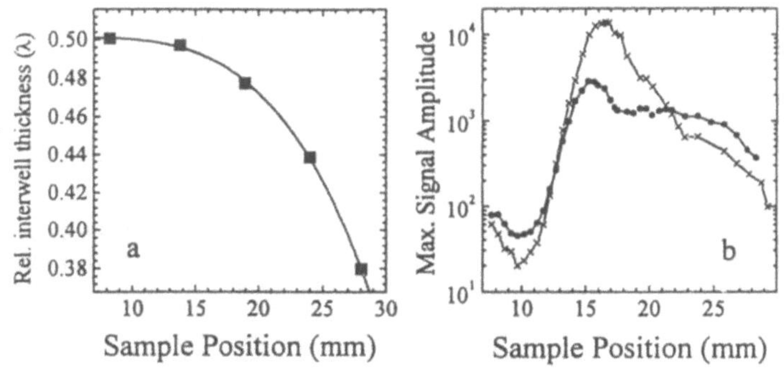

Figure 8 (a) Position-dependent barrier thickness on the tapered $N = 30$ (In,Ga)As/GaAs sample. (b) TI-DFWM signal amplitude measured in forward (dots) and backward (crosses) diffraction geometry vs. sample position.

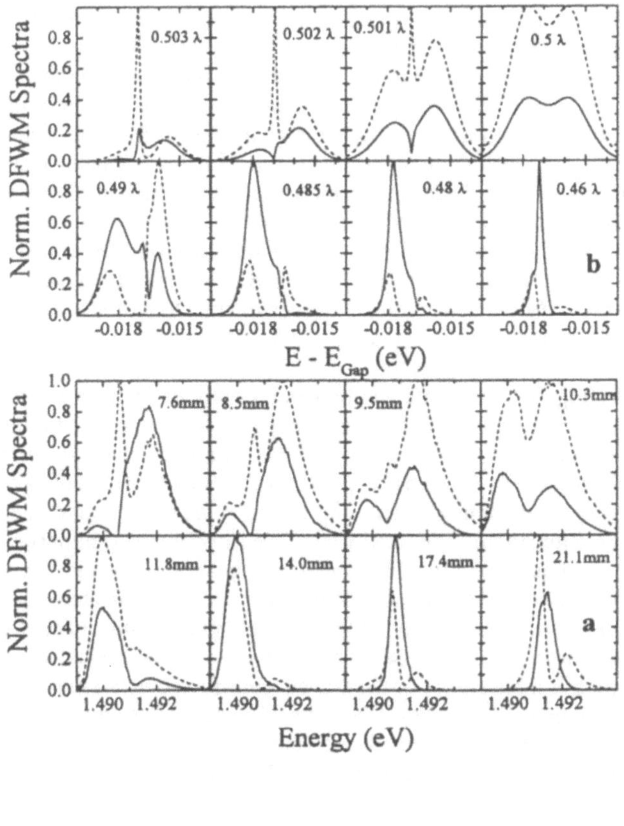

Figure 9 Normalized spectra of the TI-DFWM signals measured (a) and calculated (b) for different positions on the wedge-shaped $N = 30$ (In,Ga)As Bragg sample in forward (dotted line) and backward (solid line) diffraction geometry.

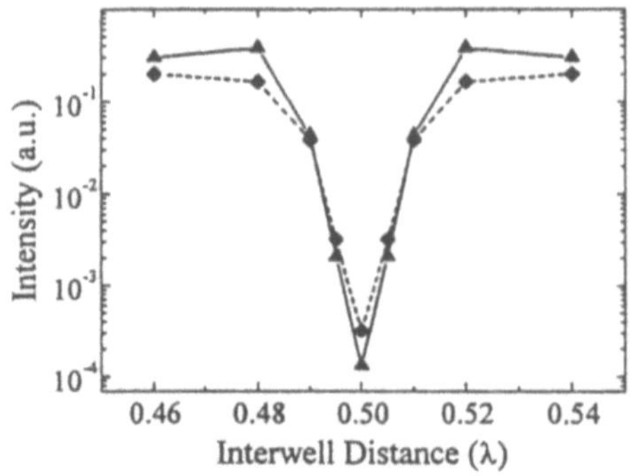

Figure 10
Calculated DFWM signal amplitude for forward (diamonds) and backward (triangles) diffraction geometry vs. interwell distance.

dip will disappear. At resonance, the signal shape is identical for the backward and forward geometry (Fig. 9b). The FWHM reaches a value of 3.2 meV in accordance with the fast decay of 0.22 ps measured in the time domain and the FWHM of the linear reflection spectrum.

Detuning of d with respect to $\lambda/2$ leads to a gradually increasing suppression of the superradiant mode and a concommittantly growing strength of the subradiant modes. It generates an asymmetry between the low and high energy wing of the broad spectrum observed for the Bragg condition and produces distinct differences between the spectra diffracted into the forward and backward direction. For $d > \lambda/2$, a continuously growing sharp spike (dip) appears on the low energy side of the forward (backward) spectrum in agreement with the theoretical model which predicts that the signal due to subradiant modes is preferently emitted in forward direction. This asymmetry manifests itself as a hole in the spectrum of the backwardly diffracted signal. The shift of the signal center frequency is restricted to the spectral range covered by the broad Bragg resonance. Larger shifts occur for detunings between d and $\lambda/2$ by more than 10% which imply a significant change of the heavy hole exciton resonance frequency.

Because of the superradiant decay of the coherent polarization, the 3rd order nonlinear response in a Bragg sample is strongly reduced. This effect has been demonstrated in Sect. 3 by the different density dependences of the DFWM signal amplitudes on the GaAs/(Al,Ga)As Bragg sample A and the SQW. The signal amplitudes measured at $\tau_{12} = 0.66$ ps for different positions on the $N = 30$ (In,Ga)As MQW vary by more than 3 orders of magnitude (Fig. 8). The signal amplitudes for both the forward and backward diffraction geometry reach their minimum at the Bragg resonance and detuning of the sample geometry to either $d > \lambda/2$ or $d < \lambda/2$ involves a rapid rise of the signal. The experimentally found variation of the signal is caused by the superposition of two effects: (i) the formation of a superradiant mode at $d = \lambda/2$ and (ii) the constructive interference of the DFWM signals excited in the individual wells. The interference is important for the backward diffraction configuration [29]. Phase-matched generation of the signal in the backward direction is restricted to a total sample thickness of approximately $t_s = 25$ nm. For larger sample thicknesses t_s, the unavoidable phase-mismatch results in an oscillation of the signal amplitude with t_s. In a Bragg sample, however, where d is equal to a multiple integer of $\lambda/2$, constructive interference of the signals contributed by all individual wells is accomplished in the backward direction. This interference leads to the strong peak at 16.5 mm. The interference effect is expected to create a signal enhancement by a factor of 900 as compared to a single QW according to the N^2 law for the signal growth with

N. This enhancement should be observable at high excitation densities when the radiative coupling is suppressed by intrawell excitation-induced dephasing. At low intensities, however, the interference effect is overcompensated by the influence of superradiance. The decay of the backwardly diffracted signal amplitude for $x > 16.5$ mm is due to growing destructive interference of the polarization from the different wells. The appearance of the pronounced maximum at $x = 16.5$ mm demonstrates remarkably different sensitivities of radiative coupling and the interference effect against detunings from the Bragg condition in agreement with the theoretical predictions of Fig. 10. The superradiant mode is almost completely supressed for detunings $\delta = (\lambda/2 - d)/(\lambda/2)$ as small as 2%. In constrast, an enhanced signal amplitude due to interference is easily detectable for $\delta \approx 10\%$. The lower minimum obtained for the backward direction signal at the Bragg resonance as compared to the forward direction is explained by reflection from the sample/air interface. Since the seperation of the top QW from the surface is $\lambda/2$, the reflected part of the backward DFWM signal is constructively superimposed to the signal diffracted in the forward direction.

7 Absorption in Radiatively Coupled MQWs

Because of radiative interwell coupling, the excitation density in a MQW Bragg structure depends critically on sample geometry. In particular it can be significantly smaller than the density generated for the same excitation conditions in a nominally identical SQW [30]. Rapid superradiant decay of the coherent exciton polarization in a Bragg sample is a polarization decay channel competing with absorption which requires fast scattering into nonradiative states. Therefore, the incoherent correlated electron/hole pair population residing in the sample after several ten ps when the phase coherence is lost is drastically reduced. This incoherent population can be detected in a three-pulse DFWM experiment where two linearly copolarized pulses with wavevectors $\vec{k}_1$ and $\vec{k}_2$ create a transient absorption and phase grating due to interference between two light beams. A third pulse incident on this grating after a variable delay τ_{23} is diffracted into the direction $\vec{k}_1 - \vec{k}_2 + \vec{k}_3$ and monitored as a function of τ_{23}. Comparison of signals recorded at different positions on the wedge-shaped $N = 10$ (In,Ga)As QW sample (Fig. 11) reveals strong

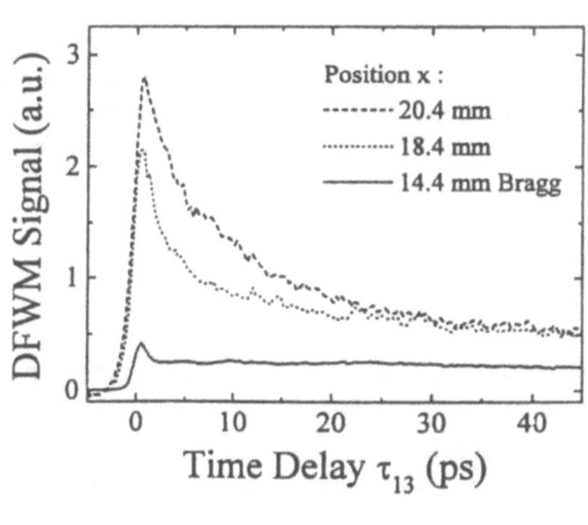

Figure 11
Three-pulse TI-DFWM signal on the $N = 10$ (In,Ga)As/GaAs MQW sample measured for the Bragg position (solid line) and positions with $d = 0.47\lambda$ (dotted line) and $d = 0.45\lambda$ (dashed line).

suppression and fast decay of the coherent polarization in the sample at the resonance position. For delays larger than 30 ps, the amplitude and dynamics are determined by the slowly decaying incoherent correlated electron/hole pair population. The data show that strong superradiant emission at the resonance position results in a drop of the effective absorption by a factor of more than 2. For the detuned position, the coherent regime lasts considerably longer since the radiative contribution to dephasing is suppressed and the residual decay (1/e decay time constant $\approx$ 8 ps) is most likely due to scattering by defects. With increasing excitation intensity the ratio of the carrier densities excited at the Bragg resonance, at the detuned position and a SQW finally approaches unity when the dynamics is completely determined by excitation-induced dephasing.

8 Conclusions

Comparison of TI- and TR-DFWM and reflectivity experiments performed with subpicosecond laser pulses on MQW Bragg structures and corresponding SQWs with solutions of the SMBE has revealed efficient superradiant emission of radiatively coupled exciton/photon modes. The formation of the superradiant mode implies phase coherence between excitons excited in different wells of the MWQ sample and thus necessitates low excitation densities and low temperature. In high-quality SQWs, radiative recombination with a lifetime as short as 10 ps provides a strong contribution to optical dephasing. In MWQs, the radiative contribution to dephasing critically depends on the number of QWs and their spacing. For a $N = 30$ (In,Ga)As MQW Bragg structure, a radiative contribution to the dephasing time is as short as ≈ 0.4 ps. The radiative lifetime varies with the thickness of the top cladding layer because of coupling to the radiation field reflected by the sample/air interface. The magnitude and dynamics of the linear and nonlinear optical properties of the coupled exciton/photon modes are strongly influenced by barrier width and QW thickness fluctuations. In particular, the presence of disorder and radiative coupling leads to a distribution of $\gamma_{\rm hom}$ implying a complex transition from FPD to PE in TR-DFWM experiments.

Acknowledgments

We gratefully acknowledge partial funding of this work by the Deutsche Forschungsgemeinschaft through the Quantenkohärenz Schwerpunkt, the Sonderforschungsbereich 383, the Leibniz Prize and AFOSR, NSF AMOP, NSF Lightwave Technology, DARPA/ARO, NSF DMR.

Bibliography

[1] *Coherent Optical Interactions in Semiconductors.*, ed. by R.T. Phillips NATO ASI Series B: Physics Vol. **330**, Plenum Press, New York – London (1993).

[2] E. Hanamura, Phys. Rev. B **38**, 1228 (1988).

[3] L. C. Andreani, F. Tassone, and F. Bassani, Sol. State Commun. **77**, 641 (1991).

[4] E. L. Ivchenko, A. I. Nesvizhskii, and S. Jorda, Phys. Sol. State **36**, 1156 (1994).

[5] T. Stroucken, A. Knorr, and S. W. Koch, Phys. Rev. B **53**, 2026 (1996).

[6] B. Deveaud et al., Phys. Rev. Lett. **67**, 2355 (1991).

[7] A. Vinattieri et al., Phys. Rev. B **50**, 10868 (1994).

[8] A. Vinattieri et al., Sol. State Commun. **88**, 189 (1993).

[9] Y. Yamamoto et al., Opt. Commun. **80**, 337 (1991).

[10] N. Ochi et al., Appl. Phys. Lett. **58**, 2735 (1991).

[11] M. Hübner et al.,Phys. Rev. Lett. **76**, 4199 (1996).

[12] M. Opher-Lipson, E. Cohen, and L. N. Pfeiffer, Phys. Rev. B **55**, 13778 (1997).

[13] A. P. Heberle, J. J. Baumberg, and K. Köhler, Phys. Rev. Lett. **75**, 2598 (1995).

[14] T. Stroucken et al. in *Festkörperprobleme/Advances in Solid State Physics* **38**, Ed. B. Kramer, Vieweg, Braunschweig/Wiesbaden (1998).

[15] B. Grote et al., phys. stat. sol. **164** a, 467 (1997).

[16] T. Yajima and Y. Taira, J. Phys. Soc. Jpn. **47**, 1620 (1979).

[17] M. Hübner et al., accepted for publication in Sol. State Commun.

[18] J. Kuhl et al. in *The Femtosecond Technology*, Eds. T. Kamiya. F. Saito, H. Yajima, and O. Wada, Springer Heidelberg, to be published 1998

[19] M. Hübner et al., Sol. State. Commun. **105**, 105 (1998).

[20] S. Haas et al., Phys. Rev. B. **57**, 14860 (1998).

[21] see the article by J. Kuhl et al. in ref. [1].

[22] M. D. Webb, S. T. Cundiff, and D. G. Steel, Phys. Rev. Lett. **66**, 934 (1991).

[23] H. P. Wagner et al., Phys. Rev. B **56**, 12581 (1997).

[24] M. Lindberg, R. Binder, and S. W. Koch, Phys. Rev. A **45**, 1865 (1992).

[25] H. Schneider, and K. Ploog, Phys. Rev. A **49**, 17050 (1994).

[26] M. Hübner et al., phys. stat. sol. **206 b**, 333 (1998).

[27] Y. Merle d'Aubigne et al., Phys. Rev. B **54**, 14003 (1996).

[28] C. Ell et al., Phys. Rev. Lett. **80**, 4795 (1998).

[29] A. Honold et al., Appl. Phys. Lett. **52**, 2105 (1988).

[30] S. Haas et al., phys. stat. sol. **164** a, 19 (1997).

Quantum Kinetics in Semiconductors: Experiments on a 10-Femtosecond Timescale

M. Wegener[1], M.U. Wehner[1], D. Steinbach[1], M.H. Ulm[1], G. Kocherscheidt[1], and D.S. Chemla[2]

[1] Institut für Angewandte Physik, Universität Karlsruhe (TH), Kaiserstr. 12, D-76128 Karlsruhe, Germany
[2] Permanent Address: Material Sciences Division of Lawrence Berkeley Laboratory, and Physics Department of the University of California, 1 Cyclotron Road MS 66, Berkeley, CA 94720, USA

Abstract: In Quantum Kinetics, a few tens of femtoseconds after optical excitation, scattering processes in semiconductors are not completed, thus relaxation has not yet become irreversible. Here we review recent experiments on bulk GaAs, GaAs quantum wells and bulk ZnSe under these conditions. We demonstrate the fascinating possibility of actually reversing or enhancing incompleted scattering events by means of coherent control.

1 Introduction

In many situations, we are using the concept of decay rates, scattering rates or relaxation times to describe the temporal behavior of physical systems which are subject to some sort of damping. Consider a simple example: an optically excited ion in a solid coupled to a lattice vibration. It is often argued as follows: the number of scattering events at some time t is proportional to the number of excited ions at the same time t – the dynamics is local in time. As a result the decay becomes exponential, in frequency space the line is Lorentzian. Why is the number of scattering events not also proportional to the number of excited ions at earlier times? In fact, in general it is! Only in the limit of vanishing correlations among scattering events the simple picture of a dynamics local in time becomes valid. How does scattering come about in quantum mechanics? Consider that we prepare a quantum mechanical system, e.g. a semiconductor, in a known state at time $t = 0$. It is clear that from thereon its wavefunction obeys the time-dependent Schrödinger equation. Thus, the evolution of its wavefunction at time t is completely determined by the state of the crystal at the same time t – the dynamics is local in time. Two aspects are worth mentioning at this point. (i) The semiconductor is such a complicated many-body system that its Schrödinger equation is not solvable. Thus, it is necessary to restrict ourselves to treat only certain degrees of freedom of the problem, for example the electrons in a conduction and a valence band. In this approach, we have to consider that this subsystem interacts with other degrees of freedom, as the vibrations of the nuclei or other electrons, which are modelled as a thermal bath. Then, the equation of motion of the subsystem becomes nonlocal in time. The evolution at time t not only depends on the state of the system at time t but also on its history. Such non-Markovian dynamics has been discussed for systems which are well described by an ensemble of uncoupled two-level systems [1–5]. In semiconductors the physics is more involved because we have bands of interacting states. The goal of quantum kinetic theory is to treat such problems [6]. (ii) Let us assume we were able to prepare the system in any arbitrary initial state. Furthermore we assume that the energy level spectrum is sufficiently dense, an assumption which is well justified for a macroscopic semiconductor. We represent the initial state wavefunction Φ in terms of the complete set of orthonormal eigenfunctions φ_n of the semiconductor

$$\Phi = \sum_n a_n \varphi_n \tag{1.1}$$

with in general complex coefficients a_n. With these assumptions we can adjust the amplitudes and phases of the a_n such that (almost) any temporal behavior for any observable results. This is closely analogous to a Fourier-transformation. In this way of thinking, any relaxation process is nothing but an interference phenomenon of wavefunctions. It is somewhat counter-intuitive that relaxation or scattering – often thought of as irreversible processes – are not intrinsic properties of the solid but also a result of the specific way we perform the experiment. It is this unusual option which we want to call coherent control of the system. We will describe corresponding experiments in Section 4.

2 Theory

The Hamiltonian H of a semiconductor excited by a laser field which is roughly resonant with its band edge is well established [6,7]. In particular it contains the Coulomb electron-electron interaction $H_{\mathrm{e-e}}$ and the electron-phonon $H_{\mathrm{e-LO}}$ interaction. So far it has not been possible to solve the complete Hamiltonian for conditions corresponding to the experiments described below. In fact, not even the special cases $H_{\mathrm{e-e}} = 0$ or $H_{\mathrm{e-LO}} = 0$ are solvable without further approximations (except for the case of linear energy dispersion for the electrons [8] or the so-called phonon staircase [9]). For $H_{\mathrm{e-e}} = H_{\mathrm{e-LO}} = 0$ the optical Bloch-equations are obtained, for $H_{\mathrm{e-LO}} = 0$ and $H_{\mathrm{e-e}}$ treated in Hartree-Fock (mean field) approximation the well-known semiconductor Bloch-equations [7,10,11] result. Let us focus on two special cases. (A) $H_{\mathrm{e-LO}} \neq 0$ and $H_{\mathrm{e-e}}$ treated in Hartree-Fock approximation [12–16]. If we neglect the small dispersion of the LO-phonons in the center of the Brillouin-zone, we have a model system in which the electrons couple to a single characteristic frequency ω_{LO} of the bath of LO-phonons. The LO-phonons in GaAs have an oscillation period of 115 fs, a value which roughly determines the time scale for memory effects in the electron-LO-phonon scattering. The effective strength of the electron-phonon interaction is usually described by the dimensionless Fröhlich-constant [6] α which can be expressed by quantities which are altogether accessible experimentally.

$$\alpha = \sqrt{\frac{E_{\mathrm{R}}}{\hbar\omega_{\mathrm{LO}}}} \left(\frac{\varepsilon(0)}{\varepsilon(\infty)} - 1 \right) \tag{2.2}$$

E_R is the exciton Rydberg energy, $\varepsilon(0)$ and $\varepsilon(\infty)$ are the static and the high frequency limits of the dielectric function, respectively. For $\alpha \ll 1$ we are in the limit of weak phonon coupling, for $\alpha \approx 1$ we enter the regime of strong coupling. Examples are $\alpha_{\mathrm{GaAs}} \approx 0.06$ ($E_R = 4.2\,\mathrm{meV}$, $\varepsilon(0) = 12.9$, $\varepsilon(\infty) = 10.9$, $\hbar\omega_{\mathrm{LO}} = 36\,\mathrm{meV}$), or $\alpha_{\mathrm{ZnSe}} \approx 0.43$ ($E_R = 19\,\mathrm{meV}$, $\varepsilon(0) = 8.8$, $\varepsilon(\infty) = 5.7$, $\hbar\omega_{\mathrm{LO}} = 31\,\mathrm{meV}$). (B) The situation for electron-electron scattering, i.e. $H_{\mathrm{e-LO}} = 0$ and $H_{\mathrm{e-e}}$ beyond the Hartree-Fock approximation, is somewhat less obvious. In a classical picture, the electrons excited by a short pulse are initially randomly distributed and form a disorder potential landscape [17–21]. It is clear that they will move and eventually be able to screen a positive test charge for long times. How long does this process take? We currently do not know for sure. It can be argued, however, that – as a rule of thumb – the characteristic time is the plasma oscillation period. For a carrier density of $10^{18}\,\mathrm{cm}^{-3}$ in bulk GaAs the plasma period is close to 100 fs. For both cases, (A) and (B), we expect new physics on a timescale well below 100 fs. Some experiments will be discussed below.

3 A Simple Model for Electron-LO-Phonon Scattering

Let us consider the following model Hamiltonian which has the same general form as the complete Hamiltonian, however, it accounts for two electronic levels and a single phonon mode, only. This allows one to follow the mathematics easily.

$$H = \hbar\Omega c_1^\dagger c_1 + \hbar\omega_{\mathrm{LO}} a^\dagger a + \hbar g \left(a^\dagger c_1^\dagger c_1 + a c_1^\dagger c_1\right) - dE \left(c_1^\dagger c_0 + c_0^\dagger c_1\right) \tag{3.3}$$

The electronic ground state 0 of the fermionic two-level system is at zero energy, the excited state 1 at energy $\hbar\Omega$. The zero point energy of the phonon mode has been subtracted. Only scattering (with coupling g) from the upper electronic level back into the upper level via emission or absorption of phonons is included. These processes obviously violate energy-conservation for the individual scattering events. Thus they would not contribute in Fermi's golden rule or the Boltzmann-equation, however, they are especially important in quantum kinetics. The light field E couples via the dipole matrix element d. It is straightforward to compute the Heisenberg-equation of motion for the density matrix

$$\rho_{kn}^{ij} = \langle c_i^\dagger c_j (a^\dagger)^k a^n \rangle \, . \tag{3.4}$$

The usual transition amplitude (polarization) is obviously nothing but the element ρ_{00}^{01}, the occupation number of the upper electronic level is ρ_{00}^{11}. We obtain the coupled and open set of equations

$$\begin{aligned} &\frac{\partial}{\partial t}\rho_{kn}^{01} + i(\Omega + (n-k)\omega_{\mathrm{LO}})\,\rho_{kn}^{01} = \\ &\qquad + i\hbar^{-1} dE \left(\rho_{kn}^{00} - \rho_{kn}^{11}\right) + ig\left(-n\rho_{k,n-1}^{01} - \rho_{k,n+1}^{01} - \rho_{k+1,n}^{01}\right) \end{aligned} \tag{3.5}$$

and

$$\begin{aligned} &\frac{\partial}{\partial t}\rho_{kn}^{11} + i(n-k)\omega_{\mathrm{LO}}\rho_{kn}^{11} = \\ &\qquad + i\hbar^{-1} dE \left(\rho_{kn}^{10} - \rho_{kn}^{01}\right) + ig\left(k\rho_{k-1,n}^{11} - n\rho_{k,n-1}^{11}\right) \end{aligned} \tag{3.6}$$

and similar equations for ρ_{kn}^{00} and ρ_{kn}^{10}. In this simple model, relaxation is nothing but the physics of coupled oscillators. We excite one oscillator, say the electronic transition amplitude ρ_{00}^{01}, which couples to the one-phonon excitation ρ_{01}^{01}, which couples to the two-phonon excitation ρ_{02}^{01} and so forth. In the complete Hamiltonian, energy spreads into many degrees of freedom which eventually looks like an irreversible decay.

How can we solve this open set of equations? If we truncate at a certain sufficiently large number of phonons involved, i.e. $0 \leq n,k \leq n_{\max} = k_{\max}$, and assure that the solution does not depend on the particular choice of this cut-off, the dynamics can be computed numerically. Alternatively we could employ different levels of Born-approximations. A simple and intuitive limit is linear optics and weak phonon coupling at $T = 0\,\mathrm{K}$. Here we have $\rho_{00}^{00} = 1$ and $\rho_{00}^{11} = 0$. Furthermore, we factorize within the second Born-approximation (e.g. $\rho_{11}^{01} = \langle c_0^\dagger c_1 a^\dagger a \rangle \approx \langle c_0^\dagger c_1 \rangle \langle a^\dagger a \rangle$. If the reservoir remains in thermal equilibrium at $T = 0$, the Bose-factor is zero, i.e. $\langle a^\dagger a \rangle = 0$), which means that we account for one-phonon processes only. We obtain the following simple form for the transition amplitude $p(t) = \rho_{00}^{01}$ and the scattering amplitude $s(t) = i\rho_{01}^{01}$

$$\frac{\partial}{\partial t}\tilde{p} + g\tilde{s} = i\hbar^{-1} d\tilde{E} \tag{3.7}$$

$$\frac{\partial}{\partial t}\tilde{s} + i\omega_{\mathrm{LO}}\tilde{s} = g\tilde{p} \tag{3.8}$$

with $E = \tilde{E}\exp(-i\Omega t)$, $p = \tilde{p}\exp(-i\Omega t)$, and $s = \tilde{s}\exp(-i\Omega t)$. Formal integration of (3.8) introduced into (3.7) results in

$$\frac{\partial}{\partial t}\tilde{p}(t) + g^2 \int_{-\infty}^{t} \mathrm{e}^{-i\omega_{\mathrm{LO}}(t-t')}\,\tilde{p}(t')\,dt' = i\hbar^{-1} d\tilde{E}\,. \tag{3.9}$$

It is because we have written an equation of motion for the electrons only that (3.9) takes this non-Markovian form, whereas the combined equations for electrons and phonons, i.e. for $\tilde{p}$ and $\tilde{s}$, are local in time. The solution of (3.9) can easily be written as a main oscillation and a phonon sideband. We will show solutions of this simple model within the second Born approximation when discussing the coherent control experiments on bulk GaAs in Section 4.

4 Experiments

The first experiments on semiconductors using sub-10 fs optical pulses have already been published several years ago [22,23]. They were performed at room temperature and showed no indication of non-Markovian relaxation. While these experiments used low repetition rate ($\approx$ 10 kHz) laser sources, recent advances in generation of ultrashort pulses directly from laser oscillators provide us with high stability and high repetition rate ($\approx$ 100 MHz) sources [24] of very short, even of sub-10 fs, pulses.

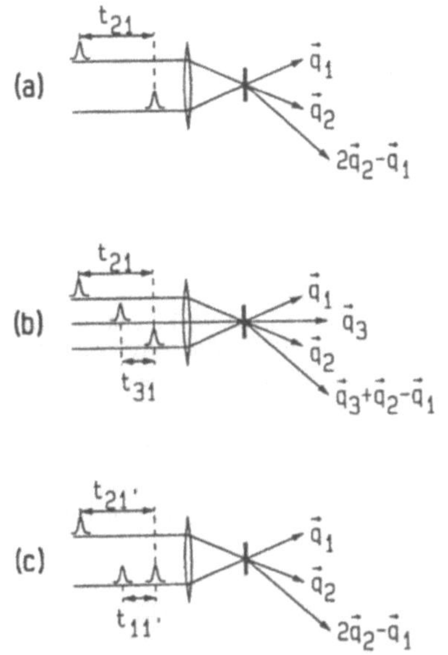

Figure 1
Schematic of the three different types of four-wave mixing (FWM) experiments discussed in this article. (a) Two-pulse experiment, (b) three-pulse experiment, and (c) coherent control configuration.

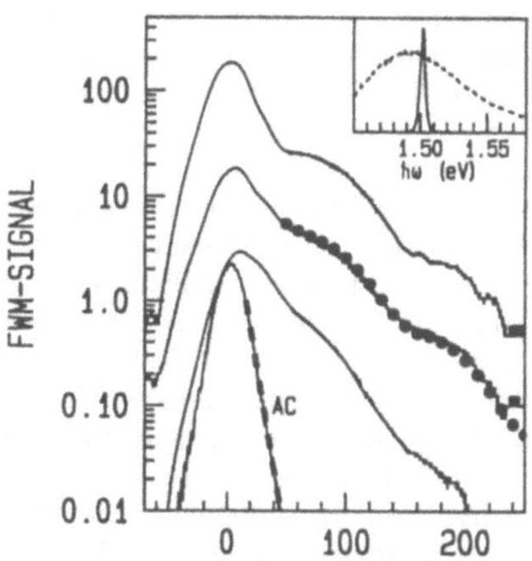

Figure 2
Spectrally integrated two-pulse FWM-signal on bulk GaAs at $T = 77\,\mathrm{K}$ for three different carrier densities (from top to the bottom: 1.2×10^{16}, 1.9×10^{16}, $6.3 \times 10^{16}\ \mathrm{cm}^{-3}$). The curves are vertically displaced for clarity. The dots are the result of the quantum kinetic theory. The curve labeled AC is the autocorrelation of the laser pulse showing 6.5 fs time constant exponential wings of the 14.2 fs pulses. The inset shows the FWM-spectrum for the lowest excitation density at zero time delay and the laser spectrum.

4.1 Two-Pulse Experiments on Bulk GaAs

First evidence for non-Markovian electron-LO-phonon scattering has been observed in two-pulse four-wave mixing (FWM) experiments (Fig. 1(a)) on bulk GaAs at $T = 77\,\mathrm{K}$ [25]. In these experiments the 14.2 fs sech^2-shaped laser pulses resonantly excite the material near its band gap (inset in Fig. 2). As shown in Fig. 2, the time integrated FWM-signal exhibits clear oscillations as a function of the time delay. From the mere fact that the oscillations occur, we can conclude that memory effects have been observed, since the decay of the signal would have been simply exponential in the Boltzman-limit [25]. The origin of the oscillations is that an electron at wavevector k in the conduction band

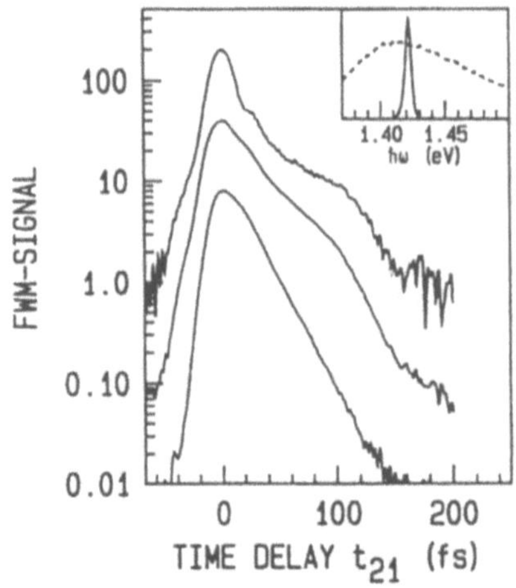

Figure 3
Two-pulse FWM-signal on bulk GaAs at $T = 300\,\mathrm{K}$. The carrier density is parameter (from top to the bottom: 1.0×10^{16}, 5.8×10^{16}, $3.5 \times 10^{17}\,\mathrm{cm}^{-3}$). The curves are vertically displaced for clarity. The inset shows the FWM-spectrum for the lowest excitation density at zero time delay and the laser spectrum.

scatters to a state k' separated by one LO-phonon energy. As long as the scattering process is not completed the electron oscillates back and forth between these two states resulting in an oscillation of the interband polarization, which then translates into an oscillation of the light emitted in the direction of diffraction $2\vec{q}_2 - \vec{q}_1$ as a function of the delay time. A simple calculation yields that the corresponding oscillation frequency is given by $\omega_{osc} = \omega_{LO}(1 + m_e/m_h)$, with the electron (heavy-hole) mass m_e (m_h). The FWM-spectra, which show no phonon satellite (inset in Fig. 2), as well as additional three-pulse experiments (Fig. 1(b)) are consistent with this picture [26] (not shown). At room temperature we have to employ a different mirror set in the laser oscillator in order to again resonantly excite the band edge. Here we obtain somewhat shorter, sech^2-shaped, transform-limited, optical pulses of $10 - 11\,\mathrm{fs}$ in duration. In Fig. 3 the phonon oscillations are somewhat less pronounced, in part due to a worse signal to background ratio. Again the oscillations disappear at elevated excitation densities, similar to $T = 77\,\mathrm{K}$. Notice, that the response at low excitation levels is purely excitonic even at room temperature (inset in Fig. 3). It is important to note that the phonon oscillations can neither be explained by so-called coherent phonons [27–30] nor by heavy-hole light-hole beating [31], a point which has been discussed in detail previously [32]. We will discuss coherent phonons later in the context of the coherent control experiments. Recent pump/probe experiments on bulk GaAs with 100 fs time resolution are consistent with our interpretation of non-Markovian electron-LO-phonon scattering [33]. They are described by the same theory [34]. Recently, also the Coulomb quantum kinetics has been discussed in pump/probe experimens on bulk GaAs [35,36] and FWM-experiments on GaAs quantum wells [37].

4.2 Two-Pulse Experiments on Bulk ZnSe

As pointed out above, the electron-LO-phonon coupling for GaAs is within the weak coupling limit ($\alpha_{GaAs} = 0.06$) which allows for the second Born approximation, that is discussion of one-phonon scattering processes only. ZnSe on the other hand has a Fröhlich-coupling constant that is almost an order of magnitude larger ($\alpha_{ZnSe} = 0.43$). Unfortunately, for large couplings, theory is yet to be developed. From Section 3, however, it is clear that the scattering amplitude which describes the simultaneous emission

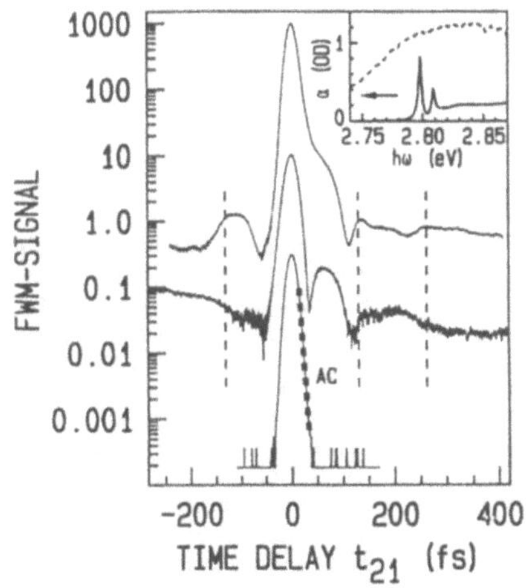

Figure 4
Spectrally integrated two-pulse FWM-signal on bulk ZnSe at $T = 77\,\mathrm{K}$ at a carrier density of $2.6 \times 10^{17}\,\mathrm{cm}^{-3}$. Collinear and perpendicular linear polarization of the incident beams are shown on the same scale. The curve labeled AC is the autocorrelation of the laser pulse showing 7.1 fs time constant exponential wings of the 13 fs pulses. The dashed lines at integer multiples of the phonon oscillation period are guides to the eye. The inset shows the linear absorption as well as the laser spectrum.

of n phonons, i.e. ρ^{01}_{0n}, gains relative weight proportional to $g^{2n} \sim \alpha^n$. Thus we expect an increased influence of multiple phonon scattering processes in ZnSe. For the experiments, we simply frequency-double the output of the laser oscillator. Group velocity dispersion is more pronounced in the blue, thus we have to compensate the optics of the experiment and the cryostat window by a prism sequence and a grating (or chirped mirror) sequence. In this fashion we can obtain blue 10.0 fs optical pulses [38] with low but sufficient average power. These are the shortest optical pulses generated in this spectral region by any means to date. The sample is a thin (100 nm) film of bulk ZnSe, epitaxially grown on a GaAs substrate. It is glued to a sapphire disk at room temperature, the GaAs substrate is etched away and the front side of the sample is antireflection coated. When cooled to low temperatures, the different thermal expansion coefficients of ZnSe and sapphire induce a strain in the ZnSe film which leads to a heavy-hole light-hole splitting of about 10.6 meV (corresponding to a beat period of 390 fs). The experiment is again closely similar to the one on bulk GaAs. Fig. 4 depicts the two-pulse FWM-signal of ZnSe for collinear and perpendicular linear polarization of the two incoming beams. The inset shows the linear absorption (optical density) as well as the laser spectrum. Here, the half width at half maximum of the heavy-hole exciton line is 1.5 meV. Eventhough the absolute carrier density seems quite large ($n_{\mathrm{eh}} = 2.6 \times 10^{17}\,\mathrm{cm}^{-3}$), the relevant quantity in the theory [7] is $n_{\mathrm{eh}}\, a_{\mathrm{B}}^3$ with the exciton Bohr radius a_{B}. Thus, a density of $n_{\mathrm{eh}} = 2.6 \times 10^{17}\,\mathrm{cm}^{-3}$ in ZnSe ($a_{\mathrm{B}} = 3.9\,\mathrm{nm}$) has to be compared with a density of $8 \times 10^{15}\,\mathrm{cm}^{-3}$ for GaAs ($a_{\mathrm{B}} = 12.5\,\mathrm{nm}$). We have verified experimentally by further attenuation that the low-density limit has indeed been reached. The observed decay is quite complicated and shows indications of a beating with approximately the phonon period (here 132 fs, see dashed lines) and also hints of twice this frequency. Notice, that either of these structures is clearly separated from the beating with the exciton Rydberg energy [39] ($E_{\mathrm{R}} = 19\,\mathrm{meV}$) which would correspond to an oscillation period of 218 fs. As for the GaAs case the exciton dominates the FWM-spectra of ZnSe and no phonon satellites are seen (not shown). No indications of propagation effects [40] are observed. We interpret the rapid structures as being due to multiple-phonon scattering processes. This shows that coherence is preserved for some time not only if one phonon is emitted (absorbed) but also if two or more phonons are emitted (absorbed).

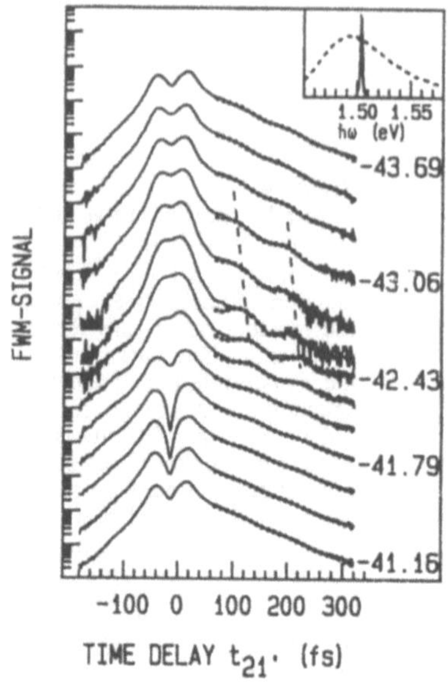

Figure 5
Coherent control of scattering processes: the shape of the decay of the FWM-signal changes as a function of the fixed time delay $t_{11'}$ between the two phase-locked pulses (from the top to the bottom from $t_{11'} = -43.69$ to -41.16 fs in steps of 0.21 fs). The curves are vertically displaced for clarity. Notice, that the phonon oscillations are almost absent for $t_{11'} = -43.69$ fs and pronounced for $t_{11'} = -42.43$ fs. Bulk GaAs, $T = 77$ K, the excited carrier density (incoherent sum) is $3.6 \times 10^{15}\,\mathrm{cm}^{-3}$. The inset shows the two-pulse FWM-spectrum at zero time delay and the laser spectrum.

4.3 Coherent Control on Bulk GaAs

Above we have argued that in quantum kinetics scattering processes are not yet completed. Thus a natural question is: can scattering actually be reversed at later times? Suppose a first pulse excites the system and scattering begins. As electrons, however, still behave as quantum mechanical waves, scattering is not yet irreversible. Thus, in principle, suitable excitation by a second pulse with a well-defined relative phase should allow to reverse or at least modify the scattering event. Obviously, such experiments demand coherent control of the semiconductor [41–47]. The required phase-locked pulses are derived from a novel scheme based on Pancharatnam's phase, the optical analogue to Berry's phase for electrons. This scheme has been discussed in detail elsewhere [48] and allows to keep the time delay of the phase-locked pulses constant with less than 100 attoseconds deviation. In addition, it makes scanning very easy and reliable. In the experiment (Fig. 1(c)) the two phase-locked pulses 1 and 1′ with time delay $t_{11'} = t_1 - t_{1'}$ collinearly propagating in direction $\vec{q}_1$ and a second pulse in direction $\vec{q}_2$ with time delay $t_{21'} = t_2 - t_{1'}$ lead to two nonequivalent diffracted beams with wavevectors $2\vec{q}_2 - \vec{q}_1$ and $2\vec{q}_1 - \vec{q}_2$, respectively. Here, we focus the discussion on the direction $2\vec{q}_2 - \vec{q}_1$ in which the phase-locked pulses enter linearly (in the $\chi^{(3)}$-limit). All beams have equal linear polarization (in the plane of incidence) and roughly equal intensity.

Results for bulk GaAs at $T = 77$ K excited by optical pulses with a duration of 15 – 16 fs and a center frequency tuned to the band gap are depicted in Fig. 5. The excitation densities are comparable to the two-pulse four-wave mixing data of Fig. 2. (Notice, that for constructive interference between pulses 1 and 1′ their incoherently summed density has to be multiplied by a factor of four.) Depending on the choice of the time delay $t_{11'}$, the oscillations in the time delay domain $t_{21'}$ can almost disappear completely (e.g. $t_{11'} = -43.69$ fs or $t_{11'} = -41.16$ fs). Here the decay time constant is especially large. For other delay times (e.g. $t_{11'} = -42.43$ fs) the initial rapid decay is more pronounced, however, pronounced oscillations occur at low absolute signal levels. Also (e.g. $t_{11'} =$

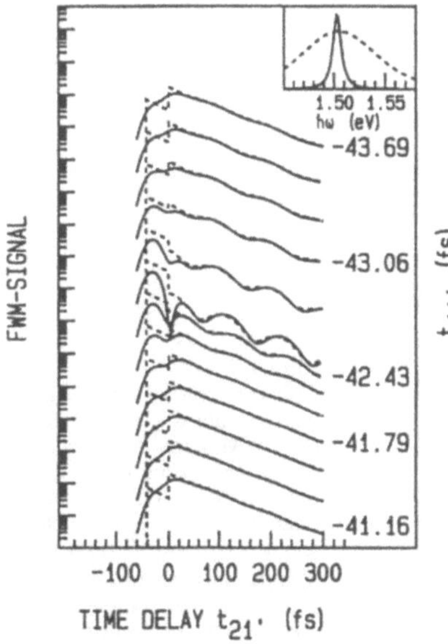

Figure 6
Coherent control of electron-LO-phonon scattering. The solutions of the simple model in the weak coupling limit ($g = 0.2\,\omega_{LO}$) and for parameters corresponding to the experiment can directly be compared with the experiment (Fig. 5). The inset shows the two-pulse FWM-spectrum at zero time delay and the laser spectrum.

−43.06 fs and −41.79 fs) a more usual decay with oscillations superimposed is observed. This scenario repeats itself until for about $t_{11'} = 80\,\text{fs}$ the disappearance of the oscillations is not observed any more. For overlapping laser fields, i.e. $|t_{11'}| < 10\,\text{fs}$, only an overall modulation of the signal amplitude is observed.

Let us focus our discussion on the non-Markovian phonon scattering contribution and for the moment ignore the initial rapid decay which has previously been attributed to exciton-electron scattering [49–52]. It is instructive to consider the analogy between the electron-photon and the electron-phonon scattering. In the Markov-limit, we can coherently control the transition amplitude $\tilde{p}$ (Section 3), that is the electron-photon scattering. An incident photon generates a coherent oscillation of the interband transition amplitude. It is only after the dephasing time that the photon is actually absorbed. In other words: at early times, the electron-photon interaction process is not yet completed and we can reverse it by coherent control [41]. The physics of the electron-LO-phonon scattering process is analogous. Here, we have a coherent oscillation of the electron-phonon scattering amplitude $\tilde{s}$ (see Section 3). As long as $\tilde{s}$ has not dephased, i.e. as long as the memory has not been lost, we can coherently control the scattering process. In the early stage we can even reverse it. For later times, the phonon is actually emitted or absorbed and scattering becomes irreversible. Corresponding solutions of the simple model introduced in Section 3 are depicted in Fig. 6 and are consistent with the experiment (Fig. 5). Even in a much more detailed comparison [53] this excellent agreement remains. It should, however, be noted that the same model essentially fails for the description of the three-pulse FWM-experiments [26], which have been performed under identical conditions, while the complete quantum kinetic theory describes these experiments well [26].

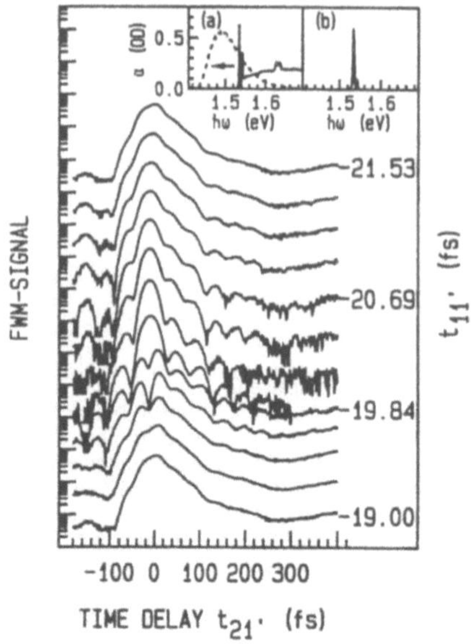

Figure 7
Coherent control on a GaAs quantum well sample at $T = 77\,\mathrm{K}$. The decay of the FWM-signal changes as a function of the fixed time delay $t_{11'}$ between the two phase-locked pulses. The excited carrier density (incoherent sum) is $2.5 \times 10^{10}\,\mathrm{cm}^{-2}$. Inset (a) shows the absorption spectrum and the laser spectrum, inset (b) the two-pulse FWM-spectrum at zero time delay.

4.4 Coherent Control on GaAs Quantum Wells

What happens to the electron-LO-phonon quantum kinetics in a (quasi-) two-dimensional system? Recently, theory predicted a smaller amplitude and a higher frequency of the resulting oscillations [54]. In the experiments we use a GaAs / $Al_xGa_{1-x}As$ ($x = 0.3$) multiple quantum well sample with 20 periods of 11.6 nm wide wells and 15.0 nm barriers, grown by molecular beam epitaxy. It is glued to a sapphire disk and its front-side is antireflection coated. The half width at half maximum (HWHM) of the heavy-hole exciton line at $T = 77\,\mathrm{K}$ is 1.5 meV (see inset in Fig. 7). The dynamics of the two-pulse FWM-signal (not shown) is rather complex and also shows up in the coherent control experiment (Fig. 7). The heavy-hole light-hole beating with about 500 fs period is well understood and coincides with the splitting in the linear absorption spectrum and with the FWM-spectrum. The rapid oscillation with a period around 45–50 fs is compatible with the energetic difference of the first and second subband (92 meV $\longleftrightarrow$ 45 fs). In a second sample (not shown) with a smaller well-width of 7.9 nm and a corresponding larger subband separation of 130 meV we indeed find a faster oscillation. The rapid oscillation vanishes for larger excitation densities without noticable change in the decay time constant (Fig. 8). Notice, however, that no satellite occurs (to better than 1/1000 of the first subband peak) in the FWM-spectrum corresponding to an oscillation with about 45–50 fs period. Hence, this oscillation is neither a simple quantum beating nor a polarization interference in the nomenclature introduced in [39]. In addition to these two oscillations, one can also identify a third oscillation around the phonon oscillation period. It is observed in the low density data (Fig. 7) and, unlike the other two oscillations, remains to be seen even at elevated densities (Fig. 8).

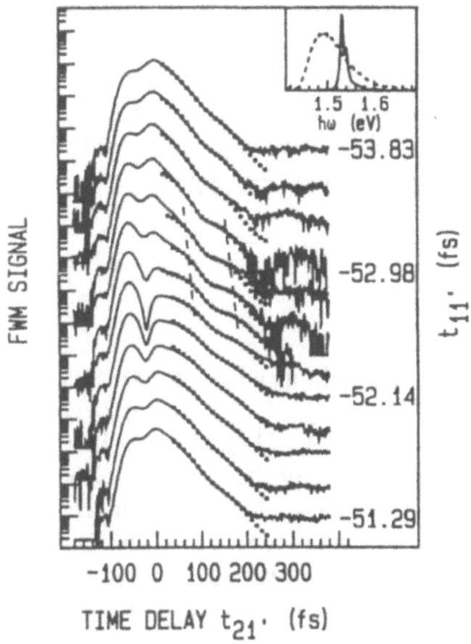

Figure 8
Coherent control experiment as Fig. 7, however, for a larger excitation density $I/I_0 = 37$. I_0 corresponds to the conditions of Fig. 7.

As in the bulk-GaAs case, the visibility of the phonon oscillation can be modulated by coherent control. This allows to distinguish experimentally between coherent phonon effects [27–30] and coherence in the phonon scattering amplitude (quantum kinetics). In a medium with inversion symmetry coherent phonons can not be generated. More simply: the phonon has no reason to start oscillating in one direction rather than in the opposite one. In a properly designed bulk-GaAs sample this practically rules out coherent phonons right away. The inversion symmetry in quantum wells, however, can be broken by asymmetric heterointerfaces or other effects which, in principle, allows for generation of coherent phonons. How would coherent phonons show up in the experiment? After excitation with a short pulse at $t = 0$ coherent phonons would lead to a time-dependent eigenfrequency $\Omega(t)$ of the electronic excitation of the form $\Omega(t) = \Omega_0 + a\Theta(t)\sin(\omega_{\text{LO}}t)$ with some amplitude a. As this amplitude is small in the low density regime, coherent phonons do not lead to an oscillation in a two-pulse FWM-experiment. At elevated densities, oscillations also appear in the two-pulse experiment. Under these conditions, however, in a coherent control experiment the signal strength A_{exp} and the phonon visibility A_{phon} oscillate *in phase*, while A_{exp} and A_{phon} oscillate *out of phase* for the quantum kinetics scenario [55]. From this comparison we can conclude that the phonon oscillations seen on quantum wells (Fig. 9) are primarily due to coherence in the electron-LO-phonon scattering amplitude (quantum kinetics) and not primarily due to coherent phonons.

5 Conclusions

We have presented evidence for non-Markovian relaxation due to electron-LO-phonon scattering. This shows up as phonon oscillations which are superimposed on the exponential decay of the coherent four-wave mixing signal. They are observed in bulk GaAs, GaAs quantum wells and bulk ZnSe. These oscillations can be magnified or suppressed using coherent control over the system on an attosecond timescale. Such coherent control experiments also demonstrate that the electron-phonon interaction shares many common features with the electron-photon interaction. The challenge for both, experiment and theory, remains to observe the corresponding effects of quantum kinetics for the case of electron-electron interaction. This question is in fact intimately related to the fundamental problem of the temporal build-up of screening in an interacting gas of charged particles.

6 Acknowledgements

We gratefully acknowledge support by the Krupp-Stiftung and by the DFG-Schwerpunktprogramm *Quantenkohärenz in Halbleitern*. The stay of D. S. C. has become possible due to funding by the Humboldt-foundation and the DFG-Graduiertenkolleg *Kollektive Phänomene im Festkörper*. His work was supported by the Director, Office of Energy Research, Office of Basic Energy Sciences, Division of Material Sciences of the U.S. Department of Energy. We thank D. Hommel, K. Ohkawa, W. Stolz, and G. Weimann for the high quality samples and the group of H. Kalt for help with the preparation of samples. Stimulating discussions with W. Schäfer, H. Haug, L. Bányai, and T. Kuhn are gratefully acknowledged.

Bibliography

[1] P. Meystre, and M. Sargeant III, *Elements of Quantum Optics*, second edition, Springer Verlag (1991)

[2] J.-Y. Bigot, M. T. Portella, R. W. Schoenlein, C. J. Bardeen, A. Migus, and C. .V. Shank, Phys. Rev. Lett. **66**, 1138 (1991)

[3] E. T. J. Nibbering, D. A. Wiersma, and K. Duppen, Phys. Rev. Lett. **66**, 2464 (1991)

[4] J. P. Lavoine, and A. A. Villaeys, Phys. Rev. Lett. **67**, 2780 (1991)

[5] B. Fainberg, Phys. Rev. A **48**, 849 (1993)

[6] H. Haug, and J.-P. Jauho, *Quantum Kinetics in Transport and Optics of Semiconductors*, Springer series in Solid-State Sciences **123**, Springer Verlag (1996)

[7] H. Haug, and S. W. Koch, *Quantum Theory of the Optical and Electronic properties of Semiconductors*, World Scientific, second revised edition (1993)

[8] K. Schönhammer, and C. Wöhler, Phys. Rev. B **55**, 13564 (1997)

[9] J. A. Kenrow, K. El Sayed, and C. J. Stanton, Phys. Rev. Lett. **78**, 4873 (1997)

[10] K. Leo, M. Wegener, J. Shah, D. S. Chemla, E. O. Göbel, T. C. Damen, S. Schmitt-Rink, and W. Schäfer, Phys. Rev. Lett. **65**, 1340 (1990)

[11] M. Wegener, D. S. Chemla, S. Schmitt-Rink, and W. Schäfer, Phys. Rev. A **42**, 5675 (1990); T. Rappen, U. Peter, M. Wegener, and W. Schäfer, Phys. Rev. B **48**, 4879 (1993)

[12] A. V. Kuznetsov, Phys. Rev. B **44**, 8721 (1991)

[13] H. Haug, Phys. Stat. Sol. B **173**, 139 (1992); L. Bányai, D. B. Tran Thoai, C. Remling, and H. Haug, Phys. Stat. Sol. B **173**, 149 (1992); D. B. Tran Thoai and H. Haug, Phys. Rev. B **47**, 3574 (1993); D. B. Tran Thoai, L. Bányai, E. Reitsamer, and H. Haug, Phys. Stat. Sol. B **188**, 387 (1995)

[14] M. Hartmann and W. Schäfer, Phys. Stat. Sol. B **173**, 165 (1992)

[15] R. Zimmermann, J. Luminescence 53, 187 (1992)

[16] J. Schilp, T. Kuhn, and G. Mahler, Phys. Rev. B **50**, 5435 (1994); J. Schilp, T. Kuhn, and G. Mahler, Phys. Stat. Sol. B **188**, 417 (1995)

[17] V. L. Gurevich, M. I. Muradov, and D. A. Parshin, Europhys. Lett. **12**,4 (1990)

[18] K. El Sayed, S. Schuster, H. Haug, F. Herzel, and K. Henneberger, Phys. Rev. B **49**, 7337 (1994); K. El Sayed, L. Banyai, and H. Haug, Phys. Rev. B **50**, 1541 (1994)

[19] M. Bonitz, D. Kremp, D. S. Scott, R. Binder, W. D. Kraeft, and H. S. Köhler, J. Phys. Condens. Matter **8**, 6057 (1996)

[20] G. S. Setlur, and Y.-C. Chang, Phys. Rev. B **55**, 1517 (1997)

[21] Q. T. Vu, L. Banyai, P. I. Tamborenea, and H. Haug, Europhys. Lett. **40**, 323 (1997)

[22] P. C. Becker, H. L. Fragnito, C. H. Brito Cruz, J. Shah, R. L. Fork, J. E. Cunningham, J. E. Henry, and C. V. Shank, Appl. Phys. Lett. **53**, 2089 (1988)

[23] J. Y. Bigot, M. T. Portella, R. W. Schoenlein, J. E. Cunningham, and C. V. Shank, Phys. Rev. Lett. **67**, 636 (1991)

[24] M. T. Asaki. C. Huang, D. Garvey, J. Zhou, H. C. Kapteyn, and M. Murnane, Opt. Lett. **18**, 977 (1993)

[25] L. Bányai, D. B. Tran Thoai, E. Reitsamer, H. Haug, D. Steinbach, M. U. Wehner, M. Wegener, T. Marschner, and W. Stolz, Phys. Rev. Lett. **75**, 2188 (1995)

[26] D. Steinbach, M. U. Wehner, M. Wegener, L. Bányai, D. B. Tran Thoai, E. Reitsamer, and H. Haug, in: *Confined Excitations in Molecular and Semiconductor Nanostructures*, eds. S. Mukamel, D. S. Chemla, Chem. Phys. **210**, 49 (1996)

[27] G. C. Cho, W. Kütt, and H. Kurz, Phys. Rev. Lett. **65**, 764 (1990)

[28] R. Scholz, and A. Stahl, Phys. Stat. Sol. B **168**, 123 (1991)

[29] R.W. Schoenlein, D.M. Mittleman, J.J. Shiang, A.P. Alivisatos, and C.V. Shank, Phys. Rev. Lett. **70**, 1014 (1993)

[30] A. V. Kuznetsov, and C. J. Stanton, Phys. Rev. B **51**, 7555 (1995)

[31] M. Joschko, M. Woerner, T. Elsässer, E. Binder, T. Kuhn, R. Hey, H. Kostial, and K. Ploog, Phys. Rev. Lett. **75**, 2188 (1997)

[32] M. U. Wehner, M. H. Ulm, D. S. Chemla and M. Wegener, Phys. Stat. Sol. B **206**, in press (1998)

[33] C. Fürst, A. Leitenstorfer, A. Laubereau, and R. Zimmermann, Phys. Rev. Lett. **78**, 3733 (1997)

[34] A. Schmenkel, L. Banyai, and H. Haug, Journal of Lumin., in press (1998)

[35] S. Bar-Ad, P. Kner, M. V. Marquezini, D. S. Chemla, and K. El Sayed, Phys. Rev. Lett. **77**, 3177 (1996)

[36] F. X. Camescasse, A. Alexandrou, D. Hulin, L. Bányai, D. B. Tran Thoai, and H. Haug. Phys. Rev. Lett. **77**, 5429 (1996)

[37] D. S. Chemla, J.-Y. Bigot, M.-A- Mycek, and S. Weiss, Phys. Rev. B **50**, 8439 (1994)

[38] D. Steinbach, W. Hügel, and M. Wegener, J. Optical Soc. Am. B **15**, 1231 (1998)

[39] M. Koch, J. Feldmann, G. von Plessen, E. O. Göbel, P. Thomas, and K. Köhler, Phys. Rev. Lett. **69**, 3631 (1992)

[40] T. Rappen, G. Mohs, and M. Wegener, Phys. Rev. B. **47**, 9658 (1993)

[41] A. P. Heberle, J. J. Baumberg, and K. Köhler, Phys. Rev. Lett. **75**, 2598 (1996)

[42] W. S. Warren, H. Rabitz, and M. Dahleh, Science **259**, 1581 (1993)

[43] M. S. Pshenichnikov, W. P. de Boeij, and D. A. Wiersma, Phys. Rev. Lett. **76**, 4701 (1996)

[44] A. Haché, Y. Kostoulas, R. Atanasov, J. L. P. Hughes, J. E. Sipe, and H. M. van Driel, Phys. Rev. Lett. **78**, 306 (1997)

[45] M. Gurioli, F. Bogani, S. Ceccherini, and M. Colocci, Phys. Rev. Lett. **78**, 3205 (1997)

[46] S. Ogawa, H. Nagano, H. Petek, and A. P. Heberle, Phys. Rev. Lett. **78**, 1339 (1997)

[47] V. Blanchet, C. Nicole, M.-A. Bouchene, and B. Girard, Phys. Rev. Lett. **78**, 2716 (1997)

[48] M. U. Wehner, M. H. Ulm, and M. Wegener, Opt. Lett. **22**, 1455 (1997)
for other approaches see the references cited therein

[49] M. U. Wehner, D. Steinbach, and M. Wegener, Phys. Rev. B **54**, R 5211 (1996)

[50] S. T. Cundiff, M. Koch, W. H. Knox, and J. Shah, Phys. Rev. Lett. **77**, 1107 (1996)

[51] M. U. Wehner, J. Hetzler, and M. Wegener, Phys. Rev. B **55**, 4031 (1997)

[52] D. Birkedal, V. G. Lyssenko, J. M. Hvam, and K. El Sayed, Phys. Rev. B **54**, 14250 (1996); K. El Sayed, D. Birkedal, V. G. Lyssenko, and J. M. Hvam, Phys. Rev. B **55**, 2456 (1997)

[53] M. U. Wehner, M. Ulm, D. S. Chemla, and M. Wegener, Phys. Rev. Lett. **80**, 1992 (1998)

[54] E. Reitsamer, L. Banyai, D. B. Tran-Thoai, P. I. Tamborenea, and H. Haug, Proceedings of the 23d International Conference on the Physics of Semiconductors, M. Scheffler and R. Zimmermann eds., World Scientific, 685 (1997)

[55] M. U. Wehner, D. S. Chemla, and M. Wegener, Phys. Rev. B **58**,3590 (1998).

Theory for Semiconductor Near-Field Optics: Selection Rules and Coherent Spatio-Temporal Dynamics

A. Knorr, B. Hanewinkel, H. Giessen, and S.W. Koch

Fachbereich Physik und Zentrum für Materialwissenschaften,
Philipps-Universität, Renthof 5, D-35032 Marburg, Germany

Abstract: A theoretical approach to the description of semiconductor near-field optics is presented. Starting from the microscopic light-matter interaction, the equations of motion for the relevant observables are derived and analyzed. Selected applications include optical near-field distributions in mesoscopic semiconductor structures, near-field selection rules and spatio-temporal wavepacket dynamics.

1 Introduction

Micro- and nanofabrication can be used to structure semiconductors on lengthscales from below to above the wavelength of visible light [1]. The idea to combine high spatial with temporal resolution for the investigation of quantum structures makes it possible not only to obtain the optical response of single quantum dots [2], wires [3] and wells [4] but also to study transport dynamics [5] as well as spatio-temporal coherence and its subsequent dephasing [6]. Experimentally, spatial resolution up to several nanometers is reached by scanning near-field optical microscopy (SNOM) [7]. SNOM increases the spatial resolution of conventional microscopy by using a probe object smaller than the wavelength of light as source or detector in the sub-wavelength vicinity (near-field region) of the sample [8]. Near-field interaction cannot be described assuming an independent external field because the radiation source and the sample influence each other. Especially important is the dielectric environment of the sample investigated [9]. Most near-field probes currently used experimentally consist of a metallized fiber tip which confines the light intensity to a fraction of the wavelength [10]. SNOM and short pulse techniques combined can be applied to simultaneously obtain spatial and temporal resolution. An example for experimental investigations of semiconductors which connect the temporal and spatial resolution in SNOM is the study of electron-hole carrier diffusion [5].

The theoretical description of spatial phenomena in the interaction of light and structured solids can be classified by the comparison of experimentally relevant length scales l with the wavelength λ of the incident radiation. In a typical experiment, l is determined by the sample shape (e.g. the linear extension l_s) as well as the observation distance (l_o). To illustrate this, two examples of scattering objects in the x-y-plane are considered:

(i) The object with a geometrical extension l_s in the x-y-plane is excited by an wave propagating in z-direction (incident wavenumber $k_l = 2\pi/\lambda$). The dispersion relation for the scattered field in vacuum, including in-plane momenta Q, is given by $\omega^2 = c_0^2 k_l^2 = c_0^2(k_z^2 + Q^2)$. A rough estimate for the scattered in-plane momentum Q is given by $Q \in [-2\pi/l_s, +2\pi/l_s]$ [11]. Corresponding to $k_z = \pm\sqrt{k_l^2 - Q^2}$, the scattered electric field contains propagating components $\exp(\pm i|k_z|z)$ for $Q < k_l$ and non-propagating (evanescent) components $e^{-|k_z|z}$ for $Q > k_l$. Clearly, evanescent fields occur if the extension l_s of the scattering object is smaller than the wavelength ($l_s < \lambda$). Thus, evanescent modes contain the information relevant for sample variations smaller than the wavelength of light and can only be observed within a distance $l_o < l_s$.

(ii) An infinitely extended boundary in the x-y plane ($l_s \to \infty$) supports the propagation of surface waves which contain evanescent components in the z-direction. Again, these surface waves can only be measured when the observation distance l_o between surface and detector is within their decay range [12].

Following these arguments, one may distinguish between different theoretical methods to analyze the optical response in the context of evanescent waves [13]

- *Geometric or macroscopic* optics describes the interaction for $\lambda \ll l = l_s, l_o$ within scalar diffraction theory where evanescent fields are neglected [14].
- In the so called *mesoscopic* regime, $\lambda \geq l$, optics includes the action of evanescent waves. Here, spatial sample details l_s below the wavelength $\lambda \geq l_s$ can be resolved, however, additionally the observation distance l_o must be in the vicinity of the sample ($\lambda \geq l_o$, "near-field optics"). A recent review on near-field theory, which requires the solution of the full Maxwells equations, can be found in [13].
- If the structure investigated varies on a *microscopic* length scale (atomic range), $\lambda \gg l = l_s, l_o$, the spatio-temporal variation of the fields on the length scale of the scattering object is of minor importance. In this case, the light matter interaction can be described in the multipole approximation [15].

Within a quantum treatment of the semiconductor crystal, the optical response is determined by the electronic wave functions which consist of a lattice periodic part (varying on the scale of the atomic unit cell, i.e. several Å) and an envelope function determined by the quantum confinement potential (varying on scales from nanometers to microns) [16]. Therefore, the treatment of semiconductor structures includes the microscopic and the mesoscopic regimes.

The resonant optical response of semiconductors is determined by interacting electron-hole pairs [17]. Via the optical excitation process, electrons which absorb the energy of the incident light are transferred into a quantum state in the conduction band. Simultaneously, positively charged holes are left in the valence band. Linear optics in high quality samples is dominated by excitons which form the bound state of an electron-hole pair. They represent a source for the electromagnetic field which requires a self-consistent treatment. Of course, the general discussion of near-field optics given above, apply for the optical excitation of excitons as well. For example, not only energy but the momentum of the light is transferred to the excitons. For near-field excitation of quantum well excitons, a broad excitonic momentum distribution is generated by a subwavelength near-field probe (i, see above). The resulting spatial distribution of the excitonic polarization, Section 3.4, contains evanescent components in the growth direction of the well (ii, see above).

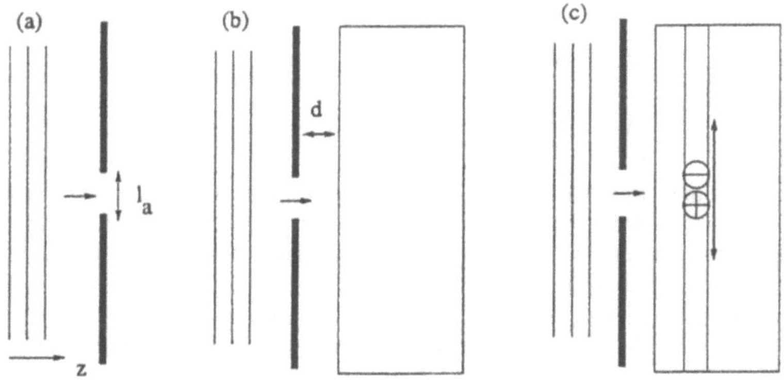

Figure 1 (a) transmission of a polarized plane wave field through a metallic circular aperture with $l_a = 60$ nm, used as a near-field probe; (b) non-resonant excitation of a semiconductor bulk material with a dielectric constant $\varepsilon = 12$, used as barrier material; (c) resonant excitation of quantum well / bulk excitons in the dielectric barrier material, where in-plane propagation of excitons is possible.

This article discusses the optical near-field excitation and response of semiconductor quantum well or bulk excitons embedded in non-resonant barrier material. The near-field is assumed to be generated by the transmission of a plane wave through a circular aperture. This aperture forms the near-field probe situated within sub-wavelength distance from the semiconductor sample. The total system, consisting of semiconductor structure and near-field probe, is described by Maxwells equations for the electromagnetic fields and the material equations for the semiconductor material.

The article is organized as follows: After this introductory section, the equations of motion are summarized in Section 2. Several applications of the theory are discussed in Section 3.

The reader who is more interested in applications may prefer to skip Section 2. The different applications/configurations are sketched in Fig. 1. To characterize the action of the near-field probe, Fig. 1a, the electric field distribution after the transmission of a plane wave through a circular subwavelength aperture ($l_a = 60$ nm) in an ideal metallic screen is calculated (Section 3.1). This computation yields information about the excitation conditions of the semiconductor sample spatially below the near-field probe. Next, the near-field response of the quantum well barrier material, Fig. 1b, is investigated (Section 3.2). This calculation shows how the field distribution is modified in the presence of an off-resonant dielectric semiconductor material. The resonant response of quantum well and bulk excitons, sandwiched in the dielectric material (Fig. 1c) is analyzed in Section 3.3. Here, we focus on the modifications of the near-field selection rules. In Section 3.4 we investigate pulsed near-field excitation. Here, the excitation of excitons in the near-field leads to the spatio-temporal propagation of wavepackets in the plane of the sample.

2 Field and Material Equations

The observables in an optical experiment are the electric field $\vec{E}$ and the magnetic field $\vec{B}$. The fields obey the macroscopic Maxwells equations [18]

$$\begin{aligned}\nabla\times\vec{E}(\vec{R},t) &= -\frac{1}{c_0}\partial_t\vec{B}(\vec{R},t)\\ \nabla\times\vec{B}(\vec{R},t) &= \frac{\varepsilon(\vec{R})}{c_0}\partial_t\vec{E}(\vec{R},t)+\frac{4\pi}{c_0}\vec{j}(\vec{R},t).\end{aligned} \tag{2.1}$$

Here, the material sources have been divided in non-resonant and resonant contributions. The non-resonant contributions, such as off-resonant band to band transitions as well as the near-field probe, are described by a dispersionless dielectric function $\varepsilon(\vec{R})$. The current $\vec{j}$ contains the electron-hole response of the resonantly excited electronic semiconductor transitions. Usually, the optical contributions to the current $\vec{j}$ are represented by the dipole density $\vec{P}$, a higher order electric tensor $\hat{Q}$ and the magnetization $\vec{M}$ of the medium [18]

$$\vec{j}(\vec{R},t)=\partial_t\left(\vec{P}(\vec{R},t)-\nabla\cdot\hat{Q}(\vec{R},t)\right)+\nabla\times\vec{M}(\vec{R},t). \tag{2.2}$$

To determine the different contributions for a multiband semiconductor microscopically, we start from the expression [19]

$$\vec{j}(\vec{R},t)=\frac{e}{m_0V}\sum_{k_1,k_2}\sum_{\lambda_1,\lambda_2}e^{-i(\vec{k}_1-\vec{k}_2)\cdot\vec{R}}\times \left\langle\lambda_1\vec{k}_1\left|\left(\vec{p}+\frac{\vec{k}_1+\vec{k}_2}{2}-\frac{e}{c_0}\vec{A}(\vec{R},t)\right)\right|\lambda_2\vec{k}_2\right\rangle\sigma^{\lambda_1\lambda_2}_{\vec{k}_1\vec{k}_2}(t). \tag{2.3}$$

Here, the current (bulk material) is represented by a sum over all density matrix elements $\sigma^{\lambda_1\lambda_2}_{\vec{k}_1\vec{k}_2}$ which account for the different possible electronic transitions between the bands λ and the wavenumbers $\vec{k}$ in a semiconductor. Furthermore, m_0 is the bare electron mass, $\vec{p}$ is the momentum and $\vec{A}$ is the vector potential. The occuring Bloch-integral

$$\langle\lambda_1\vec{k}_1|\,\vec{p}\,|\lambda_2\vec{k}_2\rangle=\int_{\Omega_0}d^3r\,u^*_{\lambda_1\vec{k}_1}(r)\,\vec{p}\,u_{\lambda_2\vec{k}_2}(r) \tag{2.4}$$

over the lattice periodic part of the Bloch-functions $u_{\lambda\vec{k}}$ results from the spatial average taken in the theoretical description of macroscopic experiments. To obtain an expression for the current similar to (2.2), the Bloch-integrals are expanded for $\vec{k}\ll a_0^{-1}$, where a_0 is the extension of an elementary cell. This expansion is carried out in a $\vec{k}\cdot\vec{p}$ perturbation theory [16]. This procedure yields electric and magnetic moments to first order and results in a simple expression for the current if a generalized polarization $\vec{\Pi}$ [20] is introduced

$$\vec{j}(\vec{R},t)=\partial_t\vec{\Pi}(\vec{R},t). \tag{2.5}$$

In contrast to the more conventional form (2.2), $\vec{\Pi}$ includes dipole, quadrupole and magnetic contributions in a single expression [20]. The systematic inclusion of contributions beyond the dipole approximation $\vec{j}=\dot{\vec{P}}$ is necessary, because for the stronger field gradients $\gg\lambda^{-1}$ (near-field optics) one can expect situations where such contributions become significantly more important than in far-field optics.

In linear optics, $\vec{\Pi}$ is conveniently expressed in an excitonic basis [16]

$$\vec{\Pi}(\vec{R},t) = \sum_{\lambda\gamma} \Big[\vec{d}_{\lambda_1\lambda_2} + \left(\vec{\nabla}_r + \xi_{\lambda_2}^{\lambda_1\lambda_2}\vec{\nabla}_R\right) \cdot \hat{B}^{\lambda_1}_{\lambda_1\lambda_2} + \hat{B}^{\lambda_2}_{\lambda_1\lambda_2} \cdot \left(\vec{\nabla}_r - \xi_{\lambda_1}^{\lambda_1\lambda_2}\vec{\nabla}_R\right) \Big] \; \varphi_\gamma^{\lambda_1\lambda_2}(\vec{r})|_{\vec{r}=0} \; P_\gamma^{\lambda_1\lambda_2}(\vec{R},t). \quad (2.6)$$

Here, $P_\gamma^{\lambda_1\lambda_2}(\vec{R},t)$ is the polarization amplitude for the excitonic state γ weighted by the corresponding wavefunction $\varphi_\gamma^{\lambda_1\lambda_2}(\vec{r})$ of the relative motion between electron and hole in real space $\vec{r}$. The relative and center of mass motion have been decoupled by introducing coordinates $\vec{r}$ and $\vec{R}$, weighted with the effective mass ratio for parabolic bands $\xi_{\lambda_1}^{\lambda_1\lambda_2} = m_{\lambda_1}/(m_{\lambda_1}+m_{\lambda_2})$ and $\xi_{\lambda_2}^{\lambda_1\lambda_2} = m_{\lambda_2}/(m_{\lambda_1}+m_{\lambda_2})$, respectively. The explicitly calculated microscopic expressions for the dipole components $d^\alpha_{\lambda_1\lambda_2}$ and quadrupole-like moments $B^{\alpha\beta\lambda}_{\lambda_1\lambda_2}$ in (2.6) read

$$d^\alpha_{\lambda_1\lambda_2} \equiv -ie\frac{\langle\lambda_1|p^\alpha|\lambda_2\rangle}{m_0(\varepsilon_{\lambda_1}-\varepsilon_{\lambda_2})} \quad (2.7)$$

$$B^{\alpha\beta\lambda}_{\lambda_1\lambda_2} \equiv e\sum_{\nu\neq\lambda}\frac{\langle\lambda_1|p^\alpha|\nu\rangle\langle\nu|p^\beta|\lambda_2\rangle}{m_0^2(\varepsilon_{\lambda_1}-\varepsilon_{\lambda_2})(\varepsilon_\lambda-\varepsilon_\nu)}. \quad (2.8)$$

Here, we restrict ourselves to non-degenerate bands. ε_λ denotes the transition energies at the bandedge and the integrals between the Bloch-functions ($\vec{k}=0$) have to be taken over one elementary cell of the solid. Note, that the coefficients $d^\alpha_{\lambda_1\lambda_2}$ and $B^{\alpha\beta\lambda}_{\lambda_1\lambda_2}$ are directly related to the expressions for one- and two-photon absorption coefficients, respectively [21]. From (2.6) it can be recognized that the leading term is the dipole current, proportional to $\vec{d}_{\lambda_1\lambda_2}$, which occurs for s-like excitons only ($\varphi_\gamma^{\lambda_1\lambda_2}(\vec{r})|_{\vec{r}=0}\neq 0$) [16]. The next term in (2.6) is proportional to the gradient of the excitonic wavefunction at $\vec{r}=0$, which is non-vanishing for p-like excitons ($\vec{\nabla}_r\varphi_\gamma^{\lambda_1\lambda_2}(\vec{r})|_{\vec{r}=0}\neq 0$) [22]. Furthermore, in a spatially inhomogeneous situation, like near-field optics, $\vec{\nabla}_R P_\gamma^{\lambda_1\lambda_2}(\vec{R},t)$ contributes. Because this contribution is connected with the spatial inhomogenity in the sample, it contains quadrupole and magnetic interaction effects.

In far-field optics, usually the quadrupole transition is important only if the dipole transition is not allowed. Because near-field excitation invokes spatial field gradients below the wavelength, the selection rules may be modified in comparison to the usual far-field situation.

To calculate the excitonic polarization, (2.6), the equations of motion for the excitonic polarization amplitudes $P_\gamma^{vc}(\vec{R},t)$ ($\lambda_1=v$ for valence band, $\lambda_2=c$ for conduction band) is determined from the Heisenberg equations of motion [16,19]

$$-i\partial_t P_\gamma^{vc}(\vec{R},t) = (-\varepsilon_\gamma - \frac{\Delta_R}{2M^{cv}})P_\gamma^{vc}(\vec{R},t) + \varphi_\gamma^*(\vec{r})|_{\vec{r}=0}\vec{d}_{vc}\cdot\vec{E}(\vec{R},t) + \Big[\left(\vec{\nabla}_r - \xi_v^{vc}\vec{\nabla}_R\right)\varphi_\gamma^*(\vec{r})\cdot\hat{B}^c_{cv} + \hat{B}^v_{cv}\cdot\left(\vec{\nabla}_r + \xi_c^{vc}\vec{\nabla}_R\right)\varphi_\gamma^*(\vec{r})\Big]_{\vec{r}=0}\cdot\vec{E}(\vec{R},t). \quad (2.9)$$

$P_\gamma^{vc}(\vec{R},t)$ is driven by the total electric field $\vec{E}(\vec{R},t)$ and its spatial derivative. This shows that not only dipole but quadrupol and magnetic effects are included in the equations of motion. Due to spatial dispersion ($\sim\Delta_R/M^{cv}$ where $M^{cv}=m_c+|m_v|$ is the exciton mass), ballistic propagation of excitons is included in our description. On the same level as (2.9), the electron density dynamics can be calculated by the conservation law [23] $N_c(\vec{R},t) = |P^{vc}_{\gamma}(\vec{R},t)|^2$. Note, that in general, the restriction to the linear dynamics is not necessary [23], however, nonlinear Bloch-equations for near-field optics are very demanding in their numerical evaluation and should even contain higher order correlations [24].

3 Applications

In this section, several applications of our theoretical approach are discussed. The analysis contains (1) the electric field distribution characterizing a near-field probe, (2) the field distribution in a dielectric semiconductor layer, (3) the resonant response of quantum well and bulk excitons and a study of near-field selection rules, (4) pulsed excitation of quantum well excitons as an example of wavepacket excitations.

To numerically investigate the simultaneous solutions of the material equation (2.9), and Maxwells equations (2.1), a finite difference time domain method (FDTD) is applied [25]. For this purpose, the equations of motion are discretized in time ($\Delta t \approx 0.01$ fs) and in space ($\Delta r \approx 5$ nm). Using an initially given field-distribution at time $t = 0$, the evolution of the field and the material polarization are calculated. This way, pulsed excitation can be investigated yielding temporally resolved information or excitation with a monochromatic wave can be studied resulting in a stationary field distribution. Since we want to focus on the excitonic response we choose the excitation energy at the exciton resonance for cw excitation. As an example of pulsed excitation we concentrate on 10 fs pulses with carrier frequency at the exciton resonance to calculate excitonic probe spectra over a broad frequency range. For all examples we choose material parameters corresponding to GaAs [26]. If not mentioned otherwise, the near-field probe is approximated by a circular hole (aperture diameter $l_a = 60$ nm) in a perfectly conducting screen which is excited by a circularly polarized plane wave, compare Fig. 1a. This assumption eliminates the dependence of the results on the parameters of the SNOM probe, allowing us to focus on the near-field response of the semiconductor.

3.1 Electric Field Distribution Below a Sub-Wavelength Aperture

To characterize the spatial excitation conditions we calculate the electric field distribution $|E^{\alpha}(\vec{R},z)|$ at a distance $z = 30$ nm below the aperture. We choose the independent vector components of the electric field as $\vec{\sigma}^z = (0,0,1)$ as well as in both circular in-plane components $\vec{\sigma}^{\pm} = \frac{1}{\sqrt{2}}(1, \pm i, 0)$. Here, z is the propagation direction of the field and the two circular polarization components are convenient since they are adopted to the selection rules for dipole transitions in semiconductors [1]. Fig. 2 shows the computed field distributions in the plane (a) and in the propagation direction (b), for σ^+ plane wave excitation of the aperture, compare Fig. 1a.

From Fig. 2a it can be recognized that all vector components of the optical near-field are generated by the subwavelength aperture. The σ^+ component of the electric field is peaked directly below the aperture. However, the peak ratio of incident to transmitted σ^+ field amplitude $|E^+(\vec{R} = 0)|/E_{\text{inc}}$ in the observation plane below the aperture is only 0.06 due to the rapid spatial decay of the evanescent field. In addition to σ^+ light, large contributions of σ^- and σ^z components occur below the metal edges. The σ^z contribution

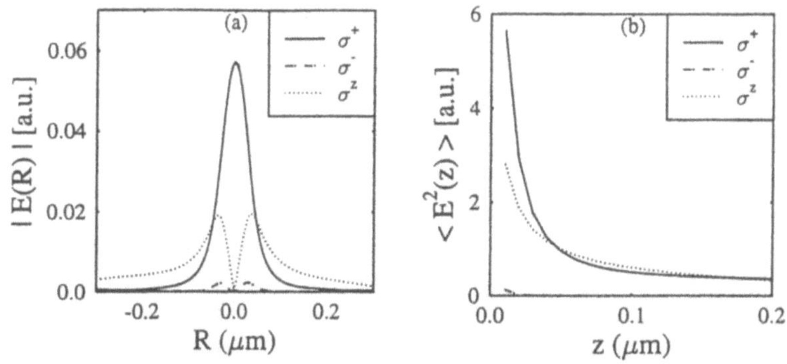

Figure 2 (a) Vector field distribution at $z = 30$ nm below a $l_a = 60$ nm aperture, where the incident wave is σ^+-polarized. (b) Transversally integrated field intensity as a function of distance below the near-field probe.

reaches the same order of magnitude as the σ^+ component which is the directly transmitted portion of the incident polarization σ^+. These results reproduce the main features already obtained in the first investigations of the field distributions below an aperture [27]. The polarization mixing results from the aperture edges which cross the incident field vector (σ^+) and do not allow the incident tangential components to be continuously transmitted. The existence of vector-field components in the transmitted field that are not present in the incident field have a strong impact on polarization sensitive spectroscopy, such as the excitation of heavy- and light holes, compare Section 3.3. Fig. 2b shows the decay of the transversally integrated field intensity, $\langle E^2(z)\rangle \equiv \int d^2\vec{R}E^2(\vec{R},z)$, over the radius of 1 μm for increasing distance z below the near-field probe. Here, the integration over the in-plane field intensity simulates the collection of light by an ideal, nondisturbing detector. After an initial fast decrease, which corresponds to the action of evanescent modes, we see in Fig. 2b that a small amount of far-field contributions propagates away from the near-field probe.

3.2 Non-Resonant Response in a Bulk Semiconductor

In this section, we investigate a non-resonantly excited semiconductor bulk sample (dielectric constant $\varepsilon = 12$), located at $z = d \equiv 10$ nm below the tip ($z = 0$), compare Fig. 1b. This setup is useful to study the role of the dielectric environment for resonantly excited quantum well excitons, because the quantum well is usually sandwiched between barrier material characterized as an off-resonant dielectric medium (Fig. 1c). Fig. 3 shows the transversal distributions of the σ^+ component of the electric field in air and in the barrier material at different positions ($z = 20$ nm, 80 nm).

The comparison reveals that the propagation in the dielectric medium leads to a broadening of the field distribution by a factor of about 1.2 in comparison to the reference distribution in vacuum. Therefore, we can conclude that the presence of the background material causes a slight decrease in resolution for embedded structures. For distances close to the aperture ($z = 20$ nm), the field in air is larger than the field in the medium. It is remarkable that this ratio changes for increasing distance ($z = 80$ nm). This reversed trend is best illustrated in the z-dependence of the σ^+-component of the field (Fig. 4a). Close to the aperture and in the surface region of the sample, a destructive interference of

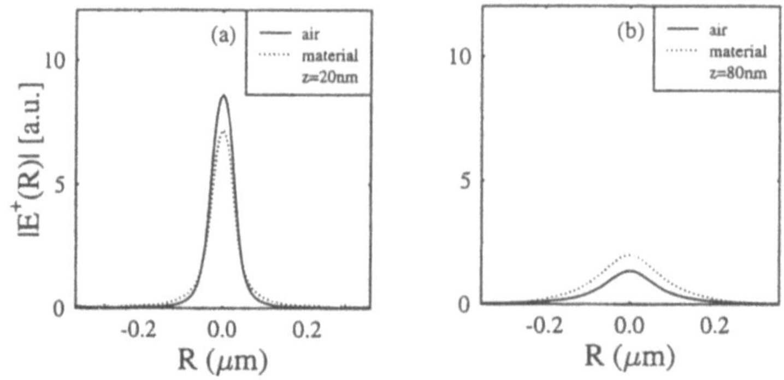

Figure 3 Transversal field distribution at different positions below the near-field probe in a semiconductor barrier material. Light confinement with and without the material are compared at different positions below the surface (a) $z = 20$ nm, (b) $z = 80$ nm.

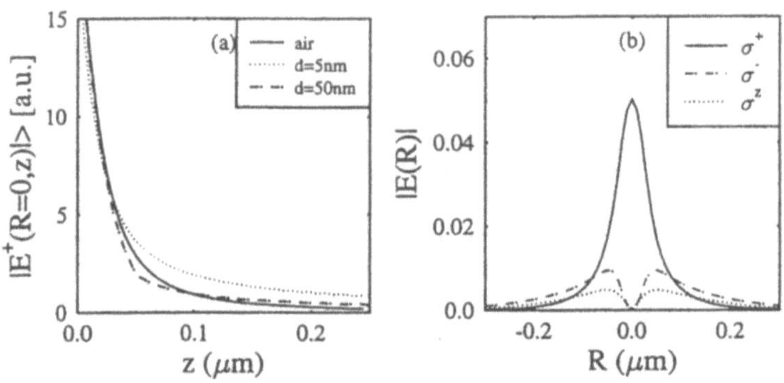

Figure 4 (a) Field distribution $E^+(R = 0,z)$ in z-direction for different distances d between near-field probe and barrier medium ($d = 5$, 50 nm) in comparison to the distribution in air. (b) Influence of the barrier medium on the polarization mixing observed in air, Fig. 2a.

the forward and backward propagated fields dominats, such that the field in air is larger than in the presence of the sample. On the other hand, the high refractive index of the sample leads to a partial transformation of evanescent into propagating modes which causes a larger far-field contribution in the material than for the case of air [12].

Next, we discuss the influence of the barrier material on the polarization mixing. Fig. 4b shows the computed field components at $z = 30$ nm for the aperture-sample distance of $d = 10$ nm. Even though significant polarization mixing is present, it can be recognized that below the surface of the barrier medium the σ^z-component of the field is strongly reduced in comparison to the corresponding distribution in air, Fig. 2a. This reduction is a consequence of two competing mechanisms. First, due to the field boundary conditions at the interface, the z-component of the electric field is reduced by the factor of the inverse dielectric constant, i.e. approximately one order of magnitude. Second, the strong localization of the σ^z-component results in an increase of the field in the medium (so called photon tunneling). The simultaneous action of both mechanisms results in an effective decrease of the σ^z contribution to about 20 percent of its value in air. The σ^- component is strongly enhanced in comparison to the distribution in air because is not screened at the boundary but enhanced due to the increased incoupling into the medium.

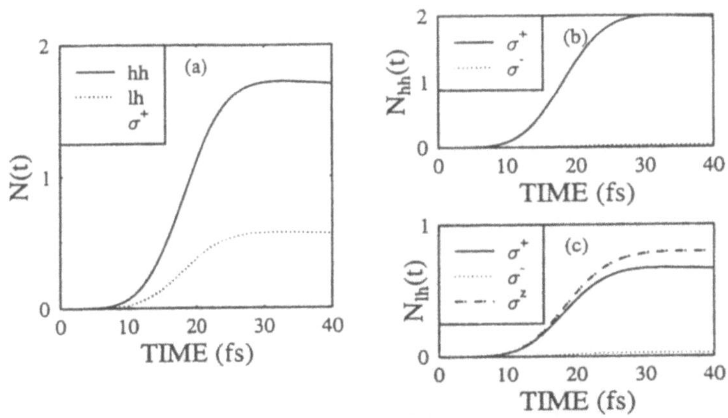

Figure 5 (a) Temporal development of the excitonic heavy and light hole density for a free-standing quantum well for far-field excitation with a σ^+ polarized 10 fs plane wave pulse propagating perpendicular to the quantum well surface at $z = 10$ nm. (b+c) Same, but excitation through the near-field probe for the *hh*-transition (b) and the *lh*-transition (c).

3.3 Resonant Excitation of Excitons

This section discusses quantum well and bulk excitons in the framework of linear near-field optics where we focus on the selection rules. A similar analysis of different subband transitions in quantum dot systems is reported in [28]. We study here the effects on a femtosecond timescale, such that excitonic propagation in (2.9) connected with spatial dispersion is of minor importance. Such propagation signatures will show up on a picosecond timescale, Section 3.4.

First, heavy- and light-hole excitons in a GaAs-quantum well [1] are studied. The material equations for a quantum well are derived in exactly the same way as in the bulk case, however, all wavenumbers are now the in-plane wavenumbers. In our analysis, we focus on the dominant dipole transitions with *s*-excitons only and neglect quadrupole and *p*-excitons in Eqs.(2.6) and (2.9). Their influence is discussed below for a forbidden band to band transition in bulk material. Since all heavy-holes (*hh*) are circularly polarized ($\sigma^{\pm}$), the dipole moments are $\vec{d}_h^+ = d_{vc}\vec{\sigma}^+$ and $\vec{d}_h^- = d_{vc}\vec{\sigma}^-$. The light-holes (*lh*) have both, circular polarization and contributions in the $z-$direction $\vec{d}_l^+ = \frac{d_{vc}}{\sqrt{3}}\vec{\sigma}^+$, $\vec{d}_l^- = \frac{d_{vc}}{\sqrt{3}}\vec{\sigma}^-$ and $\vec{d}_l^z = \sqrt{\frac{2}{3}}d_{vc}\vec{\sigma}^z$ which is twofold degenerate.

In the following, we discuss the temporally resolved generation of the excitonic transition probability $N_\gamma \equiv \int d^2\vec{R}|P_\gamma^{vc}(z,\vec{R},t)|^2$, where the integration is again over a radius of 1 μm in the transverse area at the position of the quantum well. The sum over all $\gamma = hh$ (or $\gamma = lh$) contributions yields a measure of the total excitonic density at the *hh* (or *lh*) resonance. The excitation takes place with a 10 fs Gaussian pulse. Fig. 5a shows the resulting build-up of the exciton density. For plane wave excitation (far-field optics), the ratio of hh- to lh- exciton density follows the selection rules without aperture which imply for σ^+ excitation a density ratio of 3 : 1 after the pulse. Figs. 5b,c show the excitonic density for the excitation with the near-field probe for a free standing quantum well (without barrier

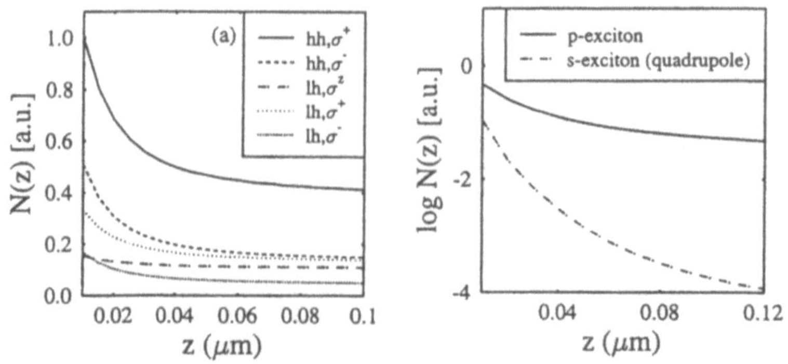

Figure 6 (a) Steady state exciton density for the heavy- and light-hole transition at different distances between probe and sample ($d = 10$). (b) Steady state p-exciton and quadrupole-exciton density excited by the optical near-field excitation at different distances between probe and sample ($d = 10$).

material). Due to the polarization mixing (Fig. 2a) not only σ^+ but also σ^- and σ^z polarization is excited. Because only the lh-transition is sensitive to σ^z light, the ratio of hh- to lh-exciton density changes in the present case from 3 : 1 to about 3 : 2.

However, the situation is different in the presence of quantum well barriers because the σ^z-component of the field at the quantum well position is screened and σ^- is enhanced, compare Fig. 2a and Fig. 4b. Fig. 6a shows the computed exciton transition amplitude for the different hh- and lh-transitions at different positions of the quantum well in the barrier material (probe-barrier distance $d = 10$ nm). It can be recognized that, for excitation with a near-field probe, the hh- and lh- exciton density (N_γ) depends strongly on the position of the near-field probe, i.e. the ratio of generated hh- and lh-excitons is changed in the presence of the aperture. The contribution of σ^z polarization increases the lh- in comparison to the hh-population. A rough estimate for a distance close to the aperture yields a decreased hh- to lh-ratio of 2 : 1 instead of 3 : 1 (the ratio without aperture).

Another modification of selection rules can be discovered for quadrupole transitions. Corresponding signatures should be observable e.g. by studying a forbidden band to band transition, where p-like transitions yield the dominant contributions. Corresponding to (2.6), the quadrupole contribution depends on the spatial gradient of the incident field distribution. To inspect this contribution, we solve (2.6) for a thin slice of bulk semiconductor in air [26] (Fig. 1c). Fig. 6b shows the comparison of the excited p-like and quadrupole-exciton density in the presence of the near-field probe for different distances of bulk material and near-field probe. It can be recognized that with decreasing distance between sample and probe, the strength of the quadrupole-transition increases in a more pronounced way than that for the p-like excitons. This enhancement of the quadrupole transition is a direct consequence of increasing spatial gradients of the field close to the sample.

3.4 Transport Dynamics: Wavepackets

In the previous sections, all spatial propagation effects due to spatial dispersion of excitons have been neglected, because excitons do not move appreciably on a femtosecond timescale. However, a rough estimate shows that ballistic transport over several microns

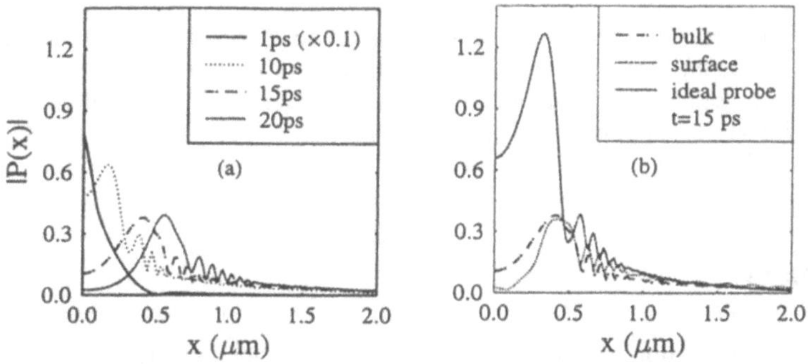

Figure 7 (a) Spatial distribution of the exciton polarization at different times after its excitation. (b) Influence of the finite barrier width and the near-field probe on the wavepacket for $t = 15$ ps.

should occur on a picosecond timescale. To discuss this phenomenon, we study a situation where a 30 fs pulse is propagated through the aperture to excite ballistic excitons. Their wavenumber (or momentum) distribution is determined by the inverse width of the aperture and results in the propagation of exciton-polariton wavepackets [6]. Because of the numerical complexity we simplify the investigations by assuming that the aperture consists of a stripe in the metallic screen. Furthermore, we include only two bands in the analysis (GaAs, *hh*). Using light which is polarized perpendicular to the stripe leads to the generation of excitons with $\vec{Q}||\vec{E}_{\vec{Q}}$ (excitonic Z-mode, [29]). To investigate the subsequent influence of the near-field probe and the quantum well barriers on the wavepacket propagation, three situations are investigated: (i) propagation of an initially near-field excited distribution in a quantum well, embedded in an infinitly extended barrier material (no presence of probe and barrier boundary after the wavepacket generation), (ii) presence of the barrier boundary ($d = 10$ nm) for the quantum well below the surface at $z = 30$ nm (no presence of the near-field probe), (iii) presence of the barrier and the ideal metallic near-field probe.

(i) Fig. 7a shows the spatial distribution for the exciton polarization $|P^{vc}(x,t)|$ as a function of the in-plane coordinate x, chosen to be perpendicular to the aperture for different times after its preparation. After the arrival of the near-field pulse ($t = 0$), a peaked exciton distribution is created in front of the near-field source ($x < 0.5$ μm), see e.g. the $t = 1$ ps trace in Fig. 7a. At larger times, the distribution expands out of the excitation region. The propagation involves a superposition of quantum well polaritons with different momentum states excited by the near-field probe [29]. Spatial and temporal profiles of the wavepacket are determined by the corresponding dispersion relation $\omega = \omega(Q)$ which includes the interplay of spatial dispersion and the back coupling of the field to excitons, (2.9), compare [6]. The dispersion relation describes polaritons with in-plane wavenumbers $Q < k_l$ which decay radiatively, as well as evanescent states $Q > k_l$ which propagate without temporal decay [29]. Therefore, all polaritons which contribute to the wavepacket on a time scale longer than the radiative decay have at least a velocity determined by the momentum $Q = k_l$. This limit determines the trailing edge of the wavepacket. The spatial oscillation on top of the wavepacket results from the interference of weakly and strongly radiation coupled excitons, in analogy to the two exciton polariton branches in bulk material [30].

(ii,iii) Fig. 7b shows the comparison of the polarization distribution after a propagation time of 15ps where the barrier surface and the near-field probe itself have been taken into account. In the presence of an ideal metallic screen the wavepacket amplitude is enlarged because the radiative damping of the polaritons is reduced [31]. The barrier surface has only a minor influence on the propagation dynamics.

4 Summary

The simultaneous solution of material and Maxwells equations yields new insights for spatially resolved spectroscopy such as semiconductor near-field optics. Near-field optics involves not only a better optical resolution but a change in the selection rules due to the mixing of the vector components of the electromagnetic field. Furthermore, pulsed near-field should lead to new wavepacket phenomena on ultrashort time and space scales. In general, barrier materials and the near-field probe have to be taken into account for a realistic modelling of quantum confined excitons in their dielectric environment.

Acknowledgements

The authors thank Anna von der Heydt, Tineke Stroucken and Peter Thomas for useful discussions. This work is supported by the Deutsche Forschungsgemeinschaft (Quantum Coherence Project) and the Leibniz Prize.

Bibliography

[1] G.Bastard, *Wave Mechanics Applied to Semiconductor Heterostructurs.* (Les Editions de Physique, Paris 1988); *Frontiers in Nanoscale Science of Micron/Submicron Devices.* edited by A.P.Jauho and E.Buzaneva (Kluwer Academic Publishers, Dordrecht-Boston-London 1996)

[2] A. Zrenner, L. V. Butov, M. Hagn, G. Abstreiter, G. Böhm, and G. Weinmann, Phys. Rev. Lett. **72**, 3382 (1994); D. Gammon, E. S. Snow, B. V. Shanabrook, D. S. Katzer, and D. Park, Phys. Rev. Lett. **76**, 3005 (1996).

[3] A.Richter, G.Behme, M.Süpitz, Ch.Lienau, Th.Elsässer, M.Ramsteiner, R.Nötzel, and K.H.Ploog, Phys.Rev.Lett. **79**, 2145 (1997).

[4] H.F.Hess, E.Betzig, T.D.Harris, L.N.Pfeiffer, K.W.West, Science **264**, 1740 (1994); W.Langbein, J.M.Hvam, S.Madsen, M.Hetterich, and C.Klingshirn, Phys.Stat.Sol.(a) **164**, 541 (1997).

[5] Ch. Lienau, A.Richter, G.Behme, M.Ramsteiner, Th.Elsässer, R.Nötzel, and K.H.Ploog, Phys.Stat.Sol.b **206**, 153 (1998).

[6] P.Thomas and A.Weller, J. Non-Cryst. Solids 97/98, 245 (1987); A.Knorr, F.Steininger, B.Hanewinkel, S.Kuckenburg, P.Thomas, and S.W.Koch, Phys.Stat.Sol.(b) **206**, 139 (1998).

[7] *Photons and Local Probes.* edited by O.Marti and R.Möller (Kluwer Academic Publishers, Dordrecht-Boston-London 1996)

[8] D. W. Pohl, Advances in Optical and Electron Microscopy, **12**, (1991); E. Betzig, J.K. Trautman, T.D. Harris, J.S. Weiner, and R.L. Kostelak, Science Vol. **251**, 1468 (1991).

[9] C. Girard, O. J. F. Martin, A. Dereux, Phys. Rev. Lett. **75**, 3098, (1995). R. X. Bian, R. C. Dunn, S. Xie, P. T. Leung, Phys. Rev. Lett. **75**, 4772, (1995).

[10] L. Novotny, D. W. Pohl, and B. Hecht, Opt. Lett. **20**, 970 (1995).

[11] J.P.Fillard *Near-Field Optics and Nanoscopy.* (World Scientific, Singapore 1988)

[12] A.Otto, Z.Phys. **216**, 398 (1964).

[13] C.Girard and A.Dereux, Rep.Pro.Phys. **59**, 657 (1996).

[14] M.Born and E.Wolf, *Principles of Optics.* (Pergamon Press, London, New York, Paris 1959)

[15] S.Mukamel, *Principles of Nonlinear Optical Spectroscopy.* (Oxford University Press 1995)

[16] H.Haug and S.W.Koch, *Electronic and Optical Properties of Semiconductors.* (World Scientific, 3rd ed. 1994)

[17] Proceedings of the conference *Nonlinear Optics and Excitation Kinetics in Semiconductors* in Phys.Stat.Sol.b 188 (1994) and 206 (1998)

[18] J.D.Jackson, *Classical Electrodynamics.* (2nd. edition, John Wiley & sons, New York 1975)

[19] G.D.Mahan, *Many Particle Physics* (Plenum Press, New York 1990)

[20] A.Knorr, A.v.d.Heydt, B.Hanewinkel, and S.W.Koch, to be published.

[21] D.Fröhlich in *Advances in Solid State Physics X.* 227, ed. by O. Madelung (Vieweg, Braunschweig 1970)

[22] P.Y.Yu and M.Cardona, *Fundamentals of Semiconductors.* (Springer, Berlin, Heidelberg, New York 1996)

[23] A.Knorr, F.Steininger, A.Girndt, T.Stroucken, S.Haas, W.W.Chow, P.Thomas, and S.W.Koch, Il Nuovo Cimento D **17**, 1265 (1995); F.Steininger, A.Knorr, T.Stroucken, P.Thomas, S.W.Koch, Phys.Rev.Lett. **77**, 550 (1996).

[24] V.M.Axt and A.Stahl, Z. Phys. B **93**, 195 (1994); M.Lindberg, Y.Z.Hu, R.Binder, and S.W.Koch Phys. Rev. B **50**, 18060 (1994).

[25] A.Taflove, *Computational Electrodynamics - The Finite-Difference Time -Domain Method.* (Artech House Inc., Norwood, 1995)

[26] dielectric constant of the barrier material: $\varepsilon = 12$; longitudinal-transversal splitting: $\Delta_{LT} = 0.08$ meV; energy gap $\omega_{gap} = 1.5$ eV; dipole moment: $d_{vc} = 3e$ Å; for the quadrupole interaction (for instance in Cu_2O), $\hat{B}$ is assumed to be approximately isotropic: $B^{\alpha\beta} = d_{cv}a_0$, $a_{\text{Bohr}} = 10$ Å.

[27] H.A. Bethe, Phys. Rev. **66**, 163 (1944), C.J. Bouwkamp, Philips Res. Rep. **5**, 401 (1950).

[28] G.W.Bryant, Appl. Phys. Lett. **72**, 768 (1998).

[29] V.M.Agranovitch and V.L.Ginzburg, *Crystal Optics with Spatial Dispersion and Excitons.* (Springer, Berlin, Heidelberg, New York 1984), C.L. Andreani, D.S.Citrin in *Confined Electrons and Photons: New Physics and Devices.*, ed. E. Burstein and C. Weisbuch (Plenum Press, New York 1994)

[30] D. Fröhlich, A. Kulik, B. Uebbing, A. Mysyrowicz, V. Langer, H. Stolz, and W. von der Osten, Phys. Rev. Lett. **67**, 2343 (1991).

[31] H.Kuhn, Jour.Chem.Phys. **33**, 101 (1970).

Near-field Scanning Optical Spectroscopy of Semiconductor Nanostructures

Christoph Lienau

Max-Born-Institut für Nichtlineare Optik und Kurzzeitspektroskopie, Rudower Chaussee 6, D-12489 Berlin, Germany

Abstract: The nanoscopic optical properties of a novel quantum-well-embedded GaAs quantum wire structure grown on patterned (311)A GaAs surfaces are mapped using near-field scanning optical microscopy at temperatures between 10 and 300 K. The local confinement potential of a single quantum wire and the embedding 2D quantum well continuum is directly extracted using near-field photoluminescence excitation spectroscopy with subwavelength resolution. Shallow asymmetric potential barriers separating quantum wire and embedding quantum well are identified and their strong influence on carrier transport and trapping into the quantum wire is analyzed in steady state and time-resolved near-field photoluminescence experiments at variable temperatures.

1 Introduction

There is currently a tremendous worldwide research effort towards the fabrication of low-dimensional semiconductor nanostructures such as one-dimensional quantum wires (QWR) or zero-dimensional quantum dots (QD). This effort arises partly from the desire to design laser devices with improved optical characteristics such as higher optical gain, ultra-low threshold current and narrow spectral linewidths [1,2]. Moreover, unique new physical properties such as narrowed densities of states, increased excitonic effects [3,4], strongly enhanced carrier mobilities [5] and modified phonon scattering rates [6] are predicted for such low-dimensional systems. The low-dimensional confinement of carriers in semiconductor nanostructures is based on spatial variations of the bandstructure, e.g. the bandgap, on a nanometer scale. The study of such confinement potentials and their influence on the carrier dynamics in nanostructures requires sophisticated experimental techniques providing very high spatial resolution. While non-optical methods such as scanning or transmission electron microscopy, scanning tunneling, ballistic electron emission [7,8] and atomic force microscopy are readily available, these dimensions are far beyond the resolution limit of classical optical techniques. Thus, so far, the optical properties of low-dimensional semiconductors have mainly been studied with ensembles consisting of up to hundreds of nanostructures. In such experiments, size fluctuations within the ensemble lead to a strong inhomogeneous broadening of the optical spectra and make the observation of the intrinsic low-dimensional behavior difficult. Furthermore, processes involving real-space transfer of carriers are in general difficult to separate from carrier relaxation phenomena. This is particularly relevant for the coupling between nanostructures

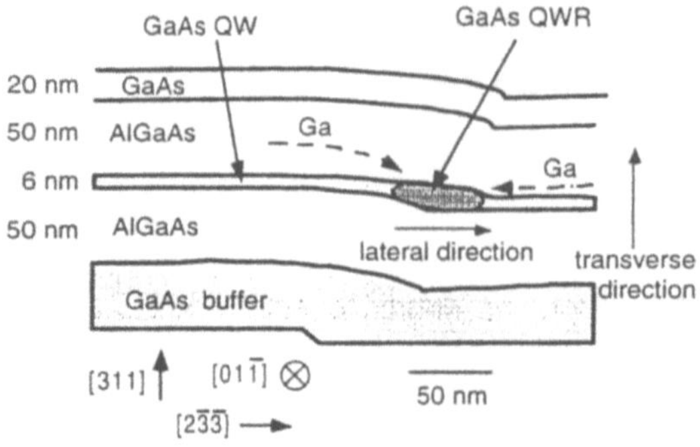

Figure 1
Scheme of the quantum wire structure. The one-dimensional confinement originates from the thickness variation of the GaAs quantum well along the lateral [2−3−3] direction.

of different dimensions, e.g. in the case of carrier capture into quantum wires or dots, where – in general – both real-space transfer and trapping of carriers are involved [9,10].

Here, optical techniques with a spatial resolution in the 100 nm range, such as near-field scanning optical microscopy (NSOM) [11] [12] not only have the potential to select single nanostructures but also to differentiate between real-space transfer and relaxation phenomena and thus to provide entirely new insight into the physical properties of such structures. This potential has been demonstrated in pioneering near-field spectroscopic experiments on quantum wells [13] and wires [14]. Very recently, the absorption properties of single quantum wires have been investigated for the first time using near-field spectroscopy [15]. The near-field technique becomes even more powerful when it is combined with time-resolved excitation and detection schemes as it offers direct acess to the spatio-temporal carrier dynamics on ultrashort time and length scales. This has very recently been demonstrated in spatially-resolved pump-probe experiments on GaAs microdisks [20] and in a study of exciton spin dynamics in magnetic heterostructures [16].

In this article, we report results of a recent experimental investigation of the nanoscopic optical properties of a novel class of quantum wires grown on patterned high-index GaAs surfaces [17]. Low-temperature near-field spectroscopy is used to map the local energetics of a single quantum wire [18]. Real-space transfer and capture of carriers into the QWR are separately resolved and their dynamics are studied by combining both stationary and picosecond time-resolved near-field photoluminescence spectroscopy over a wide temperature range [19].

2 Experimental

2.1 Samples

The QWR structure investigated in our experiments is depicted schematically in Fig. 1. The sample was grown by molecular beam epitaxy (MBE) on patterned GaAs (311)A substrates at the sidewall of 15 to 20 nm high mesa stripes oriented along [01−1] [21]. The sample consists of a nominally 6 nm thick GaAs QW layer clad by 50 nm thick $Al_{0.5}Ga_{0.5}As$ barriers. Formation of a sidewall QWR arises from the preferential migration of Ga atoms within the QW layer from both the mesa top and the mesa bottom towards the sidewall during MBE growth [22]. Cross sectional TEM images [21] indicate that – as a consequence of this migration process – the thickness of the GaAs QW near the sidewall increases from 6 up to 13 nm. This change in QW thickness results in a 1D confinement over a lateral QWR width of 50 nm. Confinement of excitons in the wires was independently inferred from the diamagnetic shift of the QWR photoluminescence (PL) [21]. For device applications it is important that these wires exhibit high uniformity, narrow PL emission lines and high PL efficiency up to room temperature. They can be vertically stacked in the growth direction without increase in interface roughness and wire size fluctuations.

2.2 Near-Field Spectroscopy

In near-field microscopy, subwavelength spatial resolution beyond the diffraction limit is achieved by transmitting light through a nanometer-sized aperture at the end of a NSOM probe tip. Such probes are made by pulling single mode optical fibers to a sharp taper and coating the taper with a 50–100 nm thick aluminum or gold layer [24]. The aperture diameter is controlled, through variation of the shape of the taper, between 50 and 200 nm. Time- and spatially resolved experiments at temperatures between 10 and 300 K are performed with a novel home-built near-field microscope (LT-NSOM) [25], while for room temperature experiments a modified commercial instrument (Topometrix Aurora) was used. In the LT-NSOM, the sample is cooled by attaching it to the cold finger of a continuous flow liquid helium cryostat, the temperature of which can be varied between 5 and 330 K (Fig. 2a). Cold finger, sample, and NSOM scan head, consisting of NSOM probe, tuning fork [26] shear-force distance regulation (Fig. 2b) [27], and coarse and fine positioning, are mounted inside a vacuum chamber. In this way, the sample is cooled while leaving the temperature of the NSOM scan head at room temperature.

The experiments reported in this article are performed in an illumination geometry [28]. Spatial resolution is provided in the excitation process by transmitting light through the NSOM fiber probe. Photoluminescence (PL) emitted by the sample is collected in the far-field by a conventional microscope objective. The luminescence is dispersed in a 0.25 m double monochromator with a spectral resolution of 1.2 nm and detected using a single-photon-counting silicon avalanche photodiode either in steady state or with a

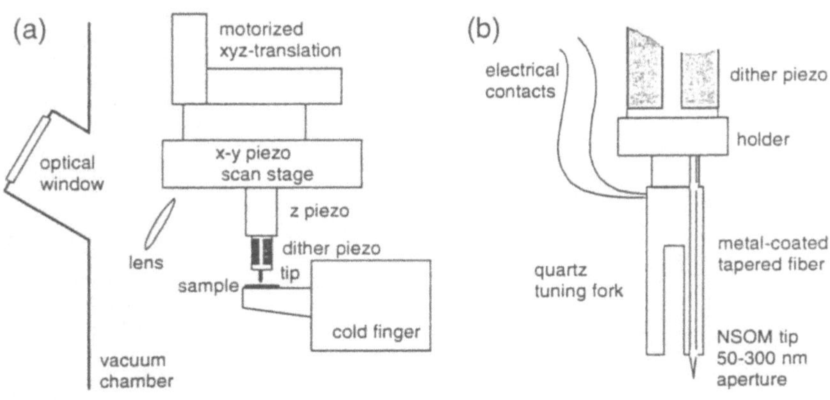

Figure 2 (a) Schematic of the NSOM scan head mounted inside a vacuum chamber. The temperature of the cold head is actively controlled between 5 and 330 K. (b) Schematic of the quartz tuning fork shear-force setup.

time resolution of 250 ps. Different tunable lasers served for steady-state or time-resolved excitation of the sample with powers between 10 and 100 nW, corresponding to a very low carrier density between 10^4 and 10^5 cm^{-1}. Experiments performed in a collection mode geometry are reported elsewhere [28]. In all experiments, PL is detected as a function of tip position while the tip is scanned relative to the sample. During the scan, the tip-to-sample distance is kept constant at about 5 nm using either an optical (room temperature NSOM) [27] or a tuning fork (LT-NSOM) [26] shear-force setup. Shear-force images recorded simultaneously with the PL images give information on the sample topography and allow for a direct correlation between optical and topographic signal.

3 Near-field Excitation Spectroscopy: Mapping of the Local Bandgap

In this section, we report on experiments combining photoluminescence excitation (PLE) spectroscopy and near-field microscopy for mapping the local energetics of the quantum wire structure in real space with subwavelength resolution. At a sample temperature of 10 K, the photoluminescence (PL) spectrum of the sample shows two prominent emission peaks [25]. QWR emission appears as a single narrow peak centered at 1.545 eV with a spectral full-width-at-half-maximum (FWHM) of 8 meV (Fig. 3(b)). QW emission is centered around 1.604 eV with a spectral width of 12 meV. Information on the local energetics of the QWR and the embedding QW structure is obtained by mapping the local absorption properties of the sample by recording spatially-resolved PLE spectra. In such an experiment, the sample is locally excited through the NSOM probe with a tunable narrowband Ti:sapphire laser and QWR luminescence is detected in the far-field. The tip is scanned along the lateral y direction perpendicular to the QWR located at $y = 0$.

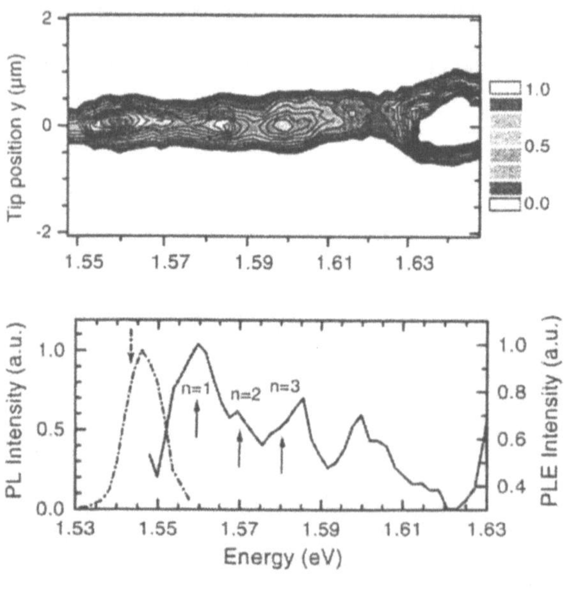

Figure 3
(a) Near-field PLE spectrum recorded for spatially resolved excitation of the QWR sample through the NSOM fiber probe at a temperature of 10 K. The intensity of the QWR luminescence at 1.544 eV is plotted as a function of excitation energy (abscissa) and of lateral distance between the QWR located at $y = 0$ and the fiber tip (ordinate). Only carriers excited at the QWR contribute to the signal. (b) Cross-section through (a) at $y = 0$ (solid line) and QWR near-field luminescence spectrum (dash-dotted line).

In Fig. 3(a), the intensity of QWR luminescence at a detection energy of 1.544 eV (spectral resolution 2 meV), i.e. slightly below the maximum of QWR emission, is plotted as a function of excitation energy (abscissa) and of lateral distance of the exciting fiber tip from the QWR position at $y = 0$ (ordinate). We find a PLE spectrum that is spatially confined to the QWR region. For excitation energies below 1.63 eV, the width of this region has a value of about 300 nm and is limited by the finite distance of 70 nm between QWR and sample surface and the finite aperture size of about 200 nm used in this experiment. A cross section through this image (Fig. 3b) exhibits several maxima below the onset of QW absorption at 1.63 eV. Such peaks are due to the quasi one-dimensional subband structure of the QWR and reflect the QWR absorption spectrum [29]. Preliminary calculations of the 1D subband structure within the adiabatic approximation [30] indicate that the first peaks are due to the $n = 1$ (1.558 eV), $n = 2$ (1.570 eV), and $n = 3$ (1.583 eV) heavy-hole to conduction band transitions (arrows in Fig. 3(b)). The splitting between the $n = 1$ and $n = 2$ transition is on the order of 12 meV, which is in good agreement with the width of this QWR structure of about 50 nm, that was estimated from TEM images [21] and observed diamagnetic shifts [23]. The strong intensities of the third peak (at 1.583 eV) and fourth peak (at about 1.60 eV) suggest a significant contribution from excitonic transitions between valence band states with predominant light-hole character and conduction band states. The role of such transitions has recently been addressed both experimentally [31] and theoretically [32,33] in investigations of valence band mixing effects on the linear optical properties of quantum wires. Currently polarization resolved experiments are underway that will lead to a definite assignment of these transitions. The small inhomogeneous broadening of the spectra of about 10 meV is limited by the structural properties of the corrugated (311)A surface. The absorption line is comprised of contributions from different localized excitons, as observed for QWs on low-index GaAs surfaces [13,36]. On (311)A surfaces the typical lateral size of monolayer-high islands is much smaller than on (100) surfaces, where these islands can extend to diameters of more than 100 nm [36]. (311)A surfaces have been shown to spontaneously form a corrugation with a lateral periodicity of 3.2 nm and a step height of 2 monolayers (0.34 nm). On a larger, 20–40 nm scale, STM images of these surfaces show mesoscopic scale roughness of few monolayers. Because of this short corrugation length, components from individual localized excitons are not resolved even on a 100 nm length scale and the NSOM PLE spectrum shows significant inhomogeneous broadening. We note, however, that attempts to resolve the NSOM PLE spectrum with reduced spatial resolution, using a confocal setup [22] with a resolution of 1.5 μm were not successful.

Similar excitation energy dependencies of the QWR luminescence are observed for detection on the high-energy side of the QWR spectrum at 1.550 eV [37] and at different spatial positions along the wire axis. This points to a predominantly inhomogeneously broadened QWR PL spectrum and to exciton localization along the wire direction. The very intense peak at energies above 1.63 eV is due to the onset of QW absorption. From its separation to the QWR first peak one derives a confinement energy of 80 meV. Unlike for QWR absorption, the spatial dependence of the QW absorption peak is not limited by the spatial resolution of our experiment. Its spatial intensity variation is reasonably described by a Gaussian profile with a width of about 800 nm. This width is mainly given by the confinement in the vicinity of the QWR wire as will be discussed below.

The spatially resolved PLE spectrum changes drastically as the sample temperature is increased to 77 K. In Fig. 4, the luminescence intensity detected near the maximum of the QWR emission at 1.533 eV is plotted as a function of excitation energy and position as the tip is scanned perpendicular to the QWR. The QWR PLE spectrum appears again as

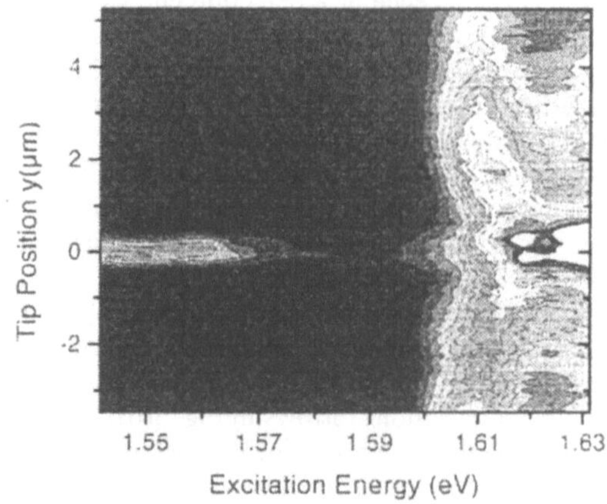

Figure 4
Near-field PLE spectrum at a sample temperature of 77 K and a detection energy of 1.533 eV, corresponding to the maximum of the QWR emission spectrum. Notice the pronounced spatial blue-shift of the excitonic QW absorption peak as the excitation tip approaches the QWR at $y = 0$.

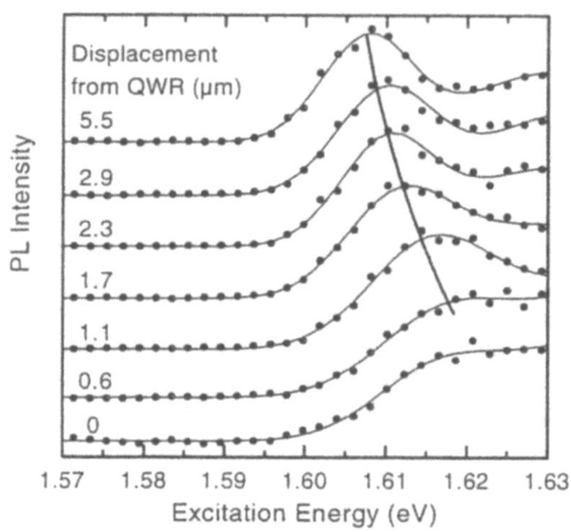

Figure 5
Cross-sections through Fig. 4. Near-field mesa top QW PLE spectra at fixed excitation position relative to the QWR location at $y = 0$ at a sample temperature of 77 K and a detection energy of 1.533 eV. Experimental data points are shown as solid circles while the solid lines represent fits to the data to (1).

a spatially narrow peak around $y = 0$, slightly red-shifted with respect to the 10 K spectrum. This red-shift reflects the decrease of the GaAs bandgap with lattice temperature. The subband-related modulation of the spectrum is washed out due to increased thermal broadening. More important and in contrast to the low-temperature spectrum, strong QWR luminescence is now detected for QW excitation at energies higher than 1.6 eV and at distances y of several μm away from the QWR location. The spectral shape of the PLE spectrum for $y \neq 0$ changes strongly with excitation position. In Fig. 5, we present cross sections through the image of Fig. 4 for different separations from the QWR on the mesa top ($y > 0$) (symbols). The peak in these spectra which is due to the excitonic enhancement of QW absorption shifts by about 14 meV to higher energies as one moves from a distance of $y = 5.5$ μm to the location of the QWR. On the mesa bottom ($y < 0$) (Fig. 4), this shift is even more pronounced with PLE maxima at 1.610 eV for $y = -3.5\mu$m and at 1.628 eV for $y = -0.4$ μm.

The data of Figs. 3 and 4 demonstrate that carriers locally generated within the QW undergo real-space transfer to the location of the QWR where they are trapped into quasi-one-dimensional QWR states and contribute to the QWR emission. The time interval during which real-space transfer occurs, is given by the recombination lifetime of carriers within the QW. Time-resolved PL measurements under the same excitation conditions give a lifetime of 1.2 ns at 77 K. This time scale is much longer than typical thermalization and cooling times of carriers, processes occurring on the femto- and picosecond time scale [38]. As a result, the carriers undergoing real-space transfer form a quasi-equilibrium distribution with a temperature close to lattice temperature.

The PLE spectra recorded on the QW are closely related to the local QW absorption spectrum at the specific excitation position with the overall intensity at a specific position being proportional to the probability for carrier trapping into the QWR. The results in Fig. 5 reveal (i) a pronounced exciton-like feature in the absorption of the QW, and (ii) a blue-shift of this maximum with decreasing separation from the QWR. To analyze this behavior in a more quantitative way, we used the Elliott formula [39] for the absorption spectrum $\alpha(E)$ of a quasi-two-dimensional semiconductor

$$\alpha(E) \propto \frac{E}{E_0}\left(\sum_{n=0}^{\infty}\frac{4}{(n+1/2)^3}\delta\left(\Delta+\frac{1}{(n+1/2)^2}\right)+\Theta(\Delta)\,\frac{\exp\left(\pi/\sqrt{\Delta}\right)}{\cosh\left(\pi/\sqrt{\Delta}\right)}\right). \tag{3.1}$$

Here, E is the excitation energy, n the quantum number of the bound exciton states, and E_0 the 3D exciton binding energy. $\Delta = (E-E_g)/E_0$, with E_g being the bandgap energy. $\Theta(\Delta)$ denotes the Heavyside step function. The dominant features are the $n=0$ exciton absorption peak and the continuum absorption, i.e. the second term in (3.1). At each excitation position, the QW absorption spectrum in Fig. 5 was fitted to (3.1) using a Gaussian profile with a width of about 10 meV instead of the δ-function to account for the inhomogeneous broadening of the lines which is probably due to interface roughness of the QW. A similar broadening was used for the step-like continuum absorption (second term in (3.1)). The solid lines in Fig. 5 represent the calculated spectra which are in good agreement with the experimental results. The calculation gives the local bandgap energy $E_g(y)$ as a function of excitation position y [40]. A plot of the lateral variation of the QW bandgap versus y is shown in Fig. 6. On each side of the wire, the potential shows two pronounced maxima, separated by approximately 1 μm. On the mesa bottom ($y<0$) the band-gap increases by as much as 18 meV over a length scale of about 2 μm as approaching the location of the QWR, while on the mesa top ($y>0$) the bandgap shift is slightly less pronounced and amounts to about 14 meV.

The change in bandgap energy originates mainly from a change in the average local thickness of the quantum well, that decreases from about 5.6 nm on the flat-area mesa top and bottom down to 4.8 nm ($y<0$), respectively 5.0 nm ($y>0$), in the vicinity of the wire. This thinning is a consequence of the Ga-atom migration towards the sidewall during the growth process and determined by the specific MBE growth parameters, in particular the substrate temperature. The asymmetry in the bandgap profile on mesa top ($y>0$) and bottom ($y<0$) is most likely related to the wet chemical etching procedure that is used to pattern the mesa structure[22], and the resulting changes of Ga migration. Based on these results, we derive the complete confinement potential of the quantum-well embedded QWR. Two different spatial regions are distinguished. In the region $|y|>100$ nm, the potential can directly be taken from the results of Fig. 6, while in the central 100 nm

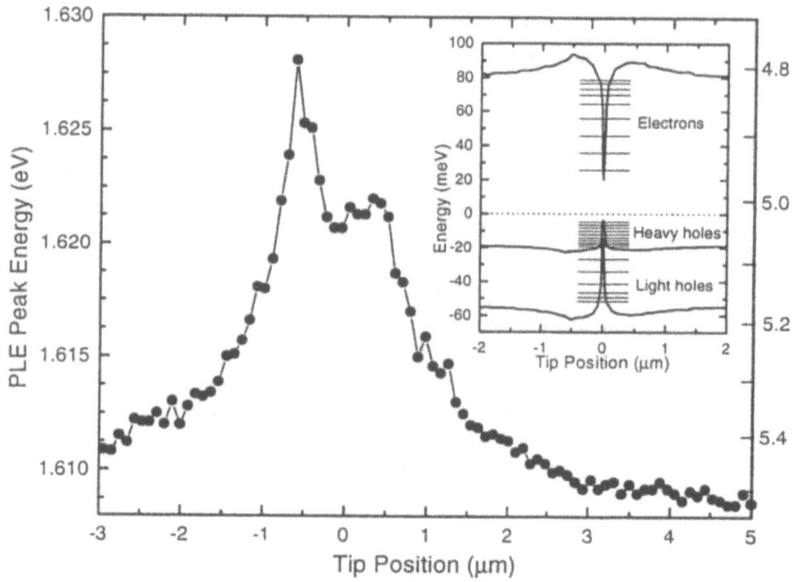

Figure 6 Lateral band gap profile as obtained from an analysis of the QW PLE spectrum shown in Fig. 4. The lateral shift of the excitonic QW absorption peak corresponds to a decrease of the average QW thickness in the region close to the QWR. Inset: QWR confinement potential as derived from the data in Fig. 5(a) and calculated energies of the 1D subbands (lines).

around $y = 0$ the local thickness variation of the QWR and thus the confinement potential is taken from the PLE spectrum at 10 K, and indepently from cross sectional TEM images [21]. In the inset of Fig. 6 the potential is plotted as a function of y for the whole interval between -2 and $+2$ μm. Using this information and a ratio of 2:1 for the conduction to valence band offset energy in the 2D GaAs/$Al_{0.5}Ga_{0.5}As$ structure, the quasi-one-dimensional subband structure and the interband transition energies of the QWR can now be calculated within the adiabatic approximation. We find a splitting of 10 meV between the two lowest conduction band subbands, in agreement with the low-temperature PLE spectrum [28]. We note that Coulomb correlation effects that affect both the position and shape of the peaks in the PLE spectrum are neglected in these calculations. Since the height of the local barriers of about 15 meV is only a small fraction of the QWR confinement energy and the barriers are separated by about 1 μm, the influence of the barriers on the energetic position of the lowest QWR subbands is rather small. Nonetheless, the presence of the barriers strongly influences the carrier transport within the sample as is evidenced from temporally and spatially resolved PL measurements discussed in the next section.

To confirm the above findings, we recorded a spatially resolved PLE spectrum at a temperature of 10 K in which the QW instead of the QWR luminescence was monitored for different separations of the excitation tip from the QWR [37]. Again the thinning of the QW for excitation near the QWR center is clearly observed as a characteristic blue shift of the first exciton peak in the spectrum. An analysis of the spectrum gives a similar spatial dependence of the local quantum well band gap as shown in Fig. 6 and thus confirms the inter-pretation of PLE spectra at 77 K. In contrast to the 77 K spectrum, the

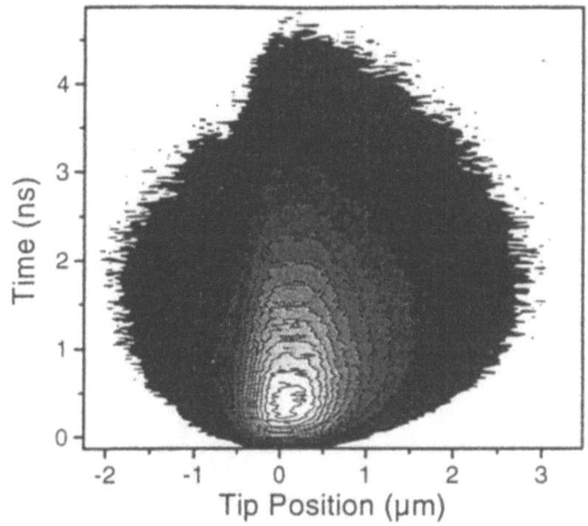

Figure 7
Spatially and temporally resolved photoluminescence of the QWR at a temperature of 100 K after localized excitation in the embedding quantum well (E_{ex} = 1.614 eV, detection energy 1.54 eV). The luminescence intensity is plotted as a function of the lateral distance y between excitation and QWR position and as a function of time.

low temperature spectra show energetic as well as spatial fluctuations of the luminescence intensity that are a signature of exciton localization within the QW [36,13].

4 Time-Resolved Luminescence Spectroscopy: Exciton Transport into a Single QWR

In this section, we combine low-temperature near-field microscopy and time-resolved luminescence spectroscopy to study the influence of the local bandstructure, described in the previous section, on the microscopic carrier transport dynamics within the nanostructure. By transmitting femtosecond excitation pulses from a mode-locked Ti:Sapphire laser through a metal-coated NSOM fiber probe with an aperture of 200 nm, we can generate a transient electron-hole pair distribution that is localized within a narrow, subwavelength spot within the GaAs QW, with dimensions that are mainly given by the aperture diameter. The dynamics of this carriers distribution, in particular their real-space transfer towards the QWR and trapping into 1D QWR states is now monitored by time-resolving the QWR luminescence using time-correlated single-photon-counting detection with a temporal resolution of 260 ps.

First, experiments at a sample temperature of 100 K are discussed. We know from the PLE spectra presented above and from stationary PL experiments [37] that in this temperature range, electron-hole pairs that are generated within the embedding QW, at distances of several μm away from the QWR are effectively trapped into the QWR. Moreover, the energy of the potential barriers separating QWR and QW is similar to the thermal energy at this temperature so that we expect a considerable influence of the barrier potential on the carrier dynamics. In Fig. 7, the intensity of the time-resolved QWR emission at a sample temperature of 100 K is shown as a function of tip position along the lateral y-axis (abscissa) and delay time (ordinate). In this experiment, the 16 meV broad spectrum of the excitation laser is centered at 1.614 eV and overlaps the excitonic absorption band of the QW. QWR PL at 1.54 eV, the maximum of the QWR PL spectrum, is collected in the far field, dispersed in a 0.25 monochromator with a resolution of 1.2 nm and time-resolved. Cross sections through Fig. 7, i.e. the time evolution of luminescence for fixed

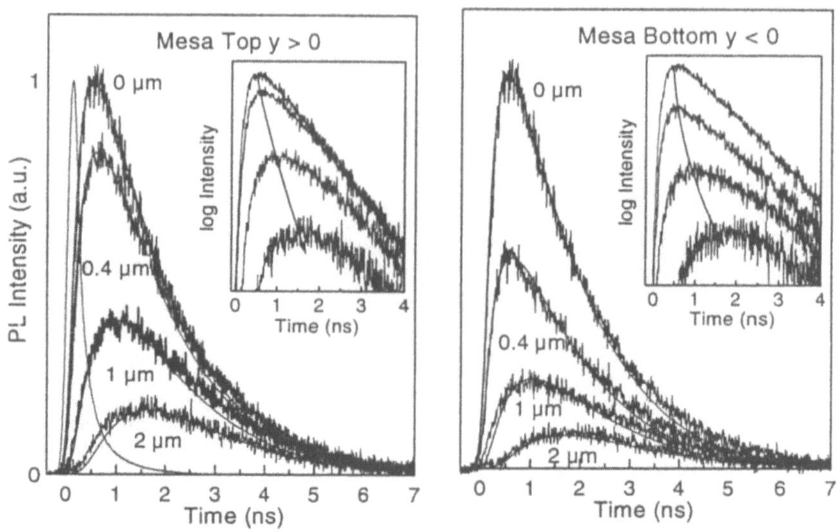

Figure 8 Time evolution of the QWR luminescence after excitation on the (a) mesa top and (b) mesa bottom of the QWR structure. The luminescence intensity at 1.54 eV is plotted versus time for different excitation positions y and shows an increasingly delayed rise with increasing y which is due to the transport of excitons from the excitation spot to the QWR. Insets: Data plotted on a logarithmic intensity scale. Solid lines: results of a numerical calculation based on a drift-diffusion model for exciton transport.

excitation position $|y| = 0$ (QWR position), 0.4, 1, and 2 μm are shown in Fig. 8(a) and (b) for excitation on mesa top and mesa bottom, respectively. In the insets, these data are plotted on a logarithmic intensity scale. For excitation at $y = 0$, we find a rise of the QWR luminescence that is limited by the temporal resolution of our experiment of 260 ps and a decay of the QWR luminescence that is single exponential over two orders of magnitude with a decay time of 1.5 ns. This temporal dependence of the QWR luminescence remains unchanged for $|y| < 320$ nm, i.e. for excitation between the barriers separating QWR and embedding QW. For a distance $|y| > 320$ nm, the onset of QWR luminescence is delayed with respect to that for excitation at $y = 0$. The rise of the QWR luminescence becomes monotonously slower with increasing $|y|$, as is highlighted in Fig. 9, where normalized QWR luminescence profiles for fixed excitation positions of $|y| = 0$, 0.9, and 1.8 μm are compared. This delayed rise manifests itself in a shift of the maximum of the PL intensity in time with increasing y.

In Fig. 10, we plot the temporal position of the maximum PL intensity, $t_{max} - t_0$, as a function of the excitation position y ($t_0 \approx 0.4$ ns: maximum position for $|y| < 0.32\mu$m). The value of $t_{max} - t_0$ increases almost linearly with increasing distance y, with a slope of 0.73 ns/μm for $y < 0$ and 0.62 ns/μm for $y > 0$. The maximum intensity of QWR luminescence after excitation on the mesa bottom ($y < 0$) is significantly smaller than for mesa top ($y > 0$) excitation, i.e. less carriers are trapped into the QWR. This is evident from the amplitude values in Fig. 8(a) and (b) and from the shape of the plot in Fig. 7 at positive and negative y.

In the following, we discuss the luminescence behavior found for different excitation positions y. For $|y| < 0.32\,\mu$m, the single exponential decay of the QWR PL gives a population lifetime of the QWR [41,42] of 1.5 ns, slightly longer than the PL decay time of the

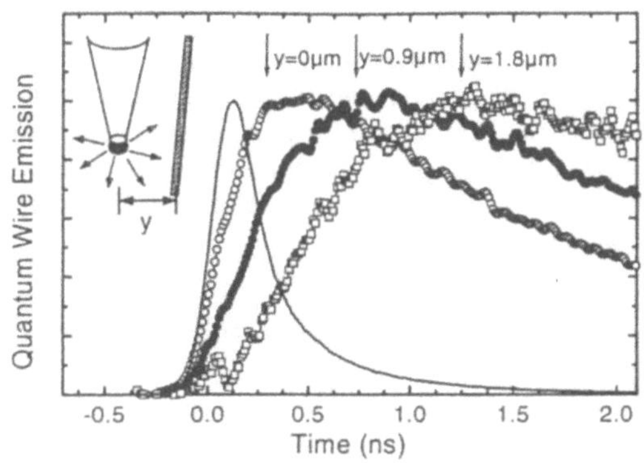

Figure 9
Temporal profiles of the QWR luminescence for fixed excitation position y for mesa top QW excitation as in Fig. 7. Exciton transport within the GaAs QW from the generation point towards the QWR location causes a delayed rise of the QWR luminescence. The maximum of the QWR emission for each excitation position has been normalized to unity. Shown as a solid line is the response function of the single photon counting detection.

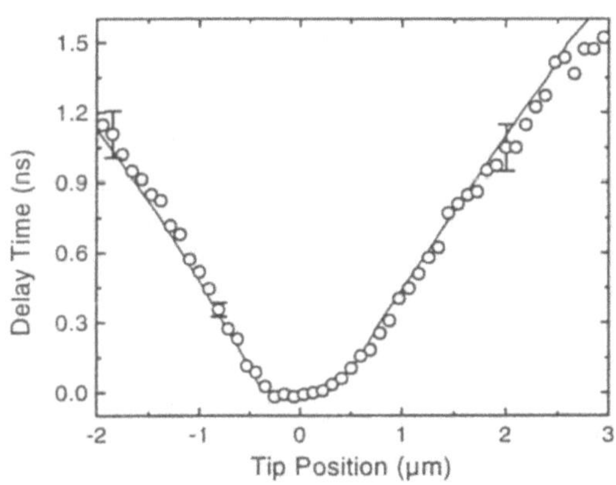

Figure 10
Temporal position of the luminescence maxima of Fig. 9 as a function of lateral separation between excitation spot and QWR (symbols). Solid line: Temporal positions calculated from the drift-diffusion model for exciton transport.

embedding QW of 1.35 ns. The fast rise of the QWR PL after excitation at $|y| < 0.32\ \mu$m demonstrates that carrier relaxation from continuum states into localized QWR states occurs within the time resolution of the experiment of 260 ps. This relaxation involves both trapping into and energy relaxation within the QWR. For our very low excitation density of $\leq 10^5$ cm^{-1}, carrier-carrier scattering is negligible and emission of longitudinal optical phonons represents the dominant process populating low-lying QWR states [9,43].

The onset of the QWR luminescence for excitation at $|y| > 0.32\ \mu$m, i.e. for photogeneration of carriers in the QW, occurs with a delay of several hundred picoseconds. This delay increases with y (Fig. 10), reflecting the traveling time of carriers from the excited QW area to the QWR location. In our experiment, the optical excitation is resonant to the heavy-hole exciton transition of the QW which is about 10 meV below the onset of the band-to-band continuum. For a lattice temperature of 100 K and weak excitation, real space transfer in the QW is dominated by excitonic transport [44]. The time scale of transport is several hundred picoseconds, much longer than the formation and energy relaxation times of excitons, i.e. excitons undergoing real-space transfer form a quasi-equilibrium energy distribution close to lattice temperature. In such a case, an isothermal drift-diffusion model is appropriate to describe the spatially resolved exciton dynamics. Exciton transport is described by the two-dimensional particle current density $\vec{j}(\vec{r},t) = \vec{j}_{\mathrm{diff}}(\vec{r},t) + \vec{j}_{\mathrm{drift}}(\vec{r},t)$ with a diffusion term $\vec{j}_{\mathrm{diff}}(\vec{r},t) = -D_{\mathrm{ex}}\vec{\nabla}n(\vec{r},t)$ induced by the gradient of the exciton concentration $n(\vec{r},t)$ ($\vec{r} = (x,y)$, D_{ex}: exciton diffusion coefficient) and a drift term $\vec{j}_{drift}(\vec{r},t) = -\mu_{ex} n(\vec{r},t)\vec{\nabla}U(\vec{r})$ induced by the action of the local band gap gradient $\vec{\nabla}U(\vec{r},)$ on the center of mass motion of the exciton (μ_{ex}: equivalent charged particle mobility of the exciton) [45]. Under the conditions of our experiment, μ_{ex} is linked to the to the diffusion coefficient D_{ex} by the Einstein relation $\mu_{ex} = eD_{ex}/kT$ (e: electron charge, k: Boltzmann constant). The spatio-temporal evolution of the exciton concentration $n(\vec{r},t)$ is described by the two dimensional continuity equation, including a generation, a diffusion, a drift, and a recombination term:

$$\begin{aligned} \frac{\partial n(\vec{r},t)}{\partial t} &= g(\vec{r},\vec{r}_0,t) + D_{ex}\Delta n(\vec{r},t) + \mu_{ex} n(\vec{r},t)\Delta U(\vec{r}) + \\ &\quad \mu_{ex}\vec{\nabla}n(\vec{r},t)\vec{\nabla}U(\vec{r}) - n(\vec{r},t)/\tau(\vec{r}) \end{aligned} \tag{4.2}$$

Here, the spatial variation of the band gap $U(\vec{r})$ along the lateral y direction, perpendicular to the wire axis, is directly taken from the confinement potential in Fig. 6. Note that $U(\vec{r})$ does not depend on the x coordinate, i.e. the direction parallel to the wire axis. The exciton lifetime $\tau(\vec{r})$ has a value of 1.5 ns for $|y| < 50$ nm (QWR region) and 1.35 ns for $|y| >$ 50 nm (QW region). For the generation term $g(\vec{r},t)$, we use a Gaussian shape in time (FWHM of 1 ps) and space (FWHM of 300 nm) centered at $\vec{r} = (0,y)$. The intensity of the time-resolved QWR luminescence is proportional to $I(t) = \int dx \int dy \partial n(\vec{r},t)/\partial t$, where the integration is performed over the length of the QWR along the x-axis and over the width of the QWR region along the y-axis. For comparison with experiment, $I(t)$ is convoluted with the temporal response function of the photodetector.

In Fig. 8, the results of such simulations for different tip positions y are compared to the experimental results. Assuming an exciton diffusion coefficient D_{ex} = 13 cm^2/s, the calculated temporal dependence of the QWR luminescence is in good agreement with experiment for excitation on both mesa top and bottom. This diffusion coefficient corresponds to an excitonic mobility $\mu_{ex} = 1\,500$ cm^2/Vs in the QW, given mainly by the 2D hole mobility and limited in this temperature range by LO phonon scattering. In particular, the model calculation correctly describes the influence of the lateral band gap variation $U(\vec{r})$ on the exciton transport. In the region outside the barriers, $U(\vec{r})$ exerts a force on the excitonic center of mass motion that opposes the diffusive real space transfer towards the QWR. This is manifested in the experimental data as *(i)* a significantly weaker QWR luminescence for mesa bottom ($y < 0$) than mesa top ($y > 0$) excitation due to the higher barrier on the mesa bottom, and *(ii)* a slight shift of the temporal position of the maximum of the QWR emission t_{max} to shorter delay times on the mesa top. Both on the mesa top and bottom, the calculated t_{max} are in good agreement with the measured values (see Fig. 10), indicating that the exciton diffusion constants are similar on both sides of the wire. The pronounced influence of the drift current relative to the diffusion current becomes even more important at lower temperatures, where the ratio between barrier height and thermal energy increases. This is demonstrated in a similar time-resolved experiment at a sample temperature of 10 K (Fig. 11). Again the sample is locally excited by transmitting fs pulses that are resonant to the excitonic QW absorption band and time-resolved QW luminescence is detected at the center of the QW emission band. At this temperature QWR luminescence is only observed for excitation in a narrow region around $y = 0$. The intensity of the time-integrated QWR luminescence is found to decrease monotonously with increasing separation of excitation tip and QWR. The spatial $|y|$ dependence of the time-integrated luminescence fits to a Gaussian profile with a full width at half maximum

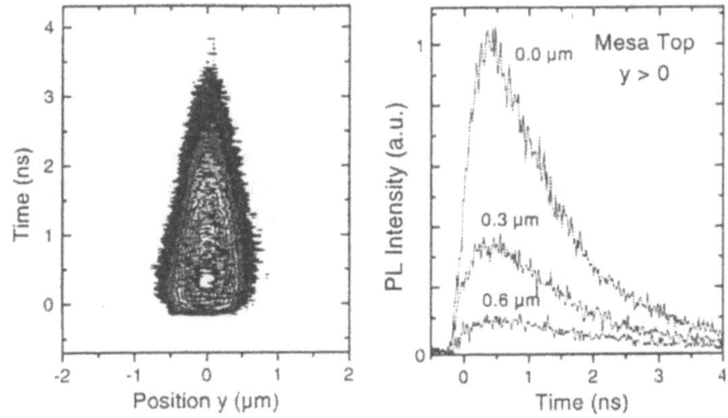

Figure 11 (a) Temporally resolved QWR photoluminescence at a temperature of 8 K after localized excitation in the QW. Excitation conditions as in Fig. 7. The luminescence intensity is plotted as a function of the lateral distance y between excitation and QWR position and as a function of time. (b) Time evolution of the QWR luminescence after excitation on the mesa top. The luminescence intensity at 1.545 eV is plotted versus time for different excitation positions.

of 570 nm [46]. At each excitation position, the temporal dependence of the QWR luminescence can be described by a single-exponential decay with a decay constant of 0.8 ns, the population lifetime of the QWR. No shift of the temporal position of the maximum PL intensity is observed within the accuracy of the experiment (±50 ps). This indicates that at low temperatures (T=10 K) exciton transfer from outside the barriers into the QWR is fully suppressed and that QWR luminescence is mainly observed for excitation in between the maxima of the energetic barriers separating QWR and QW. This behavior is also evident from the PLE spectrum in Fig. 3 recorded with steady-state excitation. The fact that we observe no temporal shift of the PL maximum suggests that at this temperature diffusive exciton transport is of minor importance for real-space transfer and that the trapping of excitons generated in the region between the barriers is greatly accelerated due to the bandgap-induced drift motion.

In conclusion, we presented a study of the nanoscopic optical properties of a novel quantum well embedded quantum wire structure. The high spatial resolution of the NSOM technique allows us to separate the optical spectra of QWR and QW and provides evidence for a quasi-one-dimensional carrier confinement. Local, subwavelength variations of the bandgap energy are quantitatively mapped and their influence on the real-space exciton transport into single GaAs quantum wires is identified by using picosecond time-resolved near-field spectroscopy. The experiments demonstrate that through combination of near-field spectroscopy with time-resolved detection and excitation schemes, direct access to the spatio-temporal carrier dynamics in single nanostructures becomes possible. In particular the extension of the temporal resolution towards the femtosecond regime offers exciting new perspectives for our understanding of transport and relaxation processes in low-dimensional semiconductors.

5 Acknowledgments

The author wishes to thank Alexander Richter, Doris Heinrich, Gerd Behme, Marko Süptitz and Monika Tischer for their invaluable contributions to this project and in particular Thomas Elsaesser for continuous support and fruitful discussions. The high-quality QWR samples have been provided by Richard Nötzel, Manfred Ramsteiner and Klaus Ploog from the Paul-Drude-Institut in Berlin. Financial support by the Deutsche Forschungsgemeinschaft (SFB 296) is gratefully acknowledged.

Bibliography

[1] For a recent overview, the reader is referred to the proceedings on the 23rd International Conference on the Physics of Semiconductors, M. Scheffler, R. Zimmermann (Eds.), World Scientific, Singapore 1996.

[2] N. Kirstaedter, N. N. Ledentsov, M. Grundmann, D. Bimberg, V. M. Ustinov, S. S. Ruvimov, M. V. Maximov, P. S. Kopev, Zh. I. Alferov, U. Richter, P. Werner, U. Gösele, and J. Heidenreich, Electron. Lett. **30** (1996) 1416.

[3] S. Benner and H. Haug, Phys. Rev. B **47**, (1993) 15750.

[4] S. Glutsch and D. S. Chemla, Phys. Rev B **53**, (1996) 15902; S. Glutsch and F. Bechstedt, Phys. Rev. B **47**, 4315 (1993); **47**, 6385 (1993).

[5] H. Sakaki, Japan. J. Appl. Phys. **19**, L735 (1980).

[6] U. Bockelmann and G. Bastard, Phys. Rev. B **42**, 8947 (1990).

[7] M. E. Rubin, G. Medeiros-Ribeiro, J. J, O'Shea, M. A. Chin, E. Y. Lee, P. M. Petroff, and V. Narayanamurti, Phys. Rev. Lett. **77**, 5268 (1996)

[8] C. Eder, J. Smoliner, and G. Strasser, Appl. Phys. Lett. **68**, 2876 (1996).

[9] J. F. Ryan, A. C. Maciel, C. Kiener, L. Rota, K. Turner, J. M. Freyland, U. Marti, D. Martin, F. Morier-Gemoud, and F. K. Reinhart, Phys. Rev. B, **53**, R4225 (1996)

[10] M. Walther, E. Kapon, J. Christen, D. M. Hwang, and R. Bhat, Appl. Phys. Lett. **60**, 521 (1992).

[11] D.W. Pohl, W. Denk, and M. Lanz, Appl. Phys. Lett. **44**, 651 (1984)

[12] E. Betzig, and J.K. Trautman, Science **257**, 189 (1992).

[13] H. F. Hess, E. Betzig, T. D. Harris, L. N. Pfeiffer, K. W. West, Science, **264**, 1740 (1994).

[14] R. D. Grober, T. D. Harris, J. K. Trautman, E. Betzig, W. Wegscheider, L. Pfeiffer, and K. West, Appl. Phys. Lett. **64**, 1421 (1994).

[15] T. D. Harris, D. Gershoni, R. D. Grober, L. Pfeiffer, K. W. West, and N. Chand, Appl. Phys. Lett. **68**, 988 (1996).

[16] J. Levy, V. Nikitin, J. M. Kikkawa, A. Cohen, N. Samarth, R, Garcia, and D. D. Awschalom, Phys. Rev. Lett. **76**, 1948 (1996).

[17] A. Richter, G. Behme, M. Süptitz, Ch. Lienau, T. Elsaesser, M. Ramsteiner, R. Nötzel, and K.H. Ploog, Phys. Rev. Lett. **79**, 2145 (1997).

[18] Ch. Lienau, A. Richter, G. Behme, M. Süptitz, D. Heinrich, T. Elsaesser. M. Ramsteiner, R. Nötzel, and K.H. Ploog, Phys. Rev. B **58**, 2045 (1998).

[19] A. Richter, M. Süptitz, D. Heinrich, Ch. Lienau, T. Elsaesser, M. Ramsteiner, R. Nötzel, and K.H. Ploog, Appl. Phys. Lett., in press (1998).

[20] J. B. Stark, U. Mohideen, E. Betzig, and R. E. Slusher, in: Ultrafast Phenomena IX, Springer Series in Chemical Physics, Berlin, 349 (1996).

[21] R. Nötzel, M. Ramsteiner, J. Menniger, A. Trampert, H.-P. Schönherr, L. Däweritz and K. H. Ploog, Jpn. J. Appl. Phys. **35**, L297 (1996).

[22] R. Nötzel, J. Menniger, M. Ramsteiner, A. Ruiz, H.-P. Schönherr, and K. H. Ploog, Appl. Phys. Lett. **68**, 1132 (1996).

[23] R. Nötzel, M. Ramsteiner, J. Menniger, A. Trampert, H.-P. Schönherr, L. Däweritz, and K. H. Ploog, J. Appl. Phys. **80**, 4108 (1996).

[24] E. Betzig, J. K. Trautman, T. D. Harris, J. S. Weiner, and R. L. Kostelak, Science **251**, 1468 (1991).

[25] G. Behme, A. Richter, M. Süptitz, and C. Lienau, Rev. Sci. Instrum. **68**, 3458 (1997).

[26] K. Karrai and R. D. Grober, Appl. Phys. Lett. **66**, 1842 (1995).

[27] E. Betzig, P. L. Finn, and J. S. Weiner, Appl. Phys. Lett. **60**, 2484 (1992).

[28] A. Richter, M. Süptitz, Ch. Lienau, and T. Elsaesser, Surf. and Interface Analysis **25**, 583 (1997).

[29] F. Rossi and E. Molinari, Phys. Rev. Lett. **76**, 3642 (1996).

[30] E. Kapon, D. M. Hwang, and R. Bhat, Phys. Rev. Lett. **63**, 430 (1989).

[31] F. Vouilloz, D. Y. Oberli, M.-A. Dupertuis, A. Gustaffson, F. Reinhardt, and E. Kapon, Phys. Rev. Lett. **78**, 1580 (1997).

[32] G. Goldoni, F. Rossi, E. Molinari, and A. Fasolino, Phys. Rev. B. **55**, 7110 (1997).

[33] G. W. Bryant, Appl. Phys. Lett. **72**, 768 (1998).

[34] R. Nötzel, N. N. Ledentsov, L. Däweritz, K. Ploog, and M. Hohenstein, Phys. Rev. B. **45**, 3507 (1992).

[35] M. Wassermeier, J. Sudijono, M. D. Johnson, K. T. Leung, B. G. Orr, L. Däweritz, and K. H. Ploog, Phys. Rev. B **51**, 14721 (1995).

[36] D. Gammon, E. S. Snow, B. V. Shanobrook, D. S. Katzer, and D. Park, Phys. Rev. Lett. **76**, 3005 (1996); Science **273**, 87 (1996).

[37] Ch. Lienau, A. Richter, G. Behme, M. Süptitz, T. Elsaesser, M. Ramsteiner, R. Nötzel, and K. H. Ploog, phys. stat. sol. (b) **206**, 153 (1998).

[38] for a review see: J. Shah, *Ultrafast spectroscopy of semiconductors and semiconductor nanostructures*, Springer, Berlin, 1996.

[39] H. Haug, and S. W. Koch, *Quantum Theory of the Optical and Electronic Properties of Semiconductors*, World Scientific, Singapore, 1997.

[40] The change in quantum well thickness near the QWR is correlated with an increase in inhomogeneous broadening. This is directly observed as an increase in the spectral width of the $n = 0$ exciton absorption peak.

[41] D. Gershoni, M. Katz, W. Wegscheider, L. N. Pfeiffer, R. A. Logan, and K. West, Phys. Rev. B. **50**, 8930 (1994).

[42] H. Akiyama, S. Koshiba, T. Someya, K. Wada, H. Noge, Y. Nakamura, T. Inoshita, A. Shimizu, and H. Sakaki, Phys. Rev. Lett. **72**, 924 (1994).

[43] A. C. Maciel, C. Kiener, L. Rota, J. F. Ryan, U. Marti, D. Martin, F. Morier-Gemoud, and F. K. Reinhard, Appl. Phys. Lett. **66**, 3039 (1996).

[44] H. Hillmer, A. Forchel, C. W. Tu, Phys. Rev. B **45**, 1240 (1992); and references therein.

[45] M. A. Tamor, and J. P. Wolfe, Phys. Rev. Lett. **44**, 1703 (1980).

[46] This width is about a factor of 2 larger than the spatial resolution in the experiment that is independently determined by mapping the QWR luminescence for near-resonant QWR excitation.

Lorentz Microscopic Investigations of Micromagnetic Structures

Josef Zweck

Institut für Experimentelle und Angewandte Physik, Universität Regensburg, D-93040 Regensburg, Germany

Abstract: Magnetic materials are characterized mainly by their response to external magnetic fields, i.e. by hysteresis loops. These hysteresis loops are an averaged signal for the specimen, and the specimen is in most cases macroscopic. For new applications in microsystem components, the lateral dimensions of magnetic functional units will soon reach 0.1 μm and less, thus being in the dimension of micromagnetic structures such as local induction fluctuations. Therefore, severe influence of the lateral dimensions on the physical (magnetic) behaviour can be expected. Lorentz microscopy is a technique which can give the necessary information on micromagnetic properties at high spatial resolution. Some experimental results will be presented, which show both the capability of Lorentz microscopy to resolve the micromagnetic structure of modern multilayered thin films and also the necessity to look at the micromagnetic structure at a microscopic scale.

1 Introduction

Until recently, magnetic materials have been characterised mainly by the measurement of their response function to an applied external magnetic field, i.e. the measurement of their hysteresis loops. From these hysteresis loops, it is fairly easy to derive the most commonly used characteristic magnetic parameters, such as coercivity field H_c, anisotropy constant H_k and permeability χ, which give a good idea of the averaged magnetic properties over the whole volume of the specimen. This approach has served very well for many years to characterize magnetic materials and to judge for which purpose they may be useful. However, this worked only well because the magnetic materials used so far have been generally larger than the micromagnetic fluctuations within the specimen. Even in the case of magnetic tapes for data storage the mean diameter of an information unit ("bit") is large compared to the size of the magnetic particles of the usually granular magnetic recording medium. Therefore, the information is stored by several magnetic particles, and the readout signal is an average over the magnetisations of these individual particles. Things become different, when modern hard disks are considered: They are continuous media (generally thin magnetic films), and the information is stored by the local orientation of the magnetisation, written into the medium as a small magnetic domain of a certain orientation. Using so-called vertical recording media, sub-micrometer dimenbions of the "bits" have been sucessfully achieved. With domains so small, a new problem turns up: Depending on the size and magnetic surroundings (i.e. neighbouring bits) as well as on the materials used, the domains may become unstable or deviate from a nice circular shape, which causes increased noise in the readout signals. Here, the importance of a micromagnetic characterisation becomes obvious. Only, when the general micromagnetic behaviour of the material can be measured and characterised, a prediction can be made with respect to the usability of the material under consideration for certain purposes. Beyond that, the improvement of magnetic functional materials can be achieved easiest, when the changes in micromagnetic behaviour due to a modification of the material (i.e. production process, composition, etc.) can be observed directly and attributed to the modification. The micromagnetic characterisation will become even more important for future micro- and nanostructured functional magnetic units in microsystems such as on-chip magnetic sensors or magnetic random access memories (MRAMs) for computer applications. These functional units will necessarily have dimensions well below 100 nm, and therefore be exactly in the range of micromagnetic inhomogeneities. A macroscopic approach to the magnetic design of such devices is very likely to fail. For example, up to now it is not completely clear, whether below a certain critical size of a magnetic particle (either genuine or nanostructured) a single domain state will be the minimum energy state or not. On the other hand, the magnetic switching behaviour will be influenced by the existence or non-existence of magnetic domain walls, because the switching will be dominated by wall movements or homogeneous rotation, respectively. Additionally, the stability of the magnetic state will also depend on the micromagnetic properties and also the interaction of the magnetic unit with surrounding magnetic structures. In the following, the use of Lorentz microscopy for the microscopic characterisation of magnetic materials will be demonstrated by several examples of modern magnetic materials such as $[Co/Cu]_N$ multilayers (GMR media) and $[Fe/Tb]_N$ or $[Fe/Cu/Tb]_N$ multilayers for vertical magnetic recording.

2 Experimental Technique for Micromagnetic Characterisation: Fresnel Imaging

With the Fresnel technique, the specimen is imaged in a defocused condition of the Lorentz lens, which acts as a long focal length objective lens with a negligible magnetic field at the specimen's location [1]. Due to the Lorentz force that acts on the electrons when passing through a specimen of (ideally homogeneous) in plane induction component B_0 over a thickness t of the specimen, the electron becomes deflected by an angle β_L, the Lorentz angle (Fig. 1a).

If now the imaging conditions are chosen in such a way that the object plane does not coincide with the real object (as it is usually done in Gaussian focus condition), a plane above or below the specimen at a distance $+\Delta f$ is being imaged, where partial waves either converge or diverge. As can be seen from Fig. 1b, the converging partial waves create basically an increase of electron density (i.e. measurable intensity), in the case of sufficient coherence accompanied by biprism interference fringes (not shown here). In the case of diverging partial waves, the corresponding areas become darker than the average intensity. Therefore, domain walls are visible as bright or black lines, which change contrast when the defocus is changed from overfocus ($+\Delta f$) to underfocus ($-\Delta f$). A detailed explanation for these features can be found in [1,2].

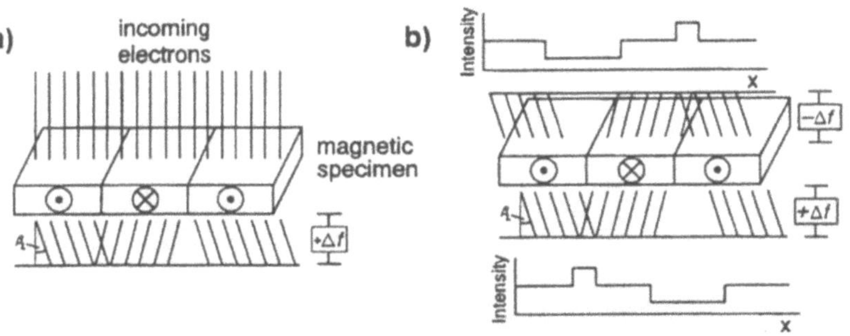

Figure 1 Schematic representation of Fresnel imaging of magnetic structures. a: The incoming electrons become deflected within the specimen by the Lorentz angle β_L. The deflected partial waves converge diverge. b: In a plane at a distance $+\Delta f$ behind the specimen's exit plane, the convergent and divergent partial waves create increased or decreased intensity compared to the average intensity, giving the locations of domain walls as "black" or "white" lines. The same is possible for a defocus value $-\Delta f$ above the specimen's plane, the contrast is, however, inverted.

3 Experimental Results

In this section, experimental results will be presented, that can be obtained for magnetic thin films of recent interest, such as thin films showing the Giant Magnetoresistance effect (GMR) or of films which serve the investigation of the coupling behaviour and the range of magnetic coupling between different atomic species such as Fe and Tb in Fe/Tb multilayers with and without a nonmagnetic spacer layer in between the magnetic layers. It will also be shown, that the electron microscope can serve as a microlab, which permits the variation of the ambient temperature and/or the external magnetic fields that influence the specimen. It is therefore ideally suited for an investigation of the micromagnetic response behaviour of magnetic thin films, where the micromagnetic effects that in total generate the macroscopic magnetic behaviour can be observed directly.

3.1 Magnetisation Reversal in $[Co/Cu]_N$ Multilayer Films

$[Co/Cu]_N$ multilayers show an either ferromagnetic or antiferromagnetic coupling, depending on the thickness of the nonmagnetic Cu spacer layer [3–5]. For example, in the case of a 1.6 nm Cu layer thickness, the coupling is ferromagnetic, while for a 2.2 nm spacer layer thickness, the coupling is mostly antiferromagnetic [8]. For other spacer layer thicknesses there are again ferromagnetic, antiferromagnetic and mixed ferro/antiferromagnetic states observeable. Since the GMR effect is found only with antiferromagnetically coupled $[Co/Cu]_N$ multilayers, the difference in magnetisation behaviour between ferro- and antiferromagnetically coupled multilayer systems is of interest. Besides that general interest, it was found that the antiferromagnetically coupled thin film system, which showed the maximum GMR, showed a distinct "kink" in the hysteresis loop at ± 86 Oe (see Fig. 2b) , which could not be explained from the hysteresis loop alone. This kink was missing in the case of the minimum GMR effect (Fig. 2a), which indicates that it is a property directly related to the GMR effect itself. Below, the magnetisation behaviour will be shown as a series of Lorentz microsccopy images for both typs of specimens.

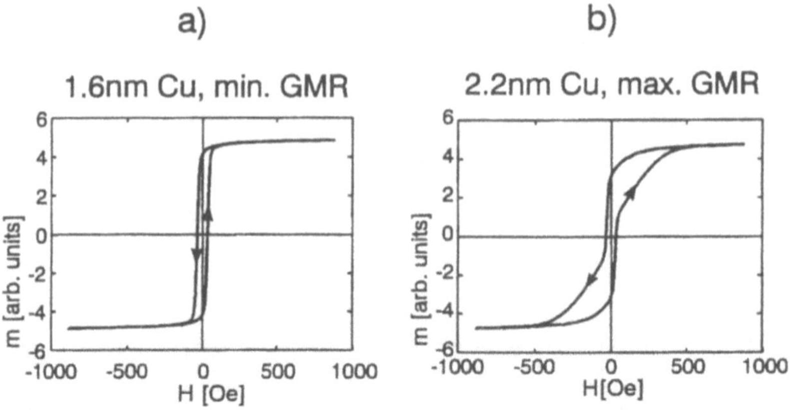

Figure 2 Comparison of the hysteresis loops for $[Co/Cu]_N$ multilayers with a Cu spacer layer thickness between the Co layers of a: 1.6 nm (minimum GMR) and b: 2.2 nm (maximum GMR).

Magnetisation Behaviour for $[Co/Cu]_N$ Multilayers in the GMR Minimum.

The magnetic hysteresis loop of the multilayer system in the GMR minimum state exhibits a standard behaviour as expected from any soft magnetic material, as shown in Fig. 3. No unusual and unexplainable details are found in the hysteresis loop. The series of Lorentz microscopic images shown in Fig. 3 illustrate the behaviour during a half cycle of a hysteresis loop for this multilayer system. Starting point is a state relatively close to saturation at 175 Oe, where the film system exhibits a fairly homogenous state of magnetisation with only light ripple structures.

The starting field is sufficiently high to have the specimen close to magnetic saturation, and consequently only the micromagnetic fine structure, the so-called magnetic ripple [7] is visible. The direction of the external magnetic field and of the local induction of the specimen is indicated by black and white arrows, respectively. When the external field is reduced to zero, i.e. the specimen is in its remanent state, slightly more pronounced fine structures become visible, with a larger mean angle between the induction directions of adjacent areas. When in succession an increasing reverse field is applied, the local induction direction rotates more and more, until at approx. -45 Oe the first 180° walls (Bloch walls) are being formed. With increasing external field, this process continues, always turning the local induction more and more parallel to the external field, until for approx. -122 Oe a nearly saturated state is reached again, with no walls present.

Magnetisation Behaviour for $[Co/Cu]_N$ Multilayers in the GMR Maximum.

As already mentioned above, the magnetic hysteresis loop of the multilayer system in the GMR maximum state shows a "kink" in the otherwise smooth function, which is located at ± 86 Oe (see Fig. 4a). This kink coincides with the maximum of the GMR effect $\frac{\Delta R}{R}$ when plotted as a function of the applied external field, as in Fig. 4b.

This indicates that the kink and the GMR are related, a clear explanation can, however, not be derived from the hysteresis loop alone. When the same hysteresis half cycle as for

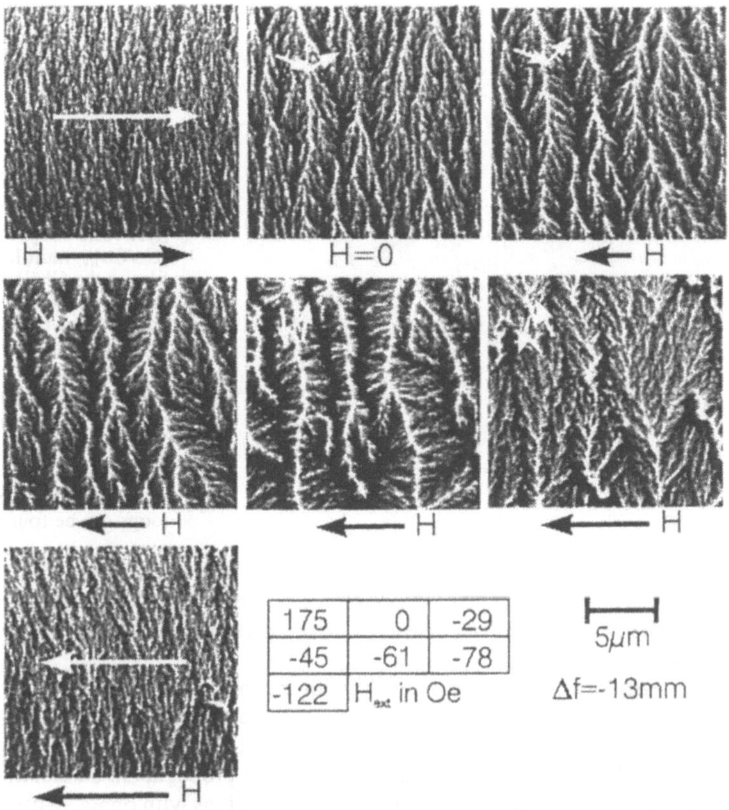

Figure 3 Fresnel images of a magnetisation reversal cycle of a $[Co/Cu]_N$ multilayer system in the minimum GMR state (Defocus: $\Delta f = -13$ mm). External fields are given in the table below the images, black arrows below the images indicate the direction and strength of the external applied field, white arrows show the local direction of induction within the specimen. The micromagnetic changes can be described by local rotation of the induction, and the result is a fairly homogeneous state in the reversed direction.

the multilayer system in the GMR minimum is looked at, the initial image is quite similar (Fig. 5, H_{ext} = +204 Oe).

Again, the starting point is a saturated specimen with a fairly homogeneous induction. Upon reduction of the external magnetic field to zero, the remanent state shows an only slight increase in non-uniformity of the induction within the specimen together with slightly larger angles between adjacent directions of local induction, similar to Fig. 3. When a reverse field is applied, at first the same appears as for the minimum GMR's specimen case in Fig. 3, namely a coherent rotation of the local induction in order to align the local induction parallel to the external field. Also, Bloch walls are being formed for approx. -37 Oe to -49 Oe. When the external field is further increased, a different effect than before can be observed. The Bloch walls condense to form loop-shaped wall structures, which turn out to be 360° walls [9]. These 360° walls are stable at external fields, at which a minimum GMR specimen is already close to saturation again. Additionally, the field values that are sufficient to create the 360° domain walls correspond well to the

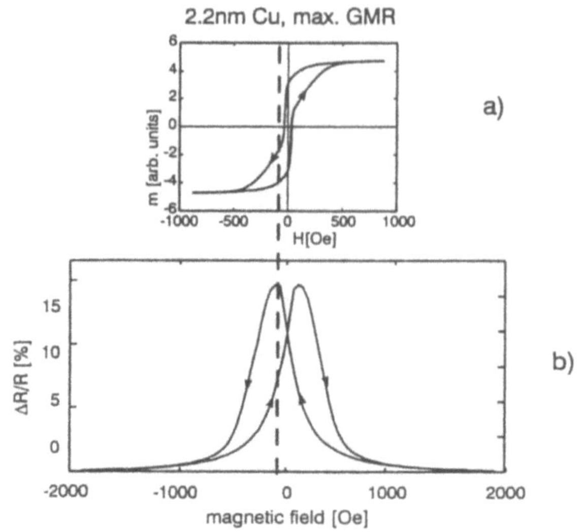

Figure 4
Comparison of a: the hysteresis loop of a $[Co/Cu]_N$ multilayer film in the GMR maximum, i.e. with a Cu spacer layer thickness of 2.2 nm, and b: the resistivity ratio $\frac{\Delta R}{R}$ vs. the applied magnetic field. The maxima of $\frac{\Delta R}{R}$ are at ±86 Oe, the same field value where the kinks in the hysteresis loops can be found.

fields, at which the maximum change of the electrical resistivity for these specimens occurs, as well as to the field values that correspond to the kinks in the hysteresis loop. This is shown in Fig. 4.

The coincidence of the maximum value of $\frac{\Delta R}{R}$ with the formation of 360° walls can be easily understood when the domain wall structure is looked at in detail. A 360° wall shows a rotation of the induction of 360° from one area within the specimen to another one, being separated by the wall [10]. For conduction electrons, which propagate in an electric field parallel to the thin film's plane, the domain walls contain regions with opposite directions of induction, which increases the possibility of spin dependent scattering, even without the necessity to change into a different magnetic layer. Therefore, an increase of the electrical resistance due to the occurrence of 360° walls results. From the contrast of the 360° walls one can assume that they extend completely through the multilayer system. When the average separation between the walls of 2.2 ± 0.1 μm and the mean width of the walls of 0.24 ± 0.01 μm is taken into account (after correction for the defocus applied [6]), a volume fraction of $11 \pm 1\%$ is occupied by the 360° walls. These regions within the specimen, where the direction of the induction changes locally by 360° may contribute considerably to the GMR effect and should be taken into account for any theories that describe the GMR for non-perfect (i.e. non-epitaxial) but application-related systems.

3.2 Micromagnetic Behaviour of a $[Fe/Tb]_N$ Multilayer System at Various Temperatures.

$[Fe/Tb]_N$ multilayer systems have been suggested as promising candidates for vertical magnetic recording with magnetooptical read/write technique. To increase the magnetic anisotropy perpendicular to the plane of the film, attempts have been made to use multilayers instead of alloys, where the density of oriented Fe-Tb pairs at the interface region of alternating layers is strongly increased compared to an alloy. The pairs are frequently considered to be the microscopic reason for the perpendicular anisotropy of these materials. The coupling behaviour between the adjacent Fe and Tb layers, and especially the

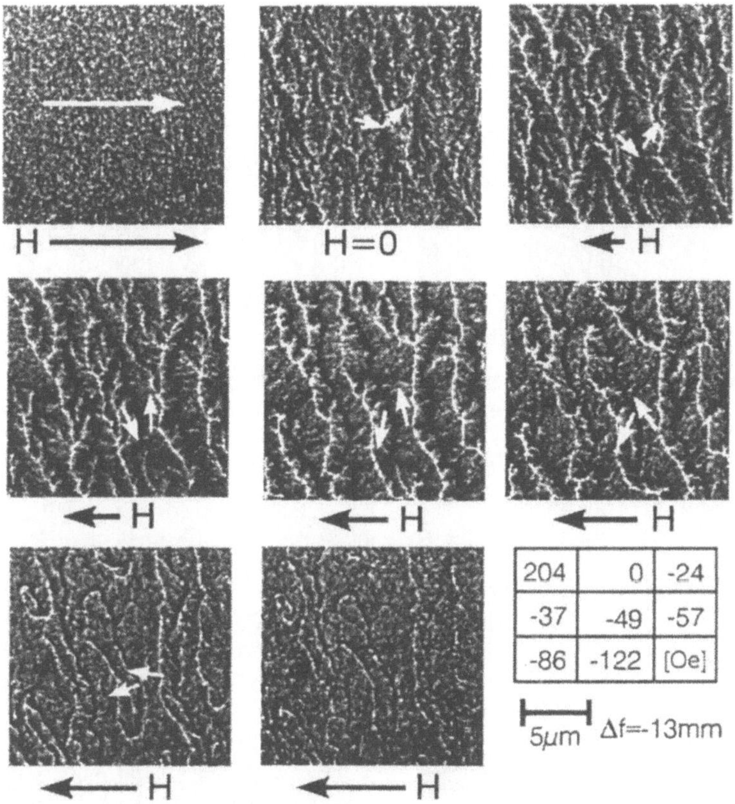

Figure 5 Fresnel images of a magnetisation reversal cycle of a $[Co/Cu]_N$ multilayer system in the maximum GMR state. (Defocus: $\Delta f = -13$ mm.) Applied external fields are given in the table below the images, black arrows below the images indicate the direction and strength of the external applied field, white arrows show the local direction of induction within the specimen. As in Fig. 3, the micromagnetic changes can initially be described by a local rotation of the induction. At a field value of -86 Oe, so-called 360° walls appear. These 360° walls are very stable, and remain even at $H_{ext} = -122$ Oe.

range of magnetic coupling between the individual layers are the subject of recent investigations. From these, it was found that a layer sequence of 1.5 nm Fe and 2.2 nm Tb gives the maximum magnetic coupling between the layers.

When there is an optimum coupling of the individual layers, an energetic balance is reached between the energy of magnetisation of the individual layers *and* their coupling energy. As can be seen below, the coupling is obviously reduced at higher temperatures. Therefore, micromagnetic imaging can help to understand the process of thermally induced decoupling of adjacent layers.

When a $[Fe/Tb]_N$ multilayer thin film is imaged by the Fresnel technique, at room temperature (298 K) regular domain walls (Bloch walls) can be found. In the images in Fig. 6, which show the same area of the specimen, but at different temperatures, certain features are marked by arrows. The left and bottom arrow point to a saw-like domain wall, which consists of Bloch walls of alternating sequence, therefore showing alternating white and black walls. The right arrow points to an island-like domain, which is limited on the

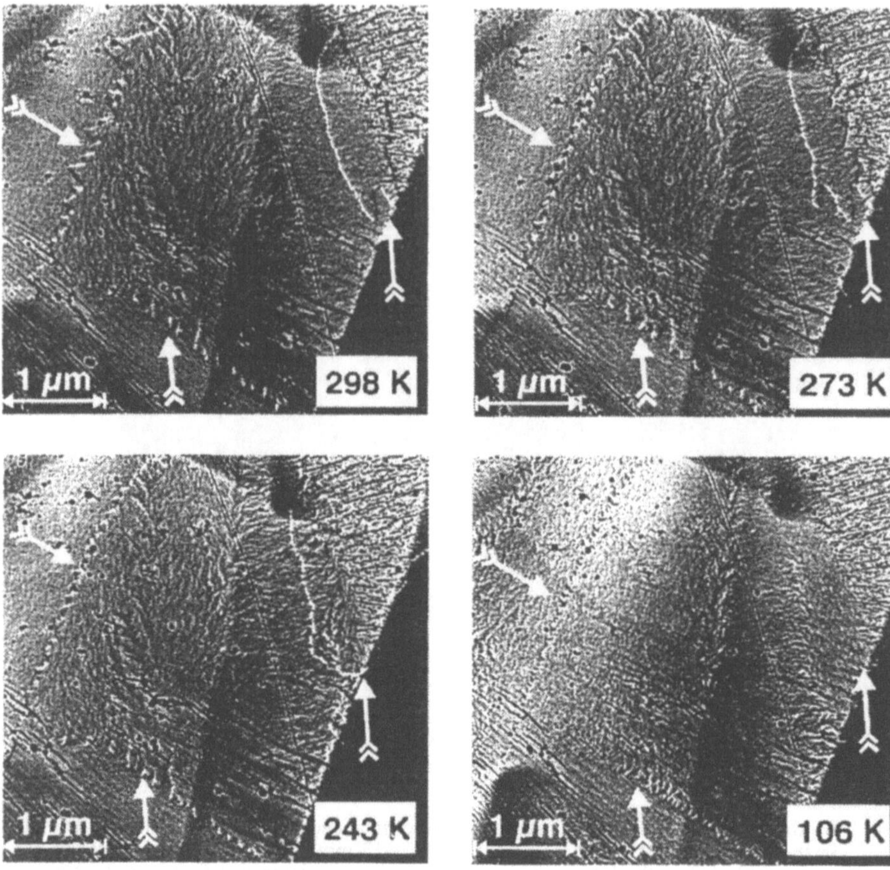

Figure 6 $[Fe/Tb]_N$ multilayer system showing various micromagnetic features at different temperatures between 298 K and 106 K. Equivalent areas in the images taken at different temperatures are marked by arrows.

left side by a white wall, and to the right side by a black wall (both Bloch type). This island-like domain shows on its lower end two finger-like features. After decreasing the specimen's temperature to 273 K, the saw-like domain wall shows less pronounced features. The island-like domain developed another, third "finger". Further decrease of the temperature leaves the saw-like domain wall basically unchanged, however the contrast is reduced. (This cannot be seen on the micrographs in Fig. 6, because for printing purposes the contrast has been enhanced.) The island-like domain now shows strong so-called magnetic ripple contrast, and the finger-like features disappeared. In the region, where initially the space between the "fingers" was located, strong ripple contrast has evolved. The black wall to the right disappeared. Finally, at 106 K, the saw-like domain wall is still visible, however again at reduced contrast. The island-like domain disappeared, while the former location of the white (left) domain wall still shows some magnetic contrast.

The changes found can be understood as follows:

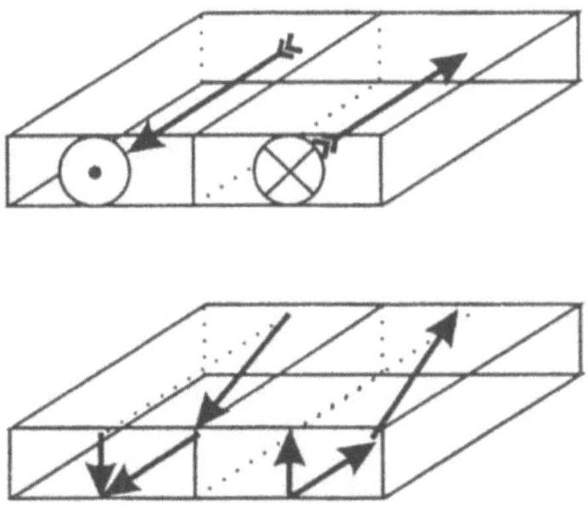

Figure 7
Schematic representation of the alignment of local adjacent moments within a multilayered thin film. The local moments are shown as an average over the film thickness, i.e. over a whole number of different layers. a: At room temperature, the moments in the individual layers are decoupled, leading to a basically in-plane induction distribution. b: At lower temperatures, the local moments tend to align themselves more and more perpendicular to the film plane, creating an out-of-plane component.

At room temperature, the magnetisation is basically in-plane, i.e. the dominant magnetic process within the multilayer stack is to align the magnetic moments in each layer individually, with basically no or only minor coupling between the adjacent layers. This means that the thermal fluctuations are stronger than (or in the order of) the coupling between the adjacent layers, resulting in an in-plane magnetisation. This is supported by the Bloch type domain walls, which show strong contrast, i.e. the components of the local induction are nearly perfectly aligned in the film plane (Fig. 7a).

When the temperature is decreased, the random thermal disturbances are reduced and the coupling will lead to an increase of the alignment of local moments perpendicular to the multilayer plane. This is faciliated by the tendency to align the magnetic moment along pairs of type Fe-Tb (Fig. 7b). Formerly well developed sections of Bloch walls (alternating black and white) along the saw-like wall become shorter and less clearly visible. Also, the contrast is reduced. This is an effect of the gradual alignment of the local moments in a direction perpendicular to the thin film's plane. If this occurs, the in-plane component of the local induction becomes reduced, leading to a reduced contrast. Additionally, the local moments seem to have a tendency to get "twisted" with each other, leading to a more diffuse image of the corresponding domain wall. This is most clearly seen in Fig. 6 (Temperatures 243 K and 106 K), where the initially present white domain wall of the island-like domain radually disappears in a ripple-like pattern, which is predominantly oriented in a plane perpendicular to the film plane.

4 Dynamic Observation of Domain Wall Movements in Individual Layers of a $[Fe/Cu/Tb]_N$ Multilayer Thin Film.

When the alternating layers consisting of Fe and Tb are separated by a 2.9 nm Cu layer, the coupling between the individual layers becomes reduced due to the larger separation. As a consequence, domain walls do not form anymore on top of each other and thus throughout the complete film, but may vary in their location from layer to layer. This can be seen in Fig. 8. Here, a series of images is shown that was taken in time intervals of 5 seconds between the individual exposures. The image taken at 0 s shows clearly several

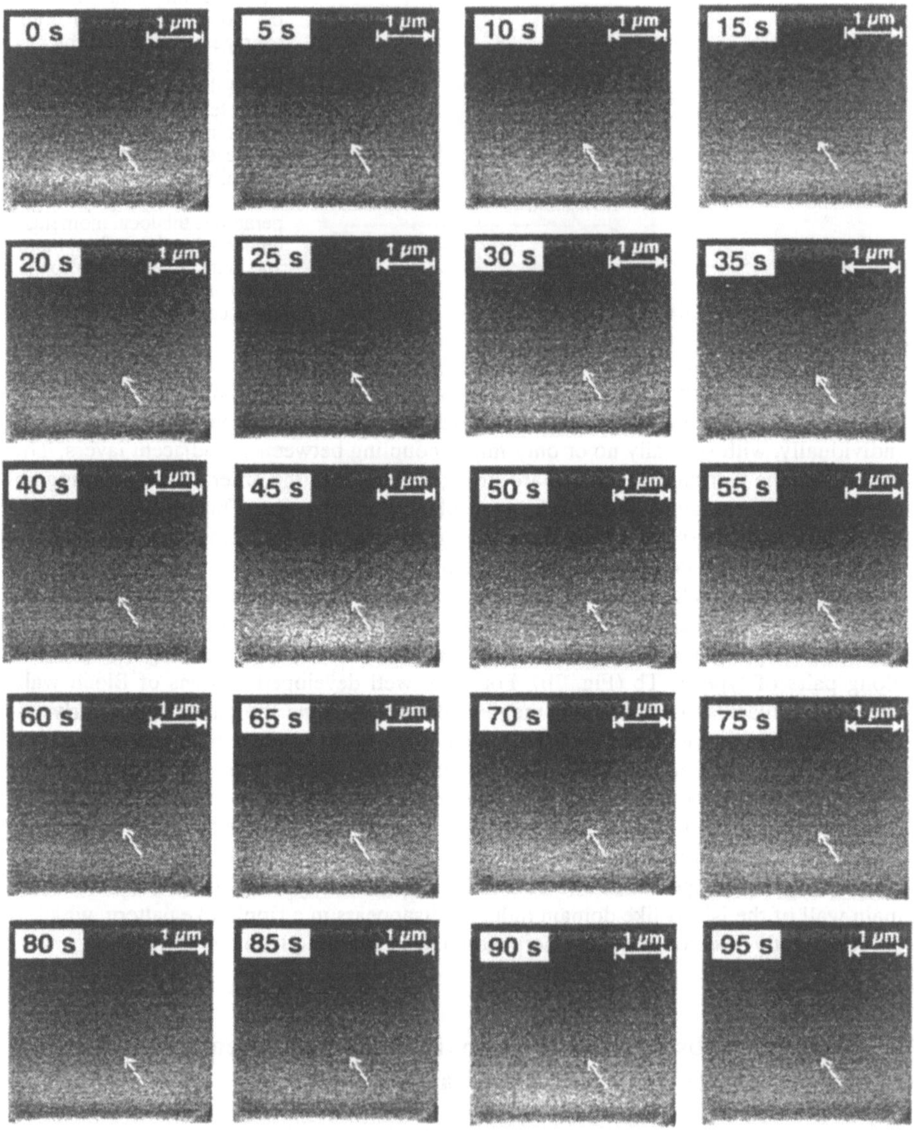

Figure 8 Series of Fresnel images taken at intervals of 5 seconds. Initially, several black walls can be seen with no white walls in between. A small external magnetic in-plane field is applied to the specimen, causing wall movement. Initially, a series of black walls move to the upper left of the image, until (65 s to 75 s) a white wall moves to the lower right.

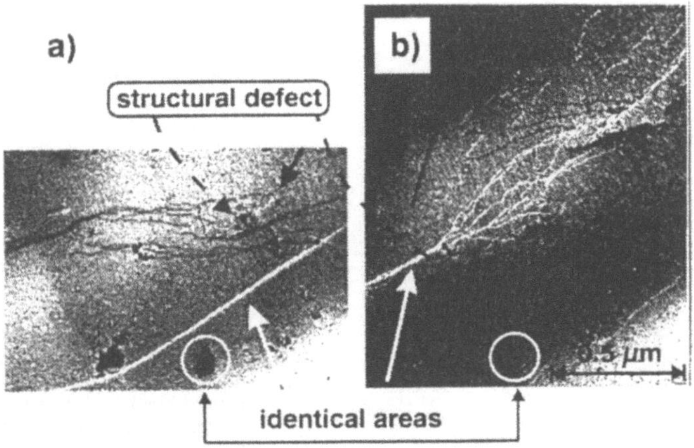

Figure 9 Identical areas of the same specimen ([$Fe_{1.5\ nm}$/$Cu_{2.9\ nm}$/$Tb_{2.2\ nm}$]$_N$), imaged by the Fresnel technique. Note the encircled identical areas in both images for comparison. a: In the upper part of the image, a split black wall can be seen. b: The black walls moved up, showing more splitting plus additional white domain wall sections. The previous solid white wall has now moved to the top, and started to split up into several white walls.

black walls (arrow), with no white wall in between them. From Fig. 1 it is clear, that this cannot be the case for a homogeneous thin film with alternating domains of magnetisation. One would find always alternating black and white walls. Therefore, the only explanation for the observation here is that in subsequent layers the domain walls are displaced with respect to each other, while every layer does have the same sequence of domains, which give rise to a black wall each. These black walls are, however, in different layers and thus in different depths of the multilayer film.

- At 0 seconds, several black walls can be seen, which move (5 s to 60 s) due to the presence of a small external magnetic field (in-plane). The initial changes are marked by an arrow. It can be seen, that the black walls move to the upper left part of the images.
- At 65 seconds, a white wall appears, which moves to the lower right (70 seconds) and then disappears.

A similar behaviour can be seen in Fig. 9: Again, a magnetic in-plane field has been applied to the specimen, causing domain wall movement. In this case, a white wall (lower part of the micrograph) can be observed, which is strong in contrast. Above the white wall, a split black wall can be seen, which starts as a single black wall in the left side of the image, then splits into different black walls close to a structural defect in the specimen (Fig. 9a). After some time, the black split wall moved upward, splitting even more severe. The black domain walls are now intersected by white domain wall sections. Also, the white wall, which was a solid white wall before, moved upward close to the structural inhomogeneity, which obviously caused a splitting of the white wall (Fig. 9b).

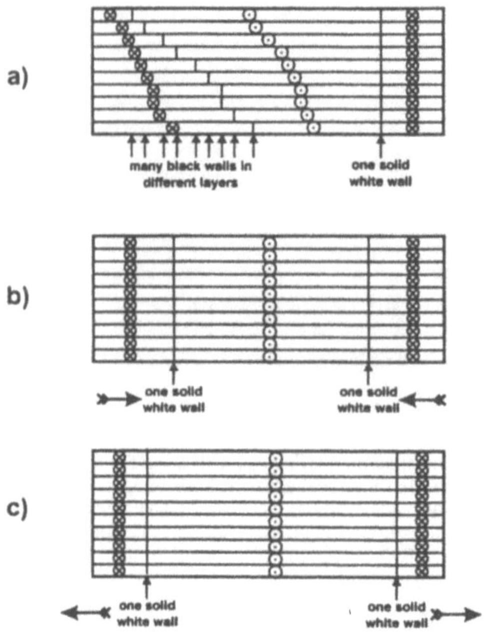

Figure 10
Schematic representation of the domain wall splitting in individual layers in a $[Fe_{1.5\ nm}/Cu_{2.9\ nm}/Tb_{2.2\ nm}]_N$ multilayer specimen. a: While the white walls are strongly coupled (right) and form one solid white wall, the black walls are less strongly coupled and form individual black walls in the corresponding layers. b: When an external magnetic field is applied, the black walls move until they meet, and form a solid black wall together. c: Further increase of the external field causes growth of the central area of the scheme, spreading the central magnetisation at the cost of the magnetisation of the neighbouring domains. This causes a movement of both the white and black walls in opposite directions, as indicated by arrows.

As long as no disturbance occurs, the walls show a tendency to stay together, although they are located in different individual layers of the multilayer sequence. However, as soon as slight disturbances occur, for example due to some defects in the thin films etc., the coupling between the individual layers and therefore also the walls is reduced and splitting occurs. The walls start to move independently, but still show a tendency to stay closely together. This is supported by the fact that the number of both the white and black walls in their split region are found to be seven. Since there are 10 layers of Fe and Tb each, nearly every individual double layer of Fe and Tb develops its own walls. While it is reasonable that the top and bottom layer show different coupling behaviour due to their asymmetric position, there seems to be one layer that exhibits a stronger coupling to the adjacent layer than the other ones. This may be due to thickness variations, as has been detected in cross sections of the specimens [10,11].

The behaviour of the system of magnetic layers, weakly coupled together, can be understood by the situation given schematically in Fig. 10.

In Fig. 10a, the white walls are strongly coupled (right) and form one solid white wall, the black walls are less strongly coupled and form individual black walls in the corresponding layers. (Compare this to Fig. 8 (0 s to 60 s).) Fig. 8b shows the situation when an external magnetic field is applied: the black walls move until the meet, and form a solid black wall together. In Fig. 8c, a further increase of the external field finally causes a growth of the central area, spreading the central magnetisation at the cost of the magnetisation of the neighbouring domains. This causes a movement of both the white and black walls outward, as indicated by arrows. (Compare this to Fig. 8, where a white wall moves away from the black walls towards the lower right due to the increase of the domain between the white and the black walls (65 s and 70 s)).

Obviously, the insertion of the non-magnetic spacer layers reduces the coupling of the individual layers and therefore the creation of domain walls of the same colour (black or white) in the individual layers, which are still weakly coupled due to the magnetic coupling of the correspondent domains.

5 Summary and Conclusions

In this article, it was shown that the imaging of micromagnetic features by high resolution Lorentz microscopy is possible under controllable external conditions such as external magnetic fields and/or temperatures. Lorentz microscopy is therefore able to offer a technique which can give direct images of the micromagnetic behaviour of modern magnetic materials under controllable experimental conditions. For a future micromagnetic understanding of the overall magnetic properties of magnetic materials, Lorentz microscopy can contribute to a more global understanding of macroscopic magnetic features as the average over an ensemble of micromagnetic local properties. In order to understand and describe the micromagnetic behaviour of magnetic media, it is no longer sufficient to describe the global properties, but to be aware of the locally varying properties. Only when this is taken into account, a detailed understanding of the behaviour of modern magnetic materials in the microscopic regime can be possible without the need for hypothetic assumptions, that give only indirect explanations instead of direct evidence as Lorentz microscopy does (as well as other highly resolving magnetic imaging techniques).

To illustrate the ability of Lorentz microscopy to image micromagnetic behaviour under controllable external conditions, several experiments have been described briefly:

- The micromagnetic behaviour of $[Co/Cu]_N$ multilayers both in the minimum and maximum of the GMR effect have been investigated. The results of the investigation were used to explain the coincidence of the maximum in the GMR curve ($\frac{\Delta R}{R}$ vs. H, compare with Fig. 4) with the creation of 360° walls, and therefore at least one microscopic reason for the increase of the field dependent resistance.
- The micromagnetic behaviour of a $[Fe/Tb]_N$ multilayer system at various temperatures was investigated. The change from in-plane magnetisation to gradual out-of-plane magnetisation for decreasing temperature could be imaged and explained.
- The splitting and movement of domain walls in individual magnetic layers was shown in a time series. The origin of the multiple black or white domain walls was explained. For the behaviour of the wall movement, a qualitative explanation could be given, too.

Bibliography

[1] J. Zweck, B. Bormans, Philips Electron Optics Bulletin **132**, 1 (1992).

[2] J.N. Chapman, J. Phys. D: Appl. Phys. **17**, 623 (1984)

[3] S.S.P. Parkin, R. Bhadra, K.P. Roche, Phys. Rev. Lett. **66**, (16) 2152 (1991)

[4] C. Dorner, M. Haidl, H. Hoffmann, J. Appl. Phys. **74**, (9) 5886 (1993)

[5] G. Binasch, P. Grünberg, F. Saurenbach, W. Zinn, Phys. Rev. B **39**, (7) 4828 (1989)

[6] J.N. Chapman, E.M. Waddell, P.E. Batson, R.P. Ferrier, Ultramicroscopy **4**, 283 (1979)

[7] H. Hoffmann, *Mikromagnetische Theorie der quasistationären Eigenschaften dünner Schichten* Thesis, München (1965)

[8] T. Zimmermann, J. Zweck, H. Hoffmann, Journ. Magn. Magn. Mater. **148**, 239 (1995)

[9] L.J. Heyderman, H. Niedoba, H.O.Gupta, I.B. Puchalska, Journ. Magn. Magn. Mater. **96**, 125 (1991)

[10] J. Zweck, T. Zimmermann, Th. Schuhrke, Ultramicroscopy **67**, 153–162 (1997)

[11] T. Schuhrke, J. Zweck, H. Hoffmann, Thin Solid Films **292**, 118–123 (1997)

Optical and Mobility Measurements of Alkali Earth Atoms and Ions in Superfluid Helium

M. Foerste, I. Baumann, U. Pritzsche, G. zu Putlitz, B. Tabbert, J. Wiebe, and C. Zühlke

Physikalisches Institut der Universität Heidelberg,
Philosophenweg 12, D-69120 Heidelberg, Germany

Abstract: This article is devoted to optical measurements of alkali earth atoms and mobility measurements of alkali earth ions in superfluid helium. Four topics serve as examples. The optical excitation spectrum of implanted Ca atoms is quantitatively explained with the bubble model. The lack of optical transitions in the recombination process of implanted ions with electrons is discussed for the case of Ba. In case of the Mg triplet system a prolonged lifetime and an unexpected emission profile of the lowest triplet states have been measured. Mobility measurements of the alkali earth ions Be^+, Mg^+, Ca^+, Sr^+ and Ba^+ are reported. These results raise the question if the defect structure of Be^+ may be a snowball like defect in contrast to the bubble like defect of the heavier alkali earth ions.

1 Introduction

The implantation of ions and atoms into superfluid helium has long been a target of different authors in order to develop a tool for the study of the interaction of point defects with the superfluid matrix. Early attempts have used radioactive sources in the superfluid and X-rays to obtain positive and negative charges [1–3]. Glaberson and Johnson utilised the heating of a coated wire to implant ions of alkali and alkali earth elements into the superfluid [4]. A seeded beam technique has been used by Gordon and Mezhov-Deglin [5]. A much more general method was developed by Reyher and Winnacker [6] in Heidelberg utilizing an evaporation device in combination with a flowing afterglow applicable for any elements which could be evaporated at reasonable temperatures. With the advent of powerful laser sources laser sputtering above the surface of the fluid or inside the fluid became more fashionable, in some cases combined with a neutralisation technique by electrons emitted from a tip in the superfluid [7–11]. All these techniques have promoted a larger number of studies of atomic and ionic defects in superfluid helium by different methods. This paper is devoted predominantly to the spectroscopic investigations of atoms in superfluid helium and the corresponding drift behaviour of the same ions. The goal of these studies is to classify the different defects in a model picture of either snowball like defects or bubble like defects, similar like the defect structure of a helium ion or an electron in the liquid [12,13]. Experiments of the Heidelberg group serve as example here. However, work of other groups in this field should be acknowledged. Their work can be found in a recent review [14].

The outline of this paper is the following: The first paragraph deals with the different experimental realizations for optical and mobility measurements in superfluid helium. The different techniques to implant alkali earth ions in the liquid will also be shown. Section 3 deals with a selection of four experimental results and their theoretical interpretation. It includes results of the optical excitation spectrum of Ca, of the recombination of ions with electrons and of the prolonged life time of the lowest magnesium triplet states. The temperature dependent mobility of alkali earth ions is described in the last section.

2 Experimental Realizations

2.1 Implantation Techniques

In the present experiments described here the material of interest can be evaporated in the focused light emitted from a powerful laser. A pulsed N_2 or Nd:YAG laser is focused onto the sample material. Due to the high temperature at the sample a plasma is produced [7,8]. The positive ions in this plasma can be drawn into the liquid with an applied electric field. At a temperature above 1.3 K helium atoms from the background gas are ionized always simultaneously with the probe ions during the sputtering process. Therefore the relative mobility of a foreign ion compared to the helium can be measured conveniently [15]. In the liquid a field emission tip at a high voltage emits a continuous stream of electrons [16]. The ions move towards the electrons. When they approach each other a recombination process takes place. The light of this recombination process can be measured with a monochromator-photomultiplier arrangement. The produced atoms stay for several hundreds of milliseconds in the optical detection region. During this time they can be investigated by laser induced fluorescence.

In the two following sections the detailed experimental set-up for optical [19] and mobility measurements [17,18] using the laser sputtering technique are discussed.

2.2 Experimental Set-Up for Optical Investigations

The experiments are carried out in a ^{4}He bath cryostat working at saturated vapour pressure in a temperature range of $T = 1.2\,\mathrm{K}$ to $4.2\,\mathrm{K}$. A cube-shaped chamber is submerged in the liquid helium bath. It contains the ion source, electrically charged grids and a field emission tip at the bottom of the cell to emit electrons (Fig. 1). The chamber is filled with helium and the helium level can be regulated by a magnetic valve at the bottom of the cell with a precision of 0.5 mm. This opens the possibility to use the laser sputtering techniques in the gas phase as well as in the liquid phase. After the laser sputtering the produced ions were drawn towards the field emission tip by an electric field. The recombination process of an ion and an electron happens if the applied electric field strength is smaller than the electrostatic attraction of the two charged particles. The light emitted by the recombination process is detected in a scanning range of 200 nm to 900 nm by an arrangement of a 50 cm grid monochromator and a photomultiplier [19]. The measured recombination spectra exhibit the common feature that the emission lines are fairly symmetric and only slightly shifted, but visibly broadened compared to the vacuum values.

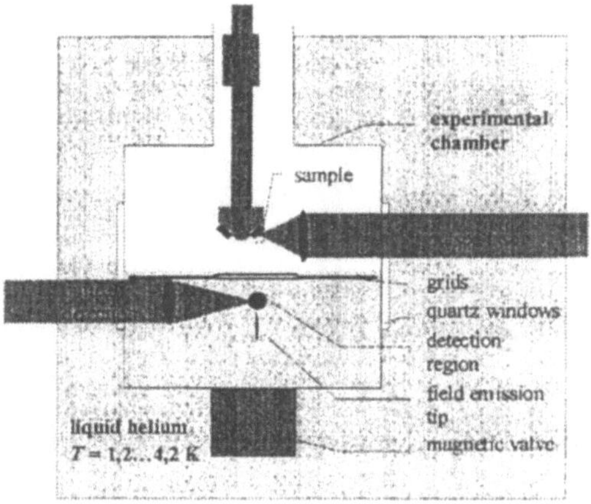

Figure 1
Experimental chamber for spectroscopic measurements in superfluid helium. The set-up allows the production of foreign ions in the gas phase as well as in the liquid phase due to a magnetic valve at the bottom of the chamber [19].

After the recombination the produced atoms were no longer influenced by the applied electric field. Hence the atoms become trapped in the liquid matrix for more than 100 ms. The only loss of atoms out of the optical detection region can be attributed to the convective flow in the liquid. Due to this long trapping time laser induced fluorescence experiments can be done with the implanted atoms. In Sections 3.1, 3.2 and 3.3 some experimental results obtained with such an experimental set-up will be shown.

2.3 Experimental Set-up for Mobility Measurements

Besides the optical experiments many attempts have been made to measure the mobility of charged particles in the superfluid helium matrix. This kind of measurements opens the chance to investigate the quantum liquid with its thermal excitations as well as the defect structures of the implanted ions. The interaction of the ions with the superfluid helium is mediated through the scattering with phonons, rotons and quantized vortices. In this experiment the impurity ions drifted in low electric fields, which means that the mobility $\mu(T) = \langle v \rangle / E_d$ is only a function of the temperature. $\langle v \rangle$ denotes the drift velocity and E_d denotes the electric field that forces the ions to move through the liquid. The mobility is governed by the temperature dependent density of scattering partners and their scattering cross sections. Consequently drift time measurements can reveal information about both the structure of the implanted ions through the known density of the scattering partners and on the interaction cross section. An extensive survey on mobility measurements is given in the review articles of Schwarz [20] and of Shikin [21].

The crude experimental set-up is very similar to the one used in the optical experiments (Fig. 2). The ions were produced with the laser sputtering technique in the gas phase. After the laser pulse the ions were drawn into the liquid by means of an electric field. Inside the liquid helium the ion cloud is controlled by an arrangement of three charged grids located at the top of the drift space. Initially, the voltages at these grids are chosen appropriately to stop the ions. To start the drift measurements the voltage of the second grid is changed by a pulse of some milliseconds and an ion cloud is released into the drift space. After passing the grids the ions are guided by a homogenous electric drift field towards the collector. Before the ion cloud reaches the collector its induced image charge leads to a

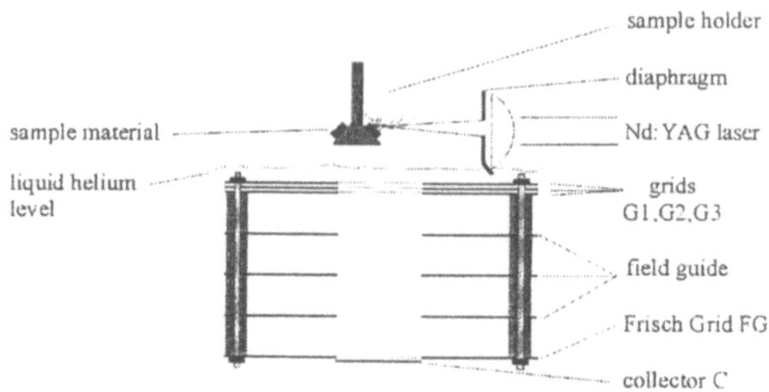

Figure 2 Experimental set-up for mobility measurements with foreign ions in superfluid helium.

premature detector signal. This systematic error is considerably reduced by mounting an additional grid in front of the collector.

The drift time t_d is determined by the middle of the gate opening pulse and the maximum of the ion peak at the collector. This time difference is related to the average drift velocity $\langle v \rangle = s/t_d$, where the drift distance $s = 4.23\,\mathrm{cm}$. The drift velocity $\langle v \rangle$ shows a linear dependence of the electric drift field E_d and a linear fit leads to the mobility μ according to the relation $\mu(T) = \langle v \rangle / E_d$.

3 Results and Discussion

3.1 Measured Optical Spectra and their Explanation with the Bubble Model

The optical spectra of the majority of atoms investigated in liquid helium exhibits the common feature that, for any detected optical transitions the absorption wavelength differs from the emission wavelength. This behaviour is characteristic for all measured spectra of atoms belonging to the alkali, alkali earth and noble metals as well as for the atoms of the group IIb, IIIb and IV elements [19]. The emission lines are fairly symmetric and slightly shifted, but visibly broadened in comparison to the vacuum transition. The measured excitation spectra are broad bands, which are displaced to higher energies. Fig. 3 shows as an example an absorption band of Ca with a blue shift of 2.65 nm in comparison to the vacuum transition and broadened to 4.75 nm (FWHM) [22].

For the alkali earth atoms the spectral features explained above, can be understood with the bubble model [24,25]. In this model the modified spectra are a consequence of the interaction between the helium atoms and the foreign particles: the impurity particles are not free but rather form a complex defect structure. The structure is similar to the bubble structure of an implanted electron in the liquid. The electron repels the surrounding helium atoms due to the repulsive Pauli interaction, an empty bubble with a radius of about 17 Å is formed. In case of the alkali earth atoms the situation is similar: there are two spatially expanded valence electrons which repel the surrounding helium atoms. This interaction is responsible for the missing helium atoms in the vicinity of the impurity particle. In order to calculate the interaction energy the pair potential $V^{\mathrm{pair}}(r)$ of a foreign particle and a helium atom [23] is integrated over the surrounding helium atoms weighted with a

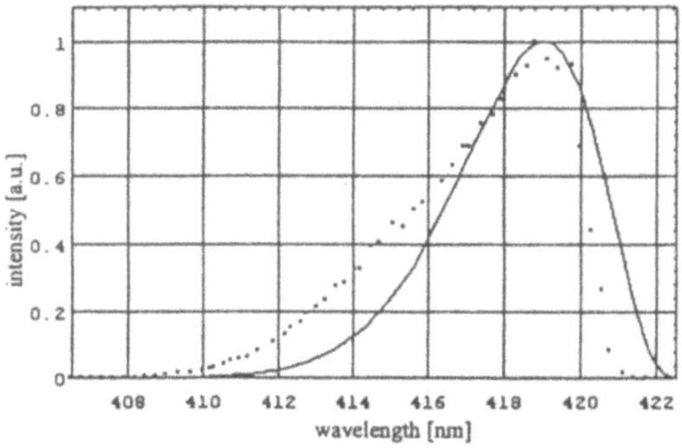

Figure 3
The $4s^2\,{}^1S_0 \rightarrow 4s4p\,{}^1P_1$ excitation spectrum of Ca atoms in superfluid helium at about $T = 1.3\,K$ under saturated vapour pressure. The points are the measured values [29] whereas the line shows the spectra calculated with the bubble model and the statistical line broadening theory in its static limit [26].

density function. This density function displays the vanishing helium density $\rho(r)$ in the neighbourhood of the atom and is expressed by the following trial function [24]:

$$\rho(r,r_0,\alpha) = \begin{cases} 0 & \text{for} \quad r \leq r_0 \\ \rho_0[1-[1+\alpha(r-r_0)]e^{-\alpha(r-r_0)}] & \text{for} \quad r > r_0 \end{cases}$$

In this equation r_0 and α denote parameters related to the bubble size and the smoothness of the shape which were determined by variational minimization of the total energy E_{tot} of the defect structure. The bubble radius r_b of the defect is defined as the center of mass of the density distribution at the bubble margin by the relation

$$\int_0^{r_b} \rho(r) 4\pi r^2\, dr = \int_{r_b}^{\infty} [\rho_0 - \rho(r)] 4\pi r^2\, dr$$

Besides the repulsive interaction the surrounding liquid will reduce the radius of the bubble. These attractive terms of the liquid E_{liq} can be described by the classical terms of the surface energy E_s, the pressure volume work E_{pv} and an additional term E_{vk}. The last term arises from the enhanced kinetic energy of the liquid at the bubble margin.

$$E_{liq} = E_{pv} + E_s + E_{vk} = p\frac{4}{3}\pi r_b^3 + \sigma 4\pi r_b^2 + \frac{\hbar}{8m_{He}} \int \frac{(\nabla\rho)^2}{\rho}\, d^3r$$

The equilibrium configuration of the defect can be determined by minimization of the total energy with respect to the size and shape parameters of the bubble.

$$E_{tot} = E_{liq} + \int V^{pair}(r')\rho(r')d^3r'$$

E_{tot} is the shift of the energy of the impurity center in comparison to the free atom. The time scale for an electronic transition (10^{-15} s) is much smaller than a possible relaxation of the surrounding helium configuration (10^{-12} s). Hence it is sensible to calculate the shift of the transition directly as the difference of the impurity-helium interaction of both states:

$$\Delta\nu = \frac{1}{\hbar} \int \rho(r)[V_f^{pair}(r) - V_i^{pair}(r)]\, d^3r$$

By this means the energy change of the initial (i) and the final (f) state of the impurity atom is determined under the assumption that the configuration of the surrounding helium atoms is not changed during the electronic transition.

In order to determine the shape of the excitation or emission bands two different approaches have often been used. One approach is to calculate the vibrations of the bubble surface [22]. For that purpose the total energy as a function of the defect size is regarded as a potential for the quantized oscillating bubble. The probability density of the vibronic wave function of the initial state is projected onto the potential curve of the final state, which obeys the Frank-Condon principle. Due to this process the width of the band is determined from the slope and the asymmetry in the shape is deduced from the curvature of $E_{tot}(r)$. The form of the spectrum produced with this method is qualitatively in good agreement with the measured line shape. However, a quantitative comparison shows that it always underestimates the observed broadening.

Another way to explain the spectra is utilizing the standard statistical line broadening theory in its static limit [27]. This method is often applied to calculate the spectra of atoms in gases under high pressure. In this theory the fluctuations of the helium atoms are assumed to be uncorrelated, in contrast to the correlated fluctuations of an oscillating bubble explained above. The optical autocorrelation time, which is proportional to the reciprocal width of the optical transition, is on the order of $\tau_{opt} \approx 10^{-14}$ s. From the experimentally obtained kinetic energy of helium atoms at a temperature of about 1.7 K [28] the atoms move in the time τ_{opt} only a distance of 0.1 Å. This distance is much less than the bubble radius and width of the smoothness of the bubble shape. Hence the problem can be treated in its static limit. With this approximation the line intensity is given by [27,28]

$$I(\nu) \propto \int_{-\infty}^{\infty} e^{i\nu\tau} \phi(\tau)\, d\tau$$

where the transition-dipole correlation function is expressed by

$$\phi(\tau) = \exp\{-\int (1 - e^{-i\Delta V(r)\tau}) \rho(r, r_0, \alpha)\, d^3 r\}$$

with the difference $\Delta V(r) = V_f^{pair}(r) - V_i^{pair}(r)$ between the final and initial-state pair potential energies and $\rho(r, r_0, \alpha)$ the equilibrium density obtained in the minimization process. The calculated excitation spectrum with the statistical line broadening theory in its static limit is shown in Fig. 3 for the $4s^2\,{}^1S_0 \rightarrow 4s4p\,{}^1P_1$ transition of Ca together with the measured values (points) [29]. The calculated curve is obtained without any fit of parameters. In the calculation the following parameters have been used: surface tension $\sigma = 0.36\,\mathrm{erg/cm^2}$, helium pressure $p = 2.937$ mbar and the helium density of $\rho_0 = 2.2 \cdot 10^{22}\,\mathrm{1/cm^3}$. The values determined by the minimization of the total energy are for the smoothness of the bubble surface region $\alpha = 1.15\,1/a_0$ and for the size of the bubble $r_0 = 9.25\,a_0$ ($1\,a_0 = 0.529 \cdot 10^{-10}$ m).

3.2 The Recombination Process of Ions and Electrons

After the laser sputtering process positive ions of the element in question are drawn into the liquid where they are neutralized with electrons emitted from a field emission tip. This recombination process normally involves the emission of light: Commonly, a series of optical emission lines can be detected. Obviously, excited levels are populated when the atoms are formed by the recombination process. However, a closer look at the measured spectra reveals the absence of any emission lines starting from the levels above a threshold at about 1.8 eV above the ionization limit of the foreign atoms investigated [19].

To find a theoretical explanation for this effect the process of recombination in liquid helium is investigated in more detail [30]. The ions and the electrons drift towards each other in the low external electric fields. If the field strength of the electrostatic interaction of the ion and electron exceeds the external fields the two species are accelerated to each other. Due to the finite mobilities in liquid helium the Coulomb potential energy is dissipated into the surrounding liquid through the creation of rotons and even vortices. At smaller distances the tunnelling probability of the electron in a bubble like defect structure to the positive ionic defect steadily increases. The most likely tunnelling distance can be calculated leading to an average energy loss of $\approx 0.7\,\text{eV}$ for a single recombination process. During the tunnelling process the surrounding helium atoms are assumed to be unchanged. The newly built atomic configuration resides in a structure which is not in an equilibrium state. After the tunnelling process the cavity around the recombined atom relaxes into a new equilibrium configuration. This is the second part of the energy dissipated during the recombination process. The energy difference can be calculated with the bubble model explained in the last section. In the case of a barium atom the relaxation energy is on the order of 1.1 eV. Together with the loss of 0.7 eV liberation energy during the approach of the ion and the electron, a total energy loss of 1.8 eV for the case of Barium is calculated, which is in good agreement with the experimental result of the absence of any emission line from states closer than $\approx 1.8\,\text{eV}$ to the ionization threshold [30].

3.3 Pecularities of the Magnesium Atom

In the preceding section it has been shown that the spectral features and the absence of some optical transitions can be understood with the bubble model of the implanted atoms in superfluid helium. But besides these rather normal features there are several unexplained experimental results. The lifetime measurements of the magnesium triplet state may serve as an example [32].

The intercombination transition $3s3p\,^3P_1 \rightarrow 3s^2\,^1S_0$ of the free atomic magnesium is possible only because of small deviations from the pure LS coupling scheme. The $^3P_{0,2}$ do not decay at all into the $^1S_0$ state by electric dipole radiation, because of the selection

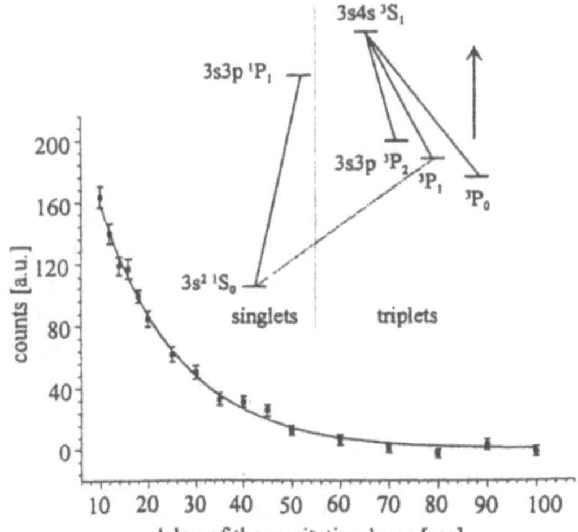

Figure 4
Time dependence of the pulsed laser induced fluorescence signal of magnesium atoms in liquid helium and the level scheme for its lowest states (Simultaneous excitation of all $3s3p\,^3P$ levels to the $3s4s\,^3S_1$ state and detection of the $^3S_1 \rightarrow {}^3P$ transition). Hence the fluorescence transition of the $^3S_1 \rightarrow {}^3P$ is of the order of some nano seconds this measurement displays the lifetime of the $^3P_{2,1,0}$ levels.

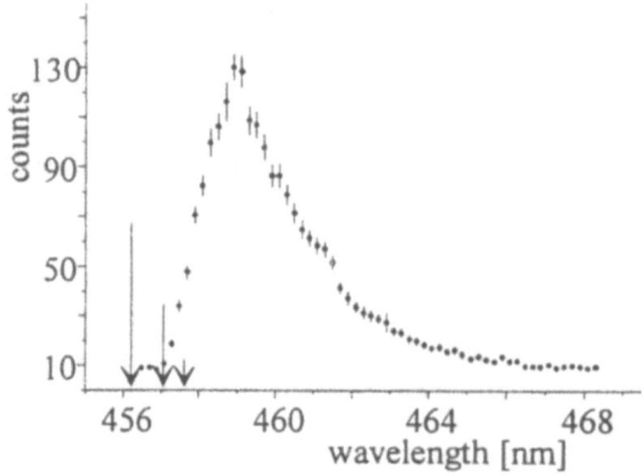

Figure 5
Emission light of the $3s3p\,^3P_1 \rightarrow 3s^2\,^1S_0$ transition of magnesium atoms. The three arrows mark the free atomic transitions. The middle one is due to the observed vacuum intercombination line.

rules $\Delta J = 0, \pm 1$; $J + J' \geq 1$. The measured lifetime of the $^3P_1$ state of free metastable magnesium atoms results in a value of 5.1 ± 0.7 ms [31]. In contrast, the findings of the measurements of magnesium atoms in liquid helium establish an equal lifetime for all three triplet p-states of about $\tau(^3P_{2,1,0}) = 15 \pm 2$ ms [32], that is three times the lifetime in vacuum. This value is obtained with two different techniques. In the first case all three $^3P_{2,1,0}$ levels are excited simultaneously into the $3s4s\,^3S_1$ state with a pulsed dye laser and the fluorescence from this state is monitored. In order to get the time constant for the population of the ^{3}P-levels, the intensity of the fluorescence signal is measured as a function of the time delay of the dye laserpulse (Fig. 4). The same decay constant was determined by the time resolved detection of the intercombination fluorescence $3s3p\,^3P_1 \rightarrow 3s^2\,^1S_0$. Besides these time resolved measurements the $3s4s\,^3S_1 \rightarrow 3s3p\,^3P_{2,1,0}$ transition of magnesium atoms was measured with spectrally resolved wavelength. The line intensities of this measurement show a population of the lowest triplet state of roughly 5 : 3 : 1 corresponding to their statistical weights [32].

A possible explanation for the measured lifetime is the existence of a mechanism which mixes the ^3P_J levels. This mixing mechanism will presumably conserve the measured population ratio despite of the constant ongoing decay of the $^3P_1$ state into the ground state. Considering the rate equations for this process a lifetime of $\tau(^3P_{2,1,0})/3 = \tau(^3P_1) = 5 \pm 0.4$ ms can be deduced which is in rather good agreement to the latest lifetime measurements in the free space. However, the lineshape recording of Fig. 5 rather suggests an emission from all three levels of the ^{3}P configuration.

Such a mixing mechanism is not fully understood but two suggestions were made for its origin: the first relies on non-central collisions of the defect structure with rotons in the superfluid which induces a spin flipping M1 transition. The other mechanism is an exciplex formation which enhances the impurity helium interaction and induces a radiationless transition. These assumptions include, that the total angular momentum is still a good quantum number even if the atom is embedded in liquid helium. However, this may not be guaranteed under all circumstances. For this reason it seems to be adequate to consider the wavelength resolved measurements of the intercombination transition, which is broader and more asymmetric than any other detected emission lines (Fig. 5). The broadening and asymmetry could be explained if it is assumed that all $^3P_{2,1,0}$ states can decay into the ground state with an equal transition probability and a statistical population of 5 : 3 : 1. This assumes of course that all three states can emit into the $^1S_0$ state and J is not a good quantum number any more. In order to understand completely the prolonged lifetime of the magnesium triplet state further investigations are necessary. For example,

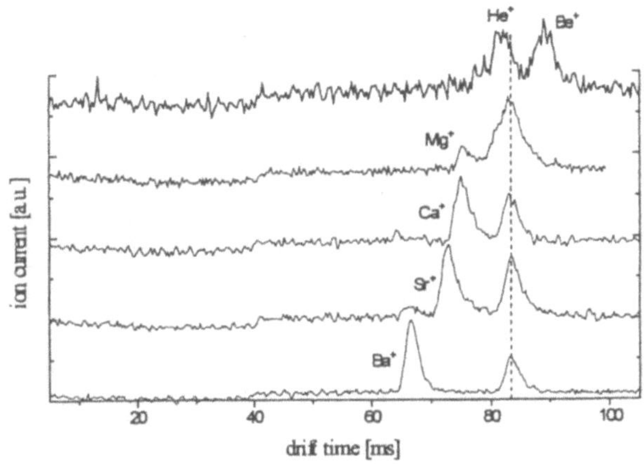

Figure 6
Time of flight spectrum of different alkali earth ions at $T = 1.361\,\mathrm{K}$ and $E_d = 72.3\ \mathrm{Vcm^{-1}}$.

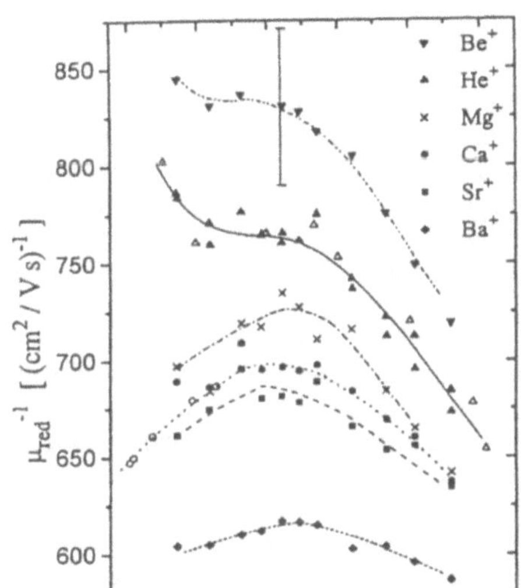

Figure 7
Inverse reduced mobilities $\mu_{\mathrm{red}}^{-1} = \mu_{\mathrm{Ion}}^{-1} \cdot \exp(\Delta(T)/kT)$ for Be^+, Mg^+, Ca^+, Sr^+, Ba^+ and He^+ ions. The Ca^+ data points are taken from [4], the He^+ values are taken from [36,37] (open symbols: $Ca^+ = \circ$, $He^+ = \triangle$). These absolute mobility values μ for the older data from [4,36,37] are increased by a factor 1.1 for the purpose of comparison because the new measurements presented here show systematic deviation of 10% compared to the previous data. The lines are drawn to guide the eyes.

it is very interesting to know how much the lifetime is influenced if the pressure of the surrounding helium is changed.

3.4 Mobility Measurements of Alkali Earth Ions

Much experimental and theoretical work has been carried out to understand the structure and the mechanism of the interaction between ions and excitations in superfluid helium. Mobility measurements reveal the bubble like structure of implanted electrons [13] and the snowball like structure of He^+ ions [12]. In the case of the positive ion it is assumed that the strong attractive polarization forces (Atkins forces) form a hard sphere core of neutral helium atoms around a He_n^+-complex with a radius of about 6 Å [12].

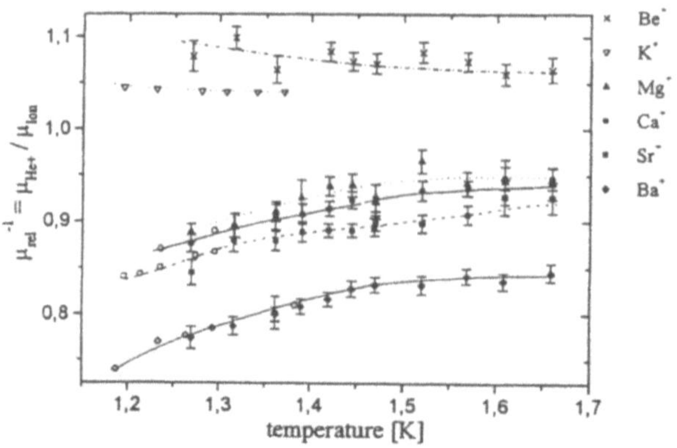

Figure 8 Inverse relative mobilities $\mu_{rel}^{-1} = \mu_{He^+}/\mu_{Ion}$ of K^+, Be^+, Mg^+, Ca^+, Sr^+ and Ba^+. The open symbols (▽ ○ □◇) were taken from reference [4]. The lines are drawn to guide the eyes.

In the following part, the results of the mobility measurements with alkali earth ions in the temperature range of 1.27 K to 1.66 K, where the most important scattering partners for the impurities in liquid helium are rotons [35], are described [18]. In the field and temperature region discussed here the inverse mobility is roughly proportional to the roton density $\mu(T)^{-1} \propto f(T) \cdot \exp(-\Delta(T)/kT)$, where $\Delta(T)$ is the temperature and pressure dependent roton energy gap [35]. The physics of interest in the interaction of the implanted ions with rotons seems to be contained in the prefactor $f(T)$, because all existing theories predict the exponential dependence of the inverse mobility [20]. The drift times t_d of the alkali earth ions at a given temperature and constant drift field decrease with increasing atomic number (Fig. 6). If one assumes that the scattering process depends only on the radius of the defects, Ba^+ should have a smaller defect radius in liquid helium than Ca^+. This is at a first glance a little bit surprising but can be qualitatively understood with the model of Cole and Bachmann [33]. In this model it is supposed that the alkali earth ions form a bubble like structure which is similar to the structure of an implanted electron in liquid helium. These ions have one valence electron and – showing an extended wavefunction similar to the single electron – they also experience a repulsive potential with the surrounding helium atoms. Beside this repulsive force the monopol-dipole and the van der Waals interaction of the core ion with the surrounding bulk helium causes an attraction. A closer examination of the repulsive and attractive forces indicates that the tendency measured for the total radius of the defects is in agreement with the calculation [33].

The most important result is a significant difference between the ionic mobilities of Mg^+ and Be^+ which cannot be explained by their slightly different polarizability and valence electron wave function. Consequently this difference may indicate a different defect structure for these two ions in liquid helium, a snowball like structure for Be^+ and a bubble like structure for the Mg^+ ion. Additionally to the different mobilities of Mg^+ and Be^+ we found a different temperature dependence of their inverse reduced moblities $\mu_{red}^{-1} = \mu_{Ion}^{-1} \cdot exp(\Delta(T)/kT)$ (see Fig. 7) and of the inverse relative mobility $\mu_{rel}^{-1} = \mu_{He^+}/\mu_{Ion}$ with respect to the He^+ ion (see Fig. 8).

For $T < 1.4$ K the inverse reduced mobility μ_{red}^{-1} of Mg^+ shows a similar temperature dependence as the Ba^+, Sr^+ and Ca^+ ions. The model of Cole and Bachmann [33] predicts a bubble like structure for these elements. If one assumes that the scattering process for a certain structure has always the same temperature dependence, then Mg^+ should exhibit a bubble like structure. On the other hand μ_{red}^{-1} of Be^+ shows a similar temperature dependence to that of He^+ and K^+, as can be seen in Fig. 7 and Fig. 8. The theory predicts a snowball like structure for K^+ and He^+ [33,12], which suggests that the Be^+ defect in liquid helium has a similar structure. As mentioned above, this interpretation relies on the hypothesis, unproved so far, that the temperature dependence is the same for the same class of defect structures.

To answer the open questions raised above more theoretical and experimental work is needed including pressure dependent mobility experiments. With this adjustable parameter it might be possible to look for a transition from a bubble to a snowball like structure of Mg^+ ions, as predicted by [34].

4 Conclusion

General methods have been developed to implant practically any atom and singly ionized ion into superfluid and solid helium [6,7,9,11,38]. This opened up the possibility to investigate the behaviour of the mobility of ions and the spectra of ions, atoms and multimers in the superfluid. The questions of interest with regard to these atomic and ionic defects are numerous:

- When does a defect form a bubble and when a snowball?
- Can these two types of defects be discriminated by their mobilities in the fluid?
- Can their spectra – line shifts, line shapes and line width – contribute to this discrimination?
- How can the absence of resonance radiation in some alkalis be explained quantitatively?
- Are there bound states in the region 1.8 eV below the ionization threshold which can emit light if they are populated, e.g. by a two step laser absorption process?
- What is the reason for emission of light from states which could not emit light in the free atom or ion according to the selection rules? Why are their lifetimes altered considerably in some cases?
- Will it be possible to select defects which undergo changes from bubble to snowball under the influence of pressure on the liquid thus allowing to investigate the intricate energy balance between attractive and repulsive forces in the defect matrix interaction?

According to our judgement all these questions can be answered in forthcoming investigations. In addition to the basic insight gained into the nature of quantum fluids the result may also apply to questions of quantum turbulence.

Bibliography

[1] R.J. Donnelly, *Experimental Superfluidity*, University of Chicago Press, Chicago (1967).

[2] A.N. Gerritsen, Physica **14**, 407 (1948).

[3] A.N. Gerritsen, J. Koolhaas, Physica **10**, 49 (1943).

[4] W.I. Glaberson, W.W. Johnson, J. Low Temp. Phys. **20**, 313 (1975).

[5] E.B. Gordon, L.P. Mezhov-Deglin, O.F. Pugachev, ZhETF Pis. Red. **19**, 103 (1973); JETP Lett. **19**, 63 (1974).

[6] H. Bauer, M. Hausmann, R. Mayer, H.J. Reyher, E. Weber, A. Winnacker, Phys. Lett. A **110**, 279 (1985).

[7] M. Himbert, A. Lezama, J. Dupont-Roc, J. Physique **46**, 2009 (1985).

[8] H. Bauer, M. Beau, J. Fischer, H.J. Reyher, J. Rosenkranz, K. Venter, Physica B **165 & 166**, 137 (1990).

[9] T. Yabuzaki, A. Fujisaki, K. Sano, T. Kinoshita, Y. Takahashi, Atomic Physics 13, ed. by H. Walther and T.W. Hänsch, North Holland, New York, (1992).

[10] M. Arndt, S.I. Kanorsky, A. Weis, T.W. Hänsch, Phys. Lett. A **174**, 298 (1993).

[11] J.H.M. Beijersbergen, Q. Hui, M. Takami, Phys. Lett. A **181**, 393 (1993).

[12] K.R. Atkins, Phys. Rev. **116**, 1339 (1959).

[13] R.A. Ferrell, Phys. Rev. **108**, 167 (1957).

[14] B. Tabbert, H. Günther, G. zu Putlitz, J. Low Temp. Phys. **109**, 653 (1997).

[15] H. Günther, M. Foerste, G. zu Putlitz, Th. Schumacher, Fiz. Nizk. Temp. **22**, 189 (1996); Low Temp. Phys. **22**, 143 (1996).

[16] P.V.E. McClintock, Phys. Lett. A **29**, 453 (1969).

[17] H. Günther, M. Foerste, M. Kunze, G. zu Putlitz, U. von Stein, Z. Phys. B **101**, 613 (1996).

[18] M. Foerste, H. Günther, O. Riediger, J. Wiebe, G. zu Putlitz, Z. Phys. B **104**, 317 (1997).

[19] B. Tabbert, M. Beau, H. Günther, W. Häußler, C. Hönninger, K. Meyer, B. Plagemann, G. zu Putlitz, Z. Phys. B **97**, 425 (1995).

[20] K.W. Schwarz, Advances in Chem. Phys. **33**, 1 (1975).

[21] V. B. Shikin, Sov. Phys. Usp. **20**, 226 (1977).

[22] H. Bauer, M. Beau, B. Friedl, C. Marchand, K. Miltner, H.J. Reyher, Phys. Lett. A **146**, 134 (1990).

[23] E. Czuchaij, F. Rebentrost, H. Stoll, H. Preus, Chem. Phys. Lett. **182**, 191 (1991) and private communication.

[24] K. Hiroike, N.R. Kestner, S.A. Rice, J. Jortner, J. Chem. Phys. **43**, 2625 (1965).

[25] W.B. Fowler, D.L. Dexter, Phys. Rev. **176**, 337 (1968).

[26] M. Foerste, Dissertation, Universität Heidelberg 1997.

[27] P.W. Anderson, Phys. Rev. **86**, 809 (1952).

[28] A.P. Hickman, W. Steets, N.F. Lane, Phys. Rev. B **12**, 3705 (1975).

[29] H. Bauer, Dissertation, Universität Heidelberg 1988.

[30] M. Beau, H. Günther, G. zu Putlitz, B. Tabbert, Z. Phys. B **101**, 253 (1996).

[31] A. Godone, C. Novero, Phys. Rev. A **45**, 1717 (1992).

[32] H. Günther, M. Foerste, C. Hönninger, G. zu Putlitz, B. Tabbert, Z. Phys. B **98**, 395 (1995).

[33] M.W. Cole, R.A. Bachmann, Phys. Rev. B **15**, 1388 (1977).

[34] M.W. Cole, F. Toigo, Phys. Rev. B **17**, 2054 (1978).

[35] R. Barrera, G. Baym, Phys. Rev. A **6**, 1558 (1972).

[36] K.W. Schwarz, Phys. Rev. A **6**, 837 (1972).

[37] B. Brody, Phys. Rev. B **11**, 170 (1975) and private communication.

[38] M. Arndt, R. Dziewior, S. Kanorsky, A. Weis, T.W. Hänsch, Z. Phys. B **98**, 377 (1995).

Disorder and Impurities in Hubbard-Antiferromagnets

M. Ulmke[1], P.J.H. Denteneer[2], V. Janiš[3], R.T. Scalettar[4], A. Singh[1,*], D. Vollhardt[1], and G.T. Zimanyi[4]

[1] Theoretische Physik III, Elektronische Korrelationen und Magnetismus, Institut für Physik, Universität Augsburg, 86135 Augsburg, Germany
[2] Lorentz Institute for Theoretical Physics, University of Leiden, P.O. Box 9506, 2300 RA Leiden, The Netherlands
[3] Institute of Physics, Academy of Sciences of the Czech Republik, CZ 18040 Praha 8, Czech Republik
[4] Department of Physics, University of California, Davis, CA 95616, USA
* On leave from Department of Physics, Indian Institute of Technology, Kanpur 208016.

Abstract: We study the influence of disorder and randomly distributed impurities on the properties of correlated antiferromagnets. To this end the Hubbard model with (i) random potentials, (ii) random hopping elements, and (iii) randomly distributed values of interaction is treated using quantum Monte Carlo and dynamical mean-field theory. In cases (i) and (iii) weak disorder can lead to an enhancement of antiferromagnetic (AF) order: in case (i) by a disorder-induced delocalization, in case (iii) by binding of free carriers at the impurities. For strong disorder or large impurity concentration antiferromagnetism is eventually destroyed. Random hopping leaves the local moment stable but AF order is suppressed by local singlet formation. Random potentials induce impurity states within the charge gap until it eventually closes. Impurities with weak interaction values shift the Hubbard gap to a density off half-filling. In both cases an antiferromagnetic phase without charge gap is observed.

1 Introduction

Antiferromagnetic spin correlations are present in many strongly correlated electron systems, notably the prototype Mott insulators NiO and V_2O_3, the parent compounds of HTSC cuprates, and heavy fermion systems such as YbP, U_2Zn_{17}, and many others. Many of those systems are intrinsically disordered, in particular upon additional homo- or heterovalent doping. The influence of impurity doping on antiferromagnetic (AF) order and electronic properties has recently been studied in a variety of systems. Doping with static scatterers like nonmagnetic impurities usually weakens antiferromagnetic order, a prominent example being Zn doping in $YBa_2Cu_3O_6$ [1]. In spin chains ($CuGeO_3$) [2] and ladder compounds ($SrCu_2O_3$) [3] doping with magnetic and nonmagnetic impurities can *induce* AF order while the pure systems show spin gap behavior. Very effective in destroying AF order are mobile carriers, e.g. hole doping in $La_{1-x}Sr_xCuO_4$ [4]. The stability of AF order strongly depends on the positions of the dopant level. While in the nickel oxides $La_{1-x}Sr_xNiO_4$ [5] and $Ni_{1-x}Li_xO$ [6] holes are supposed to be localized, in the cuprate

$La_{1-x}Sr_xCuO_4$ the hole level lies in the valence band leading to mobile scatterers. As a result, AF order is stable in $La_{1-x}Sr_xNiO_4$ up to $x = 0.5$, but is destroyed in the cuprate already at approx. 5% Sr doping.

In the present paper, we will study the influence of disorder on AF order and the Mott band gap in correlated antiferromagnets. We employ the Hubbard model in the presence of different types of disorder. While the (disordered) Hubbard model is certainly far too simple to describe real materials it already contains very rich physics including local moment formation, magnetic ordering, Mott-Hubbard transition, and Anderson localization. On the other hand, the interplay of disorder and interactions in electronic systems belongs to the most difficult problems in physics, and reliable results within simple models are still very desirable. The problem has been investigated in the past by a variety of methods, including field theoretical approaches [7], renormalization group treatments [8,9], unrestricted Hartree-Fock [10,11], dynamical mean-field theory (DMFT) [12–14], quantum Monte Carlo (QMC) [15–17], and several more (see [18] for a review). Here, we give an overview of results obtained by QMC and DMFT concentrating on the AF phase diagram and the Mott gap.

We consider the following Hubbard Hamiltonian

$$\hat{H} = \sum_{i\sigma}(\varepsilon_i - \mu)\hat{n}_{i\sigma} + \sum_{\langle ij\rangle\sigma} t_{ij}(\hat{c}^{\dagger}_{i\sigma}\hat{c}_{j\sigma} + \text{h.c.}) + \sum_i U_i(\hat{n}_{i\uparrow} - \frac{1}{2})(\hat{n}_{i\downarrow} - \frac{1}{2}). \tag{1.1}$$

In principle all parameters $\varepsilon_i, t_{ij}, U_i$ can be randomly distributed. The precise definition of the different disorder types studied in this paper will be given in the following sections. The average $t \equiv \langle t_{ij}\rangle$ sets our energy scale. We will restrict the hopping t_{ij} to nearest-neighbors hence not allowing for frustration. Longer range, random hopping amplitudes will be important in the modeling of amorphous materials such as doped semiconductors [18,19] and are not considered in the present work.

2 Methods

2.1 Determinant Quantum Monte Carlo (*d=2*)

We use a finite temperature determinant quantum Monte Carlo method [20] to obtain approximation-free results for finite lattices. The algorithm is based on a mapping of the interacting electron problem onto a $d+1$ dimensional quasi-classical problem using auxiliary Ising-type spins. It provides for calculating thermal averages of observables, $\hat{A}$, at a temperature $T = 1/\beta$,

$$\langle \hat{A}\rangle = \frac{\text{Tr}\hat{A}\, e^{-\beta\hat{H}}}{\text{Tr}\, e^{-\beta\hat{H}}}. \tag{2.2}$$

The phase space sampling over the auxiliary field configurations is performed using Monte Carlo techniques. The weight of a configuration is proportional to a product of two determinants, one for each electron spin species. In the case of half-filling without random potentials, i.e. $\varepsilon_i - \mu \equiv 0$, on a bipartite lattice the determinants always have the same sign, hence their product is always positive semi-definite, which can be shown by particle-hole transformation of one spin species $[c_{\vec{i}\downarrow} \to (-1)^{\vec{i}} c^{\dagger}_{\vec{i}\downarrow}]$. In the general situation that the product can become negative the algorithm can still be employed in principle. However, the signal to noise ratio decreases exponentially with systems size, inverse temperature, and interaction, putting severe restrictions to the applicability of the method. This so-called "minus-sign problem" is a general obstacle for all exact fermionic Monte Carlo methods as well as for spin-systems in the presence of frustration. Even without the minus-sign problem the computational effort is large because the computer time grows cubically with system size N, restricting N to the order of 100 on present supercomputers.

In the case of disorder all observables have to be averaged over the (frozen) disorder configurations. Because of the computational effort we restrict ourselves to two dimensional lattices with linear size up to $L_x = 10$ which often allows a reliable finite size scaling. Since we are interested in AF ordering we calculate the magnetic correlation functions $C(\vec{l})$ and their Fourier transforms, the magnetic structure factors $S(\vec{q})$,

$$C(\vec{l}) = \frac{1}{N}\sum_{\vec{j}} \langle m_{\vec{j}} m_{\vec{j}+\vec{l}} \rangle, \quad S(\vec{q}) = \sum_{l} C(\vec{l}) e^{i\vec{q}\vec{l}}. \tag{2.3}$$

In particular the AF structure factor $S(\pi,\pi)$ is used to obtain the ground state sublattice magnetization M by a finite size scaling Ansatz according to spinwave theory [21]:

$$\frac{S(\pi,\pi)}{N} = \frac{M^2}{3} + O(\frac{1}{L_x}). \tag{2.4}$$

2.2 Dynamical Mean-Field Theory (Limit of Infinite Dimensions)

The dynamical mean-field theory [22,23] is a local approximation in which the self energy becomes site diagonal, or momentum independent:

$$\Sigma_{ij}(\omega) = \delta_{ij}\Sigma(\omega), \qquad \hat{\Sigma}(\vec{k},\omega) = \hat{\Sigma}(\omega). \tag{2.5}$$

The one-particle Green function $G(\vec{k},\omega)$ can hence be obtained from the non-interaction Green function $G^0(\vec{k},\omega)$ by $G(\vec{k},\omega) = G^0(\vec{k},\omega - \hat{\Sigma}(\omega))$, and the local Green function is given by $G_{ii}(\omega) = 1/N\sum_{\vec{k}} G(\vec{k},\omega)$ This does not imply a simple shift of energies, like in traditional mean-field theories (e.g. Hartree-Fock), because Σ remains dynamical, i.e. frequency dependent, preserving local quantum fluctuations. The local approximation becomes exact in the limit of infinite spatial dimensionality and maps the interacting lattice model onto a self-consistent single impurity model like, for example, the Wolff model [24]:

$$\hat{H}_{\text{Wolff}} = \sum_{\vec{k}\sigma} \varepsilon_{\vec{k}} \hat{n}_{\vec{k}} + U(\hat{n}_{0\uparrow} - \frac{1}{2})(\hat{n}_{0\downarrow} - \frac{1}{2}) + \varepsilon(\hat{n}_{0\uparrow} + \hat{n}_{0\downarrow}). \tag{2.6}$$

Here, the one-particle energies $\varepsilon_{\tilde{k}}$ have to be defined such that the non-interacting local Green function of the Wolff model fulfills $G^0_{\text{Wolff}} \equiv (G_{ii} - \tilde{\Sigma})^{-1}$. In the self-consistent solution the local (interacting) Green function of the Wolff model has to be equal to G_{ii}. In the presence of disorder one has to average over all possible values of U or ε, respectively. This type of local averaging is equivalent to the "coherent potential approximation", well known from investigations of disordered alloys.

While the self-consistency is rather easily reached by iteration the solution of the single impurity problem is the hard part. There exists no analytic solution and different numerical and approximative techniques have been employed [23]. Here we again use auxiliary field QMC [25], an algorithm quite similar to the one for finite dimensional lattices sketched above. The computer time grows like L^3 where the number of Matsubara frequencies, $L \propto \beta$. Fortunately, QMC for the single band model is free from the minus-sign problem.

In the following, the non-interacting DOS is chosen as a semi-elliptic model DOS with bandwidth=8, equal to the $d = 2$ tight binding bandwidth for $t = 1$. A typical quantity under consideration is the staggered magnetic susceptibility χ_{AF} whose divergence signals the transition to an AF ordered state. One can also extend the DMFT equations to the ordered phase to obtain spin and sublattice dependent electron densities and the sublattice magnetization M. The one-particle density of states (DOS) is obtained by analytical continuation of the imaginary time Green function using the Maximum Entropy method. For details of the algorithm, the implementation of disorder averages and determination of expectation values see [23,14].

3 Random Potentials

3.1 Local Moment Quenching

Random potentials are the most frequently studied type of disorder in the context of Anderson localization. Contrary to the Hubbard interaction which at half-filling favors single occupation on each site, different local potentials lead to different occupations and hence to a quenching of local magnetic moments on sites with large absolute value of the local potential. This is seen in Fig. 1a where the average local moment squared, m^2, is plotted versus disorder strength for a flat distribution of ε values with width Δ. It is also shown that the spin-spin-correlations go in parallel with m^2. For a large width of random potentials this type of disorder is apparently very effective in the destruction of magnetic order. To study the behavior of the charge gap, the electronic compressibility, $\kappa \equiv \partial n/\partial \mu$, is calculated as a function of Δ (Fig. 1b). While disorder decreases κ in the non-interacting case, κ is enhanced by disorder at finite U. The reason is the introduction of states within the AF charge gap as will be discussed below. As mentioned above, random potentials break the particle-hole symmetry leading to a minus-sign problem even at half filling. This is the reason why only small lattices (4×4 in Fig. 1) can be studied for this type of disorder.

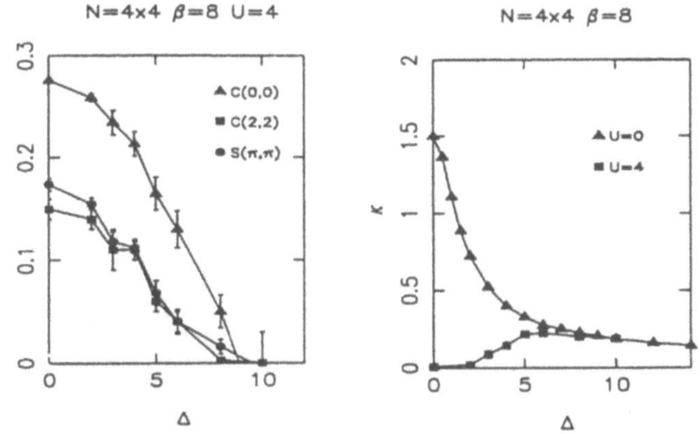

Figure 1 (a) Local magnetic moment $C(0,0)$, spin-spin correlation function at largest possible distance (2,2), and AF structure factor $S(\pi,\pi)$ as a function of disorder strength for a constant distribution of random potentials (in all cases the values at $U = \Delta = 0$ subtracted). (b) compressibility κ vs. Δ; disorder reduces κ in the non-interacting case and enhances it at $U = 4$ [16].

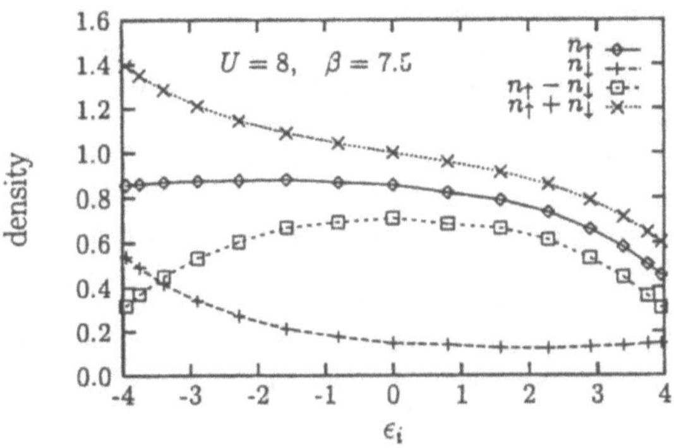

Figure 2
Local spin resolved densities in DMFT as a function of the random potential value ε_i [26]. Width of disorder distribution is $\Delta = 8$(=bandwidth).

3.2 Disorder-Enhanced Delocalization

Consider the situation deep in the AF phase, i.e. with a large staggered moment. A high value of the local potential on a given site reduces the potential barrier for a majority spin electron to tunnel to a neighboring sites. The electron is hence delocalized and the local magnetic moment is reduced. A low potential on the other hand cannot significantly further localize the majority spin because it is already almost saturated. This asymmetry of the localizing and delocalizing effects of random potentials is depicted in the DMFT results, Fig. 2. The majority spin ($\uparrow$) is strongly reduced for $\varepsilon > 0$ but almost unchanged for $\varepsilon < 0$. Note that the total local density monotonically decreases with ε, and that the net magnetization $(n_\uparrow - n_\downarrow)$ is reduced for any $|\varepsilon| > 0$.

The fact that the delocalization is not compensated can be expressed in terms of an enhanced effective hopping parameter t_{eff}. In the case $(U - \Delta) \gg t$ the effective t_{eff} can be estimated by the mapping onto the AF Heisenberg model: the AF exchange arises from virtual hopping of an electron with spin σ to a neighboring site occupied with an electron of spin $-\sigma$. The exchange energy for this and the opposite process is

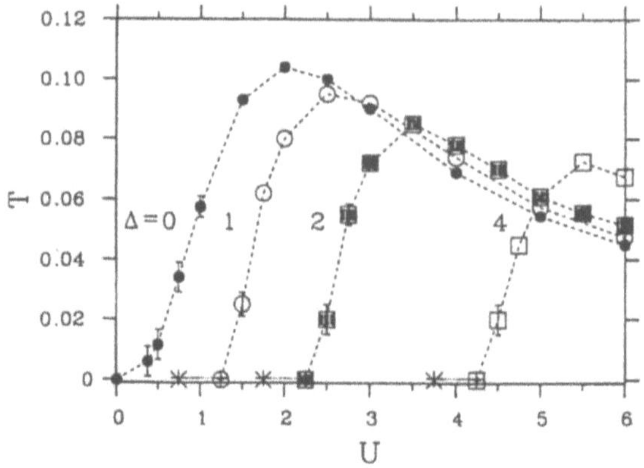

Figure 3
AF $(T-U)$ phase diagram in DMFT for a bimodal random potential distribution. Energies in units of the half bandwidth. [14]

$$J_{ij} = \frac{t^2}{U-(\varepsilon_i-\varepsilon_j)} + \frac{t^2}{U-(\varepsilon_j-\varepsilon_i)} \approx \frac{2t^2}{U}\left[1+\frac{(\varepsilon_i-\varepsilon_j)^2}{U^2}\right]. \tag{3.7}$$

J_{ij} is hence always *enhanced* from the pure case ($J_0 = 2t^2/U$), and the average exchange becomes $J_{\rm av} = J_0[1+\lambda(\Delta/U)^2]$ where λ depends only on the shape of the disorder distribution (e.g., $\lambda = 1/12$ for the constant, $\lambda = 1/2$ for a binary distribution). Furthermore, one can show that the form of the magnon propagator remains unchanged in second order in Δ/U [26], but magnons are stiffened by a factor of $(1+\lambda(\Delta/U)^2)$. The effective hopping in strong coupling can thus be expressed as $t_{\rm eff} = t[1+\lambda(\Delta/U)^2]^{1/2}$. In dimensions $d > 2$ one expects the Néel temperature T_N to be proportional to J and hence T_N, too, should be enhanced by weak disorder.

While disorder-enhanced delocalization stabilizes AF order at strong coupling it suppresses it at weak coupling. For small U the AF state is rather "spin density wave like" and an enhanced kinetic energy tends to weaken AF order and to reduce T_N. In addition, random potentials destroy the perfect nesting instability responsible for AF order at small U.

The T vs. U phase diagram within DMFT (Fig. 3) summarizes and confirms the above considerations: For small U, T_N is reduced and eventually vanishes when Δ becomes roughly equal to U. At large U, however, the T_N curves for different Δ cross each other, i.e. at a given value of U, T_N increases with disorder. The opposite effects of disorder on T_N depending on U are due to the non-monotonic behavior of the function $T_N(U)$. If the slope of $T_N(U)$ is positive (negative) an effectively reduced ratio $U/t_{\rm eff}$ leads to a suppression (enhancement) of T_N.

3.3 Closing of the Charge Gap

Local potentials can induce electronic states in the charge gap. As also found experimentally the positions of the impurity states are crucial for the stability of AF order with respect to carrier doping. Disorder-induced gap states will reduce the size of the charge gap. We now want to determine the critical disorder strength at which the gap vanishes. First we treat the disorder within the T-matrix approximation which becomes exact for a single impurity [27,26]. Since the local host Green function of the correlated, not-disordered

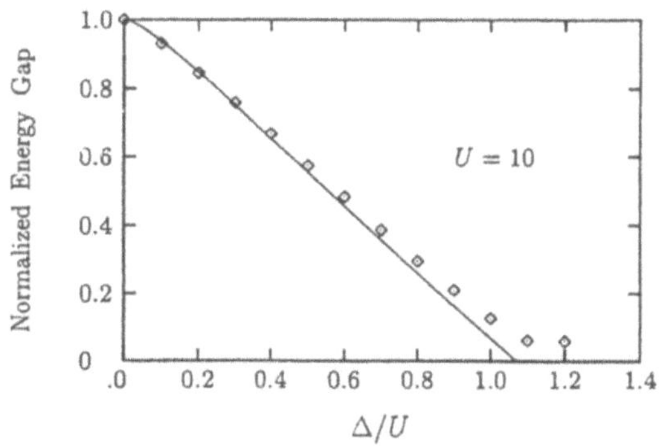

Figure 4
Normalized energy gap, $\tilde{D}/D$, in $d = 2$ versus disorder strength Δ at $U = 10t$ as obtained by the T-matrix approximation (solid line) and the UHF analysis (symbols) [26].

problem is unknown we approximate it by the Green function of the AF Hartree-Fock solution:

$$[g^0_\sigma]_{ii} = (1/N)\sum_{\vec{k}}(\omega - \sigma D)/(\omega^2 - E^2_{\vec{k}}). \tag{3.8}$$

Here $2D = mU$ is the Hubbard energy gap in the pure AF, $E_{\vec{k}} = \sqrt{D^2 + \varepsilon^2_{\vec{k}}}$ is the AF band energy, and D is obtained from the self-consistency condition $(1/N)\sum_{\vec{k}}(2E_{\vec{k}})^{-1} = U^{-1}$. With this local host Green function we can calculate the location of impurity-induced states from the poles in the T-matrix, $T_\sigma(\omega) = \varepsilon_i/(1 - \varepsilon_i[g^0_\sigma(\omega)]_{ii})$. We consider a constant distribution of random potentials between $\pm\Delta/2$. There are disorder-induced states within the gap if $|[g^0_\sigma]_{ii}| > 2/\Delta$. The energy $\tilde{D}$ up to which states are formed within the gap is given by $[g^0_\sigma(-\tilde{D})]_{ii} = 2/\Delta$. The remaining charge gap $2\tilde{D}$ is plotted versus Δ in Fig. 4. The decrease is almost linear and $2\tilde{D}$ vanishes close to $\Delta = U$. Also shown is the result from a numerical unrestricted Hartree-Fock (UHF) analysis. In this approach, the HF Hamiltonian on a finite lattice is numerically (self-consistently) diagonalized, so that disorder is treated exactly [10,11,26]. The energy gap is obtained from the energy difference between the lowest energy state of the upper Hubbard band and the highest energy state of the lower Hubbard band. Averages are taking over 100 disorder realizations on a 10×10 lattice. The saturation of the gap at $\Delta/U \sim 1$ is due to the finite system size. The agreement with the T-matrix approach is excellent for the present interaction value $U = 10t$. Deviations from the T-matrix approach are more pronounced at lower interaction strengths where the fermion states are more extended.

The closing of the charge gap is also observed within the DMFT approach. Fig. 5a shows the density of states for several disorder values. Note that all spectra shown are within the AF ordered phase, i.e. AF order is much more stable than the charge gap which closes about $\Delta = U$. A linear reduction of the charge gap is also observed in the case of a binary disorder distribution (Fig. 5b). For this stronger type of disorder both the charge gap and the AF order vanish near $\Delta = U$.

3.4 Spin Vacancies

As discussed in the previous section the one-particle excitation gap decreases with Δ and vanishes near $\Delta = U$. Upon further increase of Δ the two Hubbard bands will overlap whereby electrons from the highest levels of the lower band (with $\varepsilon_i > U/2$) will be

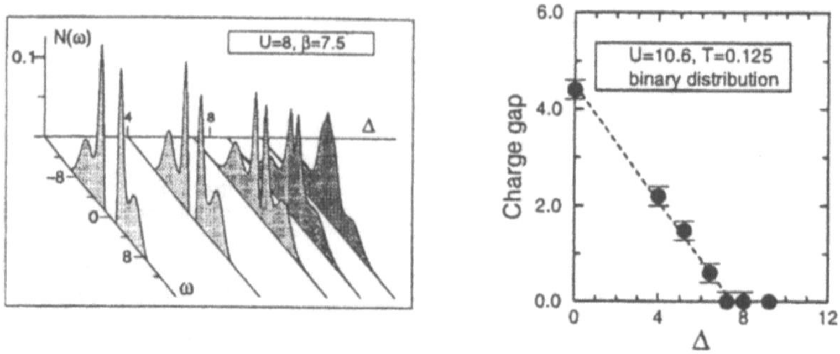

Figure 5 (a) Density of states (DOS) for a constant disorder distribution of different width in DMFT [26]; all spectra are within the AF phase. (b) Charge gap vs. Δ in DMFT for the bimodal distribution of random potentials.

transfered to the lowest levels of the upper band (with $\varepsilon_i < -U/2$). Those sites will become doubly occupied, their local moment will vanish as observed in Sec. 3.1. In the limit $t \ll U,\Delta$ the Hubbard bands at $\pm U/2$ are nearly flat with width Δ. The overlap region is $\sim (\Delta - U)$ and the fraction of nonmagnetic sites can be estimated by $x \sim (\Delta - U)/\Delta$.

In the following this situation is modeled by spin vacancies of concentration x which will lead to strong magnon scattering. Such diluted antiferromagnets have been widely studied recently mostly within spin diluted Heisenberg models [28,29] and also Hubbard models [27]. A perturbative analysis in the strong coupling limit ($O(t^2/U)$) [27,26] yields a softening of magnons by a factor of $(1-2x)$ in $d=2$. Extrapolating to large fractions x, the magnon energy scale $\tilde{J} = J(1-2x)$ hence vanishes only close to the percolation threshold $x_{\text{perc}} \approx 0.4$. In $d>2$ the Néel temperature is assumed to be proportional to $\tilde{J}$ and a linear decrease of T_N is thus expected. In experiments, e.g. on Li doped NiO [30] such a linear dependence on x is indeed observed with prefactor of 2.2. In the limit $d \to \infty$ the percolation threshold is 1.0; hence we can expect AF order to persist for arbitrarily large dilution. Within DMFT, too, a linear decrease of $T_N \propto (1-x)$ is observed for the diluted model at large U (Fig. 6a). At small U, however, AF order is remarkably robust, T_N being constant up to dilutions close to $x=1$ where T_N eventually drops to zero. This behavior

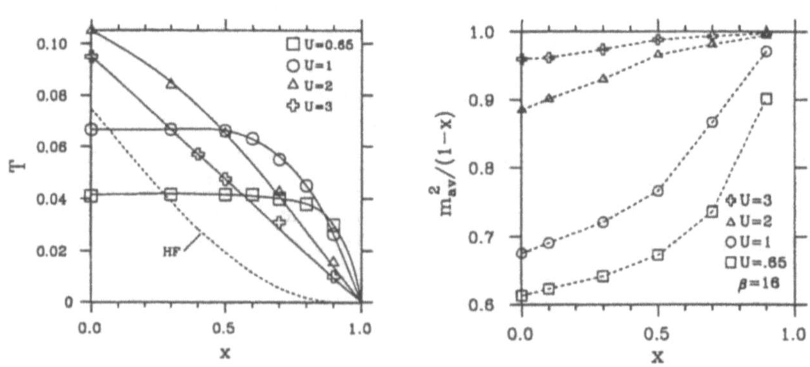

Figure 6 (a) T_N as a function of site dilution in DMFT for different values of U. Dashed line Hartree-Fock approximation for $U=0.65$ (energies in units of the half bandwidth). (b) Enhancement of local moments on the remaining fraction $(1-x)$ of the lattice at inverse temperature $\beta = 16$. [14]

of T_N can be explained by a strong enhancement of the local moment density on the remaining sites at small U (Fig. 6b). The reason for this enhancement is that with reduced average number of nearest neighbors the kinetic energy decreases, leading to a stronger localization. At large U the local moments are already almost saturated at $x = 0$ and just cannot be further enhanced. The situation in the case of vacancies is therefore quite different from the effect of weak site disorder where disorder-enhanced *delocalization* is observed.

4 Random Hopping

The case of spin vacancies discussed in the previous section can of course be regarded as a specific type of randomness in the hopping elements. The more generic case of a continuous (flat) distribution of $t_{ij} \in [1-\Delta/2, 1+\Delta/2]$ was studied in $d = 2$ using QMC. Since the hopping is still restricted between nearest neighbors on a square lattice particle-hole symmetry is preserved and no minus-sign problem occurs at $n = 1$. Random t_{ij} hardly affect the density of local moments (Fig. 7a), the slight decrease may be due to the enhanced kinetic energy which is $\propto \sqrt{\langle t_{ij}^2 \rangle}$. Nevertheless, longer-range spin-spin correlations are strongly suppressed if Δ is of the order of t (see Fig. 7a). The finite size scaling according to (2.4) yields the AF order parameter (staggered moment) M versus Δ (Fig. 7b). M vanishes at a critical disorder strength of $\Delta_c \approx 1.4$. We propose [16] that the phase boundary is determined basically by the variance of the AF exchange coupling $v = (\langle J_{ij}^2 \rangle - \langle J_{ij} \rangle^2)/\langle J_{ij}^2 \rangle$. AF order persists for $v < v_c \approx 0.4$. This criterion is consistent with the phase boundary of the bond-disordered AF Heisenberg model with a bimodal distribution of J_{ij} [31]. The reason for the vanishing of AF order for this type of disorder is supposedly the formation of local singlets. Such singlets will form first on the strongest bonds and will leave some spins which are weakly coupled to their neighbors unpaired. Those "free" spins are expected to give a Curie-like contribution to the susceptibility as observed in doped semiconductors [18,19]. The numerical results indeed show a strong enhancement of the uniform susceptibility in the disordered case [16].

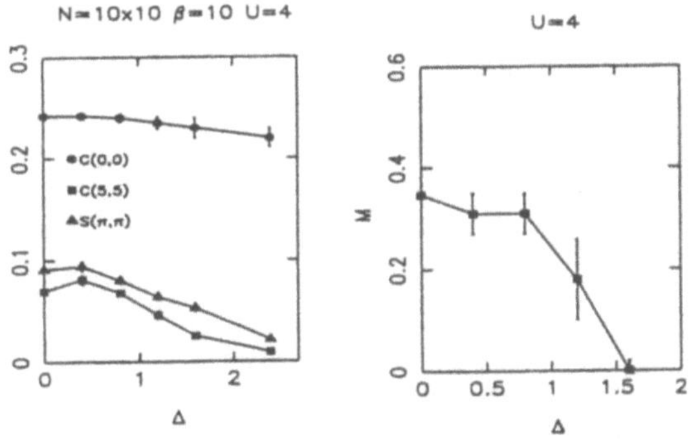

Figure 7 (a) Local moment and spin-spin correlations in $d = 2$ for random hopping (same quantities as in Fig. 1). (b) Staggered magnetization M vs. Δ as obtained by finite size scaling [16].

5 Impurities with Weak Local Interactions

The Hubbard model at $n = 1$ exhibits both AF order and the Mott-Hubbard metal-insulator transition (MIT). It is important to note that both effects are in principle independent. The Mott-Hubbard MIT occurs at intermediate interaction ($U \sim$ bandwidth) while AF order can set in at arbitrarily small U due to the Fermi surface nesting instability. It is argued that the MIT at $T = 0$ represents a quantum critical point which is however concealed by the low temperature AF phase [32], and attempts are made to suppress the AF phase to very low or zero T by different types of frustration [23]. A different possibility to separate AF order and the MIT is to shift the filling at which the Mott-Hubbard MIT occurs away from $n = 1$ by introducing impurities with a low (or zero) local interaction with a concentration f. Such impurity sites can be doubly occupied without the cost of the local repulsion and hence the MIT is expected at a density $n = 1 + f$.

5.1 Mott-Hubbard Gap

The Hubbard model with a bimodal distribution of U values ($U = 0$ on a fraction f of the lattice, and $U = 8t$ on the remaining sites) was investigated by QMC on square lattices and in DMFT. Note that for technical reasons (minus-sign problem) the choice of the $U-$disorder in model (1.1) preserves particle-hole symmetry. This corresponds to different chemical potentials on the two constituents such that at $n = 1$ the electronic density is homogeneous, i.e. independent of U_i. To detect charge gaps the electron density n is plotted versus the chemical potential μ (Fig. 8a). As expected the gap moves to a density off half filling close to $1 + f$. This agrees with the kinetic energy which shows minima at the corresponding densities (Fig. 8b). At a closer look one observes an additional gap at $n = 1$ which is due to the doubling of the unit cell in the AF ordered state. The gaps at $n = 1$ and $n = 1 + f$ can also be detected by the one-particle spectrum, obtained in DMFT [17]. Between densities 1 and $1 + f$ the systems is apparently compressible.

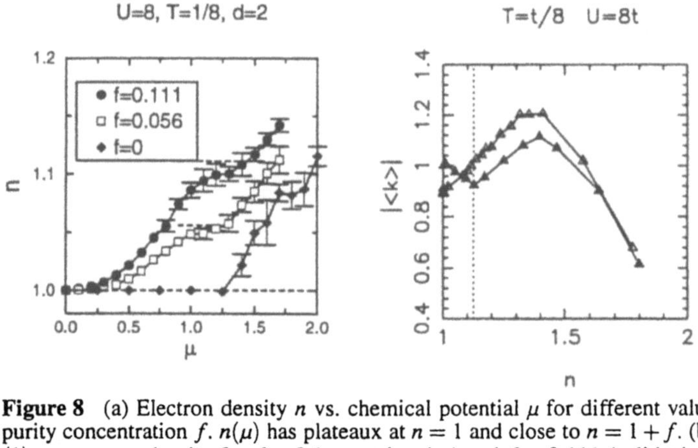

Figure 8 (a) Electron density n vs. chemical potential μ for different values of $U = 0$ impurity concentration f. $n(\mu)$ has plateaux at $n = 1$ and close to $n = 1 + f$. (b) Kinetic energy $\langle k \rangle$ vs. average density for $f = 0$ (open triangles) and $f = 0.111$ (solid triangles). $\langle k \rangle$ shows minima at $n = 1 + f$ [17].

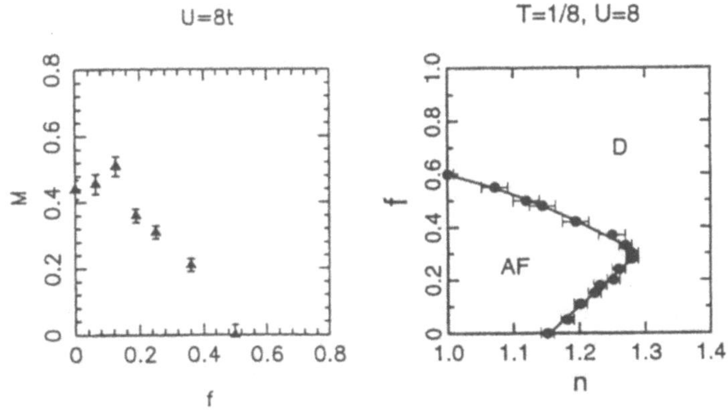

Figure 9 (a) Staggered magnetization in $d = 2$ as a function of impurity concentration f. (b) $f - n$ phase diagram at temperature $T = 1/8$ within DMFT. For small f the AF phase is stabilized against doping. [17]

5.2 Antiferromagnetic Order

Figure 9a shows the staggered moment M extrapolated to the thermodynamic limit in $d = 2$ at $n = 1$ as a function of impurity concentration f. For small f AF order is robust leading to the charge gap at $n = 1$. AF order vanishes at $f_c \approx 0.45$, i.e. close to the percolation threshold $f_{\text{perc}} \approx 0.4$.

Off half-filling the finite size scaling is no longer possible due to the minus-sign problem and the AF phase boundary is obtained within DMFT only (shown in Fig. 9b at $T = 1/8$). It is found that the $U = 0$ impurities can induce AF order at densities for which the clean model is disordered. In the clean model ($f = 0$) additional electrons are free to move and hence very effective in destroying long range order. $U = 0$ sites provide localizing centers which are energetically favorable for the additional carriers. This is why the AF phase extends to larger dopings at finite (but small) values of f. Since the mechanism of localizing mobile dopants is observed in $d = 2$, too, we expect the enlargement of the AF phase to be present in $d = 2$ at $T = 0$ as well, in spite of the fact that in the clean model in $d = 2$ the critical doping is supposed to be zero. Eventually at larger fractions f the AF phase shrinks, and the critical density approaches 1.0 for $f_c \approx 0.6$ ($f_c \approx 0.75$ in the ground state [17]). For even larger values of f there is no AF order even at $n = 1$.

Both the stabilization of AF order and the shift of the Mott gap to higher densities result from the localization of carriers at the $U = 0$ impurities. AF order is only destroyed when the density on the $U = 0$ sites saturates, i.e. it is stable *across* the Mott-Hubbard MIT. Hence the separation of Mott-Hubbard MIT and AF order is not present in the $U = 0$ impurity model, at least not for the present parameter values.

6 Summary and Conclusion

In this paper we discussed different types of disorder in correlated antiferromagnets and presented results obtained mostly by quantum Monte Carlo simulations in $d = 2$ and within dynamical mean-field theory ($d = \infty$).

Different mechanisms were identified by which disorder can enhance antiferromagnetic order: (i) disorder-enhanced delocalization at strong coupling in the case of weak disorder in the chemical potentials and (ii) localization of surplus carriers in the case of impurities with weak local interaction. In both cases compressible antiferromagnetic phases are observed. To determine if the gapless AF phase is metallic requires the calculation of transport properties which is presently in progress.

Quantum Monte Carlo simulations of electronic tight binding models are just making the transition from addressing rather abstract issues of correlation effects to making contact with real experiments. One important feature in this respect is the treatment of intrinsic disorder. For a quantitative description of experiments, however, the inclusion of a realistic band structure is mandatory. Here the DMFT will be particularly helpful because it allows for the treatment of multi-band models in a wider range of parameters and in the thermodynamic limit.

Bibliography

[1] G. Xiao, M. Z. Cieplak, A. Gavrin, F. H. Streitz, A. Bakhshai, and C. L. Chien Phys. Rev. Lett. **60**, 1446 (1988); B. Keimer et al., Phys. Rev. B **45**, 7430 (1992); A. V. Mahajan, H. Alloul, G. Collin, and J. F. Marucco, Phys. Rev. Lett. **72**, 3100 (1994).

[2] M. C. Martin, M. Hase, K. Hirota, G. Shirane, Y. Sasago, N. Koide, and K. Uchinokur, Phys. Rev. B **56**, 3173 (1997). P. E. Anderson, J. Z. Liu, and R. N. Shelton, Phys. Rev. B **56**, 11014 (1997).

[3] M. Azuma, Y. Fujishiro, M. Takano, M. Nohara, and H. Takagi, Phys. Rev. B **55**, 8658 (1997).

[4] H. Takagi et al., Phys. Rev. B **40**, 2254 (1989). S. Uchida et al., Phys. Rev. B **43**, 7942 (1991).

[5] T. Ido, K. Magoshi, H. Eisaki, and S. Uchida, Phys. Rev. B **44**, 12094 (1991).

[6] S. Hüfner, P. Steiner, I. Sander, F. Reinert, and H. Schmitt, Z. Phys. B **86**, 207 (1992); ibid. **88**, 247 (1992) F. Reinert et al., Z. Phys. B **97**, 83 (1995).

[7] M. Ma and E. Fradkin, Phys. Rev. B **28**, 2990 (1983); A. M. Finkelshtein, Zh. Eksp. Teor. Fiz. **84**, 168 (1983) [Sov. Phys. JETP **57**, 97 (1983)]; C. Castellani, C. Di Castro, P. A. Lee, and M. Ma, Phys. Rev. B **30**, 527 (1984); C. Castellani, C. Di Castro, and M. Grilli, *ibid.* **34**, 5907 (1986).

[8] A. M. Finkelshtein, Z. Phys. B **56**, 189 (1984); C. Castellani, et al. Phys. Rev. B **30**, 1956 (1984); **33**, 6169 (1986).

[9] M. Ma, Phys. Rev. B **26**, 5097 (1982); J. Yi, L. Zhang, and G. S. Canright, Phys. Rev. B **49**, 15 920 (1994).

[10] M. E. Tusch and D. E. Logan, Phys. Rev. B, **48**, 14 843 (1993).

[11] A. Singh and Z. Tešanović, Phys. Rev. B **41**, 614 (1990); **41**, 11 457 (1990), S. Basu and A. Singh, Phys. Rev. B **53**, 6406 (1996).

[12] V. Janiš and D. Vollhardt, Phys. Rev. B **46**, 15 712 (1992).

[13] V. Dobrosavljević and G. Kotliar, Phys. Rev. Lett. **71**, 3218 (1993); Phys. Rev. B **50**, 1430 (1994).

[14] V. Janiš, M. Ulmke, and D. Vollhardt, Europhys. Lett. **24**, 287 (1993); M. Ulmke, V. Janiš, and D. Vollhardt, Phys. Rev. B **51**, 10 411 (1995).

[15] A. Sandvik and D. J. Scalapino, Phys. Rev. **47**, 10090 (1993); A. Sandvik, D. J. Scalapino, and P. Henelius, Phys. Rev. **50**, 10474 (1994).

[16] M. Ulmke and R. T. Scalettar, Phys. Rev. B **55**, 4149 (1997).

[17] P. J. H. Denteneer, M. Ulmke, R. T. Scalettar, G. T. Zimanyi, Physica A **251**, 162 (1998); M. Ulmke, P. J. H. Denteneer, R. T. Scalettar, G. T. Zimanyi, Europhys. Lett. **42**, 655 (1998).

[18] D. Belitz and T. R. Kirkpatrick, Rev. Mod. Phys. **66**, 261, (1994).

[19] R. N. Bhatt and P. A. Lee, Phys. Rev. Lett. **48**, 344 (1982). M. Milovanović, S. Sachdev, and R. N. Bhatt, Phys. Rev. Lett. **63**, 82 (1989).

[20] R. Blankenbecler, D. J. Scalapino, and R. L. Sugar, Phys. Rev. **D 24**, 2278 (1981). J. E. Hirsch, Phys. Rev. **B 28**, 4059 (1983). G. Sugiyama and S.E. Koonin, Ann. Phys. **168**, 1 (1986); S. Sorella, S. Baroni, R. Car, and M. Parrinello, Europhys. Lett. **8**, 663 (1989). S. R. White, D. J. Scalapino, R. L. Sugar, E. Y. Loh, J. E. Gubernatis, and R. T. Scalettar, Phys. Rev. **B 40**, 506 (1989).

[21] D. A. Huse, Phys. Rev **B 37**, 2380 (1988).

[22] W. Metzner and D. Vollhardt, Phys. Rev. Lett. **62**, 324 (1989); D. Vollhardt, in *Correlated Electron Systems*, ed. V. J. Emery, World Scientific, Singapore, 1993.

[23] T. Pruschke, M. Jarrell, J. K. Freericks, Adv. Phys. **44**, 187 (1995); A. Georges, G. Kotliar, W. Krauth, M. Rozenberg, Rev. Mod. Phys. **68**, 13 (1996).

[24] P. A. Wolff, Phys. Rev. **124**, 1030 (1961).

[25] J. E. Hirsch and R. M. Fye, Phys. Rev. Lett. **56**, 2521 (1986).

[26] A. Singh, M. Ulmke, and D. Vollhardt, Phys. Rev. B **58** (to appear Oct. 1998); cond-mat/9803094.

[27] P. Sen, S. Basu, and A. Singh, Phys. Rev. B **50**, 10 381 (1994); P. Sen and A. Singh, Phys. Rev. B **53**, 328 (1996); A. Singh and P. Sen, Phys. Rev. B **57**, 10598 (1998).

[28] N. Bulut, D. Hone, D. J. Scalapino, and E. Y. Loh, Phys. Rev. Lett. **62**, 2192 (1989); D. Poilblanc, D. J. Scalapino, and W. Hanke, Phys. Rev. Lett. **72**, 884 (1994); G. B. Martins, M. Laukamp, J. Riera, and E. Dagotto, Phys. Rev. Lett. **78**, 3563 (1997). Y. Motome, N. Katoh, N. Furukawa, and M. Imada, J. Phys. Soc. Jpn. **65**, 1949 (1996).

[29] W. Brenig and A. P. Kampf, Phys. Rev. B **43**, 12 914 (1991); E. Manousakis, Phys. Rev. B **45**, 7570 (1992); C. C. Wan, A. B. Harris and D. Kumar, Phys. Rev. B **48**, 1036 (1993).

[30] M. Corti et al., Phys. Rev. B **56**, 11056 (1997).

[31] A. Sandvik and M. Vekić, Phys. Rev. Lett. **74** 1226 (1995).

[32] For a recent review see: F. Gebhard, *The Mott Metal–Insulator Transition*, Springer Tracts in Modern Physics, vol. 137 (Springer, Heidelberg, 1997).

Metallic Ferromagnetism: Progress in our Understanding of an Old Strong-Coupling Problem

D. Vollhardt, N. Blümer, K. Held, M. Kollar, J. Schlipf, M. Ulmke, and J. Wahle*

Theoretische Physik III, Elektronische Korrelationen und Magnetismus, Universität Augsburg, D-86135 Augsburg, Germany
* Present address: Theoretische Physik, Universität Duisburg, Lotharstr. 1, D-47048 Duisburg, Germany

Abstract: Metallic ferromagnetism is in general an intermediate to strong coupling phenomenon. Since there do not exist systematic analytic methods to investigate such types of problems, the microscopic origin of metallic ferromagnetism is still not sufficiently understood. However, during the last two or three years remarkable progress was made in this field: It is now certain that even in the one-band Hubbard model metallic ferromagnetism is stable in dimensions $d = 1$, 2, and ∞ on regular lattices and at intermediate values of the interaction U and density n. In this paper, the basic questions and recent insights regarding the microscopic conditions favoring metallic ferromagnetism in this model are reviewed. These findings are contrasted with the results for the orbitally degenerate case.

1 Introduction

What is the microscopic origin of ferromagnetism? Exactly seventy years ago, in 1928, Heisenberg addressed this question [1] after having discovered the phenomenon of quantum mechanical exchange and the corresponding exchange interaction. He formulated a spin model (the Heisenberg model), hoping to be able to answer precisely this question. However, it was pointed out by Bloch [2] that a model of localized spins cannot explain *metallic* ferromagnetism as observed in iron, cobalt, and nickel, and that a proper model would have to include the itineracy of the electrons, i.e. the band aspect. Based on the observation that the Curie temperature $T_c \sim 10^3$ K ~ 0.1 eV in these systems it is clear that the kinetic energy and the spin-*independent* Coulomb interaction, together with the Pauli principle, must ultimately be responsible for metallic ferromagnetism. Ever since one has been looking for the simplest microscopic model and mechanism explaining the origin of metallic ferromagnetism and, equally important, for analytic solutions or at least controlled approximations for these models [3]. Today we know that even with the "right" model these answers are not easily obtained since metallic ferromagnetism generally occurs only at *intermediate* to *strong* coupling and off half filling [4,5]. Thus, it belongs to the class of problems for which systematic theoretical approaches do not exist. Namely, weak-coupling theories or renormalization group approaches [6] which are so effective in detecting instabilities with respect to antiferromagnetism or superconductivity, do not work in this case. Instead, *nonperturbative* methods are required.

During the last two or three years significant progress was made in our understanding of the microscopic foundations of metallic ferromagnetism. These insights were made possible by both new analytic methods and new numerical techniques. In this paper, some of these recent developments will be reviewed. In particular, the microscopic conditions for metallic ferromagnetism in the one-band Hubbard model (Section 2) and in the case of orbital degeneracy (Section 3) are explained and the differences discussed. A conclusion (Section 4) closes the presentation.

2 The One-Band Hubbard Model

The simplest lattice model for correlated electrons, the one-band Hubbard model

$$H_{\mathrm{Hub}} = -\sum_{i,j,\sigma} t_{ij}(c_{i\sigma}^{\dagger}c_{j\sigma} + \mathrm{h.c.}) + U\sum_{i} n_{i\uparrow}n_{i\downarrow} \tag{2.1}$$

was proposed independently by Gutzwiller [7], Hubbard [8], and Kanamori [9] in 1963, with the explanation of metallic ferromagnetism in 3d transition metals in mind. Concerning the suitability of (2.1) to describe metallic ferromagnetism for general U and electron densities n the three authors came to different conclusions. In any case, the theoretical methods used at that time were not controlled enough to provide definitive conclusions. This is also true for most of the research following their original work, with a few exceptions such as Nagaoka's theorem for a single hole at $U = \infty$ [10]. We note that in the past the kinetic energy in (2.1) was usually restricted to nearest-neighbor (NN) hopping; then it is useful to divide the underlying lattices into bipartite and nonbipartite ones. About ten years ago the interest in the subject started to rise again [11]. In particular, by reducing Kanamori's [9] model density of states (DOS) of noninteracting electrons, $N^0(E)$ (Fig. 1a), to its barest minimum (Fig. 1b) Mielke [12] began to investigate the stability of ferromagnetism in systems with flat, i.e. dispersionless, bands. He [13] and Tasaki [14] were able to derive rigorous criteria for the existence of ferromagnetism in these particular systems [15]. Generalizations to nearly-flat bands are also possible [16]. Ferromagnetism is proven to exist when the lowest band is half-filled and the system is insulating, as well as close to half filling. Due to the pathological degeneracy of the ground state it is still not exactly clear whether away from half filling one really obtains *metallic* ferromagnetism [17].

A different route to ferromagnetism was taken by Müller-Hartmann [18] who investigated the t-t' Hubbard model

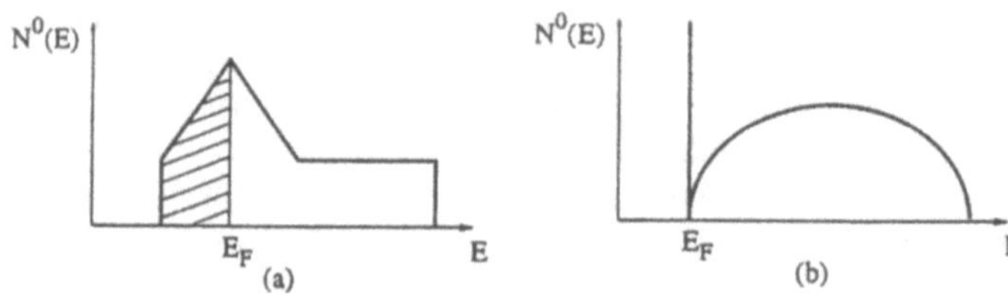

Figure 1 (a) Model DOS favoring ferromagnetism in the Hubbard model as suggested by Kanamori [9]; (b) DOS for flat-band ferromagnetism (schematic) [15].

$$H_{\text{Hub}}^{tt'} = -t \sum_{\text{NN},\sigma} (c_{i\sigma}^{\dagger} c_{j\sigma} + \text{h.c.}) - t' \sum_{\text{NNN},\sigma} (c_{i\sigma}^{\dagger} c_{j\sigma} + \text{h.c.}) + U \sum_i n_{i\uparrow} n_{i\downarrow}, \tag{2.2}$$

i.e. model (2.1) with NN and next-nearest neighbor (NNN) hopping. He found that in $d = 1$ at $U = \infty$ ferromagnetism becomes possible in the low-density limit ($n \to 0$). This scenario was extended by Pieri et al. [19] and, in particular, by Penc et al. [20] who introduced a generalized model which can be shown to have a metallic phase in $d = 1$.

In the Hubbard model the interaction term is completely independent of lattice and dimension. Therefore the kinetic energy, or dispersion, of the electrons and the underlying lattice must play an important role for the stability of metallic ferromagnetism. This is indeed seen explicitly in all of the above-mentioned investigations and is also apparent in the studies of the single spin-flip instability of the Nagaoka state for which Hanisch et al. [21] recently derived significantly improved bounds for various lattices in $d = 2$ in $d = 3$, and which was solved analytically by Uhrig [22] in the limit of $d = \infty$ for several nonbipartite lattices.

2.1 Routes to Ferromagnetism

On bipartite lattices the t'-hopping term destroys the antiferromagnetic nesting instability at small U [23]. In $d > 1$ it shifts spectral weight to the band edges and thereby introduces an asymmetry into the otherwise symmetric DOS. It will be shown below that a high spectral weight at the band edge (more precisely: the *lower* band edge for $n < 1$) minimizes the loss of kinetic energy of the overturned spins in the magnetic state and is hence energetically favorable. Therefore, frustrated, i.e. nonbipartite lattices, or bipartite lattices with frustration due to hopping (e.g. $t' \neq 0$) support the stabilization of metallic ferromagnetism. The fcc lattice is an example for a frustrated lattice in $d = 3$. The corresponding DOS of the noninteracting particles is shown in Fig. 2. Switching on an additional NNN hopping t' is seen to further increase the spectral weight at the lower band edge. For $t' = t/2$ one even obtains a square-root-like divergence.

To understand why a high spectral weight at the band edge is favorable for the kinetic energy we first consider the case $U = 0$, $n < 1$ [5]. Let us consider a flat, symmetric DOS as in Figs. 3a, b. Fig. 3a describes the paramagnetic state. The fully polarized state is obtained by inverting the spin of the down electrons, which due to the Pauli principle have to occupy higher energy states. Counting the energy from the lower band edge the Fermi energy of the polarized state, $\mu_\uparrow$, is seen to be twice that of the unpolarized state (Fig. 3b). This should be contrasted with the DOS having large spectral weight at the lower band edge shown in Figs. 3c and 3d. Here the Fermi level of the polarized state

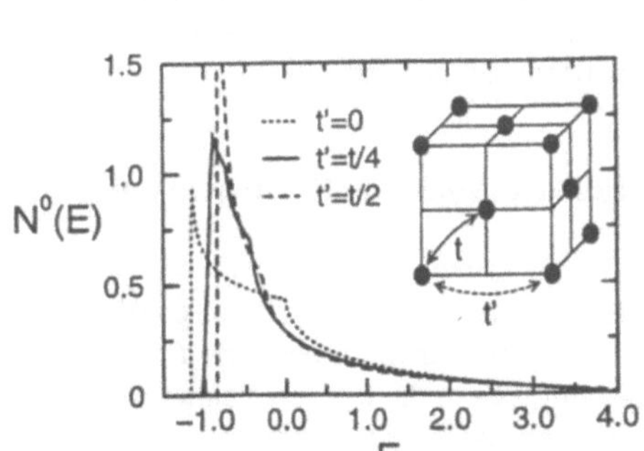

Figure 2
DOS of noninteracting electrons on a fcc lattice in $d = 3$ with and without additional NNN-hopping t'.

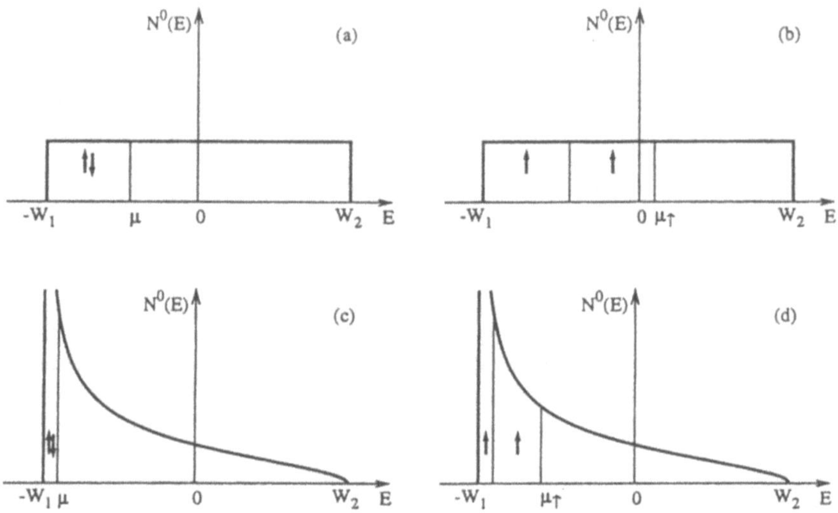

Figure 3 Flat, symmetric DOS for (a) unpolarized and (b) fully polarized electrons; (c) and (d): same as in (a), (b) but for a strongly peaked DOS.

is not so strongly shifted upwards, i.e. fewer high energy states are populated, which is clearly energetically favorable. The energy difference between the fully polarized state and the unpolarized state

$$\Delta E = \left[\int_{-W_1}^{\mu_\uparrow} - 2 \int_{-W_1}^{\mu} \right] dE\, N^0(E)\, E \tag{2.3}$$

must become negative for the ferromagnetic state to be stable. Of course, in the noninteracting case $\Delta E > 0$ [24]. Nevertheless, even for $U = 0$, ΔE attains its lowest value for a DOS with peaked spectral weight at the lower band edge for all n [5]. To show that $\Delta E < 0$ for $U > 0$ requires a good estimate of the energy of the *correlated paramagnet* – this is indeed a central problem of any correlation theory. It should be noted that the above discussion concerning the shape of the DOS goes beyond the well-known Stoner criterion which predicts an instability of the paramagnet for U equal to the inverse of the DOS precisely *at* the Fermi level.

Another possibility to stabilize ferromagnetism is to consider those interactions which are neglected in the Hubbard interaction, in particular the NN direct-exchange interaction. The effect of this and other terms will be discussed in Section 2.3.

2.2 Numerical Investigation of the Hubbard Model on Frustrated Lattices in d=1, 2, and ∞

Since metallic ferromagnetism is an intermediate coupling problem purely analytic approaches meet only with limited success, in particular in dimensions $d > 1$. In this situation the development of new numerical techniques in the last few years was of crucial importance for progress in this field. In particular, the density matrix renormalization group (DMRG), applicable mostly in $d = 1$, the projector quantum Monte Carlo method, and

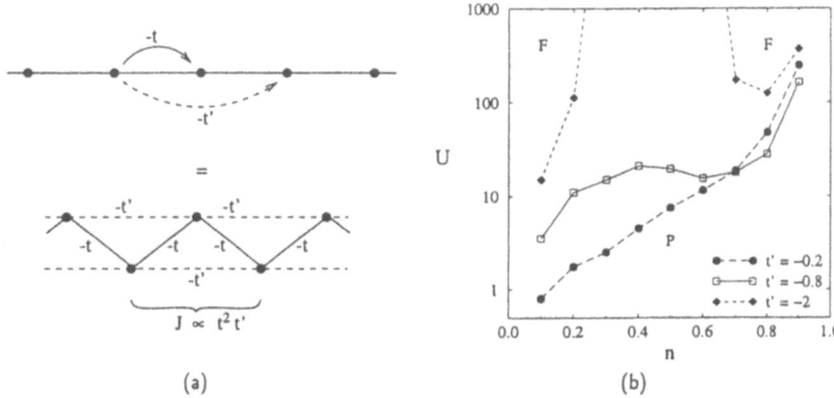

Figure 4 (a) t-t' Hubbard model in $d = 1$. (b) Ground state phase diagram U vs. n for the t-t' Hubbard model in $d = 1$ for several values of t' (after Daul and Noack [26]). P: paramagnetic phase, F: ferromagnetic phase.

the dynamic mean-field theory (DMFT), i.e. the large d limit, in connection with quantum Monte Carlo (QMC) have led to expiicit, reliable results in dimensions $d = 1,2,\infty$.

***d*=1:** In one dimension the t-t' Hubbard model may be viewed as a zig-zag chain made of triangular units (Fig. 4a). Taking one of these units by itself the effective exchange interaction between the spins of two electrons due to hopping of an electron or hole along the triangle is $J \propto t^2 t'$ [20,25]. It clearly shows that the sign of t' is crucial: only for $t' < 0$ does one obtain a ferromagnetic exchange; this seems to hold even in the extended system. Of course, $d = 1$ is a special dimension, since (i) ferromagnetic order is only possible at $T = 0$ and $t' \neq 0$ [24], and (ii) the DOS is always large at the band edges.

The t-t' Hubbard model was studied in detail by Daul and Noack [26] using DMRG for systems as large as 140 sites. The resulting magnetic phase diagram is shown in Fig. 4b. Even at small $|t'|$ a large ferromagnetic region in the U vs. n phase diagram is found to exist. As $|t'|$ increases the region of stability shrinks. This is due to the fact that for $|t'| \to \infty$ the zig-zag chain separates into two unconnected chains with t' as pure NN hopping; in this case the Lieb-Mattis theorem [24] rules out a ferromagnetic state.

***d*=2:** The t-t' Hubbard model on a square lattice was investigated by Hlubina et al. [27]. At $T = 0$, using projector QMC on systems as large as 20×20 sites and working at specific "van Hove densities" for which the Fermi energy coincides with the divergence in the noninteracting DOS, these authors found a region of metallic ferromagnetism, e.g. at $|t'| = 0.47t$.

***d*=∞:** Already in three dimensions the coordination number of a fcc lattice is $Z = 12$. It is therefore quite natural to view Z as a large number, and to consider the limit $Z \to \infty$ [28]. In this case one has to scale the hopping, e.g. as $t = t^*/\sqrt{Z}$ (in the following $t^* = 1$), and thus obtains a purely local theory where the self energy becomes $\vec{k}$ independent and where the propagator $G(\vec{k},\omega) = G^0(\vec{k},\omega - \Sigma(\omega))$ may be represented by the noninteracting propagator at a shifted frequency [29]. In this limit the dynamics of the quantum mechanical

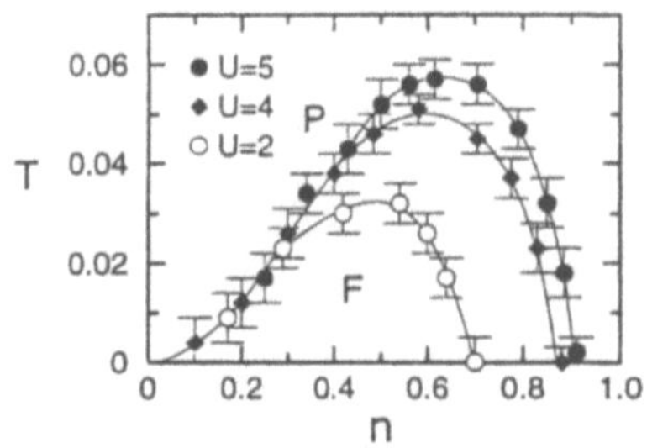

Figure 5
T vs. n phase diagram of the Hubbard model for a fcc lattice in $d = \infty$ for several values of U [31].

correlation problem is fully included, but due to the local nature of the theory there is no *short*-range order in position space. The dependence on the lattice or the dispersion is then encoded in the DOS $N^0(E)$ of the noninteracting particles. In view of these properties the $d \to \infty$ limit is now generally referred to as "dynamical mean-field theory" (DMFT) [30].

Investigations of the stability of metallic ferromagnetism on fcc-type lattices in large dimensions, obtained by solving the DMFT equations by finite-temperature QMC techniques, were first performed by Ulmke [31]. The resulting T vs. n phase diagram is shown in Fig. 5 for different values of the interaction parameter U. At $T = 0$ the critical interaction $U_c(n)$ (Fig. 5) is consistent with the analytically obtained spin-flip results by Uhrig [22]. The region of stability is seen to increase with U. By using an improved iterated perturbation theory to solve the DMFT equations Nolting et al. [32] obtained a similar phase diagram. To make contact with $d = 3$ we now use the corresponding fcc DOS shown in Fig. 2. For $t' = 0$ no instability is found at temperatures accessible to QMC. However, already a small contribution of t'-hopping (which is present in any real system) is enough to produce a large region of stability for metallic ferromagnetism in addition to an antiferromagnetic phase close to half filling (Fig. 6a) [31]. This shows the strong and subtle dependence of the stability on the dispersion and the distribution of spectral weight in the DOS. The maximal transition temperature is $T_c^{max} = 0.05 \simeq 500$ K for a band width $W = 4$ eV. This is well within the range of real transition temperatures, e.g. in nickel.

So far we only argued on the basis of the shape of the DOS of the *non*interacting electrons, $N^0(E)$. On the other hand the interaction will renormalize the band and relocate spectral weight. Therefore it is not *a priori* clear at all whether the arguments concerning the kinetic energy etc. (Fig. 3) still hold even at finite U. To settle this point we calculate the DOS of the interacting system, $N(E)$, by the maximum entropy method. In Fig. 6b we show $N(E)$ corresponding to the parameter values leading to the phase diagram in Fig. 6a. Clearly the ferromagnetic system is metallic since there is appreciable weight at the Fermi level ($E = \mu$). Furthermore, the spectrum of the majority spins is seen to be only slightly affected by the interaction, the overall shape of the noninteracting DOS being almost unchanged (the magnetization is quite large ($m = 0.56$ at $n = 0.66$) and hence the electrons in the majority band are almost noninteracting). This implies that the arguments concerning the distribution of spectral weight in the noninteracting case and the corresponding kinetic energy are even applicable to the polarized, interacting case. The spectrum of the minority spins is slightly shifted to higher energies and has a pronounced peak around $E - \mu \simeq U = 6$.

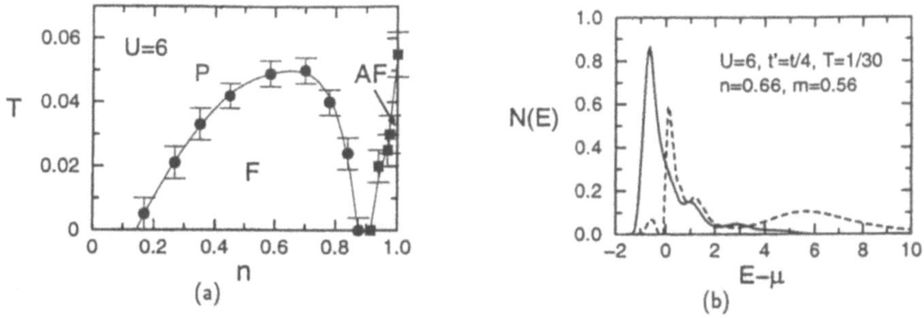

Figure 6 (a) T vs. n phase diagram of the Hubbard model as obtained within DMFT for the DOS corresponding to a three-dimensional fcc lattice with NN-hopping $t' = t/4$ (see Fig. 2) [31]; (b) DOS of the interacting electrons in the ferromagnetic phase of (a), solid line: majority spin, dashed line: minority spin.

To study the influence of the distribution of spectral weight on the stability of ferromagnetism within the DMFT systematically Wahle et al. [33] recently solved the DMFT equations with a tunable model DOS (Fig. 7),

$$N^0(E) = c\,\frac{\sqrt{D^2 - E^2}}{D + aE}. \tag{2.4}$$

with $c = (1 + \sqrt{1 - a^2})/(\pi D)$ and half-bandwidth $D \equiv 2$. Here a is an asymmetry parameter which can be used to change the DOS continuously from a symmetric, Bethe lattice DOS ($a = 0$) to a DOS with a square-root divergence at the lower band edge ($a = 1$), corresponding to a fcc lattice with $t' = t/4$ in $d = 3$ (Fig. 2). It is possible, in principle, to map any $N^0(E)$ to a dispersion $E(\vec{k})$ (although not uniquely).

The strong dependence of the stability of metallic ferromagnetism on the distribution of spectral weight is shown in Fig. 8a. Already a minute increase in spectral weight near the band edge of the noninteracting DOS, obtained by changing a from 0.97 to 0.98 (see Fig. 8b) is enough to almost double the stability region of the ferromagnetic phase. It should be mentioned that Obermeier et al. [34] found ferromagnetism even on a hypercubic, i.e. bipartite, lattice, but only at very large U values ($U > 30$).

The importance of genuine correlations for the stability of ferromagnetism is apparent from Fig. 9, where the DMFT results are compared with Hartree-Fock theory [33]. The quantum fluctuations, absent in Hartree-Fock theory, are seen to reduce the stability regime of ferromagnetism drastically. Spatial fluctuations (e.g. spin waves), absent also in the DMFT, should be expected to reduce that stability regime further.

2.3 Additional Interactions

In the one-band Hubbard model only the local interaction is retained. Thereby, several interactions which naturally arise when the Coulomb interaction is expressed in Wannier representation are neglected. Even in the limit of a single band and taking into account only NN contributions, four additional interactions appear [35,36,5]:

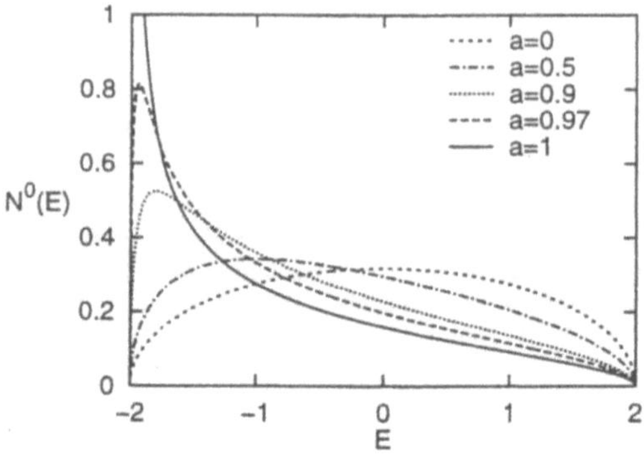

Figure 7
Model DOS, Eq. (2.4), shown for different values of the asymmetry parameter a [33].

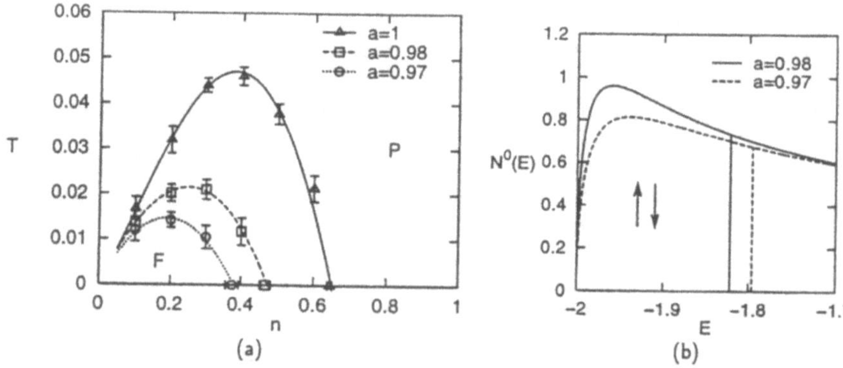

Figure 8 (a) T vs. n phase diagram of the Hubbard model as obtained within DMFT; (b) corresponding shapes of the noninteracting DOS; Fermi energies for $n = 0.3$ are indicated by vertical lines [33].

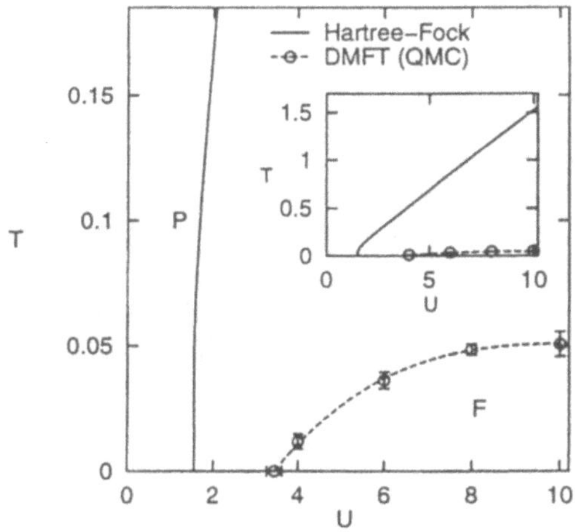

Figure 9
T vs. U phase diagram for a strongly peaked DOS ($a = 0.98$, see Fig. 8b) at $n = 0.4$ within DMFT (circles; dashed line is guide to the eyes only) in comparison with Hartree-Fock (solid line) [33].

$$V_{\text{1-band}}^{\text{NN}} = \sum_{\text{NN}} \Big[V n_i n_j + X \sum_{\sigma} (c_{i\sigma}^{\dagger} c_{j\sigma} + \text{h.c.})(n_{i,-\sigma} + n_{j,-\sigma}) \\ -2F(\vec{S}_i \cdot \vec{S}_j + \frac{1}{4} n_i n_j) + F'(c_{i\uparrow}^{\dagger} c_{i\downarrow}^{\dagger} c_{j\downarrow} c_{j\uparrow} + \text{h.c.}) \Big] . \tag{2.5}$$

Here the first term corresponds to a density-density interaction, the second term to a density-dependent hopping, and the fourth term describes the hopping of doubly occupied sites. In particular, the third term (with $F = F^*/Z > 0$)

$$H_F = -2\frac{F^*}{Z} \sum_{\text{NN}} \vec{S}_i \cdot \vec{S}_j \tag{2.6}$$

describes the direct ferromagnetic exchange between electrons on NN sites. It is this spin-type interaction which Heisenberg in his original model singled out as the main source of ferromagnetism. It should be noted, however, that this interaction is present even when the electrons are not localized but are free to move. The exchange interaction will be quite small, but nevertheless it favors ferromagnetic ordering in the most obvious way. Hirsch [36] argued that this term is the main driving force for metallic ferromagnetism in systems like iron, cobalt, and nickel. Indeed, one can show rigorously that a next-neighbor direct exchange interaction, if chosen large enough, can easily trigger the ferromagnetic instability [37,5]. To investigate the importance of the direct exchange interaction we supplement the Hubbard model by this term, $H = H_{\text{Hub}} + H_F$, neglecting all the other NN interactions. We note that within the DMFT the Heisenberg exchange reduces to the Weiss/Hartree-Fock contribution. In Fig. 10 the influence of the exchange interaction on the stability regime of ferromagnetism is depicted [33]. For $F^* = 0$ ferromagnetism is unstable down to the lowest temperatures for $U = 4$. However, by taking into account a small value of $F^* \simeq 0.15 \ll U$ at $T = 0$, the ferromagnetic phase is stabilized. Likewise, at larger values of U the critical temperature for the onset of ferromagnetism is significantly enhanced. Hence, F^* (and also the other neglected interactions) may well be important for systems on the verge of a ferromagnetic instability. Nevertheless, since we now know that the Hubbard interaction U together with a suitable kinetic energy is sufficient to trigger a ferromagnetic instability the ferromagnetic exchange interaction does not, in general, play an unrenounceable role and is thus less important than U itself.

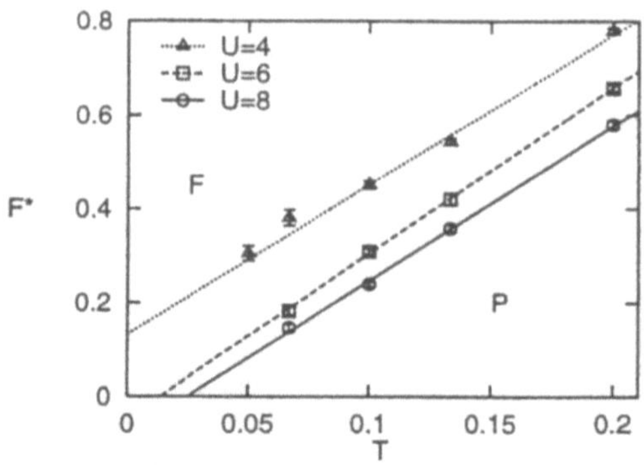

Figure 10
Direct exchange F^* vs. T phase diagram for the generalized model $H = H_{\text{Hub}} + H_F$ for different values of U in the case of a strongly peaked DOS ($a = 0.98$, see Fig. 8b) at $n = 0.6$. The linear extrapolation to $T = 0$ shows that there exists a critical value of U above which ferromagnetism is stable even in the absence of the direct exchange [33].

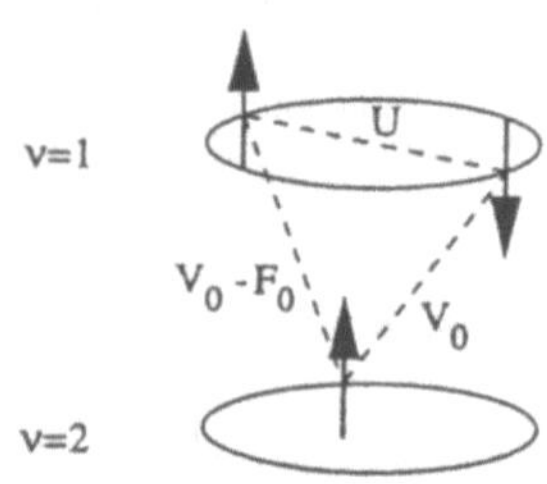

Figure 11
Illustration of the local interactions between electrons in a two-band model.

3 Orbital Degeneracy

The properties of the metallic ferromagnets iron, cobalt, and nickel are determined by 3d electrons, implying a five-fold degeneracy. Therefore it has long been speculated that band degeneracy is an essential precondition for metallic ferromagnetism. Band degeneracy leads to additional on-site matrix elements of the Coulomb interaction describing intra-atomic interactions

$$V_{\text{interband}} = \sum_i \left[\sum_{\nu<\nu';\sigma\sigma'} (V_0 - \delta_{\sigma\sigma'} F_0) n_{i\nu\sigma} n_{i\nu'\sigma'} - F_0 \sum_{\nu<\nu';\sigma\neq\sigma'} c^\dagger_{i\nu\sigma} c_{i\nu\sigma'} c^\dagger_{i\nu'\sigma'} c_{i\nu'\sigma} \right] \tag{3.7}$$

as shown in Fig. 11 in the case of a two-fold degeneracy. In particular, they imply a density-density interaction V_0 and a (ferromagnetic) exchange interaction F_0 between electrons on different orbitals. These "Hund's rule couplings" are responsible for the ferromagnetic alignment of the spins on an isolated atom. Slater [38] and van Vleck [39] suggested that this "atomic magnetism" may be transmitted from one atom to another by the kinetic energy, leading to coherent bulk order in the system. The relevant Hamiltonian is then a sum of Hubbard models for each orbital, complemented by the purely local interband coupling terms in (3.7):

$$H = \sum_\nu \left[-t \sum_{\text{NN},\sigma} c^\dagger_{i\nu\sigma} c_{j\nu\sigma} + U \sum_i n_{i\nu\uparrow} n_{i\nu\downarrow} \right] + V_{\text{interband}}. \tag{3.8}$$

This model has received wide attention [40], especially most recently [41]. Away from quarter or half filling the model is particularly difficult to treat due to the high degeneracy in the atomic limit. Quite generally ferromagnetism is found to be stabilized by Hund's

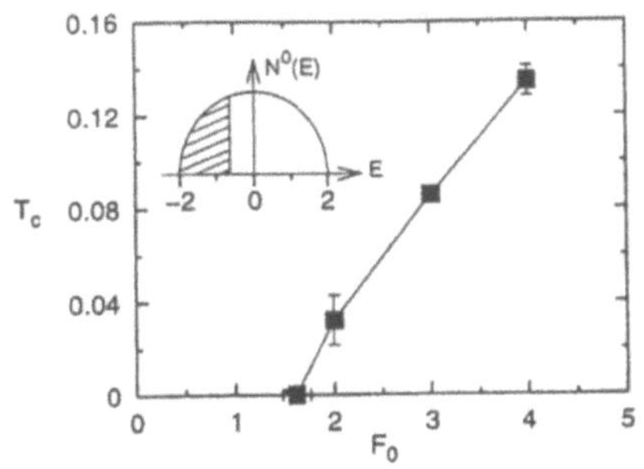

Figure 12
Curie temperature T_c vs. Hund's rule exchange coupling F_0 for a two-band Hubbard model with symmetric noninteracting DOS (see inset) as obtained within DMFT at $U = 9$, $V_0 = 5$, and $n = 1.25$ [42].

rule coupling at intermediate to strong interactions. In this regime the DMFT, solved by QMC, once more provides a powerful method for the investigation of (3.8) [42].

To identify the main mechanism responsible for ferromagnetism in the band-degenerate model and to distinguish it from that relevant for the one-band Hubbard model (namely, the strongly peaked DOS near the band edge) we here choose a featureless, symmetric Bethe-DOS as shown in the inset of Fig. 12. For such a DOS no ferromagnetic instability was found in the one-band model up to the largest U values within the DMFT [33] (see also [34]). In the following we restrict our discussion to a two-fold degeneracy. As can be seen from Fig. 12 no ferromagnetism occurs in the orbitally degenerate model even at $U = 9$ if the Hund's rule exchange interaction F_0 is absent. However, already a comparatively small value of F_0 is sufficient to make the ferromagnetic state favorable. The magnetic phase diagram T vs. n is shown in Fig. 13a for the same interaction parameters as in Fig. 12 at $F_0 = 4$ [42] (here we took into account the relation $U = V_0 + F_0$ which makes (3.8) form-invariant with respect to orbital rotations). Close to half filling ($n = 2$) the antiferromagnetic state is found to be stable, while for lower filling ferromagnetism is stable in a broad range of densities. The maximum critical temperature is $T_c^{max} \sim 0.1$ which, for a band width of 4 eV, corresponds to about 1000 K. This result should be compared with the Hartree-Fock result (Fig. 13b) which is both qualitatively and quantitatively insufficient. In particular, Hartree-Fock theory does not describe the suppression of T_c caused by the antiferromagnetic super-exchange near half filling. Furthermore, the critical temperatures are by more than an order of magnitude too high, reflecting the absence of dynamical fluctuations in this approximation. Fig. 12 clearly shows that already moderately large Hund's rule couplings are able to mediate metallic ferromagnetism even in a system with an unspecific, symmetric DOS. It is interesting to see that the magnetic phase diagrams T vs. n for the one-band model (Fig. 6a) and the band-degenerate model (Fig. 13a) look very similar, although the origin for the ferromagnetic phase is quite different.

Here we did not discuss the possibility of orbital ordering where the electron densities alternate on the two orbitals on neighboring sites. In [42] it is found that orbital ordering sets in around quarter filling ($n = 1$) when F_0 is decreased.

4 Discussion

In this paper we discussed recent developments in our understanding of the origin of metallic ferromagnetism both in the one-band Hubbard model and the band-degenerate

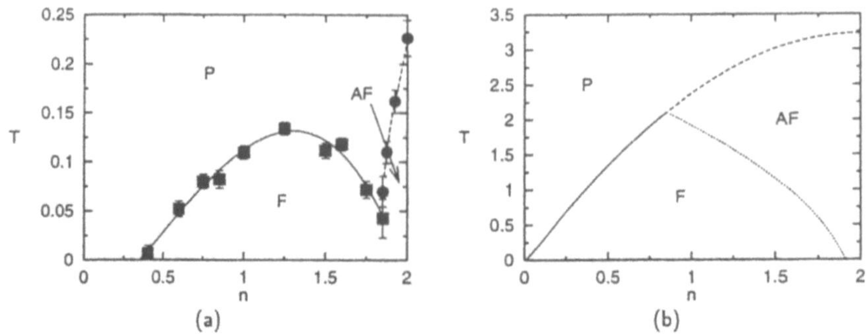

Figure 13 Magnetic phase diagram T vs. n of a two-band Hubbard model for the same U and V_0 as in Fig. 12, and $F_0 = 4$: (a) DMFT, (b) Hartree-Fock [42].

model. Analytical results for $d = 1$ and, in particular, numerical results for $d = 1, 2$, and ∞ were finally able to show convincingly that the one-band Hubbard model has a metallic ferromagnetic phase in a surprisingly large region of the on-site interaction U and density n. A stabilization of this phase at intermediate U values requires a sufficiently large spectral weight near the band edge. Such a DOS is typical for frustrated lattices which optimize the kinetic energy of the polarized state and at the same time frustrate the parasitic antiferromagnetic ordering. By contrast, the origin of metallic ferromagnetism in the band-degenerate Hubbard model need not primarily be due to a DOS effect but is rather caused by (moderate) Hund's rule couplings. In this respect the origin of ferromagnetism in the orbitally degenerate model is more straightforward than in the one-band case. In the absence of orbital ordering the resulting magnetic phase diagrams are remarkably similar.

The identification of a single main driving force for the stabilization of metallic ferromagnetism in the one-band and the band-degenerate model, respectively, helps to differentiate between different effects. In real systems these effects will tend to conspire, as is evident, for example, in nickel where an fcc lattice leads to a strongly asymmetric DOS and the band degeneracy provides for Hund's rule couplings. The combination of these effects will be investigated in the future.

Bibliography

[1] W. Heisenberg, Z. Physik **49**, 619 (1928).

[2] F. Bloch, Z. Phys. **57**, 545 (1929).

[3] D. C. Mattis, *The Theory of Magnetism I*, Springer, Berlin, 1988.

[4] E. H. Lieb, in *The Hubbard Model: Its Physics and Mathematical Physics*, edited by D. Baeriswyl et al., page 1, Plenum, New York, 1995; P. Fazekas, Phil. Mag. B **76**, 797 (1997).

[5] D. Vollhardt, N. Blümer, K. Held, M. Kollar, J. Schlipf, M. Ulmke, Z. Phys. B **103**, 283 (1997).

[6] R. Shankar, Rev. Mod. Phys. **66**, 129 (1994).

[7] M. C. Gutzwiller, Phys. Rev. Lett. **10**, 59 (1963).

[8] J. Hubbard, Proc. Roy. Soc. London A **276**, 238 (1963).

[9] J. Kanamori, Prog. Theor. Phys. **30**, 275 (1963).

[10] Y. Nagaoka, Phys. Rev. **147**, 392 (1966).

[11] E. H. Lieb, Phys. Rev. Lett. **62**, 1201 (1989).

[12] A. Mielke, J. Phys. A **24**, 3311 (1991).

[13] A. Mielke, J. Phys. A **25**, 4335 (1992); A. Mielke, Phys. Lett. A **174**, 443 (1993).

[14] H. Tasaki, Phys. Rev. Lett. **69**, 1608 (1992).

[15] A. Mielke and H. Tasaki, Commun. Math. Phys. **158**, 341 (1993).

[16] H. Tasaki, Phys. Rev. Lett. **73**, 1158 (1994); *ibid.* **75**, 4678 (1995).

[17] H. Tasaki, Prog. Theor. Phys. **99**, 489 (1998).

[18] E. Müller-Hartmann, J. Low Temp. Phys. **99**, 349 (1995).

[19] P. Pieri, S. Daul, D. Baeriswyl, and M. Dzierzawa, Phys. Rev. B **54**, 9250 (1996).

[20] K. Penc, H. Shiba, F. Mila, and T. Tsukagoshi, Phys. Rev. B **54**, 4056 (1996).

[21] T. Hanisch, G. S. Uhrig, and E. Müller-Hartmann, Phys. Rev. B **56**, 13960 (1997).

[22] G. S. Uhrig, Phys. Rev. Lett. **77**, 3629 (1996).

[23] W. Hofstetter and D. Vollhardt, Ann. Physik **7**, 48 (1998).

[24] E. H. Lieb and D. Mattis, Phys. Rev. **125**, 164 (1962).

[25] H. Tasaki, J. Phys.: Condens. Matter **10**, 4353 (1998).

[26] S. Daul and R. Noack, Z. Phys. B **103**, 293 (1997). S. Daul and R. Noack, Phys. Rev. B **58**, 2635 (1998).

[27] R. Hlubina, S. Sorella, and F. Guinea, Phys. Rev. Lett. **78**, 1343 (1997).

[28] W. Metzner and D. Vollhardt, Phys. Rev. Lett. **62**, 324 (1989).

[29] D. Vollhardt, in *Correlated Electron Systems*, edited by V. J. Emery, World Scientific, Singapore, 1993.

[30] T. Pruschke, M. Jarrell, and J. K. Freericks, Adv. Phys. **44**, 187 (1995); A. Georges, G. Kotliar, W. Krauth, and M. Rozenberg, Rev. Mod. Phys. **68**, 13 (1996).

[31] M. Ulmke, Euro. Phys. J. B **1**, 301 (1998).

[32] T. Herrmann and W. Nolting, J. Magn. Magn. Mater. **170**, 253 (1997); T. Herrmann and W. Nolting, Solid State Commun. **103**, 351 (1997); T. Wegner, M. Potthoff, and W. Nolting, Phys. Rev. B **57**, 6211 (1998).

[33] J. Wahle, N. Blümer, J. Schlipf, K. Held, and D. Vollhardt, Phys. Rev. B **58** (1998, in press).

[34] T. Obermeier, T. Pruschke, and J. Keller, Phys. Rev. B **56**, R8479 (1997).

[35] J. T. Gammel and D. K. Campbell, Phys. Rev. Lett. **60**, C 71 (1988); D. K. Campbell, J. T. Gammel, and E. Y. Loh Jr., Phys. Rev. B **38**, 12043 (1988).

[36] J. E. Hirsch, Phys. Rev. B **40**, 2354 (1989); *ibid.* **40**, 9061 (1989); S. Tang and J. E. Hirsch, Phys. Rev. B **42**, 771 (1990).

[37] R. Strack and D. Vollhardt, Phys. Rev. Lett. **72**, 3425 (1994); M. Kollar, R. Strack, and D. Vollhardt, Phys. Rev. B **53**, 9225 (1996).

[38] J. C. Slater, Phys. Rev. **49**, 537 (1936).

[39] J. H. van Vleck, Rev. Mod. Phys. **25**, 220 (1953).

[40] L. M. Roth, Phys. Rev. **149**, 306 (1966); K. I. Kugel' and D. I. Khomskiĭ, Sov. Phys.-JETP **37**, 725 (1973); M. Cyrot and C. Lyon-Caen, J. Phys. C **36**, 253 (1975); W. Gill and D. J. Scalapino, Phys. Rev. B **35**, 215 (1987); A. M. Oleś, Phys. Rev. B **23**, 271 (1981); *ibid.* **28**, 327 (1983); A. M. Oleś and G. Stollhoff, Phys. Rev. B **29**, 314 (1984); G. Stollhoff and P. Thalmeier, Z. Phys. B **43**, 13 (1981).

[41] J. Kuei and R. T. Scalettar, Phys. Rev. B **55**, 14968 (1997); J. E. Hirsch, Phys. Rev. B **56**, 11022 (1997); M. Fleck, A. M. Oleś, and L. Hedin, Phys. Rev. B **56**, 3159 (1997); T. Okabe, Prog. Theor. Phys. **98**, 331 (1997); J. Bünemann and W. Weber, Physica B **230**, 412 (1997); J. Bünemann, F. Gebhard, and W. Weber, J. Phys. Condens. Matter **9**, 7343 (1997); J. Bünemann, W. Weber, and F. Gebhard, Phys. Rev. B **57**, 6896 (1998).

[42] K. Held and D. Vollhardt, Euri. Phys. J. B (1998, in press).

Interacting Electrons in a Narrow Quantum Hall System

Tapash Chakraborty

Max-Planck Institut für Physik Komplexer Systeme
Nöthnitzer Strasse 38, D-01187 Dresden, Germany

Abstract: We discuss the electronic properties of a system of interacting electrons in a narrow channel in the quantum Hall effect regime. We demonstrate that the system undergoes phase transitions when the strength of the interaction or the geometry of the channel is varied. The resulting phase diagram contains odd-denominator quantum Hall states and also a stable $\nu = \frac{1}{2}$ state. The collective mode evaluated at the half-filled case is strikingly similar to that of an odd-denominator state. We discuss the nature of the phases with non-zero total momentum or with no energy gap. We also analyze the spin polarizations of those stable states as a function of temperature. At $\nu = \frac{1}{2}$ no two-dimensional analog exists, but for the odd-denominator states the temperature dependence closely resemble the ones in the corresponding two dimensional systems where the results have received support from experiments.

1 Introduction

One of the perhaps most spectacular demonstrations of electron correlations in nature is the fractional quantum Hall effect (FQHE) [1]. It is characterized by Hall plateaus at Landau level filling factors $\nu = N_e/N_s$ (N_e is the number of electrons and N_s is the Landau level degeneracy) corresponding to certain simple fractions with mostly odd denominators. This effect is believed to occur when a two dimensional electron system (2DES) in a strong perpendicular magnetic field and at low temperatures condenses into an incompressible liquid state – the Laughlin state [2]. The FQHE occurs when the chemical potential of the electron system is discontinuous at certain magnetic field dependent densities leading to incompressibility [3]. The energy gap structures in the excitation spectra are also well established for the incompressible states [4]. Interestingly, despite intense activities at the half-filled Landau level [5–7], such a level of understanding has not yet been achieved for the simplest even-denominator state, which is compressible and shows no sign of QHE.

In recent years there has been a great deal of interest in the study of electron correlations in systems where electrons are confined in even lower dimensions [8]. For example, ballistic quantum wires fabricated [9] with long transport mean-free path and very large subband spacings enable one to study effects of electron-electron interactions in one-dimensional systems. At high magnetic fields when the radius of the cyclotron orbit – the length scale in the system – is smaller than the width of the channel, it is plausible that the system behaves like a two-dimensional electron gas (2DEG) in the sense that it exhibits QH states. The boundaries of the sample are of course still important in that regime and therefore in theoretical models one should be able to treat the edges and bulk on the same footing.

Prior to our theoretical work presented here, there were a few earlier attempts to study QHE in one dimension (1D-QHE). Theoretical work on one-dimensional systems with hard wall boundary condition [10] indicated that as the channel width is varied there is a transition between the FQH state and the $2k_F$-charge density wave state. It was also noticed that the size of the FQHE gap is comparable for the filling factors $\nu = \frac{1}{2}$ and $\nu = \frac{1}{3}$ [10], which would mean that just like the stable $\frac{1}{3}$ QHE state (Laughlin state), $\nu = \frac{1}{2}$ should also exhibit FQHE. Following that theoretical work, observation of incompressible states at specific filling factors in a narrow channel electron system was reported a few years ago [11]. In that experiment, in addition to various odd-denominator fractions, Hall resistance also showed signatures of quantization at the half-filled Landau level. That observation brought into focus the question about the existence and nature of even-denominator states in a narrow channel together with the role of boundaries of the sample. Theoretical interest on 1D-FQHE has been revived by recent work of Yoshioka [12,13] where a parabolic confinement (see below) was used and incompressible states were found at $\nu = \frac{1}{3}$ and $\frac{2}{3}$, but that approach was somewhat different in spirit than our work described below.

2 A System of Spinless Electrons

In our model for the 1D-QHE, electrons (spin polarized) are confined by an electrostatic potential which is parabolic in the lateral direction and flat along the channel, and are also subjected to a strong perpendicular magnetic field. The electrons are considered to be in a cell of length a. The width of the cell is determined by the lateral electron density profile which is finite due to the presence of the confinement potential. We impose a periodicity condition along the length of the cell. The total Hamiltonian of the system is

$$\mathcal{H} = \mathcal{H}_0 + \mathcal{H}_{\rm int} \tag{2.1}$$

where $\mathcal{H}_0$ includes the kinetic energy of N_e electrons of effective mass m^* and the electrostatic confining potential

$$\mathcal{H}_0 = \sum_{i=1}^{N_e} \left[\frac{1}{2m^*} \left(\vec{p}_i - e\vec{A}_i \right)^2 + \frac{1}{2} m^* \omega_0^2 y_i^2 \right] \tag{2.2}$$

with $\vec{A}$ being the vector potential in Landau gauge. The interaction term of the Hamiltonian consists of Coulomb interactions and also terms corresponding to the neutralizing background

$$\begin{aligned} \mathcal{H}_{ee} &= \frac{1}{2} \sum_{i \neq j} \frac{e^2}{\varepsilon |\vec{r}_i - \vec{r}_j|}, \\ \mathcal{H}_{bb} &= \frac{1}{2} \int d\vec{R} d\vec{R}' \frac{e^2 \rho(\vec{R}) \rho(\vec{R}')}{\varepsilon |\vec{R} - \vec{R}'|}, \\ \mathcal{H}_{eb} &= - \sum_i \int d\vec{R} \frac{e^2 \rho(\vec{R})}{\varepsilon |\vec{R} - \vec{r}_i|} \end{aligned} \tag{2.3}$$

where ε is the background dielectric constant and $\rho(\vec{R})$ is the background charge density. The single-electron wave functions labeled by $\kappa = \{m,n\}$ are

$$\psi_\kappa = \left(\frac{1}{a\sqrt{\pi}\lambda}\right)^{\frac{1}{2}} \exp\left(ikx - \frac{\hat{y}^2}{2\lambda^2}\right) H_n\left(\frac{\hat{y}}{\lambda}\right) \tag{2.4}$$

where $k = (2\pi/a)m$ is the quantized wave vector and m is the momentum quantum number. The effective Larmor radius for the cyclotron motion is $\lambda = (\hbar/m^*\Omega)^{\frac{1}{2}}$, where $\Omega = \sqrt{\omega_0^2 + \omega_c^2}$ and $\omega_c = eB/m^*$ is the cyclotron frequency. The single-electron states Eq. (2.4) are localized around $\hat{y} \equiv y + 2\pi\lambda^2 m/\gamma a = 0$. Here $\gamma = \sqrt{1 + (\omega_0/\omega_c)^2}$ is a dimensionless quantity. The single-electron states for m and m' are therefore separated spatially in lateral direction by an amount $\Delta y_{mm'} = 2\pi(m-m')\lambda^2/\gamma a$. For a very strong confinement energy with respect to cyclotron energy, spatial separation of the single-electron states vanishes – the system becomes truly one dimensional. On the other hand, when the confinement becomes weaker with respect to cyclotron energy, the states begin to separate spatially and the system becomes quasi-one-dimensional. In (2.4), H_n is the Hermite polynomial of order n. We consider the limit of a strong magnetic field and therefore, in what follows we set $n = 0$ (lowest Landau level). Ignoring the constant Landau level energy the single-electron energies are given by

$$\mathcal{E}_m = \frac{\hbar^2 k^2}{2m^*} \frac{\omega_0^2}{\Omega^2}. \tag{2.5}$$

It should be pointed out that a related model where the electrons are confined in both x- and y-directions by a parabolic confinement potential, one gets a system of quantum dots or more popularly termed as *artificial atoms* [14], whose electronic properties are no less intriguing.

In the non-interacting ground state, N_e electrons occupy the lowest available single-particle states. It is then reasonable to require that the electron density is symmetric around the $y = 0$ axis, i.e., $M = \sum_j m_j = 0$. This symmetry condition holds for *odd* number of electrons if m is an *integer*. In that case the boundary condition is periodic: $\psi_\kappa(\vec{r} + a\hat{\vec{x}}) = \psi_\kappa(\vec{r})$. For *even* number of electrons m has to be a *half-odd integer*. The boundary condition is then anti-periodic: $\psi_\kappa(\vec{r} + a\hat{\vec{x}}) = -\psi_\kappa(\vec{r})$. As we turn on the inter-electron interaction electrons start to avoid each other. Increasing the interaction strength with respect to the kinetic energy it becomes energetically favorable for the electrons to occupy also the higher energy single-particle levels in order to reduce the Coulomb repulsion.

Given the single-particle states the Hamiltonian of the many-electron system, projected to the lowest Landau level, is now written

$$\mathcal{H} = \sum_m \mathcal{E}_m a_m^\dagger a_m + \sum_{m_1,\cdots,m_4} \mathcal{A}_{m_1,m_2,m_3,m_4} a_{m_1}^\dagger a_{m_2}^\dagger a_{m_3} a_{m_4} \tag{2.6}$$

where $a_m^\dagger(a_m)$ is the creation (annihilation) operator for the state m. For the Coulomb potential

$$v(\vec{r}) = \frac{1}{(2\pi)^2} \int d\vec{q} \frac{2\pi e^2}{\varepsilon q} e^{-i\vec{q}\cdot\vec{r}} \tag{2.7}$$

where $\vec{q} = (q_x, q_y)$, the interaction matrix element is given by

$$\begin{aligned}
\mathcal{A}_{m_1,m_2,m_3,m_4} &= \frac{1}{2}\int d\vec{r}_1 d\vec{r}_2 \psi^*_{m_1}(\vec{r}_1)\psi^*_{m_2}(\vec{r}_2)v(\vec{r})\psi_{m_3}(\vec{r}_2)\psi_{m_4}(\vec{r}_1) \\
&= \frac{1}{2}\frac{e^2}{\varepsilon\lambda}\exp\left[-\frac{1}{2}\left(\frac{2\pi}{\gamma a}\right)^2(m_1-m_4)^2\right] \\
&\times \int dq'_y \frac{\exp\left[i2\pi(m_3-m_1)q'_y\right]\exp\left[-\frac{1}{2}\left(\gamma a q'_y\right)^2\right]}{\sqrt{\left[\frac{2\pi(m_1-m_4)}{\gamma a}\right]^2 + \left(aq'_y\right)^2}} \\
&\times \delta_{m_1+m_2,m_3+m_4}
\end{aligned} \tag{2.8}$$

with $q'_y = q_y\lambda^2/(\gamma a)$ as the dimensionless integration variable, λ is the unit of length and $\mathcal{E}_c = e^2/\varepsilon\lambda$ is a measure of the interaction strength. It is to be noted that the diagonal term $m_1 = m_4$ $(m_2 = m_3)$ diverges as expected for the long-range Coulomb potential. While Yoshioka used a truncated Coulomb potential (section 4), we neutralize the system by embedding the quantum wire into a positively charged background with a density

$$\rho(\vec{R}) = \frac{N_e}{a\sqrt{\pi}d'}\exp\left[-\frac{Y^2}{d'^2}\right]. \tag{2.9}$$

The width of the background density profile is denoted by d' and in actual calculations we set it equal to the length of the cell. The electrostatic energy of the background charge is

$$\begin{aligned}
\mathcal{H}_b &= \frac{1}{2}\int d\vec{R}d\vec{R}'\frac{e^2\rho(\vec{R})\rho(\vec{R}')}{\varepsilon|\vec{R}-\vec{R}'|} \\
&= \frac{1}{2}N_e^2\frac{e^2}{\varepsilon\lambda}\frac{1}{a}\int dq'_y\frac{1}{|q'_y|}\exp\left[-\frac{1}{2}(\gamma a d' q'_y)^2\right].
\end{aligned} \tag{2.10}$$

The interaction of electrons with the positive background charge is taken into account by the potential $\mathcal{H}_{eb}$. In general, the electron-background interaction is a one-body operator with the diagonal matrix element

$$\begin{aligned}
\mathcal{D}_{m_1,m_2} &= \int d\vec{r}\psi^*_{m_1}(\vec{r})v_{eb}(\vec{r})\psi_{m_2}(\vec{r}) \\
&= -\frac{e^2}{\varepsilon\lambda}\frac{N_e}{a}\int dq'_y\frac{1}{|q'_y|}\exp\left[i2\pi m_1 q'_y - \frac{1}{4}(\gamma a q'_y)^2(1+d'^2)\right] \\
&\times \quad \delta_{m_1,m_2}.
\end{aligned} \tag{2.11}$$

Using the above relations, the final form of the diagonal two-body term is

$$\begin{aligned}
\mathcal{A}_{m,m',m',m} &= \frac{1}{2}\frac{e^2}{\varepsilon\lambda}\frac{1}{a}\int dq'_y\frac{1}{|q'_y|}\left\{\exp\left[i2\pi(m-m')q'_y - \frac{1}{2}(\gamma a q'_y)^2\right]\right. \\
&\left. -2\exp\left[i2\pi m q'_y - \frac{1}{4}(1+d'^2)(\gamma a q'_y)^2\right] + \exp\left[-\frac{1}{2}(\gamma a d' q'_y)^2\right]\right\}
\end{aligned}$$

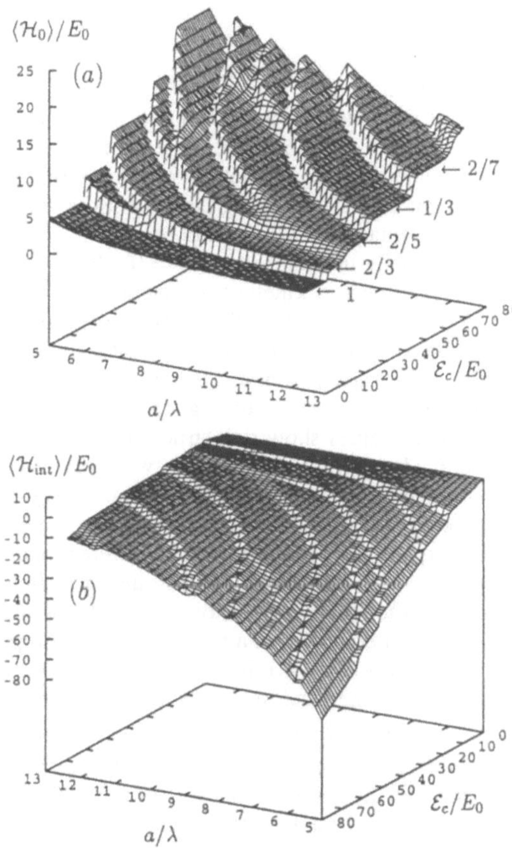

Figure 1
Expectation value of (a) kinetic energy per particle and (b) interaction energy per particle, as a function of $\mathcal{E}_c/E_0$ and length of the cell for the state M=0. The effective filling factors $\nu = \frac{1}{3}, \frac{2}{3}, \frac{2}{5}, \frac{2}{7}$ and $\nu = 1$ are also indicated.

where the divergent part of the electron-electron term is now cancelled by the positive background charge. Unlike the other theoretical models of 1D-FQHE, in our approach we can therefore include the long-range part of the potential.

2.1 Filling Factors in a Narrow Channel

We should point out that evaluation of the filling factor ν in a narrow channel is somewhat tricky. Here the single-particle states corresponding to a particular Landau "level" are not degenerate. One possible way would be to calculate the areal electron density and number of fluxes through a unit area and determine ν as the ratio of these two quantities. Alternatively, we count the number of occupied states and divide the number of electrons by that. Both methods are somewhat arbitrary: one has to choose properly either the width of the density profile in the first case (we have used below full width at half maximum) or, in the second approach, which state should be considered as occupied. We have checked that both methods agree reasonably well. The $\frac{1}{3}$ FQH state in the present system is also identified from the momentum distribution function $\langle n(k)\rangle = \langle 0|a_k^\dagger a_k|0\rangle$ by comparing it with that for a Laughlin-like wave function.

2.2 Phase Transitions

In all the numerical results that follow we have used $\omega_0/\omega_c = 0.01$. One has also to be careful about the range of a/λ, since if the ratio is too large or too small the length or width of the channel gets too large. In those situations, the system becomes truly one dimensional and the electrons prefer to line up along the bottom of the external potential or in the lateral direction respectively. For the various filling factors considered here, the range of a/λ is such that the aspect ratio remains usually around unity.

Let us first examine how the translationally invariant state ($M = 0$) changes when we change the strength of interactions with respect to kinetic and potential energies of the electrons, i.e., $\mathcal{E}_c/E_0$ (where $E_0 = (\hbar^2/2m^*\lambda^2)(\omega_0^2/\Omega^2)$ is the energy unit) and the length of the cell a. As we vary $\mathcal{E}_c/E_0$ while keeping a fixed, the expectation values of the kinetic and potential energies change very *abruptly* from one value to another. When the calculation is repeated for other fixed values of a we obtain Fig. 1a and Fig. 1b for $\langle\mathcal{H}_0\rangle$ and $\langle\mathcal{H}_{\rm int}\rangle$ respectively. These expectation values show rich structures in the parameter space spanned by $a = 5, \cdots, 12.4$ and $\mathcal{E}_c/E_0 = 0, \cdots, 80$ [15]. The two energies $\langle\mathcal{H}_0\rangle$ and $\langle\mathcal{H}_{\rm int}\rangle$ jump in opposite directions and therefore the net change in total energy does not clearly show the sudden changes in the $M = 0$ state. However, for a much longer system (at a fixed linear density) we expect sharper transitions between the different phases.

As the jump occurs in the parameter space spanned by $\mathcal{E}_c/E_0$ and a, it indicates a change in the $M = 0$ state. Yoshioka identified the filling factors $\nu = \frac{2}{3}$, $\nu = \frac{1}{3}$ and $\nu = \frac{1}{5}$ FQHE states in a system of six electrons interacting via a truncated Coulomb potential and calculating the overlap with the Laughlin-like wave functions [12]. These states are also realized in our system with real long-range Coulomb potential. The state at $\nu = \frac{1}{3}$ can also be characterized by calculating the overlap between the Coulomb-$\frac{1}{3}$ state and Haldane's pseudopotential-$\frac{1}{3}$ state [13] (see below). We have checked this overlap in our present system and found it to vary between the values 0.83 and 0.89 at $a = 9.5$.

Motivated by indications of the existence of QHE at $\nu = \frac{1}{2}$ in a narrow channel as mentioned in the introduction, we are interested in going beyond the $\nu = \frac{1}{3}$ state which seems to be indifferent to the presence of a lateral confinement. In fact, in our quest for a QHE at the half-filled Landau level we are particularly interested to know what happens in between the well established FQHE states. For example, what are the states realized in between the FQHE states $\nu = \frac{2}{3}$ and $\nu = \frac{1}{3}$? In this region there are clear jumps in both $\langle\mathcal{H}_0\rangle$ and $\langle\mathcal{H}_{\rm int}\rangle$. To get further insight on the $M = 0$ states realized in the wire, we have investigated the problem of how the electron density profile of the translationally invariant state ($M = 0$) is modified when we change $\mathcal{E}_c/E_0$ for a fixed value of a.

2.3 Electron Density

We have calculated electron density at $\vec{r}$ numerically from [15]

$$\rho(\vec{r}) = \sum_{i,j=1}^{\infty} \psi_i^*(\vec{r})\psi_j(\vec{r})a_i^\dagger a_j. \tag{2.12}$$

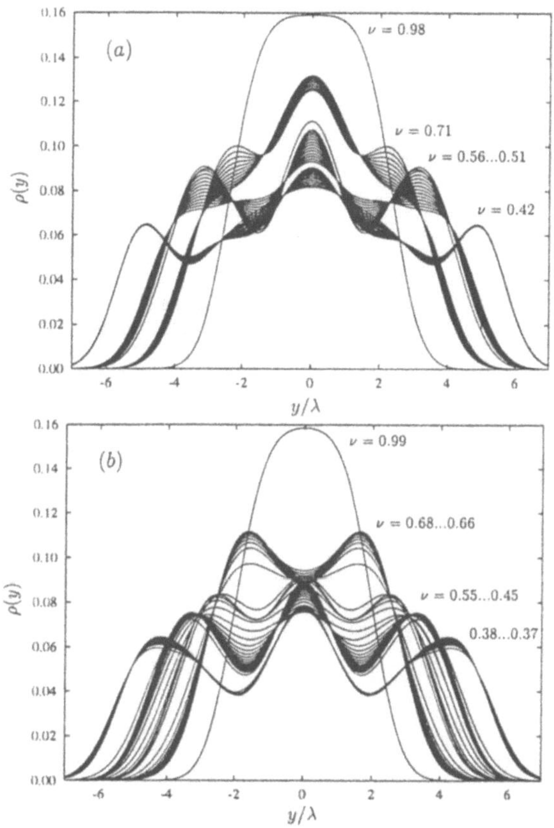

Figure 2
(a) Calculated electon densities in the lateral direction for $a = 8$ and the $M = 0$ states. (b) The results at $a = 9.5$. The effective filling factors are also indicated.

In the x-direction the charge density is constant while in the lateral y-direction it is modified because of the finite width of the system. Interestingly, as $\mathcal{E}_c/E_0$ increases the width of the charge density profile also changes *abruptly* from one value to another (Fig. 2).

The effective filling factors, calculated via $\nu = 2\pi\lambda^2 n$ (where n is the number of electrons per unit area and the width of the wire is the full width at half maximum) are found to be 0.98,0.71,...,0.51 and 0.42 which are very close to $\nu = 1, \frac{2}{3}, \frac{1}{2}$ and $\nu = \frac{2}{5}$ in Fig. 1(a) ($a = 8$) and 0.99 and 0.68,...,0.66 in Fig. 1(b) ($a = 9.5$) which suggests that these states are $\nu = 1$ and $\nu = \frac{2}{3}$, respectively. The state which has the effective filling factor 0.38,...,0.37 is identified as a $\nu = \frac{1}{3}$ state also by calculating the Coulomb-$\frac{1}{3}$ state and Haldane's pseudopotential state [13], as discussed above.

2.4 Phase Diagram

A phase diagram for the 1D-QHE states is then obtained (Fig. 3) by systematically seeking those points in the parameter space spanned by a and $\mathcal{E}_c/E_0$ where the ground state has *zero total momentum*. We then plot the energy gap between this ground state and the first excited state. In Fig. 3 the area of a filled dot is directly proportional to that gap. Clearly, the phase diagram consists of separate regions of several QHE states. Interestingly, there

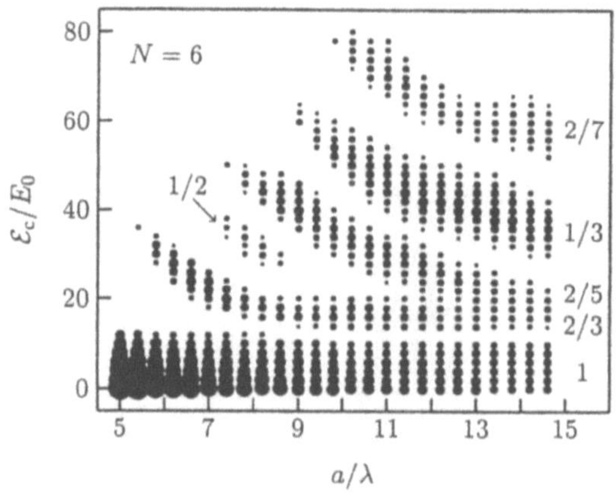

Figure 3
Phase diagram for the FQHE states derived at the effective filling factors $\nu = 1, \frac{1}{3}, \frac{2}{3}, \frac{2}{5}, \frac{1}{2}, \frac{2}{7}$.

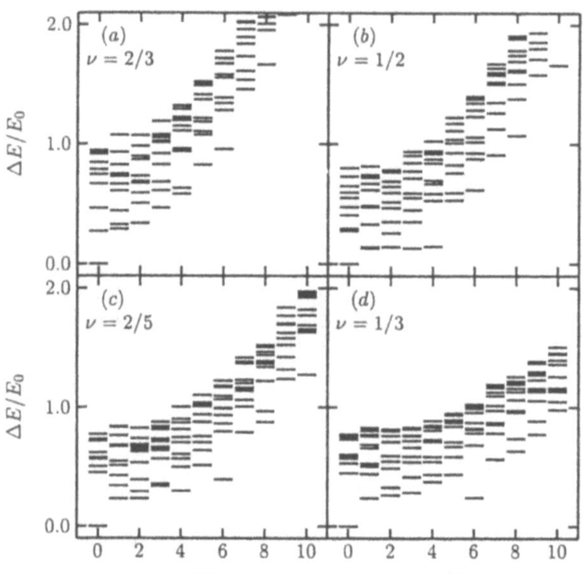

Figure 4
Energy spectra calculated at $(a, \mathcal{E}_c/E_0)$ and ν: (a) (6.8,24),$\nu = \frac{2}{3}$, (b) (7.6,36),$\nu = \frac{1}{2}$, (c) (9.2,40),$\nu = \frac{2}{5}$ and (d) (12.2,42),$\nu = \frac{1}{3}$.

is also a distinct region for the *even-denominator* state $\nu = \frac{1}{2}$ [15]. The phase diagram is dominated by two QHE states, viz., $\nu = 1$ and $\nu = \frac{1}{3}$ which are also the predominant QH states observed experimentally in a 2DEG.

2.5 Collective Modes

Figure 4 depicts the energy spectra for the states $\nu = \frac{2}{3}, \frac{1}{2}, \frac{2}{5}$ and $\nu = \frac{1}{3}$. These states are chosen from the phase diagram (Fig. 3) at the points where the gap appears to be the largest. The ubiquitous incompressible gaps in the spectra makes the analogy with those in the corresponding two-dimensional systems quite obvious. The novel result here again is, of course, the signature of incompressibility in the energy spectrum for the lowest

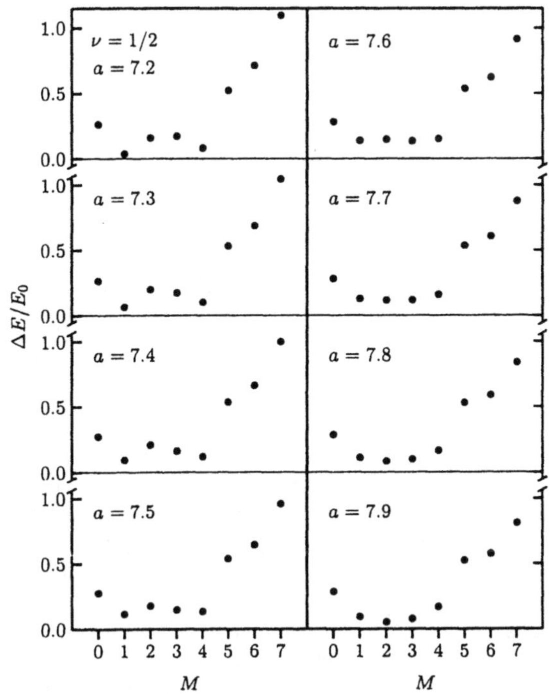

Figure 5
The lowest excitation energy at $\nu = \frac{1}{2}$ as a function of the total momentum M for various values of a.

half-filled Landau level, which is a hallmark of the fractional quantum Hall state at that filling factor [15]. The size of the gap at $\nu = \frac{1}{2}$ is comparable to that at $\frac{2}{5}$ and should be observable in experiments. There are also experimental indications [11] that the gap size at $\nu = \frac{1}{2}$ generally depends on the width of the constriction.

In order to explore the areas of the phase diagram where we do not have an incompressible state (between any two stable QH strips in Fig. 3), we looked at the energy spectra of various QH states when a is varied. In Fig. 5, we present such results at $\nu = \frac{1}{2}$. Evidently, the gap structure changes dramatically from one value of a to another and shows indications of *soft modes* at certain values of a. It therefore indicates that the symmetry changes between the ground states in different regions of the phase diagram. The precise nature of the states with $M \neq 0$ is, as yet, unclear. Similar indications of soft modes were also observed at other filling factors shown in the phase diagram.

3 Electron Spin Polarizations

So far, we have not included the role of spin into the problem. In this section we discuss the results for the electron spin polarization in a narrow quantum Hall system [16]. Here we want to describe all possible polarizations S_z from 0 to $N_e/2$. In addition, the noninteracting ground state, as discussed above, must have $M = 0$. It should also be noted that the filling factor $\nu = 1$, i.e., one electron per occupied single-particle momentum state should be realized as the ground state for *polarized* electrons. Similarly, the filling factor $\nu = 2$, i.e., two electrons per occupied state should be realized as the ground state for *unpolarized*

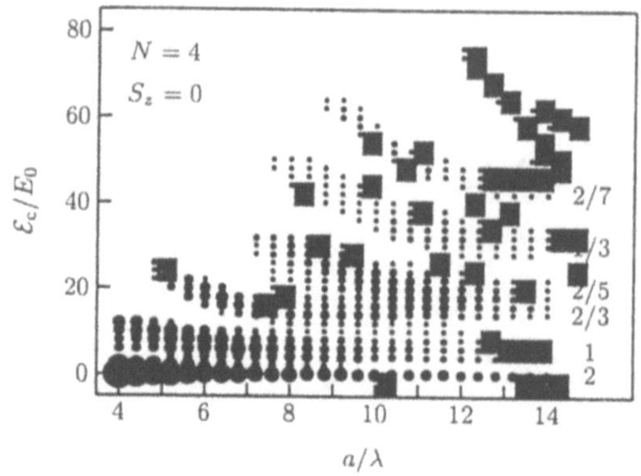

Figure 6
Phase diagram for electrons in a impurity-free narrow channel quantum Hall system with spin degree of freedom included.

electrons. As a result, the number of electrons N_e must be even. Since we have to apply antiperiodic boundary condition, we should set $N_e = 4n$ where n is an integer. This means that we have to deal with a four electron system because the next natural choice $N_e = 8$ is beyond our computational resources. If we exclude the Zeeman energy, there are no spin dependent interactions in the problem and we can set $S_z = 0$ and get simultaneously the spectrum for $S = 0, 1, \cdots, N_e/2$.

In Fig. 6, we present the results for the phase diagram, calculated for a system of four electrons with $S_z = 0$ (Zeeman energy not included) and for $\omega_0/\omega_c = 0.23$, which is appropriate for $B = 10$ Tesla and $\hbar\omega_0 = 4$ meV. As is evident in the figure, several quantum Hall states are stable with large energy gaps in the parameter range considered in this work. In this rather small system the $\nu = \frac{1}{2}$ state, though supposed to exist, cannot be resolved in the phase diagram. The $\nu = \frac{1}{2}$ states are expected to lie between $\nu = \frac{2}{3}$ and $\nu = \frac{2}{5}$ in Fig. 6. In general, the energy gaps are larger for spinless electrons (Fig. 3) because in that case there are no low-lying spin excitations available.

The temperature dependence of spin polarization for various filling factors found in the phase diagram is calculated by a method we developed earlier [18]. The spin polarization $\langle S_z(T)\rangle$ is calculated from

$$\langle S_z(T)\rangle \equiv \frac{1}{Z}\sum e^{-\varepsilon_j/kT}\langle j|S_z|j\rangle$$

where $Z = \sum_j e^{-\beta\varepsilon_j}$ is the canonical partition function and the summation is over all states including all possible polarizations. Here ε_j is the energy of the state $|j\rangle$ with Zeeman coupling included. A direct measurement of $\langle S_z(T)\rangle$ is possible through the NMR Knight-shift measurements and also via optical spectroscopic measurements [17,19]. These experiments provide unique probes of spin polarizations in the system.

Our results for $\langle S_z(T)\rangle$ vs T at $\nu = 1, 2, \frac{2}{3}, \frac{2}{5}, \frac{1}{3}$ and $\frac{1}{2}$ are shown in Fig. 7. In order to obtain the $\nu = \frac{1}{2}$ results we have employed a six-electron system. In these calculations the magnetic field was kept fixed at 10 tesla and the g-factor is varied $(0.02 - 0.52)$. At $\nu = 1$, we find the results to be similar to those for the 2DES [18] and the system is fully spin polarized even for very low values of the Zeeman energy. Qualitatively similar behavior is also seen at $\nu = \frac{1}{3}$. In the same way, $\nu = 2$ is a spin-unpolarized state even at the highest value of the Zeeman energy considered and $\nu = \frac{2}{3}$ and $\nu = \frac{2}{5}$ are spin-unpolarized states at low Zeeman energies with a non-monotonic temperature dependence as predicted for a

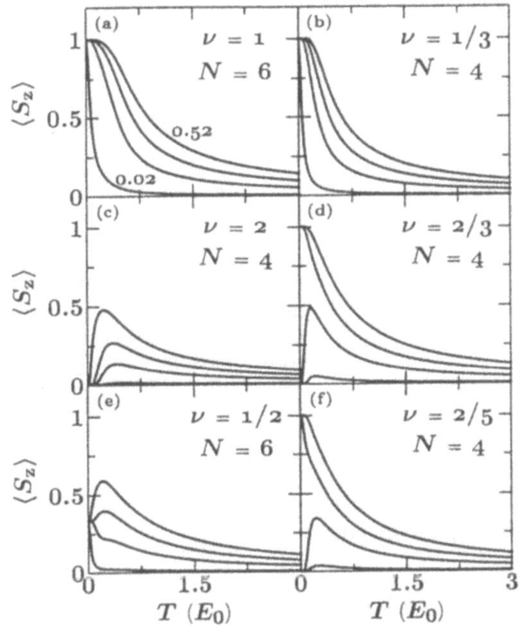

Figure 7
Spin polarization $\langle S_z \rangle$ vs T for $\nu = 1, 2, \frac{2}{3}, \frac{1}{2}, \frac{1}{3}$ and $\frac{2}{5}$ for an impurity-free system.

2DEG [18]. Such a non-monotonical behavior is observed in experiments on a 2DES [19]. Clearly, the correspondence with the results of spin polarization in a two dimensional system gives us confidence that our classification of the QH states in a narrow channel system is essentially correct. At $\nu = \frac{1}{2}$ we find a spin partially-polarized state.

4 Laughlin-Like State in a Narrow Channel

As we mentioned earlier, Yoshioka [12] considered a truncated Coulomb potential, i.e., $v(x,y) = 0$ for $|x| > a/2$, instead of the long-range Coulomb potential. He calculated the overlap between the $\frac{1}{3}$ ground state (spin polarized) and that for a model pseudopotential for which Laughlin state is the ground state [13]. The $\frac{2}{3}$-FQH state was then obtained via the electron-hole symmetry: In the $\frac{1}{3}$ FQH state, N_e electrons occupy the lowest $3N_e - 2$ states. This is the $\frac{2}{3}$ FQH state where $2(N_e - 1)$ holes occupy the lowest $3N_e - 2)$ states. The results [Fig. (8)] indicate that for a suitable choice of parameters, the Laughlin-like FQHE state for a flat external potential is a good approximation for the ground state even in the presence of the confinement potential. The Laughlin state in a narrow channel is different from the Luttinger liquid state [20] because in the former case a discontinuity exists at the Fermi points. One indeed finds a jump in the momentum distribution function from the occupied state to the unoccupied state. This is because the Laughlin state in the present system has a sharp boundary for the occupation of the single-electron state. In the $1/m$ FQHE state with N_e electrons, only the lowest $(N_e - 1)m + 1$ states are occupied. However, in an infinitely long system the momentum distribution may not have a discontinuity, and the system may behave like a Luttinger liquid.

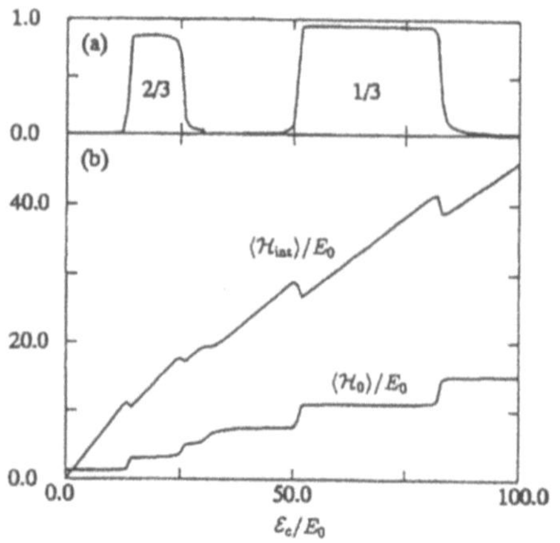

Figure 8
(a) Overlap of the ground state with the $\frac{1}{3}$ and $\frac{2}{3}$ FQH states, (b) expectation value of the kinetic energy per particle and interaction energy per particle for a truncated Coulomb potential as a function of $\mathcal{E}_c/E_0$ for the state $M = 0$ and for $a = 9.4$ [12].

The Laughlin state in a narrow channel has been also analyzed by the variational Monte Carlo method [21]. For a choice of the periodic potential

$$v(r) = \lambda \left[\left(\frac{a}{\pi} \sin \frac{\pi}{a} x \right)^2 + y^2 \right]^{-\frac{1}{2}},$$

which reduces to the usual Coulomb potential $\lambda/|r|$ in the limit $x/a \to 0$, a Laughlin-like wave function was constructed to be

$$\Psi_m(z_1, \cdots, z_{N_e}) = \prod_{j<k} \left[\sin \frac{\pi}{a}(z_j - z_k) \right]^m \prod_j \exp\left(-\frac{y_j^2}{2\mu\lambda^2} \right),$$

where $z = x + iy/\mu$ is the electron position, $\mu = \ell_0/\lambda = \sqrt{1 + (\omega_0/\omega_c)^2}$, $\ell_0 = \sqrt{\hbar c/eB}$ and m is an odd integer. In the case of $\mu = 1$, i.e., $\hbar\omega_0 \ll e^2/(\varepsilon\ell_0) \ll \hbar\omega_c$, a phase diagram at $\nu = 1/m$ was obtained for a 45-electron system. Further, a finite-size scaling of the low-lying excitation energies of the system indicated that the correlation function at long distances can be characterized by a non-universal exponent which is in sharp contrast to the standard Laughlin state.

5 Concluding Remarks

In this article, we have discussed the electronic properties of incompressible states that are expected to be present in a narrow channel electron system. In between the incompressible states there are states in the phase diagram where the excitation spectra go soft. This indicates a phase transition in going from one QH state to another. The precise nature of the states in the compressible region is, as yet, unclear. The temperature dependence of electron spin polarization for a narrow quantum Hall system shows behavior analogous to that of a 2DES at major filling factors. At the lowest half-filled quantum Hall state for which no two-dimensional analog exists, a stable spin partially-polarized ground state is expected. Optically pumped NMR experiments as well as optical spectroscopy should be the most appropriate experiments to explore the electronic properties described here.

Acknowledgments

I would like to thank Karri Niemelä and Pekka Pietiläinen for their continuing collaboration on these and other related projects. I also thank Peter Fulde for his kind hospitality.

Bibliography

[1] D.C. Tsui, H.L. Störmer, and A.C. Gossard, Phys. Rev. Lett. **48**, 1559 (1982).

[2] R.B. Laughlin, Phys. Rev. Lett. **50**, 1395 (1983).

[3] B.I. Halperin, Helv. Phys. Acta **56**, 75 (1983).

[4] T. Chakraborty and P. Pietiläinen, *The Quantum Hall Effects* (Springer, New York, 1995), second edition; T. Chakraborty, in *Handbook on Semiconductors*, vol. 1, ed. P.T. Landsberg (North-Holland, Amsterdam, 1992), Ch. 17.

[5] H.W. Jiang et al., Phys. Rev. **B 40**, 12013 (1989); R.L. Willett, et al., Phys. Rev. Lett. **65**, 112 (1990); W. Kang et al., *ibid.* **71**, 3850 (1993); R.R. Du et al., *ibid.*, **70**, 2944 (1993); *ibid.*, **73**, 3274 (1994); D.R. Leadley, et al., *ibid.*, **72**, 1906 (1994); R.L. Willett et al., *ibid.*, **75**, 2988 (1995).

[6] B.I. Halperin, P.A. Lee, and N. Read, Phys. Rev. **B 47**, 7312 (1993); B.I. Halperin, Surf. Sci. **305**, 1 (1994); B.I. Halperin, in *Perspectives in Quantum Hall Effects*, edited by S. Das Sarma and A. Pinczuk (Wiley, 1997), ch. 6.

[7] F.D.M. Haldane, Phys. Rev. Lett. **55**, 2095 (1985); G. Fano, F. Ortolani, and E. Tosatti, Nuovo Cimento **D 9**, 1337 (1987); T. Chakraborty and P. Pietiläinen, Phys. Rev. **B 38**, 10 097 (1988); T. Chakraborty, Phys. Rev. B **57**, 8812 (1998).

[8] A. Okiji and N. Kawakami, (Eds.), *Correlation Effects in Low-Dimensional Electron Systems* (Springer, 1994).

[9] S. Tarucha, T. Honda, and T. Saku, Solid State Commun. **94**, 413 (1995); A. Yacoby, H.L. Störmer, N.S. Wingreen, L.N. Pfeiffer, K.W. Baldwin, and K.W. West, Phys. Rev. Lett. **75**, 461 (1996).

[10] S.T. Chui, Phys. Rev. Lett. **56**, 2395 (1986).

[11] G. Timp, R. Behringer, J.E. Cunningham, and R.E. Howard, Phys. Rev. Lett. **63**, 2268 (1989); G. Timp, in *Nanostructured Systems*, ed. M. Reed (Academic, Boston, 1992), Ch. 3.

[12] D. Yoshioka, J. Phys. Soc. Jpn. **62**, 839 (1993).

[13] D. Yoshioka, Physica **B 184**, 86 (1993).

[14] T. Chakraborty, Comments Cond. Mat. Phys. **16**, 35 (1992); P.A. Maksym and T. Chakraborty, Phys. Rev. Lett. **65**, 108 (1990).

[15] T. Chakraborty, K. Niemelä, and P. Pietiläinen, Phys. Rev. Lett. **78**, 4829 (1997); T. Chakraborty, K. Niemelä, and P. Pietiläinen, Physica **E 1**, 80 (1997).

[16] T. Chakraborty, K. Niemelä, and P. Pietiläinen, cond-mat/9711158, and Phys. Rev. B (October 15, 1998).

[17] S.E. Barrett, et al., Phys. Rev. Lett. **74**, 5112 (1995); E.H. Aifer, et al., *ibid.*, **76**, 680 (1996).

[18] T. Chakraborty and P. Pietiläinen, Phys. Rev. Lett. **76**, 4018 (1996); T. Chakraborty, P. Pietiläinen and R. Shankar, Europhys. Lett. **38**, 141 (1997).

[19] I.V. Kukushkin, K. von Klitzing, and K. Eberl, Phys. Rev. **B 55**, 10607 (1997).

[20] J. Sólyom, Adv. Phys. **28**, 201 (1979); J.M. Luttinger, Phys. Rev. **119**, 1153 (1960).

[21] S. Tokizaki and Y. Kuramoto, J. Phys. Soc. Jpn. **64**, 2302 (1995).

Electron-Electron Interaction Effects in Single-Wall Carbon Nanotubes

Reinhold Egger[1] and Alexander O. Gogolin[2]

[1] Fakultät für Physik, Albert-Ludwigs-Universität, D-79104 Freiburg, Germany
[2] Department of Mathematics, Imperial College, London SW7 2BZ, UK

Abstract: We describe the effective low-energy theory for single-wall carbon nanotubes including the electron-electron interaction. Various experimentally relevant consequences of the correlations are discussed in detail, e.g., the phase diagram, long wavelength instabilities, anomalous conductance laws, and the slowly decaying Friedel oscillation.

1 Introduction

Nanotubes are tubular nanoscale objects which can be thought of as graphite sheets wrapped into a cylinder. Shortly after their discovery [1], an interesting prediction linking structural to electronic properties has been made [2]. The arrangement of carbon atoms on the tube surface is determined by the integer indices $0 \leq m \leq n$ of the wrapping superlattice vector $\vec{T} = n\vec{a}_1 + m\vec{a}_2$, where $\vec{a}_1$ and $\vec{a}_2$ are the primitive Bravais translation vectors of the honeycomb lattice. A necessary condition for metallicity arises because the Fermi vector should obey the transverse quantization condition $\vec{T}\vec{k} = 2\pi I$ for some integer I. The first Brillouin zone of the honeycomb lattice is a hexagon, and bandstructure calculations show that the only gapless points are the corner points of this hexagon [3]. There are two independent Fermi points denoted by K and K', and two linearly dispersing bands around each Fermi point. Imposing the transverse quantization condition for K and K' then implies $n - m = 3I$. If this condition is not fulfilled, the nanotube is insulating with the band gap $\Delta E = 2v/3R \approx 1$ eV, where v is the Fermi velocity and R the radius. Even if $n - m = 3I$, the rearrangement of local bonds due to the curvature of the cylinder can introduce a small secondary gap, $\Delta E \approx 10$ meV, which implies narrow-gap semiconducting behavior. In the case of armchair nanotubes ($n = m$), the formation of a secondary gap is prevented by the high symmetry, and therefore the armchair tube stays metallic for all n. All of these theoretical predictions have been beautifully verified in recent STM experiments [4].

Nanotubes can be fabricated using either the carbon-arc method [1] or a novel laser ablation technique [5]. The carbon-arc process usually yields multi-wall nanotubes composed of several concentric graphite sheets. On the other hand, the laser ablation method allows for the controlled fabrication of single-wall nanotubes (SWNTs). In particular, large quantities of metallic (10,10) SWNTs can be obtained using this technique [5]. In most cases, the SWNTs spontaneously crystallize in triangular-packed bundles containing

$\approx$ 20 to 100 SWNTs, but single SWNTs can also be found in this process. From a fundamental perspective, a single SWNT is very interesting, because electron-electron interactions always lead to the breakdown of Fermi liquid theory in any truly one-dimensional (1D) metal. Astonishingly, Tans et al. [6] were able to attach contacts to a single 3 μm long metallic SWNT. Their transport measurements were so far dominated by charging effects due to large lead-SWNT contact resistances. By lowering the contact resistances one can circumvent charging effects, thereby allowing one to study pronounced correlation effects directly in the current-voltage characteristics. Due to their structural stability and unique band structure, SWNTs are expected to be more reliable 1D conductors than conventional systems such as long chain molecules, edge states in the fractional quantum Hall effect, or quantum wires in semiconductor heterostructures. In these systems, the experimental verification of non-Fermi liquid behavior has been notoriously difficult.

Interacting 1D electrons often exhibit *Luttinger liquid* rather than Fermi liquid behavior characterized by, e.g., the absence of Landau quasi-particles, spin-charge separation, suppression of the electron tunneling density of states, and interaction-dependent power laws for transport quantities [7]. As shown in [8], at very low energy scales transport through a Luttinger liquid becomes fully blocked by a single arbitrarily weak impurity. The conductance vanishes as a function of temperature and/or voltage with interaction-dependent power laws. This effect can be understood in terms of the Friedel oscillation building up around a scatterer in a 1D metal. The oscillatory charge screening cloud displays an algebraically slow decay away from the impurity and thereby causes an important additional backscattering contribution [9,10].

Below we characterize in detail the nature of the non-Fermi-liquid state theoretically expected in a SWNT, and study to what extent the Luttinger liquid behavior should be realized. In Sec. 2, we describe the essential ingredients of the effective low-energy theory, and then discuss effects due to the electron-electron interaction in Sec. 3, with particular attention to their experimental relevance. Some concluding remarks are offered in Sec. 4. We skip most technical aspects of our theory which the interested reader may find in [11].

2 Effective Low-Energy Theory

The remarkable electronic properties of carbon nanotubes are due to the special bandstructure of the π electrons in graphite. There are only two linearly independent Fermi points $\alpha\vec{K}$ with $\alpha = \pm$ instead of a continuous Fermi surface. Up to energy scales ≤ 1 eV, the dispersion relation around the Fermi points is, to a very good approximation, linear. Since the basis of the honeycomb lattice contains two atoms, there are two sublattices $p = \pm$ shifted by the vector $\vec{d} = (0,d)$, with the nearest-neighbor C-C distance $d = a/\sqrt{3} = 1.42$ Å. Here a denotes the honeycomb lattice constant. As a consequence, there are two degenerate Bloch states

$$\varphi_{p\alpha}(\vec{r}) = (2\pi R)^{-1/2} \exp(-i\alpha\vec{K}\vec{r}) \tag{2.1}$$

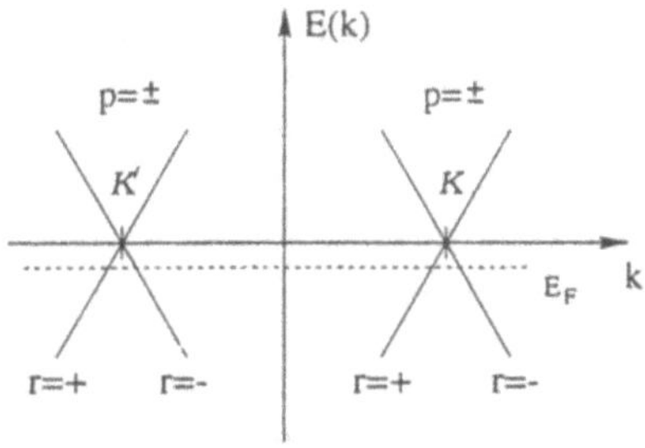

Figure 1
Schematic bandstructure of a metallic SWNT. A right- and left-moving branch ($r = \pm$) is found near each of the two Fermi points $\alpha = \pm$ corresponding to K and K', respectively. Right- and left-movers arise as linear combinations of the sublattices $p = \pm$. The Fermi energy (dashed line) can be tuned by an external gate.

at each Fermi point $\alpha = \pm$, where $\vec{r} = (x,y)$ lives on the sublattice p under consideration, and we have already anticipated the correct normalization for nanotubes. The Bloch functions are defined separately on each sublattice such that they vanish on the other. One can then expand the electron operator in terms of these Bloch functions. The resulting effective low-energy theory of a graphite sheet is the 2D massless Dirac Hamiltonian [12].

Wrapping the graphite sheet leads to the generic bandstructure of a metallic SWNT shown in Fig. 1. We shall focus on the (n,n) armchair SWNT, where the Fermi vector is $\vec{K} = (k_F, 0)$ with $k_F = 2\pi/3a$. The x-axis is taken along the tube direction, and the circumferential variable is $0 < y < 2\pi R$, with the radius $R = \sqrt{3}na/2\pi$. Quantization of transverse motion now allows for a contribution $\exp(imy/R)$ to the wavefunction. However, excitation of angular momentum states other than $m = 0$ costs the energy ≈ 10 eV$/n$. In an effective low-energy theory, we may thus omit all transport bands except $m = 0$. Evidently, the nanotube forms a 1D quantum wire with only two bands intersecting the Fermi energy. This strict one-dimensionality is fulfilled up to remarkably high energy scales (eV) here, in contrast to conventional 1D conductors.

The electron operator for spin $\sigma = \pm$ is now written as [12]

$$\Psi_\sigma(x,y) = \sum_{p\alpha} \varphi_{p\alpha}(x,y)\, \psi_{p\alpha\sigma}(x) \,, \tag{2.2}$$

which introduces slowly varying 1D fermion operators $\psi_{p\alpha\sigma}(x)$ that depend only on the x coordinate. Neglecting Coulomb interactions for the moment, the Hamiltonian is (we put $\hbar = 1$),

$$H_0 = -v \sum_{p\alpha\sigma} p \int dx\, \psi^\dagger_{p\alpha\sigma} \partial_x \psi_{-p\alpha\sigma} \,. \tag{2.3}$$

Switching from the sublattice ($p = \pm$) description to the right- and left-movers ($r = \pm$) indicated in Fig. 1 implies two copies ($\alpha = \pm$) of massless 1D Dirac Hamiltonians. By using a suitable gate, one can experimentally tune the average charge density on the nanotube and hence adjust the Fermi energy. In contrast to the half-filled band encountered in 2D graphite sheets, one is normally off half-filling in SWNT experiments ($E_F \neq 0$).

Let us now examine Coulomb interactions mediated by an arbitrary potential $U(\vec{r} - \vec{r}')$. The electron-electron interaction is described by the second-quantized Hamiltonian

$$H_I = \frac{1}{2} \sum_{\sigma\sigma'} \int d\vec{r} \int d\vec{r}'\, \Psi^\dagger_\sigma(\vec{r}) \Psi^\dagger_{\sigma'}(\vec{r}') U(\vec{r} - \vec{r}') \Psi_{\sigma'}(\vec{r}') \Psi_\sigma(\vec{r}) \,. \tag{2.4}$$

Bound electrons can be incorporated in terms of the background dielectric constant $\kappa \simeq 1.4$ [11], but free charge carriers in nearby gates could lead to an effectively short-ranged potential. For the experiments of [6], one has an externally unscreened Coulomb interaction,

$$U(\vec{r}-\vec{r}') = \frac{e^2/\kappa}{\sqrt{(x-x')^2+4R^2\sin^2[(y-y')/2R]+a_z^2}}\,, \tag{2.5}$$

where $a_z \approx a$ denotes the average distance between a $2p_z$ electron and the nucleus (the "thickness" of the graphite sheet).

The interaction (2.4) can now be reduced to a 1D interaction by inserting the expansion (2.2) for the electron field operator. The result is

$$\begin{aligned} H_{\rm I} &= \frac{1}{2}\sum_{pp'\sigma\sigma'}\sum_{\alpha_1\alpha_2\alpha_3\alpha_4}\int dx dx'\, V^{pp'}_{\{\alpha_i\}}(x-x') \\ &\times \psi^\dagger_{p\alpha_1\sigma}(x)\psi^\dagger_{p'\alpha_2\sigma'}(x')\psi_{p'\alpha_3\sigma'}(x')\psi_{p\alpha_4\sigma}(x)\,, \end{aligned} \tag{2.6}$$

with the 1D interaction potentials

$$\begin{aligned} V^{pp'}_{\{\alpha_i\}}(x-x') &= \int_0^{2\pi R} dy dy'\, \varphi^*_{p\alpha_1}(x,y)\varphi^*_{p'\alpha_2}(x',y') \\ &\times U(x-x',y-y'+pd\delta_{p,-p'})\,\varphi_{p'\alpha_3}(x',y')\varphi_{p\alpha_4}(x,y)\,. \end{aligned} \tag{2.7}$$

These potentials only depend on $x-x'$ and on the 1D fermion quantum numbers. For interactions involving different sublattices for $\vec{r}$ and $\vec{r}'$ in (2.4), i.e., $p \neq p'$, one needs to take into account the shift vector $\vec{d} = (0,d)$, see (2.7).

To simplify the resulting 1D interaction (2.6), we now exploit momentum conservation. Since an additional gate voltage tunes the average electron density on the SWNT, no lattice instabilities due to Umklapp scattering are present if one stays off half-filling. This assumption does not limit the applicability of our approach, but simplifies matters considerably. We then employ the Fermi point quantum numbers α_i to classify the allowed processes. First, there are "forward scattering" (αFS) processes, where $\alpha_1 = \alpha_4$ and $\alpha_2 = \alpha_3$. Second, we have "backscattering" (αBS), with $\alpha_1 = -\alpha_2 = \alpha_3 = -\alpha_4$.

Let us start with αFS processes. We first define the potential

$$V_0(x-x') = \int_0^{2\pi R}\frac{dy}{2\pi R}\int_0^{2\pi R}\frac{dy'}{2\pi R}\,U(\vec{r}-\vec{r}')\,. \tag{2.8}$$

For the unscreened Coulomb interaction (2.5), this can be explicitly evaluated [11]. From (2.7) and (2.1), the $\{\alpha_i\}$-independent forward scattering interaction potential reads

$$V^{pp'}_{\alpha FS}(x) = V_0(x) + \delta_{p,-p'}\delta V_p(x)\,, \tag{2.9}$$

with the correction term

$$\delta V_p(x) = \int_0^{2\pi R}\frac{dy dy'}{(2\pi R)^2}[U(x,y-y'+pd) - U(x,y-y')]\,, \tag{2.10}$$

which is only present if $\vec{r}$ and $\vec{r}'$ are located on different sublattices. Thereby important information about the discrete nature of the graphite network is retained despite the low-energy continuum approximation. Because of the periodicity of the y-integrals, Taylor expanding (2.10) in powers of d would imply $\delta V_p(x) = 0$. Since $V_0(x)$ treats both sublattices on equal footing, the resulting αFS interaction couples only the total 1D electron densities,

$$H^{(0)}_{\alpha\mathrm{FS}} = \frac{1}{2}\int dxdx'\, \rho(x)V_0(x-x')\rho(x')\,, \tag{2.11}$$

where the 1D density is $\rho = \sum_{p\alpha\sigma} \psi^\dagger_{p\alpha\sigma}\psi_{p\alpha\sigma}$. For $|x-x'| \gg a$, the argument leading to $\delta V_p = 0$ is justified. However, for $|x-x'| \leq a$, one has to be more careful. Here an additional αFS term beyond (2.11) arises due to the hard core of the Coulomb interaction. At such small length scales, the difference between inter- and intra-sublattice interactions matters. To study this term, one can put $x = x'$ and evaluate $\delta V_p(0)$ from microscopic considerations. One then finds the additional αFS contribution

$$H^{(1)}_{\alpha\mathrm{FS}} = -f\int dx \sum_{p\alpha\alpha'\sigma\sigma'} \psi^\dagger_{p\alpha\sigma}\psi^\dagger_{-p\alpha'\sigma'}\psi_{-p\alpha'\sigma'}\psi_{p\alpha\sigma}\,, \tag{2.12}$$

where $f/a \approx 0.05\,e^2/R$ for armchair SWNTs [11].

Let us next discuss the αBS contributions where $\alpha_1 = -\alpha_2 = \alpha_3 = -\alpha_4 = \alpha$ in (2.6). From (2.7) the corresponding 1D interaction potential for αBS processes reads

$$V^{pp'}_\alpha(x-x') = e^{2i\alpha k_F(x-x')}\, V_0(x-x')\,.$$

Because of the rapidly oscillating phase factor, the only non-vanishing contribution to $H_{\alpha\mathrm{BS}}$ comes from $|x-x'| \leq a$, i.e., we can effectively take a local interaction. Furthermore, it turns out that only the part of the interaction which does not distinguish among the sublattices is not irrelevant. The effective coupling constant $b > 0$ is then determined by the Fourier transform of $V_0(x)$ at $k = 2k_F$, i.e., $b = \widetilde{V}_0(2k_F)$, such that according to (2.6) the interaction contribution reads

$$H_{\alpha\mathrm{BS}} = \frac{b}{2}\int dx \sum_{pp'\alpha\sigma\sigma'} \psi^\dagger_{p\alpha\sigma}\psi^\dagger_{p'-\alpha\sigma'}\psi_{p'\alpha\sigma'}\psi_{p-\alpha\sigma}\,. \tag{2.13}$$

For the unscreened interaction (2.5), we obtain $b/a \approx 0.1\,e^2/R$, which predicts $b \approx f$. If the Coulomb interaction is externally screened, however, the qualitatively different situation $b \gg f$ could arise.

The effective low-energy theory of a metallic SWNT away from lattice commensurabilities is thus described by the fermionic Hamiltonian

$$H = H_0 + H^{(0)}_{\alpha\mathrm{FS}} + H^{(1)}_{\alpha\mathrm{FS}} + H_{\alpha\mathrm{BS}}\,. \tag{2.14}$$

This model is equivalent to two spin-$\frac{1}{2}$ fermion chains coupled by the interactions, similar to the standard two-chain problem of coupled Luttinger liquids [13,14]. Progress can then be made by employing the *bosonization* approach [7].

For that purpose, we need to bring the non-interacting Hamiltonian (2.3) into a standard form of the 1D Dirac model. This is accomplished by switching to right- and left-movers ($r = \pm$) which are linear combinations of the sublattice states $p = \pm$. In particular, we have

$$\psi_{p\alpha\sigma}(x) = \sum_r \widetilde{U}_{pr} \widetilde{\psi}_{r\alpha\sigma}(x) , \tag{2.15}$$

where the unitary matrix $\widetilde{U}$ fulfills $\widetilde{U}^\dagger \sigma_y \widetilde{U} = \sigma_z$. In this representation, the bosonization formula applies [7],

$$\widetilde{\psi}_{r\alpha\sigma}(x) = \frac{\eta_{r\alpha\sigma}}{\sqrt{2\pi a}} \exp\{iq_F r x + ik_F \alpha x + i\varphi_{r\alpha\sigma}(x)\} . \tag{2.16}$$

Here we have incorporated the spatial dependence due to the Bloch functions (2.1) into the 1D fermion operator. The chiral bosonic phase fields $\varphi_{r\alpha\sigma}(x)$ obey a commutator algebra [7] such that the fermion operator $\widetilde{\psi}_{r\alpha\sigma}(x)$ indeed shows the correct anticommutator algebra on a given branch $(r\alpha\sigma)$. To enforce anticommutator relations between different branches, we need the Majorana fermions $\eta_{r\alpha\sigma} = \eta^\dagger_{r\alpha\sigma}$ specified in [11]. The wave vector q_F, which must be carefully distinguished from the Fermi vector k_F, is related to the band filling. Relative to the half-filled case, an average excess density $\delta\rho$ gives rise to a non-zero $q_F = \pi\delta\rho/4$ and hence $E_F = vq_F$. We consider $|q_F| \ll k_F$ in the following, since otherwise the low-energy continuum approximation underlying our approach might break down.

At this stage, it is natural to introduce the standard linear combinations $\theta_{j\delta}(x)$ and their dual fields $\phi_{j\delta}(x)$ subject to the algebra

$$[\phi_{j\delta}(x), \theta_{j'\delta'}(x')] = -(i/2)\delta_{jj'}\delta_{\delta\delta'} \,\mathrm{sgn}(x-x') . \tag{2.17}$$

Written in terms of the bosonic phase fields for the total and relative ($\delta = \pm$) charge ($j = c$) and spin ($j = s$) channel, we have

$$\varphi_{r\alpha\sigma} = \frac{\sqrt{\pi}}{2}(\phi_{c+} + r\theta_{c+} + \alpha\phi_{c-} + r\alpha\theta_{c-} + \sigma\phi_{s+} + r\sigma\theta_{s+} + \alpha\sigma\phi_{s-} + r\alpha\sigma\theta_{s-}) .$$

The total electron density relative to the half-filled situation reads

$$\rho(x) = 4q_F/\pi + \frac{2}{\sqrt{\pi}}\partial_x\theta_{c+}(x) , \tag{2.18}$$

and the continuity equation then yields the current, $I = \frac{2e}{\sqrt{\pi}}\partial_t\theta_{c+}(x,t)$. The bosonized expressions for the various terms in H read

$$H_0 = \sum_{j\delta} \frac{v_{j\delta}}{2} \int dx \left[K_{j\delta}(\partial_x\phi_{j\delta})^2 + K_{j\delta}^{-1}(\partial_x\theta_{j\delta})^2\right] \tag{2.19}$$

$$H^{(0)}_{\alpha\mathrm{FS}} = \frac{2}{\pi}\int dx dx' \, \partial_x\theta_{c+}(x) V_0(x-x') \partial_{x'}\theta_{c+}(x') \tag{2.20}$$

$$\begin{aligned} H^{(1)}_{\alpha\mathrm{FS}} = \frac{f}{(\pi a)^2}\int dx \,[&-\cos(\sqrt{4\pi}\theta_{c-})\cos(\sqrt{4\pi}\theta_{s-}) \\ &- \cos(\sqrt{4\pi}\theta_{c-})\cos(\sqrt{4\pi}\theta_{s+}) \\ &+ \cos(\sqrt{4\pi}\theta_{s+})\cos(\sqrt{4\pi}\theta_{s-})] \end{aligned} \tag{2.21}$$

$$\begin{aligned} H_{\alpha\mathrm{BS}} = \frac{b}{(\pi a)^2}\int dx [&\cos(\sqrt{4\pi}\theta_{c-})\cos(\sqrt{4\pi}\theta_{s-}) \\ &+ \cos(\sqrt{4\pi}\theta_{c-})\cos(\sqrt{4\pi}\phi_{s-}) \\ &+ \cos(\sqrt{4\pi}\theta_{s-})\cos(\sqrt{4\pi}\phi_{s-})] . \end{aligned} \tag{2.22}$$

Although bosonization of (2.3) gives $K_{j\delta}=1$ in (2.19), interactions will renormalize these parameters. In particular, in the long-wavelength limit, $H^{(0)}_{\alpha\text{FS}}$ can be incorporated into H_0 by putting

$$K_{c+}=K=\left\{1+4\widetilde{V}_0(k\simeq 0)/\pi v\right\}^{-1/2}\leq 1\,, \tag{2.23}$$

while for all other channels, the coupling constant f gives rise to the (very small) renormalization $K_{j\delta}=1+f/\pi v$ away from the noninteracting value. The corresponding renormalization of K_{c+} due to $f\neq 0$ can be neglected against the large effect of (2.20). The velocities of the different modes in (2.19) are then given by

$$v_{j\delta}=v/K_{j\delta}\,. \tag{2.24}$$

Obviously, the charged $(c+)$ mode propagates with significantly higher velocity than the three neutral modes. There is a further renormalization of the values given by (2.24) due to the αBS contribution. However, this effect is very small and does not affect the power laws discussed below. We therefore neglect velocity renormalizations beyond (2.24).

For the long-ranged interaction, a logarithmic singularity in $\widetilde{V}_0(k)$ requires the infrared cutoff $k=2\pi/L$ due to the finite length of the nanotube. We then obtain

$$K=\left\{1+\frac{8e^2}{\pi\kappa\hbar v}\ln(L/2\pi R)\right\}^{-1/2}\,, \tag{2.25}$$

which predicts $K\to 0$ for $L\to\infty$. Since $\hbar c/e^2\simeq 137$, we get with $v=8\times 10^5$ m/s the estimate $e^2/\hbar v=(e^2/\hbar c)(c/v)\approx 2.7$, and therefore $K\simeq 0.18$ for the $L=3\,\mu$m tube of [6]. Quite generally, K gives the appropriate measure of the correlation strength [7], where a small value of K corresponds to strong correlations. The very small value of K found here implies that the metallic SWNT studied in [6] is a strongly correlated system that should display pronounced non-Fermi liquid effects.

It is clear from the bosonized form of H that for $f=b=0$, the nanotube constitutes a realization of the Luttinger liquid. We therefore have to address the effect of the nonlinear terms associated with the coupling constants f and b. This can be done by means of the renormalization group approach. Together with a solution via Majorana refermionization, this route allows for the complete characterization of the non-Fermi-liquid state of a nanotube [11].

Around the Gaussian fixed point $f=b=0$, the backscattering contribution is marginally relevant, while the nonlinear forward scattering term $H^{(1)}_{\alpha\text{FS}}$ is marginally irrelevant. It therefore makes sense to first consider the case $f=0$. This limiting case would also be appropriate for a well-screened interaction where $b\gg f$. For $f=0$, the $(c-)$ sector is massive, but all other branches remain massless (the $(s-)$ sector has the universality class of the 2D classical Ising model). The corresponding gap is given by

$$k_{\text{B}}T_b=D\exp[-\pi v/\sqrt{2}b]\,, \tag{2.26}$$

where $D\simeq 7.4$ eV is the bandwidth of the π band. Since the coupling constant b appears in the exponent of an exponentially small quantity, it is very difficult to give a precise estimate for T_b. A rough order-of-magnitude estimate for the setup of [6] is $T_b\approx 0.1$ mK.

Because in general $f \neq 0$, we must now reconsider the nonlinear forward-scattering contribution $H^{(1)}_{\alpha FS}$. Since it becomes relevant near the emerging strong-coupling point $b \to \infty$, the point $f = 0$ describes in fact only an intermediate fixed point. Allowing for $f \neq 0$ renders both the $(s+)$ and the $(s-)$ channel massive as well, with the gap

$$k_B T_f \simeq (f/b)\, k_B T_b\,. \tag{2.27}$$

At the emerging $T = 0$ strong-coupling fixed point, we have long-range order in the operators $\cos[\sqrt{\pi}\theta_{s+}], \cos[\sqrt{\pi}\theta_{c-}]$, and $\sin[\sqrt{\pi}\phi_{s-}]$, with exponential decay in all other operators except those of the critical $(c+)$ sector.

3 Correlation Effects

The above analysis leads to the existence of two gaps characterized by the temperature scales T_b and $T_f \leq T_b$. Therefore there are in general three different temperature regimes that need to be carefully distinguished. The intermediate strong-coupling point $(f = 0)$ should be observable on temperature scales $T_f < T < T_b$, with a crossover to a regime $T < T_f$ dominated by the true ground state. Since both T_f and T_b are well below the mK range, unless stated otherwise, we only discuss the experimentally relevant high-temperature regime $T > T_b$ in what follows. In that regime it is justified to neglect the nonlinearities associated with the coupling constants b and f, and the physical behavior is best characterized as that of a Luttinger liquid with an additional flavor index [11,15]. *Nanotubes therefore constitute a realization of the Luttinger liquid at sufficiently high temperatures $T > T_b$.*

3.1 Phase Diagram

With the strong-coupling solution briefly sketched in Sec. 2, we now examine the phase diagram of a SWNT in the $T - K$ plane, see Fig. 2. Due to the Mermin-Wagner theorem, there can be no ordered state in a 1D system. It is therefore customary to characterize the physical behavior by the most slowly decaying correlation function, which in turn indicates incipient instabilities. We have studied correlators of charge-density wave (CDW), spin-density wave (SDW), and superconducting (SC) type. For a given scaling dimension Δ and wavelength λ, the equal-time correlation function of an operator $\hat{O}$ is

$$\langle \hat{O}(x)\hat{O}(0)\rangle \sim \cos[2\pi x/\lambda]\, x^{-2\Delta}\,. \tag{3.28}$$

Depending on the temperature and the correlation strength K, the interactions lead to CDW or SDW behavior. Only at extremely low temperatures $(T < T_f)$ and well-screened interactions $(K > 1/2)$, a superconducting phase characterized by the singlet pairing operator $\hat{O}_{SC} \sim \sum_{p\alpha\sigma} \sigma \psi_{p\alpha\sigma} \psi_{p-\alpha-\sigma}$ emerges, with

$$\langle \hat{O}_{SC}(x)\hat{O}_{SC}(0)\rangle \sim x^{-1/2K}\,. \tag{3.29}$$

Despite of recent findings in conventional two-chain models [13,14], superconductivity is therefore expected to be of little practical importance in carbon nanotubes.

Notably, the interactions can induce other spatial oscillation periods than the standard wavelength $\lambda = \pi/k_F$. It follows from the dispersion relation in Fig. 1 that the wavelengths

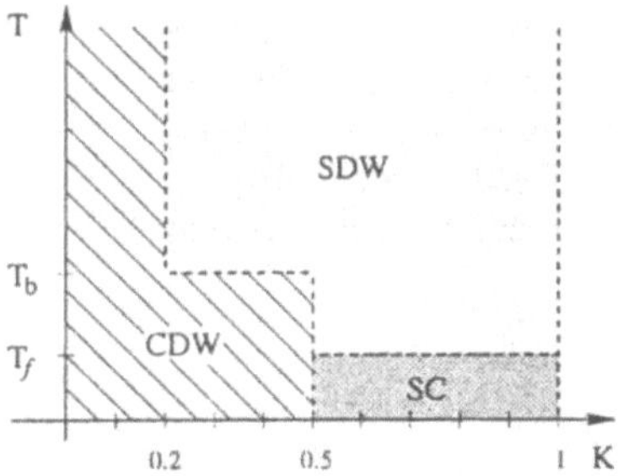

Figure 2
Phase diagram of a metallic nanotube in the $T-K$ plane.

$$\lambda = \pi/k_F, \quad \pi/q_F, \quad \pi/(k_F \pm q_F) \tag{3.30}$$

could occur in a metallic SWNT (we put $q_F > 0$ for simplicity). Without doping, only $\lambda = \pi/k_F$ is found, which corresponds to an order operator effectively involving two 1D fermions at different Fermi points ($\alpha = -\alpha'$). In the doped case, the much longer wavelength π/q_F arises if the 1D fermions are at the same Fermi point but move in opposite directions ($r = -r'$). Finally, the wavelength $\pi/(k_F \pm q_F)$ corresponds to the mixed situation ($\alpha = -\alpha', r = -r'$). The possible simultaneous occurrence of different wavelengths is a remarkable feature of doped nanotubes which has no analogue in the standard two-chain problem. The detailed scaling dimensions and wavelengths applying to each region of the phase diagram in Fig. 2 are specified in [11].

3.2 Ferromagnetic Correlations

Next we briefly discuss a peculiar feature of the SDW correlations. They represent the *dominant instability* of the nanotube for moderate correlation strength $K > 1/5$. Again, they are characterized by the simultaneous presence of the wavelengths $\lambda = \pi/q_F$ and $\pi/(k_F \pm q_F)$. For a well-screened interaction, there are strong arguments [11] that the slow wavelength is more important, and the 1D spin-spin correlation function is dominated by the $2q_F$ contribution

$$\langle s_z(x) s_z(0) \rangle \sim \cos[2q_F x]\, x^{-(3+K)/2}\,. \tag{3.31}$$

Since $\cos[2q_F x] \approx 1$ for distances x where the correlations are not already vanishingly small, the SDW correlations apparently display a pronounced *ferromagnetic* character. This reasoning could offer an explanation for the ferromagnetic tendencies observed in [6], which can be experimentally checked by increasing the doping and hence q_F. Eventually, antiferromagnetism should be recovered for sufficiently strong doping.

3.3 Tunneling Density of States

Under typical experimental conditions, the contact between a SWNT and the attached Fermi-liquid leads is not really adiabatic. For nonadiabatic contacts, the physics of the conductance is related to single-electron tunneling into the nanotube, which in turn is governed by the tunneling density of states (TDOS). A similar situation also arises in STM experiments. Here the electron tunneling occurs close to the impurity position, the role of the impurity being played by the probe itself. The presence of an open boundary, either due to the finite length of the nanotube or due to a strong impurity, can alter correlation functions and, in particular, modify the anomalous exponents [8]. This can be accomodated for by using the open boundary bosonization technique [10]. It follows that close to the boundary, i.e., for $\max(x,x') \ll vt$, the single-electron Greens function is of the form

$$\langle \Psi^\dagger(x,t)\Psi(x',0)\rangle \sim t^{-(1/K+3)/4} . \tag{3.32}$$

The boundary scaling dimension of the electron field operator is therefore $\bar\Delta = \frac{1}{8K} + \frac{3}{8}$, as opposed to its bulk scaling dimension $\Delta = \frac{1}{16}\left(\frac{1}{K}+K\right)+\frac{3}{8}$. Making use of the textbook definition of the density of states as the imaginary part of the single-electron Greens function, the TDOS is

$$\rho_{\rm TDOS}(\varepsilon) \sim \varepsilon^{2\bar\Delta-1} , \tag{3.33}$$

which is different from the bulk density of states

$$\rho_{\rm bulk}(\varepsilon) \sim \varepsilon^{2\Delta-1} . \tag{3.34}$$

Since $\bar\Delta > 1/2$ for $K < 1$, the TDOS always vanishes as the energy ε approaches zero, similar to the situation encountered in quantum wires [8]. For a Fermi liquid, we would instead have $\Delta = \bar\Delta = 1/2$, and both exponents are zero. For the setup of [6], taking our previous estimate $K \simeq 0.18$, the exponents in (3.33) and (3.34) are 1.13 and 0.46, respectively. Therefore single-electron tunneling is strongly suppressed close to the end of the nanotube. These exponents should be experimentally accessible by STM experiments. More detailed predictions for the TDOS can be found in [15].

3.4 Conductance

Since there are four transport channels ($j\delta$), a SWNT adiabatically connected to external leads exhibits the perfect conductance $G = G_0 = 4e^2/h$ in the absence of impurity backscattering. The actual conductance observed experimentally will be reduced by at least three mechanisms.

1. The *contact resistance* between the leads and the SWNT will decrease the two-terminal conductance. More importantly, a large contact resistance causes charging effects [16] implying the presence of an energy scale E_c that effectively quantizes the spectrum. In the experiments of [6], the charging energy $E_c \approx 2.6$ meV is much larger than the level spacing $\hbar v/L$ due to the finite length L of the SWNT. Transport at energy scales $\varepsilon \le E_c$ is then blocked ("Coulomb blockade").

2. At high temperatures, backscattering due to phonons or *twistons* becomes effective and leads to a linear temperature increase of the resistance [12].

3. At intermediate-to-low temperatures, *impurity backscattering* is most efficient. This case is discussed in some detail next, for simplicity focusing on the linear conductance $G(T)$. The voltage dependence of the nonlinear conductance follows the same power laws as found from the T-dependence of $G(T)$.

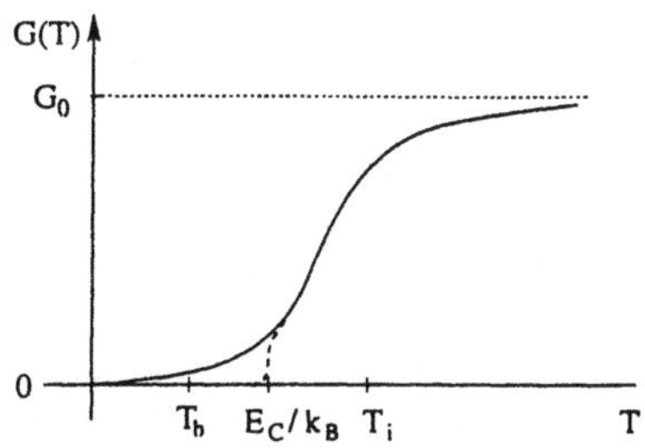

Figure 3
Schematic temperature dependence of the linear conductance in the presence of impurity backscattering. If contact resistances are large, G_0 will be strongly reduced from the perfect conductance $4e^2/h$, and transport below $T \simeq E_c/k_B$ is affected by the Coulomb blockade (dashed line). The crossover scale T_i is given in (3.36). Below $T \simeq T_b$, different exponents emerge. For a finite length L, the conductance will be constant below $T \simeq \hbar v/L$ (not shown).

Disregarding resonant tunneling effects, the generic consequences of impurity backscattering can be studied for a single impurity of strength Λ at $x = 0$,

$$H_{\rm imp} = \Lambda \cos[\sqrt{\pi}\theta_{c+}] \cos[\sqrt{\pi}\theta_{s+}] \sin[\sqrt{\pi}\theta_{c-}] \sin[\sqrt{\pi}\theta_{s-}]|_{x=0} + (\cos \leftrightarrow \sin) \,. \quad (3.35)$$

The effective temperature scale T_i generated by the impurity can be obtained from straightforward scaling arguments [8],

$$k_B T_i = D(\Lambda/D)^{4/(1-K)} \,. \quad (3.36)$$

For high temperatures $T \gg T_i$, the conductance corrections δG defined by $G = G_0 - \delta G$ can then be computed perturbatively. One finds

$$\delta G \sim (T/T_i)^{(K-1)/2} \quad (3.37)$$

to lowest order in Λ^2. The full crossover to the low-temperature regime has been obtained for arbitrary K [17], and at temperatures $T \ll T_i$, the following power law for the conductance is found,

$$G \sim (T/T_i)^{(1/K-1)/2} \,. \quad (3.38)$$

This implies a total suppression of transport through the nanotube at $T = 0$ in the presence of a single, arbitrarily weak impurity.

In practice, one has to be aware that a crossover into a temperature-independent conductance occurs at $k_B T \simeq \hbar v/L$. Furthermore, if contact resistances are large, the conductance drops to zero at $k_B T \simeq E_c$. Thereby the above anomalous conductance laws are cut off. It is thus desirable to decrease the contact resistances in order to observe Luttinger liquid transport behavior over a large temperature range. Furthermore, we have assumed here that the temperature stays large compared to the gap scales T_f and T_b. Otherwise the exponents are different due to the more complicated non-Fermi liquid states found at $T < T_b$, see [11]. The temperature dependence of the linear conductance is schematically shown in Fig. 3.

3.5 Friedel Oscillation

Finally, we study the electronic screening cloud (Friedel oscillation) building up around an impurity of strength Λ at $x = 0$, or at the end of the SWNT. In the latter case, we can alternatively speak of a strong impurity with $\Lambda = D$. The small-q response of the π electrons can be computed by linear screening. However, for the finite wavevectors in (3.30), the corresponding Friedel oscillation cannot be computed by a standard RPA-like treatment due to the breakdown of Fermi liquid theory.

Following the analysis of [9], it is straightforward to extract the power law decay of the Friedel oscillation. Remarkably, we find the simultaneous presence of the wavelengths π/q_F and $\pi/(k_F \pm q_F)$. While for $x \gg \hbar v/k_B T$ the Friedel oscillation decays exponentially on the thermal length scale $x_T = \hbar v/k_B T$, the result for $a \ll x \ll x_T$ contains first a $2q_F$ oscillatory part,

$$\begin{aligned} \langle \rho(x) \rangle &\sim \cos[2q_F x]\,(x/x_i)^{-(3+K)/4} \quad (x \gg x_i) \\ &\sim \cos[2q_F x]\,(x/x_i)^{-(1+K)/2} \quad (x \ll x_i)\,, \end{aligned} \tag{3.39}$$

and superimposed a rapidly oscillating contribution,

$$\begin{aligned} \langle \rho(x) \rangle &\sim \cos[2(k_F \pm q_F)x]\,(x/x_i)^{-(3+K)/4} \quad (x \gg x_i) \\ &\sim \cos[2(k_F \pm q_F)x]\,(x/x_i)^{-(1+K)/2} \quad (x \ll x_i)\,. \end{aligned} \tag{3.40}$$

We omit possible phase shifts here. Furthermore, $x_i = \hbar v/k_B T_i$ with T_i specified in (3.36) sets the appropriate length scale generated by the impurity. The asymptotic exponents for $x \gg x_i$ can be obtained from open boundary bosonization, while the behavior close to the impurity follows from perturbation theory in Λ. Here the important point is that the Friedel oscillation is always *slower* than the standard $1/x$ Fermi liquid result. Furthermore, the decay becomes even slower as the impurity is approached. Therefore one can expect very pronounced Friedel oscillations in a SWNT.

The power laws (3.39) and (3.40) are modified for extremely strong correlations, $K < 1/5$, due to higher-order CDW operators. Instead of (3.39) and (3.40), we then find the wavelength $\pi/4q_F$,

$$\begin{aligned} \langle \rho(x) \rangle &\sim \cos[8q_F x]\,(x/x_i)^{-4K} \quad (x \gg x_i) \\ &\sim \cos[8q_F x]\,|\ln(x/x_i)| \quad (x \ll x_i)\,. \end{aligned} \tag{3.41}$$

The only logarithmically slow decay of the Friedel oscillation close to the impurity follows from [18].

The overall behavior of Friedel oscillations in nanotubes is considerably more complicated than predicted by RPA. We find several different slow algebraic power laws in various regions of parameter space. They are all slower than the standard $1/x$ decay. Furthermore, there exist different wavelengths for the respective dominant contributions. If STM measurements of the charge screening cloud become feasible, one could directly extract the important correlation parameter K from the decay of the Friedel oscillation and observe the various regions of the phase diagram corresponding to Fig. 2.

4 Concluding Remarks

We have discussed the effective low-energy theory of a single-wall carbon nanotube, taking into account the electron-electron interaction. Nanotubes are promising candidates for revealing the Luttinger liquid behavior generally expected for one-dimensional metals. According to our theory, at temperatures above the gap scales $T_f \approx T_b \approx 0.1$ mK, a nanotube behaves as a spin-$\frac{1}{2}$ Luttinger liquid with an additional flavor index. At lower temperatures, more complicated phases characterized by gaps in the neutral modes are predicted to emerge.

A special feature of doped nanotubes is the simultaneous occurrence of different wavelengths for the various order parameter correlations. The doping is characterized by a small wavevector $q_F \ll k_F$, and one quite generally finds pronounced instabilities for the long wavelengths $\lambda = \pi/q_F, \pi/2q_F$ and $\pi/4q_F$, besides the short wavelengths $\lambda = \pi/(k_F \pm q_F)$. Unambiguous signatures of Luttinger liquid behavior include anomalous interaction-dependent power laws for the conductance, the tunneling density of states, or the decay of the Friedel oscillation. Further manifestations of the electron-electron interaction can be expected in the shot noise and in frequency-dependent transport experiments. We believe that correlation effects such as the ones discussed here can be experimentally observed in the near future.

Bibliography

[1] S. Iijima, Nature **354**, 56 (1991).

[2] R. Saito, M. Fujita, G. Dresselhaus, and M.S. Dresselhaus, Appl. Phys. Lett. **60**, 2204 (1992).

[3] P.R. Wallace, Phys. Rev. **71**, 622 (1947).

[4] J.W.G. Wildöer, L.C. Venema, A.G. Rinzler, R.E. Smalley, and C. Dekker, Nature **391**, 59 (1998).

[5] A. Thess, R. Lee, P. Nikolaev, H. Dai, P. Petit, C. Xu, Y.H. Lee, S.G. Kim, A.G. Rinzler, D.T. Colbert, G.E. Scuseria, D. Tomanek, J.E. Fischer, and R.E. Smalley, Science **273**, 483 (1996).

[6] S.J. Tans, M.H. Devoret, H. Dai, A. Thess, R.E. Smalley, L.J. Geerligs, and C. Dekker, Nature **386**, 474 (1997). See also S.J. Tans, M.H. Devoret, R.J.A. Groeneveld, and C. Dekker, Nature **394**, 761 (1998).

[7] A.O. Gogolin, A.A. Nersesyan, and A.M. Tsvelik, *Bosonization and Strongly Correlated Systems* (Cambridge University Press, 1998).

[8] C.L. Kane and M.P.A. Fisher, Phys. Rev. B **46**, 15 233 (1992).

[9] R. Egger and H. Grabert, Phys. Rev. Lett. **75**, 3505 (1995).

[10] M. Fabrizio and A.O. Gogolin, Phys. Rev. B **51**, 17 827 (1995).

[11] R. Egger and A.O. Gogolin, Phys. Rev. Lett. **79**, 5082 (1997); European Phys. Journal B **3**, 281 (1998).

[12] C.L. Kane and E.J. Mele, Phys. Rev. Lett. **78**, 1932 (1997).

[13] See, e.g., M. Fabrizio, Phys. Rev. B **48**, 15 838 (1993), and Ref. [7].

[14] H.J. Schulz, Phys. Rev. B **53**, R2959 (1996).

[15] C.L. Kane, L. Balents, and M.P.A. Fisher, Phys. Rev. Lett. **79**, 5086 (1997).

[16] H. Grabert and M.H. Devoret (eds.), *Single Charge Tunneling*, NATO-ASI Ser. B, vol. 294 (Plenum, New York, 1992).

[17] P. Fendley, A.W.W. Ludwig, and H. Saleur, Phys. Rev. B **52**, 8934 (1995).

[18] A. Leclair, F. Lesage, and H. Saleur, Phys. Rev. B **54**, 13 597 (1996); F. Lesage and H. Saleur, J. Phys. A: Math. Gen. **30**, L457 (1997).

A New Look at Low-Temperature Anomalies in Glasses

Reimer Kühn and Uta Horstmann

Institut für Theoretische Physik, Universität Heidelberg,
Philosophenweg 19, D-69120 Heidelberg, Germany

Abstract: We review a model-based rather than phenomenological approach to low-temperature anomalies in glasses. Specifically, we present a solvable model inspired by spin-glass theory that exhibits both, a glassy low-temperature phase, and a collection of double- and single-well configurations in its potential energy landscape. The distribution of parameters characterizing the local potential energy configurations can be *computed*, and is found to differ from those assumed in the standard tunneling model and its variants. Still, low temperature anomalies characteristic of amorphous materials are reproduced. More importantly perhaps, we obtain a clue to the universality issue. That is, we are able to distinguish between properties which can be expected to be universal and those which cannot. Our theory also predicts the existence, under suitable circumstances of amorphous phases *without* low-energy tunneling excitations.

Dedicated to Heinz Horner on the occation of his 60th birthday.

1 Introduction

Ever since the first measurements of Zeller and Pohl [1] revealed that the specific heat and the thermal conductivity of glassy systems at low temperatures are strikingly different from those of crystalline substances, the low temperature physics of glassy and amorphous materials has been a subject of intense research efforts, both experimentally and theoretically.

Specifically, it was found [1] that below approximately 1 K the specific heat of glassy materials scales approximately linearly with temperature, $C \sim T$ while the corresponding scaling for the thermal conductivity κ is approximately quadratic $\kappa \sim T^2$. Both findings contrast the T^3 behaviour of these quantities in crystals. Between 1 K and approximately 20 K the thermal conductivity exhibits a plateau and, then continues to rise as the temperature is further increased. The specific heat, too, changes its behaviour in the 1–20 K regime. It exhibits a bump if displayed in C/T^3 plots.

Particularly intriguing is the fact that the anomalous behaviour below 1 K appears to be universal in the sense that it is shared by virtually all glassy and amorphous materials, whereas the behaviour between 1 and 20 K seems to display a greater material dependence. The universality in the thermal conductivity data actually appears to go beyond the level of exponents, in that by a suitable rescaling with a constant depending on the Debye temperature and the the sound velocity, thermal conductivity data for various substances can be scaled onto one master-curve both below and above, though not on the plateau [2].

It should be added that there is also an extensive body of experimental data characterizing the anomalous response of glassy and amorphous materials to external probes such as ultrasound or electro-magnetic fields. The reader is invited to consult the reviews of Hunklinger et al. [5] or Phillips [6] on these matters.

The unusual material properties at low temperatures are usually attributed to the existence of a broad range of localized low-energy excitations in amorphous systems – excitations not available in crystalline materials. At energies below 1 K, these are believed to be tunneling excitations of single particles or (small) groups of particles in double-well configurations of the potential energy (DWPs). This is the main ingredient of the phenomenological so-called standard tunneling model (STM), independently proposed by Phillips [3] and by Anderson et al. [4]. As a second ingredient of the STM, it is supposed that the local DWPs in amorphous systems are random, and specific assumptions concerning the distribution of the parameters characterising them must be advanced (see below) to describe the experimental data below 1 K [3,4].

Neither Phillips [3] nor Anderson et al. [4] consider excitations other than the tunneling excitation in the DWP structures postulated by them – their local degrees of freedom are thus true two-level systems (TLSs). As a consequence their model is unable to account for the different physics that appears in the temperature range between 1 and 20 K. However, different sets of low energy excitations in amorphous systems can exist: (i) Higher excitations in the above mentioned DWPs – a possibility that is surely contained in the *physical* picture advanced by Phillips or Anderson et al., but was justly considered irrelevant for the very low temperature regime and then somehow not reconsidered when problems with the model arose at temperatures above 1 K. (ii) Localized vibrations (localized phonons) in soft anharmonic single well configurations of the potential energy. The latter have been postulated (in addition to tunneling excitations in DWPs) within the, likewise phenomenological soft-potential model (SPM) [7,8] to describe the physics above 1 K. Again, local potential energy configurations are supposed to be random, and specific assumptions concerning the distributions of the parameters characterizing them are required to account for the experimental data also above 1 K, such as the crossover to T^3-behaviour of the specific heat and the plateau in the thermal conductivity [7,8].

Both, localized soft vibrations [9] and DWPs [10] responsible for two-level tunneling systems (TLS) have been seen in molecular dynamics studies of Lennard Jones glasses. In the case of soft vibrations [9] no attempt, however, was made to determine the shape of the single-well potentials (SWPs) supporting these vibrations, so as to check the hypotheses of the SPM. On the other hand, local potential energy configurations giving rise to TLSs were analysed within the confines of a generalized SPM, [10] which assumes that locally the potential energy surface (along some reaction coordinate) can be described by certain fourth order polynomials, with coefficients distributed in a specific way. No single-well configurations were taken into account, though, to determine the statistics of the coefficients [10], as in principle they should to make full contact with the assumptions of the SPM.

Some time ago, in a lucid critical discussion of the standard theory of TLSs in amorphous systems, Yu and Leggett [11] pointed out that there was nothing in the STM that could reasonably account for the considerable degree of universality observed in amorphous systems. Analogous remarks would mutatis mutandis apply to the SPM, and it is, of course, mainly related to the phenomenological nature of these models. They have to rely on assumptions concerning in particular the distributions of parameters characterising local potential energy configurations, which – while plausible in some respects – are certainly much less so in others, and are lacking support based on more microscopic approaches, such as that of [12] for KBr:KCN mixed crystals. Neither model, to be sure, accounts for a *mechanism* that would explain *how* the required local potential energy configurations would arise, and how they would do so with the required statistics.

It is here that our model-based approach to low-temperature physics of glasses [13,14], which was started about two years ago, attempts to fill a gap.

Our approach takes as its starting point the very observation of universality of glassy low-temperature anomalies, which we translate into a strategy as follows. Since – in the light of universality – the detailed form of particle-interactions is apparently to a large extent immaterial to the phenomena that are to be modeled, we should be justified in proposing something like a *caricature-glass*. We would like this to be understood here in the sense of something as simple as possible as long as it retains essential properties of glass-forming systems. Our success will of course depend on the quality of our understanding as to what these essential ingredients might possibly be. We take them to be (i) particles moving in continuous space, subject to (ii) an interaction (*any interaction!*) that gives rise to an amorphous low temperature phase. Within these confines, our choice of the interaction has been mainly guided by the demand for simplicity, and analytic tractability. The outcome of these deliberations has been a proposal inspired by spin-glass theory [13,14] that describes a glassy material as an anharmonic interacting particle system with *random interactions* at the harmonic level, the details of which will be specified in Sect. 2 below. The interactions are chosen in such a way that the model is solvable via mean-field and replica techniques well known in the theory of spin-glasses [15].

We have organized the remainder of our material as follows. In Sect. 2 we present details of our model, and review main features of its solution within mean-field theory. We compute its glass transition temperature and exhibit its phase diagram, featuring ergodic, polarised and glassy phases. Sect. 3 is devoted to mapping out the potential energy surface of the system in (one of) its classical glassy ground-states, and to determining the statistics of its local potential energy configurations. Thermodynamic consequences at low temperatures are explored in Sect. 4, where we look at excitation spectra in local potential energy configurations, at the density of states, and at the specific heat. Dynamic consequences are briefly considered in Sect. 5. We close in Sect. 6 with a summary and with an outlook on open issues.

2 Spin-Glass Approach and a Solvable Model

We suggest to constider the following Hamiltonian as a candidate for the description of glassy low-temperature properties

$$\mathcal{H} = \sum_{i=1}^{N} \frac{p_i^2}{2m} + U_{\text{int}}(\{v_i\}) , \tag{2.1}$$

with an interaction energy given by

$$U_{\rm int}(\{v_i\}) = -\frac{1}{2}\sum_{i\neq j} J_{ij}v_iv_j + \frac{1}{\gamma}\sum_i G(v_i)\ , \tag{2.2}$$

in which we include an on-site potential of the form

$$G(v) = \frac{a_2}{2}v^2 + \frac{a_4}{4!}v^4\ . \tag{2.3}$$

For the time being we specialize to $a_2 = a_4 = 1$.

The description is in terms of *localised* degrees of freedom, i.e., the v_i may be interpreted as deviations of particle positions from a given set of reference positions – in a spirit akin to that used to describe dynamical properties of crystalline solids. Thus, we assume the system to be already in a solid state, and do not attempt to provide a faithful description of the liquid phase.[1]

We propose to model the glassy aspects by taking the interaction at the harmonic level, that is, the J_{ij}, to be random, so that the reference positions will generally be unstable at the harmonic level. This is why we have added the on-site potentials $G(v)$, namely to stabilize the system as a whole. By including in G a harmonic term, we can use the parameter γ to tune the number of modes that actually *are* unstable at the harmonic level of description. Apart from its role in this 'tuning' aspect, we have opted for the simplest choice that achieves the required stabilization.

We believe our opting for random interactions to be justified by appeal to universality (the detailed form of the interactions cannot be crucial for the emergence of glassy low-temperature anomalies), and by the observation made in recent years of the fundamental similarity between truly quenched disorder and so-called self-induced quenched disorder [16].

We choose the random interactions in such a way that the system is amenable to analysis within replica mean-field theory, well known from the theory of spin-glasses[15]. Specifically, we choose them to be Gaussian random variables with $\overline{J_{ij}} = J_0/N$ and $\overline{J_{ij}^2} - \overline{J_{ij}}^2 = 1/N$ The harmonic part of $U_{\rm int}(\{v_i\})$ is thus reminiscent of the SK spin-glass [17], except for the fact that we are dealing here with continuous degrees of freedoms rather than with classical Ising spins. Once more, by appeal to universality, and by noting that we do not intend to describe correlations at a critical point, it is not unreasonable to argue that a mean-field description should be sufficient to reveal the essential physics we are after.

Let us mention for completeness that models with an interaction energy of the form (2.2) were considered before in an entirely different field, namely in the context of analog-neuron systems and networks of operational amplifiers [18,19]. Incidentally, this connection opens up an unexpected but rather fascinating perspective of studying glassy dynamics by emulating it directly in highly integrated circuitry, albeit perhaps not in the regime of low temperatures, where quantum effects are important.

At the classical level of description, all interesting information about the thermodynamics of the system is obtained from the configurational free energy

[1] Note that we are writing down what appears to be a scalar model here. The 3-D nature of particle coordinates is, however, included in our description, if we choose to consecutively enumerate the Cartesian components of the particles, in which case N has to be read as three times the number of particles of the system.

$$f_N(\beta) = -(\beta N)^{-1} \ln \int \prod_i dv_i \exp[-\beta U_{\rm int}(\{v_i\})] \,. \tag{2.4}$$

The free energy has to be averaged over the ensemble of possible realisations of the J_{ij} configurations so as to get *typical* results. This is accomplished by means of the so-called replica method, well known in the theory of disordered systems (see, e.g. [15]). We will not here reproduce the calculations, but merely quote the result. The quenched free energy – that is, the disorder-average of (2.4) – is represented as the limit $f(\beta) = \lim_{n\to 0} f_n(\beta)$, with

$$\begin{aligned} n f_n(\beta) &= \frac{1}{2} J_0 \sum_a p_a^2 + \frac{1}{4}\beta \sum_{a,b} q_{ab}^2 \\ &- \beta^{-1} \ln \int \prod_a dv^a \exp\left[-\beta U_{\rm eff}(\{v^a\})\right] \,. \end{aligned} \tag{2.5}$$

Here

$$U_{\rm eff} = -J_0 \sum_a p_a v^a - \frac{\beta}{2} \sum_{a,b} q_{ab} v^a v^b + \frac{1}{\gamma} \sum_a G(v^a) \tag{2.6}$$

is a replicated single-site potential and the order parameters $p_a = N^{-1} \sum_i \langle v_i^a \rangle$ and $q_{ab} = N^{-1} \sum_i \langle v_i^a v_i^b \rangle$ are determined from the fixed point equations

$$p_a = \langle v^a \rangle \,, \qquad a = 1, \ldots, n \tag{2.7}$$

$$q_{ab} = \langle v^a v^b \rangle \,, \qquad a,b = 1, \ldots, n \,, \tag{2.8}$$

where $\langle \ldots \rangle$ denotes a Gibbs average corresponding to the replica potential (2.6). The limit $n \to 0$ is eventually to be taken in these equations.

Eqs. (2.5)-(2.8) were analysed in the replica symmetric (RS) [13] and the 1st step replica-symmetry breaking (1RSB) approximations [14]. In RS one assumes $p_a = p$ for the 'polarization', and $q_{aa} = \hat{q}$ and $q_{ab} = q$ for $a \neq b$ for the entries of the Edwards-Anderson matrix. These are given as solutions of

$$p = \langle \langle v \rangle \rangle_z \,, \quad C = \left\langle \frac{1}{\sqrt{q}} \frac{d\langle v \rangle}{dz} \right\rangle_z \,, \quad q = \langle \langle v \rangle^2 \rangle_z \,. \tag{2.9}$$

Here $C = \beta(\hat{q} - q)$, and $\langle \ldots \rangle_z$ denotes an average over a standard Gaussian z while $\langle \ldots \rangle$ is now a thermal average corresponding to the RS single-site potential

$$U_{\rm eff}(v) = -h_{\rm eff}\, v - \frac{1}{2} C v^2 + \frac{1}{\gamma} G(v) \,, \tag{2.10}$$

with

$$h_{\rm eff} = h_{\rm RS} = J_0 p + \sqrt{q} z \tag{2.11}$$

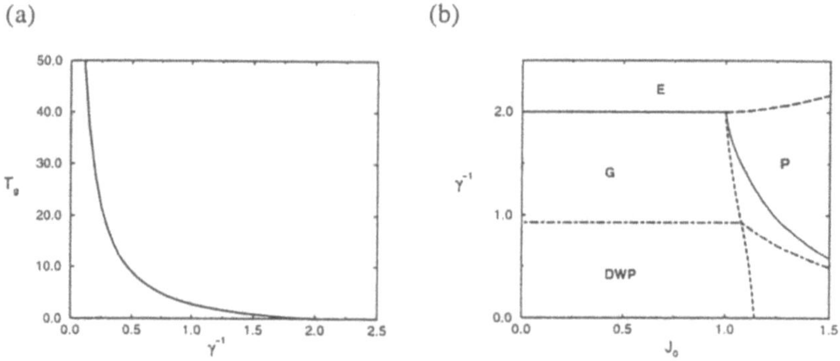

Figure 1 (a): Glass temperature T_g as a function of γ for $J_0 = 0$. (b): $T = 0$ phase diagram in RS. E denotes the $T = 0$-limit of the ergodic phase, G is the glassy phase, and P is a phase with macroscopic polarization. The full line is the AT line (see text). Below the dot-dashed line is the region with DWPs, as described in detail in Sect. 3. The short-dashed line separates the glassy phase G from the phase P with macroscopic polarization. From [14]

It turns out that our model exhibits a transition from an ergodic phase with $p = q = 0$ to a glassy phase with $q \neq 0$ at some temperature T_g which depends on J_0 and γ. For sufficiently large J_0, the transition is into a polarized phase with $p \neq 0$. The glass-transition temperature as a function of γ is shown in Fig. 1a. In Fig. 1b, we exhibit also the $T = 0$ limit of the phase diagram, which will become of importance for the discussion of of the potential energy surface of the model in Sect. 3 below.

The assumption of RS is not justified at low temperatures and large γ, where replica symmetry breaking (RSB) – an instability that breaks the permutation invariance among the n replica – is expected to occur. The location of the instability against RSB is given by the de Almeida-Thouless (AT) criterion [20] $1 = \beta^2 \left\langle (\langle v^2 \rangle - \langle v \rangle^2)^2 \right\rangle_z$. In a one-step replica-symmetry breaking (1RSB) approximation, which is believed to constitute a major step towards the full solution, $U_{\rm eff}(v)$ is of the same form (2.10), however with $h_{\rm RS}$ replaced by $h_{\rm 1RSB} = J_0 p + \sqrt{q_0}\, z_0 + \sqrt{q_1 - q_0}\, z_1$, and C by $C = \beta(\hat{q} - q_1)$, where we use standard notation [15,17] for the entries of the q-matrix. Along with a so-called partitioning parameter m, they are determined from a more complicated set of fixed point equations [15,21]. In 1RSB, z_0 is a Gaussian, whereas the distribution of z_1 is more complicated.

In the following section we will set out to demonstrate that the mean-field solution of the model we have obtained here does in fact also contain information about the potential energy surface of the model, the statistics of which is generally believed to be crucial for the emergence of glassy low-temperature anomalies.

3 Mapping Out the Potential Energy Surface

To map out the potential energy surface of the system, we observe that all that mean field theory is about, is to represent the interaction energy of the system as a sum of effective independent single site potential energies

$$U_{\rm int}(\{v_i\}) \longrightarrow \sum_i U_{\rm eff}(v_i) \; , \tag{3.1}$$

self-consistently to be determined in such a way as to get thermal averages correct. In the context of non-random systems, the most famous example is the Curie-Weiss model of ferromagnetism, where the effective single site potential is the Zeemann energy of a spin in an effective (non-fluctuating) local- or mean-field determined by its neighbours.

In our case, we obtain the sum (3.1) of independent single site potentials $U_{\text{eff}}(\nu_i)$, which contain random parameters. Replica theory achieves nothing but a self-consistent determination of the distribution of these random parameters. Within the RS approximation these effective local single-site potentials are given by (2.10),(2.11), containing a single random parameter, viz the Gaussian distributed effective fields h_{RS}, having mean $J_0 p$ and variance q. The parameters of p, q and C characterizing the Gaussian ensemble of single-site potentials $U_{\text{eff}}(\nu)$ are determined self-consistently through (2.9)–(2.11). In 1RSB, the ensemble of single site potentials is again of the form (2.10), but now with h_{RS} replaced by the more complicated h_{1RSB} and C by $C = \beta(\hat{q} - q_1)$ as noted above. Again there is only a single random parameter, a locally varying effective random field. This feature is easily seen to persist at all levels of replica symmetry breaking.

There is one additional ingredient, namely we take the the $T = 0$ limit of the theory to select one of the (possibly many) collective classical glassy groundstate configurations.

In terms of the mean field solution, we thus have a representation of the glassy potential energy landscape as an ensemble of randomly varying effective single-site potentials. Their statistics is determined by the statistics of the effective fields h_{RS} and h_{1RSB} in the RS and the 1RSB approximation, respectively.

Now note in particular, that the initial stabilising concave upward single site potential $G(\nu)$ also appears in $U_{\text{eff}}(\nu)$, but there is now an additional harmonic term, viz. $-\frac{1}{2}C\nu^2$ – entirely of collective origin – that renormalises the effective total harmonic restoring force. For

$$C - \frac{1}{\gamma} > 0 \tag{3.2}$$

the total harmonic contribution to $U_{\text{eff}}(\nu)$ becomes convex downward near the origin $\nu = 0$, so that for sufficiently small $h_{\text{RS/1RSB}}$ the effective single site potential $U_{\text{eff}}(\nu)$ attains a DWP-form, which is of collective origin! The region in parameter space where DWPs exist and its boundary are indicated in the $T = 0$ phase diagram in Fig. 1b.

We are now able to make contact with the phenomenology of glassy low-temperature anomalies, and to compare with the ideas underlying their explanation within the STM [4,3] or the SPM [7]. The main difference and, we believe, important new perspective added by our approach is that the appearance of DWPs and the statistics of local potential energy configurations need not be hypothesised from the outset, but rather that it emerges as a collective effect, originating in the frustrated interactions J_{ij}, if the temperature is sufficiently low and the parameter γ sufficiently large. Moreover, the statistics of the local potential energy configurations has been rendered part of the world of the *computable*. It follows directly from the statistics of the $h_{\text{RS/1RSB}}$.

To explore the consequences for the low temperature behaviour of glasses quantitatively, one solves the quantum mechanical problem for $U_{\text{eff}}(\nu_i)$ described be the Hamiltonian

$$H_{\text{eff}} = \frac{p^2}{2m} + U_{\text{eff}}(\nu) \tag{3.3}$$

with $U_{\text{eff}}(\nu)$ given by (2.10),(2.11) in the RS approximation, or by the corresponding expression valid for the 1RSB (or higher order RSB) ansätze.

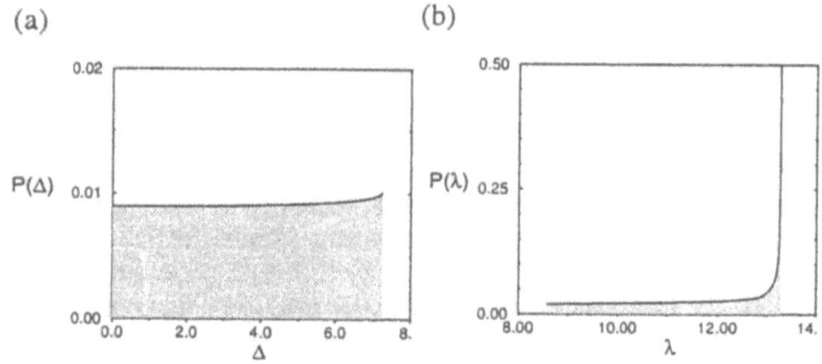

Figure 2 Distributions $\mathcal{P}(\Delta)$ (for positive Δ), and $\mathcal{P}(\lambda)$ in RS. Here $\gamma = 4$, while $J_0 = 0$. After [13]

In the STM one approximates the tunnel-splitting ε of the ground-state in the DWP case to be given in terms of the asymmetry Δ of the DWP and the tunneling-amplitude Δ_0 by the TLS-result

$$\varepsilon = \sqrt{\Delta^2 + \Delta_0^2}\,, \tag{3.4}$$

the value of Δ_0 being related with the barrier-height V between adjacent minima and the distance d between them. In WKB-approximation one has $\Delta_0 \simeq \hbar\omega_0 \exp(-\lambda)$, with $\lambda = \frac{d}{2}(2mV/\hbar^2)^{1/2}$, where ω_0 is a characteristic frequency (of the order of the frequency of harmonic oscillations in the two wells forming the double well structure), and m the effective mass of the tunneling particle. Moreover, a specific assumption is advanced concerning the distribution of asymmetries Δ, and tunneling-matrix elements Δ_0, namely, in terms of Δ and λ one assumes $P(\Delta,\lambda) \simeq$ const. Note that this assumption implies in particular also that these quantities are *uncorrelated*.

We can check these assumptions against the results from our model [13]. Obviously, since the values of both Δ and λ are determined by a single random variable, viz. $h_{\mathrm{RS/1RSB}}$, these quantities must be *perfectly correlated* in our model, in contrast to the assumptions of the STM, although their distributions separately are indeed rather flat, except for anomalies at their upper end (Fig. 2) where due to large $h_{\mathrm{RS/1RSB}}$ DWPs cease to exist.

The perfect correlation can be weakened, but *not* eliminated, by allowing *local randomness*, that is, by making the coefficients a_2 and a_4 in (2.3) vary randomly from site to site. A randomly varying a_2 is, in fact, a natural choice in our approach, since the a_2 contribution to U_{int} should be considered part of the harmonic level, which we had assumed to be random to begin with. Indeed, by allowing this kind of randomness, our results can be brought into better agreement with the experimental situation (see below). As to a_4, we have observed that taking it to be random (within limits of course) has no dramatic effect, so we adhere to it remaining non-random, serving solely, as it should, its stabilising purpose. We should add that our model remains solvable with these modifications. Outer averages in the fixed point equations simply have to be read as implying an additional average over the local randomness.

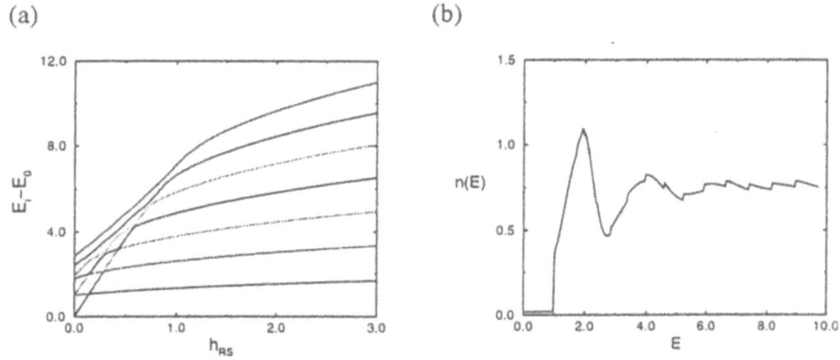

Figure 3 **a)**: Excitation spectrum as a function of h_{RS}. $\gamma = 4$ and $J_0 = 0$ are as in Fig. 2. **b)**: Density of states in RS, for parameters as in (**a**).

As a final unexpected result, note that our theory predicts the existence under suitable circumstances of an amorphous phase without DWPs, and hence without low-energy tunneling excitations (see Fig. 1b). Until very recently, this would perhaps have universally been considered a sheer absurdity. However, recent internal friction measurements in specially prepared amorphous Si:H films by Liu et al. [22] indicate that such a possibility has to be seriously taken into account.

4 Thermodynamics

Information about the thermodynamics of the system at low temperatures, where it is dominated by quantum effects, is obtained by studying the excitation spectra of the local Hamiltonians (3.3), and by using this information – averaging it over the randomness characterizing the ensemble of these local Hamiltonians, i.e. over the $h_{RS/1RSB}$-distributions (and possibly the a_2-distribution) – to compute densities of state and thermodynamic functions the usual way.

In Fig. 3a we exhibit a typical excitation spectrum, as it varies with the value of h_{eff}. The salient features are two kinds of branches, steeper and less steep ones. The former are related to tunneling (inter-well) excitations whereas the the latter are of intra-well type. That is, the latter would also exist if the wells were not communicating via a barrier of finite height and width. Note that these families of levels do not cross as might appear at first sight. Rather, there are avoided 'would-be' level-crossings precisely *due* to the tunneling mechanism. At at energies higher than the barrier or at large h_{RS} the avoided crossing-pattern disappears, because there is no longer a barrier to be tunneled through.

The corresponding density of states (DOS) (averaged over the ensemble) is displayed in Fig. 3b. Note its constant value at energies lower than that of the first intra-well excitation. It arises because, (i) apart from the immediate vicinity of its lower cutoff Δ_0, the level-splitting of the ground-state scales linearly with the asymmetry Δ (which in turn is proportional to h_{eff}, and (ii) because the distribution of h_{eff}) is virtually flat in this regime.

This energy range is to be associated with the *universal* glassy low-temperature anomalies below 1 K! Its properties are mainly driven by the distribution of effective fields h_{eff}

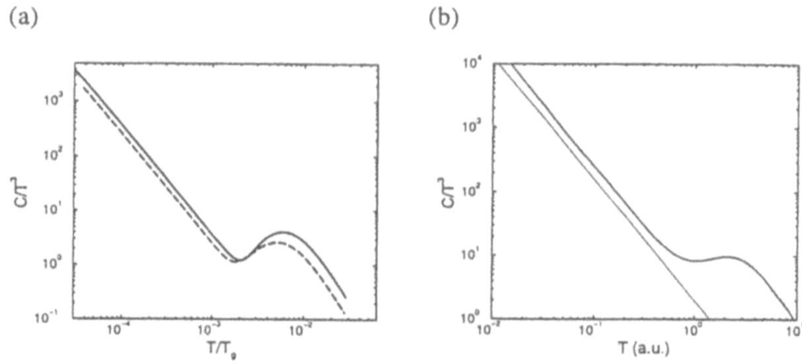

Figure 4 Specific heat at low temperatures **a):** without local disorder (full line RS, dashed line 1RSB) , **b):** with $\gamma = 1$, and a_2 a Gaussian of mean 1 and standard deviation $\sigma = 5$ in RS. Other parameters are as before. The thin line corresponds to $C \sim T^{1.1}$.

and are thus entirely of collective origin. In particular it is related to the temperature range in which the specific heat scales linearly with temperature (see Fig. 4).

By contrast, the energies of the intra-well excitations are much less sensitive to the value of the effective field, hence less steep in the excitation spectrum if plotted vs. $h_{\rm eff}$. These give rise to the strong increase of the DOS at the value of the first intra-well excitation (the contribution to the DOS being proportional to the *inverse* slope of the level as a function of $h_{\rm eff}$). The intra-well excitations depend more strongly on the shape of the *local* stabilizing potentials $G(v)$, hence they cannot be of collective origin – basically they turn out where we have chosen them to be, with or without local randomness in the $G(v)$. So the physics that depends on these features, namely the bump in $C(T)/T^3$ and the plateau in the thermal conductivity, cannot be expected to be universal.

This separation within our approach of aspects that can be expected to be universal from those which cannot is in perfect qualitative agreement with the experimental situation. It is not unreasonable to suppose that such a distinction between local and global contributions to $U_{\rm int}$ can be made for real materials as well, the contribution of the long-range global one being for instance of elastic nature (and well known to give rise to frustration).

Note that our representation of local potential energy configurations has finally turned out to be similar to that proposed within the SPM [7,8] with some marked differences in the details: (i) Within our approach the very occurrence of DWPs is of identifiable collective origin, and so is the distribution of asymmetries. (ii) *No* analogous mechanism is available for determining (distribution of) the parameters entering $G(v)$, which are mainly responsible for the distribution of barrier heights and the nature of (single- and) intra-well excitations which determine the material-specific properties. From our vantage point, therefore, one should not even attempt at proposing for these something universally valid for all glassy materials.

There is a some influence of local disorder on the behaviour of the low T specific heat also in the 'universal' regime changing it to slightly superlinear, with exponents depending on the nature of the distribution. This feature, too, agrees nicely with experimental findings, and it might be used to give some handle at fixing the distribution of local parameters for specific materials.

Concerning universality, Yu and Leggett supposed that it could result only due to a sufficiently long range interaction between TLSs, an idea which has since the been pursued in a series of papers by Burin and Kagan [23]. Here we have seen that that it can arise without assuming interactions at the level of quantized excitations, because the statistics of local potential energies is by itself already a largely collective affair.

5 Dynamics

Let us add a few brief remarks concerning dynamics. We have in mind here the computation of phonon mean free paths, of internal friction and the thermal conductivity, the latter two computable once the phonon mean free paths are known, as well as the computation of dielectric susceptibilities. All interesting properties can be computed within linear response theory. We assume a bilinear interaction of the local coordinates v_i, whose dynamics are given by $H_S = H_{\text{eff}}$ in (3.3), with the extended phonon degrees of freedom

$$H_B = \sum_{ks} \hbar\omega_{ks} b^{\dagger}_{ks} b_{ks} \tag{5.1}$$

that is mediated by the strain-field, and of the form

$$H_{SB} = v \sum_{s} \gamma_s e_s \tag{5.2}$$

with s labeling the acoustic branches of the phonon spectrum, e_s denoting the contribution of branch s to the strain field at site i and appropriate coupling constants γ_s. A Debye model is assumed for the phonon bath.

The dynamic properties of interest can be computed from the imaginary part of the dynamic susceptibility $\chi''_{vv}(\omega)$, which in turn, via fluctuation dissipation theorems, is related to the (symmetrized) centered correlation function

$$\hat{C}_{vv}(t) = (\Delta v(t)|\Delta v) = \frac{1}{2}\langle \Delta v(t)\Delta v + \Delta v \Delta v(t)\rangle \;, \tag{5.3}$$

specifically to the spectral function associated with it. Here $\Delta v = v - \langle v \rangle$. The Mori-Zwanzig Projection technique and (at this time) a weak coupling assumption are used to carry through in practice.

As of now, we have not yet produced quantitative results. So much can, however, be said at this point. In the limit of very low temperatures, our model clearly approaches the TLS-physics, and we expect a roughly quadratic temperature dependence of the thermal conductivity. The increased density of states in the 1–10 K range goes along with a new, and faster scattering mechanism: both are due to the existence of intra-well (as opposed to the slower inter-well) transitions in this energy range. As discussed by Yu and Leggett [11], the combination of these features can be expected to produce the plateau in the thermal conductivity. We shall report on these matters in greater detail in the near future [21].

6 Conclusions and Outlook

In summary, we have proposed to take a new look at glassy low-temperature anomalies from a model-based rather than phenomenological approach. The models so far considered do certainly not (yet) attempt to model the details of any specific substance. Our aim has been to formulate something very schematic that should capture the essentials of amorphous low temperature phases – a caricature in the same sense as the Ising model is a caricature description of uniaxial ferromagnets or, perhaps less boldly, the SK model a prototypic description of a spin-glass. Chances are of course, that we still haven't got the bare essentials right. But we feel that our first attempts do point in the right direction. In particular the clue we have obtained concerning the understanding as to which phenomena might be expected to be universal and which not, does look like an encouraging general qualitative result.

Our approach offers the unique possibility to study the whole range of glassy physics, from the regime of the glass transition temperature down to the low temperatures where quantum effects play a dominant role, all within a single set of model assumptions. In particular, it should be interesting to study the glassy dynamics within the type of models we have proposed at or near their respective glass transition temperature.

Another issue concerns the classification of spin-glasses into roughly two families, one exhibiting a continuous transition and infinitely many levels of RSB, the other a discontinuous transition, and only one level of RSB, the model class we have presented here belonging to the first. There are indications that at least from the point of view of dynamics near the glass transition, the second class should be more appropriate, since it exhibits dynamic freezing transitions at temperatures higher than the thermodynamic transition temperature seen in equilibrium treatments [24]. We are currently studying a modified version of the Bernasconi model [25] in this context [26]. Finally, we have started to apply our approach to the case of defect crystals like KCl:Li to study the effects of the dipolar interaction among the Li defects at higher concentration [27], a system that has the distinct advantage that the on-site potential $G(v)$ is well known.

In the present paper we have quantised our system only *after* a mean-field decoupling. In principle, the analysis should start out from a full-fledged quantum-statistical formulation, using imaginary time path integrals in conjunction with the replica method. Such a formulation has been worked out [28] and will be analysed in detail in the near future. So much can be said already at this point. Our present approach amounts to investigating the quantum-statistical physics in the so-called static approximation [29]. In the present context, this approximation amounts to ignoring the feedback of quantum fluctuations on the effective single site potentials.

Acknowledgements

This work has been supported by the Deutsche Forschungsgemeinschaft through the Graduiertenkolleg "'Physikalische Systeme mit vielen Freiheitsgraden"' (U.H.), and a Heisenberg Fellowship (R.K.). It is a pleasure to thank C. Enss, H. Horner, S. Hunklinger, M. Mézard, P. Nalbach, O. Terzidis, and A. Würger for helpful discussions and encouragement. R.K. would like to thank the Physics Department of the University of Illinois at Urbana-Champaign for its hospitality while parts of this paper were being written.

Bibliography

[1] R.C. Zeller and R.O. Pohl, Phys. Rev. B**4** 2029 (1971).

[2] J.J. Freeman and A.C. Anderson, Phys. Rev. B **34**, 5684 (1986).

[3] Phillips, J. Low Temp. Phys. **7**, 351 (1972).

[4] P.W. Anderson, B. Halperin, and S. Varma, Phil. Mag. **25**, 1 (1972).

[5] S. Hunklinger and W. Arnold, in *Physical Acoustics*, edited by W.P. Mason and R.N. Thurston (Academic, New York, 1976) Vol. XII, p. 155; S. Hunklinger and A.K. Raychauduri, in *Progress in Low Temperature Physics*, edited by D.F. Brewer (Elsevier, Amsterdam, 1986) Vol. IX, p. 26.

[6] W. A. Phillips, Rep. Progr. Phys. **50**, 1675 (1987).

[7] V.G. Karpov, M.I. Klinger, and F.N. Ignat'ev. Zh. Eksp. Teor. Fiz. **84**, 760 (1983) [Sovj. Phys. JETP **57**, 439 (1983).

[8] U. Buchenau, Yu.M. Galperin, V.L. Gurevich, D.A. Parshin, M.A. Ramos, and H.R. Schober, Phys. Rev. B**46**, 2798 (1992).

[9] H.R. Schober and B.B. Laird, Phys. Rev. B **44**, 6746 (1991).

[10] A. Heuer and R.J. Silbey, Phys. Rev. Lett. **70**, 3911 (1993).

[11] C.C. Yu and A.J. Leggett, Comments Cond. Mat. Phys. **14**, 231 (1988).

[12] E.R. Grannan, M. Randeira, and J.P. Sethna, Phys. Rev. Lett. **60**, 1402 (1988).

[13] R.Kühn, in: *Complex Behaviour of Glassy Systems*, Proceedings of the XIVth Sitges Conference, edited by M. Rubi, Springer Lecture Notes in Physics (Springer, Berlin, Heidelberg, 1996).

[14] R. Kühn and U. Horstmann, Phys. Rev. Lett. **78**, 4067 (1097).

[15] M. Mézard, G. Parisi, and M. A. Virasoro, *Spin Glass Theory and Beyond*, (World Scientific, Singapore, 1987).

[16] T. Kirkpatrick and D. Thirumalai, J. Phys. A**22**, L149 (1989); J.P. Bouchaud and M. Mézard, J. Phys. I (France) **4**, 1109 (1994): E. Marinari, G. Parisi, and F. Ritort, J. Phys. A**27**, 7615 (1994); – ibid. 7647 (1994); J.P. Bouchaud, L. Cugiandolo, J. Kurchan, and M. Mézard, preprint cond-mat/9511042.

[17] D. Sherrington and S. Kirkpatrick, Phys. Rev. Lett **35**, 1792 (1975).

[18] J. Hopfield, Proc. Natl. Acad. Sci. USA, **81**, 3088 (1984).

[19] R. Kühn, S. Bös, and J.L. van Hemmen, Phys. Rev. A **43**, 2084 (1991); R. Kühn and S. Bös, J. Phys. A **26** 831 (1993); S. Bös, PhD Thesis, Gießen (1993), unpublished.

[20] J.R.L. de Almeida and D.J. Thouless, J. Phys. A **11**, 983 (1978).

[21] U. Horstmann and R. Kühn, in preparation.

[22] X. Liu, B.E. White Jr., R.O. Pohl, E. Iwanizcko, K.M. Jones, A.H. Mahan, B.N. Nelson, R.S. Crandal, and S. Veprek, Phys. Rev. Lett. **78**, 4418 (1997).

[23] A.L. Burin and Yu. Kagan, JETP **82**, 159 (1996), and references therein.

[24] T.R. Kirkpatrick and D. Thirumalai, Phys. Rev. B **36** 5388 (1987); A. Chrisanti, H. Horner, and H.J. Sommers, Z. Phys B **92** 257 (1993).

[25] J. Bernasconi, J. Physique, **48**, 559 (1987).

[26] J. Dimitrova and R. Kühn (1998), work in progress.

[27] R. Kühn and A. Würger, in preparation.

[28] R. Kühn, unpublished.

[29] A.J. Bray and M. Moore, J. Phys. C **13**, L655 (1980).

Disorder Phenomena in Chaotic Systems

Günter Radons

Fraunhofer-Institut für Produktionstechnik und Automatisierung,
Nobelstraße 12, D-70569 Stuttgart, Germany

Abstract: The influence of static disorder on chaotic systems is investigated for two different classes. In the first case disorder is present in spatial inhomogeneities of low-dimensional extended systems. We show that chaotic diffusion is suppressed by the disorder and that such systems are simple systems exhibiting the phenomenon of aging. In the second class the randomness lies in the coupling parameters of many degrees of freedom. We demonstrate that for a system of randomly coupled limit cycle oscillators the occurrence of a first-order phase transition into a spin-glass like phase. We discuss, how these phenomena are reflected in Lyapunov spectra and other characteristic quantities of dynamical systems.

1 Introduction

Chaos in dynamical systems is nowadays a well established research field [1,2] with applications in all branches of physics, e.g. atomic physics, hydrodynamics, quantum optics, or solid state physics. The key concepts for the characterization of such systems are Lyapunov exponents, various dynamical entropies, and fractal dimensions of strange attractors in dissipative systems. Many investigations considered systems with few effective degrees of freedom living in a compact phase space. Cases where phase space is infinite were mostly treated in the presence of a discrete translational symmetry, which by reduction to a unit cell again lead to compact phase spaces, typically a high-dimensional torus. Clearly these are limiting cases, and one wonders what happens for systems with broken translational invariance, or for systems with many non-equivalent degrees of freedom. In the following we treat examples, where order in this sense is not present, but the other extreme is prevailing, namely full disorder. Disordered systems are usually considered as a branch of statistical physics or solid state physics [3] with concepts and methods very different from the ones in dynamical systems theory. The two model classes considered below, suggest that spin-glass order, aging, or disorder induced anomalous transport are phenomena that arise very frequently also in disordered dynamical systems. We will also see how the above mentioned characteristics of dynamical systems reflect these disorder phenomena.

2 Few Degrees of Freedom in Disordered Environments

The simplest disordered dynamical systems are those with one or two degrees of freedom with disorder in the environment. An example is the Hamiltonian motion of a point particle in a two-dimensional disordered potential consisting of hard discs (Lorentz gas [10]) or smoothed version thereof. The periodic counterparts of these models have recently received much experimental and theoretical attention in connection with mesoscopic transport in quantum dot lattices [11,12]. There the importance of classical phase space structures and chaotic diffusion was realized [12]. Here we will concentrate first on dissipative systems which may also show chaotic transport. A simple much studied example is the damped motion of a periodically driven particle in a periodic potential. This provides e.g. a model for superionic conductors in an external field and currently finds renewed interest in the context of ratchet physics. Simplified versions which are assumed to capture essential aspects of these systems are one-dimensional iterated maps [5–7]. While the periodicity in the equations of motion allows for the application of advanced methods such as periodic orbit theory, the thermodynamic formalism, or Levy flight statistics [9], it is clearly of much greater importance to understand the effects of static disorder in such systems. From the physics of disordered systems, it is known that static or quenched randomness may drastically alter macroscopic quantities such as transport coefficients. In the following we will first report on such an effect for dynamical systems, namely the total *suppression of normal or anomalous chaotic diffusion by quenched randomness* in the equations of motion. We will show that this can occur in both, dissipative and Hamiltonian systems. This turns out to be a non-trivial effect, since the mean-square displacement will remain finite, although chaotic transport is *not* inhibited locally. Secondly we will point out that in the presence of a global bias these systems show various *phase transitions* characterized by *anomalous chaotic transport* properties. Finally we argue that these systems can also show the phenomenon of *aging*.

2.1 Suppression of Chaotic Diffusion

Dissipative Systems

We will first concentrate on one-dimensional non-invertible maps of the type studied in [5–9]. They have the general form $x_{t+1} = f(x_t) = x_t + F(x_t)$, with $F(x)$ periodic in x. The periodicity interval, which we set equal to one, i.e. $F(x) = F(x+1)$, defines cells or half open intervals $A_i = [i, i+1), i \in \mathbb{Z}$, on the real axis. We will modify these dynamical systems by randomly changing $F(x)$ in each cell A_i to a function $F^{(i)}(x)$ resulting in

$$x_{t+1} = x_t + F^{(i)}(x_t) \tag{2.1}$$

for $x_t \in A_i$. This corresponds to a random variation in space of the driving force felt by the particle. A natural choice for $F^{(i)}(x)$ consists of random shifts of F

$$F^{(i)}(x) = F(x) + \varepsilon(i). \tag{2.2}$$

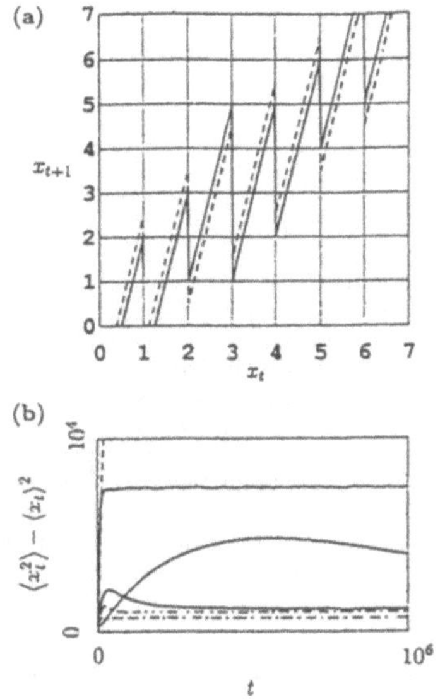

Figure 1
(a) Simple piecewise linear maps corresponding to a periodic (dashed) and a random driving 'force' (bold). The mean-square displacement (b) increases linearly for the former (dashed line) and saturates in the latter case as shown for several disorder realizations (full lines: $\sigma^2(t)$ for $\varepsilon(i) = \pm 1/2$, dot-dashed graphs: $\varepsilon(i)$ equally distributed in $(-1/2, +1/2)$). There exist environments where a first constant level is observed only after more than $t = 10^6$ iterations $(0.5 \ldots 1.0 \times 10^7$ for the second graph from the top).

In order to avoid complications connected with a global bias we assume for the moment the symmetry $F(-x) = -F(x)$, and further that the $\varepsilon(i)$ are independent, identically distributed random variables with a symmetric distribution function $p(\varepsilon) = p(-\varepsilon)$ implying $\overline{\varepsilon(i)\varepsilon(j)} \propto \delta_{ij}$ and $\overline{\varepsilon(i)} = 0$. Through the cell index i, defined as $i = [x]$, the largest integer smaller than x, the term $\varepsilon(i)$ is recognized as piecewise constant random function of x. In contrast to previous studies [5], where time-dependent noise was added to the deterministic dynamics, the random term $\varepsilon(i)$ remains constant in time, (2.1) is still deterministic, it describes a dynamical system with *quenched* randomness.

Let us now investigate the effect of this static randomness first for the simplest maps which in the absence of disorder ($\varepsilon(i) = 0$) exhibit chaotic diffusion. These are systems where $F(x)$ varies linearly in each cell, i.e. $F(x) = a\{x\} - a/2$ with $\{x\} = x - [x]$. Since the slope of $f(x)$ is $a+1$ these maps are chaotic for $a > 0$ and show chaotic diffusion for $a > 1$. The dashed graph in Fig. 1a is an example with $a = 3$. The diffusive motion for this ordered case is verified by the linear increase of the mean-square displacement $\sigma^2(t) = \langle x_t^2 \rangle - \langle x_t \rangle^2 = 2Dt$ with the correct diffusion constant $D = 1/4$ [6] (dashed line in Fig. 1b). This and the following results for $\sigma^2(t)$ were obtained numerically by iterating ensembles of 2×10^4 points (initially distributed homogeneously or inhomogeneously in one cell) for 10^6 (occasionally 10^7) time steps. An example of a map with binary disorder, $\varepsilon(i) = \pm 1/2$ in (2.2), is shown as full line in Fig. 1a. Now, with disorder $\varepsilon(i) \neq 0$, a very different behavior is observed: $\sigma^2(t)$ saturates and remains bounded for large times. As is seen from Fig. 1b this is true for discrete random variations as well as for continuously distributed random variables $\varepsilon(i)$. We emphasize that for both cases there exists no obvious reason why the spreading of the distribution $\rho_t(x)$ should be limited because the *a priori* probability for reaching one of the neighboring cells is always finite. More explicitly, *independent* of the chosen sequence $\varepsilon(i)$, a fraction $p = 1/4$ of a homogeneous distribution in some cell A_i is always transferred to the right neighboring cell A_{i+1}, the same fraction to the left cell A_{i-1}, and one quarter remains within the cell A_i. From this point of view there is no difference between the homogeneous situation ($\varepsilon(i) = 0$) and

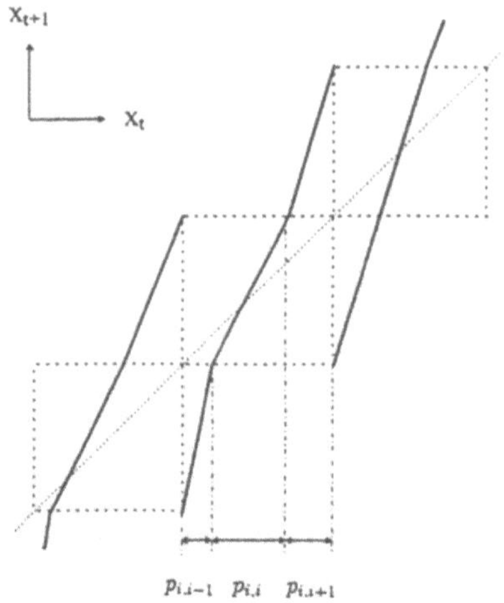

Figure 2
An example from the class of iterated maps for which the asymptotically finite mean-square displacement follows rigorously due the work of Sinai and Golosov [14] [15]. Also shown by dashed lines are the unit squares of the integer grid (see Fig. 1a) along the bisectrix. The indicated intervals p_{ii} and $p_{i,i\pm1}$ mediate the transitions from the i-th cell A_i to itself and its neighbors respectively.

the inhomogeneous case ($\varepsilon(i) \neq 0$)! The randomness affects only the last quarter, which is mapped into one or both of the next-nearest cells $A_{i\pm2}$. Note also that the degree of chaoticity as measured by the Lyapunov exponent (Fig. 1: $\lambda = \ln 4$) is not altered by the random shifts.

The explanation of this localization effect in the case of quenched randomness follows from the following connection. For the map $f(x)$ with discrete random shifts as in Fig. 1 the cells A_i define a (generating) Markov partition [13]. This implies that the evolution of piecewise constant distributions $\rho_t(x)$ (constant in the cells A_i) is fully equivalent to a Markov process, i.e. the content $\pi_i(t) = \int_i^{i+1} \rho_t(x)\,dx$ of cell A_i at time t is iterated according to

$$\pi_j(t+1) = \sum_i \pi_i(t) p_{ij}. \tag{2.3}$$

For the above piecewise linear map with $\varepsilon(i) = \pm 1/2$ (Fig. 1) the only non-zero transition probabilities p_{ij} are given by $p_{ii} = p_{i,i\pm1} = 1/4$ and $p_{i,i\pm2} = (1/2 \pm \varepsilon(i))/4$. The results of Fig. 1 were also checked by iterating (2.3) with these transition probabilities. Such a model defines a discrete random walk in a locally asymmetric random environment. The above localization effect, i.e. $\sigma^2(t)$ remaining finite for $t \to \infty$, is known as *Golosov phenomenon* in the random walk literature [4]. Inspired by Sinai's work [14] it was proven rigorously for systems with only nearest neighbor transitions by Golosov [15]. Reversing the above arguments which led us from iterated maps to random walks, it is obvious that also for the latter systems there exist realizations in terms of dynamical systems. These consist of piecewise linear chaotic maps of the form (2.1), with a typical example shown in Fig. 2. Again the cells A_i provide a Markov partition for this system. The segments of length p_{ii} and $p_{i,i\pm1}$ in each unit cell, where the map $f(x)$ is linear, correspond to the non-zero transition probabilities p_{ii} and $p_{i,i\pm1}$ of the associated Markov chain.

The asymptotically finite mean-square displacement was proven in [15] for independent random sequences p_{ii} and $p_{i,i-1}/p_{i,i+1}$ with $\overline{\ln(p_{i,i-1}/p_{i,i+1})} = 0$. The latter condition means that there is no global bias in the system and that one observes recurrent behavior with probability one.

So far we have seen that the dynamical systems defined in Fig. 1 and Fig. 2 both can be mapped to random walk models with (locally) asymmetric random transitions probabilities to next-nearest respectively nearest neighbors. Because of the short ranged correlations in the quenched disorder they belong to the same universality class [4].

An intuitive picture for the relevant physical processes is obtained from the continuum limit of these discrete random walk models, which is Brownian motion in a spatially random force field $\tilde{F}(x)$ [4]. In this limit the dynamics is governed by the Langevin equation

$$\dot{x}(t) = -\frac{\partial \tilde{V}}{\partial x}(x(t)) + \xi(t) \tag{2.4}$$

with Gaussian white noise $\xi(t)$. The important point is that the associated potential $\tilde{V}(x) = -\int^x \tilde{F}(x')dx'$ itself can be thought of as a spatial realization of a Brownian path. The resulting statistical self-similarity of the potential $\tilde{V}(Lx) \simeq L^{1/2}\tilde{V}(x)$ implies the occurrence of deeper and deeper potential wells as the particle proceeds. The work of Sinai and Golosov shows that an ensemble of initially close particles moves in a coherent fashion from one deep minimum to the next deeper potential well. In this stepwise process it is typically one minimum which dominates and therefore determines the (finite) width $\sigma^2(t)$ of the ensemble. Since the random environment in the neighborhood of these minima is the same only in a statistical sense one observes for a fixed environment still fluctuations in $\sigma^2(t)$. These fluctuations become extremely rare for large times t as follows from an Arrhenius argument [4] which says that the typical time to overcome the ever increasing relevant potential barriers increases exponentially with the barrier height, i.e. it takes a time of the order $\exp(\sqrt{x})$ for the state to travel a distance x. Solving this relation for x says that the typical distance reached in time t increases only as $\ln^2 t$. Indeed it has been shown rigorously that the mean displacement grows anomalously as $\langle x(t)\rangle = \xi(t)\ln^2 t$ with $\xi(t)$ a random function of $O(1)$ [4,14,15].

Applying this picture of a thermally activated process in a random Brownian landscape to dynamical systems presupposes the existence of a Markov partition. The results of Fig. 1 for continuous distributions of shifts $\varepsilon(i)$, however, show that the observed localization phenomenon is not bound to the existence of a Markov partition. One may also expect that the feature of piecewise linearity is not necessary for the occurrence of this effect. Indeed we have shown in [16] that even anomalously enhanced diffusion generated by certain nonlinear maps [8] is totally suppressed by the introduction of disorder of the above type.

Area Preserving Systems

One may wonder whether this sort of dynamical localization can be observed also in Hamiltonian systems or area preserving maps. This question is answered affirmatively by an explicit construction of area preserving maps, which show the same behavior. Our goal is achieved by extending the concept of homogeneous multibaker maps [17] to inhomogeneous chains by an appropriate use of the fact that every finite Markov chain can be realized by area preserving maps [18]. The construction is shown in Fig. 3 below.

It consists of an array of L rectangles called cells $A_i, i = 1,\dots,L$, of unit width in the x-direction and heights π_i. During one iteration step the inscribed rectangles within one cell A_i are mapped to the neighboring cells and into the present cell as follows: Take e.g. the black rectangle of height π_i and width $p_{i,i-1}$, which is labeled by its area $\pi_i p_{i,i-1}$ in Fig. 3. This rectangle is in one iteration (from top to bottom in Fig. 3) squeezed in the y-direction,

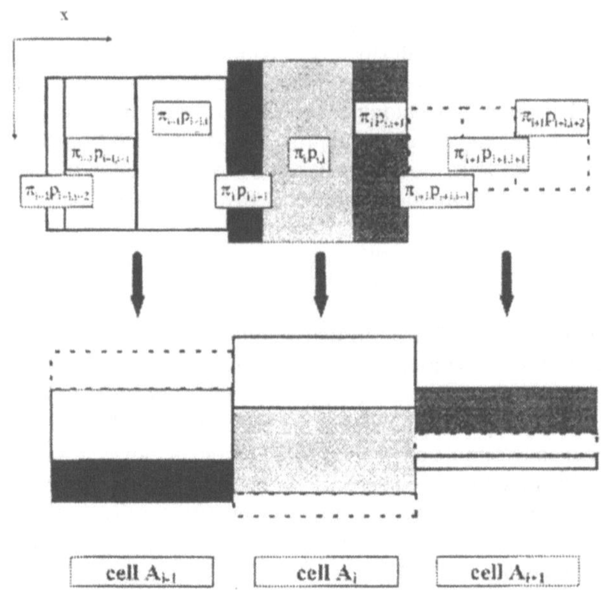

Figure 3
The construction of an area preserving inhomogeneous chain of baker maps, which in the projection on the x-axis reveals exactly the dynamics of the one-dimensional iterated map of Fig.2.

stretched in the x-direction, and transferred to the left neighboring cell A_{i-1}in an area preserving manner. In the same way the grey-shaded rectangle $\pi_i p_{i,i+1}$ is transferred to the right, and the light-grey rectangle gets squeezed and stretched, but remains in the cell A_i as shown in the figure. To make this scheme consistent we impose periodic boundary conditions $A_{i+L} = A_i$ ($L = 3$ in Fig. 3). Alternatively we could confine transport within the array by appropriate modifications of the boundary maps ($p_{1,0} = p_{L,L+1} = 0$). One observes that the x-coordinate x_t of a point (x_t, y_t) in this two-dimensional area preserving map is iterated exactly as a point x_t of the one-dimensional map of Fig. 2 (with slopes $p_{i,j}^{-1}$). This implies that the dynamical localization phenomenon of the previous section is found also in these area preserving maps!

Note, however, that in order to get globally an area preserving map, the heights π_i cannot be chosen arbitrary. This follows from the condition that in one iteration the outflow of area of a given cell, say A_i, has to be equal to its inflow,

$$\pi_{i-1} p_{i-1,i} + \pi_{i+1} p_{i+1,i} = \pi_i p_{i,i-1} + \pi_i p_{i,i+1}. \tag{2.5}$$

Adding to both sides of this equation the term $\pi_i p_{i,i}$ and using the normalization $\sum_j p_{i,j} = 1$, one finds that the π_i have to be proportional to the stationary probability distribution of a Markov chain with transition probabilities $p_{i,j}$. This is also clear from the fact that the projection of the two-dimensional stationary distribution of the inhomogeneous chain of baker maps as constructed above (which is constant in its area of definition), should yield the stationary distribution of the one-dimensional map of Fig. 2. For reflecting boundary conditions the stationary distribution can be found exactly as

$$\pi_i = \pi_1 \prod_{k=1}^{i-1} \frac{p_{k,k+1}}{p_{k+1,k}}. \tag{2.6}$$

For quenched random transition probabilities, as in the models treated here, this means that the π_i follow a random multiplicative process, which occur naturally in many branches of physics. For periodic boundary conditions a similar result is obtained [19]. A characterization of such stationary distributions in terms of its cumulants was given only recently in [20]. Of course it would be of interest if for some class of area preserving maps or corresponding continuous time Hamiltonian systems, such a distribution of chaotic areas occurs naturally. In such cases one expects that one finds the same anomalous transport properties as in the simple maps introduced here. This will be observable until the boundaries of the system are reached, which for extended systems occurs only at exponentially large times, i.e. at times of the order $O(\exp\sqrt{L})$.

2.2 Anomalous Drift and Phase Transitions

Releasing the conditions $\overline{\varepsilon(i)} = 0$ for the system of Fig. 1 or $\overline{\ln(p_{i,i-1}/p_{i,i+1})} = 0$ for that of Fig. 2 leads us from systems without global bias to such with a bias. The latter may simulate an external static field or may be attributed to systematic asymmetries of the underlying potential in more realistic models, such as driven damped particles in some potential landscape. Such aspects are clearly of great importance in the context of transport in ratchets [21,22]. Although the connection between the simplified one-dimensional maps and more realistic models is complicated and not fully understood, the investigation of the former with bias will provide some insight into the possible scenarios in the latter.

Exploiting again the connection between maps and discrete time Markov chains, allows us to apply known results from the literature to disordered iterated maps. Analytical results by Derrida and Pomeau [19] for the biased case $\overline{\ln(p_{i,i-1}/p_{i,i+1})} \neq 0$ lead to transport properties for the corresponding dynamical systems, which we summarize in Fig. 4 for a special example of the type of Fig. 2.

The example consists of setting $p_{ii} = 0$ and choosing a binary distribution for the transition probabilities $p_{i,i\pm1}$ of Fig. 2. More explicitly, we construct a chain of maps consisting of two sorts of cells (see Fig. 2), which we denote as A_+ and A_-. For the type A_+ we choose $p_{i,i+1} = a$ and correspondingly $p_{i,i-1} = 1-a$, while for type A_- we reverse the assignments, i.e. we set $p_{i,i+1} = 1-a$ and $p_{i,i-1} = a$. These cells are concatenated randomly and independently so that type A_+ is present in a concentration c and correspondingly a fraction $1-c$ of cells are of type A_-. For Fig. 4 we have chosen $a = 1/3$ so that for $c = 0$ we have an ordered system consisting only of cells A_-, which map points with larger probability, namely with probability 2/3, to the right, and with probability 1/3 to the left. In this limit one gets chaotic diffusion with a mean drift to the right, where diffusion constant D and drift velocity V are given by $D = D_0 = 2a(1-a)$ and $V = V_0 = 1-2a$, respectively. The behavior of the transport coefficients, normalized by its bare values D_0

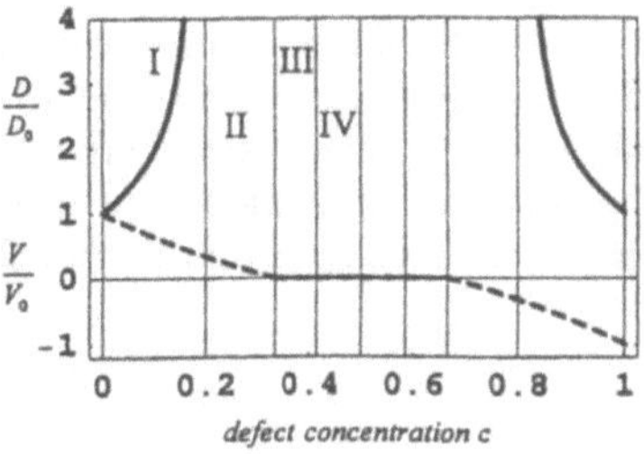

Figure 4
The dependence of the normalized transport coefficients D/D_0 (bold) and V/V_0 (dashed) is shown as fuction of the defect concentration c. In the regimes where these are finite they are self-averaging quantities and therefore these values are observed with probability one independent of the disorder realization.

and V_0, is shown in Fig. 4 as function of the concentration c. We need to discuss only the case $0 \leq c \leq 1/2$, since the rest follows by symmetry. For the other extreme of full disorder $c = 1/2$ we get the unbiased situation $\overline{\ln(p_{i,i-1}/p_{i,i+1})} = 0$, where the anomalous Sinai and Golosov results of section 2.1 hold. This means we get in this case $D = 0$ and $V = 0$. Between these extremes various transitions between dynamically different phases occur. In Fig. 4 these different regimes are numbered as I-IV. In phase I both D and V are finite and non-zero, i.e. one has normal chaotic transport as in the ordered limit. The transition to phase II occurs at concentrations where conditions $\overline{(p_{i,i-1}/p_{i,i+1})^{\pm 2}} = 1$ are fulfilled (two symmetric solutions in c) . This transition is accompanied by D becoming infinite, which holds in regimes II an III. In these phases one has therefore anomalously enhanced diffusion, i.e. the mean-square displacement grows superlinear. The transition between II an III is signaled by a vanishing drift velocity, i.e. the mean displacement grows slower than linearly in time. The transition points are given by $\overline{(p_{i,i-1}/p_{i,i+1})^{\pm 1}} = 1$. The anomalously slow growth of the displacement holds up to the value $c = 1/2$, i.e. also in regime IV. The transition from III to the latter is characterized by a crossover from superdiffusive to subdiffusive chaotic transport implying that in addition to V also D vanishes. The last transition is observed for $\overline{(p_{i,i-1}/p_{i,i+1})^{\pm 1/2}} = 1$. The fact that the qualitative changes in the drift and diffusion properties occur at different values of the concentration is quite common in disordered systems and is expected to hold also more generally. A more extensive description and a discussion of the different regimes in terms of activated Brownian motion in tilted Brownian potentials can be found in the review article [4]. Interestingly the same conclusions were reached very recently in the context of disordered ratchets by a totally different reasoning within a continuous time model [22].

2.3 Aging

The phenomenon of aging is a well-known experimental fact from glasses, spin glasses, and other complex materials [23,24]. Theoretical investigations are also mainly concerned with these systems [25], but recently aging was found also in much simpler model systems [26,27]. Aging can be defined as an anomalous behavior of response and correlation functions. Consider e.g. the correlation function $C_{AB}(t,t_w) \equiv \langle A(t_w)B(t+t_w)\rangle$ of two variables A and B, where t_w is the waiting time after the preparation of the initial state at time $t = 0$. For $t \ll t_w$ the correlation function $C_{AB}(t,t_w)$ is independent of t_w and a fluctuation-dissipation theorem is supposed to hold. Aging is present if $C_{AB}(t,t_w)$ depends strongly on t_w when t is of the same order as t_w. Furthermore for t and t_w large on expects a scaling behavior of the form

$$C_{AB}(t,t_w) = t^{-\nu} f(\frac{t}{t_w}). \tag{2.7}$$

Interestingly such a scaling behavior with $\nu = 1$ has been found numerically for various correlation functions by Marinari and Parisi [26] in the case of a random walk in a random environment of Sinai type. As pointed out above such systems are realized by dynamical systems as introduced in Fig. 2. This implies that aging occurs also in these simple disordered dynamical systems and its generalizations treated in Section 2.1. This fact has not been realized previously.

3 Many Degrees of Freedom with Random Couplings

In this Section we consider the other extreme of disordered dynamical systems, namely that of many degrees of freedom, each confined to a compact set (a circle) and coupled in a random manner. The results below are obtained for a model defined by N randomly interacting phase oscillators. Interest in such models comes from biology [28–30], e.g. from brain research [31,32], from laser physics [33,34], and other branches of physics [35]. They are a good approximation for many dynamical systems where limit cycle oscillators are the basic entities [30,35]. The model is defined by the equations of motion

$$\dot{\phi}_j(t) = \omega_j + \sum_{i=1}^{N} J_{ij} \sin(\phi_i(t) - \phi_j(t)), \tag{3.8}$$

where ϕ_i are angles, $\phi_i \in [0,2\pi)$, the ω_i are the unperturbed oscillator frequencies, and the J_{ij} are the couplings between the different limit cycle oscillators. This model has mostly been investigated for the special case of random frequencies ω_i and uniform all-to-all interactions $J_{ij} = K/N$ [35,36]. In this case, which is usually referred to as the Kuramoto model, it is well known that one observes a transition from completely incoherent motion of the oscillators for small interaction strength $K < K_c$, to partially coherent motion (phase locking) above a critical interaction strength K_c. This behavior is captured in the thermodynamic limit $N \to \infty$ by the order parameter

$$m = \lim_{t\to\infty} \langle \exp(i\phi(t)) \rangle, \tag{3.9}$$

where the brackets $\langle \ldots \rangle := \lim_{N\to\infty} 1/N \sum_{j=1}^{N} \cdots$ define an ensemble average over all oscillators. The transition is then described by a vanishing order parameter for $K < K_c$, and an increase of the "magnetization" $|m|$ as $|m| \sim \sqrt{(K-K_c)/K_c}$ above K_c. It is known that this system is not chaotic as follows from the maximal Lyapunov exponent, which is always zero, but we note that the phase locking is reflected in the full Lyapunov spectrum, which can also be calculated exactly [37].

Here we present results for the more interesting and general case, where in addition to the frequencies also the couplings are quenched random variables. They are assumed to be gaussian distributed variables with mean zero $\overline{J_{ij}} = \overline{\omega_i} = 0$ and covariances

$$\overline{\omega_i \omega_j} = \tau^{-2} \delta_{ij} \tag{3.10}$$

$$\overline{J_{ij} J_{kl}} = \frac{J^2}{N} (\delta_{ik}\delta_{jl} + \eta \delta_{il}\delta_{jk}). \tag{3.11}$$

The assumption $\overline{\omega_i} = 0$ is actually no restriction, since one can always transform to a rotating frame where this condition is fulfilled. The condition $\overline{J_{ij}} = 0$ means that the couplings are on the average neither ferromagnetic nor antiferromagnetic, but since the J_{ij} are random with positive and negative values one gets competing interactions and possibly frustration for a given oscillator. (3.10) means that the oscillator frequencies are uncorrelated and their variance can be used for fixing the time scale. In other words, setting $\tau = 1$, means that we measure the J_{ij} and J in units of τ^{-1} and time in units of τ. This leaves us with two independent parameters characterizing the system (in the limit

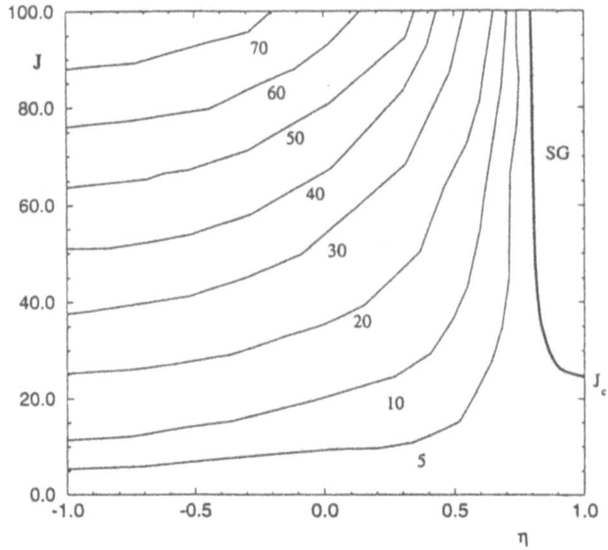

Figure 5 Value of the largest Lyapunov exponent λ_{max} in dependence on symmetry parameter η and interaction strength J represented by contour lines at levels $\lambda_{max} = 5,10,20,30,40,50,60,70$ ($N = 100$). The part of the parameter space (η,J) above $J_c(\eta)$ (bold line) where the oscillators freeze in random positions, is denoted by "SG" In this regime $\lambda_{max} = 0$. For $\eta < 0.8$ there is no spin glass order. The largest Lyapunov exponent decreases monotonically with η and for $\eta < 0.8$ grows monotonically with J.

$N \to \infty$): J with $0 < J < \infty$ is a measure for the interaction strength and $-1 \leq \eta \leq 1$ is the so-called symmetry parameter. From (3.11) follows that for $\eta = 0$ all coupling parameters are uncorrelated, while $\eta = 1$ imposes the constraint $J_{ij} = J_{ji}$ (symmetric interactions) and $\eta = -1$ implies $J_{ij} = -J_{ji}$ (antisymmetric interactions). The case of symmetric interactions has been treated previously [38].Our main results for this system are summarized in the phase diagram of Fig. 5.

As function of the parameters η and J one finds two different phases which can be regarded as a paramagnetic phase and a spin glass phase (SG in Fig. 5). The latter is characterized by a non-vanishing Edwards-Anderson (EA) parameter [3], which in our case may be defined as

$$q = \lim_{t \to \infty} \mathrm{Re} \left\langle \exp(i\phi_j(t_0)) \exp(-i\phi_j(t+t_0)) \right\rangle . \tag{3.12}$$

The spin glass phase, which exists above the full, bold line $J_c(\eta)$ in the region $0.8 \lesssim \eta \leq 1$ and for $J \geq J_c \equiv J_c(\eta = 1) \simeq 24.3$, is characterized by a vanishing "magnetization" $m = 0$ and an EA parameter $q = 1$. In addition we find in this phase that the maximal Lyapunov exponent is zero and that all other Lyapunov exponents are negative. This means that all oscillators are frozen into random directions, which is indeed a characteristic feature of a spin glass state at temperature $T = 0$. A similar freezing transition at a symmetry parameter $\eta_c \simeq 0.83$ has also been found in an asymmetric random neural network model [39], which may be related to our model at $J = \infty$. Note that in the symmetric case $\eta = 1$ the maximal Lyapunov exponent is zero for all values of J. Below $J_c \simeq 24.3$, however, not only one, but a finite fraction of zero Lyapunov exponents appears to exist which indicates that in this case a non-chaotic motion of the system on a lower dimensional manifold in phase space is still present.

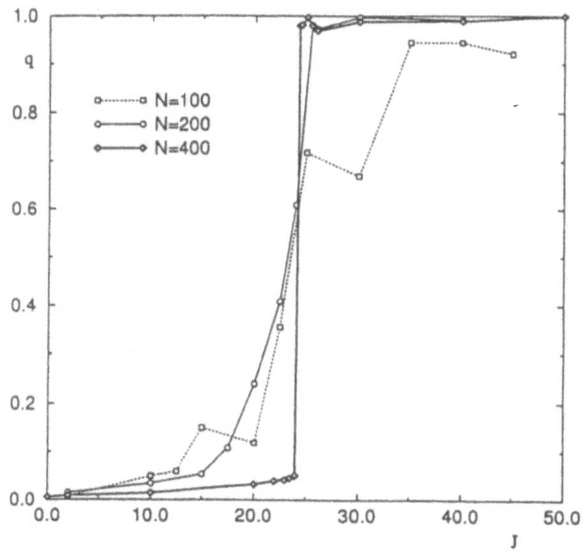

Figure 6
Order parameter q in dependence on interaction strength J for $N = 100,200,400$ oscillators. The transition from $q \approx 0$ for $J < J_c$ to $q \approx 1$ for $J > J_c$ becomes sharper with growing system size N. Between $J = 24$ and $J = 24.3$ the order parameter q changes from 0.046 to 0.982.

The situation is very different in the "paramagnetic" phase. There in addition to m also q vanishes, meaning that we are in a situation of dynamical disorder. This is also reflected in a positive maximal Lyapunov exponent implying chaotic motion, as can be seen from the level lines for the maximal Lyapunov exponent in Fig. 5. Thus we have here a certain correspondence between order parameters for disordered systems and characteristic quantities for dynamical systems.

A final point of interest, which shall be presented here concerns the nature of the phase transition at the transition line $J_c(\eta)$. Other interesting aspects of the Lyapunov spectra, behavior of correlation functions, the time-dependence of the magnetization, etc. can be found in [40]. The phase transition appears to be of first order as is demonstrated for $\eta = 1$ in Fig. 6. There the dependence of q on the coupling strength J is plotted for increasing system size. These simulations clearly suggests that the spin glass order parameter q jumps from zero to one in the thermodynamic limit $N \to \infty$ as J crosses $J_c \simeq 24.3$. The first order character of the transition is also confirmed for $0.8 \lesssim \eta \leq 1$, where interestingly we find for finite N strong intermittency phenomena in the neighborhood of the transition line [40]. A last speculative remark on the spin glass phase of this model is that one expects also aging phenomena to occur here.

4 Discussion

We have presented results for simple deterministic dynamical systems with quenched disorder. It turned out that disorder plays an extremely strong role for the macroscopic behavior of such systems. We have explicitly shown that normal and anomalous chaotic diffusion can be suppressed by disorder in dissipative and Hamiltonian systems. This may happen without any sign in Lyapunov exponents or related quantities. Interestingly, these systems show the phenomenon of aging. For biased systems one finds various phases with further anomalous transport properties. These results were obtained numerically and also rigorously. Furthermore, the underlying mechanisms appear to be universal and also robust against noise. For a high-dimensional randomly coupled dynamical system we found the occurrence of a first order phase transition into a spin-glass phase, which is also reflected in the nature of the Lyapunov spectrum. In summary the field of disordered dynamical systems provides us with many interesting problems with surprising effects to be expected also in the future.

Acknowledgement

I thank J.C. Stiller for his enthusiastic cooperation in part of this work and the Rechenzentrum der Universität Kiel for providing me with the opportunity to use the Cray Y-MP computers for the numerical computations.

Bibliography

[1] H. G. Schuster, *Deterministic Chaos*, 3rd augm. ed. (VCH, Weinheim, 1995).

[2] E.Ott, *Chaos in Dynamical Systems.* (Cambridge University Press, Cambridge, 1993).

[3] M. Mezard, G. Parisi, M.A.Virasoro, *Spin glass theory and beyond* (World Scientific, Singapore,1987).

[4] J.-P. Bouchaud and A. Georges, Phys. Rep. **195**, 127 (1990).

[5] T. Geisel and J. Nierwetberg, Phys. Rev. Lett. **48**, 7 (1982).

[6] S. Grossmann and H. Fujisaka, Phys. Rev. A **26**, 1779 (1982); H. Fujisaka and S. Grossmann, Z. Phys. B **48**, 261 (1982).

[7] M. Schell, S. Fraser, and R. Kapral, Phys. Rev. A **26**, 504 (1982).

[8] T. Geisel, J. Nierwetberg, and A. Zacherl, Phys. Rev. Lett. **54**, 616 (1985).

[9] R. Artuso, Phys. Lett. A **202**, 195 (1992); G. Zumofen and J. Klafter, Phys. Rev. E **47**, 851 (1993); X.-J. Wang and C.-K. Hu, Phys. Rev. E **48**, 728 (1993); R. Artuso, G. Casati, and R. Lombardi, Phys. Rev. Lett. **71**, 62 (1993); R. Stoop, Phys. Rev. E **49**, 4913 (1994); R. Klages and J. R. Dorfman, Phys. Rev. Lett. **74**, 387 (1995), G. Radons, Phys. Rep. **290**, 67 (1997).

[10] see e.g.: H. van Beijeren and J. R. Dorfman, Phys. Rev. Lett. **74**, 4412 (1995); and refs. therein.

[11] D. Weiss et al., Phys. Rev. Lett. **66**, 2790 (1991)

[12] R. Fleischman, T. Geisel, R. Ketzmerick, Phys. Rev. Lett. **68**, 1367 (1992).

[13] C. Beck and F. Schlögl, *Thermodynamics of Chaotic Systems* (Cambridge University Press, Cambridge, 1993).

[14] Ya. G. Sinai, Theor. Prob. Appl. **27**, 247 (1982).

[15] A. Golosov, Commun. Math. Phys. **92**, 491 (1984).

[16] G. Radons, Phys. Rev. Lett. **77**, 4748 (1996).

[17] P. Gaspard, J. Stat. Phys. **68**, 673 (1992).

[18] P. Gaspard and X.-J. Wang, Phys. Rep. **235**, 321 (1993).

[19] B. Derrida and Y. Pomeau, Phys. Rev. Lett. **48**, 627 (1982); B. Derrida, J. Stat. Phys. **31**, 433 (1983).

[20] G. Radons, J. Phys. A **31**, 4141, (1998).

[21] P. Jung, J.G. Kissner, P. Hänggi, Phys. Rev. Lett. **76**, 3436 (1996).

[22] Th. Harms, R. Lipowsky, Phys. Rev. Lett. **79**, 2895 (1997).

[23] L. Lundgren et al., Phys. Rev. Lett. **51**, 911 (1983).

[24] L.C.E. Struik, *Physical Aging in Amorphous Polymers and other Materials* (Elsevier, Amsterdam, 1978).

[25] L.F. Cugliandulo et al., Phys. Rev. Lett. **74**, 1012 (1995); L.F. Cugliandulo, J. Kurchan, F. Ritort, Phys. Rev. B **49**, 6331 (1994); H. Rieger, Physica A **224**, 267 (1996), and refs. therein.

[26] E. Marinari and G. Parisi, J. Phys. A **26**, L1149 (1993).

[27] F. Ritort, Phys. Rev. Lett. **75**, 1190 (1995).

[28] J. Buck, E. Buck, Scientific American **234**, 74 (1976).

[29] S.H. Strogatz, I. Stewart, Scientific American **269**, 102 (1993).

[30] A. T. Winfree, J. Theoret. Biol. **16**, 15 (1967).

[31] R. Eckhorn, et al., Biol. Cybern. **60**, 121 (1988).

[32] A.K. Engel, et al., Eur. J. Neurosci. **2**, 588 (1990).

[33] K. Otsuka, Phys. Rev. Lett. **67**, 1090 (1991).

[34] K. Wiesenfeld, et al., Phys. Rev. Lett. **65**, 1749 (1990).

[35] Y. Kuramoto, *Chemical oscillations, waves and turbulence.* (Springer, Berlin, 1984).

[36] H. Daido, J. Phys. A **20**, L629 (1987); Y. Kuramoto, I. Nishikawa, J. Stat. Phys. **49**, 569 (1987); S. H. Strogatz, E. Mirollo, J. Stat. Phys. **63**, 613 (1992); W. F. Wreszinski, J. L. van Hemmen, J. Stat. Phys. **72**,145 (1993).

[37] J. C. Stiller, G. Radons, Lyapunov Spectrum of the Kuramoto Model of Nonlinear Oscillators, preprint.

[38] H. Daido, Phys. Rev. Lett. **68**,1073 (1992).

[39] P. Spitzner, W. Kinzel, Z. Phys. B. **77**, 511 (1989).

[40] J. C. Stiller, G. Radons, Phys. Rev. E **58**, 1789 (1998); J. C. Stiller, Dynamik von Phasenoszillatoren, Diploma thesis, Kiel (1994).

Zero Bias Anomalies in the Tunneling Spectra of High Temperature Superconductors

Lambert Alff and Rudolf Gross

II. Physikalisches Institut, Lehrstuhl für Angewandte Physik, Universität zu Köln, Zülpicherstraße 77, D-50937 Köln, Germany

Abstract: A zero bias conductance peak (ZBCP) was often observed in the tunneling spectra of junctions with at least one electrode consisting of a high temperature superconductor (HTS). Based on general considerations, the sign change of the phase of an unconventional $d_{x^2-y^2}$-wave pairing symmetry in the HTS is expected to lead to current carrying Andreev bound states (ABS) at the Fermi level at surfaces and interfaces of HTS. In this article, the results of tunneling spectroscopy using grain boundary Josephson junctions (GBJs) formed by different hole and electron doped HTS are summarized. Including the tunneling spectra obtained by low temperature scanning tunneling spectroscopy (LTSTS) and planar superconductor - insulator - normal metal (SIN) junctions a consistent picture for the origin of the ZBCP based on the existence of ABS is developed. A further consequence of the ABS is the formation of anomalous Meissner currents for which experimental evidence is presented.

1 Introduction

A ZBCP has been observed in the tunneling spectra of HTS already shortly after the discovery of these materials. Back in the year 1992 first overviews on this phenomenon have been given by Walsh [1] and Lesueur et al. [2]. However, at that time the ZBCP was mostly interpreted by applying a theoretical model that has been developed by Appelbaum and Anderson (AA-model) for junctions formed by a transition metal, an oxide barrier, and a simple metal [3,4]. Only after it became clear that the symmetry of the order parameter of the hole doped cuprate HTS is of an unconventional $d_{x^2-y^2}$-wave, it was theoretically proposed by Hu that the physical origin of the ZBCP is related to the presence of a d-wave symmetry of the order parameter in HTS [5]. Since that time, overwhelming theoretical and experimental evidence for the $d_{x^2-y^2}$-wave symmetry of the pair potential in the HTS has been presented [6–8]. As will be shown below, with respect to possible experimental techniques for probing the pairing symmetry in HTS, tunneling spectroscopy represents a valuable method that is sensitive for both the amplitude and the phase of the order parameter. It is noteworthy that one of the milestone experiments for proving the existence of the $d_{x^2-y^2}$-wave pairing symmetry in HTS is considered the measurement of the half-integer flux quantization found by Kirtley and Tsuei et al. [8] in superconducting rings containing three HTS GBJs fabricated on tricrystal substrates. The π shift of the phase of the pair potential in orthogonal k-directions that is measured in these experiments also is the physical origin of the Andreev bound states (ABS). In our tunneling experiments presented here, HTS GBJs [9,10] are used. We note that these junctions not only form basic elements of superconducting devices but also can be used for a large variety of experiments clarifying the fascinating physics of the cuprate superconductors.

Recently, overviews on the formation of ABS at HTS-surfaces (Rainer et al. [11]), the theoretical concepts of tunneling spectroscopy in anisotropic superconductors (Kashiwaya et al. [12], Tanaka et al. [13], and Hu [14]), and the experimental evidence for ABS in tunneling measurements on HTS (Alff et al. [15]) have been given. In this article we present comprehensive experimental data obtained both from HTS GBJs and LTSTS experiments providing conclusive evidence on the nature of zero bias anomalies in the tunneling spectra of HTS. In Section 2.1 we first discuss the electrical transport properties of GBJs used in our experiments. Then, in Section 2.2 the spectroscopic results obtained from the experiments using GBJs are summarized. These experiments represent a new class of experiments to study the phase and the amplitude of the pair potential. In Section 2.3 we then give a brief comparison of the GBJ data and those obtained in experiments based on LTSTS experiments and on planar SIN junctions. In Section 2 we always will discuss the experimental results with respect to the theoretical model predictions in order to show that only the ABS-model can account in a natural way for the behavior observed in the vast majority of experiments. Finally, in Section 3 the main evidence for the ABS-model is summarized.

2 Experimental Results and Discussion

In this section the experimental results for symmetric and asymmetric [001] tilt HTS GBJs on $SrTiO_3$ bicrystal substrates with misorientation angles of 24°, 36.8° and 45° are presented. The tunneling direction in these bicrystal GBJs is along the *ab*-plane when *c*-axis oriented electrodes are used. The epitaxial thin films were fabricated by different techniques. The YBCO films were deposited by hollow cathode magnetron sputtering and pulsed laser deposition from stoichiometric targets. Two different oxygen doping concentrations were used namely $YBa_2Cu_3O_{7-\delta}$ with $\delta \approx 0.4$ and $\delta \approx 0$ resulting in T_c values of 55–60 K and 90 K, respectively. The T_c reduction in the underdoped material is due to the oxygen doping and *not* due to the presence of magnetic impurities [16]. $Bi_2Sr_2CaCu_2O_{8+x}$ (BSCCO) thin films have been fabricated by the same way slightly underdoped [17]. The T_c was about 80 K what is smaller by about 10 K than that of optimum doped samples [18]. $La_{1.85}Sr_{0.15}CuO_4$ (LSCO) and $Nd_{1.85}Ce_{0.15}CuO_{4-y}$ (NCCO) samples were prepared by reactive coevaporation from metal sources using ozone as reaction gas [19–21]. LSCO thin films have a reduced T_c as compared to single crystals of the same material. The most likely explanation is that strain is induced by the mismatch of the lattice parameters of the substrate and the film [21,22]. The NCCO thin films had T_c values about 24 K indicating optimum doping [23,24]. The tunneling spectroscopy was performed using a standard four probe lock-in technique [15]. Magnetic fields up to 12 T and low temperatures down to 100 mK were used.

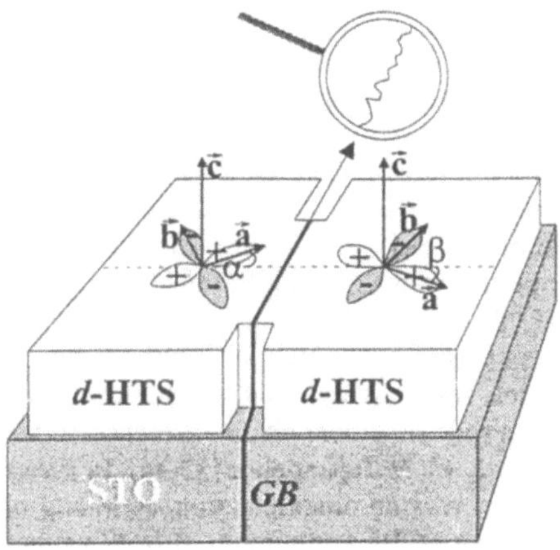

Figure 1
Sketch of the sample geometry of a symmetric [001] tilt HTS GBJ. The crystal axis $\vec{a}$, $\vec{b}$, and $\vec{c}$ are indicated. The left electrode is rotated about the $\vec{c}$-axis by α and the right electrode by β. In case of a symmetric junction $\alpha = -\beta$. For an asymmetric junction the two tilt angles are different. In the magnifying glass the facets of the grain boundary (GB) are shown. Here, the nominal misorientation angle represents the average over a wide range of different angles of the individual facets.

2.1 Grain Boundary Junctions

Transport Processes in Grain Boundary Junctions

The electrical transport and noise properties of HTS GBJs have been discussed in detail in recent reviews [9,10]. Here, we briefly summarize the main features important with respect to the interpretation of the tunneling spectra shown below. In Fig. 1 a sketch of the sample geometry is shown. The two superconducting electrodes are coupled by the grain boundary barrier to form a Josephson junction. Phenomenologically, the superconducting transport properties of GBJs can be described by the resistively and capacitively shunted junction (RCSJ) model [25]. At high temperatures GBJs represent overdamped Josephson junctions with nonhysteretic current-voltage characteristics (IVCs), whereas at low temperature some GBJs become underdamped resulting in hysteretic IVCs. A microscopic understanding of the transport processes present in GBJs is given by the intrinsically shunted junction (ISJ) model proposed by Gross et al. [26,27]. A key result of this model is a general scaling relation between the critical voltage V_c and the critical current density J_c as $V_c \propto J_c^n$ with $n \simeq 0.5$ that is found for a wide variety of junction types and different HTS materials. The ISJ-model describes the grain boundary as an insulating barrier containing a high density of localized states. Due to Coulomb repulsion the transport of Cooper pairs is dominated by direct tunneling, whereas the transport of quasiparticles is dominated by resonant tunneling via a single localized state [18,28–31]. The resonant tunneling process conserves the information about the orientation of the order parameter of the junction electrodes. Hence, the quasiparticles can carry information about the gap and the interface density of states. Note that any inelastic process as for example variable range hopping would destroy such kind of information. The localized states within the grain boundary barrier have an almost white energy distribution and, hence, do not produce specific features in the tunneling spectra.

As discussed in [15] the transmission and reflection processes depend sensitively on the angles α and β. A quasiparticle that experiences a sign change of the phase of the pair potential along its quasiclassical trajectory gives rise to the formation of ABS. This will be the case when the junction electrodes are formed by a $d_{x^2-y^2}$ HTS and α and β are suitable oriented. The presence of ABS at the Fermi level results in a ZBCP in the measured tunneling spectra.

Grain Boundary Faceting

In Fig. 1 a magnified view of the microstructure of the grain boundary is shown schematically. In the presence of faceting the nominal misorientation angle represents an average angle obtained by averaging over a large number of facets. Each facet links regions of the junction electrodes with differently oriented surfaces. The faceting of GBJs has been revealed by transmission electron microscopy (TEM) [32]. The faceting not only results in strongly inhomogeneous electrical properties of the GBJs [29] but also may cause the formation of a parallel array of so-called "0" and "π"-junctions. A "0" or "π"-junction is obtained depending on whether the sign of the pair potential locally is the same or opposite in the adjoining junction electrodes, respectively. This effect allows for locally negative and positive Josephson currents at a specific global phase difference between the junction electrodes. This leads to an irregular I_c vs. H dependence [33,34]. In general, the importance of "π"-junctions increases with increasing misorientation angle and increasing faceting. Using an asymmetric 0–45° YBCO-GBJ a minimum of I_c as a function of applied field at $H = 0$ is obtained which is a characteristic of a junction formed by d-wave superconductors [6]. At present, no exact rigorous quantitative description of the effect of faceting is available. Below some possible effects of the angle averaging due to faceting are discussed.

2.2 Tunneling Spectroscopy on Different HTS Using GBJs

In this Section, the tunneling spectra for both hole and electron doped materials are presented and discussed. In the following it is assumed that all hole doped HTS have a $d_{x^2-y^2}$-wave symmetry of the order parameter. In our analysis, the applicability of both the ABS- and the AA-model is discussed. While the ABS model is based on the unconventional d-wave pairing symmetry in the HTS [12–15], the AA-model presumes spin-dependent scattering independent on superconductivity [3,4].

Hole Doped Materials

In Fig. 2a we show the tunneling spectra for the d-wave HTS YBCO for the two doping concentrations corresponding to a T_c of 60 K and 90 K respectively, as well as for BSCCO and LSCO. The tunneling conductance is normalized by the background conductance G_n measured just above T_c, and the voltage scale is normalized by the gap voltage V_g in order to be able to compare the different materials. G_n was found to be temperature independent and its shape is mostly parabolic. We note that for s-wave BCS superconductors the gap voltage corresponds to twice the gap voltage, $2\Delta_0/e$. However, in the case of HTS it is more appropriate to take $V_g \simeq \Delta_0/e$ with the exact value depending on the detailed junction geometry [35]. With $V_g \simeq \Delta_0/e$, gap voltages Δ_0/e of 25 mV for BSCCO, 20 mV for optimum doped YBCO, 15 meV for underdoped YBCO, and 6 meV for LSCO and NCCO could be derived from our experimental data. Together with the measured T_c values, for the ratio $2\Delta_0/k_BT_c$ values ranging between 5 and 8 were obtained. As can be seen in Fig.2a all d-wave junctions qualitatively show the samer behavior. At the gap voltage no clear peak is observed in the conductance versus voltage curves. Furthermore, a reduced conductance corresponding to a density of states that is reduced by about 20 to 40% is measured at $V \ll V_g$. Close to the Fermi level ($V = 0$) a ZBCP is observed for all

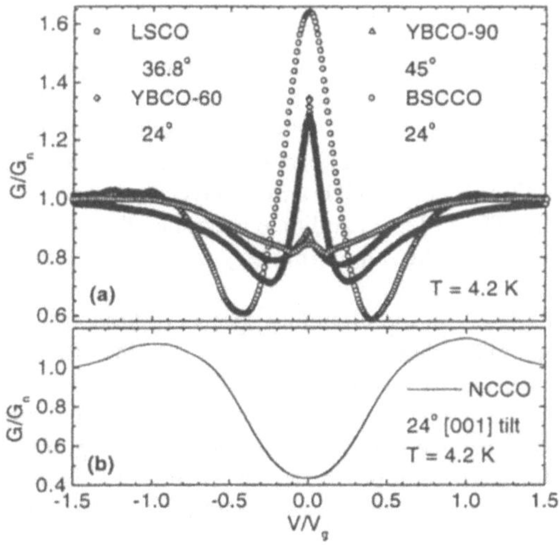

Figure 2
(a) Normalized conductance vs. normalized voltage of [001] tilt GBJs formed by YBCO (90 and 60 K phase), BSCCO, and LSCO at $T = 4.2$ K. (b) The same dependence for a NCCO-GBJ. Data taken from Ref. [37].

samples. The shape and width of this ZBCP differs considerably for the different materials and samples. This is most likely due to the sample and material dependent faceting and roughness of the grain boundary interface and the high sensitivity of the ZBCP to the microscopic structure of the interface.

Electron Doped NCCO

We now discuss the tunneling spectra obtained for the electron doped cuprate superconductor NCCO that most likely is a more or less anisotropic s-wave superconductor [38–41] without any change of sign of the order parameter in different k-directions. The NCCO-GBJs show qualitatively different tunneling spectra as compared to the d-wave superconductors. Tunneling spectra of NCCO-GBJs with misorientation angles of 10°, 24°, and 36° have been investigated. For all junctions no ZBCP was observed. A second characteristic feature of the tunneling spectra of NCCO-GBJs is a clear conductance peak at the gap voltage and a stronger reduction of the conductance (about 50% and more) at $V \ll V_g$. We emphasize that the NCCO tunneling data could not be well fitted by introducing a finite lifetime broadening according to Dynes [42] resulting in an excitation spectrum

$$D(E,\Gamma) = \Re\left\{\frac{E - i\Gamma}{\sqrt{(E - i\Gamma)^2 - \Delta_0^2}}\right\}.$$

Here, Γ is a phenomenological broadening parameter. This gives evidence that NCCO may *not* be a conventional BCS superconductor but more likely has an anisotropic s-wave symmetry of the order parameter. The latter idea is based on the fact that all NCCO spectra look similar to the spectra obtained for the materials having a strongly anisotropic d-wave symmetry of the order parameter. A pair potential without any sign change of the phase but with a strongly anisotropic amplitude is represented by the anisotropic s-wave.

The complete absence of any ZBCP for NCCO-GBJs gives strong evidence for the presence of an anisotropic s-wave order parameter. However, an extended s-wave cannot be completely ruled out since the small contributions of the angle regimes with negative sign of the pair potential may be masked by the high degree of faceting at the grain boundary. The absence of a ZBCP for NCCO junctions using a normal metal counter electrode has been reported by other groups [41,43,44]. For NCCO, an interlayer tunneling model as has been originally proposed for all HTS, still is a candidate to explain the presence of a strongly anisotropic order parameter having the same symmetry as the underlying crystal but without any sign change [45].

The presence and absence of the ZBCP in the tunneling spectra of hole and electron doped materials can be easily understood within the ABS-model due to the different kind of OP symmetry present in these materials. However, it is very difficult to understand this difference in terms of the AA-model based on magnetic impurities at the grain boundary. Since in both materials the same copper oxide planes are interrupted at the grain boundary, its is difficult to explain why for the hole doped materials there should be magnetic impurities at the grain boundary whereas these should be absent for the electron doped material.

Determination of the OP Symmetry by Tunneling Spectroscopy

Above we have discussed the clear difference between the measured tunneling spectra in hole and electron doped materials and have explained this difference by a different OP symmetry. Of course, one can change the point of view and use the measured tunneling data as strong evidence for the presence of a $d_{x^2-y^2}$-symmetry of the OP in hole doped HTS and an anisotropic s-wave symmetry for the electron doped NCCO. This result is in agreement with most other experiments that probe the pairing symmetry in the HTS [6–8]. Comparing tunneling spectroscopy to other techniques probing the symmetry of the OP we note that tunneling spectroscopy represents a technique that is sensitive for both the phase and amplitude of the OP. Hence, the conclusions that can be drawn from our tunneling spectroscopy experiments are equivalent to those of the well known tricrystal experiments [8]. Strictly speaking, the presence of a ZBCP itself represents a proof for a sign change of the order parameter in different k-directions. In combination with theoretical predictions for possible candidates of the pairing symmetry, tunneling spectroscopy certainly can provide useful evidence for distinguishing between different order parameter symmetries.

Temperature Dependence of the Zero Bias Conductance Peak

Fig. 3 shows a typical set of normalized $G(V)$ curves measured at different temperatures between 4.2 K and T_c. The data was obtained for a GBJ based on an YBCO film. In Fig. 3 the sharp conductance peak appearing directly at zero voltage due to a finite Josephson current was removed for clarity. The spectra are fully symmetric with respect to zero voltage as expected for a SIS-junction with a fully symmetric configuration. With increasing temperature the ZBCP decreases and finally disappears at the T_c of the electrode material. This behavior has been observed for all investigated samples. It is likely that the disappearance of the ZBCP below T_c which has been reported recently [43] is related to the sample and junction fabrication process. For our HTS GBJs a clear correlation between the appearance of the ZBCP and T_c has been observed. This behavior provides strong

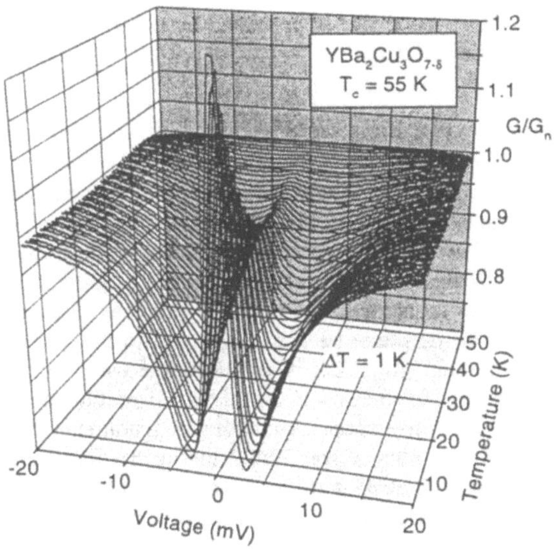

Figure 3
Normalized conductance vs. voltage for a 24° [001] tilt YBCO-GBJ for temperatures between 4 and 51 K. The sample was underdoped with $T_c \approx$ 55 K. Data taken from Ref. [15].

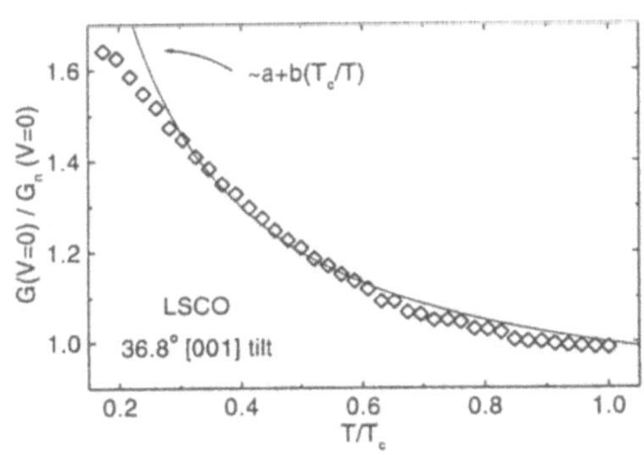

Figure 4
Normalized conductance at zero bias for a symmetric 36.8° [001] tilt LSCO-GBJ plotted vs. normalized temperature.

evidence against models that explain the ZBCP without including superconductivity, as for example the AA-model. Within the AA-model, the ZBCP is not expected to disappear at a specific temperature. Moreover, the AA-model cannot explain that the ZBCP disappears at T_c which is considerably different for the investigated materials. In contrast, the ABS-model naturally accounts for the observed correlation between the onset of superconductivity in the junction electrodes and the appearance of a ZBCP and a gap structure.

The temperature dependence of the conductance at zero bias so far has not been predicted theoretically for the full temperature range up to T_c. In Fig. 4 the temperature dependence of $G(0)/G_n(0)$ obtained for a LSCO-GBJ is shown. The conductance at zero voltage was normalized by the background conductance and plotted vs. the reduced temperature $t = T/T_c$. The functional form is consistent with a $1/T$ dependence over a wide temperature range $0.3 \leq t \leq 1$. For small t a $1/T$ dependence has been predicted within the ABS-model [36]. However, just in this temperature regime the experimental data deviates from a $1/T$ dependence. To clarify this controversy it seems to be necessary to develop more realistic theoretical models which include the effects due to the grain boundary faceting. The data shown in Fig. 4 can be used to test model predictions. We also note that the logarithmic temperature dependence that is predicted by the AA-model is not consistent with the measured experimental data.

Intrinsic Linewidth of the ZBCP

At low temperatures ($t < 0.1$) the shape of the ZBCP in the $G(V)$ curves can be better described by a Lorentzian curve than by a Gaussian curve. The transition from a Gaussian to a Lorentzian shape was observed to occur below about 10 K for YBCO and below about a few 100 mK for LSCO. This observation can be interpreted as indication of an intrinsic linewidth of the ZBCP which amounts to a few meV. One way to account for the observed width of the ZBCP is to introduce an imaginary part to the quasiparticle energy that is independent of momentum [46]. This is similar to the use of a smearing parameter Γ in the Dynes expression for the density of states. However, this method has the disadvantage of introducing a purely phenomenological lifetime parameter. Recently, Walker et al. discussed an intrinsic linewidth of the ZBCP due to the finite tunneling probability of a quasiparticle occupying an ABS into the bulk material [47]. This might provide a way to account for an intrinsic linewidth within the ABS-model.

Evidence for a Pseudogap in HTS

An interesting feature of the tunneling spectra of HTS GBJs is the temperature dependence of the conductance for $V < V_g$ which gives information on the states within the gap. It is well known that for a BCS superconductor the energy gap decreases with increasing temperature but no states within the gap are observed. In contrast, in the case of HTS a completely different behavior is usually found. With increasing temperature the states inside the gap are filled up with the gap value staying about constant. At T_c, the spectra smoothly approach the normal state conductance. This behavior of the gap can be interpreted in terms of a pseudogap that has been found for HTS using angle resolved photo emission spectroscopy [48,49]. However, up to now, in the tunneling spectra of GBJs a pseudogap was not observed above T_c. In this context we note that most of the investigated samples were optimally doped samples, for which the phase diagram does not predict a pseudogap. However, also for the underdoped YBCO (60 K phase) studied in our experiments no pseudogap above T_c was found. This is in clear contrast to other tunneling experiments using LTSTS on BSCCO surfaces, where a gap above T_c was observed by Renner et al. [50,51]. Our observation together with the observation of a pseudogap above T_c in LTSTS on HTS are important with respect to the interpretation of the pseudgap. If the superconducting gap and the pseudogap coincide below T_c and the superconducting gap develops smoothly into the pseudogap, a common physical origin of the two phenomena is likely what provides a restriction for possible theoretical explanations of the pseudogap.

Another unsettled question is the magnitude of the gap of HTS as a function of the doping level. In recent experiments using BSCCO it was observed that the gap increases with decreasing doping concentration [51,52]. In contrast to these results, for the investigated HTS GBJs using YBCO electrodes with the doping concentrations $\delta \approx 0$ and $\delta \approx 0.4$ a scaling of the gap with T_c was observed. A similar scaling behavior was found for the *s*-wave material NCCO. However, we emphasize that certainly more experimental and theoretical work is required to clarify the questions related to the pseudogap in HTS.

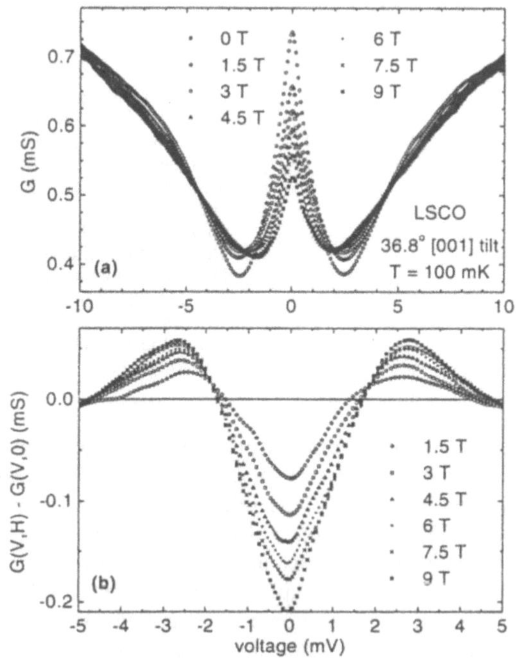

Figure 5
Magnetic field dependence of the ZBCP at 100 mK for a [001] tilt LSCO-GBJ. In (b) $(G(V,H) - G(V,0))$ is plotted for the same sample. The applied magnetic field ranged between 1 and 9 T (1.5 T steps).

Magnetic Field Dependence of the Tunneling Spectra

A typical magnetic field dependence of the tunneling spectra of GBJs is shown in Fig. 5a. The magnetic field was applied parallel to the grain boundary plane, i.e. parallel to the c-axis of the HTS thin film forming the junction electrode. The magnetic field is found to reduce the conductance around at zero voltage and to result in an increase of the conductance at larger voltages. This can be interpreted as a shift of spectral weight from zero energy and to finite energies. The spectral weight over the whole energy scale is conserved. The shift of spectral weight to finite energies leads to a clear splitting of the difference curves $G(V,H) - G(V,0)$ shown in Fig. 5b. All measurements shown in Fig. 5 were performed at 100 mK using a LSCO-GBJ. The thermal smearing at this temperature is below 10 μV. Note, that no direct splitting of the ZBCP in the $G(V)$ curves could be observed for all investigated HTS GBJs at any applied field up to 12 T. A similar behavior also was observed for SIN-type junctions fabricated from different HTS materials [43]. However, in some experiments a direct splitting with magnetic field, and even a splitting at zero field indicating a broken time reversal symmetry (BTRS) was observed [53].

Analyzing the $G(V,H) - G(V,0)$ data an indirect splitting δ_i can be defined as half the peak to peak separation in Fig. 5b. In Fig. 6, δ_i obtained for a LSCO-GBJ at 100 mK and for an oxygen reduced YBCO-GBJ at 4.2 K is plotted together with data taken from literature for SIN-type junctions [2,53]. In contrast to the GBJ, for investigated SIN-type junctions a direct splitting in the conductance versus voltage curves was observed which we denote as δ_d. Both δ_i and δ_d show a highly nonlinear magnetic field dependence. Only in the low field regime below about 1 T a linear behavior is observed. This is in clear contrast to the prediction of AA-model, where a Zeeman splitting of the impurity states results in a splitting of the ZBCP that increases linearly with increasing field. The

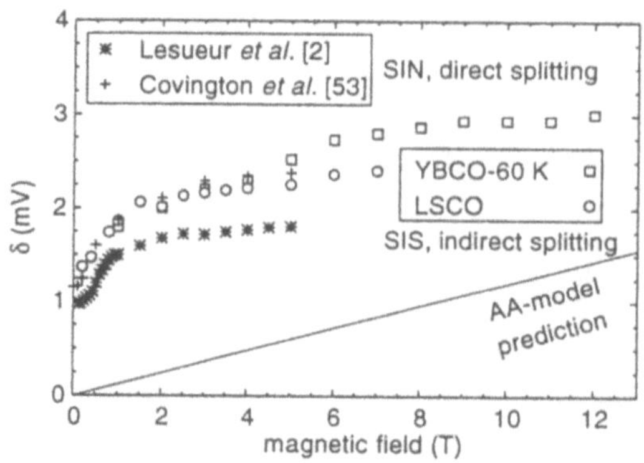

Figure 6
Indirect splitting δ_i of the $G(V,H) - G(V,H=0)$ curves plotted vs. the applied magnetic field H for a 36.8° [001] tilt LSCO-GBJ at 100 mK and a 24° [001] tilt YBCO-GBJ at 4.2 K. Also shown are data from literature for the direct splitting δ_d in the $G(V,H)$ curves of planar SIN-junctions. The straight line shows the AA-model prediction for $g = 2$.

AA-model prediction also is plotted in Fig. 6 for comparison. The AA-model predicition clearly deviates from the dependence measured for HTS junctions [54]. The tendency of the splitting to saturation at high fields is most likely related to a saturation field as predicted recently by Fogelström et al. [55]. The presence of a direct splitting of the ZBCP with a finite value at zero applied magnetic field has been interpreted as clear evidence for a surface state with BTRS. In the case of the indirect splitting as observed in GBJs, it is not yet fully established, whether the same interpretation can be applied. One reason for that controversy is that the indirect splitting δ_i by definition has to become zero for zero applied field. Hence, it is not possible to compare δ_i and δ_d directly at small magnetic fields. Also, the steep jump of δ_i that is observed already for very small applied magnetic fields is not fully understood at present. A possible reason for the presence of a shift of spectral weight without the observation of any direct splitting of the ZBCP in an applied field is the considerable faceting of the grain boundary together with impurity scattering that suppresses the splitting of the ZBCP [56]. This issue has to be further investigated both theoretically and experimentally.

Recently, Hu has discussed the magnetic field effects on the ZBCP [14]. Hu proposes that the magnetic field might be shielded or screened, thereby impeding the penetration of the magnetic field into the superconductor. However, the magnetic field behavior of GBJs has been extensively studied before. The magnetic field penetrates into the junction giving rise to a characteristic Fraunhofer diffration patter like behavior of the critical current vs. magnetic field dependence that has been measured for all of the investigated GBJs. Moreover, the magnetic field is found to even be increased by a factor of W/d, where W is the junction width and d the thickness of the junction electrodes [57]. That is, our experimental results clearly contradict the interpretation given by Hu. As discussed above a more realistic explanation for the masked splitting of the ZBCP seems to be the faceting of the grain boundary interface and impurity scattering at the grain boundary.

Anomalous Meissner Currents

HTS GBJs formed by different HTS have been successfully used to determine the relative change

$$\frac{\Delta\lambda_{ab}(T)}{\lambda_{ab}(0)} = \frac{\lambda_{ab}(T) - \lambda_{ab}(0)}{\lambda_{ab}(0)}$$

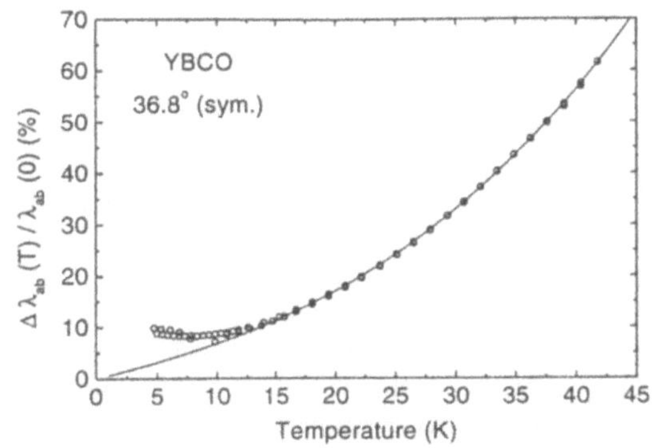

Figure 7
$[\lambda_{ab}(T)-\lambda_{ab}(0)]/\lambda_{ab}(0)$ for a 36.8° [001] tilt symmetric YBCO-GBJ. The solid line shows the temperature dependence expected for a superconductor with a $d_{x^2-y^2}$-symmetry of the order parameter. The measured upturn at low temperatures most likely is due to the presence of anomalous Meissner currents.

of the in-plane London penetration λ_{ab} depth with the high precision of only a few Å [58]. In this technique the shift of the side-lobes of the $I_c(H)$ dependence of HTS GBJs is measured as a function of temperature. Details of this precision measurement technique have been described elsewhere [58]. Recently, an anomalous temperature dependence of $\Delta\lambda_{ab}$ was measured for a symmetric 36° [001] tilt YBCO-GBJ [59] and also in ion-bombarded YBCO [60]. As can be seen in Fig. 7, for temperatures below about 10 K the $\Delta\lambda_{ab}(T)/\lambda_{ab}(0)$ dependence shows a clear upturn with decreasing temperature in contradiction to the behavior expected for a *d*-wave superconductor. Note, that this anomaly can be explained neither by assuming a pure *d*-wave nor a pure *s*-wave symmetry of the OP. In both cases $\Delta\lambda_{ab}(T)/\lambda_{ab}(0)$ is expected to monotonously decrease with decreasing temperature. However, ABS can result in an anomalous Meissner current that is flowing in opposite direction to the usual Meissner current thereby increasing the measured penetration depth. This in turn results in an upturn of the temperature dependence of the London penetration depth at low temperatures. However, a clear upturn as shown in Fig. 7 has not been observed for all investigated GBJs. At present it is not known what determines the strength of the observed upturn. Certainly, interface roughness again is expected to be an important factor. Additional experiments are required to clarify this issue.

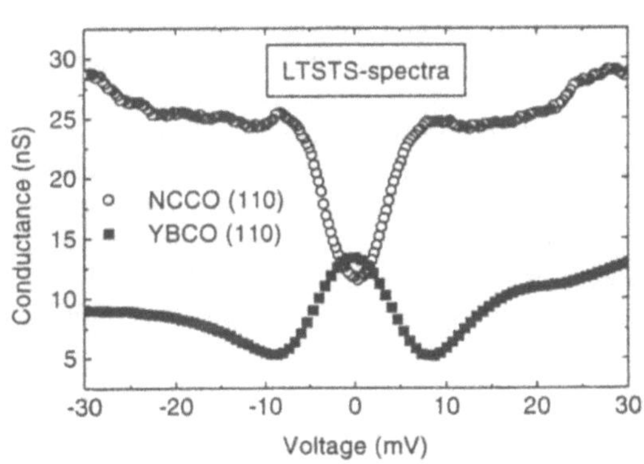

Figure 8
LTSTS spectra obtained for (110) oriented surfaces of YBCO and NCCO at 4.2 K. The data are taken from Ref. [61].

2.3 LTSTS and Other Tunneling Experiments

The experimental data presented above have been obtained using SIS-type HTS GBJs. We now briefly compare this data to that obtained by LTSTS. Using LTSTS, a ZBCP also was observed in many experiments [62–64]. For LTSTS measurements the surface quality of the sample is crucial. For BSCCO, measurements can be done by *in situ* cleaving a single crystal that breaks along well defined crystal planes. This allows to perform reliable tunneling experiments in c-axis direction. However, this is not the case for YBCO. Therefore, for most of the LTSTS measurements *ex situ* processing is required and, thus, are affected by surface contamination. Contamination in combination with surface roughness easily can lead to a masking of the ZBCP. As criteria for reliable LTSTS measurements it is required that the measured spectra are independent of position and of the distance between the surface and the tip used as the counter electrode [65]. Applying these criteria for well defined (110) oriented surfaces of YBCO, the existence of a ZBCP could be confirmed in agreement with the theoretical predictions [46]. Due to the large barrier height, the total tunneling conductance, $\sigma_T(E)$, in LTSTS experiments is calculated to [12]

$$\sigma_T(E) \approx \frac{\int_{-\pi/2}^{\pi/2} \sigma_N \rho_S(E) d\theta}{\int_{-\pi/2}^{\pi/2} \sigma_N d\theta}.$$

Here, θ is the incident angle of the quaisparticles with respect to the SI-interface, ρ_S is the surface density of states of an insulator/d-wave superconductor interface, and σ_N is the normal state conductance [46]. The experimental data obtained in LTSTS can be well fitted using this simple expression. With respect to the observed difference in the tunneling behavior of the d-wave (YBCO, BSCCO, LSCO) and s-wave HTS (NCCO), LTSTS experiments confirm the results obtained for HTS GBJs. Firstly, for NCCO a ZBCP never could be observed, as shown in Fig. 8. Secondly, for NCCO the conductance peak at the gap edges is more pronounced and the reduction of the conductance within the gap is stronger than for the hole doped HTS. In Fig. 8 the data for (110) oriented surfaces of YBCO and NCCO is shown. LTSTS measurements on (100) and (001) surfaces of both materials show that for NCCO a ZBCP never is observed [41]. However, for YBCO surfaces a ZBCP was found not only for (110) but in many cases also for (100) and (001) oriented surfaces. This behavior can be attributed to the finite surface roughness of the investigated samples.

It is also interesting to consider the background conductance in the tunneling spectra which is fully symmetric with respect to zero coltage for the HTS GBJ. In contrast, the background conductance in LTSTS measurements is found to be asymmetric with respect to the Fermi level in most experiments [65]. This asymmetry is less pronounced for NCCO than for YBCO or BSCCO. At present it is not clear, whether this effect is related to the symmetry of the pairing state or caused by the asymmetry of the junction configuration.

Table 1 Summary of the experimental situation with respect to the observation of a ZBCP in different superconducting materials. The experiments are classified into three categories according to the type of junction used for tunneling spectroscopy. The list of references is far from being complete.

Superconductor	SIS (GBJ)	LTSTS	planar SIN
LTS ($NbSe_2$, NbN)		no ZBCP BCS-like[66,67]	
YBCO ($T_c \approx 90$ K)	ZBCP [15,24,37]	ZBCP [61–64]	ZBCP [2,43,53]
YBCO ($T_c \approx 60$ K)	ZBCP [15,24,37]		
BSCCO ($T_c \approx 85$ K)	ZBCP [15,24,37]	ZBCP [68]	ZBCP [69]
LSCO ($T_c \approx 24$ K)	ZBCP [15,37]	ZBCP [68]	
$Tl_2Ba_2CaCu_2O$			ZBCP [43]
NCCO ($T_c \approx 24$ K)	no ZBCP [15,24,37]	no ZBCP [41,61]	no ZBCP [43,44]

3 Summary of the Evidence for the ABS-Model

We briefly summarize the available spectroscopic data obtained from HTS GBJs as well as from planar SIN-junctions [1,2,53,69,70] and LTSTS experiments. A comprehensive overview on the experimental situation with respect to the observation of a ZBCP in different superconducting materials using different experimental techniques is given in Table 1. Clearly, the ABS-model supplies a natural explanation for the vast majority of experimental data. The main experimental evidence for the presence of Andreev bound states at surfaces and interfaces of d-wave HTS and the validity of the ABS-model for the explanation of the available experimental facts can be summarized as follows:

- All experiments using junctions with HTS electrodes can be described naturally within a single model.
- The appearance of the ZBCP is clearly correlated to the presence of superconductivity in the junction electrodes.
- The vast majority of experimental data can be well fitted by the theoretical predictions of the ABS-model.
- All superconductors with $d_{x^2-y^2}$-symmetry of the order parameter show a ZBCP, while s-wave materials and low-temperature superconductors do not show a ZBCP (see Table 1).

- Other phenomena related to the presence of ABS such as an anomalous Meissner current are observed.
- Competing theoretical explanations as for example the AA-model fail to describe important details of the observed behavior.

4 Conclusions

It has been shown that tunneling spectroscopy on HTS can be viewed as a phase sensitive method that can be used to determine the order parameter symmetry of superconductors. A comprehensive study of various HTS GBJs gives clear evidence for the interpretation of the tunneling spectra in terms of the Andreev bound state model and against an interpretation based on magnetic scattering. Our experimental data clearly confirm that the hole doped HTS YBCO, BSCCO, and LSCO have a dominant $d_{x^2-y^2}$-wave component of the OP. In contrast, the electron doped material NCCO most likely has an anisotropic s-wave symmetry of the OP. Evidence for an anomaly of the temperature dependence of the London penetration depth that is related to the presence of ABS was presented. The question of the existence of a state with broken time reversal symmetry at the grain boundary interface still is not settled.

Acknowledgement

We would like to thank Th. Bauch, A. Beck, H. Burkhardt, M. Covington, R. Dittmann, O. M. Fröhlich, M. Fogelström, J. Halbritter, S. Kashiwaya, T. Kemen, S. Kleefisch, G. Koren, M. Koyanagi, J. Mannhart, A. Marx, M. Naito, D. Rainer, P. Richter, H. Sato, J. Sauls, U. Schoop, H. Takashima, Y. Tanaka, N. Terada, and M. Zittartz for valuable discussions. This work is supported by the Deutsche Forschungsgemeinschaft (SFB 341).

Bibliography

[1] Thomas Walsh, Int. J. Mod. Phys. **B 6**, 125 (1992).

[2] J. Lesueur, L. H. Greene, W. L. Feldmann, and A. Inam, Physica **C 191**, 325 (1992).

[3] J. Appelbaum, Phys. Rev. Lett. **17**, 91 (1966); see also Phys. Rev. **154**, 633 (1967).

[4] P. W. Anderson, Phys. Rev. Lett. **17**, 95 (1966).

[5] C. R. Hu, Phys. Rev. Lett. **72**, 1526 (1994); C. Yang and C. R. Hu, Phys. Rev. **B 50**, 16766 (1994).

[6] D. J. Van Harlingen, Rev. Mod. Phys. **67**, 515 (1995).

[7] D. J. Scalapino, Phys. Rep. **250**, 329 (1995).

[8] J. R. Kirtley, C. C. Tsuei, J. Z. Sun, C. C. Chi, Lock See Yu-Jahnes, A. Gupta, M. Rupp, and M. B. Ketchen, Nature **373**, 225 (1995).

[9] R. Gross, *Grain Boundary Josephson Junctions in the High Temperature Superconductors* in Interfaces in Superconducting Systems, S. L. Shinde and D. Rudman eds, Springer-Verlag, New York (1994), pp. 176-209.

[10] R. Gross, L. Alff, A. Beck, O. M. Fröhlich, D. Kölle, and A. Marx, IEEE Trans. Appl. Supercond. **7**, 2929 (1997).

[11] D. Rainer, H. Burkhardt, M. Fogelström, and J. A. Sauls, to appear in *Proceedings of SNS'97*, Cape Cod, September 1997; cond-mat/9712234 (1997).

[12] S. Kashiwaya, Y. Tanaka, M. Koyanagi, and K. Kajimura, Phys. Rev. **B 53**, 2667 (1996).

[13] Y. Tanaka and S. Kashiwaya, Phys. Rev. **B 56**, 892 (1997).

[14] C. R. Hu, Phys. Rev. **B 57**, 1266 (1998).

[15] L. Alff, S. Kleefisch, M. Zittartz, U. Schoop, A. Marx, and R. Gross, to be published in Eur. Phys. J. **B**; cond-mat/9806150 (1998).

[16] Yu. N. Ovchinnikov and V. Z. Kresin, Phys. Rev. **B 54**, 1251 (1996).

[17] B. Mayer, L. Alff, T. Träuble, R. Gross, P. Wagner, and H. Adrian, Appl. Phys. Lett. **63**, 996 (1993).

[18] A. Marx, U. Fath, W. Ludwig, R. Gross, and T. Amrein, Phys. Rev. **B 51**, 6735 (1995).

[19] M. Naito and H. Sato, Appl. Phys. Lett. **67**, 2557 (1995).

[20] A. Beck, O. M. Fröhlich, D. Kölle. R. Gross, H. Sato, and M. Naito, Appl. Phys. Lett. **68**, 3341 (1996).

[21] H. Sato, M. Naito, and H. Yamamoto, Physica **C 280**, 178 (1997).

[22] J. P. Locquet, Y. Jaccard, A. Cretton, E. J. Williams, F. Arrouy, E. Mächler, T. Schneider, Ø. Fischer, and P. Martinoli, Phys. Rev. **B 54**, 7481 (1996).

[23] H. Yamamoto, M. Naito, and H. Sato, Phys. Rev. **B 56**, 2852 (1997).

[24] S. Kleefisch, L. Alff, U. Schoop, A. Marx, R. Gross, M. Naito, and H. Sato, Appl. Phys. Lett. **72**, 2888 (1998).

[25] K. K. Likharev, *Dynamics of Josephson Junctions and Circuits*, Gordon and Breach, New York (1986).

[26] R. Gross and B. Mayer, Physica **C 180**, 235 (1991).

[27] R. Gross and B. Mayer, in *Advances in High Temperature Superconductivity*, D. Andreone ed, World Scientific, Singapore (1992), p. 261.

[28] J. Halbritter, Phys. Rev. **B 48**, 9735 (1993).

[29] O. M. Fröhlich, H. Schulze, A.Beck, B. Mayer, L. Alff, R. Gross, and R. P. Hübener, Appl. Phys. Lett. **66**, 2289 (1995).

[30] O. M. Fröhlich, P. Richter, A.Beck, R. Gross, and G. Koren, J. Low Temp. Phys. **106**, 243 (1997); see also IEEE Trans. Appl. Supercond. **7**, 3189 (1997).

[31] A. Marx, L. Alff, and R. Gross, IEEE Trans. Appl. Supercond. **7**, 2719 (1997).

[32] B. Kabius, J. W. Seo, T. Amrein, U. Dähne, A. Scholen, M. Siegel, K. Urban, and L. Schultz, Physica **C 231**, 123 (1994).

[33] J. Mannhart, H. Hilgenkamp, B. Mayer, Ch. Gerber, J. R. Kirtley, K. A. Moler, and M. Sigrist, Phys. Rev. Lett. **77**, 2782 (1996).

[34] H. Hilgenkamp, J. Mannhart, and B. Mayer, Phys. Rev. **B 53**, 14586 (1996).

[35] Yu. S. Barash, A. V. Galaktionov, and A. D. Zaikin, Phys. Rev. **B 52**, 665 (1995).

[36] Yu. S. Barash, A. A. Svidzinsky, and H. Burkhardt, Phys. Rev. **B 55**, 15282 (1997).

[37] L. Alff, A. Beck, R. Gross, A. Marx, S. Kleefisch, Th. Bauch, H. Sato, M. Naito, and G. Koren, to be published in Phys. Rev. **B**; cond-mat/9805162 (1998).

[38] Q. Huang, J. F. Zasadzinski, N. Tralshawala, K. E. Gray, D. G. Hinks, J. L. Peng, and R. L. Greene, Nature **347**, 369 (1990).

[39] D. H. Wu, J. Mao, J. L. Peng, X. X. Xi, T. Venkatesan, R. L. Greene, and S. M. Anlage, Phys. Rev. Lett. **70**, 85 (1993).

[40] A. Andreone, A. Cassinese, A. Di Chiara, R. Vaglio, A. Gupta, and E. Sarnelli, Phys. Rev. **B 49**, 6392 (1994).

[41] L. Alff, H. Takashima, S. Kashiwaya, N. Terada, T. Ito, K. Oka, M. Koyanagi, and Y. Tanaka, *Adv. in Supercond. IX*, S. Nakajima and M. Murakami eds, Springer-Verlag, Tokyo (1997), p. 49.

[42] R. C. Dynes, V. Narayanamurti, and J .P. Garno, Phys. Rev. Lett. **41**, 1509 (1978).

[43] J. W. Ekin, Y. Xu, S. Mao, T. Venkatesan, D. W. Face, M. Eddy, and S. A. Wolf, Phys. Rev. **B 56**, 13746 (1997).

[44] M. Naito, H. Sato, and H. Yamamoto, Physica **C 293**, 36 (1997).

[45] S. Chakravarty, A. Sudbø, P. W. Anderson, and S. Strong, Science **261**, 337 (1993).

[46] L. Alff, H. Takashima, S. Kashiwaya, N. Terada, H. Ihara, Y. Tanaka, M. Koyanagi, and K. Kajimura, Phys. Rev. **B 55**, R14757 (1997).

[47] M. B. Walker and P. Pairor, cond-mat/9803079 (1998).

[48] A. G. Loeser, Z.-X. Shen, D. S. Dessau, D. S. Marshall, C. H. Park, P. Fournier, and A. Kapitulnik, Science **273**, 325 (1996).

[49] H. Ding, T. Yokoya, J. C. Campuzano, T. Takahashi, M. Randeria, M. R. Norman, T. Mochiku, K. Kadowaki, and J. Giapintzakis, Nature **382**, 51 (1996).

[50] C. Renner, B. Revaz, J.-Y. Genoud, and Ø. Fischer, J. Low Temp. Phys. **105**, 1083 (1996).

[51] C. Renner, B. Revaz, J.-Y. Genoud, K. Kadowaki, and Ø. Fischer, Phys. Rev. Lett. **80**, 149 (1998).

[52] N. Miyakawa, P. Guptasarma, J. F. Zasadzinski, D. G. Hinks, and K. E. Gray, Phys. Rev. Lett. **80**, 157 (1998).

[53] M. Covington, M. Aprili, E. Paraoanu, L. H. Greene, F. Xu, J. Zhu, and C. A. Mirkin, Phys. Rev. Lett. **79**, 277 (1997).

[54] J. A. Appelbaum and L. Y. L. Shen, Phys. Rev. **B 5**, 544 (1972).

[55] M. Fogelström, D. Rainer, and J. A. Sauls, Phys. Rev. Lett. **79**, 281 (1997).

[56] M. Fogelström and J. Sauls, private communication.

[57] P. A. Rosenthal, M. R. Beasley, K. Char, M. S. Colclough, and G. Zaharchuk, Appl. Phys. Lett. **59**, 3482 (1991).

[58] O. M. Fröhlich, H. Schulze, R. Gross, A. Beck, and L. Alff, Phys. Rev. **B 50**, R13894 (1994).

[59] M. Zittartz, Diploma thesis, *University of Cologne* (1997).

[60] H. Walter, W. Prusseit, R. Semerad, H. Kinder, W. Assmann, H. Huber, H. Burkhardt, D. Rainer, and J. A. Sauls, Phys. Rev. Lett. **80**, 3598 (1998).

[61] L. Alff, H. Takashima, S. Kashiwaya, N. Terada, T. Ito, K. Oka, Y. Tanaka, and M. Koyanagi, Physica **C 282-287**, 1485 (1997).

[62] M. Koyanagi, S. Kashiwaya, H. Akoh, S. Kohjiro, M. Matsuda, F. Hirayama, and K. Kajimura, Jpn. J. Appl. Phys. **31**, 3525 (1992).

[63] Q. Chen and K.-W. Ng, Phys. Rev. **B 45**, 2569 (1992).

[64] S. Kashiwaya, Y. Tanaka, M. Koyanagi, H. Takashima, and K. Kajimura, Phys. Rev. **B 51**, 1350 (1995).

[65] Ch. Renner and Ø. Fischer, Phys. Rev. **B 51**, 9208 (1995).

[66] H. F. Hess, R. B. Robinson, and J. V. Waszczak, Phys. Rev. Lett. **64**, 2711 (1990).

[67] S. Kashiwaya, M. Koyanagi, A. Shoji, M. Matsuda, and H. Shibata, IEEE Trans. Magn. **MAG-27**, 837 (1991).

[68] S. Kashiwaya, private communication.

[69] Saion Sinha and K.-W. Ng, Phys. Rev. Lett. **80**, 1296 (1998).

[70] A. M. Cucolo and R. Di Leo, Phys. Rev. **B 47**, 2916 (1993).

Superconductivity in Fullerene Compounds

Michael Baenitz and Klaus Lüders

Freie Universität Berlin, Institut für Experimentalphysik
Arnimallee 14, D-14195-Berlin, Germany

Abstract: In this contribution some superconducting properties of fullerene superconductors are summarized. The dependence of the transition temperature T_C on the internal C_{60} spacing, on the external pressure and on the magnetic field is related to the density of states at the Fermi level $N(E_F)$. From low field magnetization measurements the lower critical fields of Rb_3C_{60} and K_2CsC_{60} are derived and the penetration depths are calculated. Analysing the hysterical magnetization loops in fields up to 9 T in the framework of the Bean model yields estimations of the critical current densities.

1 Introduction

The observation of superconductivity in the compound K_3C_{60} 1991 [1] with surprisingly high transition temperatures has stimulated intense further research activities of superconductivity in intercalated fullerenes [2]. Today most of the characteristics of this class of superconducting materials are known although some important details are still unknown concerning e.g. the atomic structure and the site ordering of the intercalated atoms [3,4] and the mechanisms of pinning including the existence of irreversibility lines [5].

Seven years after the discovery of superconductivity the number of compounds is quite large (Tab. 1) and they all are based on C_{60} material. No superconductivity is found in intercalated higher fullerenes or endohedral fullerenes up to now. It is remarkable that the cubic face centered alkali C_{60} compounds still exhibit the highest transition temperatures up to 32.5 K for Rb_2CsC_{60}. In contrast to graphite intercalation compounds, borocarbides and cuprates the lattice and electronic structure is isotropic. The intercalation of alkaline earth and rare earth metals in the fcc C_{60} host lattice results in the formation of other structure types with T_C values below 10 K. Superconductivity in hole doped systems like $(J_{1.9})_2C_{60}$ or $(JBr)_xC_{60}$ is not confirmed up to now.

Early attempts to prepare C_{60} compounds started in analogy to those of graphite intercalation compounds [6,7], using binary alloys like K-Tl [8,9] Rb-Tl [9], Cs-Tl, Cs-Bi, and Cs-Hg [10]. However, real ternary fullerenes containing both an alkali metal and a non-alkali metal could not be prepared. Obviously, the alloy decomposes during the preparation procedure and only the alkali metal intercalates. The second component is not involved into the synthesis or perhaps it acts as a catalyst making it possible to reduce the annealing temperature during the intercalation process. In the case of Cs-Bi alloys it was possible to prepare a sample with a superconducting volume fraction of 20% [11,12].

In this contribution some typical characteristics of the basic superconducting properties of fullerene compounds are compiled together with new magnetization investigations leading to critical field and critical current density results.

2 C_{60} Crystals

Fullerenes are a new modification of pure carbon. The ball-shaped molecules consisting of 60, 70 or more carbon atoms are able to form molecular crystals. The decisive breakthrough making solid state research possible was the discovery of a method for producing fullerenes in gram quantities by Krätschmer et al. [13]. Two rods of highly uniform graphite with typical diameters of 3–6 mm are evaporated by resistive heating applying a current of about 180 A. This takes place in a helium atmosphere with a partial pressure of 150–300 mbar. C_{60} together with about 5–10% C_{70} and higher fullerenes are then extracted by dissolving the soot in a boiling organic solvent like benzene or toluene. The separation of the fullerene mixture into its constituents is then done by several methods, e.g. liquid gravity chromatography on alumina.

2.1 Lattice Structure

The C_{60} molecules form a face-centered-cubic (fcc) lattice with a lattice constant of $a_0 = 14.152$ Å at 300 K (Fig. 1). The distance between the centres of two neighbouring molecules is 10.0 Å. With the C_{60} diameter being 7.1 Å, the resulting molecular distance is 2.92 Å which is smaller compared to the layer spacing in pure graphite (3.35 Å).

The intercalation of alkali metal ions into the interstitial octahedral and tetrahedral sites of the C_{60} fcc lattice can be done by applying the methods known from the preparation of graphite intercalation compounds. The results reported here were obtained from samples prepared exclusively by using doubly sublimed C_{60} as starting material which in a first step is intercalated to produce the saturated A_6C_{60} phase [14]. After analysing this phase is mixed intensively with the stoichiometric amount of C_{60} to get the A_3C_{60} phase. The mixture is annealed for 3 days at a temperature of 750–800 K.

2.2 Electronic Structure

The pure C_{60} crystal is a semiconductor with an energy gap of about 1.5 eV. The empty π-like t_{1u} band with a total width of 0.5–0.6 eV is able to accept six electrons per C_{60} unit. With the intercalation of alkai metals the electrons are transferred completely from the s-orbitals of the alkali atoms leading to metallic behaviour for a half-filled band in A_3C_{60}

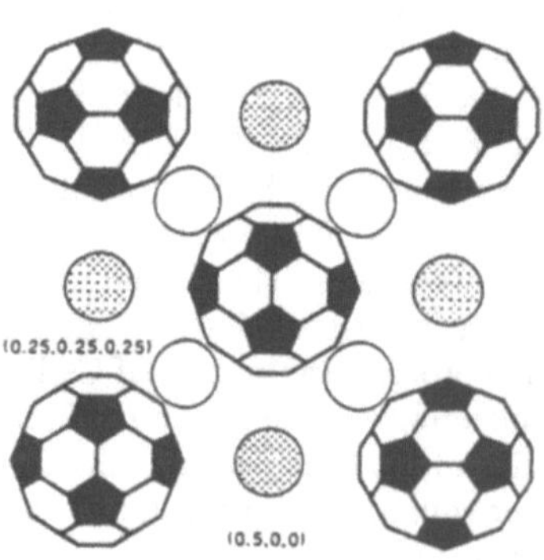

Figure 1
Crystal structure of C_{60} and A_3C_{60} (A = K, Rb). The open and the hatched spheres represent the tetragonal and octahedral interstitial sites which can be occupied with the alkali metal ions.

compounds or to the formation of an almost entirely ionic crystal A_6C_{60} by filling up the conduction band. The superconductivity in these systems is associated with the effect of band narrowing due to lattice expansion. Therefore the density of states at the Fermi level $N(E_F)$ is fairly high compared to other superconducting materials.

Caused by the divalent character of the alkaline earth or the trivalent character of the rare earth elements metallic conductivity and superconductivity in Ca_5C_{60}, Ba_4C_{60}, Sr_6C_{60}, $Yb_{2.75}C_{60}$ and Sm_3C_{60} is related to the partial population of the second band derived from the t_{1g} level of C_{60}. In contrast to alkali metal systems here the charge transfer is incomplete and an hybridization takes place. Therefore the bandwidth is enlarged and the density of states $N(E_F)$ and T_C is reduced considerably.

3 Basic Superconducting Properties

3.1 Transition Temperatures

A selection of superconducting transition temperatures of intercalated fullerenes is collected in Table 1. The largest group comprises the binary and ternary alkali metal intercalated systems with T_c values up to 30.7 K for Rb_3C_{60} and 32.5 K for Rb_2CsC_{60}. The system Cs_3C_{60} with $T_c = 40$ K could only be prepared under pressure [15]. The maximum superconducting volume fractions are about 70%. The deviation from the expected 100% value for phase pure material is a well known feature of powder samples where grain sizes are comparable to the magnetic penetration depths.

Fig. 2 shows a selection of superconducting transition curves in zero field obtained by the response of ac magnetization measurements. The respective curvatures are determined by the granular structure [28–30], i.e. the ratio of the effective particle size and the temperature dependent penetration depth, $x = r/\lambda(T)$. In the case of spherical pariticles the susceptibility is described by the following expression: $\chi = 1.5\{1-(3/x)\coth x + 1/x^2\}$. The onset of the diamagnetic transition is usually used to define T_c.

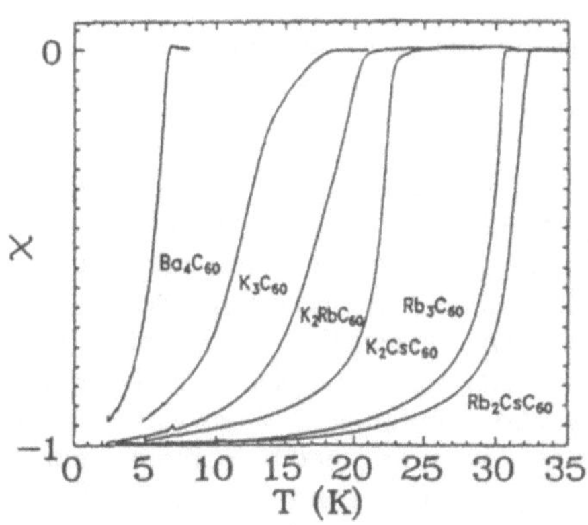

	K_3C_{60}	K_2RbC_{60}	K_2CsC_{60}	Rb_3C_{60}	Rb_2CsC_{60}
T_c / K	17,8	21,5	24,0	30,7	32,5

● Ba_4C_{60}: $T_c = 6{,}8$ K

Figure 2
Superconducting transition curves of intercalated fullerenes obtained by magnetic measurements [31]. χ' is normalized to -1 for $T \to 0$ K.

Table 1 Lattice structure and superconducting transition temperatures for the different classes of binary and ternary fullerene superconductors.

			T_C (K)	Ref.
Alkali systems				
fcc Fm3m	$Cs_3C_{60}^{\dagger}$		40	[15]
	Rb_3C_{60}		30.7	[14]
	K_3C_{60}		18	[14]
		Rb_2CsC_{60}	32.5	[16]
		K_2CsC_{60}	24	[14]
		K_2RbC_{60}	21.5	[14]
fcc Pa3		$Na_2Cs_{1-x}Rb_xC_{60}$	$3.5(x=1)-12(x=0)$	[17]
Alkali earth systems				
simple cubic	Ca_5C_{60}		8.4	[18]
bco Immm	Ba_4C_{60}		6.8	[19]
bcc Im3	Sr_6C_{60}		4	[20]
Alkali/alkaline earth systems				
fcc Fm3m		$Rb_{3-x}Ba_xC_{60}$	24 ($x=0.5$	[21]
		$K_{3-x}Ba_xC_{60}$	15 ($x=0.25$)	[21]
		$Cs_{3-x}Ba_xC_{60}$	–	[21]
bcc Im3		$K_3Ba_3C_{60}$	5.6	[22]
Rare earth systems				
orthorhombic	$Yb_{2.75}C_{60}$		6	[23]
	Sm_3C_{60}		8	[24]
Modified systems				
fcc Fm3m	$(NH_3)_4Na_2CsC_{60}$		29.6	[25]
orthorhombic	$(NH_3)K_3C_{60}^{\dagger}$		–	[26]
fcc	$N_yNa_xC_{60}$		12	[27]

† Superconductivity occurs only under pressure and the structure is not cubic.

As first pointed out by Fleming et al. [32] a relationship between the structure and the transition temperature exists. T_c increases monotonously with the lattice constant a_0. Fig. 3 demonstrates this behaviour by plotting T_c against the nearest neighbour distance of the C_{60} molecules. In the case of the alkali metal intercalated fullerenes, this distance is larger compared to that of the pure C_{60} crystal, whereas it is reduced for the intercalation with alkaline earth metals. The hatched linear region corresponds to those fullerenes with fcc structure. Depending on the preparation conditions T_c varies within this region. K_2CsC_{60}, Rb_3C_{60}, and Rb_2CsC_{60} for instance were prepared by optimized conditions, i.e. doubly sublimed C_{60} was used as starting material resulting in a shift of the linear $T_c(d_{C_{60}-C_{60}})$ relation to values higher by about 2 K. The alkaline earth systems Ba_4C_{60} and Sr_6C_{60} deviate significant from this linear behaviour whereas Ca_5C_{60} fits well to the relation. It should be mentioned, however, that the general $T_c(d_{C_{60}-C_{60}})$ relation agrees roughly with the McMillan theory with moderate coupling strenghts of $\lambda_{ep} \leq 1$ and under the assumption that the density of states, $N(E_F)$, varies with the third power of the

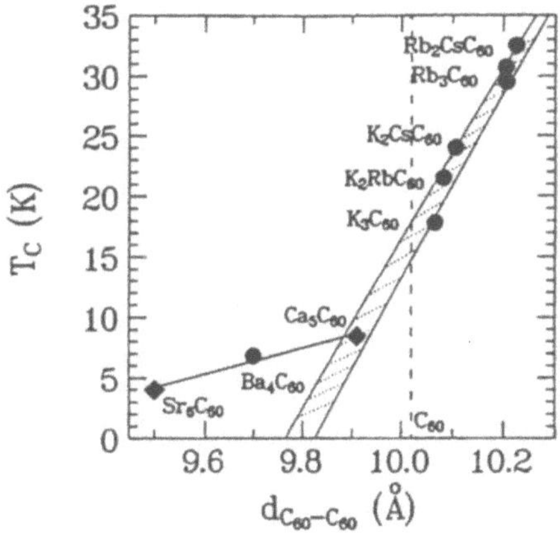

Figure 3
Relationship between the superconducting transition temperature and the intermolecular distance of C_{60} molecules in intercalated fullerenes.

intermolecular distance of the C_{60} molecules and that on ball phonon modes are responsible for the electron phonon interaction [2]. Furthermore for alkali compounds the linear relation is also confirmed by pressure experiments (see Section 3.3).

3.2 The System Ba_4C_{60}

Superconductivity in Ba intercalated fullerenes was first reported for samples with the nominal composition Ba_6C_{60} [33]. However, the stoichiometry of the superconducting phase in this system was subject to further investigations. In order to clarify this question, a series of Ba_xC_{60} samples were prepared with the stoichiometries $x = 3$, 4, 5, and 6 [19]. Only doubly sublimed C_{60} was used as starting material. X-ray diffraction patterns of the Ba samples indicate the coexistence of several Ba_xC_{60} phases. For $x = 3$ a simple cubic structure occurs while the Ba_6C_{60} phase exhibits a bcc structure. A third intrinsic phase is found which is attributed to an orthorhombic Ba_4C_{60} structure. In the samples with the nominal compositions $x = 4$, 5, and 6 this phase is present as a minority phase with the highest fraction in the $x = 4$ sample (in the following denoted as $Ba4C_{60}$). The transition temperature of this phase is $T_C = (6.8 \pm 0.2)$ K with a superconducting volume fraction of 20% which is considerably enhanced compared with 3% for Ba_6C_{60}.

The relatively large superconducting volume fraction allows the determination of the upper critical field $H_{c2}(T)$. The slope of the upper critical field near T_C is one order of magnitude smaller compared to alkali C_{60} compounds. At low temperatures a significant deviation from the theoretical prediction of Werthamer, Helfand and Hohenberg is found towards higher values, a behaviour also observed for instance in $K3C_{60}$ [34] and graphite intercalation compounds [35]. An extrapolation to $T = 0$ K by using a linear relationship also for Ba_4C_{60} leads to $\mu_0 H_{c2}(0) = (2.4 \pm 0.3)$ T resulting in a Ginzburg-Landau coherence length of $\xi(0) = (116 \pm 7)$ Å which is the largest value ever determined for fullerene compounds.

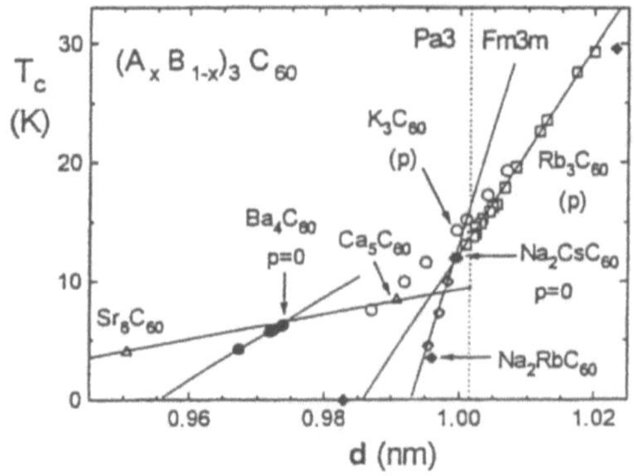

Figure 4
T_c relations obtained from pressure experiments on intercalated fullerenes (d = intermolecular distance of the C_{60} molecules) [37].

3.3 Pressure Experiments

Further characterization of Ba_4C_{60} was done by pressure studies [36]. T_C decreases linearly with pressure ($dT_C/dp = -1.9$ K/GPa) showing the same sign as was found for all alkali metal fullerenes (Fig. 4). The absolute value, however, is about four times smaller. With the assumption that Ba_4C_{60} has a comparable relative compressibility as the K_3C_{60} compounds (since almost equal ionic radii of Ba^{2+} and K^{1+}) it results that T_C of Ba^4C_{60} shows a weaker dependence on the intermolecular C_{60}-C_{60} distance than the alkali metal fullerenes. This might be attributed to the hybridization of Ba s-, d-states with C_{60} π-states. The pressure behaviour of Ba_4C_{60} which can be described by the coefficient $dT_c/dd_{C_{60}-C_{60}}$ = 360 K/nm deviates from the line connecting the zero pressure T_c values of the alkaline earth fullerenes (Fig. 4) with a slope of $dT_c/dd_{C_{60}-C_{60}}$ = 100 K/nm. This demonstrates that in contrast to the alkali metal fullerenes no uniform behaviour of T_c on d is found within this group.

3.4 STM Investigations

Scanning tunneling microscopy (STM) and spectroscopy (STS) have been successfully used to investigate surfaces of conducting solids and their electronic properties. At low temperatures these techniques are especially useful in the case of superconductors. By tunneling spectroscopy the superconducting energy gap can be determined with the high spatial resolution of STM.

Fig. 5 shows results of the application of this method to Rb_2CsC_{60} [38]. The energy gap Δ is determined by dividing the peak to peak distance in the dI/dV spectra by 2. Fig. 6 shows a histogram of the extracted gap values. Most of them are between 3.3 and 5.7 meV resulting in a mean gap value of (4.2 ± 0.6) meV as calculated from a statistical set of 44 spectra. With the transition temperature of $T_C = 32.5$ K it follows for the reduced gap value: $2\Delta/k_BT_C = 3.2 \pm 0.5$ which is close to the BCS value of 3.5 at $T = 0$ K. The spread of the data may be attributed to locally varying stoichiometry. However, some measurements yield energy gap values exceeding the BCS value by far. But these large gaps can be understood by assuming superconductor-insulator-superconductor tunneling junctions which can occur when superconducting grains are picked up by the tunneling tip. In this case the apparent energy gap is twice as large.

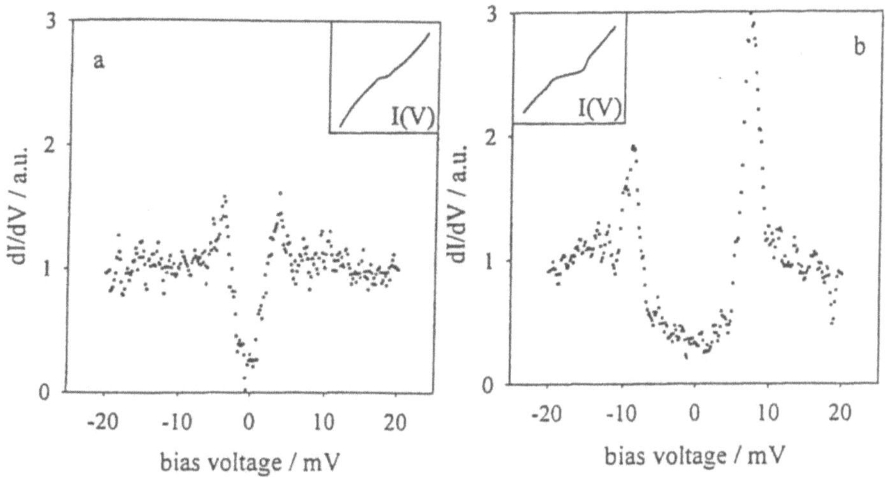

Figure 5 I/V characteristics (insets) and numerically derived dI/dV tunneling spectra acquired on Rb_2CsC_{60}. (a) Typical dI/dV characteristic exhibiting an energy gap value of 3.9 meV. (b) Unexpectedly large gap of 8.1 meV [38,39].

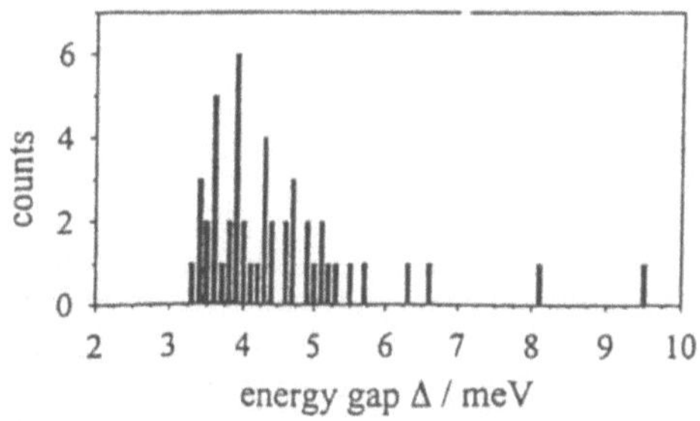

Figure 6
Histogram of energy gap values extracted from 44 dI/dV curves acquired on Rb_2CsC_{60} at 4.5 K. The mean value is 4.2 ± 0.6 meV [38,39].

4 Critical Fields and Current Densities

Ac and dc magnetization measurements provide much information about critical fields (H_{c1}, H_{c2}), critical current densities (j_c) and reversible fluxoid motion for fullerene superconductors. Most investigations are carried out on K_3C_{60} and Rb_3C_{60}. A review of these investigations is given in [40]. It has to be noted that in these measurements the intrinsic properties of the material are mostly covered by effects due to sample granularity (weak link behaviour) and size effects (where particle sizes are comparable to the penetration depth λ) [29]. In the following sections results are presented of high quality powder samples with large particle sizes (10–100 μm) and the absence of granularity where these effects can be neglected.

4.1 Upper Critical Fields

$H_{c2}(T)$ values are mostly determined by the onset of the diamagnetic transition in ac or dc magnetization measurements. Compared to graphite intercalation compounds, the influence of magnetic fields on the transition temperature is very weak. For Rb_3C_{60} and

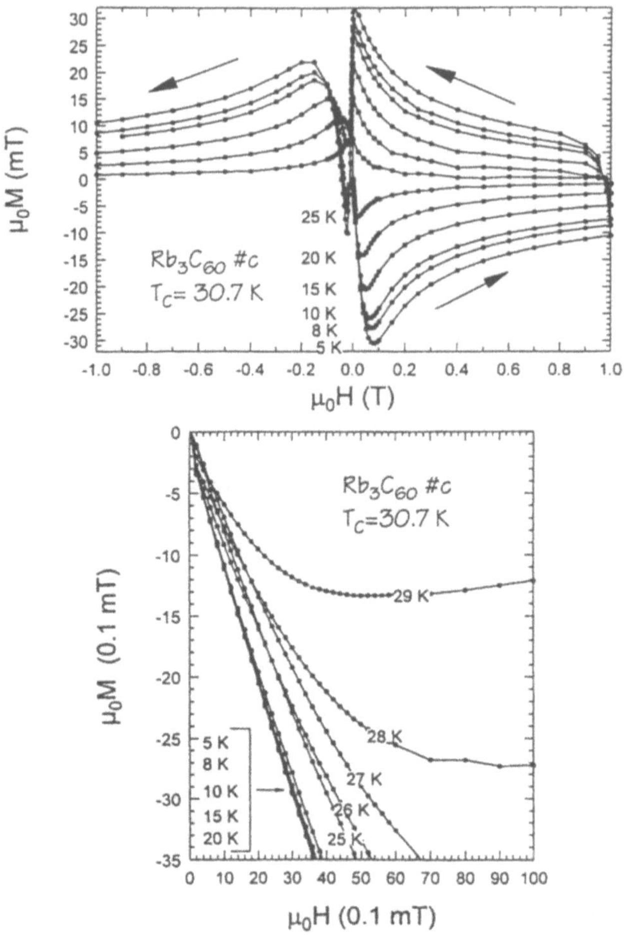

Figure 7 Magnetization of Rb_3C_{60} as a function of field for different temperatures. The arrows indicate increasing and decreasing magnetic fields.

K_2CsC_{60} we found a linear $H_{c2}(T)$ behaviour up to 9 T. The slopes of -2.2 T/K for Rb_3C_{60} and -1.7 T/K for K_2CsC_{60} are two orders of magnitude higher than values for graphite intercalation compounds.

For $T \to 0$ K the WHH-theory is applied for the determination of $\mu_0 H_{c2}(0) = 31$ T for Rb_3C_{60} and 21 T for K_2CsC_{60}. Using $\mu_0 H_{c2}(0) = \Phi_0/2\pi\xi^2(0)$ leads to $\xi(0) = 32$ Å for Rb_3C_{60} and $\xi(0) = 40$ Å for K_2CsC_{60}.

It is shown in [41] that the slope of the upper critical field is related to the density of states $N(E_F)$. The monotonous decrease of the slope for different compounds with T_c is consistent with the reduction of $N(E_F)$. Furthermore, it is very interesting that from resistivity and $H_{c2}(T)$ measurements a system independent mean free path at $T = 0$ K of approximately 22 Å is evident. This value is very close to the circumference (πd) of C_{60} and displays the on-ball charge localization in C_{60}-based superconductors.

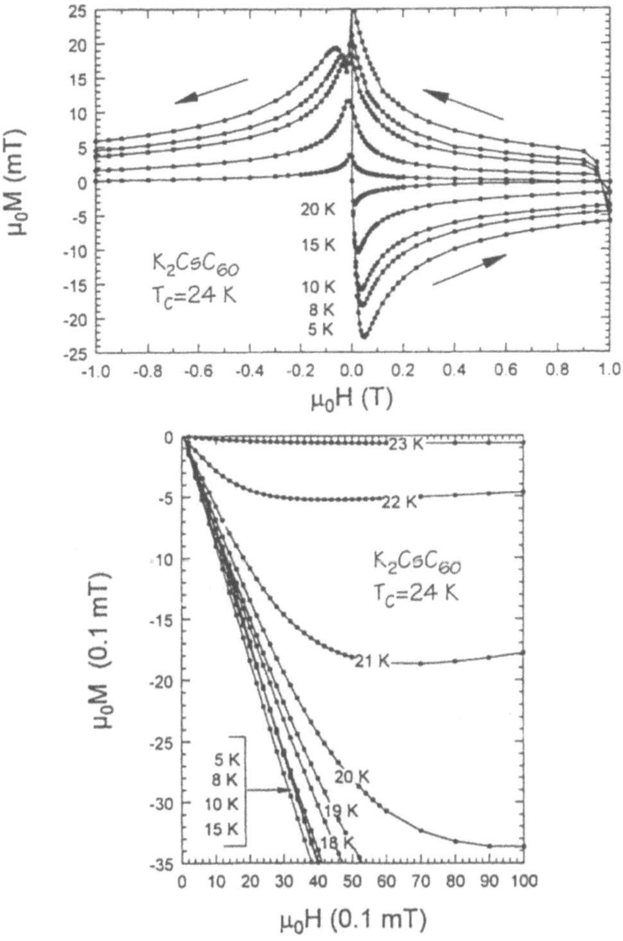

Figure 8
Magnetization of K_2CsC_{60} as a function of field for different temperatures. The arrows indicate increasing and decreasing magnetic fields.

4.2 Lower Critical Fields

In Fig. 7 magnetization loops up to 1 T for several temperatures for Rb_3C_{60} are shown. The large hysteresis indicates the presence of strong pinning centers and no reversible behaviour of the magnetization is found. On an enlarged plot the linear behaviour of the Meissner state magnetization at 5, 8, 10, 15, and 20 K is demonstrated. A deviation from the initial linear behaviour ($\chi = M/H$) of 10% is used as a criterion for the estimation of $H_{c1}(T)$. Fig. 8 shows the same curves for the ternary compound K_2CsC_{60}. Here the enlarged plot confirms linear behaviour for 5, 8, 10, and 15 K. Analysing the magnetization curves yields to $H_{c1}(T)$-curves plotted in Fig. 9. The lower critical field is described well by the parabolic behaviour $H_{c1}(T) = H_{c1}(0)(1-(T/T_c)^n)$ with n close to 2. We found $\mu_0 H_{c1}(0) = 15$ mT for Rb_3C_{60} and 17 mT for K_2CsC_{60}. For Rb_3C_{60} the ratio $H_{c2}(0)/H_{c1}(0)$ is calculated to $2 \cdot 10^3$. From the Ginzburg-Landau theory this ratio is given by $2\kappa^2/\ln\kappa$, where $\kappa = \lambda/\xi$ is the Ginzburg-Landau parameter. For a typical value of 60 for κ of Rb_3C_{60} the ratio is calculated to $1.8 \cdot 10^3$ which is very close to the

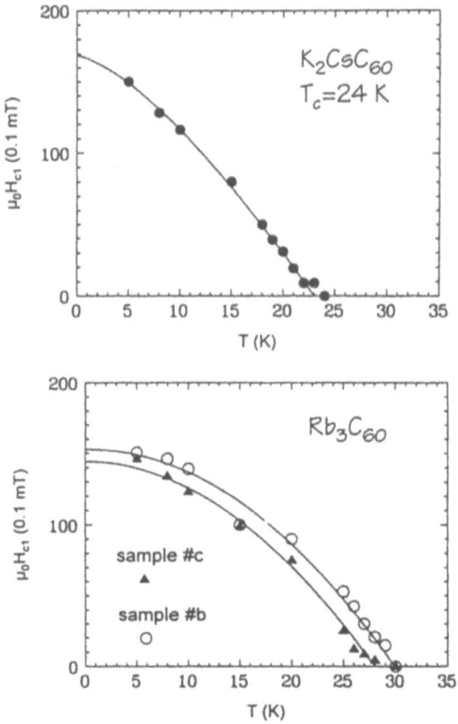

Figure 9
Lower critical field as a function of temperature for K_2CsC_{60} (top) and Rb_3C_{60} (bottom). The lines depict the behaviour $H_{c1}(T) = H_{c1}(0)(1-(T/T_c)^n)$ with $n = 2$.

estimated ratio. From $\kappa \approx 60$ a penetration depth of $\lambda(0) \approx 1860$ Å is calculated. For K_2CsC_{60} we found $\kappa \approx 50$ and $\lambda(0) \approx 2000$ Å.

Finally, it has to be mentioned that for these dirty type-II superconductors all quantities depend strongly on the sample quality. The presence of weak links for example leads to very small H_{c1} values because flux can easily penetrate into the granular material. These values are not related to the intrinsic lower critical field of the material. In the framework of the Ginzburg-Landau theory these small values are inconsistent with H_{c2} measurements.

4.3 Critical Current Densities

Due to the sensitivity of the intercalated fullerenes on air information about the critical current density is obtained from contactless magnetization measurements. One common way is to apply the Bean model for an estimation of $j_c(T,H)$ [42]. For cylindrical samples with the radius R the density j_c is related to the width of the magnetization curve $\Delta M(T,H)$ by: $j_c(T,H) = 1.5 \cdot \Delta M/R$, with $R = 10$ mm. In Fig. 10 the result of this estimation is shown at 5 K. Loop measurements in fields up to 10 T provide the determination of j_c-values over a wide field range. Generally field and temperature dependence of j_c is given by a scaling law with $j_c \propto (1-t^2)^m H^{\gamma-1}$ with $t = T/T_c$ [43,44]. At 5 K we found $\gamma \cong 0.5$ for both compounds. The extrapolated j_c(5 K, 0 T)-value of $7 \cdot 10^9$ A/m^2 is two orders of magnitude smaller than the depairing critical current density $j_{Th}(0) = 4 \cdot 10^{11}$ A/m2 calculated for this compound. Furthermore, the magnitude of the

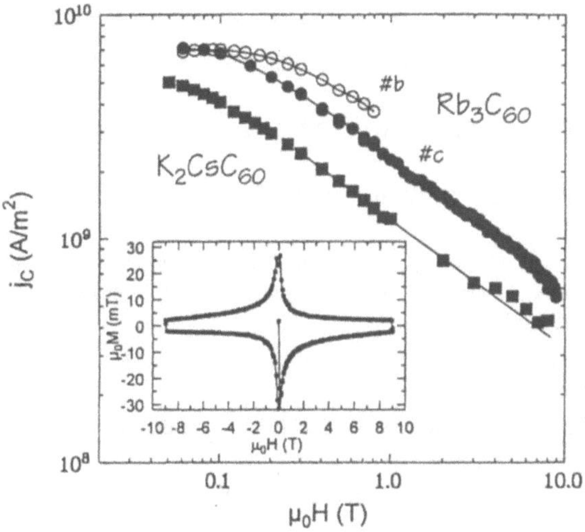

Figure 10
Critical current density for K_2CsC_{60} and Rb_3C_{60} at $T = 5$ K. The inset shows a magnetization loop in fields up to ±9 T for Rb_3C_{60} #c at $T = 5$ K.

critical current densities is comparable to that of NbTi or Nb_3Sn. From the viewpoint of magnetic investigations the class of fullerene superconductors is comparable to conventional A15 material but the transition temperature is approximately 10 K higher. Among the isotopic superconductors the fullerene based superconductors exhibit the highest transition temperatures.

Bibliography

[1] A.F. Hebard, M.J. Rosseinsky, R.C. Haddon, D.W. Murphy, S.H. Glarum, T.T.M. Palstra, A.P. Ramirez, A.R. Kortan, Nature **350**, 600 (1991).

[2] O. Gunnarsson, Rev. Mod. Phys. **69**, 575 (1997).

[3] M. Kanowski, H. Werner, R. SchlŤgl, H.-M. Vieth, K. Lüders, Appl. Magn. Reson. **11**, 285 (1996).

[4] C. H. Pennington, V. A. Stenger, Rev. Mod. Phys. **68**, 855 (1996).

[5] V. Buntar, F.M. Sauerzopf, H.W. Weber, J.E. Fischer, H. Kuzmany, M. Haluska, Phys. Rev., in press.

[6] K. Lüders, Materials Science Forum **191**, 119 (1995).

[7] K. Lüders, in "Chemical Physics of Intercalation II" (P. Bernier et al., Eds.), NATO ASI Series B **305**, Plenum Press, New York and London, p. 31 (1993)

[8] M. Kraus, J. Freytag, S. Gärtner, H.M. Vieth, W. Krätschmer, K. Lüders, Z. Phys. B - Condensed Matter **85**, 1 (1991).

[9] M. Baenitz, M. Kraus, S. Gärtner, H.M. Vieth, W. Krätschmer, K. Lüders, Springer Series in Solid-State Sciences **113**, 475 (1993).

[10] S.P. Kelty, C.-C. Chen, C.M. Lieber, Nature **352**, 223 (1991).

[11] K. Lüders. M. Baenitz, Recent Advances in the Chemistry and Physics of Fullerenes and Related Materials (K.M. Kadish, R.S. Ruoff, Eds.) **5**, 510 (1997).

[12] K. Lüders, M. Kraus-Sartori, M. Baenitz, V. Thommen-Geiser, D. Herein, to be published.

[13] W. Krätschmer, D.L. Lamb, K. Fostiropoulos, D.R. Huffman, Nature **347**, 354 (1990).

[14] M. Baenitz, M. Heinze, K. Lüders, H. Werner, R. SchlŤgl, Mat. Res. Soc. Symp. Proc. **349**, 301 (1994).

[15] T.T.M. Palstra, O.Zhou, Y. Iwasa, P.E. Sulewski, R.M. Fleming, B.R. Zegarski, Solid State Commun. **93**, 327 (1995).

[16] M. Baenitz, M. Heinze, K. Lüders, H. Werner, R. SchlŤgl, Solid State Commun. **91**, 337 (1994).

[17] T. Yildirim, L. Barbedette, J.E. Fischer, C.L. Lin, P.W. Stephens, P.E. Sulewski, in "Physics and Chemistry of Fullerenes and Derviates", H. Kuzmany, J. Fink, M. Mehring, S. Roth (eds.), World Scientific, Singapore p.379 (1995).

[18] A.R. Kortan, N. Kopylov, S. Glarum, E. M. Gyorgy, A. P. Ramirez, R. M. Fleming, F. A. Thiel, R. C. Haddon, Nature **355**, 529(1992).

[19] M. Baenitz, M. Heinze, K. Lüders, H. Werner, R. SchlŤgl, M. Weiden, G. Sparn, F. Steglich, Solid State Commun. **96**, 539 (1995).

[20] A.R. Kortan, N. Kopylov, E. Źzdas, A. P. Ramirez, R. M. Fleming, R. C. Haddon, Chem. Phys. Lett. **223**, 501(1994).

[21] T. Yildirim, L. Barbedette, J. E. Fischer, C. L. Lin, J. Robert, P. Petit, T. T. Palstra, Phys. Rev. Lett. **77**, 167 (1996).

[22] Y. Isawa, H. Hayashi, T. Furudate, T. Mitani, Phys. Rev. B **54**, 14960 (1996).

[23] E. Źzdas, A.R. Kortan, N. Kopylov, A.P. Ramirez, T. Siegrist, K.M. Rabe, H.E. Bair, S. Schuppler, P.H. Citrin, Nature **375**, 126 (1995).

[24] X. H. Chen, G. Roth, Phys. Rev. B 52, 15534 (1995).

[25] O. Zhou, R.M. Fleming, D.W. Murphy, M.J. Rosseinsky, A.P. Ramirez, R.B. van Dover, R.C. Haddon, Nature **362**, 433 (1993).

[26] M.J. Rosseinsky, D.W. Murphy, R.M. Fleming, O. Zhou, Nature **364**, 425 (1993).

[27] I.I. Khairullin, K. Imaeda, K. Yakushi, H. Inokuchi, Physica C **231**, 26 (1994).

[28] D. Schoenberg, *Superconductivity*, Cambridge Press, (1960).

[29] M. Baenitz, M. Heinze, E. Straube, H. Werner, R. SchlŤgl, V. Thommen, H.-J. Güntherodt, K. Lüders, Physica C **228**, 181 (1994).

[30] M. Baenitz, E. Straube, M. Kraus, M. Kanowski, H. Werner, R. SchlŤgl, K. Lüders, Springer Series in Solid-State Sciences **117**, 387 (1993).

[31] M. Baenitz, Thesis, Freie Universität Berlin, Fachbereich Physik, Germany, (1995).

[32] R.M. Fleming, A. P. Ramirez, M.J. Rosseinsky, D.W. Murphy, R.C. Haddon, S. M. Zahurak, A.V. Makhija, Nature **352**, 787 (1991).

[33] A.R. Kortan, N. Kopylov, S. Glarum, E. M. Gyorgy, A. P. Ramirez, R. M. Fleming,O. Zhou, F. A. Thiel, P. L. Trevor, R. C. Haddon, Nature **360**, 566 (1992).

[34] G.S. Boebinger, T.T.M. Palstra, A. Passner, M.J. Rosseinsky, D.W. Murphy, I.I. Mazin, Phys. Rev. B **46**, 5876 (1992).

[35] A. Chaiken, M.S. Dresselhaus, T.P. Orlando, G. Dresselhaus, P.M. Tedrow, D.A. Neumann, W.A. Kamitakahara, Phys. Rev. B. **41**, 71 (1990).

[36] G. Sparn, F. Laube, A. Link, F. Steglich, M. Baenitz, K. Lüders, H. Werner, R. SchlŤgl, J. Low Temp. Phys. **105**, 1703 (1996).

[37] G. Sparn, Habilitation thesis, Johann-Wolfgang-Goethe University, Fankfurt/Main, Germany, 1996

[38] P. Jess, U. Hubler, H.P. Lang, H.-J. Güntherodt, K. Lüders, H. Werner, R. SchlŤgl, Proc. Int. Winterschool on Electric Properties of Novel Materials, Fullerenes and Fullerene Nanostructures (H. Kuzmany et al., eds.) 1996, p. 426;

[39] P. Jess, U. Hubler, H.P. Lang, H.-J. Güntherodt, H. Werner. R. SchlŤgl, K. Lüders, J. Phys. Chem. Solids **58**, 1803 (1997).

[40] V. Buntar, H. W. Weber. Supercond. Sci. Technol. **9**, 599 (1996).

[41] M. Heinze, M. Baenitz, K. Lüders, H. Werner, R. SchlŤgl, Synthetic Metals, **77**, 23 (1996).

[42] C.P. Bean, Phys. Rev. Letter **8**, 250 (1962).

[43] N. Sakamoto, T. Akune, Y. Matsumoto, E. S. Otabe, T. Matsushita, M. Baenitz, K. Lüders, Jpn. J. Appl. Phys. **34**, L1267 (1995)

[44] N. Sakamoto, Y. Ide, T. Akune, M. Baenitz, K. Lüders, Proc. 16.th Cryogenic Engineering Conf. ICEC 16, Elsevier Science 1561 (1997).

Heteroepitaxy and Noise in Integrated HTS dc-SQUID Magnetometers

Meinhard Schilling

Universität Hamburg, Institut für Angewandte Physik und
Zentrum für Mikrostrukturforschung,
Jungiusstraße 11, D-20355 Hamburg, Germany

Abstract: Magnetometers based on superconducting quantum interference devices and made from high-temperature superconductors today can be realized in $YBa_2Cu_3O_7$ multilayer technology with up to five epitaxial layers. Such low-noise, integrated magnetometers can be employed in multichannel systems for medical applications. For multilayer devices, the thin-film preparation has to provide high homogeneity and reproducibility of the superconducting and morphological properties. For process control we use statistical methods of experimental design. The magnetic flux-density noise of magnetometers is ruled by fluctuations of the critical current in the Josephson junctions and by the motion of vortices in the $YBa_2Cu_3O_7$ films. The local heteroepitaxy of $YBa_2Cu_3O_7$/ $PrBa_2Cu_3O_7$/$SrTiO_3$-layers is of crucial importance for low flux noise, as demonstrated by Micro-Raman-Spectroscopy combined with noise measurements.

1 Introduction

In the 11 years since the discovery of the high-temperature superconductors (HTS) [1] the growth of thin films from $YBa_2Cu_3O_7$ has been investigated by many groups for the use in superconducting devices. Starting from grain-boundaries as barriers in HTS-Josephson junctions, techniques for the controlled preparation of artificial grain-boundaries have been developed [2]. But the use of artificial barrier materials in HTS-Josephson junctions promises a better control of their electronic properties. Taking the anisotropic nature of high-temperature superconductivity into account, *c*-axis oriented thin films are used and non-superconducting $PrBa_2Cu_3O_7$ can be used as epitaxial barrier material in Josephson junctions in ramp-edge geometry [3]. Many other suitable barrier-materials have been found.

Superconducting quantum interference devices (SQUIDs) are among the most promising applications of active superconducting electronics. Although very impressive results have been obtained for rf-SQUIDs [4], most groups work on dc-SQUID magnetometers [5]. The possible applications cover biomagnetism, nondestructive testing, geophysical measurements and many other areas. In all these applications, SQUIDs are used due to their extraordinary low-noise in the measurement of very weak magnetic signals. Most groups have used grain-boundary Josephson junctions for SQUID-magnetometers [6], but also low-noise SQUID-magnetometers using $YBa_2Cu_3O_7$ Josephson junctions with $PrBa_2Cu_3O_7$ barriers in ramp-edge geometry can be prepared [7]. The technology with

$PrBa_2Cu_3O_7$ as barrier material is currently investigated to increase the yield of low-noise Josephson junctions [8].

The most sensitive sensors are achieved, if both low noise Josephson junctions and very effective coupling of the external flux density to the flux in the SQUID are combined [9]. The most effective flux-collection antennas require epitaxial multilayer structures to connect thin films in different layers. Thus, the heteroepitaxial growth of HTS-multilayers is investigated by many groups [10].

Two main contributions to the noise of SQUID sensors can be distinguished: apparent flux noise due to fluctuations of the critical current in the Josephson junctions and flux noise due to vortex motion in the superconducting films. While the microscopic source of critical current fluctuations $\delta I_c/I_c$ is currently under investigation [11], the noise due to vortex motion has been investigated in detail [12]. Here, the influence of the epitaxial film quality on the flux noise of $YBa_2Cu_3O_7$ films is discussed, since the pinning forces for vortices are significantly lowered along grain-boundaries between a/b-axis oriented grains and *c*-axis oriented grains in the film. The influence of film quality on the noise properties of flip-chip magnetometers is shown in [10], and the local epitaxy analysis further clarified the prepositions for low-noise integrated magnetometers [13].

2 Multilayer Preparation

2.1 Preparation of $YBa_2Cu_3O_7/SrTiO_3$ Bilayers

Heteroepitaxy starts with the growth of $YBa_2Cu_3O_7$ on $SrTiO_3$ substrates. We choose $SrTiO_3$ because the crystal quality and the polishing techniques are rather advanced for this perovskite material. The lattice mismatch to $YBa_2Cu_3O_7$ is even lower for $NdGaO_3$, but not so much experience in the crystal preparation exists for this material today. We use (100)-$SrTiO_3$ substrates with dimensions $10 \times 10 \times 1$ mm^3. The twin-free substrates are specified by the manufacturer to be oriented with a tolerance below 0.5° and ±2° deviation in the orientation of the edges. They are polished for epitaxial growth on one side and specified with a flatness of 1 μm on 10×10 mm^2 with a rms-roughness of below 2.5 nm.

For film deposition we use a KrF-excimer laser deposition process. This process has been optimized by the use of statistical methods of experimental design and is described in detail in [14]. This way, we achieved a very stable process for droplet-free $YBa_2Cu_3O_7$ films with very good superconducting properties as $T_c \cong 89$ K, resistivity $\rho(100\text{ K}) \leq 60\ \mu\Omega$cm and a relative resistance ratio $R(300\text{ K})/R(100\text{ K}) \cong 3$. The density of outgrowths is low enough to realize integrated magnetometer chips of high quality.

For $YBa_2Cu_3O_7$-$SrTiO_3$ bilayers with 200 nm $YBa_2Cu_3O_7$ and 50 nm $SrTiO_3$ we achieve a rms-surface roughness of 1.0 nm which is already below one lattice constant of the $YBa_2Cu_3O_7$ film, as shown in Fig. 1. For comparison, we include in this figure an atomic force microscopy picture of a GaAs/GaAlAs heterostructure grown by molecular beam epitaxy with a rms-roughness of 0.2 nm measured with the same set-up in the same lateral resolution. Here, the roughness is about 0.35 of the lattice constant of GaAs. For the comparison one has to keep in mind, that in the case of $YBa_2Cu_3O_7/SrTiO_3$ the lattice mismatch to the $SrTiO_3$ substrate is above 1% with a Stranski-Krastanov growth mode [15]. In GaAs/GaAlAs, on the other hand, two-dimensional layer-by-layer growth is obtained due to perfect lattice matching.

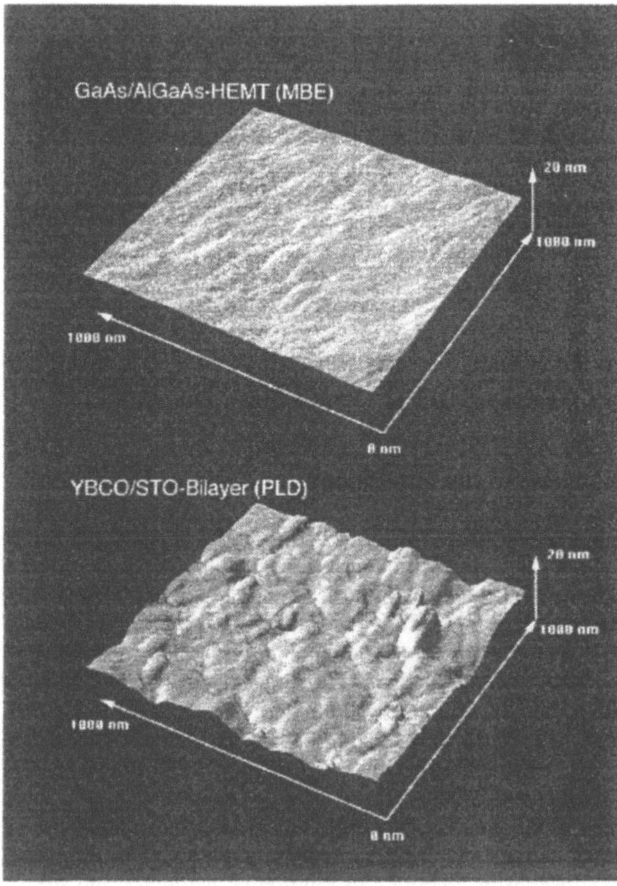

Figure 1
Comparison of the surface roughness of a GaAs/GaAlAs heterostructure grown by molecular beam epitaxy and a $YBa_2Cu_3O_7$-$SrTiO_3$ bilayer grown in our multilayer laser deposition process.

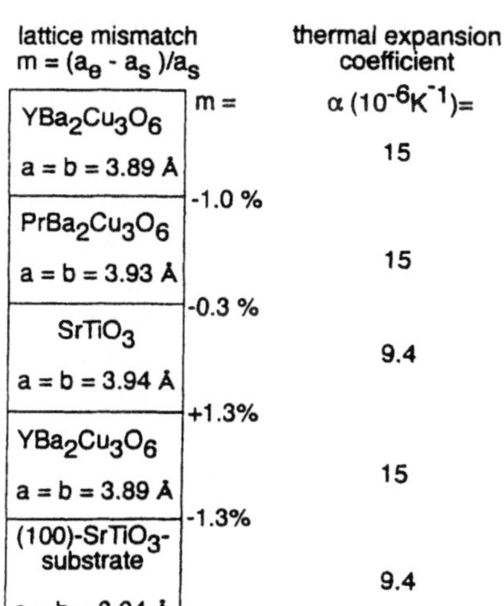

Figure 2
Heteroepitaxial conditions for the growth of multilayers at 800 °C on (100)-$SrTiO_3$ substrates.

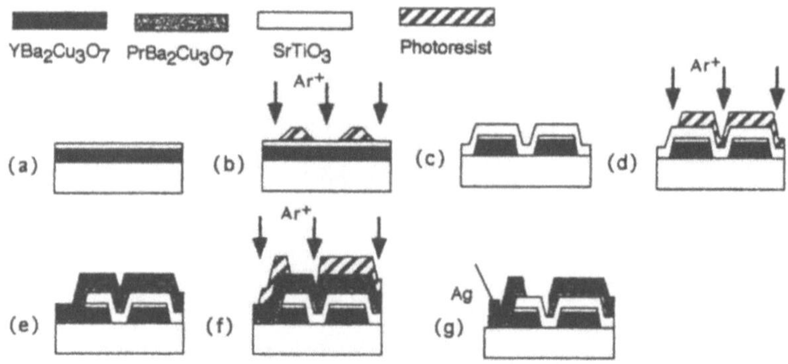

Figure 3 Multilayer process employed for the preparation of multilayer integrated magnetometers. The left hand side in each picture shows the preparation of a Josephson junction and at the right hand side the formation of an insulating crossover can be seen.

2.2 Multilayer Growth, Epitaxy and Patterning

For HTS-Josephson junctions with $PrBa_2Cu_3O_7$ as barrier material in ramp-edge geometry, the complete stack of heteroepitaxial films in our magnetometers is shown in Fig. 2. After the deposition of the first $YBa_2Cu_3O_7$ film of 200 nm thickness, $SrTiO_3$ layers are necessary to realize insulating cross-overs. Here we employ a 20 nm thin $SrTiO_3$ cap layer [10] and after patterning this bilayer by conventional photolithography and argon plasma etching a second, thicker $SrTiO_3$ insulation layer (100 nm) is deposited. On top follow the $PrBa_2Cu_3O_7$ barrier with a thickness between 10 nm and 100 nm and the counter electrode of $YBa_2Cu_3O_7$ with 100 nm thickness. Both, $PrBa_2Cu_3O_7$ and $YBa_2Cu_3O_7$ grow tetragonal at about 800 °C. The main lattice mismatch of $m = 1.3\%$ can be found at the interface between $YBa_2Cu_3O_7$ and $SrTiO_3$.

In Fig. 3, the deposition and patterning processes involved in the preparation of integrated magnetometers are schematically depicted. In this figure, both the successive steps in the preparation of the Josephson junctions in ramp-edge geometry and the insulating crossovers are shown. Starting with Fig. 3a the bottom superconducting electrode is covered by the insulating $SrTiO_3$ cap layer. In Fig. 3b the bottom lines of $YBa_2Cu_3O_7$ are defined by conventional photolithography and transferred into the superconductor by argon-ion etching, as described below. Afterwards these lines are covered (Fig. 3c) with the thicker $SrTiO_3$ insulation layer. In Fig. 3d again a photoresist edge is defined by photolithography. Etching is performed in a parallel-plate argon-plasma reactor. The angle of the photoresist edge after development and postbake influences the resulting ramp-edge angle and is transferred into the insulator and the bottom superconductor electrode. Damage can ocurr at the superconductor surface by amorphization during physical plasma etching [16]. To minimize such damage, we now employ acceleration voltages below 200 V in plasma etching. Oxygen loss at the superconductor surface can be replenished by a treatment in an oxygen plasma [17]. The ramp-edge angles of the different materials are controlled by argon-pressure and rf-power and can be described in a simple model for the edge profile of multilayers. In the parallel-plate reactor the resulting ramp-edge angle for each material α_{mat} only depends on the photoresist angle α_{pr} and the etch rates of the material r_{mat} and the photoresist r_{pr} via the relation

$$\alpha_{mat} = \arctan[\frac{r_{mat}}{r_{pr}} \tan(\alpha_{pr})].$$

Because the relative etch rates can be changed by varying the argon-pressure and rf-power, the ramp-edge angle is precisely controllable. For ion-gun etching the angle of incidence of the ions has also to be taken into account. The edge angle has to be chosen small enough to allow epitaxial, *c*-axis oriented growth of the barrier material and the top $YBa_2Cu_3O_7$ electrode [18–20]. In particular, this avoids serial grain boundaries on the ramp-edge. We produce small ramp-edge angles of $\alpha = 2 - 25°$ in the bottom $YBa_2Cu_3O_7$ electrode [8]. The ramp-edge angle can be measured ex situ by atomic force microscopy before barrier layer and top superconducting electrode are deposited. We determined a ramp-edge angle of 15° for the typical photoresist treatment and plasma-etching process. This is not done for each chip, because this ex situ measurement is expected to affect the electrical properties of the Josephson junctions.

In the next step, the barrier material is deposited. Its thickness is controlled by deposition rate and time. We assume that the higher growth velocity along the *a-/b*-direction of $PrBa_2Cu_3O_7$ does not influence significantly the growth on the ramp, since smooth overgrowth of the ramp is observed with atomic force microscopy. The microscopic growth mechanism of $PrBa_2Cu_3O_7$ on the $YBa_2Cu_3O_7$ ramp with a physically etched surface is currently under investigation.

Subsequently, the top superconducting electrode is deposited in situ. The following photolithographic and etching steps define the width of the Josephson junctions. Here, a small overlap region of the top electrode is necessary for mask alignment during photolithography. A good epitaxy in the overlap region and on the ramp is crucial, since flux trapped therein can induce an inhomogeneous current distribution in the Josephson junction [21].

Contacts to the Josephson junctions are made by soldering with indium or by bonding to evaporated metal pads. For gold-ball bonding, we observed a strong influence of the ultrasonic surface waves on patterned $YBa_2Cu_3O_7$ films which even led to the destruction of devices [22]. On the contrary, with aluminum-wedge bonding no significant changes of the junction properties have been found. Consequently, this method is now routinely employed.

For the process optimization the preparation of integrated magnetometers was devided in statistically independent subprocesses, as depicted in Table 1. Some results of the complete process optimization by experimental design methods already have been published. The optimization of the $YBa_2Cu_3O_7$ film can be found in [14] and results on the photolithography are included in [8].

3 Micro-Raman-Spectroscopy

After the first patterning process, the conditions for the epitaxy of the subsequent layers change across the chip. On the etched areas the surface is rather different from the areas which have been covered by photoresist. This implies, that the growth conditions for the overgrowth after patterning have to be a compromise to obtain the best local epitaxy as possible on the different surfaces. As a tool to characterize the *local* epitaxy in patterned multilayer devices we use micro-Raman-spectrocopy. This technique can be combined with scanning laser microscopy [23] to analyze the local superconducting properties and to compare them to the epitaxial properties in the same set-up. Also, the comparison with scanning electron microscopy reveals the very good agreement between the results of both methods [24]. An important advantage of the Raman-spectroscopy is that the

technology	*subprocesses*	*requirements*
target preparation	pressing sintering	high density, smoothness, homogeneity
film preparation	1. $YBa_2Cu_3O_7$	epitaxy, superconductivity, morphology
	$SrTiO_3$-cap	epitaxy, insulation, morphology
	$SrTiO_3$-insulation	epitaxy, insulation (edges), morphology
	2. $YBa_2Cu_3O_7$	epitaxy, superconductivity, morphology
photo-lithography	bottom layers	low angle resist ramps, dimensional stability
	insulation layer	low angle resist ramps, dimensional stability, no damage
	top layers	steep angle resist ramps. (dimensional stability)
Argon-etching	bottom layers	low angle ramps ($epitaxy SrTiO_3$, orientation)
	insulation layer	low angle ramps (epitaxy $YBa_2Cu_3O_7$, orientation)
	top layers	steep angle ramps (dimensional stability)
metallisation	Ag-evaporation	Ag-adherence, dimensional stability
contacts	bonding	no ultrasonic damage, adherence

Table 1 Contributing subprocesses in the preparation of multilayer devices.

inelastic scattered light comes from a very small volume: The laser spot is focused down to 1 μm diameter and the optical penetration depth into the $YBa_2Cu_3O_7$ film is about 70 nm [25]. By investigating the polarization dependence of the inelastically scattered light the orientation of the film can be evaluated from the relative intensities in the phonon related peaks. This is based on the extremely anisotropic Raman-tensors of the various Raman active modes being observed in $YBa_2Cu_3O_7$. We concentrate on the two strongest modes, namely the O(2)-O(3) mode at 340 cm^{-1}, which is due to the out-of-phase vibration of the oxygen atoms of the CuO_2-planes, and the O(4) mode at about 500 cm^{-1} which is related to the vibration of the apex oygen atom [25]. From these measurements three parameters can be evaluated: the amount of *c*-axis oriented material in the film δ_c, where we assume that the remaining material is purely *a*-/*b*-axis oriented, the degree of in-plane orientation with respect to the substrate for the *c*-axis oriented (Q_c) and the *a*-axis oriented fraction (Q_a) of the film. The analysis has been worked out for $YBa_2Cu_3O_7$ and $PrBa_2Cu_3O_7$ [26].

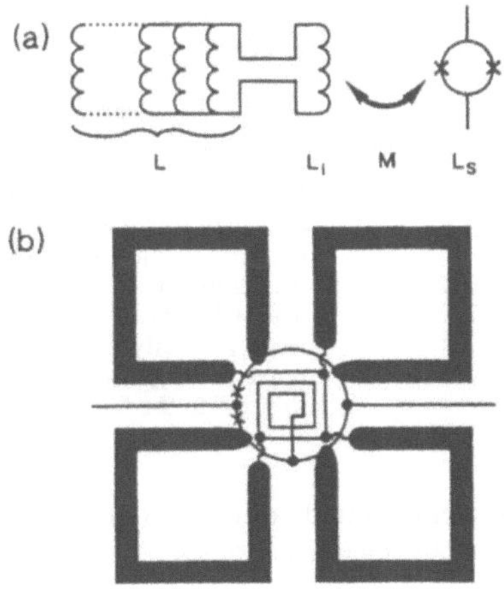

Figure 4
Scheme of the integrated multiloop pick-up coil magnetometer.

4 Magnetometer Concept

Our multilayer deposition process is employed for the fabrication of various integrated dc-SQUID magnetometers [27]. We have combined the epitaxial analysis by micro-Raman-spectroscopy with measurements of the flux-noise in integrated magnetometers to clarify the microscopic origins of the flux noise in these devices.

The magnetometer investigated here is a concept where the inductance matching between the pick-up coil and the dc-SQUID is possible by two means: the connection of several loops in parallel for the pick-up loop and the mutual inductance in the flux-transformer. To combine this way the advantages of the multiloop magnetometer [28] with the flexibility of inductance matching by a Ketchen-type magnetometer [29], we developed a new concept for a magnetometer with an integrated multiloop pick-up coil (IMPUC) [7]. In this layout, large pick-up coils are possible to couple to low inductance SQUIDs to achieve large effective areas at 77 K despite the low SQUID inductance needed for operation at high temperatures. This is shown schematically in Fig. 4.

The calculation of the inductance L of the parallel pick-up loops is analogous to the multiloop dc-SQUID [28]. In the expression $L = L_p/N^2 + L_s/N$ one has the inductance contribution L_p of the circumference, the number of parallel loops N, and the inductance contribution of the spokes L_s. The effective area of the magnetometer ($L_i = L$) is given by [7]

$$A_{\text{eff}} = A_{\text{SQ}} + \frac{k}{2}\sqrt{\frac{L_{\text{SQ}}}{L}}A_p$$

analogous to the flux transformer coupled to the dc-SQUID [29]. Due to the lowered inductance of the pick-up coils, only a few turns are needed in the input-coil of the flux-transformer allowing small area square washers. So, the parasitic inductance contribution of the slit is reduced significantly, and the LC-resonances and strip-line resonances can be shifted in frequency out of the region of the voltage modulation. Resonances due to galvanic coupling of the pick-up coil to the dc-SQUID have to be damped separately [30]. With a SQUID inductance of 90 pH and 5 turns of the input coil at 77 K we obtain an effective area of 1.4 mm^2. The dependence of the effective area on the number of loops and on the temperature has been investigated in detail [31].

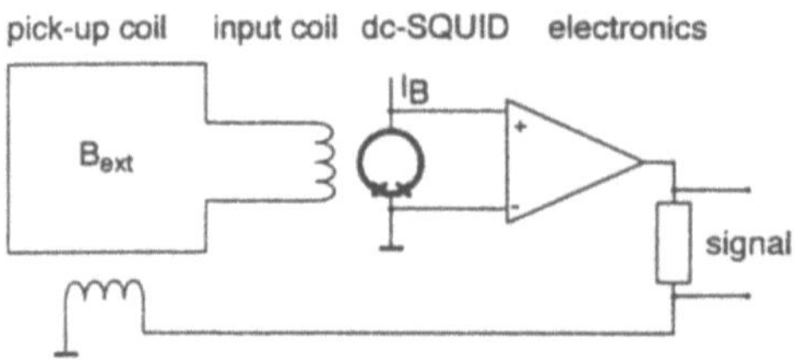

Figure 5
Principle of the operation of a dc-SQUID magnetometer in a flux-locked loop. The pick-up coil and the input coil are made from superconducting material to utilize flux quantization in superconducting rings for optimum flux transfer in to the SQUID. The voltage signal of the SQUID is galvanically coupled to a low noise preamplifier and integrator to provide the feedback current.

Fig. 5 shows such a dc-SQUID magnetometer operated in flux-locked loop to increase the dynamic range and to linearize the flux-voltage characteristics.

5 Heteroepitaxy and Noise

Noise measurements on dc-SQUIDs are typically made in a flux-locked loop. Flux modulation already suppresses the in-phase critical current fluctuations of both junctions in a dc-SQUID [32]. By using an ac-bias current also the out-of-phase critical current fluctuations are suppressed. Remaining low-frequency noise results from vortex motion in the superconducting films or is due to external disturbances. The low-frequency noise properties of two IMPUC-magnetometers with similiar electrical properties are shown in Fig. 6.We find that magnetometer R384 allows a suppression of the low-frequency $1/f$-noise by the use of ac-bias current, for R385 ac-bias current has no effect on the noise spectrum. Since external perturbations can be excluded here, the $1/f$-noise results from vortex motion in the films constituting the magnetometer R385.

A subsequent detailed analysis of the epitaxy with micro-Raman spectroscopy demonstrated the correlation of epitaxy and low-frequency noise properties [13]. The epitaxial properties of both magnetometers were measured first for the top $YBa_2Cu_3O_7$ films and then these layers were removed in our argon-ion etching process. Afterwards the epitaxy of the bottom $YBa_2Cu_3O_7$ films was determined. The results are summarized in Table 2. The micro-Raman analysis has revealed a good epitaxial quality of the bottom $YBa_2Cu_3O_7$ films of both magnetometers and a good epitaxial quality of the top $YBa_2Cu_3O_7/PrBa_2Cu_3O_7$ bilayer where it grows directly on the etched $SrTiO_3$ substrate. These results were obtained in the via-contacts, where the top $YBa_2Cu_3O_7$ grows on the etched substrate surface. But the IMPUC-magnetometer with the high $1/f$-noise exhibits a deteriorated epitaxial quality in the top $YBa_2Cu_3O_7$ layer where it grows on top of the $SrTiO_3/YBa_2Cu_3O_7$-bilayer, as obtained from the measurements of the lines of the input-coil. The different noise properties of both devices are related to the deposition parameters of the $SrTiO_3$-layer, which were varied in this experiment. Thus, for low-noise integrated magnetometers it is crucial to obtain a high and homogeneous epitaxy with high reproducability.

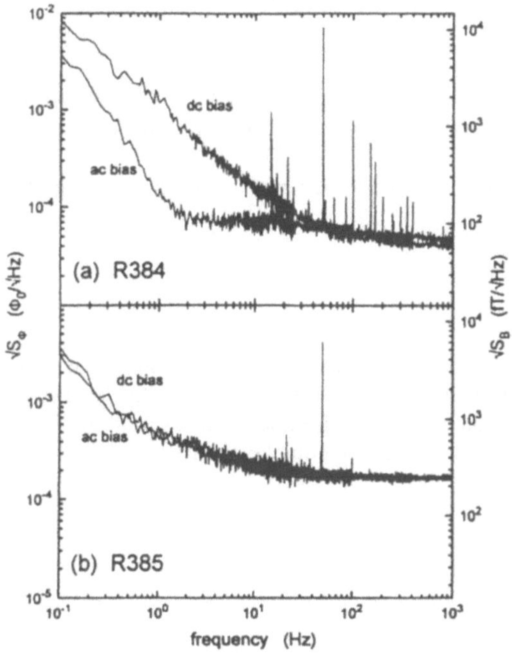

Figure 6
Noise properties at 77 K of two IMPUC-magnetometers with similiar electrical properties but different epitaxial quality of the top $YBa_2Cu_3O_7$ film.

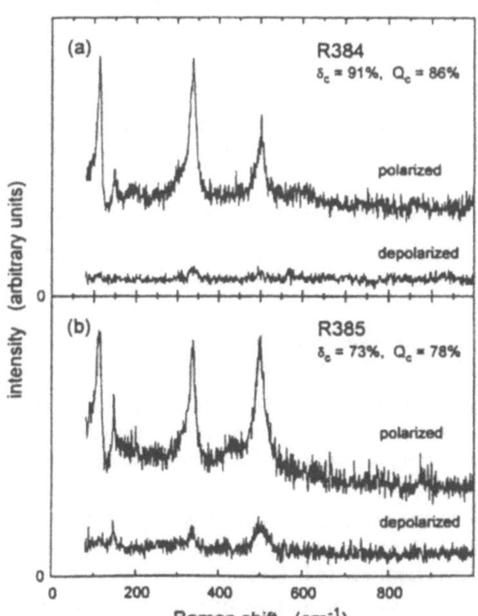

Figure 7
Micro-Raman characterization of the two IMPUC-magnetometers.

sample	p(Pa)	ac-bias	site	δ_c(bottom)	Q_c (bottom)
R384	40	$1/f$-noise	coil	91% (100%)	86% (90%)
		suppression	via	93% (94%)	84% (93%)
R385	5	no $1/f$-noise	coil	73% (95%)	78% (98%)
		suppression	via	94% (97%)	84% (93%)

Table 2 Results of the micro-Raman characterization of the two IMPUC-magnetometers.

6 Conclusions

By micro-Raman-spectroscopy in comparison to noise measurements on two IMPUC magnetometers we identified the *local* epitaxy of the top $YBa_2Cu_3O_7$ layer in integrated dc-SQUID magnetometers as a source of flux-noise due to vortex motion. For low noise properties, the epitaxy in all involved superconducting layers has to be close to perfect to avoid such flux-noise contributions.

Bibliography

[1] J.G. Bednorz and K.A. Müller, *Possible High T_cSuperconductivity in the Ba - La - Cu - O System*, Z. Phys. B. **64**, 189 (1986).

[2] R. Gross, L. Alff, A. Beck, O.M. Froehlich, D. Koelle, and A. Marx, *Physics and technology of high temperature superconducting Josephson junctions*, IEEE Trans. Appl. Supercond. **7**, 2929 (1997).

[3] J. Gao, W.A.M. Aarnink, G.J. Gerritsma, D. Veldhuis, and H. Rogalla, *Preparation and properties of all high T_c SNS-type edge DC SQUIDs*, IEEE Trans. Magn. **27**, 3062 (1991),

[4] Y. Zhang, H. Soltner, N. Wolters, W. Zander, J. Schubert, M. Banzet, and A.I. Braginski, *HTS rf SQUIDs with fully integrated planar tank circuits*, IEEE Trans. Appl. Supercond. **7**, 2870 (1997).

[5] M. Schilling, *HTS dc-SQUID magnetometers*, Applied Superconductivity 1998, in press.

[6] F. Ludwig, E. Dantsker, R. Kleiner, D. Koelle, J. Clarke, S. Knappe, D. Drung, H. Koch, N. McN. Alford, and T.W. Button, *Integrated high-T_c multiloop magnetometer*, Appl. Phys. Lett. **66**, 1418 (1995).

[7] R. Scharnweber and M. Schilling, *Integrated $YBa_2Cu_3O_7$ magnetometer with flux transformer and multiloop pick-up coil*, Appl. Phys. Lett. **69**, 1303 (1996).

[8] K.-O. Subke, S. Krey, H. Burkhardt, D. Reimer, and M. Schilling, *Process optimization for high yield of $YBa_2Cu_3O_7$ ramp-edge Josephson junctions with $PrBa_2Cu_3O_7$ barriers*, Extended Abstracts of the International Superconductive Electronics Conference, Berlin 1997, Vol 2, pp. 88 - 90 (1997).

[9] D. Drung, F. Ludwig, W. Müller, U. Steinhoff, L. Trahms, H. Koch, Y.Q.Shen, M.B. Jensen, P. Vase, T. Holst, T. Freltoft, and G. Curio, *Integrated $YBa_2Cu_3O_{7-x}$ magnetometer for biomagnetic measurements*, Appl. Phys. Lett. **68**, 1421 (1996).

[10] F. Ludwig, D. Koelle, E. Dantsker, D.T. Nemeth, A.H. Miklich, J. Clarke and R.E. Thomson, *Low Noise $YBa_2Cu_3O_{7-x}$ - $SrTiO_3$ - $YBa_2Cu_3O_{7-x}$ multilayers for improved superconducting magnetometers*, Appl. Phys. Lett. **66**, 373 (1995).

[11] A. Marx and R. Gross, *Scaling behavior of 1/f noise in high temperature superconductor Josephson junctions*, Appl. Phys. Lett. **70**, 120 (1997).

[12] M.J. Ferrari, M. Johnson, F.C. Wellstood, J.J. Kingston, T.J. Shaw, and J. Clarke, *Magnetic flux noise in copper oxide superconductors*, J. Low. Temp. Phys. **94**, 15 (1994).

[13] R. Scharnweber, N. Dieckmann, and M. Schilling, *Noise properties and epitaxial quality in low-noise integrated $YBa_2Cu_3O_7$ magnetometers*, Appl. Phys. Lett. **70**, 2189 (1997).

[14] J.K. Heinsohn, D. Reimer, A. Richter, K.O.-Subke, and M. Schilling, *Interaction of process parameters in the laser deposition of $YBa_2Cu_3O_7$ films*, Physica C 1998, in press.

[15] V. Dediu, A. Kursumovic, O. Greco, F. Biscarini, and F.C. Matacotta, *Observation of the Strnnski-Krastanov growth transition in $GdBa_2Cu_3O_{7-\delta}$ films*, Phys. Rev. B **54**, 1564 (1996).

[16] C.L. Jia, M.I. Faley, U. Poppe, and K. URBAN, *Effect of chemical and ion-beam etching on the atomic structure of interfaces in $YBa_2Cu_3O_7$ / $PrBa_2Cu_3O_7$ Josephson junctions*, Appl. Phys. Lett. **67**, 3635 (1995).

[17] M. Schilling, D. Reimer, and U. Merkt, *$YBa_2Cu_3O_7$ direct current-superconducting quantum interference devices with artificial $PrBa_2Cu_3O_7$ barriers above 77 K*, Appl. Phys. Lett. **64**, 2584 (1994).

[18] C.L. Jia, B. Kabius, K. Urban, K. Herrman, G.J. Cui, J. Schubert, W. Zander, A.I. Braginski, and C. Heiden, *Microstructure of epitaxial $YBa_2Cu_3O_7$ films on step-edge $SrTiO_3$ substrates*, Physica C, **175**, 545 (1991).

[19] E. Polturak, G. Koren, D. Cohen, E. Aharoni, and G. Deutscher, *Proximity effect in $YBa_2Cu_3O_7$ / $Y_{0.6}Pr_{0.4}Ba_2Cu_3O_7$ / $YBa_2Cu_3O_7$ junctions*, Phys. Rev. Lett. **67**, 3038 (1991).

[20] C. Stölzel, M. Siegel, G. Adrian, C. Krimmer, J. Söllner, W. Wilkens, G. Schulz, and H. Adrian, *Transport properties of $YBa_2Cu_3O_{7-\delta}$ / $Y_{0.3}Pr_{0.7}Ba_2Cu_3O_{7-\delta}$ / $YBa_2Cu_3O_{7-\delta}$ Josephson junctions*, Appl. Phys. Lett. **63**, 2970 (1993).

[21] A. Marx, K.-D. Husemann, B. Mayer, T. Nissel, R. Gross, M.A.J. Verhoeven and G.J. Gerritsma, *Distribution of the critical current density and flux trapping in $YBa_2Cu_3O_{7-\delta}$ ramp-edge Josephson junctions*, Appl. Phys. Lett. **64**, 241 (1994).

[22] L. Burmeister, D. Reimer, and M. Schilling, *Thermosonic bond contacts with gold wire to $YBa_2Cu_3O_7$ microstructures*, Supercond. Sci. Technol. **7**, 569 (1994).

[23] N. Dieckmann, A. Bock, and U. Merkt, *Spatially resolved analyses of epitaxial and electrical properties of $YBa_2Cu_3O_7$ devices*, Appl Phys. Lett. **68**, 3626-3628 (1996).

[24] N. Dieckmann, S. Friemel, A. Bock, U. Merkt, R. Gerber, R.P. Huebener, *Equivalence of LTSLM and LTSEM analyses: study of a $YBa_2Cu_3O_7$ multi-layer device*, Physica C **292**, 133 (1997).

[25] N. Dieckmann, R. Kürsten, M. Löhndorf, and A. Bock, *Epitaxial Quality of c-Axis and a-Axis oriented $YBa_2Cu_3O_7$ films: Characterization by Raman Spectroscopy*, Physica C **245**, 212-218 (1995).

[26] N. Dieckmann, M. Rübhausen, A. Bock, M. Schilling, K.-O. Subke, U. Merkt, E. Holzinger-Schweiger, *Magnetic and phononic excitations in $Y_{1-x}Pr_xBa_2Cu_3O_7$ films*, Physica C **272**, 269-276 (1996).

[27] S. Krey, K.-O. Subke, D. Reimer, M. Schilling, and R. Scharnweber, *Integrated $YBa_2Cu_3O_7$ magnetometers for biomagnetic applications*, Applied Superconductivity 1998, in press.

[28] D. Drung, S. Knappe and H. Koch, *Theory for the multiloop dc superconducting quantum interference device magnetometer and experimental verification*, J. Appl. Phys. **77**, 4088 (1995).

[29] J.M. Jaycox and M.B. Ketchen, *Planar coupling scheme for ultra low noise dc SQUIDs*, IEEE Trans. Magn. **17**, 400 (1981).

[30] K. Enpuku, T. Maruo, and T. Minotani, *Resonant characteristics of high T_c dc-SQUID caused by large dielectric constant of $SrTiO_3$*, IEEE Trans. Appl. Supercond. **7**, 3355 (1997).

[31] R. Scharnweber and M. Schilling, *A new concept for integrated* $YBa_2Cu_3O_7$ *magnetometers*, IEEE Trans. Appl. Supercond. **7**, 3485 (1997).

[32] O. Dössel, B. David, M. Fuchs, W.H. Kullmann, and K.-M. Lüdecke, *A modular low noise 7-channel SQUID-magnetometer*, IEEE Trans. Magn. **27**, 2797 (1991).

Magnetic Phase Diagram and Reentrant Superconductivity in $DyNi_2B_2C$

Klaus Winzer, Zhiqiang Peng, and Klaus Krug

I. Physikalisches Institut, Bunsenstraße 9, D-37079 Göttingen, Germany

Abstract: Detailed measurements of the resistivity ρ_{ab} as a function of temperature, magnetic field and orientation have been carried out on $DyNi_2B_2C$ single crystals. From the angular dependence of the metamagnetic transitions a phase diagram with five metamagnetic phases have been obtained. For $B||[100]$, anomalously large hysteresis effects of the metamagnetic transitions and also of the superconducting transitions were observed at low temperature ($T < 2$ K). The large hysteresis leads to a reentrant behavior of the superconductivity in small external fields. The reentrant behavior and the large hysteresis indicate a strong interplay between magnetism and superconductivity in $DyNi_2B_2C$.

1 Introduction

The newly discovered RNi_2B_2C (R = Y, Gd – Lu) [1,2] compounds have attracted attention and have been regarded as ideal materials for studies of the interplay between long range magnetic order and superconductivity because of their high magnetic ordering temperatures and comparable magnetic and superconducting condensation energies [3,4]. For R = Tm [5], Er [6], Ho [7,8], and Dy [9,10] superconductivity coexists with antiferromagnetism with $T_c \cong$ 10.8,10.5,8.5,and 6.5 K and $T_N \cong$ 1.5,5.8,5.2, and 10.3 K, respectively. The values of T_c and T_N roughly scale with the de Gennes factor across the series [9,11]. This is consistent with the coupling between the conduction electrons and the local magnetic moments via the RKKY interaction, giving rise to the long range magnetic order below T_N. The coexistence and the relatively weak coupling between superconductivity and antiferromagnetism is caused by the different degrees of localisation of electrons in the borocarbides. Band structure calculations (LAPW,PPW) show that the conduction band consists mainly of Ni 3d-electrons [12–14]. They are responsible for the superconducting transition. The magnetic properties are governed by the well localized electrons in the incompletely filled 4f-shell of R. The exchange interaction between these two electron systems is mediated by the small fraction of R 6s and 5d-electrons in the conduction band.

$DyNi_2B_2C$ is the unique member of the RNi_2B_2C system with $T_N > T_c$. At temperatures $T < T_N$, the magnetic moments of Dy are ordered ferromagnetically in each Dy-C layer of the body-centered tetragonal crystal structure. The Dy moments are aligned along the [110] direction, and the moments in each two neigboring Dy-C layers are aligned in opposite direction, forming a simple commensurate antiferromagnetic structure [4,15]. Large anisotropy of magnetic and superconducting properties were observed in $DyNi_2B_2C$ single crystals [9,10,16]. In contrast to studies on superconducting $HoNi_2B_2C$

[7,8,11], $ErNi_2B_2C$ [6,11], and $TmNi_2B_2C$ [5,11], no reentrant behavior or other anomalous features were observed in $DyNi_2B_2C$ single crystals, indicating a perfect coexistence of antiferromagnetism and superconductivity. However, all earlier measurements on $DyNi_2B_2C$ were carried out at temperatures $T > 1.8$ K.

In the first part of this paper we propose a magnetic phase diagram of $DyNi_2B_2C$ obtained from the angular dependence of metamagnetic phase transitions observed by resistivity and susceptibility measurements, respectively. In the second part we report on the observation of large hysteresis effects of the metamagnetic phase transitions and of the upper critical field in the low temperature range from 50 mK to 2 K [17].

The experiments were performed on $DyNi_2B_2C$ single crystals grown by the high temperature flux method using Ni_2B as a solvent [18]. The dimensions of the platelet like crystals are typically about $0.6 \times 0.5 \times 0.1$ mm^3 with the c-axis perpendicular to the surface of the platelet. The quality of the crystals was characterized by measurements of the residual resistivity ratio ($RRR = \rho(300K)/\rho(T_c^{\mathrm{onset}})$) and by the T_c-value and the transition width ΔT_c. For the best crystal a $RRR = 70$ and $T_c = 6.8$ K were obtained indicating a high degree of perfection. A standard ac four-probe method was used for the measurements of the resistivity ρ_{ab} with a current density of about 2 A/cm^2 in the a-b plane. The samples were orientated using the x-ray Laue back-scattering method. Angular dependent resistivity and susceptibility measurements were performed at 1.4 K in magnetic fields up to 8 T. Resistivity measurements with fixed sample orientation were performed in an adiabatic demagnetization cryostat at temperatures between 50 mK and 14 K in magnetic fields up to 5 T.

2 Results

2.1 Development of the Magnetic Phase Diagram

Fig. 1 shows the field dependence of the resistivity of a $DyNi_2B_2C$ single crystal at 1.4 K for different orientations of the magnetic field with respect to the [110] axis. The superconductivity is suppressed by a magnetic field of 0.7 T. In the normal state there is a step in the resistivity at about 1.4 T with a weak angular dependence. For $\theta \geq 20°$ and $B \geq 1.5$ T the resistivity shows a plateau followed by a steplike decrease. This decreasing

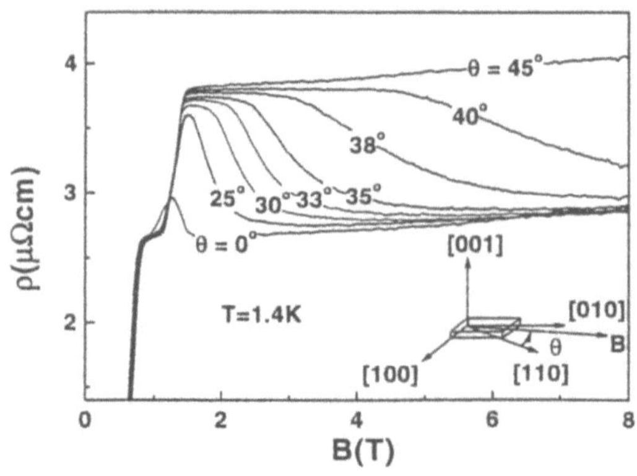

Figure 1
Magnetic field dependence of the resistivity of a $DyNi_2B_2C$ single crystal at 1.4 K for different orientations of the magnetic field with respect to the [110] axis.

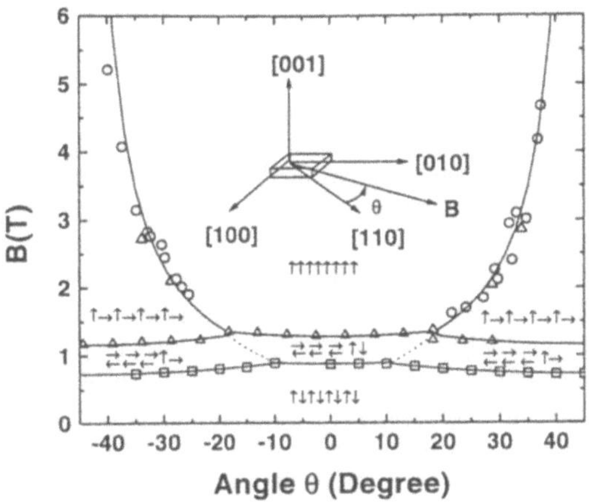

Figure 2
Angular dependence of critical magnetic fields for $DyNi_2B_2C$ at $T = 1.4$ K. Circles: $B_{3F}(\theta)$ obtained from resistivity measurements; Triangles: $B_{13}(\theta)$ and $B_{2F}(\theta)$ obtained from resistivity measurements; Squares: $B_{A1}(\theta)$ and $B_{A2}(\theta)$ obtained from magnetization measurements at $T = 2.0$ K [19].

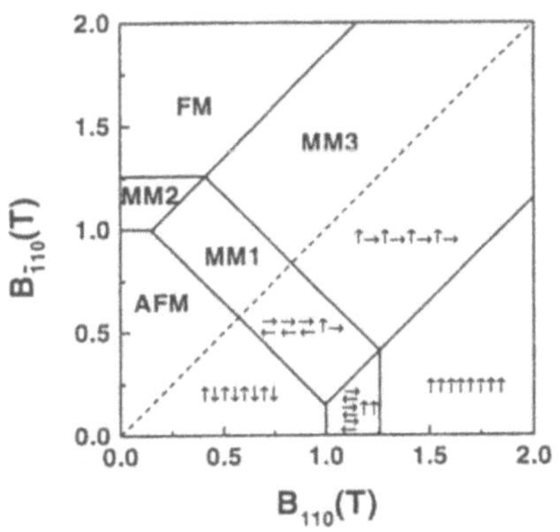

Figure 3
Magnetic phase diagram for $DyNi_2B_2C$ at $T = 1.4$ K. The B axis corresponds to the [110] and the [$\bar{1}$10] directions.

step in the resistivity shows a strong angular dependence and the corresponding transition diverges as $\theta \Rightarrow 45°$. Both steps correspond to metamagnetic transitions between phases with different net distributions of the Dy magnetic moments. A third transition was recently observed by magnetization measurements at $T = 2$ K and slightly smaller fields [19]. This transition can not be observed in our resistivity measurements since it is merged into the superconducting transition which occurs in the same range of magnetic fields at 1.4 K. The angular dependencies of the metamagnetic transitions obtained from resistivity and susceptibility measurements at 1.4 K are summarized in Fig. 2. Depending upon the angle θ between the applied magnetic field and the [110] axis, up to three transitions can be observed. The two low field transitions are bounded while the third transition diverges as $\theta \Rightarrow 45°$. From the angular dependence of these transitions a magnetic phase diagram of $DyNi_2B_2C$ can be deduced which is very simple if the components of the magnetic field $B \| \langle 110 \rangle, B \| \langle \bar{1}10 \rangle$ are choosen as variables. The phase diagram shown in Fig. 3 is constructed analogous to a recently published magnetic phase diagram for the system $HoNi_2B_2C$ [20]. The low field phase has been identified by neutron diffraction experiments [21] with the antiferromagnetic phase (AFM), which we denote as ↑↓, where ↑ is a moment along the [110] axis and ↓ is a moment along the [$\bar{1}\bar{1}$0] axis. For magnetic field directions near to the [110] axis we expect a saturated paramagnetic (or ferromagnetic (FM)) state ↑↑ in the high field limit. A transition into this FM state is connected with the decreasing step in the resistivity due to the higher degree of order of the magnetic moments. For magnetic field directions near to the [100] axis we expect a canted structure ↑→ of the magnetic moments in the easy directions, where → is a moment along the [$\bar{1}$10] axis. This expectation is experimentally confirmed by magnetization measurements [19] which show in this high field phase MM3 a magnetization $M \cong M_s/\sqrt{2}$, where $M_s = 8.7\mu_B$/Dy. The net distribution of moments in the intermediate metamagnetic states MM1 and MM2 can not be predicted definitely. The magnetization in the metamagnetic state MM2 is about $M_s/4$ [19]. The most probable net moment distribution in agreement with this magnetization is ⇄⇄⇄↑↑ in MM2 and ⇄⇄⇄↑→ in MM1, respectively. An alternative net moment distribution ↑↓↑↓↑↓↑↑ in MM2 and ↑↓↑↓↑↓↑→ in MM1 with the same magnetization is less probable with respect to the experimentally obtained magnetic field values of the metamagnetic transitions from AF to MM2 and from MM2 to FM, respectively. It should be noted that the interpretation of the magnetization data implies that the sample is a single magnetic domain in the metamagnetic states MM1 and MM2, therefore it is likely that the metamagnetic transitions from the AFM state to the metamagnetic states MM1 and MM2 consist of both a change in the magnetic structure and a spin flop of the domains. From the phase diagram of Fig. 3 the angular dependence of the phase transitions between different metamagnetic phases can easily be calculated and are given in Fig. 2 as solid lines. The magnetic-field-dependent energy difference between the low field antiferromagnetic structure AFM and the metamagnetic structure MM2 with 1/4 of the full Dy moment along the [110] axis will vary as $B\mu_{sat}\cos(\Theta)/4$. If there is a critical energy difference W_{crit}, that has to be exceeded to stabilize the metamagnetic structure, then the critical field will vary as $4W_{crit}/\mu_{sat}\cos(\Theta)$. In a similar way the angular dependencies of the phase boundaries can be obtained from the energy difference between adjacent metamagnetic phases. In Tab. 1 the angular dependencies and the experimentally obtained magnetic fields B_{MM} necessary to stabilize the four metamagnetic phases (MM1–MM3, FM) are summarized. From the last column of Tab. 1 it can be seen that some of the magnetic fields and hence the corresponding energies of adjacent metamagnetic states are very similar. For example, the difference between $B_{2F}(0)$ and $B_{13}(45°)$ is only 6.4% compared to 20.8% for the corresponding phase transitions in the $HoNi_2B_2C$ system [22]. Such similarities of magnetic energies can explain some of the large hysteresis effects observed in the $DyNi_2B_2C$ system especially at low temperatures.

2.2 Large Hysteresis Effects Below $T \cong 2$ K

Fig. 4 shows the temperature dependence of the resistivity $\rho(T)$ under various fixed magnetic fields B applied along the crystallographic [100] axis. For zero magnetic field, the $\rho(T)$ curve exhibits a drop of the resistivity below 10.3 K, which is attributed to a phase transition from paramagnetism to antiferromagnetism [15]. A sharp superconducting transition occurs at $T_c = 6.5$ K with a transition width $\Delta T_c \cong 0.3$ K, indicating that the sample has a high homogeneity. With increasing field, T_c shifts to lower temperatures with slightly broaded ΔT_c. The resistivity at the temperature just above the superconducting transition is about 2.0 ± 0.2 $\mu\Omega$cm. The residual resistivity ratio of this single crystal is $RRR = 32$. Fig. 5 shows the field dependence of the resisitivity $\rho(B,T = \text{const.})$ with increasing B from 0 to 2 T and $B \| [100]$. For the curve of $T = 4.2$ K, besides the superconducting transition occuring at $B_{c2} \cong 0.27$ T (defined as the intersection of a straight line drawn through

Table 1 Angular dependencies and experimentelly obtained magnetic fields B_{MM} for the four metamagnetic phases (MM1–MM3, FM). The values $B_{A1}(45^\circ)$ $B_{A2}(0^\circ)$ were obtained from magnetization measurements by Canfield et al. [19] by considering the different temperatures of the measurements by a factor of 1.12. This factor corresponds to the increase of $B_{A1}(T)$ from 2.0 K to 1.4 K (Fig. 7a).

Phase boundary	Range of θ	Angular dependence	$B_{MM}(\theta)$
AFM - MM1	$9^\circ < \theta < 45^\circ$	$B_{A1} = \dfrac{B_{A1}(45^\circ)}{\cos(\theta - 45^\circ)}$	$B_{A1}(45^\circ) = 0.81$ T
AFM - MM2	$0^\circ < \theta < 9^\circ$	$B_{A2} = \dfrac{B_{A2}(0^\circ)}{\cos(\theta)}$	$B_{A2}(0^\circ) = 1.00$ T
MM1 - MM2	$9^\circ < \theta < 18^\circ$	$B_{12} = \dfrac{B_{12}(0^\circ)}{\sqrt{2}\cos(\theta + 45^\circ)}$	$B_{12}(0^\circ) = 0.85$ T
MM1 - MM3	$18^\circ < \theta < 45^\circ$	$B_{13} = \dfrac{B_{13}(45^\circ)}{\cos(\theta - 45^\circ)}$	$B_{13}(45^\circ) = 1.18$ T
MM2 - FM	$0^\circ < \theta < 18^\circ$	$B_{2F} = \dfrac{B_{2F}(0^\circ)}{\cos(\theta)}$	$B_{2F}(0^\circ) = 1.26$ T
MM3 - FM	$18^\circ < \theta < 45^\circ$	$B_{3F} = \dfrac{B_{3F}(0^\circ)}{\sqrt{2}\cos(\theta + 45^\circ)}$	$B_{3F}(0^\circ) = 0.85$ T

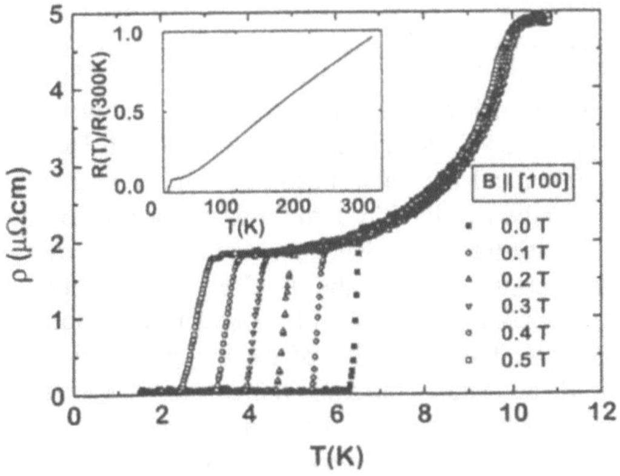

Figure 4
Temperature dependence of resistivity ρ_{ab} under various fixed fields applied along the crystallographic [100] axis for the $DyNi_2B_2C$ single crystal. Inset: Temperature dependence of zerofield normalized resistance in the temperature range between 4.2 and 300 K.

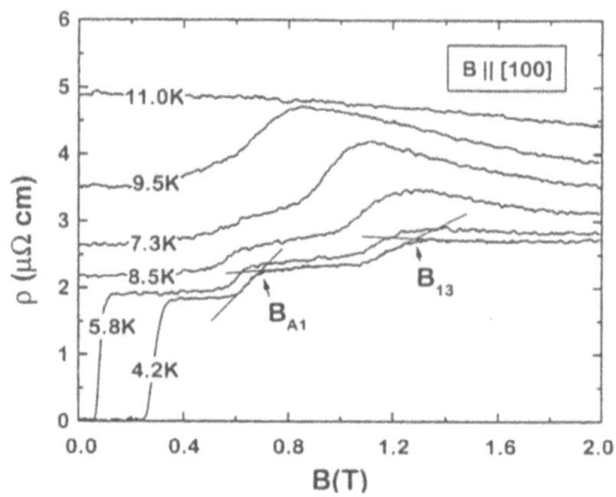

Figure 5
Resistivity $\rho_{ab}(B)$ with $B||[100]$ at various fixed temperatures. Besides the superconducting transition, two field-induced metamagnetic phase transitions at B_{A1} and B_{13} are also observed.

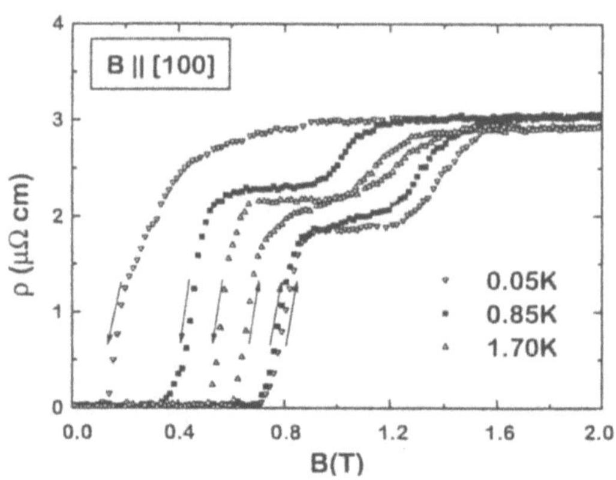

Figure 6
Field dependence of resistivity $\rho_{ab}(B)$ with $B||[100]$ at temperatures 0.05 K, 0.85 K and 1.7 K.

the steepest part of the superconducting transition with the B axis) two other steps can be observed. The first step at $B_{A1} \cong 0.7$ T corresponds to the phase transition from AFM to MM1 and the second step at $B_{13} \cong 1.3$ T corresponds to the phase transition from MM1 to MM3. Both steps shift to lower magnetic field values with increasing temperature. Fig. 6 shows the curves of $\rho(B)$ isotherms ($B||[100]$) at $T = 1.7$ K, 0.85 K and 0.05 K with increasing magnetic field from 0 to 2 T and then decreasing to zero. For $T = 1.7$ K as B increases, the superconducting transition occurs at $B_{c2} \cong 0.6$ T, followed by the metamagnetic phase transition at $B_{13} \cong 1.35$ T. With decreasing field B_{13} shifts to a lower value of about 1.25 T, and the upper critical field also becomes smaller with $B_{c2} \cong 0.5$ T. No structure associated with B_{A1} can be resolved in both field-increasing and field-decreasing curves. The metamagnetic phase transition AFM to MM1 seems to merge with the superconducting transition at temperatures below about 2.5 K. The hysteresis of the upper critical field $\Delta B_{c2} = B_{c2}(\text{field-increasing}) - B_{c2}(\text{field-decreasing})$ is about 0.1 T at 1.7 K and increases to 0.34 T at 0.85 K. The hysteresis of the metamagnetic transition ΔB_{13} is of the same order of magnitude and shows the same temperature dependence. For T = 0.05 K, an even larger hysteresis effect is observed. The field-increasing part of $\rho(B)$ shows a similar behavior as at 0.85 K, with slightly higher values of B_{c2} and B_{13}. With decreasing field, the resistivity decreases continously to zero at $B_{c2} \cong 0.14$ T, no anomalous feature associated with either metamagnetic transition can be clearly resolved above

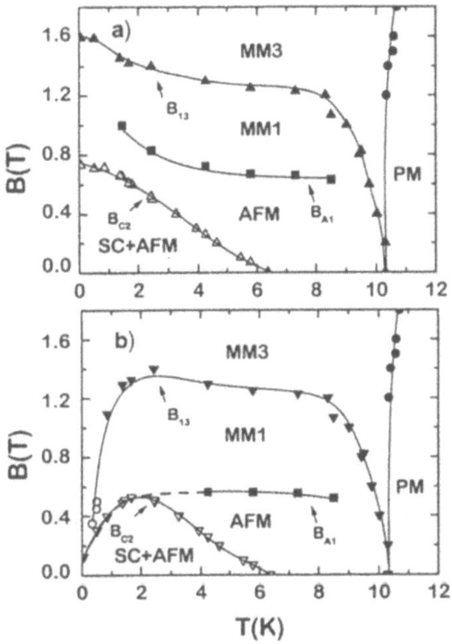

Figure 7
(a) Phase diagram for $DyNi_2B_2C$ derived from the field-increasing part of $\rho_{ab}(B)$ with $B\|[100]$; (b) phase diagram derived from the field-decreasing part of $\rho_{ab}(B)$ curves. The solid lines drawn through the data points serve as a guide to the eye. The value of $B_{A1}(T)$ at temperatures below 4 K cannot be determined by our resistivity measurements. The dashed line is proposed to be a possible boundary between AFM and MM1 (see text for details).

this field. The hysteresis of the upper critical field amounts to $\Delta B_{c2} \cong 0.6$ T at $T = 0.05$ K, hence $\Delta B_{c2}/B_{c2}(0) \cong 80\%$.

Based on $\rho(T)$ and $\rho(B)$ measurements with $B\|[100]$, the phase diagram is determined. Figures 7a,b show the results derived from the field-increasing and field-decreasing parts of $\rho(B)$ curves respectively. The magnetic phase below the B_{A1} curve is antiferromagnetic (AFM). According to the phase diagram given in Fig. 3 between curve B_{A1} and curve B_{13} we have the metamagnetic phase MM1. Above the curve B_{13} we have the metamagnetic phase MM3 with the canted structure of the magnetic moments. The phase boundary between MM3 and the paramagnetic phase PM is derived from data of $\rho(T)$. In high fields $B > 1$ T, the $\rho(T)$ curves display a clear kink around 10.5 K, whose position is almost field independent as indicated by the solid circles in Fig. 7. This kink can not be observed in the case of $B\|[110]$. In Fig. 7a, the upper critical field $B_{c2}(T)$ exhibits no local minimum or other features as observed in $HoNi_2B_2C$, $ErNi_2B_2C$ and $TmNi_2B_2C$. Near T_c the curve of $B_{c2}(T)$ displays a positiv curvature similar to that observed in other superconducting RNi_2B_2C members [23,24]. In Fig. 7b, the upper critical field B_{c2} derived from the field-decreasing part of $\rho(B)$ curves shows a parabolic-shaped behavior with a maximum of $B_{c2}^{max} \cong 0.53$ T at $T = 2$ K. Comparing Fig. 7a and Fig. 7b, one can see that the hysteresis of the superconducting transition and of the metamagnetic phase transition B_{A1} becomes larger as temperature decreases from 2.0 K to 0.05 K. The metamagnetic phase transition B_{A1} can not be resolved in the field-decreasing part of $\rho(B)$ curves at low temperatures $T < 4$ K, so we can not determine the phase boundary between AFM and MM1 at temperatures $T < 4$ K. Above 2 K, B_{c2} and B_{13} exhibit negligible hysteresis, while B_{A1} keeps a modest hysteresis effect with $\Delta B_{A1} \cong 0.15$ T at $T > 4$ K. In Fig. 7b, the upper critical field B_{c2} determined from field-decreasing parts of $\rho(B)$ displays a maximum at 2 K. Such a parabolic-shaped $B_{c2}(T)$ predicts a reentrant behavior. If we first increase B

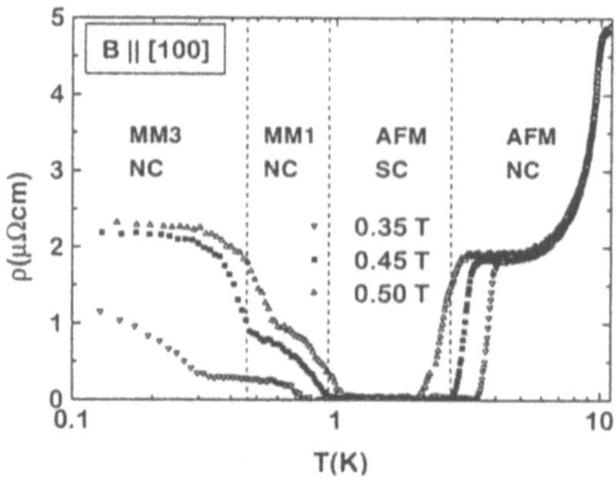

Figure 8
Measurement of $\rho_{ab}(T)$ with increasing temperature after a process with B increasing from 0 to 2 T at temperature $T = 0.05$ K, and then lowered to final fields $B_0 = 0.35, 0.45$, and 0.5 T, respectively. The dashed lines emphasize the transitions into the labeled metamagnetic phases for $B = 0.45$ T.

from 0 to 2 T at temperature $T = 0.05$ K, and than decrease B down to a field B_0 between $B_{c2}(T = 0.05\text{ K})$ and B_{c2}^{max}, the sample will remain in the normal state. Afterwards with increasing temperature, the sample shall undergo a superconducting transition. Fig. 8 shows the corresponding results with various final fields B_0. For $B_0 = 0.45$ T, with increasing temperature, the sample first undergoes a phase transition from the normal state to the superconducting state at $T_{c2} \cong 0.95$ K, and than returns to the normal state at $T_{c1} \cong 2.8$ K. With lower B_0, T_{c1} becomes larger, while T_{c2} decreases. In addition there exists a kink around 0.5 K in these $\rho(T)$ curves. As shown in Fig. 7b, the open circles corresponding to the kink positions of $\rho(T)$ curves in Fig. 8 agree well with the extrapolation of B_{13} deduced from $\rho(B)$ measurements. Hence we tend to believe that the kink observed in Fig. 8 may be associated with the phase transition from the canted structure to MM1. No hysteresis effects can be observed for $B||[001]$. In the case of $B||[110]$ the observed hysteresis effects are comparatively small, in the superconducting transition as well as in the metamagnetic transitions. The anomalous features of large hysteresis of superconductivity and the reentrant behavior indicate that a strong interplay between superconductivity and magnetism exists in $DyNi_2B_2C$. A full interpretation of such a behavior requires detailed measurements of magnetization and neutron diffraction at low temperatures $T < 2$ K. Here we note that the hysteresis of the metamagnetic phase transition from MM1 to AFM at temperatures $T < 2$ K might play an important role in determining the unusual behavior of superconductivity. Unfortunately the metamagnetic phase transition B_{A1} can not be resolved in the field-decreasing part of $\rho(B)$ at low temperatures $T < 4$ K. However, according to the magnetization measurement on the $DyNi_2B_2C$ single crystal at $T = 1.8$ K reported by Thomy et al. [16], large hysteresis of a similary field-induced phase transition from AFM to an intermediate metamagnetic phase was observed with $\Delta B_{A1} \cong 0.5$ T. In comparison with the behavior of B_{13} and B_{c2}, it is expected that the hysteresis effect of B_{A1} will also become larger with decreasing temperature. Due to this expected large hysteresis of B_{A1}, when decreasing field, the metamagnetic phase MM1 possibly extends to the low-field superconducting range. The metamagnetic phase MM1 possesses a ferromagnetic component with a value of about $3.5\mu_B$/Dy [16,25], and this strongly suppresses the superconductivity, leading to the lowering of B_{c2} in the low temperature range $T < 2$ K as shown in Fig. 7b. Comparing the results of $M(H)$ reported by Thomy et al. [16] with our $\rho(B)$ measurements in a similar temperature range, we note that in their field-decreasing part of $M(H)$ curve at $T = 1.8$ K, the metamagnetic transition occurs at $B_{A1} \cong 0.5$ T, which is nearly the same as the upper critical field $B_{c2} = 0.52 \pm 0.2$ T determined from the field decreasing part of $\rho(B)$ at 1.7 K, as shown in Fig. 5. Therefore one possibility is that the phase boundary between MM1 and AFM in the low temperature range $T < 2$ K is identical with that between the normal and the superconducting states, as indicated by the dashed line in Fig. 7b. With decreasing field, when the phase transition from MM1 to AFM occurs at B_{A1}, the suppression of superconductivity by the metamagnetic phase MM1 is also reduced, and the superconducting transition might occur at the same field. It is particular interesting that the coexistence of superconductivity with the metamagnetic phase MM1 can also not be ruled out by our resistivity measurements. In the borocarbides only $ErNi_2B_2C$ exhibits a possible coexistence of superconductivity and a weak-ferromagnetic phase of $0.33\mu_B$/Er [6,9] which is about 10 times smaller than the ferromagnetic component of the magnetic phase MM1 observed in $DyNi_2B_2C$. Detailed investigations by magnetization measurements at low temperatures $T < 1$ K are in progress.

3 Conclusion

We have carried out detailed resistivity measurements as a function of temperature and magnetic field on $DyNi_2B_2C$ single crystals. From the angular dependence of the metamagnetic transitions a phase diagram with five metamagnetic phases have been obtained. The field dependence of the resistivity $\rho(B)$ measured at temperatures $T < 2$ K with $B \| [100]$ exhibits an anomalous large hysteresis, and the upper critical field $B_{c2}(T)$ derived from the field-decreasing part of the $\rho(B)$ curves display a maximum at $T \cong 2$ K. We propose that the interaction between superconductivity and the metamagnetic phase MM1 plays a dominant role in determining the low temperature superconducting properties. Detailed magnetization and neutron scattering measurements will be needed for a complete explanation of the origin for such a low temperature behavior.

Bibliography

[1] R.Nagarajan. C.Mazumdar, Z.Hossain, S.K.Dhar, K.V.Gopalakrishnan, L.C. Gupta, C.Godart, B.D.Padalia, and R.Vijayaraghavan, Phys. Rev. Lett. **72**, 274 (1994).

[2] R.J.Cava, H.Takagi, H.W.Zandberg, J.J.Krajewski, W.F.Peck, Jr.,T.Siegrist, B. Batlogg, R.B. van Dover, R.J. Felder, K. Mizuhashi, J.O. Lee, H. Eisaki, and S. Uchida, Nature **367**, 252 (1994).

[3] L.C.Gupta, Physica B **223–224**, 56 (1996).

[4] J.W.Lynn, S.Skanthakumar, Q.Huang, S.K.Sinha, Z.Hossain, L.C.Gupta, R.Nagarajan, and C.Godart, Phys. Rev. B **55**, 6584 (1997).

[5] B.K.Cho, Ming Xu, P.C.Canfield, L.L.Miller, and D.C.Johnston, Phys.Rev. B **52**, 3676 (1995).

[6] P.C.Canfield, S.L.Bud'ko and B.K.Cho, Physica C **262**, 249 (1996).

[7] K.D.D.Rathnayaka, D.G.Naugle, B.K.Cho, and P.C.Canfield, Phys. Rev. B **53**, 5688 (1996).

[8] K.Krug, M.Heinecke, and K.Winzer, Physica C **267**, 321 (1996).

[9] B.K.Cho, P.C.Canfield, and D.C.Johnston, Phys. Rev. B **52**, 3844 (1995).

[10] C.V.Tomy, M.R.Lees, L.Afalfiz,G. Balakrishnan and D.McK. Paul, Phys. Rev. B **52**, 9186 (1995).

[11] H.Eisaki, H.Takagi, R.J.Cava, B.Batlogg, J.J.Krajewski, W.F.Peck, Jr., K.Mizuhashi, J.O.Lee, and S.Uchida, Phys. Rev. B **50**, 647 (1994).

[12] L.F.Mattheis, Phys. Rev. B **49**, 13279 (1994).

[13] W.E.Pickett, D.J.Singh, Phys. Rev. Lett. **72**, 3702 (1994).

[14] H.Kim, C.-D.Hwang, J.Ihm, Phys. Rev. B **52**, 4592 (1995).

[15] P.Dervenagas, J.Zarestky, C.Stassis, A.I.Goldman, P.C.Canfield, and B.K.Cho, Physica B **212**, 1 (1995).

[16] C.V.Tomy, M.R.Lees, G.Balakrishnan, D.T.Adroja, and D.McK.Paul, Physica B **223–224**, 62 (1996).

[17] Z.Q. Peng, K.Krug, and K. Winzer, in Phys. Rev. B **57**, R8123 (1998).

[18] M.Xu, P.C.Canfield, J.E.Ostenson, D.K.Finnemore, B.K.Cho, Z.R.Wang, and D.C.Johnston, Physica C **227**, 321 (1994).

[19] P.C.Canfield and S.L.Bud'ko, J. of Alloys and Compounds **262–263**, 169 (1997).

[20] V.A. Kalatsky and V.L.Pokrovsky, Phys. Rev. B **57**, 5485 (1998).

[21] J,W.Lynn, and S.Skanthakumar, Q.Huang, S.K.Sinha, Z.Hossain, L.C.Gupta, R.Nagarajan, and C.Godart, Phys. Rev. B **55**, 6584 (1998).

[22] P.C.Canfield, S.L.Bud'ko, B.K.Cho, A.Lacerda, D.Farrell, E.Johnston-Halperin, V.A.Kalatsky, and V.L.Pokrovsky, Phys. Rev. B **55**, 970 (1997).

[23] K.D.D.Rathnayaka, A. K.Bhatnagar, A. Parasiris, D. G.Naugle, P.C.Canfield, and B.K.Cho, Phys. Rev. B **55**, 8506 (1997).

[24] S.V.Shulga, S.L.Drechsler, G.Fuchs, K.-H.Müller, K.Winzer, M.Heinecke, and K.Krug, Phys. Rev. Let. **80**, 1730 (1998).

[25] M.S.Lin, J.H.Shieh, Y.B.You, Y.Y.Hsu, J.W.Chen, S.H.Lin, Y.D.Yao, Y.Y.Chen, J.C.Ho, and H.C.Ku, Physica C **249**, 403 (1995).

The Current-Phase Relation in HTS Josephson Junctions

E. Il'ichev, V. Zakosarenko, R.P.J. Ijsselsteijn, V. Schultze, H.-G. Meyer, and H.E. Hoenig

Institut für Physikalische Hochtechnologie, Bereich Kryoelektronik, PF 100239, D-07702 Jena, Germany

Abstract: The current-phase relation of $YBa_2Cu_3O_{7-x}$ step-edge as well as 24° and 45° grain boundary Josephson junctions has been investigated experimentally. The junctions were incorporated into a washer-shaped superconducting ring with inductance $L \sim 80$–300 pH. The ring was inductively coupled to a tank circuit with a resonance frequency 9 ... 40 MHz. The current-phase relation was obtained from the measurement of the impedance of the phase-biased junction. It is shown, that experimentally observed deviations from harmonic behavior of the apparent current-phase relation for step-edge and 24° grain boundary junctions can be explained by the influence of thermal noise. The current-phase relation of 45° grain boundary junctions was found to be extremely non-harmonic. The reasons of this unusual behavior are discussed.

1 Introduction

The density of the supercurrent j_s flowing through a weak contact between two superconducting electrodes depends on the local phase difference $\varphi = \varphi_2 - \varphi_1$ of the order parameters $\Delta_{1,2} = |\Delta_{1,2}| \exp(i\varphi_{1,2})$ in the electrodes

$$j_s(\varphi) = j_c f(\varphi) \tag{1.1}$$

where $j_c = \max(j_s(\varphi))$. For a junction of such small area, that the phase difference across the junction is independent of the coordinates in the plane of the contact, the total supercurrent can be expressed as $I_s(\varphi) = I_c f(\varphi)$, and $I_s(\varphi)$ is usually called the current-phase relation (CPR).

The CPR of the small size weak links has certain general properties: Firstly, a 2π-change of $\varphi_{1,2}$ does not change the physical state of the system, therefore, the CPR is always 2π periodic. Secondly, the CPR is an odd function

$$I_s(-\varphi) = -I_s(\varphi) \tag{1.2}$$

because interchanging the phases φ_1 and φ_2 of the electrodes leads to the change of the current direction. From (1.1) and (1.2) it follows that $I_s(\pi) = 0$.

For tunnel junctions

$$I_s(\varphi) = I_c \sin(\varphi), \tag{1.3}$$

where $I_c = \max(I_s(\varphi))$ is the critical current [1]. For other types of weak links between classical superconductors the CPR (1.3) is valid at temperatures T close to critical temperature T_c [3,5].

The CPR is one of the main characteristics of the superconducting weak links. There is considerable technical importance to know the CPR for the design of weak-link-based circuits. The performance of cryoelectronic devices can seriously degrade by a nonsinusoidal CPR as it was shown, for instance, in [4]. Apart from the technical importance the CPR is of considerable scientific interest for high-T_c Josephson junctions (HTS JJs). In particular, the CPR of the grain-boundary (GB) HTS JJs can contain additional information about the symmetry of the order parameter for HTS cuprates.

Recently, some models of HTS JJs have been discussed in the literature: (i) constriction type models, where current transport is mediated by ballistic constrictions with high transparency of interfaces [6], and (ii) resonant tunneling models assuming the presence of an insulating barrier containing a large density of localized states [7–10]. Both types of models give very similar predictions if the width of the resonant channels is larger than the superconducting energy gap [11,12] – a condition which is likely to be fulfilled in HTS JJs with intrinsic interfaces. In this case models predict a nonsinusoidal CPR at low T, similar to the classical weak links. Moreover, recently it has been revealed that the order parameter of most HTS cuprates is unconventional, dominated by a $d_{x^2-y^2}$ symmetry component [13–15]. For the CPR of various configurations of HTS JJs with $d_{x^2-y^2}$ symmetry of the order parameter, strong deviations from the harmonic dependence have been predicted [16,17]. In particular, even π-periodic CPR has been obtained theoretically [16].

In spite of its importance, the CPR is the least known weak link property, especially for high T_c superconductors. To our knowledge, the CPR was only measured for weak links prepared by ion irradiation [18] and for step-edge junctions [19] at temperatures close to T_c. In almost all of these cases nominally sinusoidal current-phase relation was observed. Thus, we have no information about the CPR of any type of HTS JJs at low temperatures. Moreover, the CPR of grain boundary HTS JJs has not yet been investigated experimentally.

For the experimental investigation of the CPR, most commonly, the weak link of interest is incorporated in a superconducting ring with a sufficiently small inductance L. This circuit is usually called a single junction interferometer. When an external flux Φ_e is applied to a superconducting ring, a circulating supercurrent is induced, and a flux Φ is admitted into the ring. In this case, the phase difference φ across the weak link must satisfy the relation

$$\varphi = 2\pi\Phi/\Phi_0 \quad \text{modulo} \quad 2\pi \tag{1.4}$$

where Φ_0 is the flux quantum. In the absence of fluctuations

$$\varphi = \varphi_e - \beta f(\varphi) + 2\pi m \tag{1.5}$$

where $\varphi_e = 2\pi\Phi_e/\Phi_0$ is the normalized external flux, $\beta = 2\pi L I_c/\Phi_0$ is the normalized critical current, and m is an integer. For the interferometer with $f(\varphi) = \sin(\varphi)$, the nonlinearity of the $\varphi(\varphi_e)$ curve depends on parameter β. At the value $\beta > 1$ the curve becomes multivalued in the vicinity of the points $\varphi_e = \pi + 2\pi m$, which causes jumps of φ and a hysteresis for a sweep of φ_e (Fig. 1A). The condition $\beta < 1$ has to be fulfilled to obtain the CPR for the complete phase range $-\pi \leq \varphi \leq \pi$. Thus, for $\beta < 1$ by monitoring φ with a superconducting flux detector as a function of φ_e, the CPR $\beta f(\varphi)$ can be obtained [20].

A

B

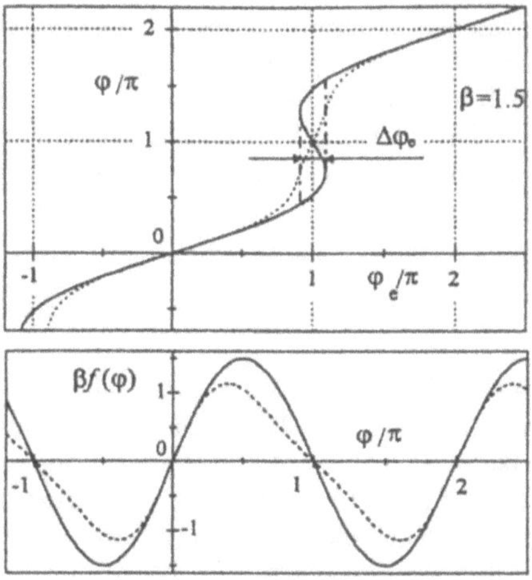

Figure 1
(A) Plot of phase φ vs normalized external flux φ_e for the interferometers with $\beta = 1.5$ (solid line) and apparent $\varphi(\varphi_e)$-dependence in the presence of noise (dashed line). (B) Sinusoidal CPR with $\beta = 1.5$ (solid line) and expected CPR in the presence of thermal noise (dashed line)

An other approach has been developed by Rifkin and Deaver [21]. The interferometer is coupled inductively to the tank circuit with the inductance L_T, the capacitance C_T, the quality factor Q, and the resonance frequency ω_0. In this arrangement the effective impedance Z_{eff} of the system of the tank circuit with the interferometer is a function of φ_e. In the case $\beta < 1$, as it has been shown [21], the CPR and the phase-dependent conductance $G(\varphi)$ both can be obtained from the measurement of Z_{eff}.

Until now the discussion has ignored the effects of thermal noise. Thermal noise modifies the internal flux in the ring [20] and, therefore, the measured signal. The estimation of this effect using a simple model is given in [22] for the case $\beta < 1$. The case $\beta > 1$ can be analyzed applying the model developed by Kurkijärvi [23]. According to (1.5), the width of the hysteresis ($\Delta\varphi_e$ in Fig. 1) is given by

$$\Delta\varphi_e = 2\left(\arccos(1/\beta) + \sqrt{\beta^2 - 1}\right). \tag{1.6}$$

In the case of finite flux fluctuations the jump can occur with a certain probability when φ_e is in a hysteresis region. The width of the distribution of the jump probability can be expressed as [23]

$$\sigma_\varphi \approx 0.26\beta \left(\frac{L}{\beta L_N}\right)^{2/3} \left(1 - (1/\beta)^2\right)^{1/6}, \tag{1.7}$$

where

$$L_N = \left(\frac{\Phi_0}{2\pi}\right)^2 \frac{1}{k_B T} \tag{1.8}$$

is so called "fluctuation-threshold inductance" [24] and $k_B T$ is the thermal energy. If the distribution width is of the order of $\Delta\varphi_e/2$, and the sweep frequency is much lower than the characteristic frequency of the flux fluctuation, the flux will jump many times during the measurements. Therefore, the hysteresis of $\varphi(\varphi_e)$ will vanish (dashed curve in Fig. 1A). Recently this effect has been considered in [26].

Thus, for $f(\varphi) = \sin(\varphi)$ in the presence of large thermal fluctuations, the common features of the apparent CPR are: for $\beta \geq 1$ the slope of the $\beta f(\varphi)$ curve at $\varphi = 2\pi m$ is larger than unity, the maximum of the $\beta f(\varphi)$ is observed at $\varphi = \varphi_{max} < \pi/2$ and the slope at $\varphi = \pi$ is less than unity (Fig. 1B). With decreasing β (increasing temperature) the phase φ_{max} approaches $\pi/2$ and for $\beta \ll 1$ the apparent CPR should have a sinusoidal form.

2 Method

As shown in [24] thermal fluctuations can be neglected if $L < L_N$, where L_N is given by (1.8). At liquid nitrogen temperatures $L_N \approx 100$ pH. This means that inductances to be used for the interferometers should be lower then L_N. In this case the application of the method proposed in [20] is technically difficult. In this respect the method developed in [21] has a great advantage. In the present work the approach [21] has been modified to achieve better accuracy.

In our experiments the interferometer was coupled to a tank circuit. For impedance measurements a driving radio-frequency (rf) current I_{rf} as well as a sweep current I_{dc} are supplied to the tank circuit. The external flux in the ring is generated by a current $I_{dc} + I_{rf}$ in the tank coil through mutual inductance M, so $\varphi_e = 2\pi(I_{dc} + I_{rf})M/\Phi_0 = \varphi_{dc} + \varphi_{rf}$, where $M^2 = k^2 L L_T$, and k is a coupling coefficient. In this case (1.5) will give

$$\varphi = \varphi_{dc} + \varphi_{rf} - \beta f(\varphi) - 2\pi\tau(\varphi)V/\Phi_0 + 2\pi m\,, \tag{2.9}$$

where $\tau(\varphi) = LG(\varphi)$ is a relaxation time of the interferometer, and V is the voltage across a weak link. We neglect a capacitance of the junction, what is usual for HTS JJs. If the quality factor $Q \gg 1$ and $k \ll 1$, we can consider the current in the tank circuit to be sinusoidal, $I_{rf} = QI_0 \exp(i\omega t)$ [25]. Here ω is a pump frequency, I_0 is the amplitude of the rf current, and i is the imaginary unit. Then the voltage across the tank circuit will be

$$U = i\omega I_{rf} L_T - \frac{d}{dt}(I_c M f(\varphi))\,. \tag{2.10}$$

Here φ is a solution of the nonlinear equation (2.9). In the small signal limit ($\varphi_{rf} \ll 1$) and in the adiabatic case ($\omega\tau \ll 1$), using the theory of linear circuits and keeping only the first-order terms, one can obtain from (2.9) and (2.10) the expression for L_{eff} and R_{eff} of the tank circuit-interferometer system

$$L_{eff} = L_T \left(1 - \frac{k^2 \beta f'(\varphi)}{1 + \beta f'(\varphi)}\right), \tag{2.11}$$

$$R_{eff} = R_T \left(1 + \frac{k^2 Q \omega\tau}{(1 + \beta f'(\varphi))^2} \frac{\omega}{\omega_0}\right), \tag{2.12}$$

were ω_0 is the resonance frequency of the unloaded tank circuit. Similar equations were obtained in [21], where the parametric inductance of the weak link was used instead of $f'(\varphi)$. According to Eqs. (2.11) and (2.12) the real as well as the imaginary part of the impedance Z_{eff} depend on φ. The phase angle α between the drive current I_{rf} and the tank voltage U is given by

$$\tan(\alpha) = \frac{1}{R_{\text{eff}}}\left(\frac{1}{\omega C_T} - \omega L_{\text{eff}}\right). \tag{2.13}$$

Substituting the dependencies (2.11) and (2.12) to (2.13) at $\omega = \omega_0$ and neglecting $\omega_0\tau$ we obtain

$$\tan(\alpha) = \frac{k^2 Q\beta f'(\varphi)}{1+\beta f'(\varphi)}. \tag{2.14}$$

From (2.9) at $\varphi_{\text{rf}} \ll 1$; $\omega\tau \ll 1$ follows, that

$$d\varphi = \frac{d\varphi_{\text{dc}}}{1+\beta f'(\varphi)}. \tag{2.15}$$

From the experimental data of $\alpha(\varphi_{\text{dc}})$ one can determine $\beta f'(\varphi)$ as function of the external magnetic flux φ_{dc} using (2.14). Then, numerical integration of (2.15) yields the needed CPR. So the phase shift α is a convenient quantity for determination of the current-phase relation of a weak link [27]. This method has the following advantages in comparison with [21]: (i) We do not deal with the parametric inductance which has singularities at $\varphi = \pm\pi/2$. (ii) The CPR is restored from a single curve (not from a set of resonance curves) which minimizes the influence of possible drifts in the measuring setup. (iii) The shift of the resonance frequency $\Delta\omega_0(\varphi_{\text{dc}})$ detected in [21] is proportional to $k^2\beta\omega_0$ [22]. The value of ω_0 and hence $\Delta\omega_0$ can be determined with an accuracy $\delta\omega_0 \sim \omega_0/Q$. Therefore $\Delta\omega_0/\delta\omega_0 \sim k^2 Q\beta$. Since ordinarily $k^2 Q \approx 1$, the measured value of $\Delta\omega_0$ is of the order of the accuracy of its measurement. In our case $\Delta\tan\alpha(\varphi_{\text{dc}}) \propto k^2 Q\beta$, *i.e.*, it can be of the order of unity and the $\alpha(\varphi_{\text{dc}})$ can be determined with the accuracy of a few percent.

In the present work we will not discuss the phase-dependent conductance problems of HTS JJs. Nevertheless, the developed method permits the determination of $G(\varphi)$ too. The quality factor

$$Q = \omega L_{\text{eff}}/R_{\text{eff}} \tag{2.16}$$

is found by standard circuit theory to be:

$$\frac{d}{d\omega}\tan(\alpha) = \frac{2Q}{\omega_0}. \tag{2.17}$$

Therefore, from the measurements of the derivation of $\tan(\alpha)$ with respect to pump frequency ω at $\omega = \omega_0$, the dependence $\tau(\varphi)$ can be restored using (2.11), (2.12) and (2.16).

3 Experimental

3.1 Samples

All layers for grain-boundary HTS JJs are grown using the pulsed laser deposition technique on (100) oriented $SrTiO_3$ substrates. An excimer laser (wave length 248 nm, laser pulse duration 20 ns), is focused on a polycrystalline, high density $YBa_2Cu_3O_{7-x}$ target. In front of the target a resistive heater with the substrate glued to it is placed in on-axis configuration. To avoid large numbers of droplets on the thin films, which may affect pattern transfer during lithography and etching, a high density (> 99%) target was used. Further droplet density reduction was achieved by polishing the target before each deposition run and using the minimum required energy density J for stoichiometric ablation of target material onto the substrate.

Other deposition parameters like deposition temperature T_{dep} , oxygen pressure P_{O_2}, and target-to-substrate distance d influence the superconducting properties and the surface-morphology and epitaxy of the $YBa_2Cu_3O_{7-x}$ layers. Highest critical temperatures T_c and critical current densities j_c are in general obtained for Cu-rich layers. However, these layers suffer from CuO precipitates, which preferentially grow at defects like grain-boundaries. Therefore, our layers were optimized with respect to both smooth surfaces and superconducting properties. This results in following deposition conditions: T_{dep} = 760 °C, P_{O_2} = 20 Pa, d = 50 mm and J = 2.0 J/cm^2. Using these conditions the edge of the visible part of the plasma just reaches the substrate and smooth $YBa_2Cu_3O_{7-x}$ with good superconducting properties can be grown. Typical values for T_c and j_c of these films are 2×10^6 A/cm^2 and 89 K, respectively. The layers are c-axis oriented with a FWHM value of the Rocking curve of the (005) XRD reflection of 0.12 to 0.14 degree.

Patterning of the layers was done using photolithography and Ar ion beam etching. AZ 5214 resist was structured using contact-photolithography. Afterwards $YBa_2Cu_3O_{7-x}$ layers were etched using Ar ions (600 eV energy, 0.5 mA/cm^2) into 8×8 mm^2 or 5×5 mm^2 square washer single-junction interferometer structures. The washer holes had a side-length of 50 μm, leading to $L \approx 80$ pH. Layer thickness and junction width were chosen such, that $\beta < 1$ resulted at the measurement temperature. Typical layer thicknesses were 40 to 70 nm, typical junction widths 2 to 4 μm.

The samples with step-edge JJs were produced at KFA Juelich. The samples were fabricated using the laser ablation process also. Films of c-axis oriented $YBa_2Cu_3O_{7-x}$, 200 nm thick, deposited on ion-milled steep steps prepared in the $LaAlO_3$ substrates. In the case of sample #1 we have used optical lithography. The samples #2 and #3 were prepared by the inhibit patterning procedure. Due to the controlled thermal diffusion of the inhibit layer into the junction area we reduced the width of junctions down to 0.3–0.5 μm [29]. The washer holes in this case had a side-length of 200 μm, leading to $L \approx 300$ pH.

3.2 Measurements

Several tank circuits with inductance of 0.3 ... 0.8 μH and resonance frequencies of 9 ... 40 MHz have been used for the measurements. The unloaded quality factor has been measured for all tank circuits at various temperatures and values $80 < Q < 200$ were obtained. A driving rf current I_{rf} was supplied to the tank circuit and a voltage U across the circuit was measured using an amplifier at room temperature with a high input impedance. The amplitude and phase of the voltage were detected by a vector-analyzer and recorded as a function of either an amplitude of I_{rf} or I_{dc} . The coupling factor k was extracted from the period ΔI_{dc} of the $\alpha - I_{dc}$ plot according to $M\Delta I_{dc} = \Phi_0$. The value of k was between 0.04 and 0.1. The interferometer was placed inside of a five-layer magnetic shielding in a gas-flow cryostat at a temperature in the range of $4.2 < T < 90$ K. Some measurements were performed also in double magnetic shielding in liquid nitrogen at 77 K and in liquid helium at 4.2 K.

The small signal limit and adiabatic case were proved experimentally. First, from $U(I_{rf})$ curves we defined the amplitude of I_{rf} needed to generate Φ_0 in the interferometer. Then, all measurements were carried out at the amplitude of I_{rf} corresponding to the flux (0.05 ... 0.15)Φ_0 ensuring the small signal limit.

The adiabatic case has been established in two ways. (*i*) The resistance of the junctions, fabricated under identical conditions as the junctions in interferometer, have been measured separately. In all cases a resistance $R_n > 1\ \Omega$ has been obtained leading to the estimation $\omega\tau < 0.01$. (*ii*) The value of $\omega\tau$ can be estimated from the change of quality factor by coupling the sample to the tank circuit. Here the quality factor was measured at a high pumping level ($\varphi_{rf} \gg 1$). The change of Q is not only due to the presence of the weak link in the ring, but also due to the losses in the washer material. Thus, from this measurements we can give only an upper limit for $\omega\tau$, which corresponds to $\omega\tau < 0.15$ for the interferometers. These estimations show that the relaxation time of the interferometer $\tau = L/R$ is short ($\tau \ll 1/\omega_0$) and hence the quasiparticle current (fourth term in (2.9)) is negligible.

4 Results

4.1 24° Grain-Boundary Junctions

An example of the measured dependence of the phase shift α on the I_{dc} for 24° GB JJ is shown in Fig. 2. The phase shift due to the electronics α_0 has been determined from the condition

$$\int_{-\pi}^{\pi} \tan(\alpha - \alpha_0)\, d\varphi_{dc} = 0 \tag{4.18}$$

which can be obtained from (2.9), (2.14) and the condition that $f(0) = f(\pi) = 0$.

Figure 3 shows the CPR corresponding to the experimental data of Fig. 2.

As was discussed in the introduction, the deviation observed for the CPR from harmonic behavior can be explained by the influence of thermal noise. Moreover, as shown in [28], taking into account thermal noise, the apparent CPR can be fitted assuming a reasonable amplitude for the flux noise. The solid lines in Figs. 2 and 3 correspond to the fits with sinusoidal CPR with $\beta = 0.9$ and thermal fluctuation $\Phi_n = 0.5\Phi_0$ [28]. The total noise flux due to the Johnson noise in the RSJ model [2] (classical limit) for this sample corresponds to $\sim 0.2\Phi_0$. This is lower than the noise required to fit the data. The situation is typical because for real SQUIDs, as a rule, the noise is higher than the classical limit. The reason is the intrinsic noise coming from the loop material and from the critical current fluctuations in the weak link.

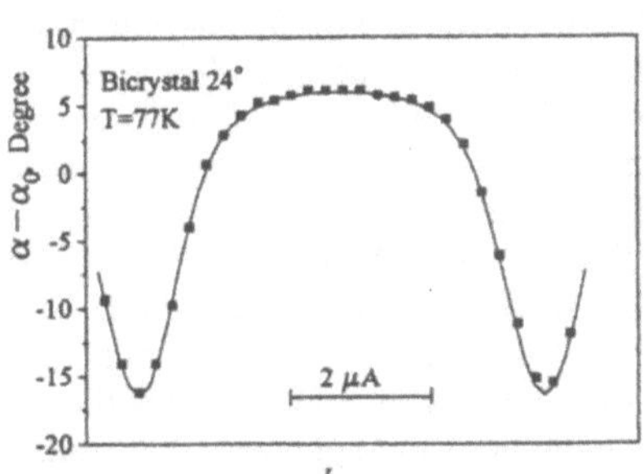

Figure 2
The phase shift α as a function of the dc current in the tank coil coupled to the superconducting ring with 24° grain boundary junction at 77 K. α_0 is the the phase shift of electronics (see text). Solid line shows the fit with sinusoidal CPR in the presence of thermal noise.

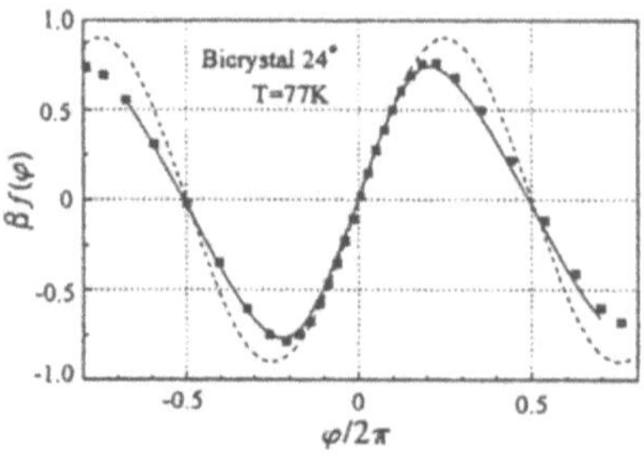

Figure 3
The CPR for 24° GB JJs at $T = 77$ K (points). The function $\beta\sin(\varphi)$ with $\beta = 0.9$ (dashed line) and fit with harmonic CPR taking into account the thermal noise (solid line) is shown for comparison.

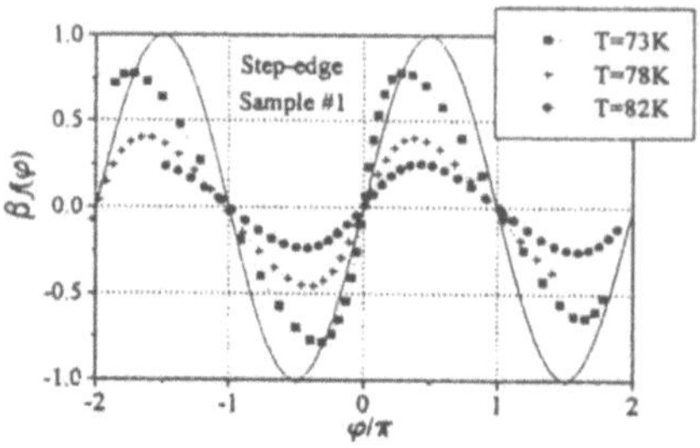

Figure 4
The CPR for sample #1 at various temperatures (symbols). $\sin\varphi$ (solid line) is shown for comparison.

4.2 Step-Edge Junctions

The same deviations from harmonic dependence have been observed for step-edge junctions (Fig. 4). It is clear that the apparent CPR is nonsinusoidal. The maximum of the current in the interferometer is observed at the phase $\varphi_{max} < \pi/2$, and the slope of the CPR curve at 73 K near $\varphi = 0$ is larger than unity. With increasing temperature φ_{max} shifts to $\pi/2$, the slope at $\varphi = 0$ decreases, and at $T = 82$ K the CPR can be described by a sinusoidal dependence. At temperatures lower than 73 K the sample #1 has characteristics typical for a hysteresis interferometer with $\beta > 1$.

The curve at 73 K can also be explained by the influence of noise on the measured characteristics. The characteristic frequency of the flux fluctuations is of the order of $1/\tau$ [23]. Therefore, in the adiabatic limit we have slow sweep. The estimation from (1.6) and (1.7) shows, that $\sigma_\varphi > \Delta\varphi_e/2$ is fulfilled for $\beta < 2$ at liquid nitrogen temperature for the given value of the inductance L. In this case, the apparent time-averaged $\varphi(\varphi_e)$ dependence has a finite slope at the points close to $\varphi = \pi + 2\pi n$ instead of the hysteresis and the slope larger than unity at $\varphi = 2\pi n$. The same has been obtained for sample #2.

Experimental results for the sample #3 are shown in Fig. 5. The measured CPR at $T = 4.2$ K is nearly sinusoidal, $\varphi_{max} \approx \pi/2$, and the slope of $\beta f(\varphi)$ at $\varphi = 0$ is 0.92, hence $\beta < 1$. A small deviation is found near $\varphi = \pi$. When β is close to unity, even the small thermal fluctuations and the rf flux φ_{rf} can distort the $\varphi(\varphi_e)$ dependence and hence the apparent CPR, because for $\beta \to 1$ the derivation $d\varphi/d\varphi_e$ diverges in the vicinity of $\varphi = \pi$ according to (1.5). At $T = 21$ K (Fig.5), the CPR deviation from a sinusoidal dependence is more pronounced than at 4.2 K, because the value of I_c at $T = 21$ K is nearly the same as at $T = 4.2$ K, but thermal fluctuations are much higher. At higher temperatures the apparent CPR becomes sinusoidal, as it was found for the samples #1 and #2. Therefore, for all samples, the deviations of the apparent CPR from harmonic behavior can be described by the influence of thermal noise in the ring.

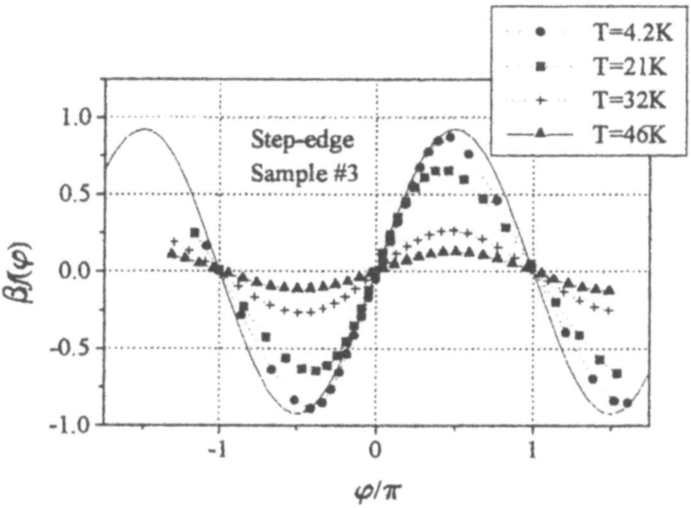

Figure 5 The CPR for sample #3 at various temperatures (symbols). $0.92 \sin \varphi$ (solid line) is shown for comparison.

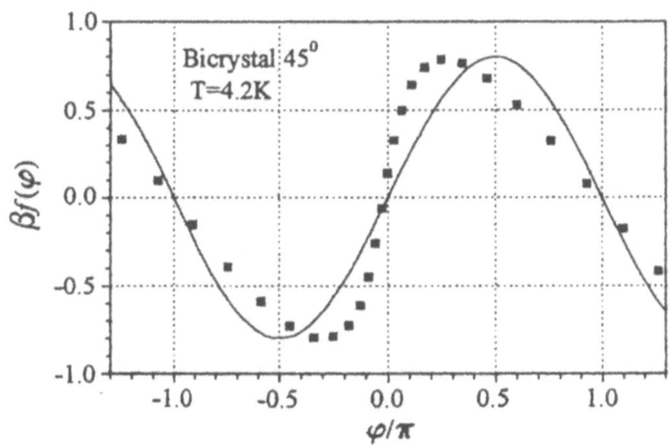

Figure 6 The CPR for 45° grain-boundary junction at 4.2 K.

4.3 45° Grain-Boundary Junctions

The CPR for 45° GB JJs at $T = 4.2$ K is shown in Fig. 6. Three samples fabricated on different substrates and measured with several tank coils followed closely the same behavior. It is clear from the Fig. 6, that the apparent CPR is strongly non-harmonic. It can not be explained in the same way as for the sample #1 at 73 K (Fig. 4). The slope of the curve $\beta f'(0) \approx 3.5$ is too large and the temperature is too low. In the next section we discuss the possible explanations of the observed anomalous CPR.

5 Discussion

For symmetric 24° bicrystal JJs as well as for step-edge JJs at high temperatures, harmonic CPR has been obtained. This result can be expected because short weak links should have harmonic CPR at T close to T_c [3]. Harmonic CPR obtained for step-edge HTS JJs at low temperatures is not so trivial. As it was discussed above, constriction type models [6] and resonant tunneling models [7–10] predict a nonsinusoidal CPR at low T, similar to the classical weak links. Since experimentally this is not the case, this gives evidence that the Cooper pairs do not use these channels, i.e. supercurrent flow is mediated by direct tunneling through interfaces of low transparency. Such a statement is consistent with many other data for HTS JJs and agrees with the phenomenological intrinsically shunted tunnel junction model, which assumes that the quasiparticle current is dominated by resonant tunneling whereas the Cooper pairs can only tunnel directly [11]. A microscopic origin for such a behavior may come from the Coulomb repulsion or from a potential barrier at the junction interface. This issue requires further study.

The experimentally observed deviations from a sinusoidal dependence for the current-phase relations of the 45° bicrystal junctions are startling. It is important to emphasize, that the estimated effective Josephson penetration depth

$$\Lambda_J = \left(\frac{\Phi_0}{4\pi\mu_0 \langle j_c \rangle \lambda} \right)^{1/2} \tag{5.19}$$

is approximately 5 μm. This value is larger than the width of the junction b (narrow junction limit). Here λ is the London penetration depth. Although several mechanisms are known to cause non-sinusoidal current-phase relations for narrow junctions fabricated from conventional superconductors, all of these mechanisms fail to account for the anomalous dependencies presented here.

First, one potential source of such deviations is thermal noise. At T=4.2 K $L \ll L_N$, therefore thermal fluctuations are negligible.

Second, for weak links with a high current density, current-induced suppression of the order parameter in the electrodes close to the weak link, can give rise to non-harmonic current-phase relations [32]. It is unrealistic that this effect is the cause for the deviations presented here, as the intragrain critical current density exceeds 10^7 A/cm^2 at 4.2 K and is therefore much larger than the grain boundary $\langle j_c \rangle < 4000$ A/cm^2 at the same temperature.

Third, several additional mechanisms, described in [5], lead also to deviations from harmonic CPR. In all these cases, the slope of $f(\varphi)$ at $\varphi = 0$ is smaller than at $\varphi = \pi$, which is in contrast to our results. Therefore also these mechanisms cannot cause the current-phase relations observed.

If the flux in the junction region is not negligible (junction is not narrow), but for some reason the total current through the junction can not be large: $\langle j \rangle \ll j_{loc}$.

(i) It may occur, e.g., if Abricosov vortices are frozen in the vicinity of the junction. Let us express the current density as

$$j(x) = j_c(x) \sin(\varphi(x)) \; . \tag{5.20}$$

Here $0 < x < b$ holds along the grain boundary. The phase profile $\varphi(x)$ is caused by the local magnetic field, and $j_c(x)$ is determined by the local properties of the boundary. When $\Phi_e = m\Phi_0$ the total current is zero, so

$$\int_0^b j_c(x)\sin\varphi_{loc}(x)dx = 0\,. \tag{5.21}$$

If $\Phi_e = m\Phi_0 + \Delta\Phi$, the total supercurrent will be

$$I = t\int_0^b j_c(x)\sin\left(\varphi_{loc}(x) + \Delta\varphi(x)\right)dx\,, \tag{5.22}$$

where t is the film thickness and $\Delta\varphi(x)$ is the additional phase difference due to the flux $\Delta\Phi$ in the loop. For our case the self shielding effect can be neglected ($\Lambda_J > b$), therefore $\Delta\varphi(x)$ does not depend on x: $\Delta\varphi(x) = \varphi$. Then from (5.22) using (5.21) one can obtain:

$$I = t\sin\varphi\int_0^b j_c(x)\cos\left(\varphi_{loc}(x)\right)dx = I_c\sin\varphi\,, \tag{5.23}$$

i. e., the apparent CPR is harmonic.

(ii) Another reason to have $\langle j\rangle \ll j_{loc}$ can be unconventional order parameter. There is considerable evidence that the symmetry of the order parameter in $YBa_2Cu_3O_{7-x}$may be a $d_{x^2-y^2}$ -wave symmetry. In this case the strong facetting of the grain boundary leads to the appearance of a parallel array of usual and so-called π-junctions. In particular, a spontaneous generation of magnetic flux in the 45° [001] tilt grain boundary junction has been observed experimentally [30]. The model describing a novel state with a self-generated magnetic flux in JJs has been developed recently [31]. Following the approach of [31], the critical current density $j_c(x)$ is a random function taking positive and negative values with a typical length-scale l,

$$j_s(x) = \langle j_c\rangle\left(1 + g(x)\right)\,, \tag{5.24}$$

where

$$\langle j_c\rangle = \frac{1}{b}\int_0^b j_c(x)dx \tag{5.25}$$

is the average value of the critical current density ($b \gg l$) and $\langle g(x)\rangle = 0$. Also $\Lambda_J \gg l$. The phase difference $\varphi(x)$ along the grain boundary $0 < x < b$ has been written as

$$\varphi(x) = \psi(x) + \xi(x) \tag{5.26}$$

where $\psi(x)$ is smooth function and $\xi(x)$ is rapidly alternating function and $|\xi(x)| \ll 1$. If these two conditions are valid, the CPR is given by

$$\langle j\rangle = \langle j_c\rangle(\sin\psi - \frac{\gamma}{2}\sin 2\psi) \tag{5.27}$$

where $\gamma > 0$. The dependence obtained is different from the measured CPR at any values of parameter γ. So this model also can not explain the observed CPR.

There are two possibilities to explain the obtained CPR. Firstly the CPR is locally non-sinusoidal. Secondly one can suppose, that the $\xi(x)$ is not small and external magnetic field can cause a drastic change of the $\xi(x)$. This situation is similar to the spin-glass behavior of superconducting clusters, where the free energy has many minima corresponding to various spatial distributions of the supercurrent [33]. It leads to a frustration of the system. In this case by increasing φ we should observe dependence (5.23) with a large slope (an effectively large I_c) up to some threshold, and then the deviation from (5.23).

For the first case an additional model should be considered, for the second case dominated by a $d_{x^2-y^2}$ symmetry component of the the order parameter should be supposed. We do not exclude the case with s-wave symmetry of the order parameter and locally harmonic CPR with other mechanisms which are not known at present.

Acknowledgements

The authors thank A.I. Braginski, V. Glyantsev, H. Hilgenkamp, J. Mannhart, and A. Golubov for fruitful discussions. We are grateful to A. Chwala and H.-J. Koeler for their assistance in the temperature-dependence measurements. Financial support by the DFG (Ho 461/1-1) is gratefully acknowledged.

Bibliography

[1] B.D. Josephson, Phys. Lett. **1**, 251 (1962), and Rev. Mod. Phys. **36**, 216 (1964).

[2] A. Barone, G. Paterno, *Physics and Application of the Josephson effect*, Wiley-Interscience Publication John Wiley and Sons, (1982)

[3] L.G. Aslamasov and A.I. Larkin, Pis'ma Zh. Eksp. Teor. Fiz. **9**, 87 (1969).

[4] L.D. Jackel, T.D. Clark, and R.A. Buhrman, IEEE Trans. Magn. **MAG 11**, 723 (1975).

[5] Likharev, Rev. Mod. Phys. **51**, 101 (1979).

[6] A.A.Golubov, V.M.Krasnov, and M.Yu.Kupriyanov, IEEE Trans. Appl. Supercond. **7**, 3204 (1997).

[7] R.Gross and B.Mayer, Physica C **180**, 235 (1991).

[8] J.Halbritter, Phys.Rev. **B46**, 14861 (1992).

[9] L.I.Glazman and K.A.Matveev, JETP Lett. **49**, 659 (1989).

[10] I.A.Devyatov and M.Yu.Kupriyanov, JETP Lett. **59**, 200 (1989).

[11] R.Gross, L.Alff, A.Beck, O.M.Froelich, D.Koelle, and A.Marx, IEEE Trans. Appl. Supercond. **7**, 2929 (1997).

[12] A.A.Golubov, I.A.Devyatov, M.Yu.Kupriyanov, G.J.Gerritsma, and H.Rogalla, Inst. of Physics Conference Series No 158 Bristol, Philadelphia, 463-466 (1997).

[13] C.C. Tsuei, J.R. Kirtley, C.C. Chi, Lock See Yu-Jahnes, A. Gupta, T. Shaw, J.Z. Sun, and M.B. Ketchen, Phys. Rev. Lett. **73**, 593 (1994).

[14] D.J. Van Harlingen, Rev. Mod. Phys. **67**, 515 (1995).

[15] D.J. Scalapino, Phys. Rep. **250**, 329 (1995).

[16] W. Zhang, Phys. Rev. **B52**, 3772 (1995).

[17] Y. Tanaka and S. Kashiwaya, Phys. Rev. **B53**, R11957 (1996).

[18] S.S Tinchev, Physica **C222**, 173 (1994).

[19] V. Polushkin, S. Uchaikin, S. Knappe, H. Koch, B. David, D. Grundler, IEEE Trans. on Appl. Supercond. **5**, 2790 (1995).

[20] L.D. Jackel, R.A. Buhrman, and W.W. Webb, Phys. Rev. **B10**, 2782 (1974).

[21] R. Rifkin and B.S. Deaver, Phys. Rev. **B13**, 3894 (1976).

[22] E.V. Il'ichev, V.M. Zakosarenko, R.P.J. IJsselsteijn, and V. Schultze, J. Low Temp. Phys. **106**, Nos. 3/4, 503 (1997).

[23] J. Kurkijärvi, Phys. Rev. **B6**, 832 (1972).

[24] J. Clarke and R.H. Koch, Science **242**, 217 (1988).

[25] T. Ryhänen, H. Seppä, R. Ilomoniemi, J. Knuutila, J. Low Temp. Phys **76**, 386, (1989).

[26] B. Cheska, J. Low Temp. Phys, to be published.

[27] E.V. Il'ichev, V. Zakosarenko, V. Schultze, H.-G. Meyer, H.E. Hoenig, V. N. Glyantsev, A. Golubov , Appl. Phys. Lett., **72**, 731 (1998).

[28] V. Zakosarenko, E.V. Il'ichev, R.P.J. IJsselsteijn, V. Schultze, IEEE Trans. on Appl. Supercond. **7**, 1057 (1997).

[29] V.N. Glyantsev, Y. Tavrin, W. Zander, J. Schubert, and M. Siegel, Supercond. Sci. Technol. **9**, A105 (1996).

[30] J. Mannhart, H. Hilgenkamp, B. Mayer, Ch. Gerber, J.R. Kirtley, K.A. Moler, M. Sigrist, Phys. Rev. Lett., **77**, 2782 (1996).

[31] R.G. Mints, to be published.

[32] M.Yu. Kupriyanov, Pis'ma Zh. Eksp. Teor. Fiz. **56**, 414 (1992) (JETP Lett. **56**, 399, (1992)).

[33] C. Ebner, D. Stroud, Phys. Rev. **B31**, 165 (1985).

Cherenkov Radiation from Josephson Fluxons

A. V. Ustinov[1], E. Goldobin[2], G. Hechtfischer[1], N. Thyssen[2], A. Wallraff[1], R. Kleiner[1], and P. Müller[1]

[1] Physikalisches Institut III, Universität Erlangen-Nürnberg, D-91058 Erlangen, Germany
[2] Institut für Schicht- und Ionentechnik, Forschungszentrum Jülich, D-52425 Jülich, Germany

Abstract: Josephson fluxons can generate Cherenkov radiation if they move with a velocity larger than the lowest phase velocity of Josephson plasma waves in the junction. This condition can not be realized in conventional one-dimensional long Josephson junctions. Nevertheless, Cherenkov radiation can occur from fluxons moving in more complex Josephson structures such as stacked Josephson junctions. We discuss the physics of this interesting phenomenon and present experimental data obtained with two different superconducting systems, namely artificially prepared Nb-Al-AlO_x-Nb stacks and naturally layered $Ba_2Sr_2CaCu_2O_{8+y}$ single crystals. These experiments, supported by numerical calculations using the model of inductive coupling between the junctions, well agree with theoretical predictions.

1 Introduction

Josephson fluxons, often also called Josephson vortices or Josephson solitons, appear in various contexts within nonlinear physics, superconductivity, and high-frequency device applications [1–3]. A fluxon in a long Josephson junction accounts for a circular supercurrent which yields the magnetic flux of one flux quantum $\Phi_0 = h/2e = 2.07 \times 10^{-15}$ Wb located between two superconducting electrodes of the junction. In many cases fluxon dynamics is similar to the dynamics of a relativistic particle, for which the role of the velocity of light plays the maximum velocity of electromagnetic wave propagation in the junction, also called Swihart velocity. Typically, this velocity is by about two orders of magnitude smaller than the velocity of light in vacuum.

The appearance of a fluxon in a long Josephson junction can be understood from Fig. 1. This figure sketches the cross view of the junction in the plane perpendicular to the external magnetic field H. Josephson tunnel barrier is a thin layer of insulator (I) between two superconducting electrodes (S). Due to the Meissner effect, the external field is screened by circulating supercurrents and it penetrates inside a bulk superconductor to the distance known as the London penetration depth, λ_L. Typically, λ_L is of the order of 100 nm. In the region of the Josephson barrier the screening effect is weakened, thus the magnetic field penetration distance is larger. This distance is called the Josephson penetration depth λ_J. Its value depends on the strength of the Josephson coupling (determined by the thickness of the tunnel barrier) and typically is of the order of some μm. We will be interested here

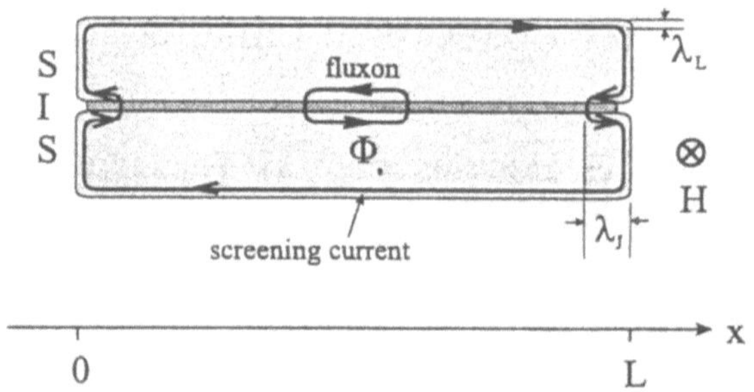

Figure 1
Schematic cross section of a long Josephson junction with a magnetic field H applied perpendicular to the plane of the picture; dimensions are not to scale.

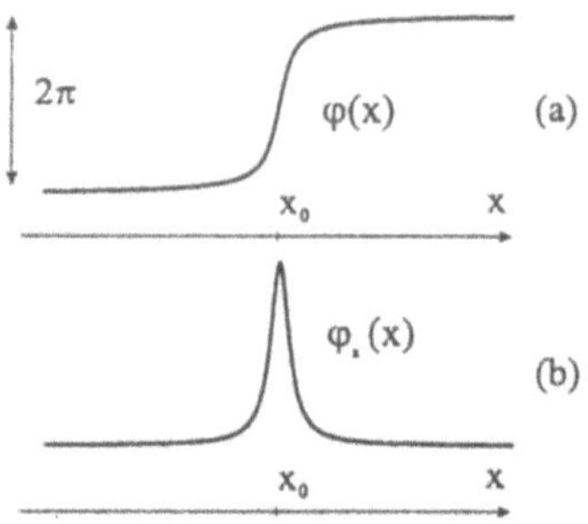

Figure 2
(a) Fluxon in a long Josephson junction accounts for a 2π kink in the phase difference $\varphi(x)$ with (b) the self-generated magnetic field proportional to $\varphi_x(x)$.

in the case when the junction is "long", so that the Josephson barrier extends over the distance L considerably larger than λ_J and the junction width W is made smaller than λ_J.

Due to the influence of the bias current flowing across the Josephson junction, the screening current tangle at the junction edge may become unstable and form a closed loop which enters the interior of the junction. This circulating supercurrent is often called Josephson vortex, and the magnetic flux generated by the vortex is equal to Φ_0.

Mathematically, the fluxon corresponds to a 2π kink of the quantum-mechanical phase difference φ between the two superconducting electrodes of the junction. The perturbed sine-Gordon equation which describes the quasi-one-dimensional dynamics of this system [4], written in the normalized form, is

$$\varphi_{xx} - \varphi_{tt} - \sin\varphi = \alpha\varphi_t - \beta\varphi_{xxt} - \gamma. \tag{1.1}$$

Here, the subscripts denote the derivatives in x and t. Time t is measured in units of ω_0^{-1}, where ω_0 is the Josephson plasma frequency, the spatial coordinate x is measured in units of λ_J, α is a dissipative term due to quasi-particle tunneling, β is a dissipative term due to surface impedance of the superconductors, and γ is a normalized bias current density. The derivation of (1.1) is based on the Maxwell and Josephson equations and can be found, e.g., in [4]. The characteristic velocity for electromagnetic waves propagating in the junction is called the Swihart velocity, $\bar{c} = \lambda_J\omega_0$.

A quasi-linear solution of (1.1) with zero right hand side exists in the limit of small amplitudes $\varphi_0 \ll 0$ where (1.1) can be linearized. This solution corresponds to the small-amplitude waves $\varphi_{pl} = \varphi_0 e^{i(kx-\omega t)}$ with the frequency ω and the wavenumber k given by the plasma-type dispersion relation $\omega^2 = 1 + k^2$. These quasi-linear waves are usually called *Josephson plasma waves*.

An important nonlinear solution to (1.1) with zero right hand side is the *soliton*

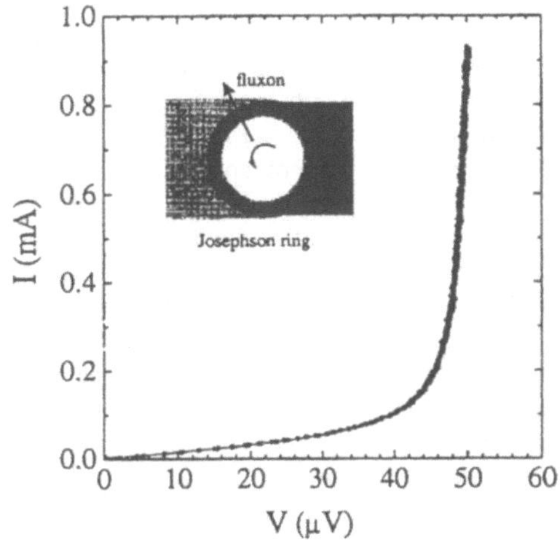

·V curve)
diameter

ar) junc-
stigation

Figure 3
Experimentally measured current-voltage characteristics (I–V curve) of a single fluxon in an annular Josephson junction. Junction diameter $D \approx 130\,\mu$m, temperature 5.7 K.

$$\varphi = 4\,tan^{-1}\left(exp\frac{x - x_0 - ut}{\sqrt{1-u^2}}\right) . \tag{1.2}$$

This solution describes a single fluxon as sketched in Fig. 2(a). It corresponds to a 2π-kink moving with a velocity u and located at $x = x_0$ for $t = 0$. The velocity u is measured in units of $\bar{c}$ and may assume values $0 \leq u < 1$. A unique property of the most of Josephson tunnel junctions is that the parameters α, β, and γ are small. Thus, the solution of the perturbed (1.1) can be well approximated by (1.2). To get a fluxon moving in a long Josephson junction, a dc bias current γ through the junction can be applied. The bias current acts with a Lorentz force F_γ on the fluxon and provides the energy input which is sufficient to compensate the friction force F_α) which corresponds to the α and β terms in (1.1). In the following, in the order to simplify our discussion, we will neglect the latter assuming $\beta = 0$.

According to the perturbation theory of McLaughlin and Scott [4], the soliton velocity u is determined by a balance between the losses, governed by α term, and the energy input due to the bias γ:

$$u = \frac{1}{\sqrt{1 + \left(\frac{4\alpha}{\pi\gamma}\right)^2}} \tag{1.3}$$

For low values of the bias current $u \propto \gamma/\alpha$, while for large values of γ/α the normalized velocity u asymptotically approaches unity, i.e., a fluxon behaves like *a relativistic particle* with respect to the limiting velocity $\bar{c}$. As $u \to 1$, the magnitude of the local magnetic field φ_x in the center of the fluxon increases and the fluxon width decreases, according to the well-known relativistic effect of the Lorentz contraction.

After the fluxon has passed a given point x_0 inside the junction, the phase difference $\varphi(x_0)$ shifts by 2π. A periodic fluxon motion will be accompanied by a dc voltage $V = \Phi_0/\langle \Delta T \rangle$, where $\langle \Delta T \rangle$ is the average time interval between the passing fluxons.

As an example, let us consider a special case of a ring-shaped (annular) junction geometry. An annular junction serves as an excellent model for investigation of soliton dynamics, which can be studied here under periodic boundary conditions. Due to the magnetic flux quantization in a superconducting ring, the number of fluxons initially trapped in the annular junction is conserved. An example of experimentally measured current-voltage characteristics (I–V curve) of a single fluxon trapped in an annular Josephson junction is given in Fig. 3. A circular motion of the fluxon under the influence of a current passing through the junction induces a dc voltage proportional to its average velocity. When increasing the bias current, the fluxon velocity increases and approaches the Swihart velocity which corresponds to the dc voltage $V_1 = \Phi_0\bar{c}/(\pi D) \approx 50\,\mu$V, with D being the junction diameter. In Fig. 3 one can see that the critical current of the annular junction with the trapped fluxon is very small, it means that we need to apply a very small force (critical current) to start the fluxon motion in this almost pinning-free system.

In conventional long junctions with open boundaries a sufficiently large external magnetic field H leads to the so-called flux-flow regime. In this mode fluxons are created at one boundary of the junction, move through the junction under the influence of the bias current, and annihilate at the other boundary. The spacing between the moving fluxons is inversely proportional to H. At large bias current, the fluxon motion with the velocity close to the Swihart velocity $\bar{c}$ is manifested experimentally by the *flux-flow step* (FFS) in the I–V characteristics [5]. For a single-barrier junction, FFS appears at the dc voltage $V_{FF} = (2\lambda_L + t)\bar{c}H$. The flux-flow regime is discussed below in relation to the measurements of $Ba_2Sr_2CaCu_2O_{8+y}$ (BSCCO) crystals.

2 Stacked Junctions

Fluxons in stacked Josephson junctions have recently become a subject of intensive theoretical and experimental investigations. The discovery of the intrinsic Josephson effect in some HTS such as BSCCO convincingly showed that these materials are essentially natural superlattices of Josephson junctions formed on the atomic scale [6–10]. The spatial period of such a superlattice is only 1.5 nm, so the Josephson junctions are packed extremely dense. The superconducting electrodes are formed by the copper oxide bilayers as thin as 0.3 nm and are separated by the non-superconducting BiO layers.

Modern thin film technology allows to grow high-quality superconducting multilayers with many Josephson tunnel barriers. Artificially prepared low-T_c stacked junctions can serve as model systems for layered high-T_c superconductors. The best multi-junction low-T_c stacks are made using Nb-Al-AlO_x-Nb junctions [11,12].

Superlattices with many Josephson layers can naturally be expected to show very complex dynamics. Therefore, it is important at first to understand in detail the dynamics of stacked junctions with very few layers.

2.1 Two Coupled Junctions

For the first time fluxon dynamics in two inductively coupled long Josephson junctions was considered theoretically by Mineev et al. [14]. The perturbation approach for small coupling has been further explored by Kivshar and Malomed [13] and Grønbech-Jensen et al. [15]. A very important step towards quantitative comparison with real experiments was made by Sakai et al. [16] who derived a model for arbitrary strong coupling between the junctions. According to that model, two stacked junctions are described by a system of coupled perturbed sine-Gordon equations

$$\begin{aligned}
&\frac{1}{1-S^2}\,\varphi^{\mathrm{A}}_{xx}-\varphi^{\mathrm{A}}_{tt} \quad -\sin\varphi^{\mathrm{A}}-\frac{S}{1-S^2}\,\varphi^{\mathrm{B}}_{xx}=\alpha\varphi^{\mathrm{A}}_{tt}+\gamma^{\mathrm{A}}\,;\\
&\frac{1}{1-S^2}\,\varphi^{\mathrm{B}}_{xx}-\varphi^{\mathrm{B}}_{tt} \quad -\sin\varphi^{\mathrm{B}}-\frac{S}{1-S^2}\,\varphi^{\mathrm{A}}_{xx}=\alpha\varphi^{\mathrm{B}}_{tt}+\gamma^{\mathrm{B}}\,.
\end{aligned} \tag{2.4}$$

Here $\varphi^{\mathrm{A}}(x,t)$ and $\varphi^{\mathrm{B}}(x,t)$ are the superconducting phase differences across the stacked junctions A and B, respectively, and γ^{A} and γ^{B} are the bias currents. The coupling coefficient $S=-\left[\left(\frac{t}{\lambda_{\mathrm{L}}}+\coth\frac{d}{\lambda_{\mathrm{L}}}+\coth\frac{d_{\mathrm{e}}}{\lambda_{\mathrm{L}}}\right)\sinh\frac{d}{\lambda_{\mathrm{L}}}\right]^{-1}<0$ can be calculated from experimental parameters such as the tunnel barrier thickness t, the middle electrode thickness d, the thickness of the top and bottom electrodes d_{e}. Obviously, the coupling parameter S vanishes for $d \gg \lambda_{\mathrm{L}}$. It is associated with screening currents in superconducting electrodes which are shared by fluxons belonging to different layers. A typical experimental value for S lies in the interval from -0.2 to -0.9.

Equations (2.4) lead to two different modes, one with in-phase and another with out-of-phase oscillations in the two junctions. Recently both in-phase and out-of-phase modes were observed experimentally [17]. The similar modes have been predicted many years ago by Ngai [18] for the Josephson junction coupled with superconducting transmission line. The corresponding wave propagation velocities are $\bar{c}_- = \bar{c}/\sqrt{1-S}$ (out-of-phase) and $\bar{c}_+ = \bar{c}/\sqrt{1+S}$ (in-phase) [19].

When applying a magnetic field H parallel to the Josephson barrier, one finds the Fiske-steps in the I–V curve of stacks. Fiske-steps are caused by the resonances of linear waves on the length of the junction. In two coupled junctions one may observe two families of Fiske-steps corresponding to different modes of plasma waves. By measuring the voltage spacings between neighboring Fiske-steps $\Delta V_- = \bar{c}_-\Phi_0/(2L)$ and $\Delta V_+ = \bar{c}_+\Phi_0/(2L)$ the characteristic velocities $\bar{c}_-$ and $\bar{c}_+$ can be measured experimentally for double-junction stacks with different thickness d of the common superconducting layer [19]. With decreasing d, the coupling is increasing, thereby increasing the difference between $\bar{c}_-$ and $\bar{c}_+$. A detailed analysis of experimental data is found to be in good quantitative agreement with theory [19].

Fluxon configurations in two-stacked junction can be noted $[N|M]$ meaning N fluxons located in one LJJ and M fluxons in the other LJJ ($N,M<0$ describe anti-fluxons). Using numerical simulations [16], it has been demonstrated that two solitary fluxons form the stable bound state $[1|1]$ with identical phases in two junctions $\varphi^{\mathrm{A}}(x,t)=\varphi^{\mathrm{B}}(x,t)$. This state has been analytically shown to be stable in the velocity range $\bar{c}_- < v < \bar{c}_+$ [20]. The coherent fluxon-antifluxon state ($[1|-1]$) which is stable up to the limiting velocity $\bar{c}_-$ has been reported in experiments by Carapella et al. [21]. The asymmetric $[1|0]$ fluxon mode in two-fold stack leads to Cherenkov radiation which will be discussed in Section 3.

2.2 Multi-Layer Stacked Junctions

Josephson superlattices consisting of many stacked tunnel junctions were discussed in the literature long before they first became available for experiments [22–24]. A multi-junction stack consists of alternating superconducting and isolating layers. In a magnetic field H applied parallel to the layers, fluxons penetrate into different Josephson junctions and, under the influence of the bias current may move coherently due to the interaction of their screening currents flowing in the inner superconducting layers.

The equations for such a stacked junction system may be obtained as a generalization of the single junction sine-Gordon equation. Josephson junction stacks consisting of N junctions ($i = 1, \ldots, N$) contain $N+1$ superconducting layers. The system of coupled sine-Gordon equations which describe the Josephson phase dynamics in the stack can be written in the following form [16]

$$\frac{\Phi_0 C_i}{2\pi}\frac{\partial^2\phi_i}{\partial t^2} + \frac{\Phi_0 G_i}{2\pi}\frac{\partial\phi_i}{\partial t} = I_B - j_{c,i}\sin\phi_i + \frac{\Phi_0}{2\pi\mu_0}\sum_{j=1}^{N} F_{i,j}\frac{\partial^2\phi_j}{\partial x^2} \tag{2.5}$$

where $i = 1, 2, \ldots, N$ is the junction index, ϕ_i is the superconducting phase difference on the junction number i, Φ_0 is the magnetic flux quantum. C_i, G_i, $j_{c,i}$ and I_B are the junction capacitance, quasiparticle conductance, critical current and bias current densities, respectively. The coupling between junctions i and j is defined by matrix element $F_{i,j}$ [16]. The major parameter which determines the strength of coupling between the junctions i and $i-1$ is

$$S \equiv \frac{s_i}{d'_{i,i-1}} = -\frac{\lambda_L}{\sinh\frac{d_i}{\lambda_L}\left(t + \lambda_L\coth\frac{d_{i-1}}{\lambda_L} + \lambda_L\coth\frac{d_i}{\lambda_L}\right)}\ , \tag{2.6}$$

where d_i is the thickness of the superconducting layer i between the junctions. t is the thickness of the tunnel barrier, the London penetration depth λ_L we assume both to be the same in all the layers. The strongest coupling ($|S| \to 0.5$) is achieved for $d_i \ll \lambda_L$.

Resonant modes for superlattices have been first calculated by Kleiner [25]. Similar to Fiske modes of conventional single-barrier junctions, Kleiner obtained general expressions for eigenfrequencies corresponding to standing wave patterns in the directions normal to the layers and along the layers. For N–junction stacks there exist N different modes [25,19]. Kleiner modes are characterized by phase shifts between Josephson oscillations in different layers. In fact, the above mentioned $\bar{c}_-$ and $\bar{c}_+$ Fiske modes of two-fold stacks are nothing else but the lowest and the highest Kleiner modes: For a two-junction stack there are only two modes possible, the symmetric mode and the antisymmetric mode.

3 Cherenkov Radiation in Stacked Junctions

In a one-dimensional Josephson junction, the phase velocity of Josephson plasma waves $u_{ph} = \omega/k$ can never be smaller than 1 (Swihart velocity in normalized units), while the fluxon velocity u can only be smaller than 1. Thus, the Cherenkov radiation condition $u_{ph} = u$ for a fluxon moving as a relativistic particle, cannot be satisfied in usual long Josephson junctions.

The idea of possible Cherenkov radiation by a fluxon moving in some more complex cases has been discussed in several theoretical papers [13,26,27]. Very recently, we reported first experimental evidences for the Cherenkov radiation [28,31]. These experiments, performed on two very different Josephson systems are discussed below. The mechanism of the phenomenon is very general: Cherenkov radiation can be generated if the fluxon velocity u becomes equal to the phase velocity $u_{ph} = \omega/k$ of linear electromagnetic waves in the system. In contrast to single-barrier long Josephson junctions, this condition can be satisfied in stacked structures.

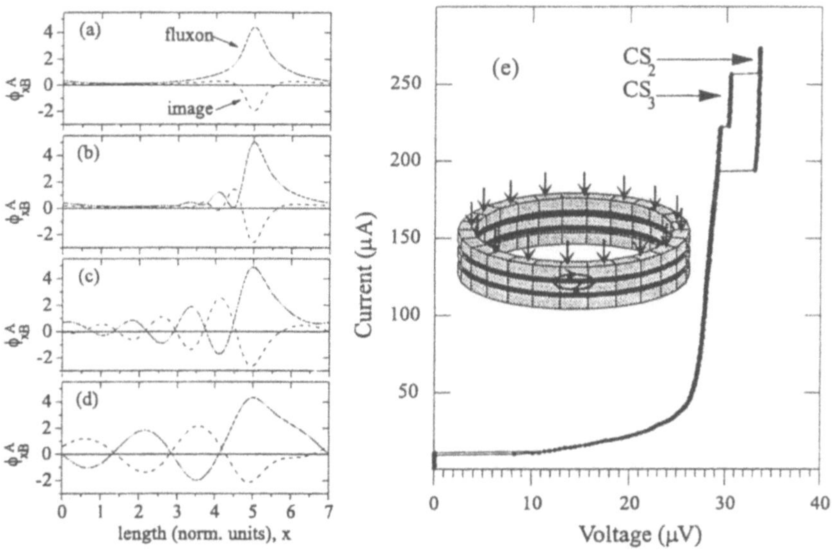

Figure 4 Simulated profiles of $\varphi_x^{A,B}(x)$ in [1|0] state of two stacked annular junctions [28]: (a) fluxon at velocity $u < c_-$, and (b), (c), and (d) with steadily increasing of u above c_-. (e) Experimental I–V characteristic of such stack in [1|0] state [28]. Cherenkov resonances are marked $CS_{2,3}$. The inset shows schematically the sample geometry.

3.1 Two Junctions: Nb-Al-AlO$_x$-Nb Stacks

Very clean evidence for Cherenkov radiation by a moving soliton has been found in a system of two stacked annular Josephson junctions [28]. Numerical simulations of the so called [1|0] configuration (1 soliton moving in junctions A and no soliton in junction B) shown in Fig. 4a–d demonstrate that the soliton in junction A moves together with is its image in junction B. As soon as the soliton velocity u exceeds $\bar{c}_-$, an oscillating wake corresponding to Cherenkov radiation arises behind the moving soliton and its image, as shown in Fig. 4b-d. The emitted radiation belongs to the out-of-phase mode as can be seen from Fig. 4b-d. With increasing the soliton velocity, the wavelength of the radiation increases while the amplitude and length of the wake quickly grow.

As the velocity of soliton increases, the Cherenkov radiation wake grows. If its length becomes comparable with the junction length, the radiation can form a running wave in which an integer number of Cherenkov radiation wavelengths fit into the circumference of the junction. The interaction of the running Cherenkov wave with the fluxon leads to the resonances on I–V characteristic of the system. The experimentally measured single-soliton step with Cherenkov resonances is shown in Fig. 4e. Resonances CS_2 and CS_3 are observed at $u > \bar{c}_-$ where $\bar{c}_-$ corresponds to the junction voltage of about 29μV. Numerical simulations of the I–V curve using experimental parameters demonstrated excellent agreement with measurements [28].

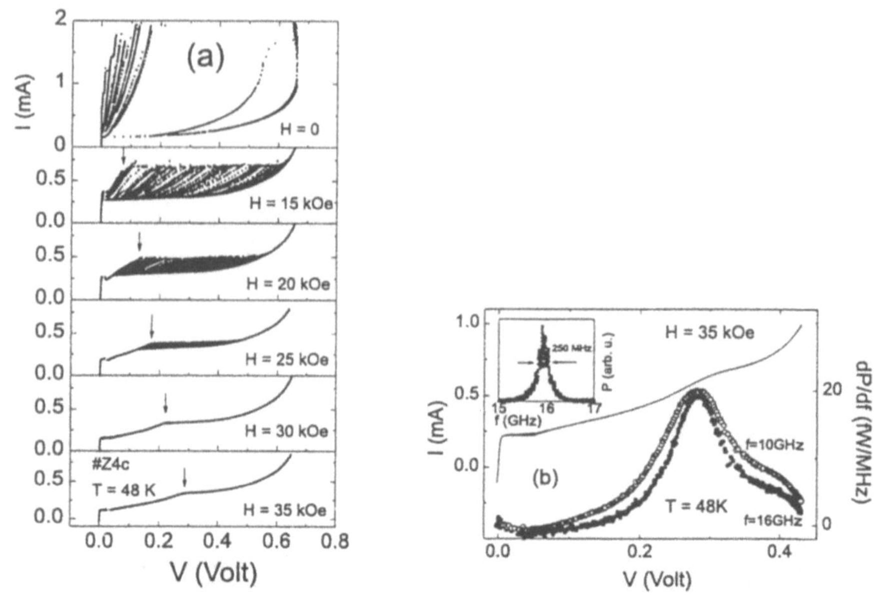

Figure 5 (a) *I–V* characteristics of a naturally grown BSCCO stack at various magnetic fields [31]. The arrow denotes the top of the flux-flow branch, i. e. the point of maximum fluxon velocity. (b) *I–V* characteristic and ***non-Josephson*** microwave emission vs. *V*. The form of the emission peak is almost the same for different frequencies, suggesting that the non-Josephson signal is extremely broadband [31]. The inset shows a typical spectrum of the conventional ***Josephson*** emission taken at 3 mV bias.

3.2 Many Junctions: $Ba_2Sr_2CaCu_2O_{8+y}$ Natural Stacks

Single crystals of BSCCO and some other materials [29] show properties expected for multi-layer stacks of Josephson junctions. Among these properties are the multi-hysteretic switching of individual tunnel junctions to gap voltages [7,8], the magnetic field dependence of the c–axis critical current [9] and flux-flow voltage [10,30], and Josephson radiation emission detected in several frequency bands [30]. Since it is rather difficult to measure the voltage on a single Josephson junction of atomic scale, natural stacks are usually measured in series.

An example of BSCCO stack *I–V* characteristics for magnetic fields between 20 kOe and 35 kOe is shown in Fig. 5a [31]. In zero magnetic field, the *I–V* characteristic of an intrinsic junction stack with N junctions consists of N branches which differ by the number of junctions switched to the gap state. The magnetic field primarily increases the slope of the flux-flow branch where no junction is switched to the gap [10,30]. The arrows in Fig. 5a show the characteristic flux-flow voltage which reaches 0.3 V at 35 kOe. As the stack consists of about 95 Josephson junctions this voltage corresponds to a Josephson frequency of 1.8 THz.

We observed [31] a very unusual broadband non-Josephson radiation emission from BSCCO samples shown in Fig. 5b. At the higher magnetic fields, an additional microwave

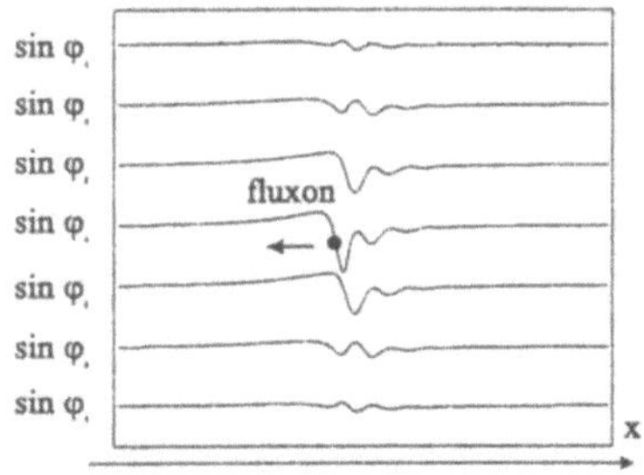

Figure 6
Spatial distribution of the supercurrents in a strongly coupled stack of seven Josephson junctions with periodic boundary conditions. A single fluxon indicated by a circle is moving in the middle junction at a velocity of 0.816 $\bar{c}$ which is higher than the lowest two mode velocities, 0.721 and 0.765 $\bar{c}$. Cherenkov radiation in the form of a trailing wave and waves induced in the neighboring junctions can clearly be seen.

emission is found at voltages much higher than the voltage of the Josephson signal illustrated in the inset of Fig. 5b. The intensity of this signal is increasing towards the top of the flux-flow branch and is by almost one order of magnitude higher than that of the Josephson signal.

Based on the magnetic field dependence of the radiation power and numerical simulations, we argue [31] that the emission is due to the Cherenkov radiation by Josephson fluxons moving in the multilayered stack. For a stack with N junctions, there are N different linear mode velocities. For strong coupling, the lowest mode velocity is about $\bar{c}/\sqrt{2}$ [30]. As soon as the fluxon velocity u rise above $\bar{c}/\sqrt{2}$ Cherenkov emission should appear. A numerical illustration of this effect in a system of 7 stacked Josephson junctions is shown in Fig. 6 [31]. Simulations are based on the coupled sine-Gordon model as described above and in [25,30]. To eliminate the influence of the boundaries, periodic boundary conditions were used. Figure 6 shows a single fluxon steadily moving in the middle junction with a velocity of 0.816 $\bar{c}$, which is above the lowest two linear mode velocities. The waves trailing the fluxon are caused by Cherenkov radiation: these waves are only observed when the lowest mode velocity of 0.721 $\bar{c}$ is exceeded! Cherenkov emission in intrinsic multi-junction stacks with large number of layers can be expected to occur simultaneously at different but close frequencies.

The non-Josephson signal was detected at frequencies that are by two orders of magnitude lower than the Josephson frequency. The occurrence of a low frequency signal can be understood as a downconverted mixing product of all the plasma waves exited by Cherenkov mechanism. The mixing of the different modes is expected in a complex and highly nonlinear system as the stacked Josephson junctions. To demonstrate this process, a stack of 21 Josephson junctions with the damping parameter $\alpha = 0.14$ and open boundary conditions was calculated numerically for the same coupling parameter as in Fig. 6. The magnetic field was chosen to give a distance between fluxons of $1.15\lambda_J$, similar to the experimental conditions. The voltage over the whole stack, taken at the boundary, was calculated for discrete time steps and was Fourier transformed to evaluate the microwave emission spectrum. Five of these spectra for different bias currents are shown in Fig. 7. For the lower two normalized bias currents, fluxons are moving slower than the lowest mode velocity. The spectrum contains mainly the Josephson signal and its second harmonic. The additional small signal at a normalized frequency of 0.1 is the fundamental cavity mode. For bias currents higher than $i = 0.560$, the fluxons in some junctions exceed the lowest mode velocity. As can be seen from Fig. 7, the Josephson signal and its harmonic are broadened, and an additional emission signal at very low frequencies appears. These simulation results, indeed, suggest that the observed non-Josephson radiation is due to the downcoversion of the multi-frequency Cherenkov emission.

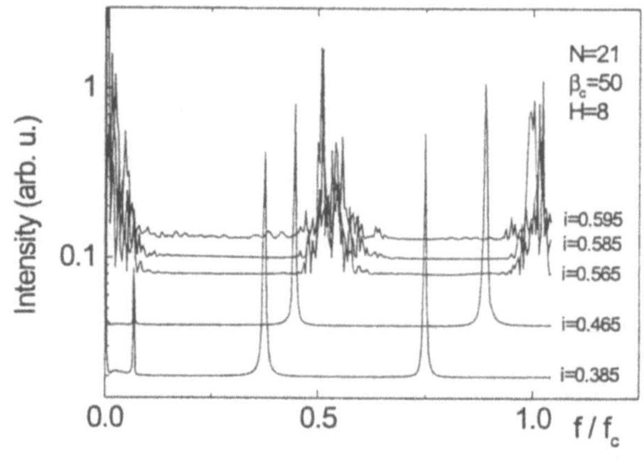

Figure 7 Fourier transforms of the voltage over a stack of 21 Jospehson junctions for five different values of normalized bias current and a normalized field of $H = 8\Phi_0/[\mu_0 B \times 1.5\,\text{nm}]$. For the two lower values, fluxons are slower than the lowest mode velocity. For current higher than about 0.560, fluxons move faster than the lowest mode velocity.

4 Conclusion

We presented observations of Cherenkov radiation by Josephson fluxons moving in multilayer tunnel junctions. Experiments have been made with two different superconducting systems, namely artificially prepared Nb-Al-AlO_x-Nb stacks and naturally layered $Ba_2Sr_2CaCu_2O_{8+y}$ crystals. In the first case, Cherenkov radiation in two stacked annular Josephson junctions leads to resonances which result from interaction of Cherenkov radiation wake with a Josephson fluxon turning around the junction. Experimental data are in good quantitative agreement with the proposed model and numerical simulations. In the second case, large number of wave propagation modes, equal to the number of intrinsically stacked junctions, leads to a broad band non-Josephson emission which has been detected experimentally. These results suggest that rather unusual fluxon dynamics may take place in intrinsic junctions of high temperature superconductors.

Finally, we would like to note that Cherenkov radiation from moving fluxons may appear in various complex Josephson structures, including planar Josephson junction arrays, two-dimensional junctions and other complex systems.

Bibliography

[1] N. F. Pedersen, in: *Solitons*, S. E. Trullinger, V. E. Zakharov, V. L. Pokrovsky, eds. (Elsevier, Amsterdam, 1986), p. 469.

[2] R. D. Parmentier, in: *The New Superconducting Electronics*, H. Weinstock, R. W. Ralston, eds. (Kluwer, Dordrecht, 1993), p. 221.

[3] A. V. Ustinov, *Solitons in Josephson junctions*, to appear in Physica D (1998).

[4] D. W. McLaughlin and A. C. Scott, Phys. Rev. A **18**, 1652 (1978).

[5] T. Nagatsuma, K. Enpuku, F. Irie, and K. Yoshida, J. Appl. Phys. **54**, 3302 (1983); **56**, 3284 (1984); **58**, 441 (1985); **63**, 1130 (1988).

[6] R. Kleiner, F. Steinmeyer, G. Kunkel and P. Müller, Phys. Rev. Lett. **68**, 2394 (1992).

[7] R. Kleiner and P. Müller, Phys. Rev. B **49**, 1327 (1994).

[8] A. Yurgens, D. Winkler, N. Zavaritsky, and T. Claeson, Phys. Rev. B **53**, R8887 (1996).

[9] Yu. I. Latyshev, J. E. Nevelskaya, and P. Monceau, Phys. Rev. Lett. **77**, 932 (1996).

[10] J. U. Lee, J. E. Nordman, and G. Hohenwarter, Appl. Phys. Lett. **67**, 1471 (1995).

[11] R. Monaco and A. Oliva, Appl. Phys. Lett. **64**, 3042 (1994).

[12] H. Kohlstedt, F. König, P. Henne, N. Thyssen, and P. Caputo, J. Appl. Phys. **80**, 5512 (1996)

[13] Yu. S. Kivshar and B. A. Malomed, Phys. Rev. B **37**, 9325 (1988).

[14] M. B. Mineev, G. S. Mkrtchjan, and V. V. Schmidt, J. Low Temp. Phys. **45**, 497 (1981).

[15] N. Grønbech-Jensen, M. R. Samuelsen, P. S. Lomdahl, and J. A. Blackburn, Phys. Rev. B **42** 3976 (1990); N. Grønbech-Jensen, O. H. Olsen, and M. R. Samuelsen, Phys. Lett. A **179A**, 27 (1993).

[16] S. Sakai, P. Bodin, and N. F. Pedersen, J. Appl. Phys. **73** (1993) 2411.

[17] A. V. Ustinov, H. Kohlstedt, M. Cirillo, N. F. Pedersen, G. Hallmanns, and C. Heiden, Phys. Rev. B **48** (1993) 10614.

[18] K. L. Ngai, Phys. Rev. **182** 555 (1969).

[19] S. Sakai, A. V. Ustinov, H. Kohlstedt, A. Petraglia, and N. F. Pedersen, Phys. Rev. B **50**, 12905 (1994).

[20] N. Grønbech-Jensen, D. Cai, and M. R. Samuelsen, Phys. Rev. B **48**, 16160 (1993).

[21] G. Carapella, G. Costabile, A. Petraglia, N. F. Pedersen, and J. Mygind, Appl. Phys. Lett. **69**, 1300 (1996).

[22] P. R. Auvil and J. B. Ketterson, J. Appl. Phys. **61**, 1957 (1987).

[23] A. F. Volkov, Pis'ma Zh. Exsp. Teor. Fiz. **45**, 299 (1987) [JETP Lett. **45**, 376 (1987)]; A. F. Volkov, Pis'ma Zh. Exsp. Teor. Fiz. **50**, 127 (1989) [JETP Lett. **50**, 139 (1989)].

[24] B. A. Malomed, I. B. Khalfin, and B. Ya. Shapiro, *Sol. St. Commun.* **87**, 223 (1993).

[25] R. Kleiner, Phys. Rev. B **50**, 6919 (1994).

[26] R. G. Mints and I. B. Snapiro, Phys. Rev. B **52**, 9691 (1995).

[27] V. V. Kurin and A. V. Yulin, Phys. Rev. B **55**, 11659 (1997).

[28] E. Goldobin, A. Wallraff, N. Thyssen, and A. V. Ustinov, Phys. Rev. B **57**, 130 (1998).

[29] P. Müller, in: *Festkörperprobleme / Advances in Solid State Physics*, vol. 34, ed. by Helbig (Vieweg, Braunschweig), p.1 (1995).

[30] G. Hechtfischer, R. Kleiner, K. Schlenga, W. Walkenhorst, P. Müller, and H. L. Johnson, Phys. Rev. B **55**, 14638 (1997).

[31] G. Hechtfischer, R. Kleiner, A. V. Ustinov, and P. Mueller, Phys. Rev. Lett. **79**, 1365 (1997).

SO(5) Symmetry in *t-J* and Hubbard Models

W. Hanke, R. Eder, E. Arrigoni, A. Dorneich, S. Meixner, and M.G. Zacher

Institut für Theoretische Physik, Universität Würzburg,
Am Hubland, D-97074 Würzburg, Germany

Abstract: Numerical and analytical results are reviewed, which support SO(5) symmetry as a concept unifying superconductivity and antiferromagnetism in the high-temperature superconductors. Exact cluster diagonalizations verify that the low-energy states of the two-dimensional *t-J* and Hubbard models, widely used microscopic models for the high-T_c cuprates, form SO(5) symmetry multiplets. Apart from a small standard deviation ($\sim J/10$), these multiplets become degenerate at a critical chemical potential μ_c (transition into doped system). As a consequence, the d-wave superconducting states away from half-filling are obtained from the higher spin states at half-filling through SO(5) rotations. Between one and two dimensions, using weak-coupling renormalization, a rather general ladder Hamiltonian including next-nearest-neighbor hopping can be shown to flow to an SO(5) symmetric point. Experimental tests and consequences such as the existence of a π-Goldstone mode both in the insulator and superconductor and, in particular, the relationship between the photoemission spectra of the insulator and superconductor, are emphasized.

1 Introduction

As their most prominent universal feature, high-temperature superconductors (HTSC) always display antiferromagnetism (AF) and d-wave superconductivity in close proximity, in their phase diagram. Recently, a unifying theory has been proposed, according to which these two at first sight radically different phases are "two faces of one and the same coin". They are unified by a common symmetry principle, the SO(5) symmetry [1]. Numerical simulations, i.e. exact cluster diagonalizations, have verified that this symmetry principle is obeyed for widely used microscopic models for the high-T_c cuprates, namely the (two-dimensional) *t-J* and Hubbard models, to within an accuracy which is better than the average superconducting energy gap [2–4]. These microscopic models are known to reproduce salient experimental results of the normal state of the high-T_c cuprates, including in particular the antiferromagnetic state. SO(5) symmetry then implies that these microscopic models also display $d-$wave superconductivity. This gives rise to a microscopic description of the complex phase diagrams of the HTSC from the insulating antiferromagnetic phase over to the metallic normal state and, finally, to the d-wave superconducting phase. These results were further corroborated between one and two dimensions by weak-coupling renormalization-group calculations, which demonstrated that rather general ladder Hamiltonians, including longer-ranged interactions [5,6] and hoppings [5], flow to an

SO(5)-invariant fixed point [7]. In this paper, the basic numerical and analytical calculations are summarized, and the microscopic principles behind SO(5) symmetry [8] are reviewed.

2 Antiferromagnetism and Superconductivity

In the HTSC, the dominant charge-carrier dynamics takes place in the two-dimensional (2D) CuO_2-planes [16]. Each CuO_2 unit cell contains an effective magnetic moment of spin $\frac{1}{2}$, essentially due to the Cu ion. Neighboring Cu-spins form singlets – the energy win due to the singlet formation, the magnetic exchange J is relatively large $\sim$ 120 meV $\sim$ 1 400 K. On the other hand, the temperatures for the transition into both low-temperature phases, the AF and the SC phases, $T_{\text{Néel}}$ and T_c, are both significantly lower and of similar magnitude ($\sim$ 250 K for $T_{\text{Néel}}$ and $\sim$ 100 K for T_c). Already this order of magnitude suggests that the mechanism of superconductivity does not directly result from the singlet formation, but is instead related to the mechanism, which results in AF in the undoped, insulating situation.

Indeed, both low temperature phases of the cuprates are "ordered": in the undoped case, i.e. in the insulator, we have AF order, in the doped case, the phase coherence of the superconductor and both phases are in the HTSC in immediate vicinity (the "spin glass phase", which occurs in some cuprates, is likely due to disorder). Therefore, it is tempting to unify these low-temperature phases, despite the fact that at glance, they appear dramatically different: on the one hand, the insulator and, on the other hand, the superconductor.

Let us consider first the insulator: at high-temperatures $\sim$ 1 000 K, the singlet pairs are completely disordered. This is a spin liquid. How does one arrive from this disordered state at high temperatures, at an ordered Néel state at low temperatures? To solve this problem, we shall consider a dynamical principle, which gives a particularly simple and transparent demonstration of the key feature of SO(5) theory, namely the one-to-one correspondence of antiferromagnetism and superconductivity: According to this dynamical principle [8], the ordered AF state can be considered as a kind of Bose-Einstein condensation of triplet excitations. The SC state, on the other hand, corresponds to a Bose-Einstein type of condensation of Cooper pairs, namely in the cuprate materials of hole pairs. Triplet excitation and hole-pairs excitation are "two faces of one and the same coin" in similarity to other unifying concepts such as the isospin theory of proton and neutron in nuclear physics [18]. The condensation energy yields then a new lower temperature scale $T_{\text{Néel}}$ and T_c.

Here, SO(5) symmetry enters in detail: Like a "magnifying glass", this symmetry principle allows to differentiate the low-energy physics of the order of T_c and, in particular, to unify the, in principle, competing AF and SC phases (specifically, this symmetry principle "rotates" triplet excitations into hole-pair excitations and vice versa). Unifying principles via symmetry are, of course, known from many fields of physics, such as quarks in high-energy physics, the predictions of which were inspired by the SU(3) classification of hadronic spectra. How does such a unification take place in the HTSC, i.e. in solid-state physics?

The order parameter of the AF is the sublattice magnetisation, a real 3-dimensional vector; if this vector is different from zero, we have AF order. Consider now two fixed neighboring sites in the 2-dimensional AF, let's say in the configuration (↑↓). This fixed spin configuration may be viewed as being due to an superposition of the singlet (↑↓ − ↓↑) with the $(S_z = 0)$−triplet (↑↓ + ↓↑). To create the macroscopic 2-dimensional AF in the CuO_2 plane, therefore, we have to mix triplet excitations already at high temperatures into all possible singlet configurations of the spin liquid. The AF states then result as a kind of "condensation" of the triplet excitations into the lowest possible energy state [8]. The three components of the triplet correspond to the three possible orientations of the AF and the density of the "condensed triplets" corresponds to the magnitude of the sublatice magnetisation. This dynamical principle illustrates the relation between AF and SC states rather clearly: If, in the AF state (we shall see below that it corresponds to a coherent state) the triplet excitation operator is replaced by a hole pair creation operator, we obtain a coherent state which creates a macroscopic number of Cooper pairs, i.e. a SC state. The "rotation" AF→SC, therefore, is described by an operator, the $\hat{\pi}$−operator of SO(5) theory [1], which replaces triplets by hole pairs.

In the following, we give a particularly simple illustration for this key-feature of SO(5) theory, which can be worked out for a ladder [8], and discuss the equivalence of antiferromagnetism and superconductivity for this example. The ground state of the ladder models is actually a resonating valence bond (RVB) type of vacuum without AF long-range-order [15]. However, for illustrative purposes, let us now construct an AF ordered state (which is in general not an eigenstate of the Hamiltonian) by condensing magnons into the RVB ground state. One can express the operator of staggered magnetization in z-direction as

$$M_s = \sum_n e^{i\pi n}(\, P_n(\uparrow\downarrow) - P_n(\downarrow\uparrow)\,), \tag{2.1}$$

where e.g. $P_n(\uparrow\downarrow)$ projects onto states where the nth rung has the configuration ↑↓ (see Fig. 4). It is easy to see that

$$\begin{aligned} (\, P_n(\uparrow\downarrow) - P_n(\downarrow\uparrow)\,)s_n^\dagger &= t_{n,z}^\dagger, \\ (\, P_n(\uparrow\downarrow) - P_n(\downarrow\uparrow)\,)t_{n,a}^\dagger &= \delta_{a,z}\, s_n^\dagger, \end{aligned} \tag{2.2}$$

where $s_n^\dagger(t_{n,\alpha}^\dagger)$ creates a singlet(α-tiplet) on rung n. Therefore

$$M_s = \sqrt{\frac{N}{2}}[\, t_z^\dagger(q=\pi) + t_z(q=\pi)\,]. \tag{2.3}$$

If we now form the coherent state

$$|\Psi_\lambda\rangle = \frac{1}{\sqrt{n}} e^{\lambda\sqrt{N} t_z^\dagger(q=\pi)}|\Omega\rangle, \tag{2.4}$$

which corresponds to z-like triplets condensed into the $k = \pi$ state, and treat the $\vec{t}$ as ordinary Bosons, we obtain

$$\langle\Psi_\lambda|M_s|\Psi_\lambda\rangle = \sqrt{2}\lambda N. \tag{2.5}$$

This calculation shows that by starting from a spin liquid – an RVB vacuum – an antiferromagnetically ordered state with M_S in z-direction can be generated by condensing z-like triplet-excitations into the $k = \pi$ state. At this point, we can invoke the SO(5) symmetry of the model [8], which tells us that since the z-like triplet with momentum π and the hole pair with momentum 0 are two different components of a 5-vector, they are dynamically indistinguishable. This means that the AF state, with condensed triplets, can be SO(5)-rotated into a state with condensed hole pairs. It follows that, if the AF state was the ground state at half-filling (which is the case for 2D materials and physical ladder systems), we can replace all z-like triplets by hole pairs with momentum 0 and by SO(5) symmetry automatically obtain the ground state in the doped case. The latter then consists of hole-pairs along the rungs condensed into the $k = 0$ states and is necessarily superconducting. In other words: both the AF and the superconducting state may be viewed as some kind of condensate 'on top of' the rung-RVB state. The SO(5) symmetry then simply implies that the condensed objects are combined into a single vector, whence unification of antiferromagnetism and superconductivity follows in a most natural way.

The above derivation makes sense only in a strong coupling limit, where a ground-state description starting out from rung-singlets is appropriate. One might expect, however, that similar considerations will apply also for cases with a weak coupling within the rungs [5,6] (see also Section 4).

In two dimensions (2D) the above interpretation of SO(5) symmetry hinges crucially on two points: first, the excitation spectrum of a 2D spin-liquid or RVB state must consist of Bosonic triplets (rather than e.g. Fermionic 'spinons'), whose condensation into the minimum of their dispersion at (π,π) leads to antiferromagnetic ordering. This would imply that the AF state in 2D also could be interpreted as a condensate of triplet Bosons. Work along this line is in progress and promising [19].

Second, these triplet-like Bosons with momentum (π,π) must be dynamically equivalent to a $d_{x^2-y^2}$ Cooper pair. Having established this equivalence one could immediately conclude that the AF phase (viz 'condensate of triplets') is identical to the d-wave superconducting phase (viz 'condensate of hole pairs') in the same sense as two nuclei belonging to the same Isomultiplet are 'identical'. Since the $\hat{\pi}$-operator in 2D [1] precisely converts a triplet with momentum (π,π) into a d-wave hole pair (see Section 3), the latter requirement is equivalent to $[H,\hat{\pi}] = \omega_0\hat{\pi}$. This commutation relation, which will be discussed in detail in the next Section, tells us whether the dynamics of charge carriers respects SO(5) symmetry, i.e. the Hamiltonian is in accord with the SO(5) rotation $\hat{\pi}$. Ideally, we should then have zero for the commutator. This will happen for a critical chemical potential μ_c, right where the transition into the doped system takes place. Further changing μ gives rise to a precession frequency, or energy to perform the AF $\rightarrow$ d-SC rotation $\omega_0 \neq 0$. The energy shift ω_0 here would correspond to the mass difference of proton and neutron in the Isospin algebra. The validity of this eigenoperator relation has already been verified numerically [3].

3 SO(5) Symmetry and Microscopic Models for HTSC

The dynamical principle of the previous Section suggests to unify the triplet excitations $\vec{t}^{\dagger}$ (corresponding to a vector) and hole-pair excitations Δ (described by (Re(Δ)), Im(Δ)) into a 5-dimensional vector, the so-called superspin vector. More precisely, Zhang suggested to group the AF order parameter $\vec{S}(\vec{Q})$ with $\vec{Q} = (\pi,\pi)$ and d-wave SC order parameter Δ into a single five-component vector, the superspin [1]

$$n_a = \left(\Delta^{\dagger} + \Delta, \vec{S}\left(\vec{Q}\right), -i\left(\Delta^{\dagger} - \Delta\right)\right), \tag{3.6}$$

where $\Delta = (i/2)\sum_{\vec{p}}(\cos p_x - \cos p_y)c_{\vec{p}}\sigma_y c_{-\vec{p}}$ denotes the $d_{x^2-y^2}$ superconducting order parameter. $\vec{S}\left(\vec{Q}\right) = \sum_{\vec{p}} c^{\dagger}_{\vec{Q}+\vec{p}}\vec{\sigma}c_{\vec{p}}$ stands for the AF Néel vector and σ_α are the Pauli spin matrices. The transition from AF to d-SC is then viewed as a kind of "superspin flop" transition as a function of the chemical potential or doping, where the direction of the superspin changes abruptly. This transition or SO(5) rotation is formally described by the so-called $\hat{\pi}$−operator.

The "superspin flop" transition is analogous to the problem of a spinning top in a uniform gravitational field, or a magnetic moment in a uniform magnetic field, as in nuclear magnetic resonance (NMR). As we shall see, in the presence of the symmetry-breaking field (which, in our case is the chemical potential or doping), the order parameter is forced to precess, and this explicit breaking of the SO(5) invariance induces rotations between AF and d-SC states (with frequency ω_0) and governs the competition between these two phenomena.

Mathematically, the $\hat{\pi}$−operator can straightforwardly be constructed [1]: as a rotation from AF to d-SC, it has to patch up the differences in the quantum numbers of the corresponding order parameters. Since Δ has spin $S = 0$, whereas $\vec{S}\left(\vec{Q}\right)$ has $S = 1$ (which is obvious from our triplet excitation picture in the last Section), $\hat{\pi}$ has to carry $S = 1$, i.e. it must be a triplet operator. $\vec{S}\left(\vec{Q}\right)$ has no charge, Δ has charge ± 2, therefore, π must create charge ± 2. Finally, $\vec{S}\left(\vec{Q}\right)$ has momentum $\vec{Q} = (\pi,\pi)$, Δ instead has $\vec{Q} = 0$, thus, $\hat{\pi}$ must have momentum $\vec{Q}$. Combination of these requirements fixes the operator uniquely up to a form factor, which is given by d-wave symmetry [38]. This results in the $\hat{\pi}$ operators (we just give one of them)

$$\hat{\pi}^{\dagger} = \sum_{\vec{p}}(\cos p_x - \cos p_y)c^{\dagger}_{\vec{p}+\vec{Q}\uparrow}c^{\dagger}_{-\vec{p}\uparrow}. \tag{3.7}$$

Charge conjugation and spin lowering operation gives five other similar operators. In real-space representation, the $\hat{\pi}\left(\hat{\pi}^{\dagger}\right)$ operator does precisely what we discussed in the previous Section, namely it replaces triplets oriented in x and y planar directions by $d_{x^2-y^2}$ hole (electron) pairs. These $\hat{\pi}$ operators were first introduced by Demler and Zhang [20] to explain the resonant neutron scattering peaks in the YBCO superconductors. Together with the total spin S_α, which is the generator of the SO(3) spin rotation forming a subgroup of SO(5), and the charge operator Q generating the $U(1)$ charge symmetry (which is also a subgroup of SO(5)), the six $\hat{\pi}$ ($\hat{\pi}^{\dagger}$) operators form the ten generators L_{ab} of the SO(5) algebra (for definition of L_{ab}, see [1]). As constructed, the $\hat{\pi}$ operators indeed rotate the AF order parameter into the d-SC order parameter

$$\left[\hat{\pi}^{\dagger}_{\alpha}, S\left(\vec{Q}\right)_{\beta}\right] = i\delta_{\alpha\beta}\Delta^{\dagger}, \tag{3.8}$$

and vice versa.

We have apparently accomplished the task of unifying AF with SC: the corresponding order parameters are grouped into a five dimensional object, and SC is "nothing" but AF viewed in some rotated coordinates and vice versa! This construction looks a bit similar to the unification of $\vec{E}$ and $\vec{B}$ by the Lorentz group. So far, this is only a mathematical construction, we haven't asked if "Mother Nature" approves the SO(5) construct or not.

In the high$-T_c$ problem, "Mother Nature" is very complicated, but we can check the SO(5) symmetry within some microscopic models, which are known to reproduce salient features of the phase diagram of HTSC. Such microscopic Hamiltonians, i.e. Hubbard and t-J models, successfully model the Mott-Hubbard insulator to metal transition [20,21], which is driven by the Coulomb correlation $U \sim 10$ eV, i.e. by "high-energy" physics. They also model prominent features of the magnetic interactions on an energy scale of order $J \sim 0.1$ eV [13–15,20–24]. However, their low-energy content of order of the average SC gap ($\sim J/5 - J/10$) has so far eluded both analytical and numerical investigations [1]. We shall demonstrate that SO(5) symmetry overcomes this major obstacle; it clarifies the role of competing orders and gives a microscopic description of the transition from AF to d-SC ground states, as the chemical potential is varied.

One can check the SO(5) symmetry by evaluating the commutator between the Hamiltonian with the $\hat{\pi}-$operators. In particular, numerical [2,3] works, summarized below, show that the $\hat{\pi}$ operators are approximate eigenoperators of the Hubbard Hamiltonian, in the sense that

$$[H, \hat{\pi}^{\dagger}_{\alpha}] = \omega_0 \hat{\pi}^{\dagger}_{\alpha}, \tag{3.9}$$

where the eigen-frequency ω_0, which is the energy to perform the AF $\rightarrow$ d-SC rotation, is of the order of J, and proportional to the number of holes. This relation (3.9) reminds us of the commutation relation between the transverse spin components S_+ and S_-in a magnetic field and the Zeeman Hamiltonian, with ω_0 being proportional to the $\vec{B}-$field. In our case, the SO(5) symmetry is broken explicitly by the chemical potential, i.e. ω_0 is proportional to the hole count or doping. Thus, the pattern of explicit symmetry breaking is simple and familiar, and, therefore, easy to handle.

In the following, we employ a kind of "computer spectroscopy" to test whether the dynamics of the charge carriers, i.e. the Hamiltonian, respects SO(5) symmetry and (3.9) is fulfilled. Ever since the early days of quantum dynamics, group theoretical interpretation of spectroscopy revealed deep symmetry and profound unity of Nature. Atomic spectra can be fitted into irreducible representations (irreps) of SO(3), and the regular patterns which emerged from this classification offered fundamental understanding of the periodic table. After the discovery of a large number of hadrons, the "embarrassment of riches" was removed by the classification of hadronic spectra into irreps of SU(3) and this hidden regularity inspired the predictions of quarks, the fundamental building block of the universe. In our quest for understanding the fundamental design of Nature, the importance of symmetry can never be over-emphasized.

In our work, we used a different kind of spectroscopy and its classification into a different kind of symmetry. The spectroscopy is performed on a computer, which numerically diagonalizes microscopic Hamiltonians, i.e. Hubbard and t-J models widely believed [20,21] to model high-T_c superconductors.

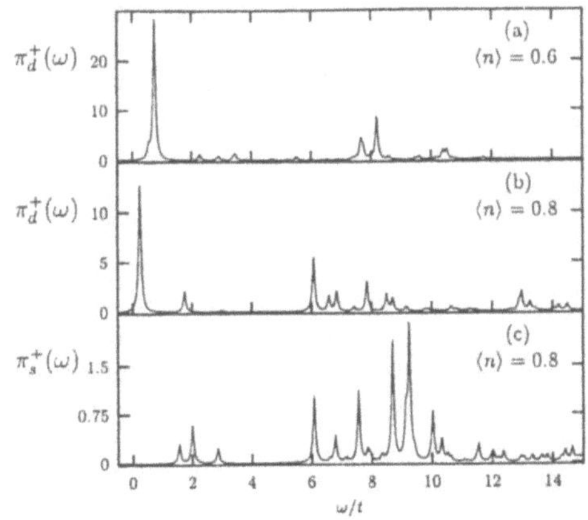

Figure 1
Dynamic correlation functions of the $\sqrt{10}\times\sqrt{10}$ Hubbard model with $U = 8t$:
(a) $\pi_d^\dagger(\omega)$-spectrum at $\langle n\rangle = 0.6$,
(b) $\pi_d^\dagger(\omega)$-spectrum at $\langle n\rangle = 0.8$,
(c) $\pi_s^\dagger(\omega)$-spectrum at $\langle n\rangle = 0.8$.

As the simplest 2D lattice model for correlated electrons, the one-band Hubbard model is defined as

$$H = -t \sum_{\langle i,j\rangle,\sigma} \left(c_{i\sigma}^\dagger c_{j\sigma} + \text{h.c.}\right) + U \sum n_{i\uparrow} n_{i\downarrow}, \tag{3.10}$$

with nearest-neighbor hopping t and Coulomb correlation U. On the other hand, the t-J model has the Hamiltonian

$$H = P\left[-t \sum_{\langle i,j\rangle,\sigma} \left(c_{i\sigma}^\dagger c_{j\sigma} + \text{h.c.}\right) + J \sum_{\langle i,j\rangle} \vec{S}_i \vec{S}_j\right] P. \tag{3.11}$$

As in the Hubbard model, $\langle ij\rangle$ denotes a summation over nearest-neighbors on the 2D square lattice and P projects onto the subspace with no doubly occupied sites. The latter constraint reflects the strong correlations in the $U/t \to \infty$ limit of the Hubbard model.

First numerical evidence for the approximate SO(5) symmetry of the Hubbard model came recently from exact diagonalizations of small-sized (10 sites) clusters [3], studying dynamic correlation functions involving the AF $\to$ d-SC rotation $\hat{\pi}$ operator. We observe that (3.9) is nothing but the ladder operator relation familiar from standard Quantum Mechanics problems such as the harmonic oscillator or the spin rising (S_+) and lowering (S_-) operators. If this equation is fulfilled, then the equal level distance ω_0 between the "rungs" of the ladder should appear – like in NMR or optical spectroscopy – as a sharp peak in the $\hat{\pi}-\hat{\pi}$ correlation function, well separated from a higher-energy incoherent background. This indeed is verified in Fig. 1a,b, which displays a typical result for the $\hat{\pi}_d$ correlation function in a ten-site Hubbard cluster with $U/t = 8$ for dopings of one hole-pair ($\langle n\rangle = 0.8$) and two hole-pairs ($\langle n\rangle = 0.6$) [3]. We note that ω_0 is a small energy, scaling with the hole count away from half-filling ($\langle n\rangle = 1$), i.e. $\omega_0 \cong J/2(1-\langle n\rangle) - 2\mu$.

The approximate relation (3.9) is highly nontrivial. One could ask if a similar relation would exist for a modified $\hat{\pi}$ operator which rotates AF into s-wave SC order parameters. The answer is negative [2,3], as seen in Fig. 1c which demonstrates that a $\hat{\pi}$–rotation with s-wave symmetry just generates an incoherent background and no sharp "π–resonance". Therefore, there is only an approximate symmetry between AF and d-wave SC (and not s-wave SC) near half filling.

In the next step, we use a most general and direct recipe for checking microscopic Hamiltonians for SO(5) symmetry, i.e. the concept of "superspin multiplets" [2]. We consider the t-J model, which, because of its more limited Hilbert space (no double occupancies), allows the exact diagonalization of larger systems (18, 20 sites). Since the t-J model explicitly projects out the states in the upper Hubbard band, some of the questions [25,26] raised recently about the compatibility between the Mott Hubbard gap and SO(5) symmetry can also be answered explicitly. In particular, if there is an approximate SO(5) symmetry of the microscopic model, the low-energy states of this model should fall into irreducible representations (irreps) of SO(5). In a given quantum mechanical system, the direction of the SO(5) superspin vector is quantized in a way similar to an ordinary SO(3) spin, and the classically intuitive picture of the precession of the SO(5) superspin vector under the influence of the chemical potential [1] can be identified with the equal level-spacing between the members of SO(5) multiplets carrying different charge. Therefore, numerically identifying the low-lying states of the microscopic model with the SO(5) irreps can lead to detailed understanding of the one-to-one correspondence and the level crossing between the excited states of the AF and the d-SC states, and thereby lead us to the microscopic mechanism by which the AF changes into the d-SC state. While finite-size calculations cannot generally be used to prove the existence of long-range order in infinite systems, the spectroscopic information about the SO(5) symmetry can be used as input for the effective field theory [1,27,28] which captures the low-energy and long-distance physics of the problem.

Exact diagonalizations (e.g. [15]) commonly study ground-state correlations, but their spatial decay is often inconclusive as a test of order due to small system size. Yet it is possible that the (excited) eigenstates show a well-defined structure characteristic of a particular symmetry; this provided the convincing evidence for long-range order in the spin-1/2 triangular lattice AF [9]. In our work, we have pursued exactly such a program in exact diagonalizations of the t-J model.

Consider as a simplest example the precession of a spin-1/2 system in a homogeneous $\vec{B}-$field in $z-$direction. $\vec{B}$ breaks the SO(3) spin-rotation symmetry and the spin (expectation value) precesses with Larmor frequency around the spatially fixed $\vec{B}-$direction. This can be read off in a spectroscopic experiment from the multiplet structure of the possible spin states $\left(|\pm\frac{1}{2}\rangle\right)$, the degeneracy of which is lifted by the $\vec{B}-$field (Zeeman effect). In this simplest case, the multiplet structure is one-dimensional, extending in S_z-"direction".

In our "computer spectroscopy", we employ a formally analogous recipe for checking the microscopic Hamiltonian for SO(5) symmetry:

In SO(5), the multiplet structure is two-dimensional, i.e. spanning both S_z-direction and Q-direction [2]. $Q = L_{15}$ is the total charge, and, in our t-J calculation, it stands for the number of doped hole pairs, and, thus, for the transition from AF to d-SC states. Formally, the multiplets are constructed by observing that $\{S_z, Q, C\}$ form a set of commuting operators with their quantum numbers labeling states of an SO(5) invariant Hamiltonian [2]. S_z is the $z-$component of the total spin ($S_z = -L_{23}$) and C the Casimir operator $\sum_{a<b} L_{ab}^2$, which is a natural generalization of the total spin operator $\vec{S}^2$. Like $\vec{S}^2$ in the familiar SO(3) spin-rotation symmetry, C fixes the level ν of an irreducible representation. Instead of $S(S+1)$ in SO(3), it takes the value $\nu(\nu+3)$ for an SO(3) level ν irreps. Fig. 2 shows the first four ($\nu = 0$ to $\nu = 3$) irreps of SO(5) with the low-lying states of an 18-site t-J model with $J/t = 0.5$, which is a typical J value [15].

What is the physical meaning of these multiplets? In fact, there is a definite physical meaning behind them, which is closely related to our coherent-state description of the

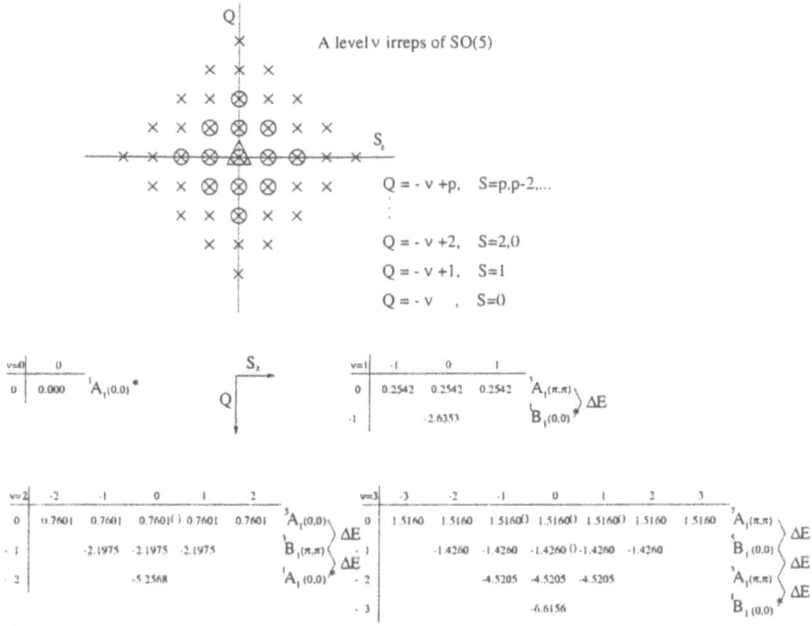

Figure 2 The upper diagram illustrates general level ν irreps of SO(5). Every state can be labeled by Q and S_z. The maximal charge is $Q = \pm\nu$. The states labeled by a × form the shape of a diamond, while states inside the nested diamonds are labeled by ∘ and △. Overlapping states with same Q and S_z are distinguished by their S quantum numbers. The lower diagrams are for $\nu = 1,2,3$ irreps of SO(5). The figure shows the energies of some low energy states for the 18-site cluster with $J/t = 0.5$. The states are grouped into different multiplets and are labeled by the spin, point group symmetry, and total momentum. A_1 denotes the totally symmetric, B_1 the $d_{x^2-y^2}$-like representation of the C_{4v} symmetry group. The () symbol denotes as yet unidentified members of the respective multiplet.

AF→d-SC transition in Section 2. The AF state is constructed from the linear superposition of the $Q=0$, $S_z=0$ states in each of the level ν multiplets. At a given level ν this state contains ν magnons or triplet excitations, which correspond to the ν−th term in the power-series expansion of the coherent-state operator, i.e. $e^{\lambda t^\dagger} = 1 + \lambda (t^\dagger)^{\nu=1} + \frac{\lambda^2}{2}(t^\dagger)^{\nu=2}\ldots$

Correspondingly, the d-SC state is constructed from a linear superposition of the lowest corner states of each level ν irreps, and may be viewed as stemming from the power-series expansion, $e^{\lambda\Delta} = 1 + \lambda\Delta^{\nu=1} + \frac{\lambda^2}{2}\Delta^{\nu=2}\ldots$

From Fig. 2 we note that the low-lying states indeed fit into the irreps of SO(5): all the different quantum numbers of the states are naturally accounted for by the quantum numbers of the superspin. However, most importantly, the levels with different charge Q are nearly equally spaced! This is explicitly indicated by the symbol ΔE in Fig. 2. More precisely, the mean-level spacing within each multiplet (up to $Q=-2$) is -2.9886 with a standard deviation of 0.0769. This standard deviation is much smaller ($\sim J/8$) than the natural energy scale J of the t-J model and comparable to or even smaller than the average SC gap. If one now adds the chemical potential term, $H_\mu = -2\mu Q$, then at a chemical potential μ_c comparable to the mean-level spacing, the superspin multiplets are nearly degenerate. In other words, (each of the coherent-state contributions to) the AF

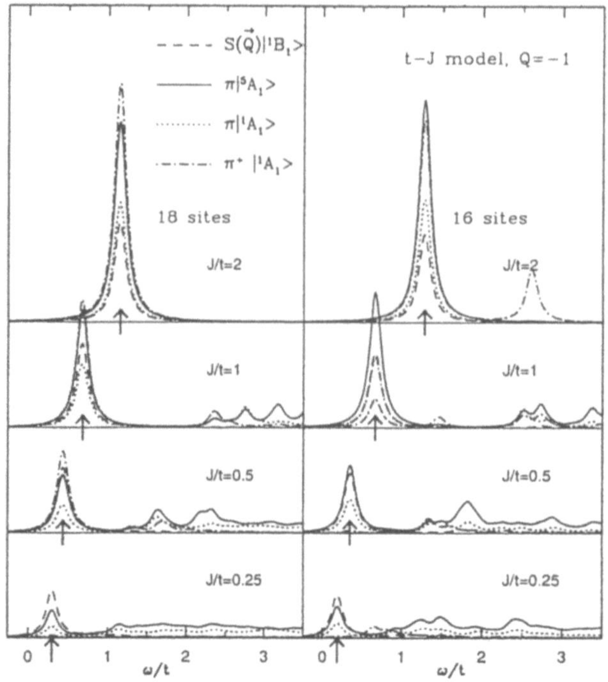

Figure 3 Spectral functions with final states in the $Q = -1$ subspace: dynamical spin correlation function for momentum transfer $\vec{Q}$, calculated for the ${}^1B_1(0,0)$ ground state; spectrum of the $\pi^\dagger$ -operator, calculated for the ${}^1A_1(0,0)$ ground state in the $Q = -2$ sector; spectra of the π-operator, calculated for the half-filled ${}^1A_1(0,0)$ ground state and the lowest half-filled ${}^5A_1(0,0)$ state.

state can freely be rotated in (the corresponding contribution to) the SC state. In Fig. 3, this is verified in terms of another "diagnostic tool", which is the spectral function

$$A(\omega) = \Im \frac{1}{\pi} \langle \psi | \hat{O}^\dagger \frac{1}{\omega - (H - E_{\text{ref}}) - i0_+} \hat{O} | \psi \rangle. \tag{3.12}$$

We see that, if we apply as the operator $\hat{O}$ the $\hat{\pi}^\dagger$ operator to the "d-SC" state in the $\nu = 2$ irreps (the $Q = -2, S_z = 0$ state ${}^1A_1(0,0)$), we end up in precisely the same final state (to within computer accuracy) as if we apply the $\hat{\pi}$ operator to the "AF" state ($Q = 0, S_z = 0$; ${}^5A_1(0,0)$). In other words, two successive $\hat{\pi}$–rotations rotate us from the AF state to the d-SC state. The energy required to perform this rotation, ω_0, is again seen to scale with J (in contrast to a false argument in [25]).

Finally, we note that the dynamical spin correlation function – where we take as the operator $\hat{O}$ in (3.12) the magnon operator $\vec{S}(\vec{Q})$ – has a peak at precisely the same position as the π–resonance. This confirms the earlier conjecture by Demler and Zhang [20] to interpret a well-known spin resonance detected in neutron-scattering experiments in HTSL, as a "π–excitation". More generally, this fingerprint of SO(5) symmetry can be described as follows: Away from μ_c, the chemical potential, which controls the changes in the total charge, breaks SO(5) symmetry. As already mentioned, the effect of this chemical potential is formally analogous to the effect of the $\vec{B}$–field on a spin. In the presence of this symmetry-breaking field, the order parameter is forced to precess and this explicit breaking of the SO(5) invariance induces transitions between AF and SC states, and governs the competition between these two phenomena. The SO(5) theory – in good accord with the experimental data [31] both concerning doping and temperature dependence – then identifies this precession frequency ω_0 with a resonance in neutron scattering.

4 SO(5) Symmetry in Ladders

4.1 Introduction

High-T_c materials are AF Mott insulators displaying long-range AF order at half filling. The AF phase is rapidly destroyed upon doping and is replaced by the superconducting phase. Below optimal doping (the doping with maximum T_c) and above the superconducting temperature there are clear experimental indications for the opening of a spin gap [30,32]. This phase is termed spin-liquid with properties which are quite difficult to reproduce on the theoretical level for a two-dimensional system. On the other hand, at and close to half filling this spin-gap variety of Mott insulators is obtained quite well in half-filled ladder systems: Upon doping the ladder, the spin gap persists and the system exhibits quasi-long-range d-wave superconducting correlations, which become dominating in some parameter range [10]. In this sense, ladder systems show properties very similar to the phase diagram of cuprate materials, the main difference being the fact that correlations are "quasi-long-ranged", i.e. they show power-law behavior, since they cannot be truly "long-ranged" because of one-dimensionality. Moreover, there also exist copper-oxides with CuO_2 planes containing line defects, which result in ladder-like arrangements of Cu-atoms [33]. These systems can be described in terms of coupled two-leg ladders exhibiting a spin gap and thus belong to the spin-liquid Mott-insulator variety. Also the related "stripe phases" of the 2D CuO_2 planes in the cuprate superconductors [11] can be mapped onto ladder systems.

It is thus interesting to study the occurrence of SO(5) symmetry in ladder systems. There is in fact a natural way to construct an SO(5) symmetric model for a two-leg ladder, which has only local interactions on a rung of the ladder [7]. We can thus use the ladder system as a theoretical laboratory to check some ideas of the SO(5) theory. On the other hand, we want to study whether *generic* models, which are *not* SO(5) invariant at the bare starting (microscopic) level may show SO(5) symmetry in their low-energy regime. An appropriate tool to study this low-energy regime, starting from a microscopic Hamiltonian, is the renormalization group (RG). This has been proven to be particularly suited for systems between one and two dimensions, for example ladder systems [34–36].

In the following, we consider generalized Hubbard-type ladders with Hamiltonians of the form

$$H = -t \sum_{\langle i,j\rangle,\sigma} \left(c^\dagger_{i,\sigma} c_{j,\sigma} + \text{h.c.} \right) - t_\perp \sum_{i,\sigma} c^\dagger_{i,\sigma} c_{i\pm\hat{y},\sigma} + U \sum_i n_{i,\uparrow} n_{i,\downarrow} + V \sum_{\langle i,j\rangle} n_i n_j + \frac{V_\perp}{2} \sum_i n_i n_{i\pm\hat{y}} + \frac{J_\perp}{2} \sum_i \vec{S}_i \vec{S}_{i\pm\hat{y}} . \quad (4.13)$$

(see Figure 4). The sum over i in (4.13) runs over the sites of a two-chain ladder, as before, $\langle i,j\rangle$ covers pairs of nearest-neighbor sites on the same chain and $i \pm \hat{y}$ is the nearest-neighbor site to i on the other chain.

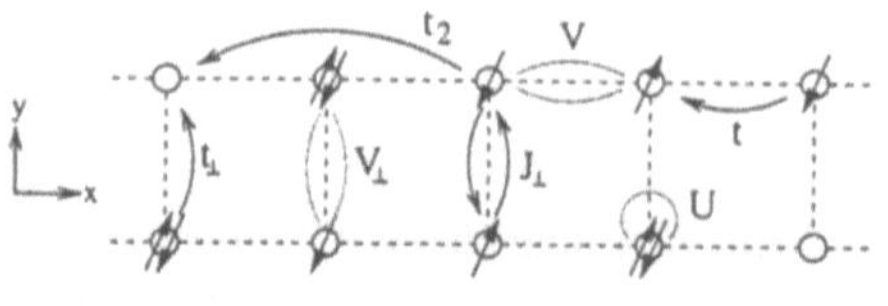

Figure 4
A generalized Hubbard-type ladder model: the Hamiltonian consists of nearest-neighbor electron hopping terms (t), on-site and off-site Coulomb interactions (U, V), and spin-spin exchange terms (J).

4.2 Exact SO(5) Ladder

In this paragraph we present "numerical spectroscopy experiments" for a recently proposed [7] exactly SO(5) symmetric ladder model based on a simplified version of the more general Hamiltonian (4.13): The non-local Coulomb and spin-spin interactions (the V and J terms in (4.13)) are restricted to act within the rungs, but not along the legs of the ladder, and hopping is only allowed between nearest-neighbor sites (with hopping t along the legs and $t_\perp$ within the rungs) (see Fig. 4). Setting $J_\perp = 4(U + V_\perp)$, the elementary magnetic and "SC" excitations of a half-filled single rung, namely the formation of a triplet and the formation of an electron (or hole) pair, become degenerate in energy. In this case, one obtains an exactly SO(5) symmetric Hamiltonian [7]. In a recent work [8], this ladder model was studied in detail and related to the triplet-excitation picture and coherent-state description of Section 2. Furthermore, this work gives the most general construction of the SO(5) irreducible representations not only for even numbers of holes (or electrons) as in Section 3, but also for odd numbers, encountered for example in direct or inverse photoemission (PE). Exact diagonalization was used to extract the practical application of the selection rules for photoemission implied by SO(5) symmetry. The classification of SO(3) symmetric states into spin multiplets, e.g. a triplet with $S = 1$, is well known. For SO(5) symmetry a similar classification holds, but to characterize a multiplet one needs two integer quantum numbers (p,q) [2,8]: within a given (p,q) multiplet the states are characterized both by S_z and the charge Q (instead of only S_z in SO(3)). Figure 5 displays some of the lowest SO(5) multiplets as well as some allowed PE transitions between them (note that for $p = q$, one gets the diamond-like structures shown in Fig. 2). These theoretical assertions can be checked numerically using the Lanczos technique [15]. The method permits to calculate ground state properties like energy and spin expectation values, but also Green's functions (cf. (3.12)), and, in particular, the photoemission and inverse photoemission spectra (here for a spin-down electron):

$$
\begin{aligned}
A_{\mathrm{PES}}(\vec{k},\omega) &= \frac{1}{\pi}\Im\langle 0|c^\dagger_{\vec{k}\downarrow}\frac{1}{\omega + H - \varepsilon_0 - i0^+}c_{\vec{k}\downarrow}|0\rangle, \\
A_{\mathrm{IPES}}(\vec{k},\omega) &= \frac{1}{\pi}\Im\langle 0|c_{\vec{k}\downarrow}\frac{1}{\omega + H - \varepsilon_0 - i0^+}c^\dagger_{\vec{k}\downarrow}|0\rangle.
\end{aligned}
$$

$|0\rangle$ is an initial energy-eigenstate found by Lanczos-diagonalization and ε_0 its energy. With this "computer experiment" one can study the direct and inverse PE within a multiplet structure and observe crucial selection rules. The SO(5) multiplets are easily identified, because the energies of states belonging to one multiplet are degenerate (i.e. identical to computer accuracy of ca. 10^{-14}). This allows to study the evolution of the single-particle spectral function as we pass from one multiplet (p,q) to the other, and within one multiplet through the different doping and spin levels (Q,S_z). The dotted line in Fig. 6a shows the single-particle spectrum for the half-filled ground state $^1(0,0)_0$ (where

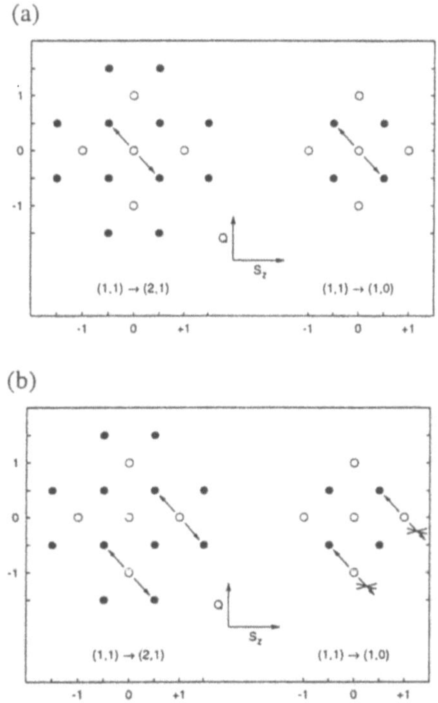

Figure 5
SO(5) multiplets connected by allowed photoemission (PES) and inverse photoemission (IPES) transitions. As we inject/remove a $\downarrow$-electron, PES corresponds to an arrow pointing south-east and IPES to one pointing north-west. (a) PES/IPES from a half-filled state with $S_z = 0$ in the $(1,1)$ multiplet (open circles). SO(5) selection rules only allow transitions to the $(2,1)$ and the $(1,0)$ multiplets (black circles), which are superimposed to the initial $(1,1)$-multiplet. (b) PES/IPES from a spin-polarized half-filled state ($Q = 0$, $S_z = +1$) and from a doped state ($Q = -1, S_z = 0$). Both states are members of the $(1,1)$ multiplet like the initial state on the left side. Note that there is no allowed photoemission transition to the $(1,0)$ multiplet for either of the two states.

the group theoretical notation is a short hand for $^{degeneracy(=S)}(momentum)_Q$); this is an RVB-state of rung-singlets (see Section 1) which actually forms a one-dimensional (0,0) multiplet. Final states can only belong to the 4-dimensional $(1,0)$ irreps [8]. Despite the fact that we are using very strong interaction parameters, there is just one single electron-like band in PES, whose cosine-type-of dispersion closely follows the dispersion of a non-interacting electron. The center of gravity of this band is given by the energy difference between a rung-singlet and a singly occupied rung [8]. The solid line in Fig. 6a represents the PES for the half-filled $^3(\pi,\pi)_0$ state (with $S_z = 0$), which carries one triplet excitation (magnon) and which belongs to the fivefold degenerate $(p = 1, q = 1)$ multiplet, i.e. the open circles in Fig. 5. From Fig. 5a we expect in this case final states belonging to both the $(2,1)$ and the $(1,0)$ irreps. Indeed, in addition to the band seen in the ground state spectra, which remains practically unchanged, there appears a "sideband" close to μ both in PES an in IPES; thus we have precisely the 4 bands expected from Fig. 5a. The physical interpretation is as follows: the initial $^3(\pi,\pi)_0$ state carries a single triplet-like Boson in the "RVB-vacuum". The "main bands" result from a creation of the photohole in a singlet rung; then the final state, consisting of the initial triplet plus a propagating singly-occupied rung carrying the entire momentum transfer, belongs to the $(2,1)$ irreps. As the two excitations, triplet and hole, can scatter from each other, the main band becomes broadened. But the photohole can also be created in the initial triplet, which results in a singly-occupied rung propagating in the RVB-vacuum, thus the new state belongs to the $(1,0)$ irreps. Since in this second case the photohole has to absorb the momentum of the initial triplet, (π,π), the sidebands' dispersion is shifted by (π,π) with respect to the main bands. Proceeding to a $^3(\pi,\pi)_0$ initial state with $S_z = 1$ (instead of $S_z = 0$) (Fig. 6b,

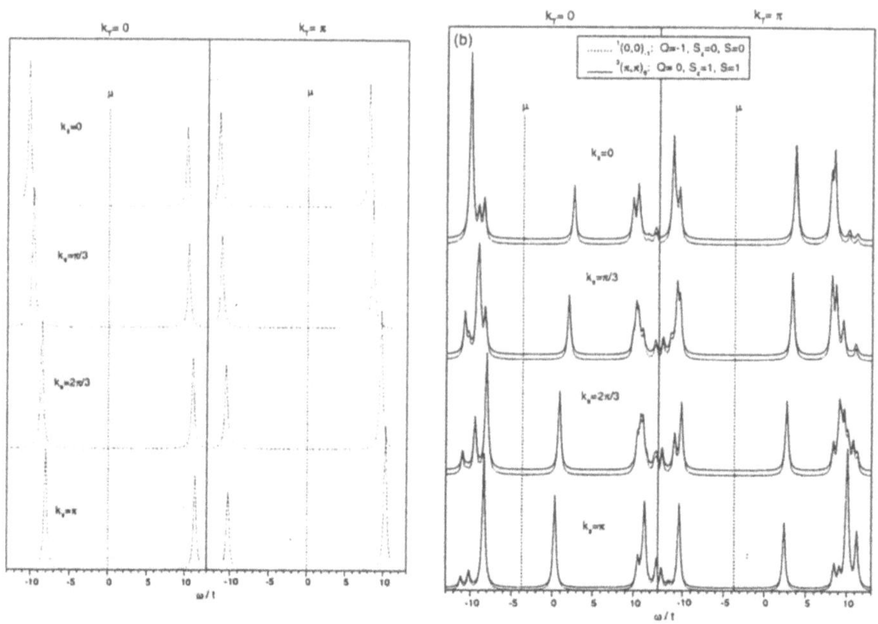

Figure 6 PES/IPES spectra of a 6-rung ladder (removal/injection of a $\downarrow$-electron. μ is the Fermi energy, defined as the average of first ionization and affinity energy. Parameter values are $U/t = 8$, $V_\perp/t = -6$, $J_\perp/t = 8$, $t_\perp/t = 1$. (a) PES/IPES from the half-filled ground state ($S_{tot} = 0$,dotted line) and the half-filled $^3(\pi,\pi)_0$ state with $S_z = 0$, $S_{tot} = 1$ (full line). (b) PES/IPES from the two-hole ground state $^0(0,0)_{-1}$ (dotted line) and the half-filled $^3(\pi,\pi)_0$ state with $S_z = S_{tot} = 1$ (full line)

full line), we get a PES/IPES spectrum similiar to the one for $S_z = 0$, but the sideband in PES has disappeared, whereas the one in IPES has gained some additional weight. The interpretation follows from Fig. 5b: As the initial state belongs to the $(1,1)$ irreps, a PES transition into the $(1,0)$ multiplet is now impossible. At this point, we remember that the SO(5) symmetry of our model implies that spin polarization and hole doping are equivalent in that the empty rung is the "SO(5) partner" of the triplet rung. Consequently, Fig. 6b also shows the spectra originating from the two-hole ground state $^0(0,0)_{-1}$ (dotted line). Both states belong to the $(1,1)$ irreps. Accordingly, their energies agree to computer accuracy. Since also the allowed final states of these two initial states belong to one and the same multiplet, namely $(2,1)$, we expect their PES/IPES amplitudes to be directly related. A more detailed consideration [8] shows that their amplitudes are in fact identical – just as Fig. 6b demonstrates. The physical reason for this identity is that the photohole cannot be created in the triplet rung, because the latter only contains two spin-*up* electrons. In summary our numerical "experiments" demonstrated, that SO(5) symmetry implies that the single particle spectra in the doped ground state are identical to those of certain higher-spin states at half-filling.

4.3 Recovering SO(5) Symmetry at Low Energies

The renormalization-group (RG) route has been proven to be particularly suited to describe the low-energy behavior of systems between one and two dimensions [34–36]. For this reason, we have carried out a RG study of a rather general ladder model of the Hubbard type [5]. The main result is that the *effective* Hamiltonian of the system, i.e. the Hamiltonian which describes excitations below a certain energy ω^*, becomes SO(5) invariant for a wide range of *bare* models. These include interactions with quite arbitrary values of U, V, and $V_\perp$ (cf. Fig. 4), provided they are weak, as well as a moderate next-nearest-neighbor hopping t_2. It is, thus, not necessary to introduce unphysical values for the parameters by hand in the model, since the exact SO(5) ladder discussed in Section 4.2 will be eventually recovered at low energies. This is remarkable, since it shows that quite general Hubbard-like models, which are relevant for the description of the cuprate materials between 1D and 2D, although not explicitly SO(5) invariant on the *microscopic* level, display an *exact* SO(5) symmetry when observed on a *macroscopic* length scale $\propto E_F/\omega^*$ (in units of the lattice spacing) [5] (for $t_2 = 0$ [6]). Physically, this means that the "standard deviation" between the multiplet splitting (cf. Fig. 2) goes to zero in these models for low energies or large length scales. An important issue here is the introduction of a next-nearest-neighbor term t_2, which is known to be essential in order to correctly describe AF correlations and Fermi-surface (FS) topology in cuprate materials.

Specifically, we have considered two coupled chains with total low-energy Hamiltonian $H = H_0 + H_I$, where H_0 represents the non-interacting part and H_I the interaction. This Hamiltonian describes interacting Fermions (creation and destruction operators $c_{\vec{k}\sigma}$ and $c^\dagger_{\vec{k}\sigma}$) close to the FS. The FS consists of four points: two bands $k_\perp = 0$ and $k_\perp = \pi$, each one with two Fermi points corresponding to right- and left-moving Fermions. Since we are interested in low-energy properties very close to the FS, the Fermion dispersion can assumed to be linear around the FS with (in general) band-dependent velocities v_F.

The idea of the RG is to divide the electronic excitations within the Brillouin zone into high-energy and low-energy modes, the latter being modes restricted within an energy $\omega \ll E_F$ from the Fermi energy. One then eliminates the high-energy modes by integrating them out and constructs an effective Hamiltonian which is restricted to the low-energy excitations. The parameters of the new effective low-energy Hamiltonian thus depend on the energy cutoff ω. Also the total spectral weights at the FS (quasiparticle weights) $Z_{k_\perp \omega}$ of the two bands are reduced due to the reduction of the cutoff and depend on ω. For this reason, in order to recover the *canonical* Fermi operators within the low-energy subspace, one should reabsorb $Z_{k_\perp \omega}$ into the definition of the Fermi operators and transform to $c_{\vec{k}\sigma} = \sqrt{Z_{k_\perp \omega}}\tilde{c}_{\vec{k}\sigma}$. This transformation has to be done in order to preserve the sum rules for the total integrated spectral weight $Z = 1$ within the restricted subspace. The $\tilde{c}_{\vec{k}\sigma}$ now acquire the meaning of *canonical* operators with the correct anticommutation relations within the low-energy subspace.

In practice, the integration of the high-energy modes is carried out by decreasing ω via infinitesimal steps, starting at $\omega = E_F$ which corresponds to the bare (microscopic) Hamiltonian (see, e.g. [37]). Due to the restriction of the modes to a small "shell" around the FS, the scattering amplitudes in the interaction H_I can also be considered as dependent only on the Fermi momenta closest to where the corresponding processes take place as well as on the spin of the scattered particles [34,36,37].

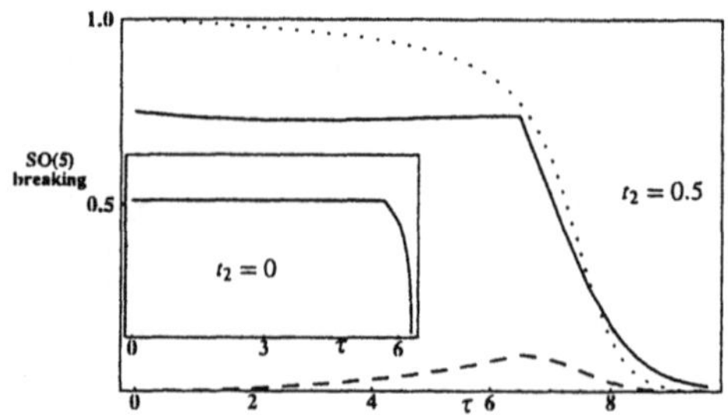

Figure 7
RG flow of the SO(5)-breaking terms of the Hamiltonian (in arbitrary units) $H_{\rm I}^{\rm (breaking)}$ (full), $H_{\rm I}^{\rm (PH-breaking)}$ (dashed), $\Delta v_{\rm F}$ (dotted) as functions of the RG flow $\tau = -\log\omega/E_{\rm F} =$ for $U = 1.$, $t = t_\perp = 1$, $t_2 = 0.5$ and half filling. The inset shows the $t_2 = 0$ case where only $H_{\rm I}^{\rm (breaking)}$ is nonvanishing.

The SO(5)invariant part $H_{\rm I}^{\rm (SO(5))}$ of $H_{\rm I}$ can conveniently be written as a sum of products of SO(5) scalar contractions in the form $\Psi^\dagger_{\vec{k}_1}\Psi_{\vec{k}_2} \times \Psi^\dagger_{\vec{k}_3}\Psi_{\vec{k}_4}$ where $\Psi_{\vec{k}}$ are SO(5) four-spinor Fermi operator (see Refs. [38,5] for details). At $\omega \approx E_{\rm F}$, the bare Hamiltonian is *not* SO(5) invariant in general and $H_{\rm I}$ consists of two parts $H_{\rm I} = H_{\rm I}^{\rm (SO(5))} + H_{\rm I}^{\rm (breaking)}$, where $H_{\rm I}^{\rm (breaking)}$ is a symmetry-breaking term. On the other hand, for $t_2 = 0$, the non-interacting part H_0 is $SO(5)$ symmetric, whereas a finite t_2 breaks particle-hole (PH) (and thus SO(5)) symmetry also in H_0 through a difference $\Delta v_{\rm F}$ between the Fermi velocities of the two bands.

The striking result of our calculation is that, by integrating away the high-energy modes, from some energy scale $\omega = \omega^*$ on, the SO(5)-breaking part of the Hamiltonian $H_{\rm I}^{\rm (breaking)}$ eventually vanishes with respect to the SO(5)symmetric part. In the inset of Fig. 7, we show the RG flow for the $t_2 = 0$ case (here the SO(5)-breaking term appears only in the interacting part of the Hamiltonian). In the bulk of the same figure, the flow for the $t_2 \neq 0$ case is displayed, for which there are three types of SO(5)-breaking terms, namely, $\Delta v_{\rm F}$, a PH-breaking term $H_{\rm I}^{\rm (PH-breaking)}$ and a plain SO(5)-breaking term $H_{\rm I}^{\rm (breaking)}$. All three terms go to zero at ω^*. This result is of extreme importance, since it means that even though the system is not SO(5) invariant at the *bare* level, i. e. at $\omega = E_{\rm F}$, the SO(5) invariant part of the Hamiltonian eventually dominates with respect to the symmetry-breaking part at energies smaller than ω^*. We have checked that this occurs for very general values of the Hamiltonian, including on-site (U) and nearest-neighbor (V and $V_\perp$) interactions, provided they are weak.

An interesting result is that for the $t_2 \neq 0$ case, the quasiparticle weights $Z_{k_\perp \omega}$ for the two bands renormalize differently. These quasiparticle weights have to be reabsorbed into the definition of the canonical Fermi operators through the transformation to the $\tilde{c}_{\vec{k}\sigma}$ variables, as explained above. This has the important consequence that the new *low-energy* SO(5) invariant Hamiltonian $H_{\rm I}^{\rm (SO(5))}$ is now invariant under a *renormalized* SO(5) symmetry in terms of *new* $\tilde{\pi}$ operators (cf. (3.7)), whereby the Fermi operators $c_{\vec{k}\sigma}$ are replaced with $\tilde{c}_{\vec{k}\sigma}$. This extended concept of SO(5) symmetry makes it possible for this *generalized* symmetry to occur in a *larger* and more *generic* class of physical systems than the ordinary SO(5). Moreover, a SO(5) theory unifying antiferromagnetism and superconductivity in terms of this *generalized* representation can admit possible asymmetries between the AF and superconducting phase, like for example the difference in $T_{\rm c}$ or in the order parameter [1,39].

In conclusion, the renormalization-group study shows that the effective low-energy Hamiltonian of a quite generic Hubbard-like ladder with weak interaction is SO(5) symmetric. This holds true also with the inclusion of a next-nearest-neighbor hopping t_2, provided it is *written in terms of the appropriate canonical Fermi operators* $\tilde{c}$ *for the low-energy subspace.*

Acknowledgements

The authors would like to express their deep gratitude to S. C. Zhang for sharing his deep insights and enthusiasm on SO(5) theory and D. J. Scalapino for a very fruitful interaction on the exact SO(5) ladder.

Bibliography

[1] S. C. Zhang, Science **275**, 1089 (1997).

[2] R. Eder, W. Hanke, and S. C. Zhang, Phys. Rev. B, **57**, 13781 (1998).

[3] S. Meixner, W. Hanke, E. Demler, and S. C. Zhang, Phys. Rev. Lett. **79**, 4902 (1997).

[4] E. Demler, S. C. Zhang, S. Meixner, and W. Hanke, Phys. Rev. Lett. **79**, 4937 (1997).

[5] E. Arrigoni and W. Hanke, cond-mat/9712143.

[6] H. H. Lin, L. Balents, and M. P. A. Fisher, Phys. Rev B **58**, 1794 (1998).

[7] D. Scalapino, S. C. Zhang, and W. Hanke, Phys. Rev. B **58**, 443 (1998).

[8] E. Eder, A. Dorneich, M. Zacher, W. Hanke, and S. C. Zhang, cond-mat/9805120.

[9] B. Bernu et al., Phys. Rev. B **50**, 10048 (1994).

[10] E. Dagotto and T. M. Rice, Science **271**, 618 (1996).

[11] J. M. Tranquada, B. J. Sternlieb, J. D. Axe, Y. Nakamure, and S. Uchida, Nature **375**, 661 (1995).

[12] D. Poilblanc, T. M. Rice, Phys. Rev. B **39**, 9749 (1989); J. Zaanen and O. Gunnarson, Phys. Rev. B **40**, 7391 (1989); H. Eskes, R. Grimberg, W. van Sarloos, and J. Zaanen, Phys. Rev. B **54**, R724 (1996).

[13] R. Eder, Y. Ohta, and S. Shimozato, Phys. Rev. B **50**, 3350 (1994).

[14] R. Preuss, W. Hanke, and W. von der Linden, Phys. Rev. Lett. **75**, 1344 (1995).

[15] E. Dagotto, Rev. Mod. Phys. **667**, 63 (1994).

[16] D. S. Marshall et al., Phys. Rev. Lett. **76**, 4841 (1996); A. G. Loeser et al., Science **273**, 325 (1996); H. Ding et al., Nature **382**, 51 (1996) and Z. X. Shen, private communication.

[17] R. Eder et al., to be published.

[18] W. Heisenberg, Z. Physik **77**, 1932.

[19] R. Eder et al., to be published.

[20] E. Demler and S. C. Zhang, Phys. Rev. Lett. **750**, 4126 (1995).

[21] P. W. Anderson, Science **235**, 1196 (1987).

[22] D. Scalapino, Phys. Rep. **250**, 329 (1995).

[23] J. R. Schrieffer, X. G. Wen, and S. C. Zhang, Phys. Rev. B **39**, 11663 (1989).

[24] D. Pines, Physica C **235**, 113 (1994).

[25] M. Greiter, Phys. Rev. Lett. **79**, 4898 (1997).

[26] G. Baskaran and P. W. Anderson, cond-mat/9706076.

[27] C. Burgess and C. Lutken, Phys. Rev B **57**, 8642 (1998).

[28] D. Arovas, J. Berlinsky, C. Kallin, and S. C. Zhang, Phys. Rev. Lett. **79**, 2871 (1997).

[29] D. Duffy, S. Haas, and E. Kim, cond-mat/9804221.

[30] B. Batlogg and V. J. Emery, Nature **382**, 20 (1996).

[31] J. Rossat-Mignod et al., Physica (Amsterdam) **235C**, 59 (1944); J. B. Keimer et al., Physica C **282-287**, 232 (1997).

[32] N. Ong, Science **273**, 321 (1996).

[33] M. Azuma, Z. Hiroi, and M. Takano, Phys. Rev. Lett. **73**, 3463 (1994).

[34] M. Fabrizio, Phys. Rev. B **48**, 15838 (1993).

[35] H. J. Schulz, Phys. Rev. B **53**, R2959 (1996).

[36] L. Balents and M. P. A. Fisher. Phys. Rev. B **53**, 12133 (1996).

[37] J. Sòlyom, Adv. Phys. **28**, 201 (1979).

[38] S. Rabello,H. Kohno, E. Demler, and S. C. Zhang, Phys. Rev. Lett. **80**, 3586 (1998).

[39] C. L. Henley, Phys. Rev. Lett. **80**, 3590 (1998).

HTS Cables for Electric Power Transmission: Basic Properties – State of the Art – Prospects

M. Leghissa, J. Rieger, J. Wiezoreck, and H.-W. Neumüller

Siemens AG, Corporate Technology,
P.O.Box 3220, D-91050 Erlangen, Germany

Abstract: Superconducting cables can contribute to cover the power consumption of densely populated areas in an economic and environmentally-friendly way. Using high-temperature superconductors cables with higher transmission capacities, smaller cross sections and lower transmission losses can be built. In this paper we review basic properties of superconducting cables, the status of the Siemens cable project and outline the future prospects for HTS power transmission cables.

1 Introduction

Electric power is usually transmitted using conventional three-phase overhead lines and underground cable systems. Overhead lines are still the most economic way of power transmission, but in the vicinity of built-up areas insufficient space and environmental reasons require the installation of cable systems. The power capacity of a three-phase cable system is given by

$$P = 3I_{\mathrm{n}}U_0 = \sqrt{3}I_{\mathrm{n}}U_n \tag{1.1}$$

with I_{n} the nominal current, U_0 the phase voltage (between phase and ground) and U_n the nominal voltage (between two phases). Since the nominal current is limited by the heat generated due to ohmic losses, high voltages are required for high power ratings. Conventional cables with oil-filled paper insulation or XLPE (cross-linked polyethylene insulation) provide a maximum transmission capacity of approximately 500 MVA (400 kV/720 A) per system. Using forced water cooling the capacity can be raised to 1 000 MVA. For comparison, the power consumption of the largest german city (Berlin, 3.5 million inhabitants) is 2 500 MW. However, there are some major disadvantages of conventional cables.

- Conventional cables have loss factors of 0.01 – 0.03 %/km. For instance, for a 20 km long 1 000 MVA system the losses are 35 GWh per year, corresponding to a CO_2-emission of 21 000 t.
- Conventional cables have limited transmission capacities at given diameters.
- High power capacities usually require installing several cable systems in parallel, leading to larger space consumption and higher installation costs.

Feasibility studies have shown that using superconducting cables these drawbacks can be reduced. Depending on the cable design, the transmission capacity and the annualy load factor the total losses can be reduced by up to 40 – 80 % [1,2]. High power ratings can be achieved at lower voltage levels, i.e., 66 kV to 110 kV rather than 220 kV to 400 kV. This leads to smaller cable diameters and the possibility of retrofitting conventional pipe-type cables, i.e., replacing conventional cables by superconducting cables with higher capacities of the same diameter [1,3,4].

From the mid 1960's through the 1980's – facing a steadily increasing demand for electric power – a couple of cable projects using conventional superconductors (Nb, Nb_3Sn, Nb_3Ge) have been started in Europe, United States and Japan [5–9]. However, studies have shown that the economic break-even point for superconducting three-phase ac cables is within the range of 3 000 to 5 000 MVA. This is well above the maximum power ratings required nowadays [5]. With the discovery of high-temperature superconductivity (HTS) the prospects are much better. Based on the much cheaper and robust liquid-nitrogen cooling technology it is believed that HTS three-phase ac cables will be an economic alternative to conventional cables even for power ratings below 1 000 MVA.

2 Superconducting Cable Concepts

2.1 Warm-Dielectric and Cold-Dielectric Designs

Superconducting cables for power transmission can be classified into designs using conventional high-voltage insulations at room temperature and designs with high-voltage insulations cooled to low temperatures [4]. These two basic concepts are shown in Fig. 1. For simplicity only single phase configurations are shown. The warm-dielectric design has the superconducting core within a cryogenic envelope. The electric insulation is placed outside the cryogenic envelope at ambient temperature. Superconducting cables with the cold-dielectric design (also called coaxial design) house the complete cable core within the cryostat. Each phase has a superconducting return conductor (shield) and an electric insulation working at low temperatures.

Roughly speaking, the cold-dielectric cable (+) has low losses, (+) allows higher power ratings, (+) lacks any electromagnetic stray fields, (−) has technical challenges due to cold dielectric insulation and (−) doubles the amount of superconductor. The warm-dielectric design (+) requires less superconductor, (+) has less technical challanges due to the conventional dielectric insulation, (+) is suitable for retrofitting existing pipe-type cables, (−) generates higher losses, (−) has lower transmission capacities and (−) induces electromagnetic stray fields.

Cable systems of either design consist of several components: superconductor, electric insulation, thermal insulation (cryostat), cooling system, terminations, joints, vacuum system [10]. We will concentrate on the transmission losses and the HTS conductor as one major component of a superconducting cable.

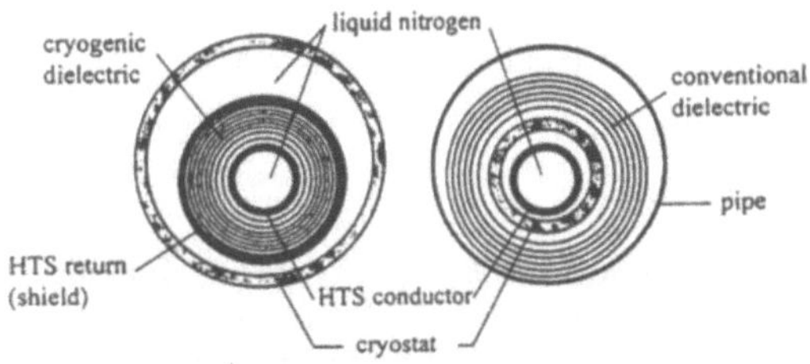

Figure 1
Comparison of the two basic concepts for superconducting cables: cold-dielectric design (left side) and warm-dielectric design (right side).

2.2 Estimation of Transmission Losses

Apart from ohmic losses due to cable joints and current leads in terminations there are three major sources of transmission losses: thermal losses of the cable cryostat, dielectric losses of the insulation and ac losses of the superconductor. The following loss values have been calculated for a HTS 110 kV / 1 000 MVA three-phase cable ($I_n = 5\,250$ A).

Thermal losses of the cable cryostat depend on the type of thermal insulation. In order to minimize these losses a vacuum superinsulation has to be used. Depending on the inner and outer cryostat diameters (D_i and D_o), the thermal conductivity $\lambda \approx 0.1 - 0.2$ mW/(m·K) of the thermal insulation and the temperature difference $\Delta T \approx 220$ K the thermal losses of a superconducting cable are in the range[1]

$$P_{therm} = \frac{2\pi \cdot \lambda \cdot \Delta T}{\ln(D_o/D_i)} \approx 1 \ldots 2 \text{ W/m} \tag{2.2}$$

Since thermal losses are independent of current and voltage they occur also in the case of a stand-by.

The dielectric losses are caused by the ac electric field applied between conductor and shield. They can be calculated for each phase by

$$P_{diel} = 2\pi f C U_0^2 \tan\delta; \quad C = \frac{2\pi \cdot \varepsilon \cdot \varepsilon_0}{\ln(r_o/r_i)} \tag{2.3}$$

with f the power frequency, U_0 the phase voltage, $\tan\delta$ the dielectric loss factor and C being the capacitance of the cable (per meter) determined by the dielectric constant ε and the outer and inner radii r_o and r_i of the insulation. Using typical values of $\varepsilon \approx 2$ and $\tan\delta \approx 1.5 \cdot 10^{-5}$ (cold-dielectric design) dielectric losses are

$$P_{diel} < 0.01 \text{ W/m} \cdot \text{phase.} \tag{2.4}$$

AC losses of superconductors operating at power frequencies are induced by the ac magnetic field generated by the current flowing in the conductor. Depending on the design of superconducting tapes these losses can be subdivided into hystersis losses, coupling losses and eddy current losses. It was already shown that the losses using multifilamentary Bi2223 tapes can be considered to be purely hysteretic as long as there is no interaction between the phases [11–13]. However, for the warm-dielectric design the interaction of the three phases yields higher losses due to the so-called "dynamic resistance" effect [14].

Within a first approximation the cable conductor can be described as a hollow superconducting cylinder (SC) with a total critical current I_c and the inner and outer diameters d_i and d_o [11]. Using the Bean model for flux penetration into a superconductor the losses can be calculated

[1] With the symbol P we will always denote losses in units [W/m].

$$P_{\mathrm{ac}}^{\mathrm{SC}} = \frac{\mu_0}{2\pi}\frac{fI_c^2}{h^2}\{(2-Fh)Fh+2(1-Fh)\ln(1-Fh)\} \tag{2.5}$$

$$\approx \frac{f\mu_0 h}{6\pi I_c}\hat{I}^3 = \frac{2f\mu_0}{3\pi^2 d_0^2 j_e}\hat{I}^3 \tag{2.6}$$

with $F = \hat{I}/I_c$ the normalized current amplitude and $h = (d_0^2 - d_i^2)/d_0^2$. Equation (2.5) represents the exact solution and (2.6) the first term of its Taylor series.

A fingerprint of hysteresis losses is the linear frequency dependence and the cubic current dependence whereas eddy current losses show $\hat{I}^2 f^2$ behaviour. In order to minimize ac losses high engineering current densities j_e and large conductor diameters d_0 are necessary. It should be mentioned that large conductor diameters cause larger cryostat diameters and therefore higher thermal losses (see 2.2)). Therefore, there will always be an optimum cable size minimizing the sum of ac losses and thermal losses. Assuming an engineering critical current density of $j_e = 25\,\mathrm{kA/cm^2}$ hysteresis losses are

$$P_{\mathrm{ac}} \approx 1\ldots2\,\mathrm{W/m}\cdot\mathrm{phase}. \tag{2.7}$$

As we will show below there is a way to lower the ac losses by using an optimized conductor layout.

For a three-phase cable system total losses of approximately 4 W/m at low temperature can be achieved. These losses have to be penalized by the efficiency of the refrigeration system, typically 10 %. Thus the cold losses have to be multiplied by a factor 10. Therefore the effective losses are 40 W/m compared to 200 W/m for a conventional 1 000 MVA copper cable with forced water cooling.

3 HTS-Tapes for Power Transmission Cables

Applications of superconductivity in the field of power engineering always require superconducting wires or tapes tailored for specific requirements. This holds also to HTS-tapes for power transmission cables. These requirements can be summarized to the following four categories:

- superconductivity: high critical currents and current densities as well as low ac losses at temperatures $T \geq 77\,\mathrm{K}$;
- availability: batch lengths of 100 m to 1 000 m;
- mechanics: flexibility and strength,
- financial aspects: reasonable specific costs (per meter and per Ampere).

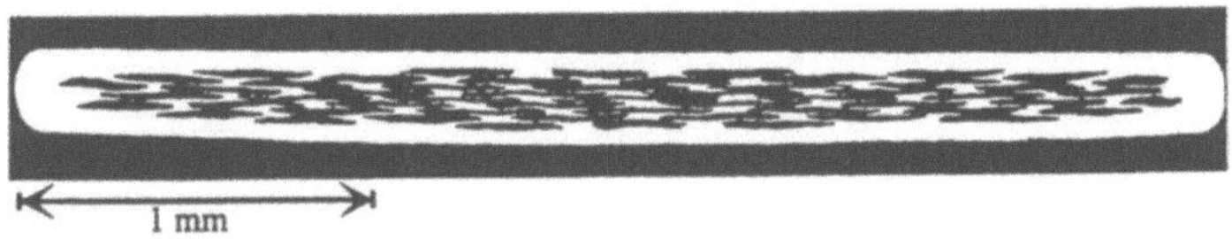

Figure 2 Cross section of a 55-filament Bi2223 tape fabricated by Vacuumschmelze and Siemens.

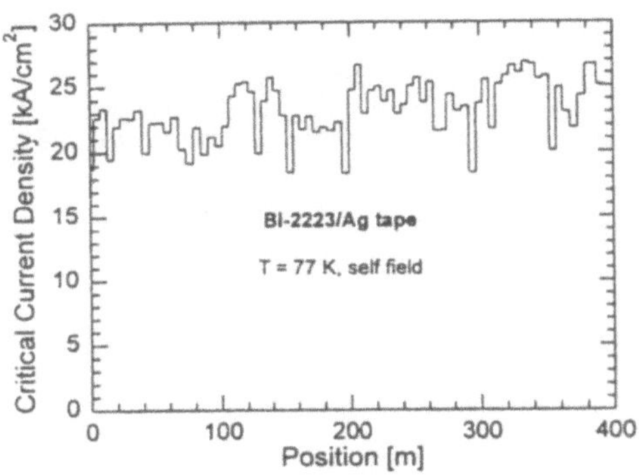

Figure 3 Homogeneity of the critical current of a 400 m batch Bi2223 tape measured with a spatial resolution of 5 m.

Up to now, all the available HTS tape systems are still more or less far from satisfying all these conditions: YBCO tapes have high critical current densities but are only available as short samples. Bi2212 tapes have too small critical current densities at and above 77 K. The same holds for Tl- and Hg-based conductors. Only Bi2223 multifilamentary tapes have reached a performance level enabling the fabrication and test of HTS-cables. As a consequence, all cable projects in Europe, USA and Japan are based on multifilamentary Bi2223 tapes [15–18].

Figure 2 shows the cross section of a 55-filament Bi2223 tape produced by Siemens and Vacuumschmelze. The superconducting filaments are embedded in a Ag- or Ag-alloy sheath. The tapes we used for the cable development typically had $I_c \approx 40$ A and $j_c \approx 20\,\mathrm{kA/cm^2}$ at 77 K. The performance depends on the length of the sample. Due to inhomogeneities within the superconducting filaments and macroscopic defects, the critical current (density) of a long tape is considerably lower than the best short-sample results [19]. In Fig. 3, the fluctuations within a 400 m long batch are shown. Due to the non-linear I-V curves, parts with low local I_c do over-proportionally contribute to the avarage I_c. Therefore, special care has to be taken to avoid inhomogeneities and defects during tape fabrication and due to handling during the cable fabrication under factory conditions.

Power transmission cables belong to the low-field applications of superconductors. The current within the cable generates a self-field below 30 mT. Usually the magnetic field is oriented parallel to the tape surface, i.e., parallel to the mean direction of the ab-planes of the Bi2223 crystallites. Using cold-dielectric coaxial conductor designs perpendicular field components generated by the other conductors are completely avoided. The working temperature of a HTS-cable will be in the range of 77 K to 90 K. In Fig. 4, the field dependence of critical current at two different temperatures is shown. Obviously, only a small I_c-reduction ($< 10\,\%$) is expected to occur due to the parallel self-field of the conductor – even at 85 K where I_c is only 65 % of its value at 77 K. For comparison the curves for perpendicular field orientation are also shown. The strong field dependence for this case underlines the necessity to screen perpendicular components.

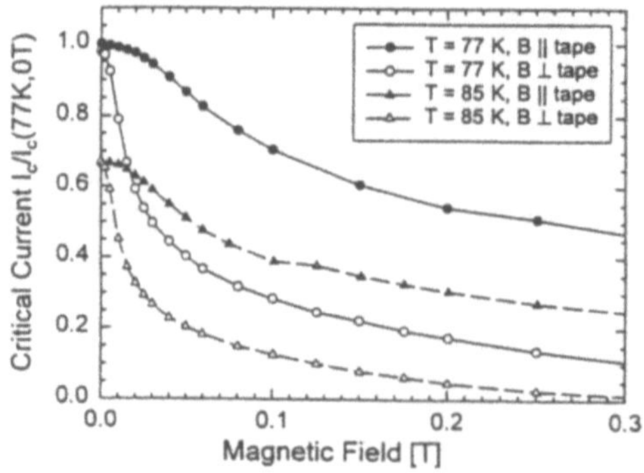

Figure 4
Magnetic field dependence of the critical current in Bi2223 tapes at 77 K and 85 K.

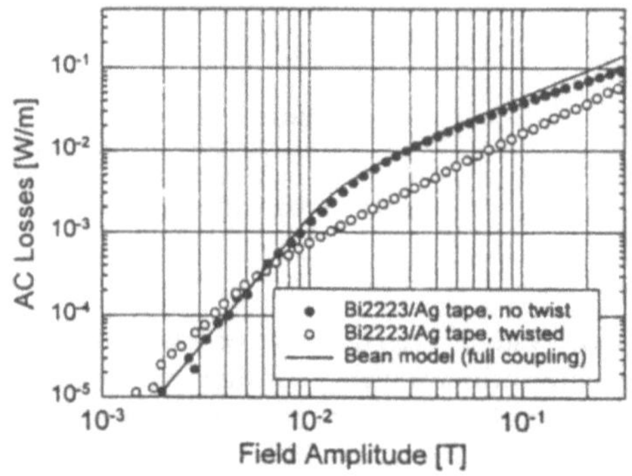

Figure 5
AC losses of non-twisted and twisted Bi2223 tapes in parallel field at a frequency of 50 Hz.

The ac losses of Bi2223 tapes at 77 K due to alternating magnetic fields are shown in Fig. 5. For magnetic fields below the full penetration field B_p (20 mT for the sample without filament twist) the losses show a B^3-dependence. For higher fields the losses increase only linearly. As long as the filaments are not twisted the tape can be described as a homogeneous superconductor despite the subdivision into individual filaments. The experimental results are compared to the theoretical prediction of the Bean model for a slab in parallel field [20,21]. By twisting the filaments losses can be reduced due to decoupling of the filaments.

4 Fabrication and Test of Cable Conductor Models

4.1 Design of the 10 m Model Conductor

In order to investigate the suitability of Bi2223 tapes for cable applications and the ac losses for our cable design a 10 m long cable conductor shown in Fig. 6 has been fabricated with a laboratory-scale winding machine. In total 145 Bi2223 tapes have been wound in four layers on a flexible former. Every layer k is characterized by its number of tapes N_k, its pitch angle α_k and its radius r_k $(k = 1 \dots 4)$. The four layers act as concentric windings. Due to their self-inductances L_k and their mutual inductances M_{kj} the total ac current I flowing in the conductor is distributed into these layers. The current distribution I_k is found by solving the equations [17,22]

Figure 6 A 10 m HTS-cable conductor fabricated and tested at the Siemens Corporate Technology Laboratories.

$$E = [R_{c,k} + R_{ac,k}(I_k)]I_k - L_k\frac{dI_k}{dt} - \sum_{j\neq k} M_{kj}\frac{dI_j}{dt} \qquad (k = 1\ldots4) \tag{4.8}$$

The contributions due to the small non-linear ac resistance $R_{ac,k}(I_k)$ and the joint resistance $R_{c,k}$ can be neglected as long as the current amplitude of each layer is lower than the critical current of the layers. By adjusting the parameters, L_k and M_{kj} an uniform current distribution can be achieved. This is the unique feature of the Siemens conductor design.

In order to test the dc and ac properties of the cable conductor, a cooling system using pressurized forced-flow liquid nitrogen (LN2) has been designed and constructed [19]. The system consists of a 10 m long vacuum insulated pipe with a cold radiation shield and two LN2-vessels with current leads and heat exchanger operating at ambient pressure. The coolant is pressurized and pumped along the cable conductor. Due to the ac losses of the conductor a temperature gradient is observed along the cooling channel. The thermal losses of the cryostat are seperately intercepted via the radiation shield. The pressure of 2 bar enables single-phase operation within a temperature range 77 – 84 K. By measuring the temperature gradient dT/dx due to the ac current and the LN2 mass flow dm/dt the ac losses can be determined calorimetrically

$$P_{ac}^{cal} = c_p \cdot \frac{dm}{dt} \cdot \frac{dT}{dx} \tag{4.9}$$

with c_p the specific heat of liquid nitrogen. The resolution of 0.1 W/m is determined by the resolution of the temperature and mass flow measurements. As a second independent method the ac losses are determined by an electric technique measuring the voltage drop component E in phase with the current along the conductor using a lock-in amplifier yielding

$$P_{ac}^{el} = E \cdot I. \tag{4.10}$$

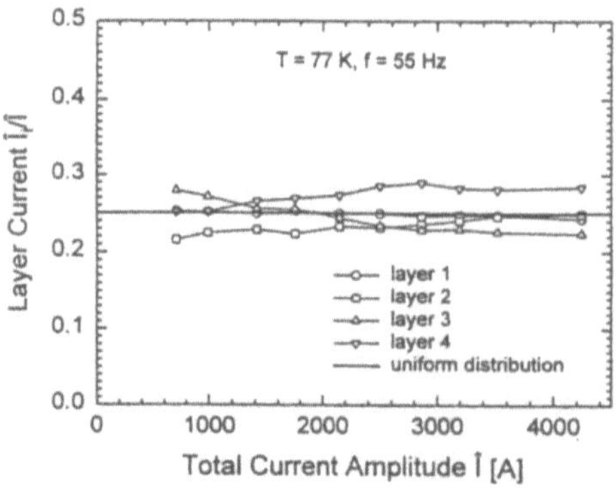

Figure 7
Measurement of the currents flowing in the different layers of the cable conductor.

The resolution of 0.001 W/m is by two orders of magnitude better than for calorimetric measurements. However, this technique is very ambitious because the results can be influenced by several factors, e.g., phase errors, ground loops, magnetic fields generated by the return line of the current and the positions of the voltage contacts [12]. For the loss measurements signals of several contacts positioned at the circumference of the conductor have been avaraged.

4.2 Measurement of the Current Distribution

The currents I_i flowing in the four layers have been measured by Rogowski-coils placed between the layers. Each Rogowski-coil gives a signal proportional to the magnetic field generated by the currents flowing in the underlaying layers. From these signals, I_i can be calculated. The current distribution is plotted in Fig. 7 as a function of the total current flowing in the conductor. As expected from our conductor design, the current distributes almost uniformly in the four layers. Measurements performed at different frequencies yield identical results.

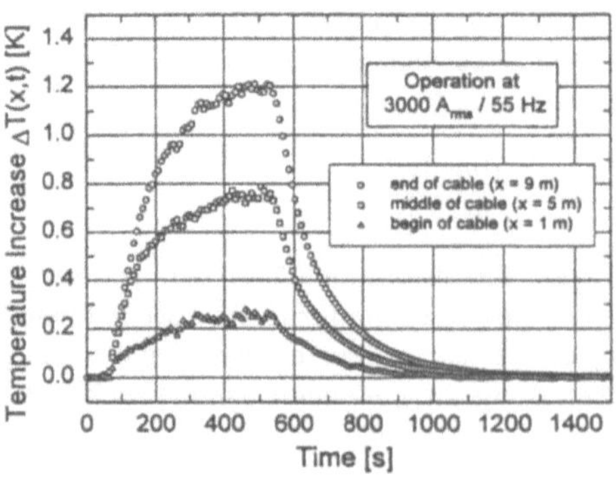

Figure 8
Typical temperature increase versus time plots at different positions of the conductor recorded at a LN2 pressure of 2 bar. Only the temperature difference $T(\hat{I}) - T(0)$ is plotted ($T(0) \approx 77$ K).

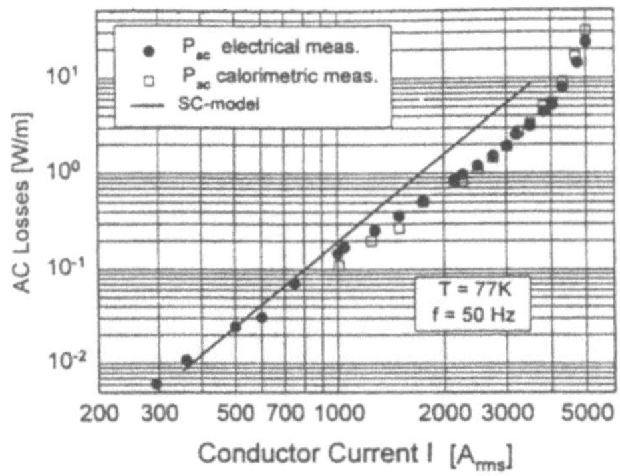

Figure 9
Comparison of the ac losses of the cable conductor measured by two independent methods with the prediction of the SC-model.

4.3 Measurement of the AC Losses

In Fig. 8, the temperature distribution along the conductor is shown for an ac current of 3 000 A_{rms} and 55 Hz. At $t = 70$ s the current is switched on. Subsequently the temperature starts to increase and after approximately 500 s a stationary temperature gradient is formed. Then, the current is switched off and the conductor cools down again. From the stationary temperature gradient the ac losses can be determined as described above. The measurement system has been thoroughly checked by using different mass flows and by calibrating using a heater with well-known power loss. For this type of measurements it is important to keep the nitrogen within the single-phase region of the phase diagram in order to rule out any evaporation of the coolant.

In Fig.9, the measured ac losses are compared to the theoretical values calculated by the superconducting cylinder model. We have observed perfect agreement between the calorimetric and the electric method. However, the experimental values are below the line predicted by the model. At low currents, the experimental losses increase with a cubic amplitude dependence as expected from (2.5). However, for larger currents the amplitude dependence is weaker than expected from the SC-model. At 3 500 A_{rms} corresponding to a current amplitude of 5 000 A a dramatic increase of the losses occurs announcing the transition to a truely resistive state. For a current of 2 000 A_{rms}, losses of only 0.8 W/m have been measured. To our knowledge this is the lowest loss value achieved up to now by a HTS-cable conductor.

4.4 Discussion of the AC Losses

The key factor to reduce the ac losses is a conductor design ensuring a homogeneous current distribution between the different layers. In the SC-model the current is assumed to flow only parallel to the cable axis and the inductances due to the axial field components generated by the solenoidal windings are neglected leading to an unbalanced current distribution. Therefore, the current penetrates from the surface into the conductor as predicted by the Bean model. By increasing the current amplitude, the penetration depth of the magnetic field increases. At the critical current the penetration depth equals the thickness of the superconducting layers.

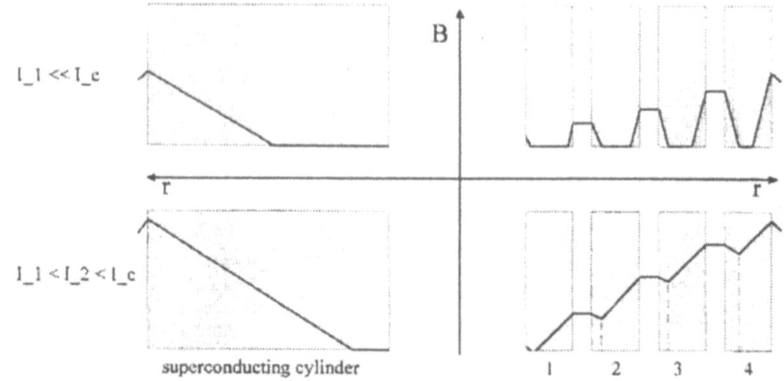

Figure 10 Magnetic field patterns for a cable conductor in the framework of the SC- (left side) and the UCD-model (right side). The upper patterns correspond to low current amplitudes, the lower one to currents close to the critical current.

On the other hand, due to the optimized winding scheme the different layers are electromagnetically coupled and the current is forced to distribute into the layers rather than to be confined within one sheath. As a consequence, the magnetic field profile differs from the prediction of the SC-model. The situation is usually refered to as the "uniform current distribution" (UCD-model) [17].

It is evident that both the current and field distributions differ from the picture expected within the SC-model. Figure 10 displays schematically the field amplitudes perpendicular to the tape axis (parallel to the surface of the tapes) for both models. On the left hand side the field profile in the SC-model is shown. For currents well below the critical current the magnetic field and therefore also the current is only flowing in the outer part of the conductor, the inner part is not contributing to the current transmission. By approaching the critical current the field penetrates throughout the conductor thickness yielding high ac losses. Due to the electromagnetic coupling of the layers in the UCD-model the transport current is distributed uniformly into the layers at any current amplitude below the critical current. The distance of flux penetration within the HTS material is reduced compared to the SC-model leading to lower losses.

Within the UCD-model the ac losses are calculated by summation of the losses of the individual layers, thus

$$P_{\text{ac}}^{\text{UCD}} = \sum_k P_{\text{ac},k}^{\text{UCD}} . \tag{4.11}$$

For each layer k, magnetic field amplitudes acting on the inner surface B_i and on the outer surface B_o as well as the influence of the transport current on the flux penetration have to be considered. A slightly modified version of the Bean model for a superconducting slab in a parallel field [20,17] has to be used. The losses of the layer k (radius r_k, thickness t_k) are given by

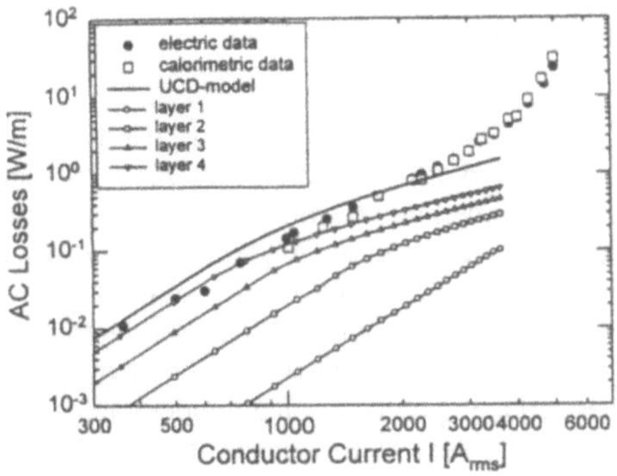

Figure 11
Comparison of the ac losses of the cable conductor with the prediction of the UCD-model using an engineering current density of $j_e = 4$ A/cm^2.

$$P_{\mathrm{ac},k}^{\mathrm{UCD}} = \begin{cases} \dfrac{4\pi f r_k}{3\mu_0^2 j_e} \cdot (B_{\mathrm{i},k}^3 + B_{\mathrm{o},k}^3) & \text{for } B_{\mathrm{i},k} + B_{\mathrm{o},k} < \mu_0 j_e t_k \\ 4\pi j_e f r_k \Big[t_{i,k}^2 (B_{\mathrm{i},k} - \frac{2}{3}\mu_0 j_e t_{\mathrm{i},k}) + \\ \qquad + t_{\mathrm{o},k}^2 (B_{\mathrm{o},k} - \frac{2}{3}\mu_0 j_e t_{\mathrm{o},k}) \Big] & \text{for } B_{\mathrm{i},k} + B_{\mathrm{o},k} \geq \mu_0 j_e t_k \end{cases} \tag{4.12}$$

$$t_{\mathrm{i},k} = \frac{t}{2}(1 - F_k) \qquad t_{\mathrm{o},k} = \frac{t}{2}(1 + F_k) \tag{4.13}$$

Similar to F being the normalized current amplitude in (2.6) F_k is defined as the normalized current amplitude $\hat{I}_k / I_{\mathrm{c},k}$ of the layer. The first line in (4.12) corresponds to the case of incomplete flux penetration into the layers (upper part of Fig. 10), the second one to full flux penetration (lower part of Fig. 10, layers 2, 3 and 4).

The ac losses for our 10 m cable conductor model have been calculated using this modell with the engineering current density j_e as the only free parameter. The comparison with the experimental value is shown in Fig. 11. We have obtained a fairly good agreement using $j_e = 4$ kA/cm^2, corresponding to a critical current of 5 000 A for the conductor. Calculations are showing that the losses are predominantely generated in the outer layers, where the highest magnetic field is acting.

From the number of tapes (145) and a critical current of 40 A/tape a "design" value of $I^* = 5760$ A is calculated. The critical current $I_{\mathrm{c,ac}} = 5000$ A determined from ac losses corresponds to 86 % of I^*. Two different effects account for the current degradation of 14 %:

- The self-field at 5 000 A is 40 mT. From the field dependence of the critical current a reduction of 7 % has to be expected.
- Axial and bending strains acting on the tapes during handling and winding of the conductor are expected to lower the critical current. This effect should be responsible for the remaining 7 %.

Due to limitation of the power supply the dc I-V curve has only been measured up to 3 000 A. We have observed a voltage drop of 0.015 μV/cm at 3 000 A which is by two orders of magnitude lower than the critical current criterion of 1 μV/cm [19]. From these findings we can conclude that $I_{\mathrm{c,dc}} \approx 5000$ A.

5 Prospects for HTS Power Transmission Cables

Superconducting power engineering applications usually benefit from reducing size, weight and losses. Superconducting cables also offer the possibility of power transmission at lower voltage levels. Furthermore environmental reasons, i.e., the absence of electromagnetic stray-fields, insulation oil and the reduction of CO_2-emission by saving energy are arguments for installing HTS devices instead of conventional solutions.

HTS power transmission cables are the most promising solution for the transmission of high power ratings ($\geq$ 1 000 MVA) in densely populated areas. They can be installed in existing underground tunnels, in pipes replacing conventional copper cables of the same diameter or in trenches much narrower than necessary for conventional cables. In all these cases intallation costs for the cable systems can be lowered. Although HTS-cables are superior at high power ratings, economic success of HTS-cable technology will depend on entering lower power ratings (300 – 1 000 MVA) which constitute the major part to the turnover by high-voltage cables.

Market forecasts by CONECTUS – a group of major european companies dedicated to superconductivity – have estimated a world wide market for HTS-cables of 2 billions US $ in 2020 [23]. These figure cleary reflects long-term strategies connected to HTS power engineering devices. It should be mentioned that lifetimes of conventional cables are at least 40 years.

Superconducting components will have to fit into the existing utility networks for power transmission and distribution. Since the supply with electric power is a basic need in industrialized countries compatibility, reliability and high quality standards are necessary. Before conquering large market segments, superconducting devices will have to proof these quality standards by long-term operation of prototypes and pilot plants. Several companies like Siemens (Germany), Pirelli (Italy/USA), BICC (GB), Southwire (USA), Sumitomo and Furukawa (both Japan) have started R&D projects on HTS transmission cables. Within a period of approximately 5 years the development advanced from meter-scale laboratory demonstrators to cable systems with length scales of 50 m.

The economical prospects for HTS-cables strongly depend on improvement of the HTS tapes. A reduction of the specific conductor costs can only be achieved by both reducing the fabrication costs and raising the critical current densities to a level of 100 kA/cm^2. Using magneto-optical [24] and magnetisation studies [25] critical current densities of this order of magnitude have been found on short lengths scales in multifilament Bi2223 tapes. These findings demonstrate that Bi2223 tapes are capable of attaining the performance level necessary for cable applications.

Conventional cable systems are based on a mature technology developed within a periode of more than one century. There is only little potential for new developments leaving only slim margins for fabricating these products in highly industrialized countries. Superconducting cables are representing a new and innovative technology and very likely this technology will help creating new jobs.

Acknowledgments

The authors want to acknowledge contributions from several colleagues from Siemens Corporate Technology, Siemens Power Transmission and Distribution Group, Vacuumschmelze GmbH Hanau, IFW Dresden, Univ. Tübingen, Univ. Erlangen, MPI für Metallforschung (Stuttgart) and FZ Karlsruhe within the joint project "Superconductivity for Power Engineering" funded with by the German BMBF under grant no. 13N6481. We have to give special credit to N. Prölß, H.-P. Krämer and P. van Haßelt for assistance in the experimental work, G. Ries for his contributions to ac losses and C. Reimann and M. Oomen for performing measurements on Bi2223 tapes.

Bibliography

[1] S.P. Ashworth, P. Metra, R.J. Slaughter, ETEP Vol. 4, 293 (1994).

[2] J. Oestergaard, IEEE Trans. on Appl. Supercond. **7**, 719 (1997).

[3] T. Hara, K. Okaniwa, N. Ichiyanagi, S. Tanaka, IEEE Trans. on Power Delivery **7**, 1745 (1992).

[4] D.W. Von Dollen, P. Metra, M. Rahman, Proceedings of the American Power Conference (Chicago 1993), 1206 (1993).

[5] G. Bogner, P. Penczynski, F. Schmidt, Kerntechnik 20 (6), 262 (1978).

[6] E. Bochenek, IEEE Trans. on Magnetics **MAG 11**, 366 (1975).

[7] P.A. Klaudy, IEEE Trans. on Magnetics **MAG 19**, 656 (1983).

[8] E.B. Forsythe, D.A. Thomas, Cryogenics **26**, 599 (1986).

[9] Y. Furuto, *Proc. of the Fifth Int. Cryogenic Eng. Conf.*, Kyoto, 180 (1975).

[10] O. Miura, S. Tanaka, K. Miyoshi, N. Ichiyanagi, Y. Tanaka, H. Ishii, T. Hara, Cryogenics **36** 589 (1996).

[11] G. Vellego, P. Metra, Supercond. Sci. Technol. **8**, 476 (1995).

[12] J. Wiezoreck, M. Leghissa, G. Ries, H.-W. Neumüller, M. Lindmayer, Inst. Phys. Conf. Ser. No. **148**, 595 (IOP Publishing Ltd 1995).

[13] J. Fujikami, N. Saga, K. Ohmatsu, T. Shibata, S. Isojima, K. Sato, H. Ishi, T. Hara, *Advances in Superconductivity VIII*, 1319 (Springer Verlag, Tokyo, 1996).

[14] D.E. Daney, H.J. Boenig, M.P. Maley, D.E. McMurry, B.G. DeBlanc, IEEE Trans. on Appl. Supercond. **7**, 310 (1997).

[15] M. Leghissa, B. Fischer, B. Roas, A. Jenovelis, J. Wiezoreck, S. Kauth, H.-W. Neumüller, C. Reimann, R. Nanke, P. MAller, IEEE Trans. on Appl. Supercond. **7**, 355 (1997).

[16] A. Bolza, P. Metra, M. Nassi, M.M. Rahmann, IEEE Trans. on Appl. Supercond. **7**, 339 (1997).

[17] S. Mukoyama, K. Miyoshi, H. Tsubouti, M. Mimura, N. Uno, N. Ichiyanagi, Y. Tanaka, M. Ikeda, H. Ishii, S. Honjo, Y. Sato, T. Hara, Y. Iwata, IEEE Trans. on Appl. Supercond. **7**, 1069 (1997).

[18] K. Sato, K. Hayashi, K. Ohmatsu, F. Fujikami, N. Saga, S. Isojima, S. Honjo, H. Ishii, T. Hara, Y. Iwata, IEEE Trans. on Appl. Supercond. **7**, 345 (1997).

[19] M. Leghissa, J. Rieger, J. Wiezoreck, H.-P. Krämer, B. Roas, B. Fischer, K. Fischer, H.-W. Neumüller, Inst. Phys. Conf. Ser. No. **158**, 1191 (IOP Publishing Ltd 1997).

[20] C.P. Bean, Phys. Rev. Lett. **8**, 250 (1962).

[21] M. Oomen, J. Rieger, M. Leghissa, H.H.J. ten Kate, Physica C **290**, 281 (1997).

[22] H. Noji, Supercond. Sci. Technol. **10**, 552 (1997).

[23] H. Piel, Inst. Phys. Conf. Ser. No. **158**, 1 (IOP Publishing Ltd 1997).

[24] Th. Schuster, H. Kuhn, A. Weißhardt, H. Kronmüller, B. Roas, O. Eibl, M. Leghissa, H.-W. Neumüller, Appl. Phys. Lett. **69**, 1954 (1996).

[25] C. Reimann, O. Waldmann, P. Müller, M. Leghissa, B. Roas, Appl. Phys. Lett. **71**, 3287 (1997).

Basic Science and Challenges in Process Simulation

J. Dąbrowski[1], H.-J. Müssig[1], M. Duane[2], S. T. Dunham[3], R. Goossens[4], and H.-H. Vuong[5]

[1] Institute for Semiconductor Physics,
Walter-Korsing-Straße 2, D-15230 Frankfurt (Oder), Germany
[2] Advanced Micro Devices, Inc., Austin, Texas, USA
[3] Electrical and Computer Engineering Dept., Boston University, Boston, MA 02215, USA
[4] Semiconductor Research Corporation, National Semiconductor Corporation
[5] Bell Laboratories, Lucent Technologies, Murray Hill, NJ 07944, USA

Abstract: We intend to turn the reader's attention towards CMOS (Complementary Metal-Oxide-Semiconductor), the mainstream semiconductor technology. Within a decade, silicon CMOS devices wil! be 200 atoms long and 50 atoms deep. Of 50 billion devices on a chip only 64 may fail. This calls for extreme control of atomistic processes. We review research topics in basic materials science which are likely to match the needs of CMOS. It is noted that a hierarchy of models which adopt methods ranging from quantum mechanics through classical and Monte Carlo schemes to easily solvable continuum equations must be constructed.

1 Introduction

Atomic-scale studies of solid state systems mark major achievements of materials science. The mainstream microchip technology, Complementary Metal-Oxide-Semiconductor (CMOS), may also profit from them. Stimulated by CMOS miniaturization (Fig. 1), atomistic research and modeling can enhance competitiveness of production by testing *technological* ideas on *fundamental* level.

A technologist often finds himself at a crossroad. Physical insight can guide one in the choice of direction to head, helping in early development phases. The same insight positions a scientist to supply efficient simulation models. This can be beneficial in failure and sensitivity analysis within the existing technologies.

Starting from an overview of CMOS, we discuss the simulation environment, TCAD tools, crystal growth, wafer processing, implantation, diffusion, oxides, silicides, back end processes, gettering, and prospective metrologies and theories. This summarizes the 182nd Heraeus-Seminar "International Workshop on Challenges in Predictive Process Simulation", ChiPPS'97 [1,2], which was primarily focused on diffusion-related problems viewed from the perspective of the whole fabrication process. These issues are posted for ranking on the Internet [2].

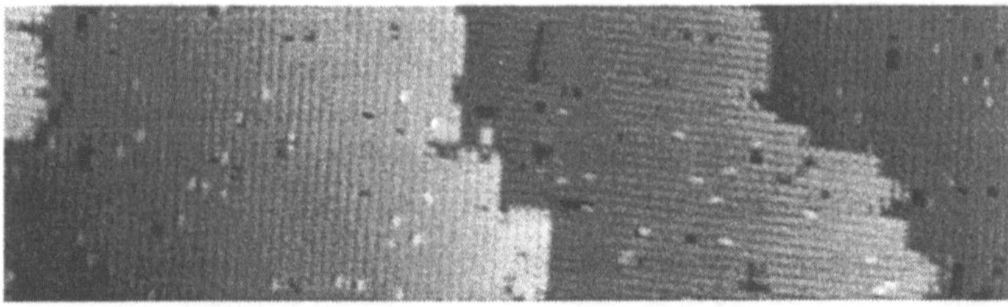

Figure 1 Miniaturization drives technology into the atomic regime. In this Scanning Tunneling Microscope (STM) image of the clean Si(001), atomic details as dimer rows, defects, and monatomic steps are clearly visible. The length of the scanned area is 0.1 μm. Within one decade, this will be the minimum feature size of state-of-the-art CMOS devices.

2 MOS technology

The workhorse of MOS technology is the Field Effect Transistor (MOSFET) (Fig. 2). Basically, it is a switch controlled by the gate potential. MOSFETs are named after the type of carriers which flow between the source (S) and the drain (D): nMOSFETs are built on p-doped material, with n^+-doped source and drain (S/D), while pMOSFETs are built on n-doped substrates, with p^+-doped S/D.

Circuits must be faster, cost less per bit, use less power. They are cheaper with interconnects on chip. Hence miniaturization [4]: shorter channels, thinner oxides, shallower junctions. Gate oxide thickness t_{ox}, now $\sim$ 8 nm, will be $\sim$ 3 nm by 2001, with roughness approaching the atomic step height [3]. Estimates from 1997 indicated that t_{ox} can be scaled down to 2 nm [5]; in the same year a 0.06 μm MOSFET with t_{ox} of only 1.2 nm (12 Å) was demonstrated [6]. Such transistors work fast but have high standby power losses due to tunneling through the oxide.

Maintaining an increasing performance of sub-μm MOSFETs may be expensive [7]. Also costs of production lines escalate: a current generation "fab" (factory) costs around a billion dollars. Only a fraction of the revenue may thus come from cutting-edge technologies (today 0.25 μm linewidth, or the gate oxide length, which somewhat exceeds the channel length). Even 95% of the products shipped in 1997 by a prospering microprocessor company may have been made with "old" 0.5 and 0.8 processes. Indeed, *though tomorrow has its challenges, innovations which make yesterday's technologies cheaper may be welcome today.*

A typical integrated circuit needs several hundred fabrication steps which apply deposition, lithography, oxidation, implantation, etching, polishing. The wafer is structured by local doping and isolation. Then gate oxides are grown and covered by gate polysilicon, contacts are made, interconnects are added. It remains to break the wafer into chips and seal them for protection. The steps needed to make the devices are named Front End Of the Line (FEOL) processes, while the steps needed to wire them are dubbed Back End Of the Line (BEOL). Most of the process steps are in BEOL and they are the primary yield-limiters. But the road from physics to technology is shorter in FEOL.

Inplanted dopant atoms are activated into substitutional sites by annealing. Standard dopants are B, P, and As; new designs may employ In and Sb. Implants are shot through a protective SiO_2 layer, sometimes covered by Si_3N_4. This may reduce channeling. To make shallower profiles, B is often implanted as BF_2.

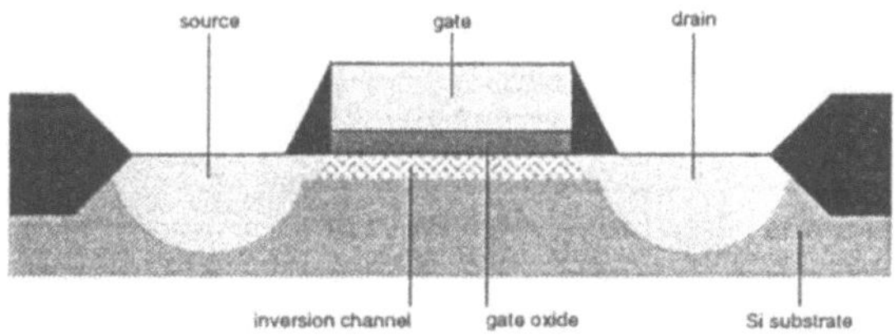

Figure 2 A schematic cross-section of a MOSFET [8]. Black areas on the sides mark the isolation oxide ("field oxide" or FOX). Triangular terminations of the FOX symbolize "bird's beaks", side-effects of LOCOS oxidation. Also in black is shown the gate spacer (SiO_2 or Si_3N_4) which encompasses the gate. The gate is a heavily doped polySi strip; the other electrodes, source (S) and drain (D), are also highly doped. The substrate between S and D is doped oppositely to S/D. A gate potential greater than the threshold voltage, V_T, opens a conducting "channel" between S and D. The electrodes are contacted by silicide (not shown).

CMOS combines nMOS and pMOS on the same wafer, usually pre-doped with B. The n-type wells ("tubs") for pMOS are implanted. Low-energy implants are then used for junctions and adjustments of channel profiles. Gates can be deposited already doped, but they are usually implanted because device parameters are optimal when the gate and the channel have the same work function.

The active areas are isolated by thermal oxidation of the pad oxide in O-containing gas. This LOCal Oxidation of Silicon (LOCOS) scheme [7] produces "bird beaks" (Fig. 2), so Shallow Trench Isolation (STI), which adds etch and deposition steps, may be needed in the future [10]. High-quality gate oxides are grown by high-temperature dry oxidation after striping the remaining pad oxide.

After gate deposition and etch, contacts are made by self-aligned silicidation (salicidation) with Ti, in the future maybe Co. The metal attacks Si, but not SiO_2. The unreacted film is removed, leaving well-defined contacts.

The circuit is wired with planes of horizontal Al lines ("metallization levels") joined by vertical W "plugs". The future material is Cu [11]. Boron Phosphorus Silica Glasses (BPSG) are used for isolation, but their high dielectric constants delay the propagation of electric signals (parasitic capacitances, RC) and facilitates cross-talk between wires. New dielectrics will be needed and even Cu may have to be replaced.

3 Simulation Environment and TCAD Tools

Modeling of technological processes is increasingly attractive to the industry. First, it can save time. Second, it can reduce processing costs of silicon wafers.

New generations are developed by scaling the existing ones; problems are solved by matrix experiments. A technology is often tuned during production, by varying the process for some wafer lots ("split lots"). If the change is too drastic, wafers can be lost. By the year 2010, a lot of 50 wafers will cost nearly a million dollars (Fig. 3) and this way may prove too expensive or too slow.

A semiconductor fab acts as a computer which may need a few months to run a batch (Fig. 3). Simulation can give results within a few days, so virtual experiments would be

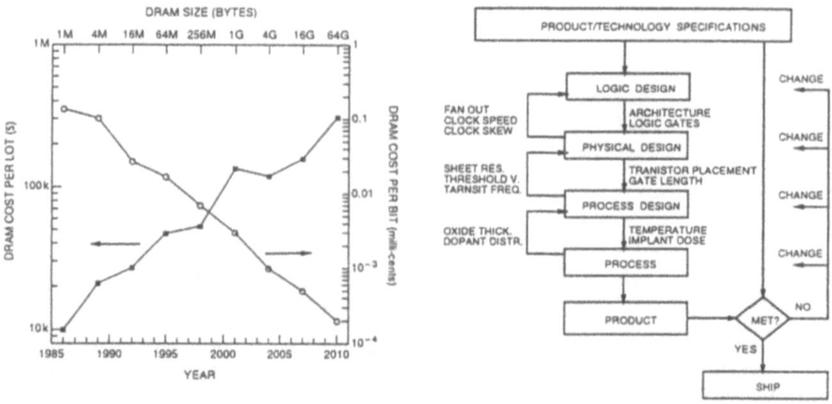

Figure 3 Left: Price per bit (circles, right ordinate) and cost per lot of 25 wafers (squares, left ordinate) for DRAM generations versus the year of introduction [9]. Right: Product realization. Logic design translates the product specifications into gates, registers, etc., assuming that parameters as clock speed are within ranges guaranteed by the physical layout. The latter outlines the physical structure (transistor placement, linewidths) which should fulfill these requests. The process design casts the physical design targets onto process conditions.

optimal. But this may work only if all design tools are *predictive*. This may be difficult (Tab. 1), but any progress is certainly welcome.

Simulation for CMOS includes *equipment modeling* (crystal growth, CVD, rapid thermal and plasma processing), *process modeling* (diffusion, implantation, oxidation, epitaxy, etching), and *device modeling* (physics, design, interaction of devices, interconnects, degradation, burn-out). These activities need an extensive database coupled to physically sound models. Process modeling may be assisted by atomistic theoretical tools employing quantum mechanics and molecular dynamics. Diffusion or interfacial physics appear particularly suited for them. But in many cases direct measurements or empirical models will dominate. As examples, one may quote reaction rates and sticking coefficients.

The key atomistic issues in modeling for *crystal growth and wafer processing* are about defects and homogeneity. The main challenges in the *front end* are related to dopant profiles in shallow junctions and short channels, an excellent playground for atomistic physics. It is intriguing that combination of models (e.g., for high-concentration regime and for defect-mediated diffusion) remains problematic. Promising for atomic-scale physics in *back end* are surface processes during deposition and etching. The simulation methods exist, but key parameters are difficult to get and little physics is built into BEOL models.

Technology Computer Aided Design (TCAD) can be used on many problems (Tab. 1). It must be efficient; physics is just *one of the means* to achieve this.

In early development phases, simulation gives insight into technology directions and elucidates interactions between optional solutions. A high return on investment can be obtained by getting *approximate but early* answers to the *right questions*. The earlier a problem is caught, the easier it is to correct. In the late development phase, TCAD is useful for process optimization, sensitivity analysis, and diagnosis [16]. Such quantitative applications are more demanding (Tab. 1). But even insight alone helps. E.g, understanding of existing processes enables predictions of process reliability and of process-induced variations.

Table 1 TCAD applications. The top three items get most attention, while much useful work is done in the bottom three classes. See also [3,12,13,15].

Application	Accuracy needed	Comments
predictive TCAD	very high	probably an elusive goal ...
advanced process control	high	provide macro models
process centering	high	optimize for a mature product
inverse modeling	high	use to extract coefficients
early exploration	medium	less splits in processing of lots
failure analysis	medium	test probable causes
learning and insight	low	high return on investment

A great potential for modeling is in new materials. There is a *lot* of data on existing technologies, so extrapolations usually work. The big questions come with new materials, as Cu instead of Al, or replacements for SiO_2. Another key area where basic physics may help is interfaces. *Any insight that physics can provide early in the development of new technologies can have a large impact.*

Practical TCAD uses computationally cheap, continuum models. This traditional method is reaching limits in defect-mediated profile evolution, chemical and physical reaction modeling, predictive equipment and topography modeling, and practical non-equilibrium transport. It is unclear when it will break down, but technology advances fast and this may happen before we realize it. Discussions between technologists and physicists are needed to smoothen the transition to discrete models. Atomistic physics is not easily integrable into the existing tools and will have to be first translated to continuum models.

Consider two examples. First, electrical characteristic of tiny transistors are affected by two- and three-dimensional (2D and 3D) structural details [14]. The 3D narrow-channel effect becomes comparable to the 2D short-channel effect. Since 2D dopant profiles cannot be properly measured (Section 10), diffusion models which would reliably extrapolate 1D data are needed. When adopted to the continuum formalism, they can be used in modern simulators which allow to define equations. Second, applications beyond 0.1 μm will need discrete description of defects and dopants, with stochastic analysis of fluctuations [4].

Optimally, TCAD should yield correct electrical characteristics given any process flow. But the models have numerous, often obscure parameters. They must be *calibrated* by fits to experiments not always doable under conditions close to those in the production. The models may also employ wrong assumptions or contain no physics. The predicivity depends thus on the choice of test structures, interpretation of the experiment, and intuition and experience of the user.

Calibration is also difficult because typically a whole process is simulated. If the threshold voltage from the subsequent device analysis does not match the measured value, one must check assumptions in the process and device simulations, the test structure layout, and electrical measurements. Knowledge in many areas is required (Tab. 2). Physics can help by systematizing this knowledge.

Models with physically meaningful parameters and with known windows for their variation are thus favored. Calibration strategies are important. A hierarchy of models from

Table 2 Knowledge needed for TCAD calibration. The wafer temperature T_0 during implantation can affect the point defect generation and thereby the diffusion later [17]. T_0 during Rapid Thermal Annealing (RTA) is difficult to measure [18], but 1 volt variation in a 220 V AC line changed T_0 by 6 ° [19]. The ion beam energy can affect Secondary Ion Mass Spectroscopy (SIMS) profiles [20]. Spreading Resistance Profiling (SRP) depends on probe pressure [21]. Poly linewidth and oxide thickness are difficult to measure; a 1% change affects the saturation current by ~1%.

Specialization	Knowledge needed for TCAD calibration
process/device	must know complete process flow, process/device physics
fab equipment	T_0 in implantation affects diffusion; local T_0 in RTA
electrical tests	electrical versus optical oxide thickness
analytical	SIMS knock-on; SRP probe pressure
metrology	uncertainty in poly length and oxide thickness
simulation	model limitations; grid dependence

atomistic to continuum is needed. Some basic models, e.g., for segregation, must be re-examined in light of advances in metrology, physical understanding, and process conditions over the last 20 years. To avoid confusion, implementation of the models must be backward compatible; their physical interpretation would assists users and vendors, adding value to TCAD tools.

4 Models for Crystal Growth and Wafer Processing

Crystal growers must produce Si base material with homogeneous distribution of O and dopants. Modeling can be used to predict the transport of the impurities in the melt and their incorporation into the crystal. Given increasing crystal diameters (a 300 mm technology fab is being built in Dresden, Germany), costs of development will increase and simulation will gain importance. But melt convection is turbulent and three dimensional, and thus difficult to handle.

Epitaxy modeling should consider heat transfer by convection, conduction, and radiation. 3D gas convection and reaction of a gas mixture ($SiHCl_3$, H_2) with the surface must be included. Accuracy of heat transfer and convection models needs improvement [22]. Models for reaction kinetics are in the development phase.

Grown-in defects (octahedral voids) in as-grown Si reduce the yield (the percentage of working chips) and must be contained. Their distributions will have to be simulated by taking into account dynamics of point defect (self-interstitials, vacancies, foreign atoms) and the formation of extended defects [23].

Finally, modeling tools for O precipitation during thermal treatments, including influence of impurities (e.g., boron), gettering of metals (Section 9), stress in Si wafers (including thermal stress), defects and homogeneity in epitaxy, and molecular simulation for cleaning and polishing of wafers must be developed.

Table 3 Shallow dopants. Note the absence of fast-diffusing Al; it would deteriorate the dopant profile of a submicrometer device. Ga diffuses fast in SiO_2 [28].

species	type	atomic mass	importance	research	diffusivity	diffusion vehicle
B	acceptor	5	high	mature	high	Si_I
P	donor	15	high	advanced	high	Si_I
As	donor	33	high	advanced	medium	Si_V, Si_I
In	acceptor	49	medium	evolving	low	Si_I , Si_V
Sb	donor	51	medium	evolving	low	Si_V

5 Models for Ion Implantation

A major challenge is the formation of shallow, heavily doped junctions [3]. Junction depths continueously decrease: in the 0.1 μm technology (year 2007), they will be 150 to 450 Å (15 to 45 nm). The difficulty lies in channeling effects and in the diffusion out of the implanted region during subsequent processing. For reviews on implantation and its rclation to diffusion [31,33].

The modeling provides initial conditions for simulation of dopant diffusion. Standard TCAD yields the spatial distribution of dopants by a sophisticated extrapolation of SIMS data. This may run out of steam for near-0.1 μm structures.

The ions create Si interstitials (Si_I) and vacancies (Si_V). Since dopants diffuse by point-defect mediated mechanisms (Tab. 3), the distribution of Si_I, Si_V, and their clusters (seeds of extended defects) is needed for further modeling of diffusion. Appropriate atomistic simulators have been presented [24–26], but more data is needed on the intermediate cluster configurations and on thermal effects on defect migration and recombination [25,27]. Though TCAD models can account for channeling and defect-mediated diffusion (Section 6.1), they are not good enough for heavily doped junctions shallower than 100 nm.

The basic theoretical methods which can help to obtain the missing data are:

1. *Quantum mechanics.* The most exact but the most CPU-intensive; hence only small systems can be examined. Static calculations dominate, dynamic simulations are needed. Can validate/calibrate higher-level methods [29].

2. *Classical molecular dynamics (MD).* Successful for low-energy impacts [24,26]. Problems with computational resources and physical intuition. Hits its limits at amorphization and defect evolution for $t > 1$ ns.

3. *Binary collisions (BC).* The leading atomistic approach [25,30–32]. Monte-Carlo based. Successful for ballistic processes, yields ion range profiles. Better coupling to MD and experiment will improve modeling of damage.

A hierarchy of models ranging from the lowest (ab initio) to the highest level (which may adopt Monte Carlo for self-annealing, diffusion, and clustering of defects [25]), with consistent and transferable parameters, would greatly enhance the predictability. Self-annihilation and thermal effects (dependence of implantation and diffusion on temperature [17]) could be then properly included.

Mostly simulated are the species of Tab. 3. It would be useful to learn more how species like N and Si affect channeling and diffusion. Si can be implanted to amorphize the crystal in order to reduce channeling and speed up silicidation in narrow stripes. N implants help to control diffusion. These are important examples where simulation plays a *minor* role. Simulation tends to be a "step behind" technology; the issue is how to get "in-step" with technology.

Briefly, in order to model shallow junctions, TCAD must include interaction between implantation and diffusion. This illustrates the key issue in FEOL: the main challenge is in making modules *interact*, not in improving each module.

6 Models for Dopant Diffusion

Research on diffusion in Si has provided many detailed data, including atomistic models and the influence of various process conditions [34]. But dopant profiles for new designs can hardly be predicted without short loop experiments for calibration. "Ramping" (increasing or decreasing of the wafer temperature, at around 100 K/s) is an extra complication. Dopants may diffuse during ramping already above 600 °C. The local temperature is difficult to measure[18,19].

The traditional approach to diffusion is to use coupled differential equations [34]. Dynamics of diffusion vehicles (Si_I, Si_V [35–37], Tab. 3) is limited to equilibrium. Recent tools provide non-equilibrium dynamics of point defects and allow to account for inhomogeneous distribution of extended defects which are sources/sinks of Si_I. The key technical details to be worked out are:

1. Diffusion parameters for Si_I and Si_V: formation energies, migration barriers, pre-exponentials, dependence on the Fermi-level, chemical composition, and strain. Experimental [27,38] and theoretical (ab initio [39], tight-binding [29]) data exist, but are often contradictory. Diffusivities of Si_I from metal diffusion and from the depth dependence of TED (Section 6.1) differ by orders of magnitude. Trapping of Si_I has been suggested as an explanation [40].

2. 3D solvers and statistical analysis for standard TCAD. Beyond 0.1 μm, statistics of defect and dopant distribution will play a role [4].

3. Methods to measure 2D and 3D profiles with sufficient accuracy (Section 10).

A number of basic problems remain unsolved. The reaction paths of defect-mediated diffusion are still doubtful. This includes diffusion in heavily doped materials. Transient diffusion effects need a complete description. Boundary conditions (notably, at SiO_2/Si(001) interfaces) must be properly defined. Diffusion in polySi is difficult to handle. These (and other) issues are addressed below.

6.1 Diffusion in Si Bulk

Dopant diffusion is commonly described by pair models: dopants and point defects are assumed to form mobile pairs. They can explain why dopants diffuse with a lower activation energy than native point defects (Si_I, Si_V). But the errors resulting from the simplification of interactions have not been estimated.

It is difficult to develop a consistent analysis of experimental data without a consistent knowledge about the paths by which point defects mediate the diffusion. Not only the form of diffusion equations but also such parameters as cross sections for defect-dopant reactions depend on the reaction path. Advanced atomistic theories for the interstitial [36,41] and vacancy mediated [36,42,43] mechanisms have been attempted; they must be verified and expanded.

Dopant diffusion is enhanced at concentrations above $2 \cdot 10^{20}$ cm^{-3}. To sort out the right explanation, quantum-mechanically verified potentials for defect-dopant interactions [42] are needed. Models including rigorously the kinetics of extended defects or Si_I injection at SiO_2/Si are missing for such concentrations.

In practice, only atoms in *isolated substitutional* sites act as shallow dopants. One must thus consider their activation, deactivation (formation of atomic-size clusters or/and precipitation of a silicide), and diffusion. Useful TCAD models exist, but they have many limitations and a lot of room for research remains.

Essential to the fabrication of shallow junctions is that diffusion is affected by excess intrinsic point defects. Depending on the migration mechanism (Tab. 3), such transient effects may enhance or retard diffusion. For example, Si interstitials enhance diffusion of B and P and retard diffusion of Sb.

The classic Transient Enhanced Diffusion (TED) is caused by Si_I produced in implantation, stored in End Of Range (EOR) defects after amorphizing and in {113} defects after non-amorphizing implants, and re-emitted. Its main features are described by the theory of cluster evaporation [27], but the initial conditions are unclear (Section 5). Description of vertical and lateral TED extensions is not reliable, particularily due to uncertainties in parameters for Si_I dynamics. TED at low concentrations (near to steady-state conditions) is well understood, but its modeling at high concentrations (far from steady-state) is problematic. Models built on the basis of a few easy to measure parameters are needed.

TED is suppressed by substitutional C [27]; first predictive models are available. The pressing question is the alleged enhancement of junction leakage by C. Some other impurities, notably F, have also been reported to affect TED [44].

An oxidizing interface ejects Si_I. The oxidation enhanced diffusion (OED [34]) is under control. Cl in the oxidizing ambient significantly reduces the number of ejected Si_I, probably due to creation of Si_V when the silicon is etched.

Another source of Si_I is deactivation of As [45]. Activation is efficiently done by Rapid Thermal Annealing (RTA) [46]: infrared lamps briefly heat the wafer to a high temperature. The concentration of the activated As exceeds solubility limits at temperatures at which As is mobile. In consequence, As atoms tend to cluster and precipitate. Even 50% of dopants can be clustered. Small As clusters contain Si_V, and their formation leads to emission of Si_I. This recent discovery [45] calls for more studies. The mechanism of the Si_I emission is not clear. Si_I may be ejected from the clusters, or the clusters may consume Si_V while Si_I are produced in the volume in Frenkel pairs. Other dopants, notably B, cluster as well. B clusters with Si_I and is deactivated; the complexes act as Si_I storage bins.

Modeling of TED, activation, and deactivation needs data on the structure (ratios of point defects to clustered impurities) and energetics (cluster sizes, emission/capture rates) of stable and intermediate clusters. Even groups of two dopant atoms [41] are important. On the other extreme, dopant interactions with extended defects must be studied. The case of polySi is addressed in 6.3.

Also silicidation injects point defects (Si_V in the case of Ti and Co). There are indications that this is *not* a major issue. When SIMS artifacts are absent, both Ti and Co silicidation are measured [47] to inject Si_V between 0.5 and 1 times the equilibrium Si_V concentration at temperatures between 800 and 890 °C. Silicidation should thus reduce diffusivities of B and P by a factor of two or less.

Finally, the concentration of Si_V is enhanced by nitridation [48]. The mechanism is unclear: it may be a direct insertion at the interface, or enhanced Si_V formation in Si volume due to the strain induced by the Si_3N_4 film.

6.2 Boundary Conditions

A dopant profile is determined by the initial distribution of dopants and point defects (Section 5) and by the physical borders (the bulk and an interface). The bulk is a reservoir of point defects. Transient processes usually dump point defects into it, but the bulk can be a source of vacancies for As deactivation.

Shallower junctions imply fine profiling of dopants closer and closer to the SiO_2/Si(001) interface, e.g., as close as 50 nm [49]. Three classes of boundary processes affect the dopant profile under the oxide. Dopant atoms either segregate to the oxide (B) or are ejected from it (P, As). Dopant atoms segregate to the SiO_2/Si(001) interface. Interstitials and vacancies are annihilated at non-oxidizing interfaces and interstitials are ejected from oxidizing interfaces.

Take P as an example. As the oxide grows into the silicon, it "shovels" P atoms, and P segregates to the interface at temperatures high enough to enable diffusion (around 600 °C). This P pile-up impacts the dependence of the threshold voltage V_T on the gate length, adding to Reverse Short Channel effect [50]. The profile can change so much that V_T obtained by standard TCAD can be off by 0.5 eV. Models for dopant profiles close to SiO_2/Si(001) exist and work for present CMOS devices (e.g.,[49]). But further detailed measurements and more comprehensive modeling are needed to give predictions for new designs. A simple but physically correct description of the interfacial processes and a reliable estimate of the physically allowed range of the model parameters would be of significant help to designers. This includes segregation coefficients and their dependence on temperature, ambient, dopant concentration, and oxide trap density.

The electronic structure of segregated dopants affects the activation, the noise figure, and the gate leakage current. The segregated atoms are deactivated, which causes a dose loss. Even 50% of the implants may "disappear" [51]. The interface can collect more than $2\cdot10^{14}$ dopant atoms per cm^2, nearly as many as there are interfacial lattice sites ($7\cdot10^{14}/cm^2$). The loss is enhanced when the annealing ambient contains O, needed to heal defects in SiO_2. An inert ambient may be critical at very high temperatures (above 1 000 °C), as during activation.

Similarly large amounts of dopants may be lost to the pad oxide after implantation. The resulting increase of series resistance becomes a problem for channel lengths below 0.25 μm. One might try to avoid this by implanting more dopants, but the atoms migrate in *all* directions and a higher dose means more diffuse profiles. Special capping layers and annealing schemes should thus be studied. The problem of dose loss in the context of silicidation is addressed in Section 7.

Ultrathin oxides are prone to B penetration from heavily doped gates (Section 6.3). Oxides which contain N are much more resistant, and nitridized oxides are a promising technological candidate. Since N tends to accumulate at the SiO_2/Si interface, the influence of N on the interfacial dopant atoms should be clarified.

6.3 Other Problems

Below, we address diffusion of dopants in gate oxides and in polySi, precipitation of oxygen, stress effects on diffusion in silicon, and diffusion in SiGe.

Stress may arise from field oxidation, different thermal expansion coefficients of layers (SiO_2, Si_3N_4), lattice misfit in SiGe heterostructures, and extended defects. The stress can be on the limit of dislocation formation, particularly during oxidation. While the influence of stress on oxidant diffusion and reaction in SiO_2 are well assessed, its effect on diffusion in Si is less studied [48].

Band-gap engineering by SiGe may become important in upcoming technologies. Quantitative models for the diffusion of Si, Ge, and dopants in SiGe structures, and for the segregation to bulk Si, are scarce [52]. There is a need to pinpoint the role of the composition and strain in the diffusion mechanisms.

Doping of polySi gates invokes important problems. First, gate oxides will be soon as thin as 3 nm (30 Å) [3]. B can diffuse through such a thin SiO_2. Atomistic models exist [53], but diffusion of B in gate oxides is not well understood. Quantitative data on the dependence of the diffusion on the B concentration and on the presence of F and other impurities is insufficient and inconsistent. Modeling of F diffusion in polySi gates and in gate oxides is in its infancy.

Second, the doping must reach 10^{20} cm^{-3}, with uniformity better than 3% [3]. It is difficult to activate even $5 \cdot 10^{19}$ cm^{-3} dopants. E.g., dielectric reflow deactivates As and can increase the contact resistance by a factor of three [54].

Third, diffusion models for polySi build on local homogenization. Poly thickness and grain sizes compare to gate lengths, so diffusion via single boundaries must be included. This is needed to estimate the effect of dopant distribution on the oxide penetration. Reliable models for diffusion and activation in polySi and for diffusion in SiO_2 should help to find a trade-off between high thermal budgets for doping homogeneity and low thermal budgets to prevent B penetration.

An area of potential interest is the interaction between impurities (such as C and dopants) and precipitation of O. O precipitates act as getters keeping the active regions clean from such "killer" impurities like Fe (Sections 4 and 9).

7 Silicide and Oxide Films

The key *silicide* in CMOS is $TiSi_2$. Silicidation is done in an inert ambient (N_2) at low temperatures, such as 600 °C. This produces a defected, high-resistivity phase of $TiSi_2$, labeled C49. Annealing transforms it into the low resistivity phase, C54. The transformation is nucleation-controlled and slows down when the silicide strip is scaled. Alternative silicides, notably $CoSi_2$, are thus studied, mostly experimentally. Standard TCAD tools lack the database for Co processes.

Silicidation is well studied [55], but its complexity hinders development of physically-based modules [56]. E.g., it is affected by dopant distribution, which must be computed in changing volumes of coexisting phases. Silicides are polycrystalline, with fast diffusion paths along boundaries [57]. Diffusion in the grains is usually anisotropic and may depend on stress. Segregation of dopants into the silicide depletes the silicon at the interface, increasing the contact resistance. Modeling should estimate V_T changes due to the dose loss and to a possible counterdoping, obtain the effect of silicidation on S/D resistance, and suggest a process window for the film thickness and the energy and dose of implantation.

Silicidation impedes scaling. The interface roughness facilitates electrical punch-through and limits junction depths. Since S/D junctions are made before silicidation, also Si consumption in the Ti process limits their depths. A junction must be at least 100 nm deep to assure low leakage after silicidation. Continued use of silicidation will soon require raised S/D structures [58].

Atomistic studies might help to solve the problems related to the dose loss and to the interfacial roughness. But to be successfull, atomistic theories must decouple the involved physical effects from the complex silicide formation.

Oxidation is troubled by oxide thickness (t_{ox}) variations and B penetration (Section 6.3). Methods to simulate t_{ox} variations, susceptibility to B penetration, and concentrations and generation kinetics of defects in the oxide (including micropores) and at SiO_2/Si (including SiO_2/polySi) are needed. Dependence on temperature and chemistry of the oxidant (O_2, O_2+N_2, NO_2, NO) and of the annealing ambient (O_2+N_2, O_2+Ar, Ar) is of interest. Physical models for device reliability would be welcome. New dielectrics, as Si_3N_4, are a prospective field.

Models for gate oxidation are largely empirical; the underlying physics is unclear. Corrections to the Deal-Grove model [59] work well for $t_{ox} > 10$ nm, but gate oxides are now thinner. As t_{ox} approaches the few-monolayer regime, atomistic schemes should be considered (Section 10). Mechanism of inhomogeneous oxide growth (as for isolation) is controversial. Parameters governing elastic, viscous, viscoelastic or plastic deformation of the oxide, and describing the effects of stress on oxidant diffusion and reaction await identification. Innovations like Rapid Thermal Oxidation [46] are not addressed in standard models.

8 Models for BEOL

Complex physics and chemistry are typical in BEOL. Practical simulations often use nearly geometrical models with much calibration. Commonly simulated process steps are deposition by sputtering, CVD, PECVD, plasma etching. The materials are SiO_2, TiN (diffusion barrier, wetting layer, and etch stop), Al, Ti, and W [60].

The goal is mainly to test the viability of increasing the aspect ratio of interconnect features, e.g., filling of trenches and vias [61], and to explore the scaling limits of present processes [62]. As these limits are reached, new materials and processes are being introduced rapidly. This includes low dielectric constant ("low-k") dielectrics to reduce delay and cross-talk [63]. Another emerging material is Cu. IBM has just announced [11] the first Cu technology (0.2 μm).

Plasma etching and deposition is treated with a limited set of chemical equations. Incoming fluxes are well simulated by equipment models. Advancements include Monte Carlo simulation for angle and energy distributions. Beyond 100 nm, molecular dynamics will be needed for the surface/plasma interaction.

The list of wishes for BEOL simulation includes [64]:

1. Decomposition paths, deposition rates: Cu, Ti, W, Al, polySi, WSi_2, $TiSi_2$.
2. Surface mechanisms (including surface diffusion) in deposition and etching.
3. Plasma processing: radical/ion interaction in plasma and with surfaces. Cross sections for excitation, total and partial ionization, dissociation, attachment, dissociative attachment. CF_x (x=1–3) interacting with Si and SiO_2, Cl etch of Al (including BCl_2 cross section, Cl_2 data exists), F interaction with Si.
4. Viscosity, thermal conductivity, diffusivity: $SiCl_x$ (x=2,3), SiH_xCl_2 (x=1,2), $SiHCl_3$, SiH_3Cl, Cl^+, BCl_2^+, CF_x^+ (x=1–3), SF_x^+ (x=1–6), WF_x^+ (x=1–6).
5. Ar^+ sputtering of Al, Cu(111), Ti. Energy, angle of sputtered atoms, and yield as a function of incoming ion energy and angle.

Another challenge is Chemical Mechanical Polishing (CMP). CMP is widely used in BEOL (dielectric planarization, metal etch back) and FEOL (STI) [63].

9 Models for Gettering of Impurities

The ultimate goal is here to improve the yield (the percentage of working chips). Simulation should pinpoint the steps most sensitive to contamination and suggest changes to reduce this sensitivity and widen the process window. This requires quantitative understanding of surface chemistry and gettering. Basic mechanisms are established, but there are many gaps in the details, such as surface and bulk segregation coefficients under various process conditions. Good surface chemistry data exist for some contaminants, especially for Fe.

The critical impurities are transition metals, mostly Fe, Cu, and Ni. Iron [65] is well-studied, ubiquitous, relatively fast diffusing. It remains electrically active (interstitial) after fast cooling. FeB pairs [67] usually determine the carrier diffusion length and lifetime, especially after processing. A single Fe atom segregated to the gate oxide can kill a 0.1 μm-channel transistor! Another frequently studied metal, Au, is not troublesome and will not become so the foreseeable future.

Progressively restrictive contamination thresholds call for new gettering methods [66]. Well-understood gettering driven by supersaturation (relaxation-type internal gettering by SiO_2 precipitates) will loose importance as supersaturation is ever more difficult to achieve. Proximity gettering (segregation gettering close to the device by p/p$^+$ epi or implanted layers) is very promising. Most processor structures use it (Intel: 100%). Quantitative calculations of proximity gettering require segregation coefficients; their values are often not well established.

10 Prospective Metrologies and Atomistic Theories

Diffusion simulations are verified mostly by comparison of electrical measurements with device simulations. This assumes at least that the latter are adequate. Below we benchmark methods which might supply direct two-dimensional (2D) data: location of the junction with accuracy better than 10 nm, and dopant and carrier profiles with vertical and lateral spatial resolution of 10 nm or better. Metrology requirements have been recently re-examined in Ref. [77].

2D profiling is too tough for traditional methods. Though the conventional Spreading Resistance Profiling (SRP) yields one-dimensional (1D) vertical profiles for junction depths below 50 nm, the required lateral resolution cannot be even approached. SRP with a conducting Atomic Force Microscopy (AFM) probe and an ohmic back contact offers some improvements. Secondary Ion Mass Spectroscopy (SIMS) can yield excellent 1D profiles, but also fails for 2D.

New methods build on electron and scanning probe microscopies. Samples are cleaved and scanned. Since a unique relation between the electronic structure of the *surface* and the dopant profile in the substrate is needed, the common problem is that the sensitivity is hampered by difficulties with surface preparation.

Selective etching [68] uses the dependence of the etch rate on the doping level. The topography is measured by AFM [69], Transmission Electron Microscopy (TEM [70]), or Scanning Electron Microscopy (SEM [70]). Stereo SEM improves sensitivity and resolution. The main problems are reproducibility, calibration, and a strong rise of the etch rate at concentrations above $5 \cdot 10^{19}$ cm^{-3}.

Chemical staining [71] detects metal layers electroplated on silicon. The main problem is low sensitivity of electroplating to the substrate Fermi level.

Scanning Tunneling Microscopy (STM [72]) detects the doping level through the dependence of the current on the depleted region in Si. The resolution is limited by the surface roughness and incomplete surface passivation.

Scanning Capacitance Microscopy (SCM) detects tip-sample capacitance changes [73]. Tip preparation methods limit its resolution to around 20 nm.

Scanning Kelvin Probe Microscopy (SKPM) measures changes in surface potential [74]. Spatial resolution, limited by the tip, is comparable to that in SCM. Surface preparation limits resolution in measurement of the potential.

TCAD must know geometrical factors such as oxide thickness, t_{ox}, and gate length, l. Essential electrical parameters, like the saturation current I_{dsat}, depend in the first-order on t_{ox} and l. For 0.25 μm gate lengths with 5 nm gate oxides, 1% accuracy for I_{dsat} implies a measurement error of only 2.5 nm (25 Å) for the poly length. Even more impressive is the condition for t_{ox}: 0.05 nm (0.5 Å), much less than a monolayer. The desired control on I_{dsat} is below $\pm 10\%$ (3σ). Errors add up, so 1% from a single input data is a conservative requirement.

STM can measure currents through ultrathin oxides [75]. This might be used to access local oxide thickness variations, interfacial states, defect states in the oxide, and to study hot-electron damage. A conducting AFM [76] may decouple the information about the oxide thickness and the interface roughness.

The ultimate goal of *atomistic simulations* for TCAD is to yield data needed to write down simple time-dependent equations. Parameters in these equations should have a well-defined physical meaning, so that they can be measured with a minimum of interpretation bias. Another value comes from the paradox of simulation: in order to simulate something, one must understand it well, but in the process of understanding it, one may reduce the need for simulation [13].

Ab initio schemes [78] are nearly arbitrarily accurate. Their assumptions can be strictly controlled. But they are CPU-intensive, working best for systems with less than thousand atoms. Even assisted by high performance computing and new algorithms, they will be most helpful in sorting out the right physics or validating and calibrating higher-level methods. Computation for TCAD seems practical only in such cases as for atomic configurations or binding energies.

Similar limitations apply to quantum-mechanical methods which use model potentials derived ab initio or from experiment. But much larger systems can be treated and molecular dynamics becomes much cheaper [79]. Disorder, complex interfaces, or dislocations become accessible with less effort, though with less confidence. These techniques are an indispensable extension of ab initio tools.

Classical methods are even faster. They introduce approximate interatomic interactions: Stillinger-Weber [80] or Tersoff [81] potentials, or "valence forces" [82]. But direct links to the electronic structure are lost. Unphysical effects can dominate the outcome. Examples of failures are known particularily for structures in which dangling bonds or unusually coordinated atoms play a role [29,83]. Critical configurations must be cross-checked against lower-level methods.

Strictly atomistic methods are limited by the period of atomic vibrations, $\sim 10^{-13}$ s: there are 13 decades from the atomic to the "technological" time scale! Monte Carlo schemes get around this by describing atomic dynamics through energy barriers and hopping frequencies. The time scale is thus increased to the hopping time, by two to six decades, and the density of mobile species drops from the atom density (10^{23}/cm^3) to the point defect density (10^{10}–10^{20}/cm^3).

Quantum-mechanical methods may have an impact through calculations for (1) dopant diffusion, notably transient effects, (2) dopants near interfaces, in particular pileup and dose loss, (3) surface chemistry during metalization, (4) chemical reactions during etching. They must be linked to higher-level methods (classical potentials, Monte Carlo, continuum equations) in virtually all cases.

11 Conclusions

We conclude noting that *as microchip technologies enter the ultrasubmicron era, physical understanding gains importance. This calls for improved process simulation supported by atomistic studies.* To secure this support, as physical insight and as input data, links between technologists and scientists must grow stronger.

Acknowledgements

We are indebted to A. Agarwal, S. Biesemans, A. Bourenkov, L. Colombo, E. Dornberger, M. Duane, R. W. Dutton, H.-J. Gossmann, E. Kaxiras, E. Langer, J. Lorenz, T. Ma, H. Marquardt, M. Meyyappan, A. Ourmazd, S. T. Pantelides, P. Pichler, J. M. Poate, M. Posselt, H. Rücker, H.-H. Vuong, E. R. Weber, and E. T. Yu for their help in organizing the work, sorting out and ranking the essential issues, and for reviewing subsequent versions of the manuscript.

Bibliography

[1] H.-J. Müssig and J. Dąbrowski, Physikalische Blätter **11**, 1132 (1997).

[2] J. Dąbrowski and H.-J. Müssig, `http://www.ihp-ffo.de/chipps/chipps.html` (1997).

[3] The National Technology Roadmap for Semiconductors, Semiconductor Industry Association (1997).

[4] C. A. Warwick et al., AT&T Techn. Journ. **72**, 50 (1993).

[5] A.-H. Lo et al., IEEE Electron Device Letters **18**, 209 (1997).

[6] R. Wilson, `http://techweb.cmp.com/eet/news/97/982news/bell.html` (1997).

[7] D. J. Foster, in Electronic Materials ed. by L. S. Miller and J. B. Mullin, (Plenum Press, New York 1991) 173.

[8] S. M. Sze, Physics of Semiconductor Devices, 2nd edition (Wiley, 1981).

[9] H.-J. Gossmann and J. M. Poate, in Proceedings of the ICPS-23, ed. by M. Scheffler and R. Zimmermann (World Scientific, 1996) p. 2569.

[10] S. Nag, A. Chatterjee, Solid State Technology (September) **40**, 129 (1997).

[11] IBM web pages at `http://www.ibm.com/News/1997/09/ls9709222.html` (1997).

[12] TCAD Roadmap: A Supplement to the National Technology Roadmap for Semiconductors, Sematech Technology Transfer 95012696A-TR (1995).

[13] M. Duane, `http://www.prismnet.com/ñaomi/tcad/road97.html` (1998).

[14] J. Lorenz, B. Baccus, and W. Henke, Microelectronic Engineering **34**, 85 (1996).

[15] M. R. Pinto et al., 1993 IEDM Digest, p. 701 (1993).

[16] P. Lloyd et al., Microelectronics Journal **26**, 79 (1995).

[17] K. Jones et. al., 11th Int. Conf. on Ion Implantation Technology, Austin, Texas, June 1996, p. 618 (1996).

[18] Solid State Technology, December 1996, p. 34 (summary of the 4th Int. Conf. on Advanced Thermal Processing of Semiconductors, RTP'96, Boise, Idaho).

[19] J. Nakos, SPIE Vol. 1804, Rapid Thermal and Laser Processing, p. 24 (1993).

[20] A. C. Diebold et al., J. Vac. Sci. Techn. B **14**, 196 (1996); L.A. Heimbrook et al., *ibid*, p. 202. See also other papers in this volume.

[21] T. Clarysse et al., J. Vac. Sci. Techn. B **14**, 358 (1996).

[22] S.Kommu and G. Wilson, ECS Proc. **97-25**, 222 (1997); H. Habuka et al., J. Cryst. Growth **151**, 375 (1995).

[23] E. Dornberger and W. v. Ammon, J. Electrochem. Soc. **143**, 1648 (1996); T. Sinno et al., J. Electrochem. Soc. **145**, 302 (1998); E. Dornberger et al.., Electrochem. Soc. Proc **97-22**, 40 (1997).

[24] T. Diaz de la Rubia and G. H. Gilmer, Phys. Rev. Lett. **74**, 2507 (1995).

[25] M. Jaraiz et al., Appl. Phys. Lett. **68**, 409 (1996).

[26] K. M. Beardmore, N. G. Jensen, `http://bifrost.lanl.gov/~reed/reed.shtml` (1998).

[27] P.A. Stolk et al., J. Appl. Phys. **81**, 6031 (1997).

[28] Properties of Silicon, ed. by T. H. Ning, C. Hilsum, (INSPEC, 1988).

[29] M. J. Tang et al., Phys. Rev. B **55**, 14279 (1997).

[30] M. Posselt, `http://www.ihp-ffo.de/chipps/97/Djpg/Dposters/posselt.html` (1997).

[31] E. Chason et al., J. Appl. Phys. **81**, 6513 (1997).

[32] E. Leitner et al., in 3-Dimensional Process Simulation ed. by J. Lorenz (1995) p. 136; M. T. Robinson and O. S. Oen, Phys. Rev. **132**, 2385 (1963).

[33] S. Moffatt, Nuclear Instruments and Methods in Physics Research B **96**, 1 (1995).

[34] P. M. Fahey, P.B. Griffin, and J. D. Plummer, Rev. Mod. Phys, **61**, 289 (1989).

[35] R. Car et al., Phys. Rev. Lett. **54**, 360 (1985).

[36] C. S. Nichols et al., Phys. Rev. B **40**, 5484 (1989).

[37] N. E. B. Cowern et al.., Phys. Rev. Lett. **69**, 116 (1992).

[38] H. Bracht, N. A. Stolwijk, and H. Mehrer, Phys. Rev. B **52**, 16542 (1995).

[39] P. E. Blochl et al., Phys. Rev. Lett. **70**, 2435 (1993).

[40] N. E. B. Cowern, Appl. Phys. Lett. **64**, 2646 (1994).

[41] J. Zhu et al., Phys. Rev. **B 54**, 4741 (1996).

[42] S. T. Dunham, C. D Wu, J. Appl. Phys. **78**, 2362 (1995).

[43] M. Ramamoorthy and S. T. Pantelides, Phys. Rev. Lett. **76**, 4753 (1996).

[44] H.-H. Vuong et al., J. Appl. Phys. **77**, 3056 (1995).

[45] P. M. Rousseau et al., Appl. Phys. Lett. **65**, 578 (1994); U. Myler et al., J. Vac. Sci. Techn. B **15**, 757 (1997).

[46] R. Singh, J. Appl. Phys. R **63**, 59 (1988).

[47] S. B. Herner et al., Appl. Phys. Lett. **68**, 2870 (1996).

[48] S. Chaudhry and M. E. Law, J. Appl. Phys. **82**, 1138 (1997).

[49] H.-H. Vuong et al., SISPAD'97, 85 (1997); H. Sakamoto and S. Kumashiro, ibid, 81.

[50] H.-H. Vuong et al., IEEE Transactions on Electron Devices **43**, 1144 (1996).

[51] P.B. Griffin, S. W. Crowder, and J. M. Knight, Appl. Phys. Lett. **67**, 482 (1995).

[52] N. E. B. Cowern et al., "Diffusion in strained SiGe" Phys. Rev. Lett. **72**, 2585 (1994).

[53] R. B. Fair, Journal of the Electrochemical Society **144**, 708 (1997).

[54] W. Taylor, SRC Topical Research Conference on TED, Boston 1995, p. 33 (1995).

[55] K. Maex, Mat Sci. Eng. **R 11**, 55 (1993).

[56] P. P. Apte et al., Mat. Res. Soc. Symp. Proc. **402**, 233, (1995).

[57] F. M. D'Heurle, Journal de Physique IV C **6**, 29 (1996).

[58] J.-Y. Tsai et al., IEEE Electron Dev. Lett. **17**, 331 (1996).

[59] R. W. Dutton and Z. Yu, Technology CAD (Kluwer Academics Publishers, 1993).

[60] F. H. Baumann, in Semiconductor Characterization, Present Status and Future Needs, ed. W. M. Bullis, D. G. Seiler, and A. C. Diebold, (AIP Press, 1995).

[61] D. S. Bang et al., IEDM Digest, 97 (1995).

[62] F. H. Baumann et al., VMIC Proceedings, 412 (1993).

[63] H.-H. Vuong, `http://www.ihp-ffo.de/chipps/97/Djpg/Dposters/vuong.html` (1997).

[64] M. Meyyappan. `http://www.ipt.arc.nasa.gov` (1998).

[65] S. A. McHugo et al., Appl. Phys. Lett. **66**, 2840 (1995).

[66] E. R. Weber and D. Gilles, in Semiconductor Silicon 1990, ed. by H. R. Huff, K. G. Barraclough, and J. I. Chikawa (The Electrochemical Society, 1990) p. 585.

[67] P. A. Stolk et al., Appl. Phys. Lett. **68**, 51 (1996).

[68] R. Subrahmanyan, J. Vac. Sci. Technol. B **10**, 358 (1992).

[69] M. Barrett et al., IEEE Electron Device Lett. 16, 118 (1995).

[70] D. M. Maher and B. Zhang, J. Vac. Sci. Technol. B **12**, 347 (1994).

[71] R. Subrahmanyan et al., Appl. Phys. Lett. **52**, 2145 (1988).

[72] E. T. Yu et al., J. Appl. Phys. **79**, 2115 (1996).

[73] C. C. Williams et al., Appl. Phys. Lett. **55**, 1662 (1989).

[74] M. Nonnenmacher et al., Ultramicroscopy **42-44**, 268 (1992).

[75] M.E. Welland and R. H. Koch. Appl. Phys. Lett. **48**, 724 (1986).

[76] S. J. O'Shea et al., J. Vac. Sci. Technol. B 13, 1945 (1995).

[77] M. Duane, `http://www.prismnet.com/ naomi/tcad/index.html` (1998).

[78] R. O. Jones and O. Gunnarsson, Rev. Mod. Phys. **61**, 689 (1989).

[79] L. Colombo, Ann. Rev. Comput. Phys., Vol. IV (World Scientific, 1996) p. 147.

[80] F. Stillinger and T. Weber, Phys. Rev. **B 36**, 1208 (1987).

[81] J. Tersoff, Phys. Rev. **B39**, 5566 (1989).

[82] H. Rücker and M. Methfessel, Phys. Rev. **B 52**, 11059 (1995).

[83] C.P. Toh et al., Surf. Sci. **292**, 114 (1993); Q.-M. Zhang et al., Phys. Rev. Lett. **75**, 101 (1995); T. Yamasaki et al., Phys. Rev. Lett. **76**, 2949 (1996).

Adaptive Cooling on the Basis of Pulse Tube Refrigerators

Günter Thummes and Christoph Heiden

Institut für Angewandte Physik, Justus-Liebig-Universität Giessen, D-35392 Giessen, Germany

Abstract: Adapting the cryocooler to the particular requirements of a given application is a task that is gaining more and more attention. Due to several inherent features of the pulse tube cooler, this type of refrigerator appears to be a good candidate for this purpose. This is illustrated in this paper for the case of single and multi-stage pulse tube refrigerators which can be used for the operation of high- and low-T_c superconducting devices.

1 Introduction

Soon after the commercial availability of SQUIDs and SQUIDsystems it became clear that for a more widespread application of such devices a proper cooling technique would be desirable. Such a cooling technique would not rely on the supply of liquid helium but use appropriate cryocoolers instead. Considerable work towards this goal was initiated and performed by J.E. Zimmerman, who was able to build low power plastic multi-stage Stirling cryocoolers with which SQUIDs made of niobium could be operated [1]. With the discovery of high-T_c superconductors in 1986 the working temperature was raised to the temperature range of liquid nitrogen. Nevertheless, cooling with liquid nitrogen, requiring to a more or less extent skilled operators to perform the task, appeared not to be the final option. Instead the demand for a cooling technique without liquid cryogens became more urgent. It may appear as a lucky coincidence, that at about the same time, a significant progress was made in the development of cryocoolers, in particular with regard to pulse tube refrigerators.

This type of refrigerator is a good candidate for meeting the requirements that have to be fulfilled by an adaptive cryocooler. Adaptive cryocoolers are refrigerators, which can be tailored to the particular needs of a given application. They are well suited for system integration. The requirements that such coolers have to meet can be grouped in (a) economical and (b) technical factors.

(a) Economical requirements:

- Reasonable investment cost (e.g. 10% of overall system costs).
- Reasonable cost of operation (i.e. unattended operation, sufficient COP).
- Long life and long MTBF (i.e. low maintenance costs).

(b) Technical requirements:

- Sufficient net cooling power at given temperature.
- Quiet operation (i.e. low mechanical vibrations and electromagnetic interference).
- Reasonable tolerance to geometrical restrictions (e.g. cooler can be integrated into a 19" rack unit).

Cryocoolers of the Stirling- or the Gifford-McMahon type [2] are among the most widely used cooling devices that meet many of the listed criteria. However they suffer from mechanical vibrations caused by the reciprocating displacer in their cold stage. This excludes them from applications for which such vibrations are critical. The situation is further aggravated by magnetic interference signals that can result from the displacer motion. Another draw back is the possible limitation of MTBF (Mean Time Between Failures) due, e.g., to slow changes of the displacer seal.

2 Development of Pulse Tube Refrigerators

The problems inherent to Stirling- and Gifford-McMahon-coolers could be solved, if one could eliminate the displacer. This in fact happened in the development of the pulse tube refrigerator, which began with the basic pulse tube cooler concept in the mid sixties [3,4]. After this a quiet period of two decades passed before an essential improvement in cooler performance was achieved by connecting a buffer volume via an orifice valve to the warm end of the pulse tube [5]. The resulting phase shift between mass flow and pressure oscillation of the working fluid increased the cooling efficiency leading to a lowest temperature of 60 K with only one stage [6]. A further significant improvement was attained by adding a second orifice between the warm end of the pulse tube and the main gas inlet [7] creating the so-called double inlet pulse tube refrigerator (DIPTR). This second inlet reduces the mass flow through the regenerator producing at the same time an additional phase shift. A lowest no load temperature of a single stage DIPTR near 30 K thus could be achieved [8].

Fig. 1 shows the scheme of such a single stage double inlet pulse tube refrigerator. There are some features that deserve to be pointed out: (1) As already mentioned, there are no moving mechanical parts in the cold stage. Vibrations of the cold head thereby are reduced by many orders of magnitude. At the same time the reliability of this stage is enhanced without increasing the cost. (2) At the (tolerable) expense of some cooling power the cold stage can be connected to the compressor and the rotary valve producing the pressure waves by means of a comparatively long and flexible line. Possible interference signals from the compressor at the cold stage thereby are reduced to a minimum. All these features are likely to make the pulse tube refrigerator a good candidate when choosing a cooler for a given application.

Instead of using a passive second inlet, the mass flow rate at the hot end of the pulse tube can also be controlled either by using an auxiliary piston [9] or, in the case of a valved compressor, by using a second flow reversing valve arrangement [10]. In the latter cases no buffer volume is needed.

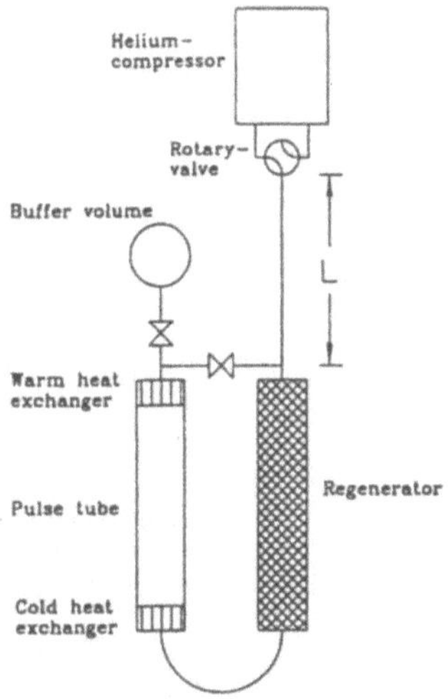

Figure 1
Scheme of a single stage double inlet pulse tube refrigerator.

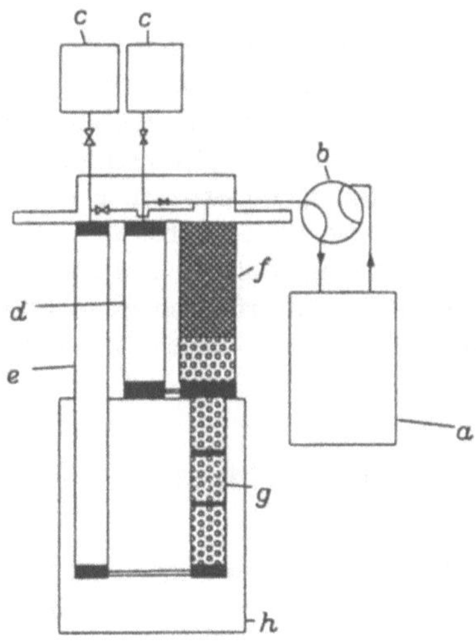

Figure 2
Scheme of a two stage double inlet pulse tube cooler: (a) compressor, (b) rotary valve, (c) reservoirs, (d) 1st pulse tube, (e) 2nd pulse tube, (f) 1st regenerator, (g) 2nd regenerator, (h) radiation shield.

The development of pulse tube refrigerators went on with cascaded designs in order to reach even lower temperatures. With three stages [11] in DIPTR configuration it was possible to reach temperatures below 4 K employing a regenerator material in the coldest stage that undergoes a magnetic phase transition [12] below 10 K. Meanwhile it has been possible to reduce the number of stages required for reaching liquid ^{4}He temperatures to only two [13,14]. Fig. 2 shows the schematic of such a two stage double inlet pulse tube cooler. Part of the mass flow is used in the first stage for precooling the gas entering the second stage regenerator containing in its coldest part a magnetic rare-earth compound. The first stage also cools the radiation shield surrounding most of the second stage (The warm end of the second stage pulse tube is at room temperature!).

Initially the coefficient of performance (COP = net cooling power/input power of compressor) of pulse tube coolers was quite low. Following systematic improvements this situation has changed now for both, single- and two-stage pulse tube coolers. Fig. 3 shows

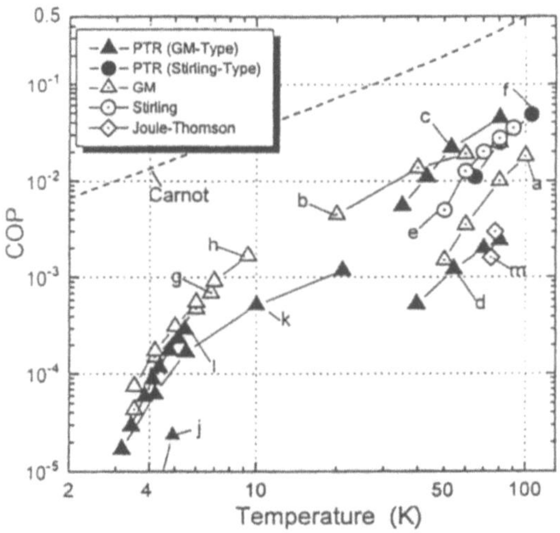

Figure 3
Coefficient of performance (COP) of various low-power cryocoolers; a: [15], b: [16], c: [17], d: [18], e: [19], f: [20], g: [21], h: [22], j: [11], k: [13], l: [23], m: [24].

the COP for different pulse tube coolers in the temperature range from 3 K to 100 K. Included are GM-type and Stirling-type pulse tube refrigerators, i.e. coolers whose pressure waves are produced by a GM-valved compressor or by a reciprocating piston compressor as used in Stirling coolers. Also included for comparison are COP values for some GM-, Stirling-, and Joule-Thomson-coolers. The figure, although not exhaustive with regard to the data base of the present available literature, illustrates clearly, that the COP of pulse tube refrigerators now approaches that of GM- and Stirling-coolers over the whole investigated temperature range. The gap between the ideal Carnot performance and the achieved COP on the other hand indicates the challenge for future improvements in cooler design.

3 Selected Applications of Pulse Tube Refrigerators

In the following, some examples of applications of pulse tube refrigerators are listed that have shown recently the potential of this refrigerator type for adaptive cooling.

3.1 Operation of high-T_c SQUIDs

One of the first demonstrations in which the application potential of pulse tube coolers could be shown was the operation of a high-T_c SQUID gradiometer, which was mounted on a single stage coaxial PTR. A flexible plastic connecting line of several meters length was used to reduce hum and noise from the compressor to a negligible level [25,26]. Fig. 4 shows a recording of the spectral flux noise density at a frequency range, in which interference signals from the compressor usually are visible. Apart from the 50 Hz line associated with the ubiquitous mains interference no signals from the compressor and the cooler can be seen in the intrinsic white noise floor of the SQUID. Thus, if a cryocooler is to be used for operating a high-T_c SQUID or SQUID gradiometer, the PTR obviously is a good choice for this purpose.

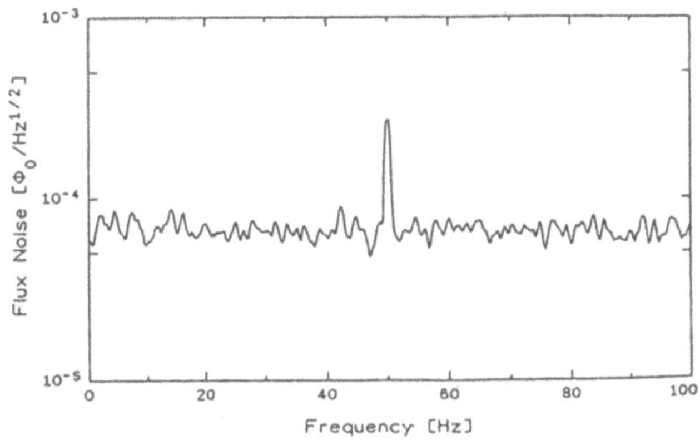

Figure 4
Low frequency spectral flux noise of a $YBa_2Cu_3O_7$ RF-SQUID gradiometer cooled by a single stage pulse tube refrigerator to a temperature near 50 K.

3.2 Cooling a HTS-Antenna for Magnetic Resonance Imaging

Another example demonstrating the potential of the pulse tube refrigerator as a quiet cryocooler is the operation of a thin film $YBa_2Cu_3O_7$ antenna coil for magnetic resonance imaging at low fields [18]. Here a single stage cooler was used, whose cold stage with the antenna was placed at the appropriate location in the magnet. Again a flexible connecting line to the rotary valve and the compressor was used. The results show a clear improvement of signal to noise ratio in the obtained images.

3.3 Small Scale ^{4}He Liquefaction

Using a two stage DIPTR the liquefaction of ^{4}He on a small scale recently could be demonstrated [27]. No precooling by liquid nitrogen was employed. A vacuum-insulated copper vessel (volume: 0.77 litres) for collecting the liquid helium was thermally attached to the cold platform of the second stage. A mass flow meter and controller served to adjust and monitor the flow of He gas that was supplied from a standard 50 litres He-cylinder at ambient temperature. Precooling of the incoming He gas was achieved by means of heat exchangers attached to the cold end of the first stage (60 K) and to the outer wall of the second stage regenerator tube. Condensation occured in a third heat exchanger at the cold end of the second stage. Fig. 5 shows the temperature variation of the liquid helium pot and the mass flow rate of He gas entering under normal pressure of 1 013 mbar. After an initial cool down period of 140 minutes, a steady mass flow rate of 0.264 g/min is obtained, corresponding to a liquefaction rate of 0.13 litres/h at 4.2 K. After a time of 380 minutes, temperature and mass flow rate decrease steadily. Since then, the liquid helium level reaches the entry of the feeding line into the vessel.

Using the two-stage pulse tube cooler for recondensation of evaporated cold helium gas should lead to higher liquefaction rates. This in turn will be of interest for applications such as cooling conventional superconducting magnets in cryostats with low evaporation rate as in magnetic resonance imaging systems [28]. Here, the two pulse tube stages could become an integral part of the whole cryostat, using the first stage for cooling radiation shields and the second stage for ^{4}He-recondensation. The cooler thus serves at the same time for refrigeration and as mechanical support structure for the suspension of the shields and the liquid helium vessel. Systems for magnetic resonance spectroscopy

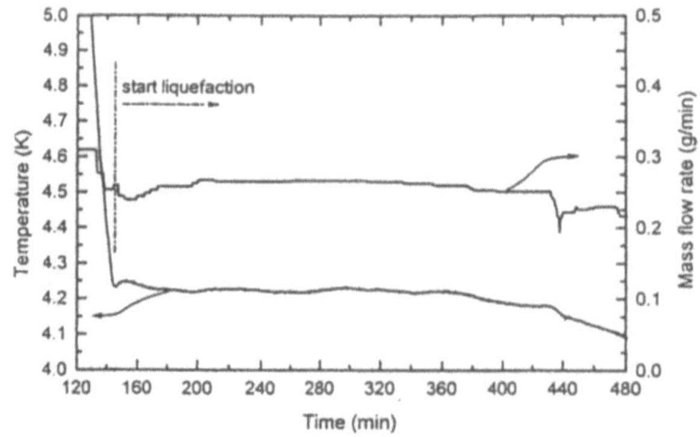

Figure 5
Temperature of liquid helium vessel and mass flow rate of ^4He gas during helium liquefaction by use of a two-stage pulse tube refrigerator.

are also likely to benefit from this development. The use of a combination of a two-stage pulse tube refrigerator and cryostat for, among others, laboratory test purposes and for cooling conventional SQUIDs or Josephson-tunnel-arrays (Voltage standard) also should to be of particular interest.

3.4 Cryogen Free Cooling of Superconducting Magnets

Pulse tube refrigerators could also be employed to conductively cool conventional superconducting magnets, thereby completely avoiding the use of cryogenic liquids. This has recently been demonstrated for a small sized superconducting magnet that was successfully operated by means of a two-stage pulse tube cooler [29]. Fig. 6 shows a photograph of the cooler with the magnet thermally anchored to the cold end of the second stage pulse tube. The magnet is made of Nb_3Sn-wire wound on a stainless steel bobbin with a clear bore of 11 mm. The end faces of the bobbin are clamped between two copper plates by means of copper bolts. The part of the current leads from room temperature to the first stage cold head consist of copper strands. From there high-T_c BiSrCaCuO (2212) rods extend to the second stage cold head. The rods are connected to the magnet terminals via a combination of copper- and NbTi-wires. After a cooling time of 6 hours the second stage cold head and the magnet surface attained stationary temperatures of 3.4 K and 4.1 K, respectively. The temperature of the first stage cold head was then 55 K. The magnet coil was successfully charged with a current of 120 A, corresponding to a central field of 2.8 T.

4 Concluding Remarks

Despite the progress which has been achieved in the performance of pulse tube refrigerators there still remain some problem areas that have to be addressed. Two effects are mentioned:

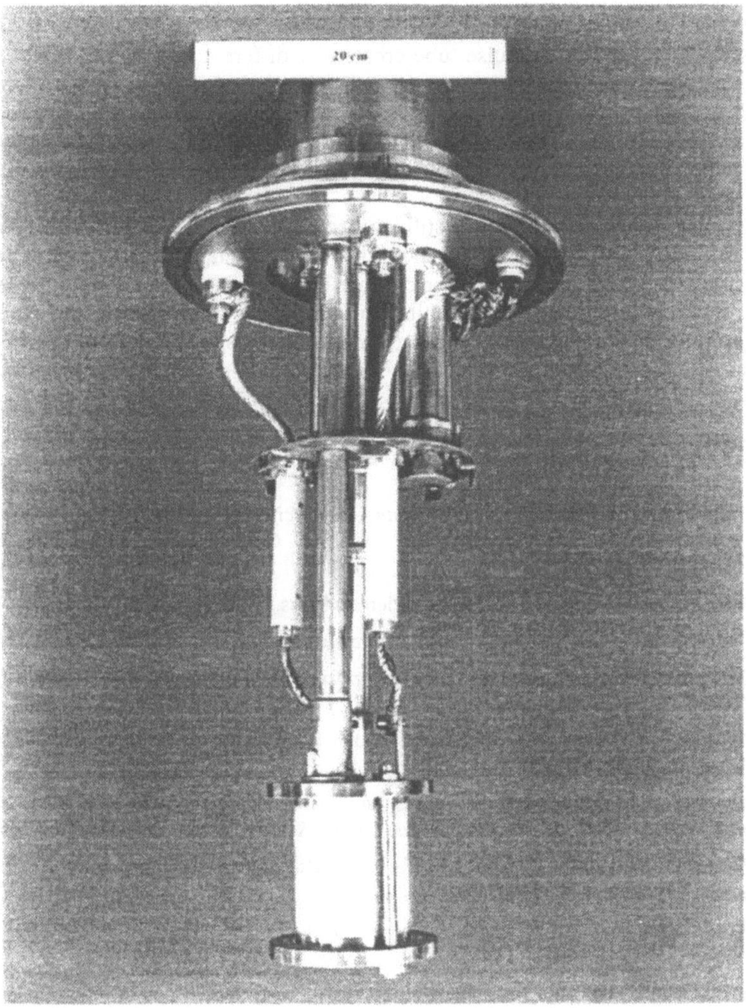

Figure 6 Photograph of two-stage pulse tube cooler with Nb_3Sn-magnet attached to the cold end of the second stage pulse tube.

(A) Dependence of performance on orientation. In particular for low pulse frequencies it is observed that the cooling power exhibits a more or less strong dependence on the orientation of the pulse tube with regard to the gravitational axis. This feature is more pronounced for low net cooling powers, say of one watt or less, than for more powerful coolers. With increasing pulse frequency it becomes less severe. At frequencies of 50 Hz or 60 Hz for instance, the effect is barely noticeable. The effect has its origin in gravity driven convection in the pulse tube by which heat is carried to the cold stage of the cooler [30].

(B) Losses due to secondary mass flow. Parasitic heat input by convection, as just mentioned, is one example for losses due to secondary mass flow. Another example is dc-flow through the pulse tube created by different effects like asymmetric flow characteristics in the various components, in particular the second-inlet orifice valve, or the particular shape of the pressure wave. Efforts to reduce the detrimental effect of dc-flow are on the way and resulted recently in first results with improved COP when operating single and two stage pulse tube refrigerators [23].

There are other losses, some of which still have to be identified. The question, whether the COP of pulse tube refrigerators in general will reach the same level, or even a higher one, in comparison to GM- or Stirling-type cryocoolers, remains open for the time being. However it certainly will be a strong stimulus for continued active research and development both on pulse tube refrigerators and on the competing conventional cryocoolers.

Bibliography

[1] J.E. Zimmerman D.B. and Sullivan, *Study of design principles for refrigerators for low-power cryoelectronic devices*, NBS Technical Note 1049, U.S. Government Printing Office, Washington D.C. 20234, (1982).

[2] G. Walker, *Cryocoolers part 1 and part 2*, (Plenum Press, New York, 1983)

[3] W.E. Gifford and R.C. Longsworth, *Pulse tube refrigeration*, Trans. ASME **86**, 264 (1964)

[4] W.E. Gifford and R.C. Longsworth, *Pulse tube refrigeration progress*, Adv. Cryog. Eng. **10 B**, 69 (1965)

[5] E.I. Mikulin, M.P. Tarasov, and M.P. Shkrebyonoch, *Low temperature expansion pulse tubes*, Adv. Cryog. Eng. **29**, 629 (1984)

[6] R. Radebaugh, J.E. Zimmerman, R.S. David, and L. Beverly, *A comparison of three types of pulse tube refrigerators: new methods of reaching 60 K*, Adv. Cryog. Eng. **31**, 779 (1986)

[7] S.W. Zhu, P.Y. Wu, and Z.Q. Chen, *Double inlet pulse tube refrigerator: an important improvement*, Cryogenics **30**, 514 (1990)

[8] A. Ravex, P. Rolland, and J. Liang, *Experimental study and modelisation of a pulse tube refrigerator*, Cryogenics **32**, 9 (1992)

[9] Y. Ishizaki and E. Ishizaki, *Experimental performance of modified pulse tube refrigerator below 80 K down to 23 K*, Proc. 17th Int. Cryocooler Conference, Santa Fee (1992), p. 140

[10] M. Thürk, H. Brehm, J. Gerster, G. Kaiser, R. Wagner, and P. Seidel, *Intrinsic behaviour of a four valve pulse tube refrigerator*, Proc. 16th International Cryogenic Engineering Conference, Kitakyushu, Japan (1996),(Elsevier Science 1997), p. 259

[11] Y. Matsubara, and J.L. Gao, *Novel configuration of three-stage pulse tube refrigerator for temperatures below 4K*, Cryogenics **34**, 129 (1994)

[12] T. Hashimoto, M. Ogawa, and R. Li, *Recent advances in magnetic regenerator material*, Cryogenics **30**, 192 (1990)

[13] C. Wang, G. Thummes, and C. Heiden, *A two-stage pulse tube cooler operating below 4 K*, Cryogenics **37**, 159 (1997

[14] G.B. Chen, L.M. Qiu, J.Y. Zhen et al., *Experimental study on a double-orifice two-stage pulse tube refrigerator*, Cryogenics **37**, 271 (1997).

[15] B.J. Huang and S.C. Chang, *System performance analysis of Gifford-McMahon cooler*, Cryogenics **35**, 117 (1995)

[16] A. Fiedeler, J. Gerban, and H.-U. Häfner, *Efficient single stage Gifford-McMahon refrigerator operating at 20 K*, Cryogenic Engineering Conference, Portland, Oregon (1997), to appear in Adv. Cryog. Eng. **43** (1998)

[17] S.W. Zhu, Y. Kakimi, K. Fujioka, and Y. Matsubara, *Active-buffer pulse tube refrigerator*, Proc. 16th International Cryogenic Engineering Conference, Kitakyushu, Japan (1996), (Elsevier Science 1997), p. 291
Y. Kakimi, S.W. Zhu, T. Ishige, K. Fujioka, and Y. Matsubara, *Pulse tube refrigerator and nitrogen liquefier with active buffer system*, Cryocoolers 9, (Plenum Press, New York 1997), p. 247

[18] K. Klundt, C. Lienerth, G. Thummes, F. Steinmeyer, M. Vester, W. Renz, and C. Heiden, *Use of a pulse tube refrigerator for cooling a HTS-antenna for magnetic resonance imaging*, Cryogenic Engineering Conference, Portland, Oregon (1997), to appear in Adv. Cryog. Eng. **43** (1998)

[19] P.J. van den Bosch, H.J.M. ter Brake, H.J. Holland, H.A. de Boer, F.F.C. Verbene, and H. Rogalla, *Cryogenic design of a high-T_c SQUID-based heart scanner cooled by small Stirling cryocoolers*, Cryogenics **37**, 139 (1997)

[20] C.K. Chan, C. Jaco, and T. Nguyen, *Advanced pulse tube cold head development*, Cryocoolers 9, (Plenum Press, New York 1997), p. 203

[21] T. Kuriyama, M. Takahashi, H. Nakagome, T. Hashimoto, H. Nitta, and M. Yakubi, *Development of 1 Watt class 4 K GM refrigerator with magnetic regenerator materials*, Adv. Cryog. Eng. **39**, 1335 (1994)

[22] A. Lang, H.-U. Häfner, and C. Heiden, *Systematic investigations of regenerators for 4.2 K refrigerators*, Cryogenic Engineering Conference, Portland, Oregon (1997), to appear in Adv. Cryog. Eng. **43** (1998)

[23] C. Wang, G. Thummes, and C. Heiden, *Performance study on a two-stage 4 K pulse tube cooler*, Cryogenic Engineering Conference, Portland, Oregon (1997), to appear in Adv. Cryog. Eng. **43** (1998)

[24] R. Hohmann, M.L. Lucia, H. Soltner et al., *Integration of HTS SQUIDs with portable cooling devices for the detection of materials defects in non-destructive evaluation*, Cryocoolers 9, (Plenum Press, New York 1997), p. 925

[25] C. Heiden, *Pulse tube refrigerators: a cooling option for high-T_c SQUIDs, in: SQUID Sensors: Fundamentals*, Fabrication and Application, H. Weinstock, ed. (Kluwer Academic Publishers, The Netherlands 1996), p. 289

[26] G. Thummes, R. Landgraf, M. Mueck, K. Klundt, and C. Heiden, *Operation of a high-T_c SQUID gradiometer by use of a pulse tube refrigerator*, Proc. 16th International Cryogenic Engineering Conference, Kitakyushu, Japan (1996), (Elsevier Science 1997), p. 283

[27] G. Thummes, C. Wang, and C. Heiden, *Small scale ^4He liquefaction using a two-stage pulse tube cooler*, Cryogenics **38**, 337 (1998)

[28] G. Klipping, *Cryogenic management of MR imaging systems and trends in their development*, presented at the 16th National Symposium on Cryogenics, Dec. 10-12, 1997, at the IIT, Kharagpur, India, Proceedings to be published

[29] C. Wang, G. Thummes, K.-J. Best, B. Oswald, and C. Heiden, *Cryogen free operation of a niobium-tin magnet using a two-stage pulse tube cooler*, 17th International Cryogenic Engineering Conference, Bournemouth, U.K. (1998), to be published

[30] G. Thummes, M. Schreiber, R. Landgraf, and C. Heiden, *Convective heat losses in pulse tube coolers: effect of pulse tube inclination*, Cryocoolers 9, (Plenum Press, New York 1997), p. 393

Current Issues in Mechanics of Layer Systems for Sensors and Actuators

Eberhard Blank and Johannes Michler

Département des matériaux, Ecole Polytechnique Fédérale de Lausanne, CH-1015 Lausanne, Switzerland

Abstract: Successful processing and functional reliability of layered microsystems depend much on the mechanical properties of the constituent layers and their geometrical arrangement. After a presentation of the basic notions of internal stress formation and measurement techniques for layer systems, the case of diamond films on silicon substrates is discussed. It is shown that the relaxation of membrane stresses in structured surfaces and at free edges leads to stress concentrations at the interface which can be problematic for adhesion. Micro-Raman spectroscopy together with finite element modelling constitute a suitable combination for stress and strain field analysis in the domain of silicon and diamond technology. The last chapter emphasises the use of indentation techniques for materials characterisation and reports recent results on adhesion, strength and load bearing capability of DLC and diamond layers.

1 Introduction

In sensors and actuators, materials with different thermo-mechanical properties are joined to form devices with sometimes complex architectures. Processing technologies are similar to IC technology but the variety of materials is still larger and the architecture generally is different, conforming to the respective device functions. Frequently, sensors and actuators operate at high temperatures and in aggressive environments, like certain gas sensors, or have to withstand external forces and contact pressure, like pressure sensors or micro-electromechanical systems. From a point of view of mechanics and materials science, the functional reliability and efficiency of such systems are not obvious. Internal stresses and stress concentrations inherently related to their composite structure and dimensional discontinuities as well as to their fabrication process may become a problem during manufacture and lead to malfunctioning and mechanical failure during service.

As a result of ongoing miniaturisation in integrated silicon technology and thin film properties which sensitively depend on processing conditions, the classical methods of mechanical testing which traditionally have focused on bulk materials had to be reviewed [1]. Bending of cantilever beams fabricated by lithographic patterning and surface micromachining has become a versatile tool to determine mechanical thin film properties [2,3]. Moreover, nano-indentation techniques now are used currently to measure stress-strain relationships in the elastic and plastic deformation range [4]. Most investigations of mechanical thin film properties assume a planar or biaxial stress state, i.e. through-the-thickness-stresses are neglected. Such simplifying assumptions are required to interprete the results of mechanical testing because the aforementionned techniques do not give access to individual components of the stress and strain tensors. These techniques also are inadequate to deal with the generally complex architectures of sensors and activators. Microstructure size scales in such devices typically range from 0.1 to several micrometers in thickness with lateral dimensions between 10 and 100 μm. Microstructure elements may be separated by lateral and vertical gaps of typically 1 μm and can be one- or two-dimensional or have high-aspect ratios.

In correspondance with the growing importance of thin film technology, the mechanical behaviour of layer systems has become the subject of intense investigation [1,5–8]. Instigated by the impressive progress in numerical simulation of mechanical problems using analytical techniques or finite element calculations as well as by new developments in experimental stress analysis by spectroscopic and micromechanical techniques, the present paper reviews some of the fundamental concepts of internal stress formation. Emphasis is placed on the deposition of diamond layers on silicon wafers. In the second part, the problem of adhesion as related to internal stress is put forward, with particular attention to stress concentrations at edges and flaws at the interface of bimaterials. The final part, which focuses on the efficiency of mechanical substrate protection by thin coatings, illustrates how interactive nano-indentation and finite element calculations can be advantageously used in materials and device analysis.

2 Internal Stress in Bilayers

2.1 Formation of Internal Stress

The concept of internal stress in bilayer systems with isotropic properties has been treated thoroughly by Menčik [5]. In the central part of a bilayer, there are basically two modes of stress relaxation, one where free bending of the bilayer is allowed and another one, to be treated first, without bending (in this latter case the free ends of the bilayer should be imagined to be clamped to a rigid reference plane). If the bi-layer is prevented from bending, the stresses in both layers are homogeneous although their values are different and of opposite sign. The stress values are as follows [5]

$$\sigma_1 = \sigma_{1,\max} \frac{1}{1 + \frac{E_1^* h_1}{E_2^* h_2}} \tag{2.1a}$$

$$\sigma_{1,\max} = E_1^* (\alpha_1 - \alpha_2)(T_0 - T) \tag{2.1b}$$

$$\sigma_2 = -\sigma_1 \frac{h_1}{h_2} \tag{2.1c}$$

Here, $E_{1,2}^*$ are Youngs moduli, $h_{1,2}$ layer thicknesses, $\alpha_{1,2}$ the thermal expansion coefficients, T_0 and T the temperatures prior to and after cooling. For two-dimensional layers, E_i^* represents the biaxial modulus given by $E_i^* = E_i/(1-\nu_i)$; in case of a thin one-dimensional line and if stresses along that line are under consideration, E_i^* is the normal Young's modulus. Layer 1 cannot relax if layer 2 is very stiff, i.e. if the ratio of tensile stiffnesses $E_1^* h_1 / E_2^* h_2$ in (2.1a) approaches zero, as in case of a very thick substrate. The stress in layer 1 then is given by the thin film formula (2.1b) which sets an upper bound for thermal stress. It should be noticed that the addition of a thin interlayer will not change

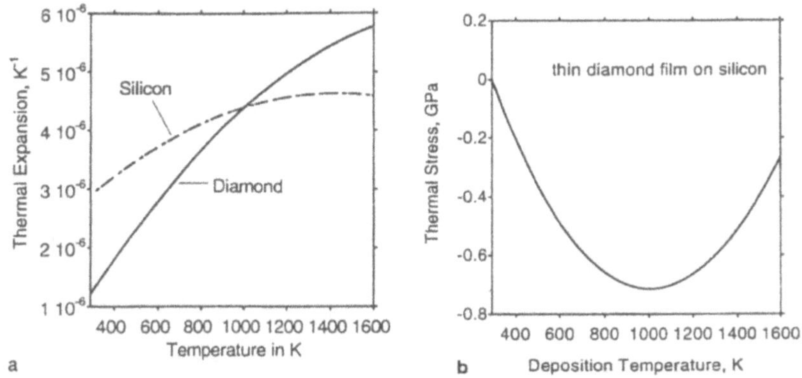

Figure 1 Thermal stress in thin diamond films on silicon; (a) temperature dependence of thermal expansion coefficients; (b) thermal stress according to (2.1b).

the stress state in layer 1 as long as the substrate remains dominant with respect to the final overall strain of the multilayer.

If bending is permitted, stress and strain become a function of z, the coordinate normal to the layers, but stresses are still considered to be biaxial. The mathematical solution for this case is given in Menčiks book (equations 3.21–3.26 in Ref. [5]). Equation (2.1b) allows to calculate the maximum stress in layer 1, as before. Bending however adds another mode of stress relaxation, in addition to elongation strain of the substrate, as illustrated by the following example : Stress in layer 1 relaxes to about 90% of its maximum value for a tensile stiffness ratio of 1/30 whereas a value of as much as 1/10 is needed if bending does not occur, i.e. the substrate must be more compliant in this latter case. As long as the thin film approximation (2.1b) applies, stress can only be influenced by changing the deposition temperature or by choosing substrates exhibiting similar thermal expansion. Generally, thermal expansion coefficients depend on temperature. For diamond and silicon, there is a crossover at about 1 000 K (Fig. 1a). Cooling to room temperature from stress-free deposition at 1 000 K leads to the highest thermal stress, as can be seen from Fig. 1b. In cooling from higher temperatures, the stress at room temperature is smaller again.

Fig. 2a illustrates the effect of growing film thickness on thermal stresses in diamond on silicon. The results of the analytical expression for bending [5] are in perfect agreement with finite element calculations. The stress values depend on the thickness and moduli ratios of the two layers rather than on their absolute values. The same graph thus would have been obtained by normalising layer thickness with respect to substrate thickness. The radius of curvature, which is also shown in Fig. 2a, varies strongly in the range of relatively thin films, as do stresses. Its value is proportional to the total thickness of the two layers.

Thermal stresses are just one example for internal stresses. Very similar stress states, compressive or tensile, are induced by the deposition process itself. They are called growth stresses. Ion bombardment in PVD generates interstitials in the deposit and leads to swelling. In amorphous carbon films, this stress moreover depends on the hydrogen content [9]. Regions of less than equilibrium atomic density in crystalline and amorphous materials have been made responsible for tensile strains [10]. Coherency strains cause compressive or tensile stresses, depending on the sign of the lattice mismatch [11,1].

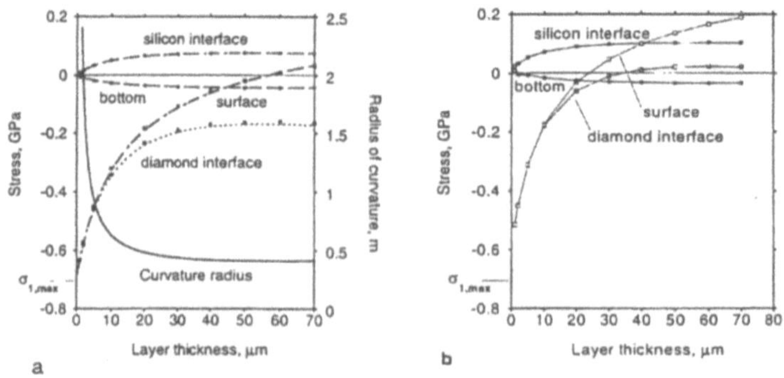

Figure 2 Calculated thermal stress and curvature radius as a function of film thickness (deposition temperature 820 °C, substrate thickness 275 μm); (a) comparison between the analytical expression given by Mencik [5] and FEM calculations; (b) results of FEM if a radial temperature gradient of 100 °C is added (the central part is hotter)

2.2 Measurement of Internal Stress

The curvature of a wafer can be determined very precisely and in a straightforward way by measuring the position of a reflected laser beam which depends on the angle of reflexion at the wafer surface. If this surface is curved and scanned by the laser beam, the wafer curvature can be calculated from the displacement of the reflected beam. Curvature measurements reveal only average stresses without spatial resolution and the results depend strongly on the assumptions made to model the stress state.

If the substrate or the thin film are single-crystalline as in case of silicon wafers and epilayers, X-ray diffraction (XRD) or Raman spectroscopy (RS) are more instructive. In both methods the stress state of the film is assumed initially and a mathematical relationship with the shift of a XRD or RS line is established. In layer systems, one operates in backreflection. In certain cases, substrate stress is measured in transmission through the thin film.

XRD is extremely versatile since each set of crystal planes can act as an in-situ strain gauge. Techniques of biaxial stress determination in polycristalline and textured thin films by XRD have been reported [12,13]. The method also applies to anisotropic materials. Relative lattice spacings variations of less than 10^{-4} can be measured [14]. A similar set-up has been used to measure residual strains in graded $Si_{1-x}Ge_x$ epilayers via reciprocal space mapping [15].

RS has the advantage of combining spatial and stress resolution [16,17] but applies to Raman-active materials only. In silicon and diamond, the frequencies of three optical phonons are measured which, without stress, form a triplet. Stress leads to frequency shifts, higher frequencies being associated with compressive, smaller ones with tensile stresses. Depending on the stress state, the triplet splits into singlets or a combination of singlets and doublets which can be identified using polarised RS [16–19]. The spectral resolution with a high resolution spectrometer is around 0.1 cm^{-1}. This corresponds to about 50 MPa for silicon [16] and 64 MPa for diamond under uniaxial stress in the [001] direction [17]. Sensitivities change with the stress state and compare well with XRD sensitivities. Depth profiling of stress in silicon is feasible by using different wavelengths while confocal micro-RS is more appropriate in high-quality diamond.

The effect of crystallographic texture on RS is not well established yet. Line broadening due to small grain size can hide the effect of stress in XRD and RS. Very careful analysis therefore is required to separate both effects. Line broadening becomes important for domain sizes below 1 μm in XRD and less than 10 nm in RS. Domain size in this context signifies a single-crystalline domain and does not necessarily correspond to the morphological grain size.

2.3 Internal Stress Analysis in Diamond Films on Silicon

On cooling diamond films on silicon substrates from deposition temperature one expects compressive stresses in the diamond film because of the higher expansion coefficient of silicon. Correspondingly, the bilayer should display convex curvature on the film side. Surprisingly, concave curvature is observed. As this curvature is preserved once the diamond film has been removed (film curvature gets even stronger), the substrate must have been plastically deformed during deposition at high temperature, driven by tensile forces exerted by the film. Transmission electron microscopy of a transverse section of the substrate indeed reveals numerous dislocations entering the substrate from the interface (Fig. 3). The dislocation density and distribution is in satisfying agreement with the substrate curvature. High temperature plastic deformation of the substrate allows film growth stresses to relax, at least partially, and one may expect the internal stress at room temperature to be dominated by the thermal stress.

In a separate experiment, the effect of growing film thickness on the evolution of the internal film stress has been investigated. A Raman head for light emission and collection in backreflection was hooked to the deposition reactor. The Raman shift was measured after interruption of the deposition process and cooling of film and substrate to room temperature [20]. It can be seen from Fig. 4 that thermal stress alone is inappropriate to explain the Raman shift, as only the general trend of a decreasing stress with increasing film thickness is reproduced. There is no unique effect which would allow to explain the observed discrepancies:

- The calculated shift corresponds to a mean stress between the film interface and surface. Taking into account that the film surface facing the light source always dominates in Raman scattering because of light absorption in the defective film, the calculated shift for thick films should be corrected to somewhat smaller valued (about 0.2 cm^{-1} at 20 μm thickness). Thin films would not be affected.

- In the present example, the film has developed a strong $\langle 001 \rangle$ fiber texture and the assumption of a polycrystalline film structure only applies to thin films, up to 4 – 5 μm at most. This effect additionally would lower the calculated shift for thick films.

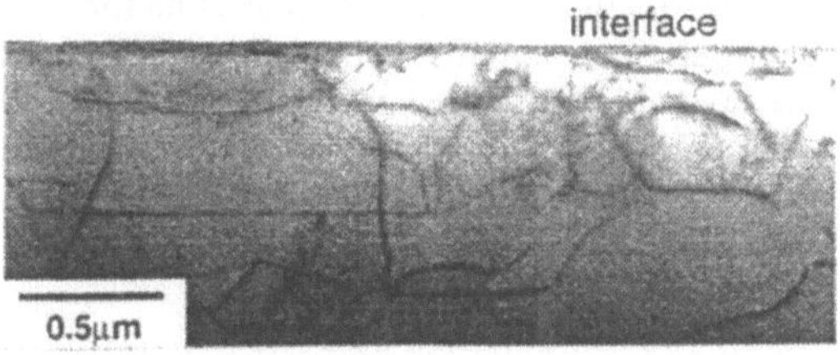

Figure 3
EM transverse section of a silicon substrate (275 μm) after growth of a heteroepitaxial diamond film. Dislocations extend from the interface (top side) into the substrate, confirming plastic deformation ($\langle 110 \rangle$ zone axis, bright field image)

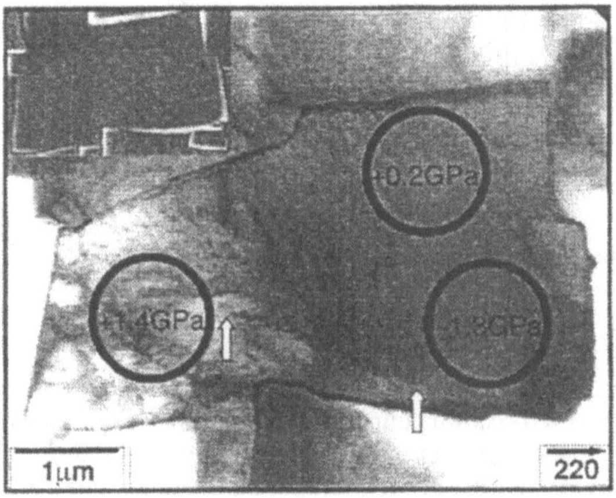

Figure 4
Raman shift of diamond film on silicon versus film thickness. For comparison, the effect of thermal stress calculated for a deposition temperature of 850 °C and polycrystalline microstructure has been plotted.

- A much more dramatic effect concerns the occurrence of a compressive stress maximum at about 1 μm film thickness which appears to be general in diamond deposition [21]. Diamond deposition occurs by island growth during the early growth stage and the stress maximum coincides with the formation of a continuous diamond film by grain coalescence [21,22].
- Finally, the effect of a radial substrate temperature gradient on thermal stress has been investigated. A comparison of Fig. 2 a and b shows that a radial temperature gradient of 100 °C (hotter in the centre, as in plasma CVD) contributes significantly to the reduction of compressive film stress, in addition to the film thickness. The colder outer part of the substrate prevents the centre from contraction, i.e. the film centre is compressed less.

Thermal stresses in diamond films should be elastic and disappear if film and substrate are separated by cleavage. Stress analysis by polarised confocal micro-Raman spectroscopy in backscattering from the (001) surface and from a (110) cross section of a free-standing heteroepitaxial diamond film reveals the presence of compressive growth stresses in the order of the thermal stress [17]. At the local scale of the probe size of about 1 μm^3, stresses can be even one order of magnitude higher. Concerning the average stress, Fig. 5 shows that the assumption of a biaxial stress state in the film plane only applies to the first 20 μm from the interface. Beyond that thickness, there is a three-dimensional stress state. Right at the film surface, the assumption of a uniaxial stress in [110] direction is in perfect agreement with the observed Raman shifts. Such a stress state would again be compatible with biaxial stress in the (001) plane. The locally fluctuating stresses are retained in thin foils for TEM and have been associated with structural defects, called disclinations [23] (Fig. 6). For adjacent grains of a few micrometer size and misorientations in the order of one degree or less it is energetically more favorable for the grains to coalesce by formation of disclinations than by forming subgrain boundaries. It is not understood yet how this mechanism relates to the three-dimensional stress state mentionned before.

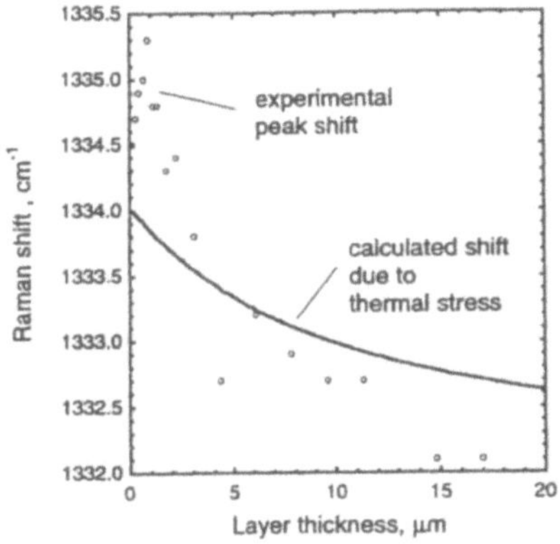

Figure 5
Stress profile in a free-standing diamond film of 50 μm thickness as measured by confocal, polarised micro-Raman backscattering from the (001) film surface (rhomboids) and a transverse (110) section (squared symbol: singlet; elliptical symbol: doublet). The measurements indicate biaxial stresses in the film plane during the first 20 μm from the interface and a three-axial stress state beyond.

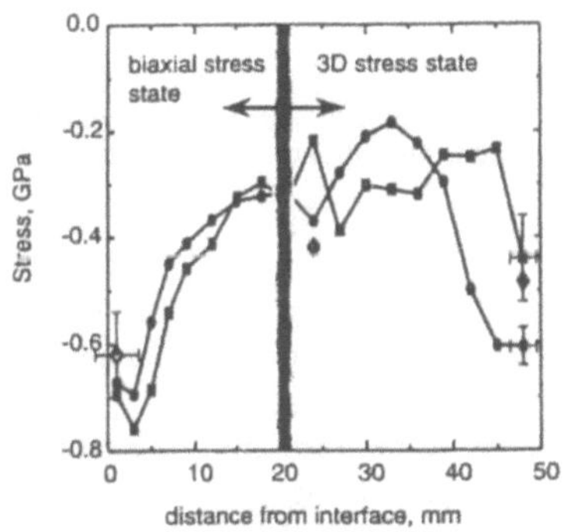

Figure 6
Stress inhomogeneities in blended grains of a heteroepitaxial diamond film which are associated with the formation of disclinations.

3 Edges, Corners and Structured Surfaces

3.1 Stress State and Stress Distribution

Close to free edges and surfaces of a bilayer or steps in one of the layers, the normal stresses, also called membrane stresses, disappear and are balanced by interface shear stresses. The reader is referred to [5,24] for a detailed discussion of the problem. As an example, the stress distribution near the surface of cylindrical silicon pin of 40 μm diameter, 100 μm height and covered with a 5 μm thick diamond layer has been calculated by FEM for the case of cooling from 820 °C (Fig. 7): The interface shear stress which is zero right at the free surface in the absence of external forces, reaches a maximum value within less than one tenth of the layer thickness and then decays exponentially over a distance of typically 3 – 5 layer thicknesses (Fig. 7a). A second shear stress maximum forms in the layer close to the free edge. Concomitantly with the decrease of the interface shear stress in moving away from the free surface, the membrane stress builds up (Fig. 7b). One would expect the membrane stress to be compressive. This is indeed the case over more than two thirds of the film section facing the substrate. However, the changing sign of the shear stress near the free surface induces a moment which is responsible for the bending of the diamond film shown by Fig. 7c (the base of the substrate remains flat). Bending causes the membrane stress to vary in z-direction. In the present example, the membrane stress varies from compressive at the interface to tensile at the surface. A tensile maximum is attained

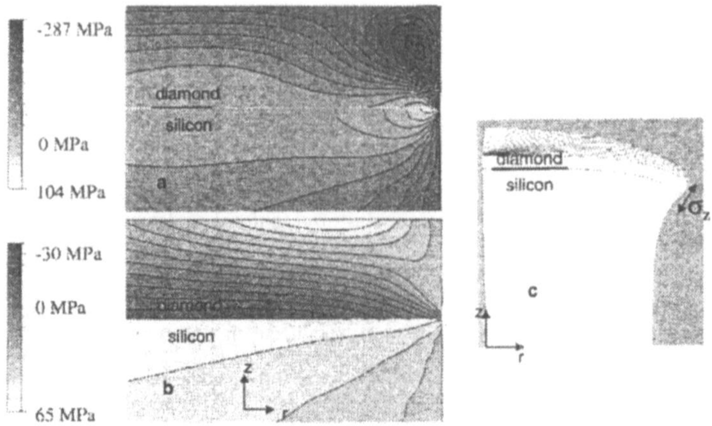

Figure 7 Thermal stress in a silicon pin (diameter 40 μm, height 100 μm) coated with a diamond layer (thickness 5 μm) (FEM, initial temperature 820 °C); (a) shear stress; (b) membrane stress; (c) elastic deformation (view under an oblique angle, deformation exaggerated 200×).

at about 1.5 layer thicknesses away from the free edge. Still another stress component related to bending is the so-called peeling stress in z-direction. It reaches a maximum at the cylinder surface right at interface between film and substrate.

Stress concentrations at free edges and steps sometimes are quite unmanageable because they can cause interface cracks. They depend strongly on the modulus difference at the transition between film and substrate. Interlayers with a comparitively small shear modulus have a very beneficial effect on stress concentrations near free surfaces because they cause the shear zone to extend over a larger distance from the free surface, thereby reducing the stress maxima [25]. Edges or layer discontinuities created by trenches are common in integrated circuit, sensor and actuator technology where they always cause stress concentrations. Metallic contact lines or the architecture of complementary metal-oxide semiconductor (CMOS) transistors are typical examples. Edges do not need to be free-standing, it is sufficient for the elastic modulus to change more or less abruptly. Illustrative examples of how interactive finite element modeling and polarised Raman spectroscopy can be used for stress analysis near steps and trenches in microelectronic devices have been given recently [16]. A spatial resolution of 0.35 μm has been achieved.

3.2 Adhesion and Internal Stress

Stress concentrations near edges, steps and corners give rise to crack formation if the local stress exceeds some critical strength in the material. Crack propagation occurs if the elastic energy necessary to drive a crack into the material can be supplied by the strained film/substrate system. This energy is necessary to break bonds at the crack tip, for crack branching and for plastic deformation in the surroundings of the crack tip.

The energy release rate, G, for crack growth is equal to the derivative of the stored elastic energy with respect to an infinitesimal increase of the crack surface area. No matter if the crack propagates perpendicularly through the layer or from edges along the interface, G is always of the type

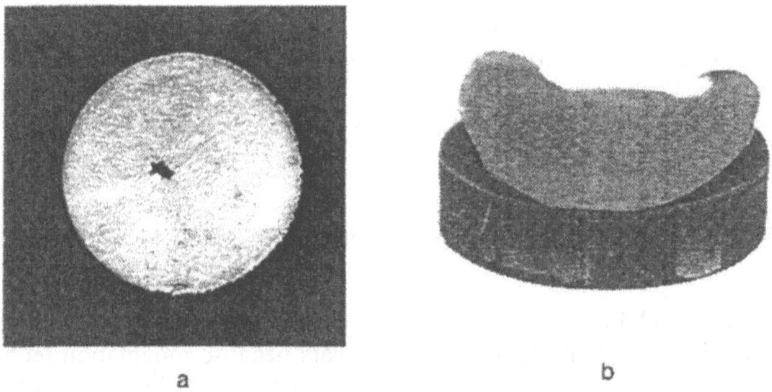

Figure 8 Two examples for delamination of diamond films deposited on steel, (a) very thin films ($< 1\ \mu$m): buckling; (b) by crack propagation from the free edge (Courtesy from [26])

$$G = k\frac{1-\nu_1}{E_1}\sigma_1^2 h_1 \tag{3.2}$$

where k represents a factor of proportionality varying between 0.5 and 1, and σ_1 a uniformely distributed stress in layer 1 [5]. The danger of thin film delamination with increasing thickness, h_1, is a well known experimental fact which can be explained by the ease of interface crack propagation according to (3.2). A very efficient way to avoid propagation of cracks driven by thermal stress consists in lowering the film deposition temperature because the energy release rate is proportional to the square of the temperature interval defined in (2.1b).

A somewhat particular situation arises if delamination begins remote from edges. Uniform tensile stresses in the film do not provide any contribution to the growth of local delaminations because no stress concentration appears at the tip of such interface cracks. By contrast, the delaminated part may buckle in case of compressive film stresses. The critical film stress for buckling, $\sigma_{\mathrm{crit},1}$ is proportional to the square of the ratio between film thickness and diameter of the delaminated area. Once buckling has occurred, there are stress concentrations at the delamination edges favoring growth of the delaminated area, with an energy release rate [5]

$$G \sim \frac{1-\nu_1}{E_1}\left[\sigma_1^2 - \sigma_{\mathrm{crit},1}^2 \left(\frac{a_0}{a}\right)\right)^4\right]^2 h_1 \tag{3.3}$$

which is smaller than in (3.2) since the compressive stress of the film is unable to relax completely as long as the buckled zone is small. Here, a_0 represents the critical and a the actual flaw size. On the other hand, the energy release rate for crack propagation in the buckling mode increases rapidly with growing size of the interface flaw in order to approach the value for crack growth from edges (3.2).

The adhesion of diamond films on steel constitutes a specially tough problem because of the large difference in thermal expansion coefficients by more than $10^{-5}\,^{\circ}\mathrm{C}^{-1}$. Figure 8 illustrates the two modes of delamination discussed just before: Wrinkling occurs if the diamond films are very thin, typically a few hundred nanometers (Fig. 8a). On a local scale, wrinkling correponds to buckling which takes place everywhere over the film surface. It is not possible to estimate the density of delamination sites at the interface from the typical wavelength for wrinkling because the accordion-type appearance of the film surface may correspond to the buckling mode of film rather than correspond each time to a new delamination site. With increasing thickness the film becomes stiffer against bending and delamination by crack growth from the edges is favored (Fig. 8b) [26].

4 Materials and Device Testing by Indentation

The issue of contact mechanics which is closely related to mechanical testing of layered media has received much attention in recent years because indentation techniques can be applied over a wide range of size scales (10^{-9} to 10^{-1} m). The interpretation of indentation data frequently relies upon FEM because of the complexity of the induced deformation field. The principles of elastic contact mechanics were developed by Hertz. An elasto-plastic analysis of the indentation of non-layered and layered systems is given by [27]. Concerning the fracture of brittle solids the reader is referred to [28]. For a review, see also [5].

Indentation depths can be controlled very precisely. Very shallow indentations of less than a tenth of the layer thickness allow to determine layer properties like the Youngs modulus without interference with substrate properties. Analysis of the stress-displacement field produced by ball indentation provides access to the treshold loads for failure of the protective film, the interface and/or the substrate.

In the coating industry, the detachment of a film from the substrate in the area surrounding a Rockwell "C" indentation has been known for many years as a means to determine adhesion of brittle films. Recently, the mechanics of this technique and a similar one based on wedge indentation have been analysed and a method to determine the interface toughness has been developed [29,30]. Plastic deformation of the substrate forces the film to be displaced radially, inducing compressive stresses in the film. The energy release rate as defined by (3.2) then drives the delamination crack away from the edge of the indentation crater as long as the critical stress for propagation of the interface crack is exceeded. The assumption here is that the film thickness is much smaller than plastic deformation field.

In the following, the results of an elasto-plastic indentation analysis for a 3.3 μm diamond-like (DLC) coating on a steel (13Cr) are reported. Both materials have very similar Youngs moduli of 210 GPa and 225 GPa for the substrate and coating, respectively. A ball indenter with a tip radius of 10 μm has been used for testing. The experimental load-displacement curve has been reproduced very accurately by FEM. At the extreme load of 820 mN, a ring crack forms in the DLC coating with a radius identical to the contact radius. The radial stress distribution at the moment of cracking is depicted by Fig. 9. There are two locations in the film with extremely high tensile stresses: Right under the indenter at the interface between coating and substrate the stress amounts to 24 GPa whereas about half of this value is reached in the shoulder of the coating surface, just outside the contact area. Both values are close to the theoretical strength of about one tenth of Youngs modulus [31].

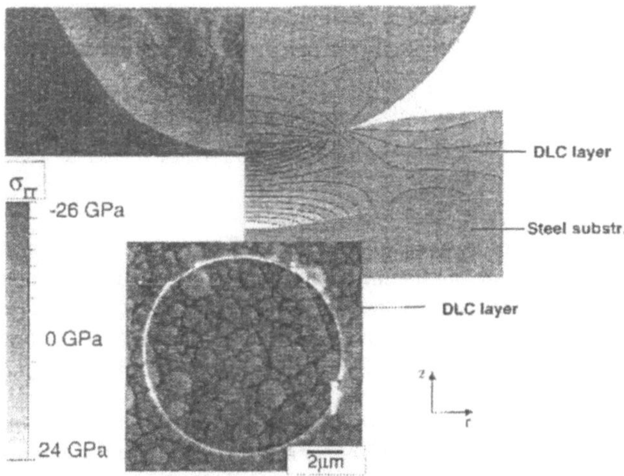

Figure 9 Numerical simulation by FEM of the indentation of a DLC coating on steel. The stress contours correspond to the radial stress. The bottom photograph shows a ring crack in the DLC coating as induced by the indenter.

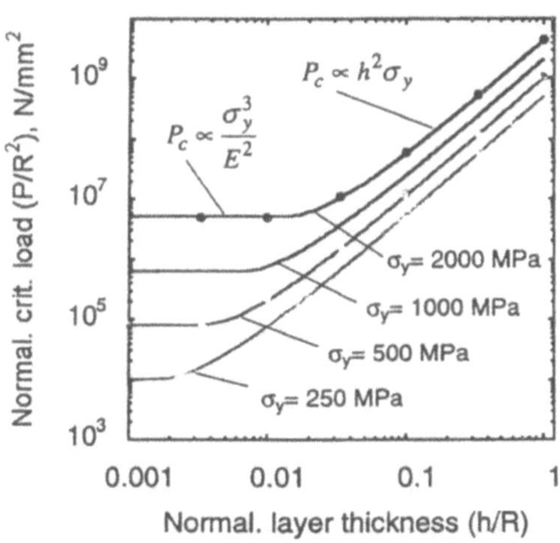

Figure 10
Load bearing capacity of DLC coatings on steel (σ_Y denotes the substrate yield strength, R the indentor radius, h the coating thickness; this plot has been normalised with respect to the indentor radius).

DLC films are applied as protective coatings for magnetic hard-discs and other devices where surfaces with low adhesive and frictional forces are required. Depending on the application, one may be interested in the load bearing capability of such coatings. In the following example, the substrate yields far below the critical load for coating failure. In order to improve the load bearing capacity either the substrate or the film thickness must increase. In such situations, FEM may be helpful for system optimization. It can be seen from Fig. 10 that very thin DLC coatings have no effect on the treshold for substrate plasticity which solely is controlled by the Young's modulus and the yield strength of the substrate. With increasing coating thickness, the influence of the coating on the critical load becomes dominant while that of the substrate is reduced to a linear proportionality.

Similar selection maps can be established for other substrate and coating materials provided the failure mode, which in the example shown here is the yield strength of the metal substrate, has been identified.

5 Conclusions

The present selection of topics on the mechanics of layered media intends to confront the reader with some basic concepts in thin film mechanics, including their inherent simplifications, and the situation when answers to the real situations have to be found. In regard to the relatively restricted, yet necessary information provided by micro-experiments and the increasing power of FEM, the combination of the two techniques is a very suitable way of materials analysis.

In future, new analytical models will be required. As interconnection linewidths in multilevel structures decrease below the micrometer scale, interface and surface stresses become more and more important [32]. Micro-electromechanical systems become vulnerable to adhesion upon contact of adjacent surfaces which may occur during micromachining as a result of release of internal stress [33].

Concerning silicon technology, the miniaturisation below the limits of spatial resolution of Raman spectroscopy puts serious constraints on the experimental determination of stress and strain. Convergent-beam electron diffraction seems to be particularly promising to measure lattice parameters with high accuracy (10^{-4}) while achieving high spatial resolution in the nanometer range [34].

Bibliography

[1] W.D. Nix, Met. Trans. **20A**, 2217 (1989).

[2] S. Hong, T.P. Weihs, J.C. Bravman, and W.D. Nix, Mater. Res. Symp. Proc. **130**, 93 (1989).

[3] T.P. Weihs, S. Hong, J.C. Bravman, and W.D. Nix, J. Mater. Res. **3**, 931 (1988).

[4] M.F. Doerner and W.D. Nix, J. Mater. Res. **1**, 601 (1986).

[5] J. Mencik, *Mechanics of Components with Treated or Coated Surfaces,* Kluwer Academic Publishers, Doordrecht (Netherlands), ISBN 0-7923-3700-X, (1996).

[6] N. Huber, D. Munz, and Ch. Tsakmakis, J. Mater. Res. **12**, 2459 (1997).

[7] J. Mencik, D. Munz, E. Quandt, and E.R. Weppelmann, J. Mater. Res. **12**,2475 (1997).

[8] J. Woirgard amd J.C. Dargenton, J. Mater. Res. **12**, 2455 (1997).

[9] X.-M. Tang, J. Weber. Y. Baer, C. Müller, W. Hänni, and H.E. Hintermann, Phys. Rev. **B 48**, 10124 (1993).

[10] E.S. Machlin, *Materials science in microelectronics,* GIRO PRESS, ISBN 1-878857-07-X, (1995).

[11] J. W. Cahn, Acta Met. **10**, 179 (1962).

[12] F. Gisen, R. Glocker, and E. Osswald ; Z. Techn. Physik **17**, 145 (1936).

[13] B.M. Clemens and J. A. Bain, MRS Bulletin July (1992) pp.46–51

[14] W.J. Bartels, J. Vac. Sci. Technol. **B 1**, 338 (1983).

[15] J.H. Li, E. Koppensteiner, G. Bauer, M. Hohnisch, H.-J. Herzog, and F. Schäffler, Appl. Phys. Lett. **67**, 223 (1995).

[16] I. De Wolf, Semicond. Sci. Technol. **11**, 139 (1996).

[17] Y. von Kaenel, J. Stiegler, J. Michler, and E. Blank, J. Appl. Phys. **81**, 1726 (1997).

[18] E. Anastassakis, A. Pinczuk, E. Burnstein, F.H. Pollak, and M. Cardona, Solid State Commun. **8**, 133 (1970).

[19] J. W. Ager III and M. D. Drory, Phys. Rev. B **48**, 2601 (1993).

[20] Notice : The stress measurements by Raman scattering were carried out by M. Mermoux, INPG, CNRS D 1213, Grenoble (France).

[21] M. Mermoux, INPG, CNRS D 1213, Grenoble (France), private communication.

[22] J. Stiegler, Y. von Kaenel, M. Cans, and E. Blank, J. Mater. Res. **11**, 716 (1996).

[23] J. Michler, Y. von Kaenel, J. Stiegler, and E. Blank, J. Appl. Phys. **83**, 187 (1998).

[24] E. Suhir, J. Applied Mechanics **56**, 595 (1989).

[25] W.T. Chen and C.W. Nelson, IBM J. Res. Develop. **23**, 179 (1979).

[26] M.D. Drory, Appl. Diam. Films and Rel. Mater., 3rd Int. Conf., 313 (1995).

[27] K. Komvopoulos, J. Tribology **111**, 430 (1989).

[28] B.R. Lawn, *Fracture of Brittle Solids, 2nd edition,* Cambridge University Press, Cambridge, UK (1993).

[29] M.D. Drory and J.W. Hutchinson, Proc. Roy. Soc. **452**, 2319 (1996).

[30] J.J. Vlassak, M.D. Drory, and W.D. Nix, J. Mater. Res. **12**, 1900 (1997).

[31] A. Kelly, *Strong Solids*, Clarendon Press, Oxford (UK), (1966).

[32] J.M.E. Harper and K.P. Rodbell,J. Vac. Sci. Technol. **B 15**, 763 (1997).

[33] R. Maboudian and R.T. Howe, J. Vac. Sci. Technol. **B 15**, 1 (1997).

[34] R. Wittmann, C. Parzinger, and D. Gerthsen, Ultramicroscopy **70**, 145 (1998).

The Crystalline Silicon Thin-Film Solar Cell – The High Temperature Approach

C. Hebling, A. Eyer, F.R. Faller, A. Hurrle, R. Lüdemann, S. Reber, and W. Wettling

Fraunhofer-Institut Solare Energiesysteme, ISE
Oltmannsstr. 5, D-79100 Freiburg, Germany

Abstract: The crystalline silicon thin film solar cell combines, in principle, the advantage of the crystalline Si wafer-based solar cell with the advantage of thin film technologies. In comparison to the other thin film solar cells the crystalline silicon thin film cell is still in a laboratory stage of development, however substantial progress has been made recently. This paper describes in some detail the processing steps of the crystalline silicon thin film solar cell based on a high-temperature technology and presents the status of its present stage of development.

1 Introduction

The fabrication of photovoltaic (PV) cells, modules and systems has seen a steady growth over the past years and an increasing number of applications. While the increase in market volume for PV modules amounted to about 15% per year in the past decade, the growth was almost 40% in 1997 with a world production of about 125 MWp [1] (1 MWp = 1 MW in standard sunlight). Photovoltaics is thus established as a small industry but with good prospects for the future. Indeed, a recent study of possible future energy scenarios up to the year 2060 that was published recently by the Shell company predicts a multigigawatt energy production by renewable energies including photovoltaics [2]. In order to realize such a development it is essential that the costs of PV electricity can be reduced substantially. In today's PV module production, which is mainly based on crystalline silicon solar cells, more than 50% of the module cost is due to the silicon wafer. In order to reduce these costs several strategies are followed in present R & D:

- Improvement of cell and module production yield in all steps of module fabrication
- Improvement of silicon usage (reduction of sawing losses by thinner and larger wafers and by silicon foil growth, improvement of cell efficiencies)
- Development of thin film solar cells on foreign substrates.

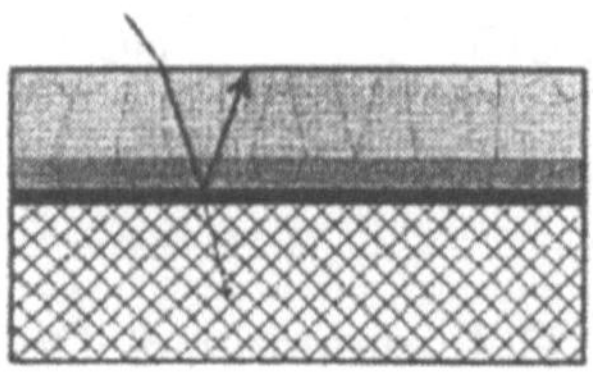

Figure 1 The basic components of a crystalline silicon thin-film solar cell.

As far as the last point is concerned, a large variety of semiconductors has been investigated as potential solar cell materials with the aim to replace crystalline silicon. All of them are direct semiconductors that absorb (and convert) the sunlight spectrum within a layer thickness on the order of 1 μm in contrast to crystalline silicon for which, as an indirect semiconductor, a thickness of 50 to 100 μm is essential to absorb most of the light. The most important thin film cells are amorphous silicon (a-Si), cadmium telluride (CdTe), and copper indium (gallium) diselenide (CI(G)S). Whereas a-Si is already produced in multi MWp quantity, CdTe and CI(G)S are in the state of pilot production. The technological status of the various technologies is discussed in several papers, e.g. [3] and all details can be found in conference proceedings, e.g. [4]. The cost reduction potential for the thin film solar cell technologies was calculated in a recent study assuming a future production on a 100 MWp scale [5]. So far, none of the thin film technologies has proven its superiority over the others. With regard to this another solar cell concept has gained a lot of interest in recent years, namely the crystalline silicon thin film solar cell (CSiTFC). It combines, in principle, the advantage of the wafer-based crystalline Si cell, (i.e. abundance and non toxicity of silicon, maturity of the technology thanks to microelectronics), with the advantage of thin film cells (i.e. low material consumption, large area deposition with integrated module production). Crystalline silicon is an indirect semiconductor, and therefore the thickness of the CSiTFC must be larger than the thickness of the other thin film cells, with typical values between 5 and 50 μm. Fig. 1 shows the main components of a CSiTFC. The substrate consists either of silicon or a foreign material and is needed for mechanical support. An intermediate layer can be applied as a diffusion barrier against impurities in the substrate and as a back side reflector in order to trap the light in the active Si-layer. A highly doped p^{++}-layer provides an excellent back surface field (BSF) which is very useful to prevent the charge carriers from the recombination at the back side. Finally there is the active Si-layer which is grown either epitaxially on a Si-substrate or on a p^{++}-layer with sufficiently large crystal grains.

The basic principles of the CSiTFC and the work done up to 1994 were reviewed by Werner et al. [82]. The activities in this field can be categorized in many ways. Generally, they are subdivided depending on the temperatures which are involved during the preparation of the layer systems or the solar cells. The high temperature approach uses temperatures up to 1250 °C for the Si-deposition, 1420 °C for the recrystallization of the Si-layer and around 800 °C for the emitter diffusion during the solar cell manufacturing. The low temperature approach is restricted to 600 °C in all process steps employed in order to be able to use glass, a very inexpensive but temperature sensitive substrate material. An overview of the activities in the field of low temperature processing on glass is given by Bergmann [13]. In the paper presented here, an overview of the high-temperature approach in CSiTFC R&D is presented. The most important aspects of the technology are discussed in Sections 2–6 that summarize the present status of activities on thin Si-layers which led to solar cells.

2 Substrate Requirements and Materials

The development of an appropriate substrate is one of the fundamental issues which are being pursued for the CSiTFC. Under investigation are either conductive substrates which are processed in the classical way or non-conductive substrates which can be series-interconnected. In the following some of the most critical requirements concerning the substrate are listed:

Low cost: This is the most stringent requirement. It is agreed that the cost of $1\,m^2$ has to be below 50 US$.

Thermal stability: The substrate has to be stable up to temperatures of 1420 °C if a recrystallization step is performed and up to 1250 °C for Si-deposition at high deposition rates.

Thermal expansion coefficient (TEC): The TEC of the substrate has to be very near ($\pm 10\%$) to that of silicon ($4 \cdot 10^{-6}\,K^{-1}$) in order to avoid crack-off of the recrystallized silicon layer during the cooling-down phase.

Mechanical stability: Mechanical loads arise during solar cell processing e.g. during screen printing of the contacts and during module integration. At least 150 MPa bending strength are required.

Surface flatness and roughness: Screenprinting or masking steps need macroscopical flatness. The microscopical roughness has to fulfill a compromise between good backside diffusivity and optimal recrystallization properties.

Possible materials for CSiTFC substrates are:

Silicon: Si as a substrate would be the most appropriate candidate, especially when it can be re-used several times as suggested in [17,81,33]. Another approach is the realization by either ingot casting, or even better, by ribbon growth using metallurgical-grade (mg) Si and applying, for example, the SSP (silicon sheets from powder) or the RGS (ribbon grown sheets) technique. This approach is presently being pursued using upgraded mg-Si the cost of which is about 30–50% of today's electronic-grade Si. Another option is tape casting and sintering of silicon powder ribbons.

Carbon/Graphite: Different types of C are under discussion. Sheets cut from graphite blocks fulfil most of the requirements but are probably too expensive because the impurity content has to be reduced by expensive cleaning procedures or by barrier layers. If conductive substrates are required, then a conducting SiC interlayer is the most appropriate.

Mullite: The TEC can be tailored to silicon very precisely, tape casting is well developed and costs can be within the target range for large scale production. The major draw-back is thermal shock resistance and that only insulating substrates can be made. Similar or even better properties and cost perspectives show mullite/ZrO_2 composites and Zirconia ($ZrSiO_4$) but additional purity questions arise from the Zr content.

SiC: It is one of the most promising systems, since the TEC is adapted very well to Si, it shows good thermal and mechanical properties, tape casting is demonstrated and electrical conduction can be achieved by doping. Starting with SiC particles, there are several ways to produce this material, e.g. sintering and reaction bonding. Solar cells with up to 9.3% efficiency have already been demonstrated at Fraunhofer ISE.

Others: There are a variety of interesting composite materials in a very early stage of development. Of special interest are Si based ceramics with Al or SiAlON as additives [75] or furthermore Al_2O_3/TiC, Si_3N_4/TiN, SiC/TiC, AlN/Al/TiB_2 (or ZrB_2) or Mullite/SnO_2. The costs and the interaction of the active silicon layer with these materials at high temperature is unclear.

3 Diffusion Barriers

Cost restrictions do not permit substrates whose purity level is comparable to that of electronic-grade silicon. On the other hand, simulations show that the impurity concentration level remains comparable to that of the substrate during the layer growth, especially for fast diffusing metals like iron or nickel – with harmful effects on the solar cell efficiency. Therefore, in order to avoid impurity diffusion from the substrate into the active silicon layer, an effective diffusion barrier has to be formed.

3.1 Requirements for the Diffusion Barrier Layer

A good barrier layer has to fulfil several physical and chemical requirements. The most important are:

- Good adhesion to the substrate and the Si layer during all high temperature steps and compensation of differences in the TECs. The barrier layer must also have a good wettability for liquid silicon
- Mechanical and chemical stability at process temperatures up to 1420°C
- Cost effective and fast deposition on large areas
- Effective diffusion barrier for impurities from the substrate

3.2 Possible Diffusion Barrier Materials

Diffusion barriers have been applied in the field of semiconductor devices for several decades. SiO_2 has been used both as a masking oxide against unwanted dopants and as an insulating intermediate layer. Metallic layers like Ti [72], Nb, Ta, Ni, Co and compounds like TiN or W-, Ta-, Mo-, Ti-SiN have been investigated [21] and employed in order to prevent diffusion of contact metals, mainly copper. SiN and nitrided SiO_2 [27] are used either as barrier against sodium, copper and moisture or as an etching mask. SiC is used against moisture and against Al diffusion. In principle, all barrier materials listed above are applicable to high temperature CSiTFCs, even various metal silicides could be possible. However, only SiO_2, SiN and SiC, for which the deposition can be scaled up and made cost-effective, have been investigated for solar cell purposes.

Metal	Cr	Cu	Fe	Ni	Ti
diffusion behaviour	slow diffusor	very fast diffusor $D_{300°C} \simeq 10^{-11} cm^2/s$,	slow diffusor	slow diffusor	very slow diffusor reaction with SiO_2
Reference	[62]	[48]	[62]	[62]	[48]

Table 1 Diffusion behaviour of transition metals in SiO_2

3.3 Properties of SiC, SiN and SiO_2

SiC can be made p- or n- conducting after doping with B/Al or N and therefore allows conventional cell processing. The chemical stability of SiC films is excellent, and the diffusion of elements like N, B, Be and Al through the SiC film occurs only at temperatures higher than 1800°C [40]. Knowledge about SiC as a diffusion barrier in the field of the CSiTFC is very limited.

Plasma enhanced CVD-SiN is, like SiC, chemically very stable, especially after high-temperature treatments. Unfortunately, the wettability of SiN with liquid silicon is worse than for SiC or SiO_2. PECVD-SiN is mainly used as anti-reflection coating since the refraction index can be adjusted according to the H_2-content. It is also used as surface passivation or as an etching mask in solar cell technology. Up to now, only few high temperature diffusion barrier properties of SiN combined with SiO_2 are known concerning transition metals [36]. A stack of PECVD deposited SiO_2-SiN-SiO_2 barrier layers was used at Fraunhofer ISE, realizing a cell with 9.3% efficiency on an industrially available very impure Si-infiltrated SiC ceramic substrate [36].

SiO_2-layers are thoroughly investigated and applied in the CSiTFC-technology. They are stochiometric, microscopically dense and provide an excellent buffer to reduce mechanical stress during temperature treatments. Due to the widespread application of SiO_2 in the semiconductor device technology quantitative diffusion data are available only for Ag [48], As [47], B [59], P [47] and Cu [48]. Table 1 shows that only a limited amount of knowledge is available concerning the SiO_2 diffusion behavior of transition metals.

Fraunhofer ISE is presently performing investigations concerning the high temperature region ($T > 1200$°C) where no experiments have been carried out previously.

4 Deposition of Silicon Layers

A large variety of methods to deposit thin silicon layers exists today. Most of them can be divided into two main groups: deposition from the liquid phase and deposition from the gaseous phase.

In the liquid phase deposition, the substrate is brought into contact with a solution that is saturated with silicon. By lowering the temperature of the melt supersaturation occurs and silicon is deposited on the substrate. The acronym for this process is LPE for Liquid Phase Epitaxy; however, it is also used when the growth is not epitaxial. In the gaseous phase deposition, or Chemical Vapor Deposition (CVD), silane or chloro-silane is decomposed, e.g., at the hot surface of the substrate, and silicon is deposited. Depending on the type of solar cell to be produced and on the characteristic properties of the substrate chosen, the following features are important to consider for a specific deposition method:

- temperature of the substrate during deposition,
- growth type (amorphous, poly-crystalline, epitaxial),
- layer quality (purity, grain size and orientation, defects, morphology),
- ability to control the doping level and switch between n- and p-type doping.

Other criteria concerning the economics and feasibility for an industrial production are

- growth rate, throughput and the possibility for up-scaling,
- complexity of equipment and simplicity of process control,
- consumption, utilization and costs of materials and energy.

Many of the different approaches for CSiTFCs require two (or even more) independent deposition steps, e.g., the deposition of a seed-layer for further recrystallization and a following epitaxial deposition.

Some distinct features of the most important deposition methods are described as follows: LPE [57,65] uses metal melts (Cu, Al, Sn, In) in which silicon is dissolved. The substrate temperature lies within the range of 800–1 000 °C and deposition rates vary from a few μm/h to up to 10 μm/h. Better layer qualities are obtained at lower growth rates. LPE growth occurs at thermal equilibrium and epitaxial layers show a very low defect density. Disadvantages include the presence of the solvent as a main impurity, the variation of the doping level during the growth process caused by lowering the temperature, and, in the case of poly-crystalline substrates, the strong unisotropic growth for different crystal orientations.

Several disadvantages of LPE can be overcome by a modified solution growth method. Rapid-LPE (RLPE) [38] as well as Temperature-Difference-Method-LPE (TDM-LPE) [74] work at a constant substrate temperature. The driving force for the deposition is a temperature gradient at the substrate-solvent interface. Homogeneous doping levels, a quasi-continuous process and very high deposition rates of up to 4 μm/min [38] are possible. It is worth mentioning that some research groups (e.g. [88]) have given up LPE in favor of CVD, which seems to have a higher potential for scaling up towards mass production.

Thermal CVD is a well-established method within microelectronics. A mixture of H_2 and the precursor SiH_4, SiH_2Cl_2 or $SiHCl_3$ is decomposed thermally at the hot substrate surface. A large variety of reactor designs, pressure- and temperature-regimes is in use [50,18,25]. In low pressure CVD (LPCVD) the relatively low deposition rates (some μm/h at about 1 000 °C) are counterbalanced by the large number of substrates (> 50) that can be coated simultaneously. A problem is the necessity of a pumping system which needs to handle the highly aggressive and reactive exhaust gases. Epitaxial growth is reported at temperatures as low as 550 °C [18].

Atmospheric pressure CVD (APCVD) is one of the fastest deposition methods. Rates up to 10 μm/min at temperatures between 1 100 and 1 300 °C are feasible and, depending on the type of substrate, the layers grow epitaxially or poly-crystalline. Excellent thickness and doping homogeneity, as well as crystal quality are achieved in today's CVD systems in microelectronics by means of very high gas-flows but extremely low chemical efficiency. This means that only a few percent of the silicon contained in the precursor is deposited. It is a challenge to design reactors that combine sufficient thickness and doping homogeneity while maintaining an economical gas utilization of $> 30\%$ [25]. At a substrate temperature lower than 600 °C, as is required for the deposition on glass, the deposition rates of thermal CVD are very low. Several methods have been developed to increase the deposition rate at low temperatures. Plasma enhanced CVD (PECVD) [84], ion-assisted deposition [53] or hot-wire CVD (HWCVD) [32] are used to deposit amorphous or micro-crystalline layers at temperatures between 170 and 600 °C with deposition rates of 0.2 – 3 μm/h. Recently, very high rates of up to 0.6 μm/min were reported for an optimized PECVD-process from $SiHCl_3$ at 500 °C substrate temperature [58].

Little is known until now about the economical aspects of the different deposition methods when used in large scale production of solar cells since no commercial equipment is yet available for this purpose. In the low-temperature range the deposition techniques may be similar to those used for amorphous silicon, in which PECVD is an established method. In the high-temperature region APCVD certainly has the potential to reach the cost goals since it makes use of the same principle as the well known SIEMENS-process for the large-scale production of high-purity poly-silicon.

5 Upgrading of Si-layers by Recrystallization

First zone melting experiments were performed more than four decades ago with a platinum wire recrystallizing luminescent material on insulating substrates [42]. Electron beams were first used in 1961 for growing large germanium grains, since this technology is very effective in providing large temperature gradients. Since the late 70s, a renewed interest has arisen due to the high device quality of remelted Si films which are applicable for the SOI technology. The advent of the Si recrystallization technique into photovoltaics took place with the growing interest of Si thin-film solar cells on various low-cost substrates. Grain boundaries in their unpassivated state can be very effective recombination centers which reduce the diffusion length of the minority carriers drastically. On the other hand, it is well known from simulations that the diffusion length should exceed the thickness of the active Si-layer at least two times in order to achieve high conversion efficiencies. Therefore, whenever the Si films deposited on intermediate layers or foreign substrates are too fine-grained, an upgrading by recrystallization is recommended. Various methods have been used, depending on the procedure used to couple the energy into the Si-layer. If only a low temperature treatment is allowed, the solid phase recrystallization (SPC) is the preferred method, since a grain enlargement occurs at temperatures as low as 600 °C. Unfortunately, the resulting average grain sizes are only some microns in dimension, even after several hours at these temperatures [13].

The most common recrystallization mechanisms which occur via the liquid phase are laser [6], electron-beam [55], strip heater [51] or halogen lamp recrystallization [36]. They are distinguished according to the form and size of the liquid zone, the grain size and defect density in the resulting Si-layer, the pulling speed of the Si-grains, the melting depth,

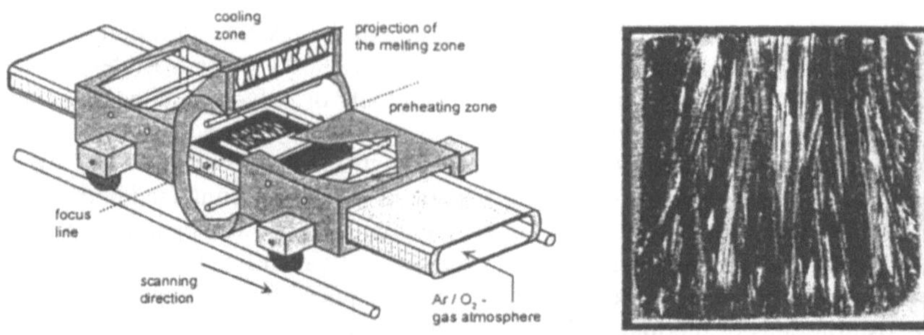

Figure 2 Diagram of an optically heated zone melting furnace developed at Fraunhofer ISE (left) and a recrystallized Si-layer on a Si-infiltrated SiC substrate out of which solar cells with 9.3% efficiency were realized (right) [36].

the three-dimensional temperature gradient and the scalability of the apparatus. Each of these methods has its distinct advantage and disadvantage, depending on the demands of the substrate and Si-layer. Generally, the higher the scanning speed, the higher the defect density and the smaller the grains are [51]. Additionally there is a correlation between layer thickness and defect density on flat substrates, but very little is known about the solidification-front morphologies on rough surfaces as they exist on low-cost substrates. It could be shown by Mitsubishi Corp. that large grain sizes and good crystallographic quality can be achieved by means of zone melting recrystallization with a strip heater resulting in 16.45% efficiency when Cz-Si is used as substrate [70]. Fraunhofer ISE realized 11.0% and 9.3% on graphite and Si-infiltrated SiC-substrates, respectively, using an optically heated zone melting furnace as shown in Fig. 2 [36]. Further investigations are needed for a better understanding in oriented recrystallization on various low cost substrates resulting in high quality Si-layers. Additionally, reliable cost-evaluations must be performed in order to be able to weigh up the gains in efficiency due to this upgrading step in comparison to the arising costs.

6 Solar Cell Design and Technology

In comparison to the well developed technology for thick crystalline Si solar cells, the fabrication of CSiTFCs sets new tasks for the processing technology: Cost-effective substrates call for a suitable technology and new contact designs and thin Si layers call for light-trapping and passivation.

6.1 Basic Solar Cell Design Considerations

In contrast to conventional solar cells, the base of a CSiTFC can be contacted from the rear side only if the substrate is conducting and no insulating intermediate layer is applied (e.g. [45,11]). Otherwise, both contact grids have to be applied on the front side (Fig. 3, left). This access to the base can be achieved either by a selective emitter, or by trenches etched through the homogeneous emitter [45]. Insulating substrates or intermediate layers make an integrated series-interconnection possible, leading to a monolithic solar cell module on a single substrate [37] (Fig. 3, right).

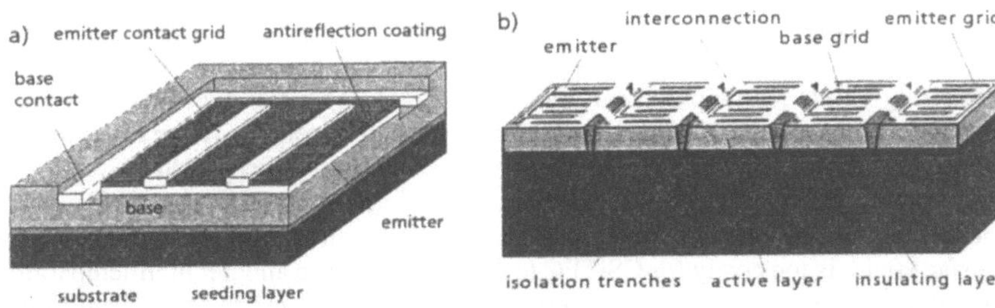

Figure 3 (a) Design of a CSiTFC on an insulating substrate, (b) Monolithic solar cell module on a single substrate with insulating intermediate layer.

6.2 Cell Technology for CSiTFCs on Foreign Substrates

Substrates from materials other than silicon are usually unwanted in the silicon processing technology. Even if the substrate material itself is chemically inert, the generally high content of impurities is critical to a clean and reliable process. Furthermore, suitable substrate materials like graphite or some ceramics are porous. Thus, wet-chemical treatments must be avoided because of contamination in succeeding furnace steps. Only two concepts avoid these problems: The use of highly pure and chemically inert substrates, i.e. high-quality glasses [82], or the complete encapsulation of the substrate by pure and inert layers or layer systems of silicon compounds or even silicon.

Recently, a dry-chemical technology for the CSiTFC has been developed [44]. It substitutes all cleaning and wet-etching steps by reactive ion etching (RIE). Thus, a completely dry solar cell process, which avoids problems because of soaked substrates and which is less susceptible to contamination, is realized. Because of the physical etch-component in RIE, high etch rates and the use of shadow masks for pattern transfer are possible, and even layers which are hard to etch in wet chemicals can be removed. Anisotropic etching, trenching, and even random surface texturing may be realized. Trenching is very important, because laser or saw cutting is often not suitable with foreign substrates.

6.3 Cell Technology for CSiTFCs on Silicon-Based Substrates

In principle, design and technology of CSiTFCs on silicon-based substrates do not differ from conventional processing of thick c-Si solar cells. Challenges arise from the impurity content of low-grade silicon substrates. The needed diffusion barriers, as discussed in Section 3, set two tasks: First, some of the concepts of upgrading the deposited thin silicon film discussed in Section 5 require seeding holes through the intermediate layer to the Si substrate. The design of these via holes is crucial. It has to be optimized and transferred correctly. Photolithographical masking is needed if the desired pattern is realized by wet-etching. However, reactive ion etching allows the use of shadow masks. If no via holes are etched into the intermediate layer, contacting the solar cell requires the same design as in the case of a foreign insulating substrate.

6.4 Gettering

Gettering stands for the process of transportation of impurities out of the active region of a device. In the case of solar cells, the impurities must be removed from the entire volume. Most suitable for CSiTFCs is phosphorus gettering [61], which could be combined with the emitter diffusion. It strongly increases the performance of CVD layers [8]. Aluminum gettering, conventionally applied to mc-Si in combination with the formation of a back surface field, is too costly for CSiTFC technology, because another high-temperature step and the etching of the Al would necessarily have to be applied from the front side.

6.5 Surface Passivation

The influence of the surface passivation to the solar cell performance increases with decreasing thickness of the cell [71]. Oxides with low defect densities at the Si/SiO_2 interface, which are grown thermally at around 1 000 °C, reduce the surface recombination strongly [28]. Alternatively, different methods with less thermal budget, like rapid thermal oxidation [20] or low-temperature oxidation [41] are under investigation. Another more suitable technique of surface passivation is based on the depletion of positive charge carriers at the surface. This could be achieved by a SiN layer deposited by a low-temperature PECVD which additionally acts as an antireflection coating [7].

6.6 Bulk Passivation by Hydrogen

Hydrogen passivation takes advantage of the ability of hydrogen to deactivate impurities and defects in silicon, and to passivate grain boundaries. Different kinds of hydrogen incorporation into Si are possible [60,56,43,66]. Microwave induced remote plasma hydrogen passivation (RPHP) methods developed recently [22,68,43], turned out to be most efficient. Thin-film solar cells from CVD-Si layers have shown a very strong increase in performance (up to 2.0% absolute in their efficiency) after RPHP treatment [78,46].

7 Present Status of the CSiTFC

The first considerations concerning thinner Si-wafer for solar cells were made by M. Wolf [83] and J. Lofersky [67] while simulating the ideal parameters for record high efficiencies. They pointed out that with decreasing cell thickness, the open circuit voltage increases due to the reduced saturation current which again is a result of a decreasing geometry factor. In these papers, the benefits of thinner Si-layers were quantified for the first time, also demonstrating the importance of low surface recombination velocities and a good optical confinement to make use of these advantages. The first theoretical work about light trapping in thin Si-layers was performed by Goetzberger [30] who suggested a Lambertian back reflector as a simple but efficient structure. The wide range of technological advantages incorporating the potential for a significant cost reduction as well as suggestions for their realization were described first also by Goetzberger [31]. After these fundamental works, it took more than a decade until an increasing number of groups singled out the various technological problems and realized first test solar cells to overcome these problems. Nowadays, there is a wide range of worldwide activities which have demonstrated the efficiency potential of the CSiTFC, but the cost advantage has not yet been converted into production lines. Generally, it must be noted, that most realized cells were still made on Si-wafers as substrates. Such test structures which were made under ideal conditions using the best understood materials available aim to prove the respective concept and to study the influence of different boundary conditions on the cell performance. These cells usually attain high efficiencies on a fairly small laboratory scale using some technologies which will probably not be suitable for large scale manufacturing. Nevertheless, they are valuable tools in order to compare the manifold measured parameters with the predicted values of the simulations. The insights gained from these ideal systems have to be transferred subsequently to cost-efficient systems. An overview of the solar cell results which were obtained by the various institutes and utilities on Si-layers of max. 100 μm thickness is given in Tab. 2, underlining the variety of the different concepts as well as the technologies used. In the following, some selected approaches which are being pursued will be discussed briefly.

The only thin Si-cell concept which is about to make it to a commercially available product is the Silicon Film™ technology developed by Astropower [11]. In this approach, the 100 μm thick active layer is deposited on a micro-grooved, conducting ceramic substrate which is covered by an intermediate layer as back reflector and diffusion barrier. The grains are about 2 mm in dimension and the electrical contact to the substrate was made through via holes in the dielectric layer. The materials used as well as the production method are unpublished. Efficiencies of up to 16.6% were demonstrated on 1 cm^2 and cells having a width of 15 cm and a variable length are about to go into production.

The University of New South Wales obtained 21.5% conversion efficiency on a thinned FZ-wafer of 47 μm thickness demonstrating the potential of such thin layers. An excellent optical confinement was achieved applying inverted pyramids at the front and a reflecting SiO_2 layer on the back and the good back surface passivation could be reached by introducing a rear surface floating junction structure [86]. The main emphasis however is put on a multilayer solar cell using a stack of CVD layers of alternating polarity. The particular advantage of this system is that fine-grained Si-material of very poor quality and hence low lifetime can be used for the epitaxy, still having the potential for considerable conversion efficiencies. Up to 17.6% efficiency was obtained on 6 n-p-n-p-n-p epitaxially grown layers on a Cz-substrate, but no results are reported on a low cost system. Critical seems to be the influence of defects causing junction recombination which limits the optimal number of device junctions [69].

The approach of Mitsubishi Corp. relies closely on the knowledge of the SOI-technology (silicon on insulator) and focuses on the formation of a high quality mc-Si layer on an insulator. A 3 μm thick LPCVD layer deposited on a SiO_2 layer was zone-melted in a strip heater resulting in grains of cm-length and defect densities on the order of some 10^6 cm^{-2}. After the epitaxial growth of the 60 μm thick active Si-layer, the supporting Si-wafer was etched back by a patterned mask for back electrode formation. The 16.45% conversion efficiency is the highest value obtained by means of recrystallizing a Si-layer [70]. This concept was further applied to the VEST-cell (Via-hole Etching for the Separation of the Thin Film) and resulted in 16% efficiency. The main features in this concept are that the substrate is separated from the SOI-structure by etching the SiO_2 intermediate layer through via-holes in the Si-layer and that the substrate is supposed to be re-used subsequently. Both emitter and base electrodes are located on the back side. The high conversion efficiency is a result of an excellent front texturing in spite of poly-Si as active Si-layer and of an upgrading by hydrogen implantation for defect passivation. Finally it is interesting to mention, that the 80 μm thick self-supporting cell had been processed

Institute/ Company	Substrate	Concept	Area [cm^2]	W [μm]	η [%]	V_{oc} [mV]	I_{sc} [$\frac{mA}{cm^2}$]	FF [%]
ASE/ISFH [8]	Graphite, SiC	CVD-Si, ZMR, CVD-Epi-Si	1.25	30	**8.3**	561.0	20.1	73.6
AstroPower [11]	unknown	Silicon-Film™,Process unpubl.	0.98	100	**16.6**	608.0	33.5	81.5
AstroPower [11]	unknown	Silicon-Film™,Process unpubl.	240	100	**12.2**	581.9	27.4	76.5
AstroPower [11]	Steel	Silicon-Film™, Product 1	0.12	-	**9.6**	536.0	26.0	68.5
Au. Nat. U. [10]	mc-Si p^+	LPE-Si	4	30	**15.4**	612.0	28.0	76.0
Au. Nat. U. [14]	mc-Si-Wafer	LPE-Si, Substrate thinned	4	28	**17.0**	651.0	32.6	80.3
Daido Hoxan [49]	Carbon-Fiber	Plasmasprayed Si, recryst.	1.05	500	**10.7**	527.9	31.1	68.5
ETL [39]	Al_2O_3,Si_3N_4-IL	CVD-Si, Laser-ZMR	0.013	4.2	**6.5**	480.0	25.5	53.0
Everg. Solar [80]	Silicon	String Ribbon, no epitaxy	1	100	**15.1**	588.0	33.7	76.1
Fraunh. ISE [34]	SIMOX-Wafer	CVD-Epi-Si p^+/p	4	46	**19.2**	667.6	37.1	77.7
Fraunh. ISE [24]	Cz-Si p^+	CVD-Epi-Si	4	40	**17.4**	658.1	32.0	82.6
Fraunh. ISE [23]	mc-Si p^+	RTCVD-Epi-Si	4	40	**12.1**	607.3	25.7	77.3
Fraunh. ISE [25]	SSP-Si p^+	RTCVD-Epi-Si	4	15	**8.0**	552.7	21.8	66.8
Fraunh. ISE [89]	SSP-Si, SiO_2-IL	Large Area Recryst.	1	30	**11.2**	563.7	27.1	75.3
Fraunh. ISE [34]	Cz-Si, SiO_2-IL	Large Area Recryst.	1.3	30	**9.3**	568.0	22.4	73.1
Fraunh. ISE [79]	Cz-Si p^+	LPE-Si(In), conc. 30 suns	0.25	30	**15.9**	705.0	-	-
Fraunh. ISE [36]	Si-infiltrated-SiC	CVD-Si, ZMR, CVD-Epi-Si	1	72	**9.3**	567.0	21.6	76.1
ISE / ASE [45]	Graphite, SiC	CVD-Si, ZMR, CVD-Epi-Si	1	30	**11.0**	570.0	25.6	75.5
Georgia Tec. [52]	Dendritic Web	Si-Ribbon, no epitaxy	4	100	**17.3**	613.0	35.4	79.2
Hitachi [76]	Corrug. Fz-Si	V-grooves on thinned Si-Wafer	-	-	**12.3**	556.0	33.5	66.0
IMEC [78]	mc-Silso-Si p^+	APCVD-Epi-Si, H^+-passiv.	4	20	**13.9**	615.0	27.5	78.4
IMEC [77]	RGS-Si	APCVD-Epi-Si, H^+-passiv.	4	30	**10.4**	558.0	24.5	75.8
Kaneka [85]	Glass	PECVD poly-Si	1	3.5	**9.4**	480.0	26.1	74.8
Mitsubishi [70]	Cz-Si, SiO_2-IL	CVD-Si, ZMR, substr.etch.	4	60	**16.5**	608.0	35.1	77.1
Mitsubishi [33]	Cz-Si p^+	VEST	95.8	77	**16.0**	588.7	35.6	76.3
Mitsubishi [51]	Cz-Si, SiO_2-IL	CVD-Si, ZMR	100	-	**14.2**	608.0	30.0	78.1
MPI-FKF [82]	Fz-Si p^+	LPE-Si(In)	1	16.8	**14.7**	659.0	27.2	80.2
MPI-FKF [15]	Fz-Si	thinned wafer	1	47	**20.6**	683.0	37.4	80.8
MPI-FKF [82]	Cz-Si p^+	CVD-Epi-Si	4.1	50	**17.3**	655.0	32.5	81.1
NREL [19]	Cz-Si	LPE-Si(Cu/Al), substr. dipp.	1	5	**15.3**	573.1	32.6	81.6
Sanyo [9]	Metal	Solid Phase Crystallization	1	5.2	**9.2**	553.0	25.0	66.4
Tonen [73]	Glassy Carbon	DC-RF plasma sprayed Si	-	-	**4.3**	426.0	16.7	64.3
U. Neuchat. [63]	Glass	PECVD-Si (micromorph)	0.33	-	**12.0**	1280	13.5	69.2
UNSW [86]	Fz-Si	thinned wafer	4	47	**21.5**	698.5	37.9	81.1
UNSW [64]	Cz-Si p^+	LPE-Si, att. to glass, thinn.	4	25	**15.6**	649.0	29.9	80.2
UNSW [87]	Cz-Si p^+	6 layers n-p CVD-Epi-Si	4	32	**17.6**	660.7	32.8	81.4

ZMR:	Zone Melting Recrystallization	LPE:	Liquid Phase Epitaxy
CVD:	Chemical Vapor Deposition	RTCVD:	Rapid Thermal CVD
APCVD:	Atmospheric Pressure CVD	PECVD:	Plasma Enhanced CVD
LPCVD:	Low Pressure CVD	IL:	Intermediate layer

Table 2 Worldwide activities in crystalline silicon thin-film solar cells.

with screen printed electrodes regardless of the thousands of via-holes which additionally make the thin layer very fragile.

The Fraunhofer ISE is investigating three different approaches to reach the goal of cost-efficient CSiTFCs: In the first approach, highly doped SSP pre-sheets are used as substrates and a direct epitaxial growth in a special RTCVD-reactor [25] provides the active Si-layer. First solar cells on 15 μm thick Si layers resulted in efficiencies of 8.0% [24] up to now. The second approach uses highly doped Si-substrates of low quality which are covered by a SiO_2 layer with via holes having an area coverage of 0.5%. The CVD-deposited Si-layer on top is exposed to a large area recrystallization step in order to enlarge the grain size. Cells on such systems had efficiencies of up to 9.3% [34]. In the third approach, encapsulated foreign substrates such as ceramics or graphite are used and the upgrading of the fine-grained Si-layer occurs with an zone melting recrystallization step. An interdigitated front contact scheme was developed for such SOI-structures which allows the emitter and base electrode formation at the front. An efficiency of 11.0% was realized on an SiC-covered graphite substrate [45,8] as well as 9.3% on an SiO_2 covered Si-infiltrated SiC-ceramic [36]. In order to study the potential of such a front-contacted SOI structure, epitaxially grown CVD layers on SIMOX-wafers were processed to solar cells which resulted in a maximum of 19.2% efficiency [35]. Additionally, 24 series-connected cells of 1 cm^2 size each, using such an interdigitated front grid achieved an open circuit voltage of 15.2 V [29] (see Fig. 3, right). The characterization of such an ideal system and the comparison with the results from the modeling provided valuable insights for the realization of different kinds of cost-effective SOI solar cells.

8 Conclusions

The CSiTFC is still in an early stage of development as compared to the other thin film cell technologies that have been investigated much longer and, thus, are more advanced, to date. However large progress has been made in recent years as was described in this paper. The concept of a CSiTFC has been demonstrated with very high efficiencies on small areas using crystalline silicon as a substrate or using monocrystalline thin silicon layers. On foreign substrates as graphite or silicon carbide efficiencies between 9 and 11% have been processed. At one place (Astropower) large area cells have been fabricated. But most important, the number of laboratories worldwide that are engaged in the development of CSiTFCs has increased considerably during the past years. With progress continuing like at present it is to be expected, that this type of solar cell will reach the status of a pilot production in 5 to 8 years. It will then be able to demonstrate its technical but also economical and environmental advantage in comparison with other solar cell technologies.

Bibliography

[1] P.D. Maycock ed., Photovoltaic News, Vol 17 No2 , p. 2 (1998).

[2] Erneuerbare Energie, Communication of Deutsche Shell AG, 22297 Hamburg (1997).

[3] H.W. Schock, A. Sah, 14th EC-PVSEC, 2000 (1997).

[4] Proceedings of the 14th EC-PVSEC (1997), 2 volumes, 2700 pages.

[5] T.M. Bruton, 14th EC-PVSEC, 11 (1997).

[6] G. Andrä, J. Bergmann, F. Falk, and E. Ose, 14th EC-PVSEC, 1400 (1997).

[7] A.G. Aberle, R. Hezel, Progess in Photovolt. Vol5 (1997).

[8] R. Auer, J. Zettner, J. Krinke, G. Polisski, T. Hierl. R. Hezel, M. Schulz, H.-P. Strunk, F. Koch,D. Nikl, and H.v. Campe, Proc. 26th IEEE Photovoltaic Specialists Conf. (1997).

[9] T. Baba, M. Shima, T. Matsuyama, S. Tsuge, K. Wakisaka and S. Tsuda, 13th EC-PVSEC, 1708 (1995).

[10] G. Ballhorn, K.J. Weber, S. Arnand, M.J. Stocks and A.W. Blakers, 14th EC-PVSEC, 1011 (1997).

[11] A.M. Barnett, Progress in Photovolt. Vol.5, 317 (1997).

[12] R.B. Bergmann, G. Oswald, M. Albrecht, V. Gross, Sol. Mat. and C. 46, 147 (1997).

[13] R.B. Bergmann, J. Köhler, R. Dassow, C. Zaczek and H.J. Werner, Phys. Stat. Sol. (a) **166** (1998).

[14] A.W. Blakers, K.J. Weber, M.F. Stuckings, S. Armand, G. Matlakowski, A.J. Carr, M.J. Stocks, A. Cuevas, and T. Brammer, Prog. in Photovolt. Vol3, 193 (1995).

[15] R. Brendel, M. Hirsch, R. Plieninger and J. Werner, 13th EC-PVSEC, 432 (1995).

[16] R. Brendel, 13th EC-PVSEC, 436 (1995).

[17] R. Brendel, 14th EC-PVSEC, 1354 (1997).

[18] A. Chin, B. C. Liu, W. J. Chen, Appl. Phys. Lett. **69**, 1617 (1996).

[19] T.F. Ciszek, T.H. Wang, and X. Whu, J. of Cryst. Growth **128**, 314 (1993).

[20] P. Doshi, J. Moschner, A. Rohatgi, R. Singh, S. Narayanan, 26th IEEE-PVSC, 87 (1997).

[21] A. M. Dutron, E. Blanquet, C. Bernard, A. Bachli, R. Madar, Applied Surface Science, 91 (1-4), 277 (1995).

[22] H.E. Elgamel, C. Vinckier, M. Caymax, M. Ghanam, J. Poortmans, P. De Schepper, J. Nijs, R. Mertens, 12th EC-PVSEC, 724 (1994).

[23] F.R. Faller, N. Schillinger, A. Hurrle, A. Eyer, 25th IEEE-PVSC, 645 (1995).

[24] F.R. Faller, N. Schillinger, A. Hurrle, C. Schetter, 14th EC-PVSEC, 784 (1997).

[25] F.R. Faller, A. Hurrle, N. Schillinger, MRS Spring Meeting, San Francisco (1998).

[26] H. Föll, Appl. Phys. A53, 8 (1991).

[27] M.R. Frost, C.W. Magee, Applied Surface Science **104**, 379 (1996).

[28] S.W. Glunz, A. Sproul, W. Warta, W. Wettling, J. Appl. Phys. **75** (3), 1611 (1994).

[29] S.W. Glunz, J. Knobloch, C. Hebling, W. Wettling, 14th EC-PVSEC (1997).

[30] A. Goetzberger, 15th IEEE-PVSC, 867 (1981).

[31] A. Goetzberger, J. Knobloch, B. Voss, Tec.Dig. Int'l PVSEC-1, 517 (1984).

[32] J. Guillet, A.R. Middya, J. Huc, J. Perrin, B. Equer, J.E. Bourée 14th EC-PVSEC, 1475 (1997).

[33] S. Hamamaoto, H. Morikawa, H. Naomoto, Y. Kawama, A. Takami, S. Arimoto, T. Ishihara, K. Namba, Proc. 14th EC-PVSEC (1997).

[34] C. Hebling, R. Gaffke, P. Lanyi, H. Lautenschlager, C. Schetter, B. F. Wagner and F. Lutz, 25th IEEE-PVSC, 649 (1995).

[35] C. Hebling, S. W. Glunz, J. O. Schumacher and J. Knobloch, 14th EC-PVSEC (1997).

[36] C. Hebling, S. Reber, K. Schmidt, R. Lüdemann, F. Lutz, 26th IEEE-PVSEC (1997).

[37] C. Hebling, R. Lüdemann, W. Wettling, Vacuum Solutions, 17 (1998).

[38] J. Hötzel, K. Peter, G. Willeke, E. Bucher, 14th EC-PVSEC, 1421 (1997).

[39] K. Ishii, H. Nishikawa, T. Taka. and Y. Hayashi, Jpn.J. Appl. Phys. **32**, 770 (1993).

[40] Landolt-Börnstein, *Num. Data and Func. Rel. in Sc. and Tech., New Series, Group III, Vol17, Semic.* O. Madelung, M. Schulz, H. Weiss (ed.), Springer Verl. Berlin (1984).

[41] G. Langguth, M. Schöfthaler, T. Sames., J. Werner, 13th EC-PVSEC, 1530 (1995).

[42] E. Leitz, British Patent 691, 335 (1953).

[43] R. Lüdemann, S. Schäfer, U. Schubert, H. Lautenschlager, 14th EC-PVSEC, 131 (1997).

[44] R. Lüdemann, R.R. Bilyalov, C. Schetter, 14th EC-PVSEC, 780 (1997).

[45] R. Lüdemann, S. Schaefer, C. Schüle, C. Hebling, 26th IEEE-PVSC, 159 (1997).

[46] R. Lüdemann et al., 2nd WCPEC, Vienna, Austria (1998) (to be published).

[47] I. Matsuura, J. Murota, N. Mikoshiba, I. Kawashima, T. Sawai, J. of the Electrochemical Soc. **138** No. 11, 3474 (1991).

[48] J. D. McBrayer, R. M. Swanson, T. W. Sigmon, J. of the Electrochemical Soc. **133** No. 6, 1242 (1986).

[49] T. Mishima, S. Itoh, G. Matuda, M. Yamamoto, K. Yamamoto, H. Kiyama, and T. Yokohama, Techn. Digest 9th PVSEC, 243 (1996).

[50] R. Monna, D. Angermeier, A. Slaoui, J. C. Muller, 25th IEEE-PVSC, 701 (1996).

[51] H.Naomoto, S. Hamamoto, A. Takami, S. Arimoto, T. Ishihara, Solar Energy Materials and Solar Cells 41/42, 261 (1997).

[52] S. Narashima, G. Crotty, T. Kryg., A. Rohatgi and D. Meier, 26th IEEE-PVSC, 235 (1997).

[53] S. Oelting, D. Martini, H. Köppen, D. Bonnet, 13th EC-PVSEC, 1681 (1995).

[54] M. Pauli, J. Müller, K. Hartkopf, T. Barth, Rev. Sci. Instrum. **63**, 2288 (1992).

[55] M. Pauli, T. Reindl, W. Krühler, F. Homberg, J. Müller, Solar Energy Materials and Solar Cells 41/42, 119 (1996).

[56] M. Pirzer, R. Schindler, Springer Proceedings in Physics 35, 122 (1989).

[57] J. A. M. van Roosmalen, C.J.J. Tool, R. C. Huiberts, R.J.G. Beenen, J. P. P. Huijsmans, W. C. Sinke,Proc. 25th IEEE-PVSC, 657 (1996).

[58] M. Rostalsky, T. Kunze, N. Linke, J. Müller, 26th IEEE-PVSC, 743 (1997).

[59] Y. Sato, K. Ehara, K. Saito, J. of the Electrochemical Soc. **136** No. 6, 1777 (1989).

[60] R. Schindler, M. Kaiser, 21st IEEE-PVSC, 691 (1990).

[61] H. Lautenschlager, F. Lutz, C. Schetter, R. Schindler, 26th IEEE-PVSC, 7 (1997).

[62] P. F. Schmidt, Journal of the Electrochemical Society **130**, 196 (1983).

[63] A. Shah, J. Meier, P. Torres, U. Kroll, D. Fischer, N. Beck, N. Wyrsch, and H. Keppner, 26th IEEE-PVSC, 569 (1997).

[64] Z. Shi and S.R. Wenham, Prog. in Photovolt. **2** 153 (1994).

[65] I. Silier, M. Konuma, A. Gutjahr, E. Bauser, 25th IEEE-PVSC, 681 (1996).

[66] B.L. Sopori et al., Solar Energy Materials and Solar Cells 41/41, 159 (1996).

[67] M. Spitzer, J. Shewchun, E. S. Vera and J.J. Lofersky, 14th IEEE-PVSC, 375 (1980).

[68] M. Spiegel, S. Keller, P. Fath, G. Willeke, E. Bucher, 14th EC-PVSEC, 743 (1997).

[69] M.J. Stocks and A.W. Blakers, 14th EC-PVSEC, 305 (1997).

[70] A. Takami, S. Arimoto, H. Morikawa, S. Hamamoto, T. Ishihara, H. Kumabe and T. Murotani, 12th EC-PVSEC, 59 (1995).

[71] H. Takato, T. Sekigawa, Jpn. J. Appl. Phys. **34** (12A), 6358 (1995).

[72] M. Takeyama, A. Noya, K. Sakanishi, H. Seki, K. Sasaki, Ja J. of Appl. Phys. Part 1, **35** (7), 4027 (1996).

[73] F. Tamura, Y. Okayasu, K. Kumagai, Tec. Dig. Inter'l PVSEC-7, 237 (1993).

[74] B. Thomas, G. Müller, P. Heidborn, H.Wawra, 14th EC-PVSEC, 1483 (1997).

[75] C.J.J. Tool, J.A.M. van Roosmalen, S. Schiermeier, R.C. Huiberts, G.M. Christie, W.C. Sinke, 14th EC-PVSEC, 1014 (1997).

[76] T. Uematsu, M. IDA, K. Hane, S. Kokkunai and T. Saitoh, IEEE Transactions on Electron Devices, **37** No. 2 (1990).

[77] T. Vermeulen, O. Evrard, W. Laureyys, J. Poortmans, M. Caymax, J. Nijs, R. Mertens, C. Vinckier, H.U. Höfs, 13th EC-PVSEC, 1501 (1995).

[78] T. Vermeulen, J. Poortmans, M. Caymax, J. Nijs, R. Mertens, C. Vinckier, 14th EC-PVSEC, 728 (1997).

[79] B.F. Wagner, C Schetter, O.V. Sulima, and A. Bett, 23rd IEEE-PVSC, 356 (1993).

[80] R. Wallace, J. Hanoka, S. Narashima, S. Kamra, A. Rohatgi, 26th IEEE-PVSC, 99 (1997).

[81] K.J. Weber, K. Catchpole, M.Stocks, and A.W. Blakers, 26th IEEE-PVSC, 107 (1997).

[82] J.H. Werner, R. Bergmann, and R. Brendel, Festkörperprobleme/Advances in Solid State Physics, Ed. R. Helbig (Vieweg, Braunschweig), **34**, 115 (1994).

[83] M. Wolf, IEEE Transactions on Electron Devices ED-**27**, 751 (1980).

[84] K. Yamamoto, A. Nakajima, N. Nishio, M. Izumina, 1st WCPEC, 1575 (1994).

[85] K. Yamamoto, T. Suzuki, and A. Nakajima, Jpn. J. Appl. Phys. **36** L 569 (1997).

[86] J. Zhao, A. Wang, S.R. Wenham an M.A. Green, 13th EC-PVSEC, 1566 (1995).

[87] M.A. Green, J. Zhao, G. Zheng, 14th EC-PVSEC, 2324 (1997).

[88] G.F. Zheng, A.B. Sproul, S.R. Wenham and M.A. Green, Proc. 25th IEEE-PVSC, 465 (1996).

[89] W. Zimmermann, S. Bau, F. Haas, K. Schmidt, A. Eyer, 2nd WCPEC, Vienna Austria (1998), to be published.

Thin Films for Gas Sensors in New Application Fields – MBRS studies and semiempirical calculations.

Th. Ochs, W. Geyer, Ch. Krummel, M. Fleischer[1], H. Meixner[1], and D. Kohl

Institute of Applied Physics, Justus-Liebig-University,
Heinrich-Buff-Ring 16, D-35392 Gießen, Germany,
[1] Siemens AG, ZT KM2, Otto-Hahn-Ring 6, D-81730 München, Germany

Abstract: Reducing gases in air are detectable by a conductance increase of Ga_2O_3 films. In molecular beam relaxation spectroscopy (MBRS) Ga_2O_3 films are exposed to gas pulses of methane. The velocity constants of the desorption of the reaction products H_2O and CO_2 are observed. The desorption energies and the frequency factors are calculated (H_2O: E_a=0.2 eV, $f_a = 2 \times 10^3\ s^{-1}$; CO_2: $E_a = 1.0$ eV, $f_a = 100 \times 10^3\ s^{-1}$). Semiempirical Hartree-Fock calculations with the MNDO-PM3 method give information about the adsorption process. The energy gain of physisorption amounts to −0.27 eV. The HoF (Heat of Formation) of the chemisorbed CH_3 group and the H atom is 4.05 eV lower in comparison to the value of the physisorbed methane. After the $CO_{latt.}$ group formation the energy gain is 7.33 eV. The binding energy of the formed $CO_{latt.}$ is calculated to 2.07 eV. The charge transfer is estimated using the mulliken charges. A reaction scheme for the catalytic oxidation of methane by lattice oxygen is proposed. The conductance increase of Ga_2O_3 on exposure to methane can be explained by the creation of oxygen vacancies and by the formation of absorbed hydrogen, both acting as surface donors. Ga_2O_3 layers with operating temperatures between 600 and 900 °C allow new applications for semiconductor gas sensors in harsh environments, e. g. the monitoring of combustion gases.

1 Introduction

Gallium(III)-oxide in its thermodynamically stable β-modification exhibits n-type conductivity. A change of conductivity caused by gas solid interactions can hence be used for the detection of a gas. The suitability of Ga_2O_3 as material for the detection of methane and other oxidisable gases at operating temperatures between 600 and 900 °C has been shown [1]. By means of MBRS the surface reactions of methane on polycrystalline Ga_2O_3 films were investigated (Fig. 1). Additionally the reaction of methane on a (100) plane of Ga_2O_3 is calculated using the semiempirical MNDO-PM3 method.

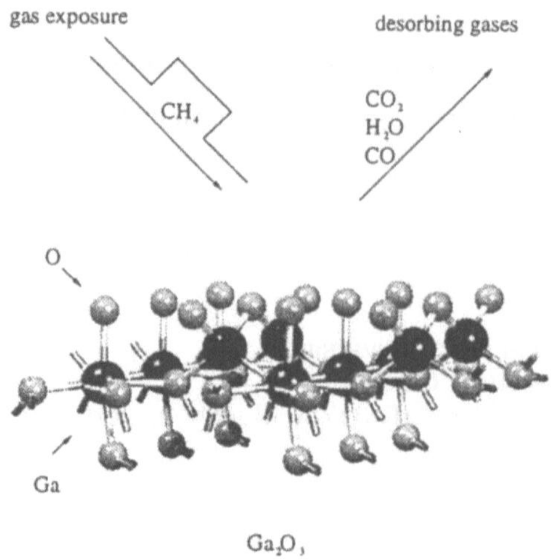

Figure 1
Scheme of the MBRS method.

2 Methods

2.1 Experimental

The reaction products of CH_4 on Ga_2O_3 surfaces are studied applying a Time-of-Flight Mass spectrometer (ToF-MS). The ToF-MS is part of a vacuum system, consisting of a preparation chamber with a gas inlet, a main UHV chamber with a solenoid gas valve in a differentially pumped system and a differentially pumped chamber for the ToF-MS. During the experiments a pressure of less than 10^{-7} Pa is maintained by means of a titanium sublimation pump and an ion pump. The compact ToF-MS with an insertion length of 45 cm is equipped with a special storing ion source and an electrostatic ion reflector [2]. The storing ion source increases sensitivity as against conventional ToF instruments by a factor of 1 000. In contrast to ion mirrors with potential-defining metal grids which limit the overall transmission to about 20%, the electrostatic ion reflector causes no reduction in transmission. The mass spectra can be acquired with a maximum repetition rate of 16 kHz, recorded by a digital storage oscilloscope with a 16 bit resolution. Usually a repetition rate of 1 – 2 kHz was used during the measurement. The gas pulses are produced by a solenoid gas valve. This valve does not bounce, is bakeable up to 200 °C and is resistant against aggressive gases. It generates gas pulses with rise and fall times of less than 2 ms. In the experiment pulses with a length of 10 ms are chosen. The gas inlet is differentially pumped with a turbo molecular pump. The pressure in front of the valve during the experiment is typical 10 – 20 Pa. The beam intensity is about 10^{16} molecules$\cdot$cm^{-2}s^{-1}. Assuming that approximately 10^{15} molecules$\cdot$cm^{-2} are equivalent to a coverage of 1 ML, the exposure dose of one pulse amounts to less than 1/10 ML.

The samples are prepared by applying a suspension of β-Ga_2O_3 powder (Johnson-Matthey; 32102; 99.99%) in bi-distilled water to aluminium oxide substrates and are dried at room temperature. The thickness of the film amounts to 200 μm. The aluminium oxide substrates with a platinum heating structure and a platinum thermoresistor (100 Ω at 293 K) on the back side of the substrate are used to reach a sample temperature up to 1 000 K. Higher temperatures are realised by using a pyrolytic Bor-Nitride substrate with a graphite heating structure inside. The graphite contacts are covered with a platinum layer to protect the graphite structure. This heater type makes sample temperatures up to 1 500 K possible. The temperature of the substrate is monitored by using a Ni/Ni-Cr thermoresistor on the backside. To provide a clean and stochiometric surface an oxygen treatment of the sample at 933 K and 20 000 Pa partial oxygen pressure for 15 min in the preparation chamber is performed.

The oxidised sample is positioned in front of the gas nozzle. After the shutting of the gas valve the gas flux intensity decreases exponentially:

$$f(t) = \phi_0 \exp(-k_r t)$$

The velocity constants k_r of the desorption are calculated by using an exponential fit to the declining part of the resulting MBRS graphs. The relation between the velocity constants and temperature T are shown in the Arrhenius equation:

$$k_r = f_a \exp(-E_a/kT)$$

The logarithm of the velocity constant as a function of temperature is sketched as an Arrhenius plot. The desorption energy and the frequency factor can be evaluated by using linear regression.

2.2 Model Calculations

Semiempirical Molecular Orbital (MO) calculations of the heat of formation (HoF) were performed by the semiempirical Hartree-Fock method MNDO-PM3 (Modified Neglect of Diatomic Overlap - Parametric Method 3) [3]. This method based on the Hartree-Fock theory was proposed by Stewart in 1989. Instead of calculating all the three- and four-centre integrals, the integrals with the so-called diatomic differential overlap, 18 parameters for each atom are used. These parameters are determined by an optimisation process using empirical data. The parametrisation as it is implemented in MOPAC93 [4] is used for the calculations. The way the parameters are optimised is described for O and H [5] and for gallium [6]. Gallium oxide exhibits a C2/m symmetry [7] with two different sites of gallium atoms. One is centred in the middle of a tetrahedron, the others are centred in the middle of an octahedron. We simulated the (100) plane of Ga_2O_3. This plane is one of the surfaces produced by cleavage of Ga_2O_3 and is the growth axis of single crystals [8]. Applying the program Solids Builder of MSI (Molecular Simulations Inc.) we simulated the crystal by using data of the asymmetric unit and the crystal structure. The optimisation of the structure was stopped when the values known from literature were reached.

We simulated the (100) plane by a cluster of 90 atoms. The cleaved bonds on the edges of the cluster are saturated with hydrogen atoms. These H atoms are sketched as open bonds in Figs. 6, 7, and 8. The (100) plane consisting of only oxygen atoms was not saturated with hydrogen, in order to provide adsorption sites for the adsorbates.

The geometry of the adsorbed methane as well as the geometry of the dissociated products CH_3 and H were obtained by a geometry optimisation. The adsorption geometries of the surface products H_2CO and CO were calculated, too. The geometry optimisation was carried out in a cyclic fashion as follows: each optimisation of angle was followed by variation of the C-O bond length. These variations were carried out until a structure was found, whose energy represented a global minimum. The convergence criteria of the SCF calculations were satisfied if the change in total energy on successive iterations was less then 4.18×10^{-4} kJ/mol.

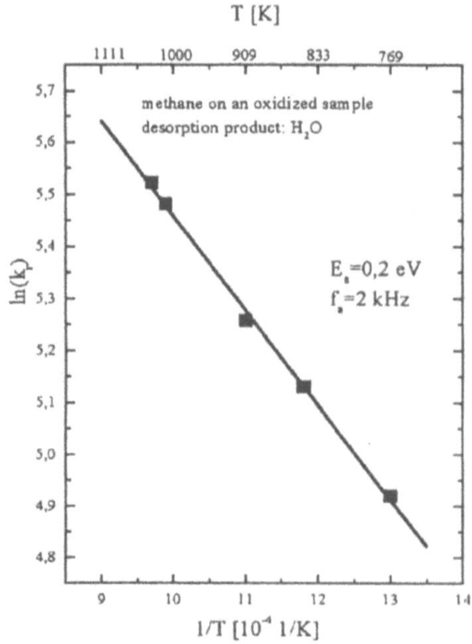

Figure 2
Arrhenius Plot of the velocity constants of H_2O.

3 Results of Experiments

During the MBRS-measurements the partial H_2O pressure has been observed at temperatures of 769, 847, 909, 1 010 and 1 030 K while exposing the samples to methane pulses. The velocity constants have been calculated using exponential regression. The Arrhenius plot (Fig. 2) of these values yields a desorption energy of 0.2 eV and a frequency factor of $2 \times 10^3\ s^{-1}$. The desorption flux ascended from 0.1% of the primary gas pulse at 769 K up to 1% of the primary gas flux at above 1 000 K.

The velocity constants of the desorption of CO_2 were measured at temperatures of 1 270, 1 420 and 1 520 K. The resulting desorption energy amounts to 1 eV with a frequency factor of $100 \times 10^3\ s^{-1}$ (Fig. 3). As a result of interfering effects, the velocity constant of CO was not observed.

Additionally, the desorption flux of CO, CO_2 and H_2O has been observed during TDS studies after exposure of the samples to methane (partial pressure: 100 Pa). The temperature of the samples has been varied between 290 K and 1 030 K with a heating rate of about 8.6 K/s. The TDS-spectrum for H_2O shows two peaks with the maximums at 470 K and 920 K. The CO peak exhibits a maximum at 870 K and the CO_2 peak has a maximum at 970 K (Fig. 9).

4 Results of Calculations

The estimated heat of formation (HoF) of the pure Ga_2O_3 cluster was -67.32 eV and of the free methane -0.56 eV. The HoF of the physisorption system (Fig. 6) has a value of -68.16 eV. Comparing these values leads to an energy of physisorption of -0.27 eV. The HoF of the dissociation products, the chemisorbed CH_3 and H, was calculated to be

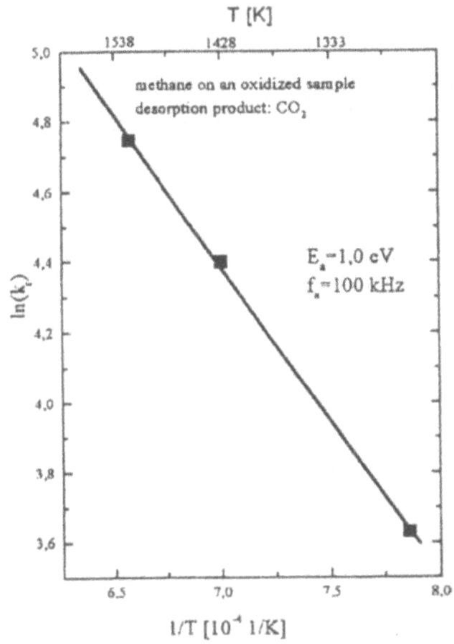

Figure 3
Arrhenius Plot of the velocity constants of CO_2.

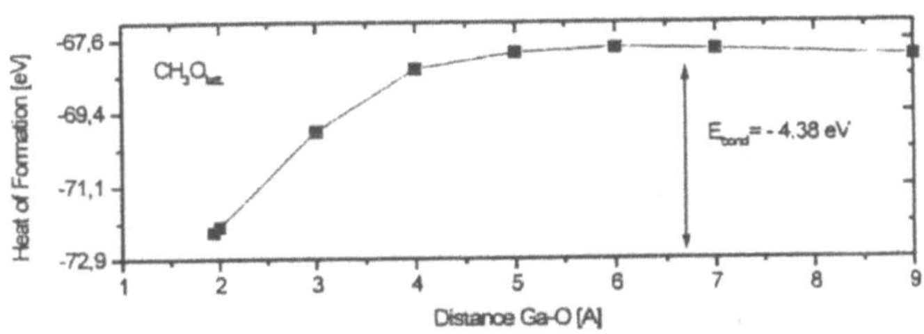

Figure 4 Heat of formation vs. the distance of the Ga-O bond of chemisorbed CH_3.

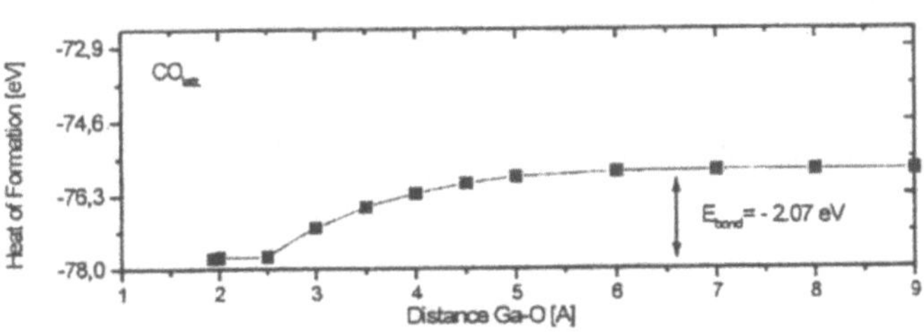

Figure 5 Heat of formation vs. the distance of the Ga-O bond of adsorbed C.

4.05 eV lower compared to the energy of physisorption. This value was estimated to be −72.21 eV. The final energy gain after the formation of the CO_{latt}. was 7.33 eV, comparing the values of 79.94 eV for the CO_{latt}. system and of the chemisorbed position of CH_3. The calculation of the binding energy of the adsorbate on the cluster was performed by calculating the HoF as a function of the Ga-O bond length. Comparing the HoF of the original system and the system with the distant adsorbate yields the binding energy. The activation energy for decomposition of the CH_3 group was calculated to be −4.38 eV

Table 1 Calculated Mulliken charges of the C and the H and bond energies of the C-Olatt. bond using the PM3 method.

	Mulliken Charge					Charge Transfer	Bond Energy of C–O
	C	H(I)	H(II)	H(III)	H(IV)		
Adsorption of CH_4	$-0.38e$						
Chemisorption of CH_3	$0.13e$	$0.08e$				$0.21e$	182 kcal/mol
Formation of H_2CO	$0.19e$	$0.12e$	$0.15e$			$0.46e$	212 kcal/mol
Formation of CO	$0.21e$	$0.14e$	$0.20e$	$0.14e$	$0.14e$	$0.83e$	326 kcal/mol

using the data of Fig. 4. After the $CO_{latt.}$ group is formed at the surface the calculation of the bond energy yields a value of -2.07 eV using the data of Fig. 5.

4.1 Mulliken-charge

Using the PM3 method we calculated the mulliken charges of the atoms listed in Tab. 1 and the bond energies of the C-O bond.

5 Discussion

We now want to propose a reaction scheme for the catalytic oxidation of methane. As a first step an intermediate state can be suggested considering the calculated physisorption energy of 0.3 eV.

$$CH4_{gas} \rightarrow CH4_{phys} \qquad \Delta H_0 = -0.3 \text{ eV}.$$

The calculated binding energy of 0.3 eV can be explained by the formation of three hydrogen bonds between the adsorbate and the surface (Fig. 6). On exposure to methane at temperature above 680 K, a dissociative adsorption occurs:

$$CH4_{phys} \rightarrow CH3_{ads} + H_{ads} \qquad \Delta H_0 = -4.05 \text{ eV}.$$

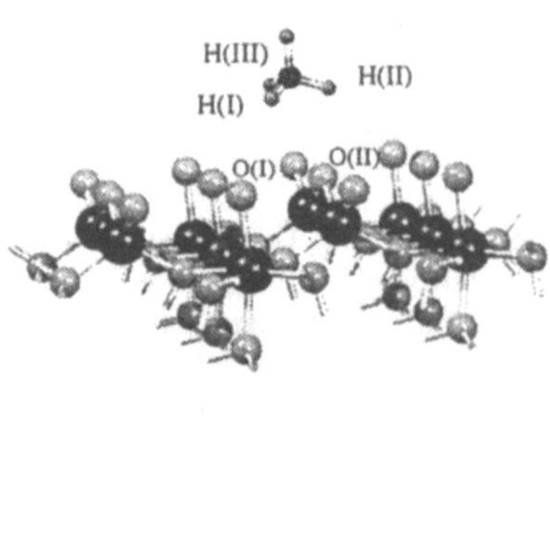

Figure 6
Position of physisorbed methane. The H(I)-O(I) distance is 2.80 Å, the H(II)- O(II) distance 2.93 Å and the H(III)-O(II) distance 2.94 Å.

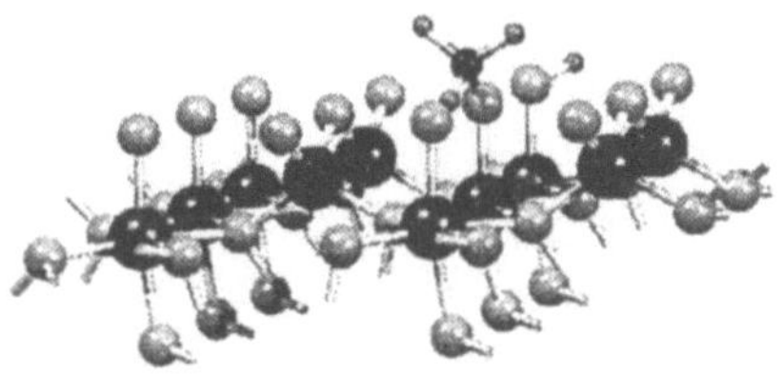

Figure 7
Position of the chemisorbed CH_3 group and the H atom.

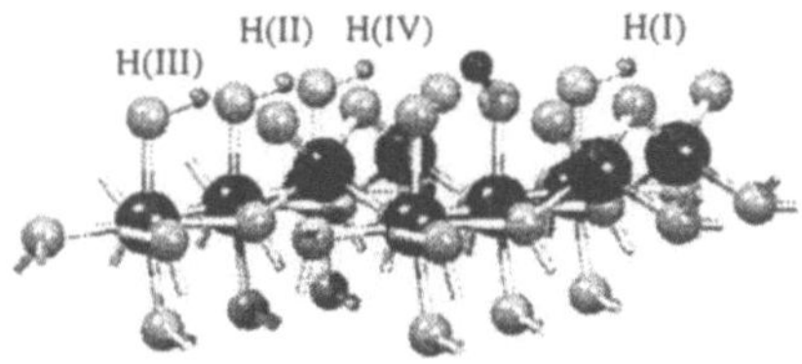

Figure 8
Position of the adsorbed C atom and the H atoms.

The high value of ΔH_0 makes any desorption of CH_4 unlikely. CH_3 adsorbs on the (100)-plane with a C-O bondlength of 1.34 Å and a binding angle of 39° between the surface and the methyl group (Fig. 7). A dissociation of the remaining H atoms takes place by subsequent separation of the hydrogen atoms:

$$CH_{3ads} \rightarrow C_{ads} + 3H_{ads} \qquad \Delta H_0 = -7.33 \text{ eV}$$

The remaining carbon shows a C-O bondlength of 1.16 Å (Fig. 8). The bond between the CO_{latt}. molecule and the lattice is weakened to a value of 2.07 eV. CO desorption was observed during TDS investigations on polycrystalline Ga_2O_3 after exposure to CH_4 [9].

$$C_{ads} + O_{latt.} \rightarrow CO_{gas} \qquad E_a = 2.07 \text{ eV}$$

The calculated bond energy of CO is consistent with the observed desorption maximum in the TDS spectra (Fig. 9). CO_2 and H_2O have been observed in desorption. The desorption energies have been evaluated during the MBRS investigations.

$$2H_{ads} + O_{latt.} \rightarrow H_2O_{gas} \qquad E_a = 0.2 \text{ eV} \qquad f_a = 2 \times 10^3 \text{ s}^{-1}$$
$$C_{ads} + 2O_{latt.} \rightarrow CO_{2\,gas} \qquad E_a = 1.0 \text{ eV} \qquad f_a = 100 \times 10^3 \text{ s}^{-1}$$

The MBRS results are confirmed by the position of the desorption maximums of the TDS spectra using the following relation [10]:

$$E_a/RT_{max} = \ln\left(\frac{f_a T_{max}}{\beta - 3.64}\right).$$

The desorption energy E_a and the frequency factor f_a are results of the MBRS measurements. The temperatures of the peak maximums have been evaluated using TDS results with a specific heating rate β. The resulting adsorption energy using this relation is 1.08 eV for CO_2 and 0.25 eV for H_2O.

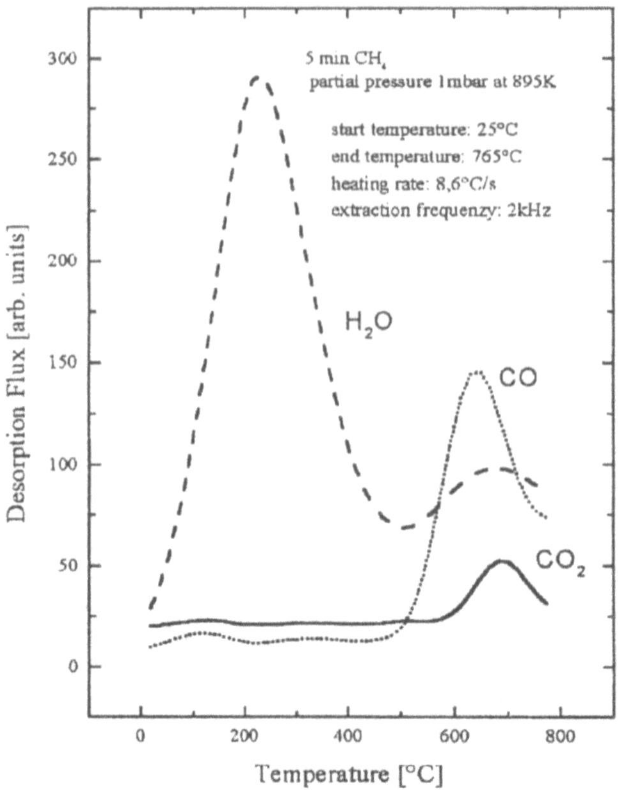

Figure 9
TDS spectra of an CH_4-treated sample.

5.1 Charge Transfer

During the chemisorption of the CH_3 group and the H atom we calculated a charge transfer of $0.21e$ to the surface by using mulliken charges with the MNDO-PM3 method. The following surface reaction steps lead to another charge transfer of $0.46e$ after the formation of the CH_2 group and two H atoms at the surface and to a charge transfer of $0.83e$ after the formation of the C atom and the 4 H atoms. Similar mulliken charges were obtained using an ab initio method for the calculations.

The charge transfer is responsible for a change in conductance. As discussed above the desorption of CO results in a formation of an oxygen vacancy at the surface. A hint of the utilisation of surface oxygen is the desorption of Ga_2O as described in [9]. This process can lead to a change in conductance, too, as a second process beside the electron transfer.

Bibliography

[1] Fleischer, M. and H. Meixner, *Sensing reducing gases at high temperatures using long-term stable* Ga_2O_3 *thin films*, Sensors and Actuators B, **6**, 257–261 (1992).

[2] Grix, R., R. Kutscher, G. Li, U. Grüner and H. Wollnik, *A Time-of-flight Mass Analyzer with High Resolving Power*, Rapid Communications in Mass Spectrometry, Vol. 2, No. 5, (1988).

[3] Stewart, J.J.P., Journal of Computational Chemistry, **10**, 209–220 (1989).

[4] Stewart, J.J.P., MOPAC Manual (7th Edition).

[5] Stewart, J.J.P., *Optimisation of Parameters for Semiempirical Methods*, Journal of Computational Chemistry, Vol. 10, No. 2, 209–220 (1989).

[6] Dewar, M.J.S. and C. Jie, Organometallics, **8**, 1544 (1989).

[7] Geller, S., *Crystal Structure of* β-Ga_2O_3, The Journal of Chemical Physics, **33**, 676–684 (1960).

[8] Chase, A.B. , *Growth of* β-Ga_2O_3 *by the Verneuil Technique*, Journal of the American Ceramic Society, 47/9, 470 (1964).

[9] Becker, F., C. Krummel, A. Freiling, M. Fleischer and C. Kohl, *Decomposition of methane on polycrystalline thick films of* Ga_2O_3 *investigated by thermal desorption spectroscopy with a mass spectrometer*, Fresenius J. Anal. Chem. **358**, 187– 189 (1997).

[10] Henzler, M. and W. Göpel, *Oberflächenphysik des Festkörpers*, Teubner StudienbΑcher (Stuttgart 1994).

Contents of Volumes 33–37

MIX
Papier aus verantwortungsvollen Quellen
Paper from responsible sources
FSC® C105338

FSC
www.fsc.org

If you have any concerns about our products,
you can contact us on
ProductSafety@springernature.com

In case Publisher is established outside the EU,
the EU authorized representative is:
Springer Nature Customer Service Center GmbH
Europaplatz 3, 69115 Heidelberg, Germany

Printed by Libri Plureos GmbH
in Hamburg, Germany